Physics

FOR THE IB DIPLOMA

SECOND EDITION

John Allum and
Christopher Talbot

Orders: please contact Bookpoint Ltd, 130 Milton Park, Abingdon, Oxon OX14 4SB. Telephone: (44) 01235 827720. Fax: (44) 01235 400454. Lines are open from 9.00 - 5.00, Monday to Saturday, with a 24 hour message answering service. You can also order through our website www.hoddereducation.com

First edition published in 2012

by Hodder Education, An Hachette UK Company

Carmelite House, 50 Victoria Embankment, London EC4Y 0DZ

www.hoddereducation.com

This second edition published 2014

Impression number 5 4 3

Year 2019 2018 2017

Cover photo Linden Gledhill

Illustrations by Pantek Media, Barking Dog Art, Aptara

Typeset in Goudy Oldstyle Std 10/12pt by Aptara inc.

Printed in India

A catalogue record for this title is available from the British Library

ISBN: 978 147 1829048

Contents

Additional higher level (AHL)

Options

Available on the website accompanying this book: www.hoddereducation.com/IBextras

Introduction

Welcome to the second edition of *Physics for the IB Diploma*. The content and structure of this second edition has been completely revised to meet the demands of the 2014 *IB Diploma Programme Physics Guide*.

Within the IB Diploma Programme, the physics content is organized into compulsory topics plus a number of options, from which all students select one. The organization of this resource exactly follows the IB *Physics Guide* sequence:

- **Core:** Chapters 1–8 cover the common core topics for Standard *and* Higher Level students.
- **Additional Higher Level (AHL):** Chapters 9–12 cover the additional topics for Higher Level students.
- **Options:** Chapters 13–16 cover Options A, B, C and D respectively. Each of these is available to both Standard and Higher Level students. (Higher Level students study more topics within the same option.)

Each of the core and AHL topics is the subject of a corresponding single chapter in the *Physics for the IB Diploma* printed book.

The Options (Chapters 13–16) are available on the website accompanying this book, as are useful appendices and additional student support (including Starting points and Summary of knowledge): www.hoddereducation.com/IBextras

There are two additional short chapters offering physics-specific advice on the skills necessary for **Graphs and data analysis** and **Preparing for the IB Diploma Physics examination**, including explanations of the **command terms**. These chapters can be found on the accompanying website.

Special features of the chapters of *Physics for the IB Diploma* are described below.

- The text is written in **straightforward language**, without phrases or idioms that might confuse students for whom English is a second language.
- The **depth of treatment** of topics has been carefully planned to accurately reflect the objectives of the IB syllabus and the requirements of the examinations.

Nature of Science

- The **Nature of Science** is an important new aspect of the IB Physics course, which aims to broaden students' interests and knowledge beyond the confines of its specific physics content. Throughout this book we hope that students will develop an appreciation of the processes and applications of physics and technology. Some aspects of the *Nature of Science* may be examined in IB Physics examinations and important discussion points are highlighted in the margins.
- The **Utilizations** and **Additional Perspectives** sections also reflect the *Nature of Science*, but they are designed to take students *beyond* the limits of the IB syllabus in a variety of ways. They might, for example, provide a historical context, extend theory or offer an interesting application. They are sometimes accompanied by more challenging, or research-style, questions. They do *not* contain any knowledge that is essential for the IB examinations.
- Science and technology have developed over the centuries with contributions from scientists from all around the world. In the modern world science knows few boundaries and the flow of information is usually quick and easy. Some international applications of science have been indicated with the **International Mindedness** icon.
- **Worked examples** are provided in each chapter whenever new equations are introduced. A large number of **self-assessment questions** are placed throughout the chapters close to the relevant theory. **Answers** to most questions are provided at the end of the book.
- It is not an aim of this book to provide detailed information about experimental work or the use of computers. However, our **Skills** icon has been placed in the margin to indicate wherever such work may usefully aid understanding. A number of key experiments are included in the *IB Physics Guide* and these are listed in Chapter 18: Preparing for the IB Diploma Physics examination, to be found on the website that accompanies this book.

- A selection of **IB examination-style questions** is provided at the end of each chapter, as well as some past IB Physics examination questions.
- Links to the interdisciplinary **Theory of Knowledge (ToK)** element of the IB Diploma course are made in all chapters.
- Comprehensive **glossaries** of words and terms for Core and AHL topics are included in the printed book. Glossaries for the Options are available on the website.

Using this book

The sequence of chapters in *Physics for the IB Diploma* deliberately follows the sequence of the syllabus content. However, the *IB Diploma Physics Guide* is not designed as a teaching syllabus, so the order in which the syllabus content is presented is not necessarily the order in which it will be taught. Different schools and colleges should design a course based on their individual circumstances.

In addition to the study of the physics principles contained in this book, IB science students carry out experiments and investigations, as well as collaborating in a Group 4 Project. These are assessed within the school (Internal Assessment), based on well-established criteria.

The contents of Chapter 1 (Physics and physical measurement) have applications that recur throughout the rest of the book and also during practical work. For this reason, it is intended more as a source of reference, rather than as material that should be fully understood before progressing to the rest of the course.

Author profiles

John Allum

John has taught pre-university physics courses as a Head of Department in a variety of international schools for more than 30 years. He has taught IB Physics in Malaysia and in Abu Dhabi, and has been an examiner for IB Physics for many years.

Christopher Talbot

Chris teaches TOK and IB Chemistry at a leading IB World School in Singapore. He has also taught IB Biology and a variety of IGCSE courses, including IGCSE Physics, at Overseas Family School, Republic of Singapore.

Authors' acknowledgements

We are indebted to the following teachers and lecturers who reviewed early drafts of the chapters: Dr Robert Smith, University of Sussex (Astrophysics); Dr Tim Brown, University of Surrey (Communications and Digital Technology); Dr David Cooper (Quantum Physics); Mr Bernard Taylor (Theory of Knowledge, Internal Assessment and Fields and Forces); Professor Christopher Hammond, University of Leeds (Electromagnetic Waves); Professor Phil Walker, University of Surrey (Nuclear Physics); Dr David Jenkins, University of York (Nuclear Physics) and Trevor Wilson, Bavaria International School e.V., Germany.

We also like to thank David Talbot, who supplied some of the photographs for the book, and Terri Harwood and Jon Homewood, who drew a number of physicists.

For this second edition, we would like to thank the following academics for their advice, comment and feedback on drafts of the chapters: Dr Robert Smith, Emeritus Reader in Astronomy, University of Sussex, Dr Tim Brown, Lecturer in Radio Frequency Antennas and Propagation, University of Surrey, Dr Alexander Merle, Department of Physics and Astronomy, University of Southampton, Dr David Berman, School of Physics and Astronomy, Queen Mary College, Professor Carl Dettmann, School of Mathematics, University of Bristol and Dr John Roche, Linacre College, University of Oxford. We would like to thank Richard Burt, Windermere School, UK for authoring Option A Relativity (Chapter 13).

Finally we would also like to express our gratitude for the tireless efforts of the Hodder Education team that produced the book you have in front of you, led by So-Shan Au and Patrick Fox.

1 Measurements and uncertainties

ESSENTIAL IDEAS

- Since 1948, the Système International d'Unités (SI) has been used as the preferred language of science and technology across the globe and reflects current best measurement practice.
- Scientists aim towards designing experiments that can give a 'true value' from their measurements but, due to the limited precision in measuring devices, they often quote their results with some form of uncertainty.
- Some quantities have direction and magnitude, others have magnitude only, and this understanding is the key to correct manipulation of quantities.

1.1 Measurements in physics *– since 1948, the Système International d'Unités (SI) has been used as the preferred language of science and technology across the globe and reflects current best measurement practice*

Fundamental and derived SI units

To communicate with each other we need to share a common language, and to share numerical information we need to use common **units of measurement**. An internationally agreed system of units is now used by scientists around the world. It is called the **SI system** (from the French 'Système International'). SI units will be used throughout this course.

Nature of Science

Common terminology

For much of the last 200 years many prominent scientists have tried to reach agreement on a metric (decimal) system of units that everyone would use for measurements in science and commerce. A common system of measurement is invaluable for the transfer of scientific information and for international trade. In principle this may seem more than sensible, but there are significant historical and cultural reasons why some countries, and some societies and individuals, have been resistant to changing their system of units.

The SI system was formalized in 1960 and the seventh unit (the mole) was added in 1971. Before that, apart from SI units, a system based on centimetres, grams and seconds (CGS) was widely used, while the imperial (non-decimal) system of feet, pounds and seconds was also popular in some countries. For non-scientific, everyday use, people in many countries sometimes still prefer to use different systems that have been popular for centuries. Confusion between different systems of units has been famously blamed for the failure of the Mars orbiter in 1999 and has been implicated in several aviation incidents.

The fundamental units of measurement

There are seven **fundamental (basic) units** in the SI system: kilogram, metre, second, ampere, mole, kelvin (and candela, which is *not* part of this course). The quantities, names and symbols for these fundamental SI units are given in Table 1.1.

They are called 'fundamental' because their definitions are not combinations of other units (unlike metres per second, for example). You do not need to learn the definitions of these units.

Table 1.1 Fundamental units

Quantity	Name	Symbol	Definition
length	metre	m	the distance travelled by light in a vacuum in 1/299 792 458 seconds
mass	kilogram	kg	the mass of a cylinder of platinum-iridium alloy kept at the International Bureau of Weights and Measures in France
time	second	s	the duration of 9 192 631 770 oscillations of the electromagnetic radiation emitted in the transmission between two specific energy levels in caesium-133 atoms
electric current	ampere	A	that current which, when flowing in two parallel conductors one metre apart in a vacuum produces a force of 2×10^{-7}N on each metre of the conductors
temperature	kelvin	K	1/273.16 of the thermodynamic temperature of the triple point of water
amount of substance	mole	mol	an amount of substance that contains as many particles as there are atoms in 12 g of carbon-12

Nature of Science

Improvement in instrumentation

Accurate and precise measurements of experimental data are a cornerstone of science, and such measurements rely on the precision of our system of units. The definitions of the fundamental units depend on scientists' ability to make very precise measurements and this has improved since the units were first defined and used.

Scientific advances can come from original research in new areas, but they are also driven by improved technologies and the ability to make more accurate measurements. Astronomy is a good example: controlled experiments are generally not possible, so our rapidly expanding understanding of the universe is being achieved largely as a result of the improved data we can receive with the help of the latest technologies (higher-resolution telescopes, for example).

Derived units of measurement

All other units in science are combinations of the fundamental units. For example, the unit for volume is m^3 and the unit for speed is $m\,s^{-1}$. Combinations of fundamental units are known as **derived units**.

Sometimes derived units are also given their own name (Table 1.2). For example, the unit of force is $kg\,m\,s^{-2}$, but it is usually called the newton, N. All derived units will be introduced and defined when they are needed during the course.

■ **Table 1.2** Some named derived units

Derived unit	Quantity	Combined fundamental units
newton (N)	force	$kg\,m\,s^{-2}$
pascal (Pa)	pressure	$kg\,m^{-1}\,s^{-2}$
hertz (Hz)	frequency	s^{-1}
joule (J)	energy	$kg\,m^2\,s^{-2}$
watt (W)	power	$kg\,m^2\,s^{-3}$
coulomb (C)	charge	$A\,s$
volt (V)	potential difference	$kg\,m^2\,s^{-3}\,A^{-1}$
ohm (Ω)	resistance	$kg\,m^2\,s^{-3}\,A^{-2}$
weber (Wb)	magnetic flux	$kg\,m^2\,s^{-2}\,A^{-1}$
tesla (T)	magnetic field strength	$kg\,s^{-2}\,A^{-1}$
becquerel (Bq)	radioactivity	s^{-1}

Note that students are expected to write and recognize units using superscript format, such as $m\,s^{-1}$ rather than m/s. Acceleration, for example, has the unit $m\,s^{-2}$.

Occasionally physicists use units that are not part of the SI system. For example, the electronvolt, eV, is a conveniently small unit of energy that is often used in atomic physics. Units such as this will be introduced when necessary during the course. Students will be expected to be able to convert from one unit to another. A more common conversion would be changing time in years to time in seconds.

ToK Link

Fundamental concepts

As well as some units of measurement, many of the ideas and principles used in physics can be described as being 'fundamental'. Indeed, physics itself is often described as the fundamental science. But what exactly do we mean when we describe something as fundamental? We could replace the word with 'elementary' or 'basic', but that does not really help us to understand its true meaning.

One of the central themes of physics is the search for *fundamental particles* – particles that are the basic building blocks of the universe and are not, themselves, made up of smaller and simpler particles. It is the same with *fundamental laws and principles*: a physics principle cannot be described as fundamental if it can be explained by 'simpler' ideas. Most scientists also believe that a principle cannot be really fundamental unless it is relatively simple to express (probably using mathematics). If it is complicated, maybe the underlying simplicity has not yet been discovered.

Fundamental principles must also be 'true' everywhere and for all time. The fundamental principles of physics that we use today have been tested, re-tested and tested again to check if they are truly fundamental. Of course, there is *always* a possibility that in the future a principle that is believed to be fundamental now is discovered to be explainable by simpler ideas.

Consider two well-known laws in physics. *Hooke's law* describes how some materials stretch when forces act on them. It is a simple law, but it is not a fundamental law because it is certainly not always true. The *law of conservation of energy* is also simple, but it is described as fundamental because there are no known exceptions.

Scientific notation and metric multipliers

Scientific notation

When writing and comparing very large or very small numbers it is convenient to use scientific notation (sometimes called '*standard form*').

In scientific notation every number is expressed in the form $a \times 10^b$, where a is a decimal number larger than 1 and less than 10, and b is a whole number (integer) called the exponent. For example, in scientific notation the number 434 is written as 4.34×10^2; similarly, 0.000 316 is written as 3.16×10^{-4}.

Scientific notation is useful for making the number of significant figures clear (see the next section). It is also used for entering and displaying large and small numbers on calculators. $\times 10^x$ or the letter *E* is often used on calculators to represent 'times ten to the power of…'. For example, 4.62*E*3 represents 4.62×10^3, or 4620.

The worldwide use of this standard form for representing numerical data is of great importance for the communication of scientific information between different countries.

Table 1.3 Standard metric (SI) multipliers

Prefix	Abbreviation	Value
peta	P	10^{15}
tera	T	10^{12}
giga	G	10^9
mega	M	10^6
kilo	k	10^3
deci	d	10^{-1}
centi	c	10^{-2}
milli	m	10^{-3}
micro	μ	10^{-6}
nano	n	10^{-9}
pico	p	10^{-12}
femto	f	10^{-15}

Standard metric multipliers

In everyday language we use the words 'thousand' and 'million' to help represent large numbers. The scientific equivalents are the prefixes kilo- and mega-. For example, a kilowatt is one thousand watts, and a megajoule is one million joules. Similarly, a thousandth and a millionth are represented scientifically by the prefixes milli- and micro-. A list of standard prefixes is shown in Table 1.3. It is provided in the *Physics data booklet*.

ToK Link

Effective communication needs a common language and terminology

What has influenced the common language used in science? To what extent does having a common standard approach to measurement facilitate the sharing of knowledge in physics?

There can be little doubt that communication between scientists is much easier if they share a common scientific language (symbols, units, standard scientific notation etc. as outlined in this chapter). But are our modern methods of scientific communication and terminology the best, or could they be improved? To what extent are they just a historical accident, based on the specific languages and cultures that were dominant at the time of their development?

Significant figures

The more precise a measurement is, the greater the number of significant figures (digits) that can be used to represent it. For example, an electric current stated to be 4.20 A (as distinct from 4.19 A or 4.21 A) suggests a much greater precision than a current stated to be 4.2 A.

Significant figures are all the digits used in data to carry meaning, whether they are before or after a decimal point, and *this includes zeros*. But sometimes zeros are used without thought or meaning, and this can lead to confusion. For example, if you are told that it is 100 km to the nearest airport, you might be unsure whether it is approximately 100 km, or 'exactly' 100 km. This is a good example of why scientific notation is useful. Using 1.00×10^3 km makes it clear that there are three significant figures. 1×10^3 km represents much less precision.

When making *calculations*, the result cannot be more precise than the data used to produce it. As a general (and simplified) rule, when answering questions or processing experimental data, the result should have the same number of significant figures as the data used. If the number of significant figures is not the same for all pieces of data, then the number of significant figures in the answer should be the same as the least precise of the data (which has the fewest significant figures). This is illustrated in Worked example 1.

Worked example

1 Use the equation:

$$P = \frac{mgh}{t}$$

to determine the power, P, of an electric motor that raises a mass, m, of 1.5 kg, a distance, h, of 1.128 m in a time, t, of 4.79 s. ($g = 9.81\,\mathrm{m\,s^{-2}}$)

$$P = \frac{mgh}{t} = \frac{1.5 \times 9.81 \times 1.128}{4.79}$$

A calculator will display an answer of 3.4652..., but this answer suggests a very high precision, which is not justified by the data. The data used with the least number of significant figures is 1.5 kg, so the answer should also have the same number:

$P = 3.5\,\mathrm{W}$

'Rounding off' to an appropriate number of significant figures

'Rounding off', as in Worked example 1, should be done at the *end* of a multi-step calculation, when the answer has to be given. If further calculations using this answer are then needed, *all* the digits shown previously on the calculator should be used. The answer to this calculation should then be rounded off to the correct number of significant figures. This process can sometimes result in small but apparent inconsistencies between answers.

Orders of magnitude

Physics is the fundamental science that tries to explain how and why everything in the universe behaves in the way that it does. Physicists study everything from the smallest parts of atoms to distant objects in our galaxy and beyond (Figure 1.1).

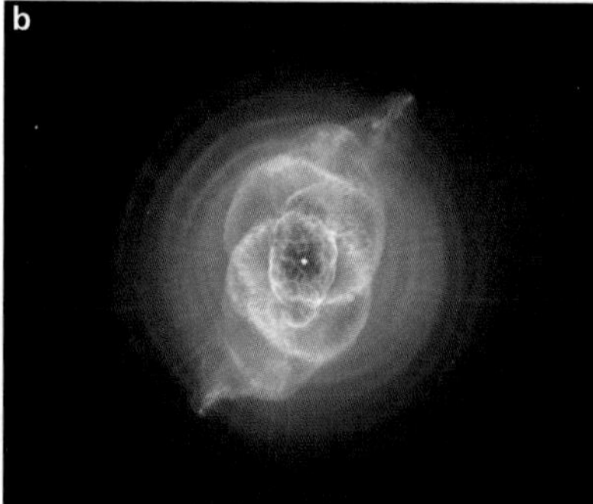

Figure 1.1
a The behaviour of individual atoms in graphene (a material made from a single layer of carbon atoms) can be seen using a special type of electron microscope
b Complex gas and dust clouds in the Cat's Eye nebula, 3000 light years away

Physics is a quantitative subject that makes much use of mathematics. Measurements and calculations commonly relate to the world that we can see around us (the macroscopic world), but our observations may require microscopic explanations, often including an understanding of molecules, atoms, ions and sub-atomic particles. Astronomy is a branch of physics that deals with the other extreme – quantities that are very much bigger than anything we experience in everyday life.

The study of physics therefore involves dealing with both very large and very small numbers. When numbers are so different from our everyday experiences, it can be difficult to appreciate their true size. For example, the age of the universe is believed to be about 10^{18} s, but just how big is that number? The only sensible way to answer that question is to compare the quantity with something else with which we are more familiar. For example, the age of the universe is about 100 million human lifetimes.

When comparing quantities of very different sizes (magnitudes), for simplicity we often make approximations to the nearest power of 10. When numbers are approximated and quoted to the nearest power of 10, it is called giving them an order of magnitude. For example, when comparing the lifetime of a human (the worldwide average is about 70 years) with the age of the universe (1.4×10^{10} y), we can use the approximate ratio $10^{10}/10^{2}$. That is, the age of the universe is about 10^{8} human lifetimes, or we could say that there are eight orders of magnitude between them.

Here are three further examples:

- The mass of a hydrogen atom is 1.67×10^{-27} kg. To an order of magnitude this is 10^{-27} kg.
- The distance to the nearest star (*Proxima Centauri*) is 4.01×10^{16} m. To an order of magnitude this is 10^{17} m. (Note: log of $4.01 \times 10^{16} = 16.60$, which is nearer to 17 than to 16.)
- There are 86 400 seconds in a day. To an order of magnitude this is 10^5 s.

Tables 1.4 to 1.6 list the ranges of mass, distance and time that occur in the universe. You are recommended to look at computer simulations representing these ranges.

■ **Table 1.4** The range of masses in the universe

Object	Mass/kg
the observable universe	10^{53}
our galaxy (the Milky Way)	10^{42}
the Sun	10^{30}
the Earth	10^{24}
a large passenger plane	10^{5}
a large adult human	10^{2}
a large book	1
a raindrop	10^{-6}
a virus	10^{-20}
a hydrogen atom	10^{-27}
an electron	10^{-30}

Distance	Size/m
distance to the edge of the visible universe	10^{27}
diameter of our galaxy (the Milky Way)	10^{21}
distance to the nearest star	10^{16}
distance to the Sun	10^{11}
distance to the Moon	10^{8}
radius of the Earth	10^{7}
altitude of a cruising plane	10^{4}
height of a child	1
how much human hair grows by in one day	10^{-4}
diameter of an atom	10^{-10}
diameter of a nucleus	10^{-15}

■ **Table 1.5** The range of distances in the universe

Time period	Time interval/s
age of the universe	10^{18}
time since dinosaurs became extinct	10^{15}
time since humans first appeared on Earth	10^{13}
time since the pyramids were built in Egypt	10^{11}
typical human lifetime	10^{9}
one day	10^{5}
time between human heartbeats	1
time period of high-frequency sound	10^{-4}
time for light to travel across a room	10^{-8}
time period of oscillation of a light wave	10^{-15}
time for light to travel across a nucleus	10^{-23}

■ **Table 1.6** The range of times in the universe

Estimation

Sometimes we do not have the data needed for accurate calculations, or maybe calculations need to be made quickly. Sometimes a question is so vague that an accurate answer is simply not possible. The ability to make sensible estimates is a very useful skill that needs plenty of practice. The worked example and questions 2–5 below are typical of calculations that do not have exact answers.

When making estimates, different people will produce different answers and it is usually sensible to use only one (maybe two) significant figures. Sometimes only an order of magnitude is needed.

Worked example

2 Estimate the mass of air in a classroom. (density of air = $1.3\,kg\,m^{-3}$)

A typical classroom might have dimensions of $7\,m \times 8\,m \times 3\,m$, so its volume is about $170\,m^3$.

mass = density × volume = $170 \times 1.3 = 220\,kg$

Since this is an estimate, an answer of 200 kg may be more appropriate. To an order of magnitude it would be $10^2\,kg$.

1 Estimate the mass of:
  a a page of a book
  b air in a bottle
  c a dog
  d water in the oceans of the world.

2 Give an estimate for each of the following:
  a the height of a house with three floors
  b how many times a wheel on a car rotates during the lifetime of the car
  c how many grains of sand would fill a cup
  d the thickness of a page in a book.

3 Estimate the following periods of time:
  a how many seconds there are in an average human lifetime
  b how long it would take a person to walk around the Earth (ignore the time not spent walking)
  c how long it takes for light to travel across a room.

4 Research the relevant data so that you can compare the following measurements. (Give your answer as an order of magnitude.)
  a the distance to the Moon with the circumference of the Earth
  b the mass of the Earth with the mass of an apple
  c the time it takes light to travel 1 m with the time between your heartbeats.

1.2 Uncertainties and errors – *scientists aim towards designing experiments that can give a 'true value' from their measurements, but because of the limited precision in measuring devices, they often quote their results with some form of uncertainty*

Nature of Science

Certainty

Although scientists are perceived as working towards finding 'exact' answers, an unavoidable uncertainty exists in any measurement. The results of *all* scientific investigations have uncertainties and errors, although good experimentation will try to keep these as small as possible.

When we receive numerical data of any kind (scientific or otherwise) we need to know how much belief we should place in the information that we are reading or hearing. The presentation of the results of serious scientific research should always have an assessment of the uncertainties in the findings, because this is an integral part of the scientific process. Unfortunately the same is not true of much of the information we receive through the media, where data are too often presented uncritically and unscientifically, without any reference to their source or reliability.

No matter how hard we try, even with the very best of measuring instruments, it is simply not possible to measure anything *exactly*. For one reason, the things that we can measure do not exist as perfectly exact quantities; there is no reason why they should.

This means that *every* measurement is an approximation. A measurement could be the most accurate ever made, for example the width of a ruler might be stated as 2.283 891 03 cm, but that is still not perfect, and even if it was we would not know because we would always need a more accurate instrument to check it. In this example we also have the added complication of the fact that when measurements of length become very small we have to deal with the atomic nature of the objects that we are measuring. (Where is the edge of an atom?)

The **uncertainty** in a measurement is the range, above and below a stated value, over which we would expect any repeated measurements to fall. For example, if the average height to which a ball bounced when dropped (from the same height) was 48 cm, but actual measurements varied between 45 cm and 51 cm, the result should be recorded as 48 ± 3 cm. The uncertainty is ± 3 cm, but this is sometimes better quoted as a percentage, in this example ± 6%. Obviously, it is desirable that experiments should produce results with low uncertainties – such measurements are described as being **precise**. But it should be noted that sometimes results can be precise, but wrong!

The more precise that a measurement is, the greater the number of significant figures (digits) that can be used to represent it.

If the correct ('true') value of a quantity is known, but an actual measurement is made that is not the same, we refer to this as an experimental **error**. That is, an error occurs in a measurement when it is not exactly the same as the correct value. For example, if a student recorded the height of a ball's bounce as 49 cm, but careful observation of a video recording showed that it was actually 48 cm, then there was an error in the measurement of +1 cm.

All measurements involve errors, whether they are large or small, for which there are many possible reasons, but they should not be confused with mistakes. Errors can be described as either *random* or *systematic* (see below), although all measurements involve both kinds of error to some extent.

The words *error* and *uncertainty* are sometimes used to mean the same thing, although this can only be true when referring to experiments that have a known correct result.

ToK Link

Scientific knowledge is provisional

'One aim of the physical sciences has been to give an exact picture of the material world. One achievement of physics in the twentieth century has been to prove that this aim is unattainable.'

Jacob Bronowski

Can scientists ever be truly certain of their discoveries?

The popular belief is that science deals with 'facts' and, to a large extent, that is a fair comment, but it also gives an incomplete impression of the nature of science. The statement is misleading if it suggests that scientists typically believe that they have uncovered certain universal 'truths' for all time. Scientific knowledge is provisional and fully open to change if and when we make new discoveries. More than that, it is the essential nature of science and good scientists to encourage the re-examination of existing 'knowledge' and to look for improvements and progress.

Different kinds of uncertainty

The uncertainty in experimental measurements discussed in this chapter is a consequence of the limitations of scientists and their equipment to obtain 100% accurate results. However, we should also consider that the act of measurement, in itself, can change what we are attempting to measure. For example, connecting an ammeter in an electric circuit must affect the current it is trying to measure, although every effort should be made to ensure this effect is not significant. Similarly, putting a cold thermometer in a warm liquid will alter its temperature.

'Uncertainty' also appears as an important concept in modern physics: the Heisenberg uncertainty principle deals with the behaviour of sub-atomic particles and is discussed in Chapter 12 (Higher Level students). One of its core ideas is that the more precisely the position of a particle is known, the less precisely its momentum can be known, and vice versa. But it should be stressed that the Heisenberg uncertainty principle is a fundamental feature of quantum physics and has nothing to do with the experimental limits of current laboratory technology.

Random and systematic errors

Random errors

Random errors cannot be avoided because exact measurements are not possible. Measurements can be bigger or smaller than the correct value and are scattered randomly around that value.

Random errors are generally unknown and unpredictable. There are many possible reasons for them, including:

- limitations of the scale or display being used
- reading scales from wrong positions
- irregular human reaction times when using a stopwatch
- difficulty in making observations that change quickly with time.

The reading obtained from a measuring instrument is limited by the smallest *division* of its scale. This is sometimes called a **readability** (or **reading**) **error**. For example, a liquid-in-glass thermometer with a scale marked only in degrees (23°C, 24°C, 25°C, etc.) cannot reliably be used to measure to every 0.1°C. It is usually assumed that the error for **analogue** (continuous) scales, like a liquid-in-glass thermometer, is half of the smallest division – in this example ±0.5°C. For **digital** instruments the error is assumed to be the smallest division that the meter can display. Figure 1.2 shows analogue and digital ammeters that can be used for measuring electric current.

A common reason for random errors is reading an analogue scale from an incorrect position. This is called a **parallax error** – an example is shown in Figure 1.3.

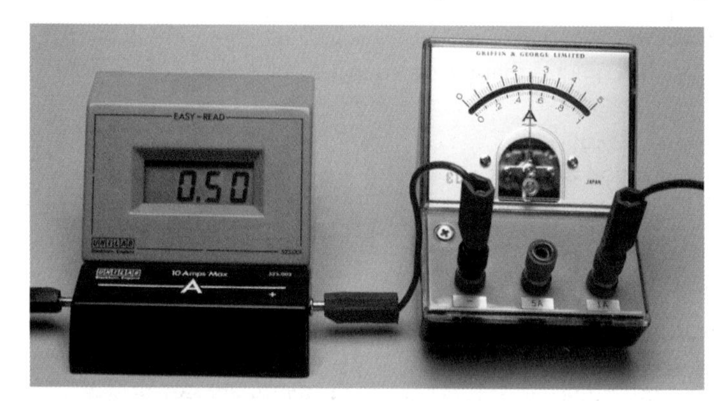

Figure 1.2 Digital and analogue ammeters measuring the same electric current

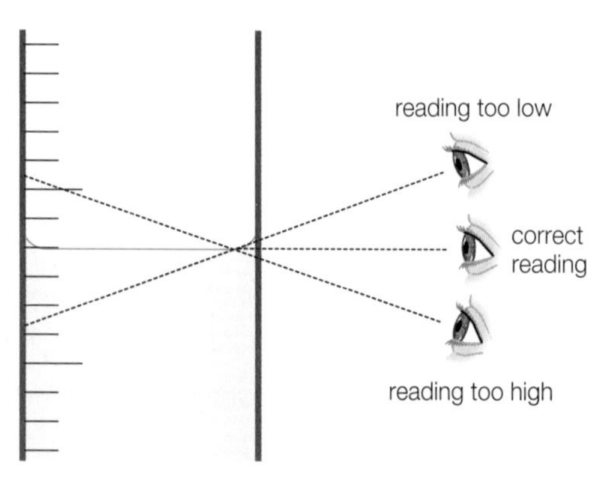

Figure 1.3 Parallax error when reading the level of liquid in a measuring cylinder

Systematic errors

A **systematic error** occurs because there is something *consistently* wrong with the measuring instrument or the method used. A reading with a systematic error is always either bigger or smaller than the correct value by the same amount. Common causes are instruments that have an incorrect scale (wrongly **calibrated**), or instruments that have an incorrect value to begin with, such as a meter that displays a reading when it should read zero. This is called a **zero offset error** – an example is shown in Figure 1.4. A thermometer that incorrectly records room temperature will produce systematic errors when used to measure other temperatures.

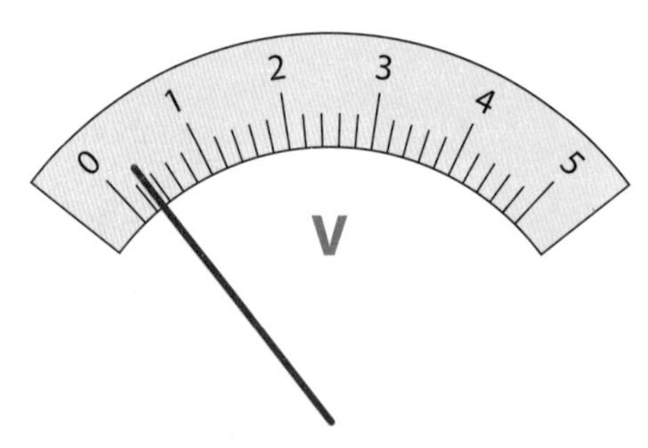

Figure 1.4 This voltmeter has a zero offset error of 0.3 V, so that all readings will be too large by this amount

Accuracy

A measurement that is close to the correct value (if it is known) is described as **accurate**, but in science the word *accurate* also means that a *set* of measurements made during an experiment have a small, systematic error. This means that an accurate set of measurements are approximately evenly distributed around the correct values (whether they are close to it or not), so that an average of those measurements will be close to the true value.

In many experiments the 'correct' result might not be known, which means that the accuracy of measurements cannot be known with any certainty. In such cases, the quality of the measurements can best be judged by their **precision**: can the same results be repeated?

The difference between precise and accurate can be illustrated by considering arrows fired at a target, as in Figure 1.5. The aim is precise if the arrows are grouped close together and accurate if the arrows are approximately evenly distributed around the centre of the target. The last diagram shows both accuracy *and* precision, although in everyday conversation we would probably just describe it as accurate.

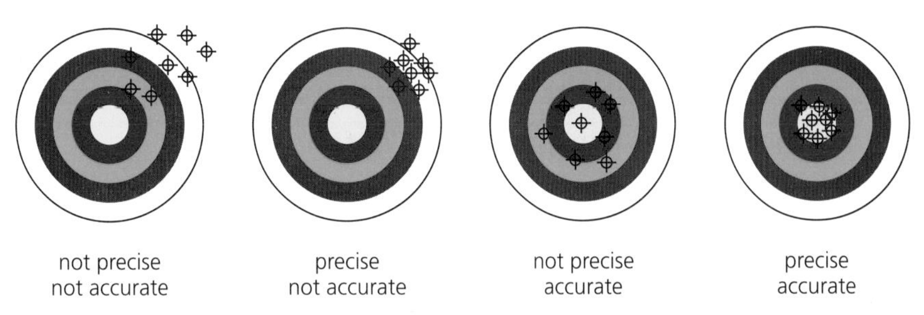

■ **Figure 1.5** Difference between precision and accuracy

A watch that is always 5 minutes fast can be described as precise but not accurate. This is an example of a systematic zero offset error. Using hand-operated stopwatches to time a 100 m race might give accurate results (if there are no systematic errors), but they are unlikely to be precise because human reaction times will produce significant random errors.

Identifying and reducing the effects of errors

If a single measurement is made of a particular quantity, we may have no way of knowing how close it is to the correct result; that is, we probably do not know the size of any error in measurement. But if the same measurement is repeated and the results are similar (low uncertainty, high precision), we will gain some confidence in the results of the experiment, especially if we have checked for any possible causes of systematic error.

The most common way of reducing the effects of *random errors* is by repeating measurements and calculating a mean value, which should be closer to the correct value than most, or all, of the individual measurements. Any unusual (**anomalous**) values should be checked and probably excluded from the calculation of the mean.

Many experiments involve taking a range of measurements, each under different experimental conditions, so that a graph can be drawn to show the pattern of the results. (For example, changing the voltage in an electric circuit to see how it affects the current.) Increasing the number of pairs of measurements made also reduces the effects of random errors because the line of best-fit can be placed with more confidence.

Experiments should be designed, wherever possible, to produce large readings. For example, a metre ruler might only be readable to the nearest half a millimetre and this will be the same for all measurements that are made with it. When measuring a length of 90 cm this error will probably be considered as acceptable (it is a percentage error of 0.56%), but the same sized error when measuring only 2 mm is 25%, which is probably unacceptable. The larger a measurement (that is made with a particular measuring instrument), the smaller the percentage error should be. If this is not possible, then the measuring instrument might need to be changed to one with smaller divisions.

It is possible to carry out an experiment carefully with good quality instruments, but still have large random errors. There could be many different reasons for this and the experiment may have to be redesigned to get over the problems. Using a stopwatch to time the fall of an object dropped from a hand to the floor, or measuring the height of a bouncing ball, are two examples of simple experiments which may have significant random errors.

The effects of *systematic errors* cannot be reduced by repeating measurements. Instruments should be checked for errors before they are used, but a systematic error might not even be noticed until a graph has been drawn of the results and a line of best-fit found not to pass through the expected intercept, as shown in Figure 1.6. In such a case it might then be sensible to adjust all measurements up or down by the same amount if the cause of the systematic error can be determined.

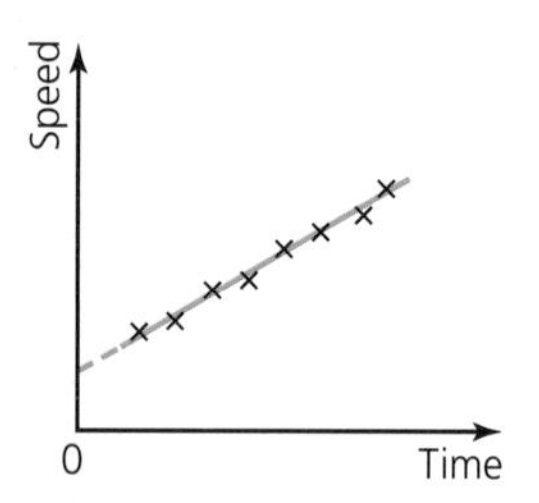

Figure 1.6 The best-fit line for this speed–time graph for a trolley rolling from rest down a slope does not pass through the origin, so there was probably a systematic error

Absolute, fractional and percentage uncertainties

Uncertainties in experimental data

Uncertainties in experimental data can be expressed in one of three ways:

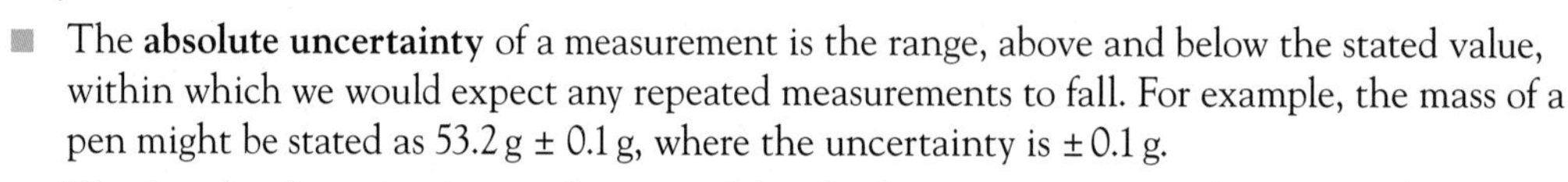

- The **absolute uncertainty** of a measurement is the range, above and below the stated value, within which we would expect any repeated measurements to fall. For example, the mass of a pen might be stated as 53.2 g ± 0.1 g, where the uncertainty is ±0.1 g.
- The **fractional uncertainty** is the ratio of the absolute uncertainty to the measured value.
- The **percentage uncertainty** is the fractional uncertainty expressed as a percentage.

Uncertainties expressed in percentages are often the most informative. Experiments that produce results with uncertainties of less than 5% may be desirable, but are not always possible.

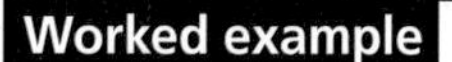

Worked example

3 The mass of a piece of metal is quoted to be 346 g ± 2.0%.
a What is the absolute uncertainty?
b What is the range of values that the mass could be expected to have?
c What is the fractional uncertainty?

a 2.0% of 346 g is ± 7 g (to the nearest gram, as provided in the data in the question)

b 339 g to 353 g (to 3 significant figures)

c 2% is equivalent to $\frac{1}{50}$

Ideally uncertainties should be quoted for all experimental measurements, but this can be repetitive and tedious in a learning environment, so they are often omitted unless being taught specifically.

It is usually easy to decide on the size of an uncertainty associated with taking a single measurement with a particular instrument. It is often assumed to be the readability error, as described earlier. However, the overall uncertainty in a measurement, allowing for all experimental difficulties, is sometimes more difficult to decide. For example, the readability error on a hand-operated stopwatch might be 0.01 s, but the uncertainty in its measurements will be much greater because of human reaction times.

The amount of scattering of the readings around a mean value is a useful guide to random uncertainty, but not systematic uncertainty. After the mean value of the readings has been calculated, the random uncertainty can be assumed to be the largest difference between any single reading and the mean value. This is shown in the following worked example.

Worked example

4 The following measurements (in cm) were recorded in an experiment to measure the height to which a ball bounced: 32, 29, 33, 32, 37 and 28. Estimate values for the absolute and percentage random uncertainties in the experiment.

The mean of these six readings is 31.83 cm, but it would be sensible to quote this to two significant figures (32 cm), as in the original data. The measurement that has the greatest difference from this value is 37 cm, so an estimate of the uncertainty is 5 cm, which means a percentage uncertainty of (5/37) × 100 = 14%.

Note that if the same data had been obtained in the order 28, 29, 32, 32, 33, 37, it would be difficult to believe that the uncertainties were random, and another explanation for the variation in results would need to be found.

Uncertainties in calculated results

When making further calculations based on experimental data, the uncertainty in individual measurements should be known. It is then important to know how to use these uncertainties to determine the uncertainty in any results that are calculated from those data.

Consider a simple example: a trolley moving with constant speed was measured to travel a distance of 76 cm ± 2 cm (± 2.6%) in a time of 4.3 s ± 0.2 s (± 4.7%).

The speed can be calculated from distance/time = 76/4.3 = 17.67…, which is 18 m s^{-1} when rounded to two significant figures, consistent with the experimental data.

To determine the uncertainty in this answer we consider the uncertainties in distance and time. Using the largest distance and shortest time, the largest possible answer for speed is 78/4.1= 19.02…. Using the smallest distance and the longest time, the smallest possible answer for speed is 74/4.5 = 16.44…. (The numbers will be rounded at the end of the calculations.)

The speed is therefore between 16.44 cm s^{-1} and 19.02 cm s^{-1}. The value 19.02 has the greater difference (1.35) from 17.67. So the final result can be expressed as 17.67 ± 1.35 cm s^{-1}, which is a maximum uncertainty of 7.6%. Rounding to two significant figures, the result becomes 18 ± 1 cm s^{-1}.

Uncertainty calculations like these can be very time consuming and, for this course, *approximate methods are acceptable*. For example, in the calculation for speed shown above, the uncertainty in the data was ± 2.6% for distance and ± 4.7% for time. The percentage uncertainty in the final result is approximated by adding the percentage uncertainties in the data: 2.6 + 4.7 = 7.3%. This gives approximately the same value as calculated using the largest and smallest possible values for speed. Rules for finding uncertainties in calculated results are given below.

Rules for uncertainties in calculations

- For quantities that are added or subtracted: *add* the *absolute* uncertainties. In the *Physics data booklet* this is given as:

$$\text{If } y = a \pm b \text{ then } \Delta y = \Delta a + \Delta b$$

- For quantities that are multiplied or divided: *add* the individual *fractional* or *percentage* uncertainties. In the *Physics data booklet* this is given as:

$$\text{If } y = \frac{ab}{c} \text{ then } \frac{\Delta y}{y} = \frac{\Delta a}{a} + \frac{\Delta b}{b} + \frac{\Delta c}{c}$$

- For quantities that are raised to a power, n, the *Physics data booklet* gives:

$$\text{If } y = a^n \text{ then } \frac{\Delta y}{y} = \left| n\left(\frac{\Delta a}{a}\right) \right|$$

- For other functions (such as trigonometric functions, logarithms or square roots): calculate the highest and lowest *absolute* values possible and compare with the mean value, as shown in the following worked example. But note that although such calculations can occur in connection with laboratory work, they will *not* be required in examinations.

Worked example

5 An angle, θ, was measured to be 34° ± 1°. What is the uncertainty in the slope of this angle?

tan 34° = 0.675 tan 33° = 0.649 tan 35° = 0.700

Larger absolute uncertainty = 0.675 − 0.649 = 0.026
(0.700 − 0.675 = 0.025, which is smaller than 0.026)

So, tan θ = 0.67 ± 0.03 (using the same number of significant figures as in the original data).

5 A mass of 346 ± 2 g was added to a mass of 129 ± 1 g.
 a What was the overall absolute uncertainty?
 b What was the overall percentage uncertainty?

6 The equation $s = \frac{1}{2}at^2$ was used to calculate a value for s when a was 4.3 ± 0.2 m s^{-2} and t was 1.4 ± 0.1 s.
 a Calculate a value for s.
 b Calculate the percentage uncertainty in the data provided.
 c Calculate the percentage uncertainty in the answer.
 d Calculate the absolute uncertainty in the answer.

7 A certain quantity was measured to have a magnitude of (1.46 ± 0.08). What is the maximum uncertainty in the square root of this quantity?

Using computer spreadsheets to calculate uncertainties

Computer spreadsheets can be very helpful when it is necessary to make multiple calculations of uncertainties in experimental results. For example, the resistivity, ρ, of a metal wire can be calculated using the equation $\rho = R\pi r^2/l$, where r and l are the radius and length of the wire, and R is its resistance. Figure 1.7 shows the raw data (shaded green) of an experiment that measured the resistance of various wires of the same metal. The rest of the spreadsheet shows the calculations involved with processing the data to determine resistivity and the uncertainty in the result. A computer program can then be used to draw a suitable graph of the results, and this can include error bars (see page 13).

RESISTANCE		RADIUS			LENGTH		RESISTIVITY		
Resistance, R / Ω ±0.2 Ω	Percentage uncertainty in R	Radius, r /mm ±0.01 mm	Percentage uncertainty in r	Percentage uncertainty in r^2	Length, l /cm ± 1 cm	Percentage uncertainty in l	Resistivity, $\rho = R\pi r^2/l$ / Ωm	Percentage uncertainty in ρ	Absolute uncertainty in ρ / Ωm
9.4	2.1	0.15	6.7	13.3	44	2.3	0.0000015	18	0.0000003
6.2	3.2	0.22	4.5	9.1	67	1.5	0.0000014	14	0.0000002
6.2	3.2	0.25	4.0	8.0	80	1.3	0.0000015	12	0.0000002
5.2	3.8	0.30	3.3	6.7	99	1.0	0.0000015	12	0.0000002
5.0	4.0	0.35	2.9	5.7	128	0.8	0.0000015	10	0.0000002
3.8	5.3	0.43	2.3	4.7	149	0.7	0.0000015	11	0.0000002
3.4	5.9	0.51	2.0	3.9	175	0.6	0.0000016	10	0.0000002
2.4	8.3	0.62	1.6	3.2	198	0.5	0.0000015	12	0.0000002

Figure 1.7 Using a spreadsheet to calculate uncertainties in a resistance experiment

8 a Use a computer spreadsheet to enter the same raw data as shown in Figure 1.7.
 b Use the spreadsheet to confirm the results of the calculations shown.
 c What difference would it make to the results if the radius of the wire could only be measured to the nearest half a millimetre?

Nature of Science

Uncertainties

'All scientific knowledge is uncertain…'

Richard P. Feynman (1998), *The Meaning of It All: Thoughts of a Citizen-Scientist*

It is not only measurements that have uncertainties. All scientific knowledge is uncertain in the sense that good scientists understand that anything we believe to be true today, may have to be changed in the light of future discoveries or insights. This doubt is fundamental to the true nature of science. At any time, past or present, in the development of science there is an accepted body of knowledge, and the greatest advances come from those who question and doubt the status quo of existing knowledge and thinking.

Representing uncertainties on graphs

Graph drawing skills are discussed in detail in *Graphs and data analysis* on the free accompanying website.

The range of random uncertainty in a measurement or a calculated result can be represented on a graph by using crossed lines to mark the point (instead of a dot).

Error bars

Figure 1.8 shows an example – a graph of distance against time for the motion of a train. Vertical and horizontal lines are drawn through each data point to represent the uncertainties in the two measurements. In this example, the uncertainty in time is ±0.5 s and the uncertainty in distance is ±1 m. These lines, which usually have small lines to indicate clearly where they end, are called error bars (perhaps they would be better called *uncertainty* bars). In Figure 1.8 the space outlined by each error bar has been shaded for emphasis – it is expected that a line of best-fit should pass somewhere through each shaded area.

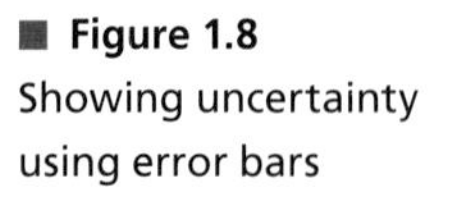

Figure 1.8 Showing uncertainty using error bars

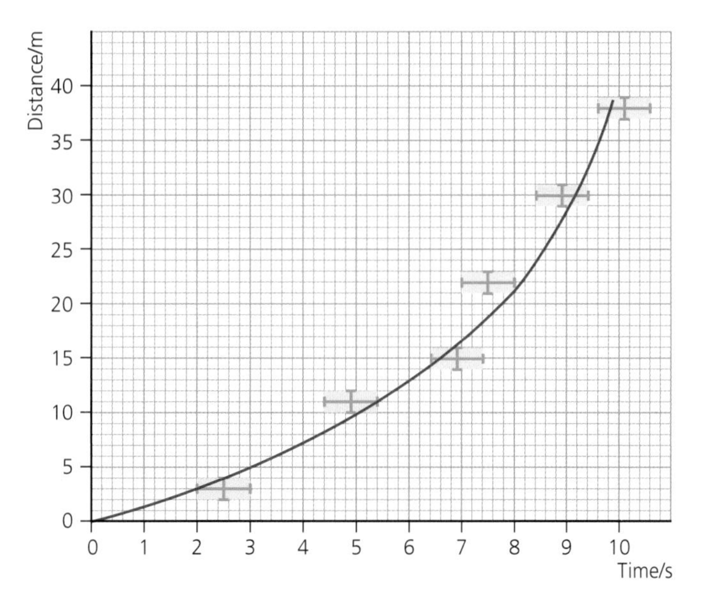

In some experiments the error bars are so short and insignificant that they are not included on the graph. For example, a mass could be measured as 347.46 ± 0.01 g. The uncertainty in this reading would be too small to show as an error bar on a graph. (Note that error bars are not expected for trigonometric or logarithmic functions.)

Uncertainty of gradients and intercepts

If the results of an experiment suggest a straight-line graph, it is often important to determine values for the gradient and/or the intercept(s) with the axes. However, it is often possible to draw a range of different straight lines, all of which pass through the error bars representing the experimental data.

We usually assume that the best-fit line is midway between the lines of maximum possible gradient and minimum possible gradient. Figure 1.9 shows an example (for simplicity, only the first and last error bars are shown, but in practice all the error bars need to be considered when drawing the lines).

Figure 1.9 shows how the length of a metal spring changed as the force applied was increased. We know that the measurements were not very precise because the error bars are long. The line of best-fit has been drawn midway between the other two. This is a **linear graph** (a straight line) and it is known that the gradient of the graph represents the force constant (stiffness) of the spring and the *x*-intercept represents the original length of the spring. Taking measurements from the best-fit line, we can make the following calculations:

$$\text{force constant} = \text{gradient} = \frac{90 - 0}{6.6 - 1.9} = 19\,\text{N cm}^{-1}$$

$$\text{original length} = x\text{-intercept} = 1.9\,\text{cm}$$

To determine the uncertainty in the calculations of gradient and intercept, we need to consider the range of straight lines that could be drawn through the error bars. The uncertainty will be the maximum difference between values obtained from graphs of maximum and minimum possible gradients and the value calculated from the best-fit line. In this example it can be shown that:

force constant is between $14\,\text{N cm}^{-1}$ and $28\,\text{N cm}^{-1}$

original length is between 1.1 cm and 2.6 cm.

The final result can be quoted as:

$$\text{force constant} = 19 \pm 9\,\text{N cm}^{-1}$$

$$\text{original length} = 1.9 \pm 0.8\,\text{cm}.$$

Clearly, the large uncertainties in these results confirm that the experiment lacked precision.

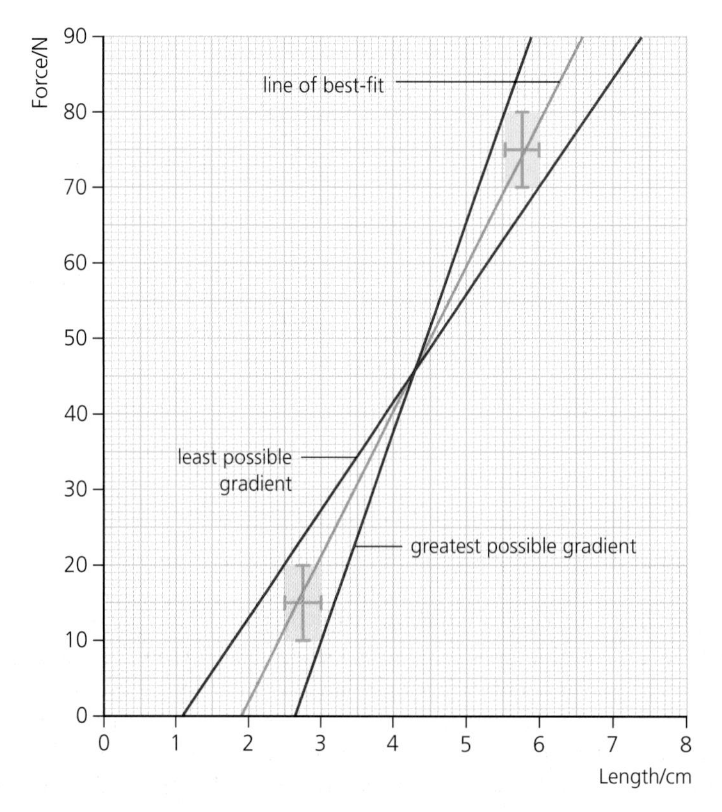

■ **Figure 1.9** Finding maximum and minimum gradients for a spring-stretching experiment

1.3 Vectors and scalars – *Some quantities have direction and magnitude, others have magnitude only, and this understanding is the key to correct manipulation of quantities*

Nature of Science

Models in three dimensions

Spatial awareness and an appreciation that the principles of science apply to three-dimensional space can easily be overlooked when studying the two-dimensional pages of a book or a screen. Knowing the directions of some physical quantities (in two or three dimensions) is important for understanding their effects. Such quantities are called vectors. Mathematical treatment of vector quantities in three dimensions (vector analysis) began in the eighteenth century.

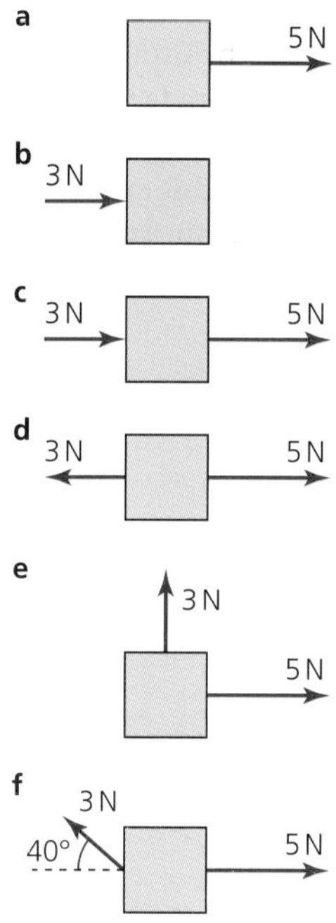

a 5 N
b 3 N
c 3 N 5 N
d 3 N 5 N
e 3 N 5 N
f 3 N 40° 5 N

■ **Figure 1.10** Forces are vector quantities

Vector and scalar quantities

The diagrams in Figure 1.10 show the force(s) acting on an object. In Figure 1.10a the object is being pulled to the right with a force of 5 N. The length of the arrow represents the size of the force and the orientation of the arrow shows the direction in which the force acts. The length of the arrow is proportional to the force. In Figure 1.10b there is a smaller force (3 N) pushing the object to the right. In both examples the object will move (accelerate) to the right.

In Figure 1.10c there are two forces acting. We can add them together to show that the effect is the same as if a single force of 8 N (= 3 + 5) was acting on the object. We say that the *resultant* (net) force is 8 N.

In Figure 1.10d there are two forces acting on the object, but they act in different directions. The overall effect is still found by 'adding' the two forces, but also taking their direction into account. This can be written as +5 + (−3) = +2 N, where forces to the right are given a positive sign and forces to the left are given a negative sign. The resultant will be the same as if there was only one force (2 N) acting to the right. In Figures 1.10e and 1.10f there are also two forces acting, but they are not acting along the same line. For these forces, the resultant can be determined using a scale drawing or trigonometry (see page 16).

Clearly, force is a quantity for which we need to know its *direction* as well as its *magnitude* (size).

Quantities that have both magnitude and direction are called **vectors**.

Everything that we measure has a magnitude and a unit. For example, we might measure the mass of a book to be 640 g. Here 640 g is the magnitude of the measurement, but mass has no direction.

Quantities that have only magnitude, and no direction, are called **scalars**.

Most quantities are scalars. Some common examples of scalars used in physics are mass, length, time, energy, temperature and speed. However, when using the following quantities we need to know both the magnitude and the direction in which they are acting, so they are *vectors:*

- displacement (distance in a given direction)
- velocity (speed in a given direction)
- force (including weight)
- acceleration
- momentum and impulse
- field strength (gravitational, electric and magnetic).

The symbols for vector quantities are sometimes shown in bold italic (for example, ***F***). Scalar quantities are shown with a normal italic font (for example, *m*).

In diagrams, all vectors are shown with straight arrows, pointing in the correct direction, which have a length proportional to the magnitude of the vector (as shown by the forces in Figure 1.11). In this course vector calculations will be limited to two dimensions.

The importance of vectors is easily illustrated by the difference between distance and displacement. The pilot of an international flight from, say, Istanbul to Cairo needs to know more than that the two cities are a distance of 1234 km apart. Of course, the pilot also needs to know the 'heading' (direction) in which the plane must fly in order to reach its destination.

Similarly, in order to draw accurate maps or make land surveys, the distance *and* direction of a chosen position from a reference point must be measured.

Combination and resolution of vectors

Adding vectors to determine a resultant

When two or more scalar quantities are added together (for example masses of 25 g and 50 g), there is only one possible answer (resultant): 75 g. But when vector quantities are added, there is a range of different resultants possible, depending on the directions involved.

To determine the resultant of the two forces shown in Figures 1.10e or 1.10f there are two possible methods: by drawing (graphical method) or by trigonometry.

Graphical method

The two vectors shown in Figure 1.10f are drawn carefully to scale (for example, by using 1 cm to represent 1 N), with the correct angle (140°) between them. A parallelogram is then completed. The resultant is the diagonal of the parallelogram (see Figure 1.11). Remember that the magnitude and the direction should *both* be determined from the diagram. In this example the resultant force is represented by the line drawn in red. Its length is 3.4 cm, which represents 3.4 N, at an angle of 36° to the 5.0 N force.

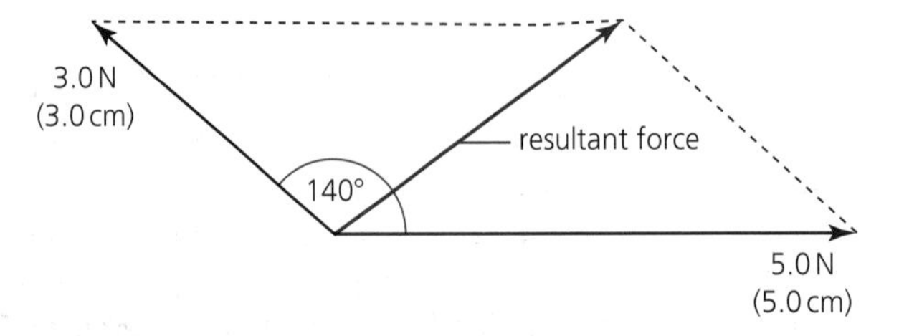

Figure 1.11 Using a parallelogram to determine a resultant force

Trigonometric method

The forces in Figure 1.11e are at right angles to each other. This means that a parallelogram drawn to represent these forces will be a rectangle (Figure 1.12) and the magnitude of the resultant of the forces, *F*, can be found using Pythagoras's theorem:

$$F^2 = 3.0^2 + 5.0^2 = 34$$

$$F = 5.8\,\text{N}$$

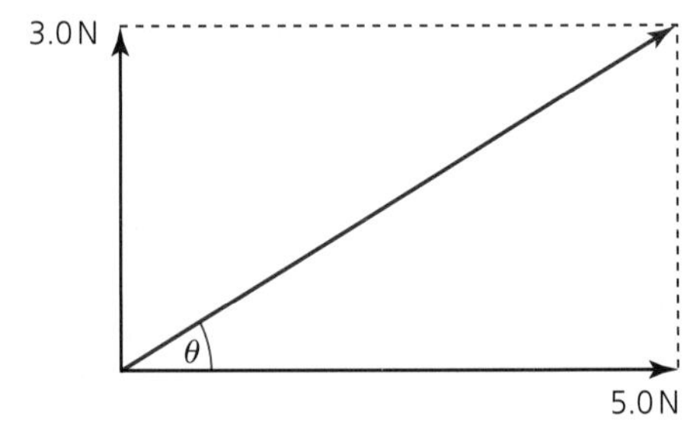

Figure 1.12

The direction of this force can be determined by using trigonometry:

$$\tan\theta = \frac{3.0}{5.0}$$ (θ is the angle that the resultant makes with the direction of the 5.0 N force)

$$\theta = 31°$$

You will *not* be expected to determine trigonometrical solutions if the parallelogram is not a rectangle.

Subtracting vectors to find their difference

We may need to know the *difference* between two vectors when we are considering by how much a vector quantity has changed. This is determined by *subtracting* one vector from the other. A negative vector has the same magnitude, but opposite direction, as a positive vector, so when finding the difference between vectors **P** and **Q** we can write:

$$\mathbf{P} - \mathbf{Q} = \mathbf{P} + (-\mathbf{Q})$$

Figure 1.13 shows how vectors are subtracted graphically. The red line represents the difference when a particular vector changed in magnitude and direction from **P** to **Q**.

a

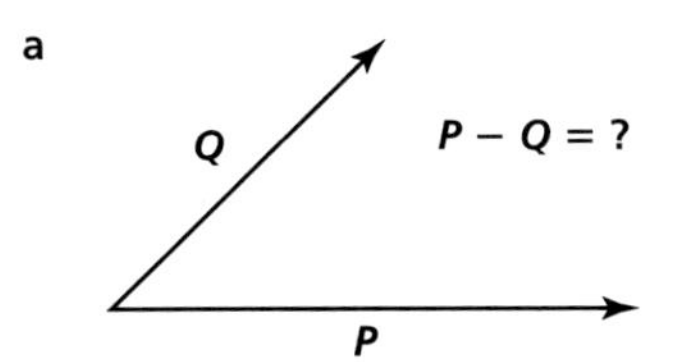

b

−*Q*
P − *Q*
P

■ **Figure 1.13** If we want to know the difference between ***P*** and ***Q*** (diagram **a**) we add ***P*** to −***Q*** (diagram **b**)

Multiplying and dividing vectors by scalars

If a vector **P** is multiplied or divided by a scalar number k, the resultant vectors are simply $k\mathbf{P}$ or $\mathbf{P}/k$. If k is negative, then the resultant vector becomes negative, meaning that the direction is reversed.

Resolving a single vector into two components

We have seen that two individual vectors can be combined mathematically to find a single resultant that has the same effect as the two separate vectors. This process can be reversed: a single vector can be considered as having the same effect as two separate vectors. This is called resolving a vector into two components. Resolving can be very useful because, if the two components are chosen to be perpendicular to each other (often horizontal and vertical), they will then both be independent of each other, so that they can both be considered totally separately.

Figure 1.14 shows a single vector, **A**, acting at an angle θ to the horizontal. If we want to know the effects of this vector in the horizontal and vertical directions, we can resolve it into two components:

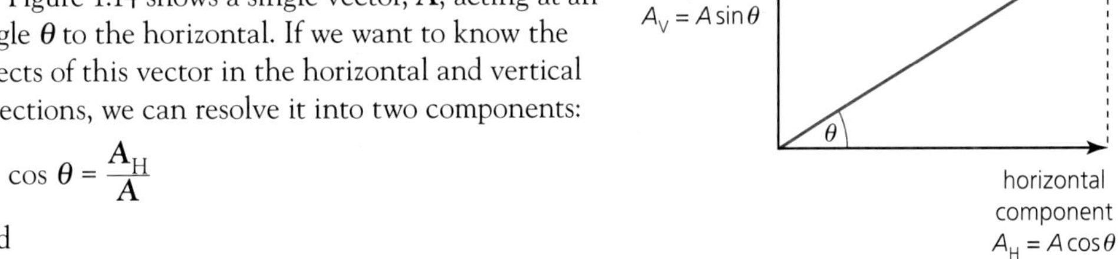

■ **Figure 1.14** Resolving a vector into two perpendicular components

$$\cos\theta = \frac{\mathbf{A}_H}{\mathbf{A}}$$

and

$$\sin\theta = \frac{\mathbf{A}_V}{\mathbf{A}}$$

so that

$$\mathbf{A}_H = \mathbf{A}\cos\theta$$

and

$$\mathbf{A}_V = \mathbf{A}\sin\theta$$

Both of these equations and the associated diagram are given in the *Physics data booklet*.

Worked example

6 Figure 1.15 shows a box resting on a sloping surface (an 'inclined plane'). The box has a weight of 585 N. What are the components of weight:

a down the slope?

b perpendicularly into the slope?

a component down the slope = $585 \sin 23°$
= 230 N

b component into the slope = $585 \cos 23°$
= 540 N

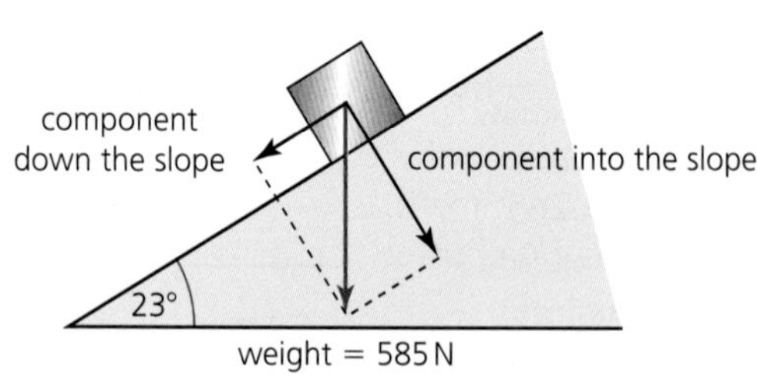

■ **Figure 1.15**

ToK Link

Physics and mathematics

What is the nature of certainty and proof in mathematics*?*

Science is mostly based on knowledge gained from experimentation and measurement, although it has been made very clear in this chapter that absolute accuracy and certainty in the gathering of data is not possible. In contrast, the essential theories and methods of pure mathematics seem to deal with certainty. Mathematics is an indispensible tool for a physicist for many reasons, including its conciseness, its lack of ambiguity and its usefulness in making predictions. Most important principles in physics can be summarized in mathematical form.

Examination questions – a selection

Paper 1 IB questions and IB style questions

Q1 The diameter of a wire was measured three times with an instrument that has a zero offset error. The results were 1.24 mm, 1.26 mm and 1.25 mm. The average of these results is:
A accurate but not precise
B precise but not accurate
C accurate and precise
D not accurate and not precise.

Q2 The approximate thickness of a page in a textbook is:
A 0.02 mm
B 0.08 mm
C 0.30 mm
D 1.00 mm.

Q3 Which of the following an approximate conversion of a time of 1 month into SI units?
A 0.08 y
B 30 d
C 3×10^6 s
D all of the above

Q4 The masses and weights of different objects are independently measured. The graph is a plot of weight versus mass that includes error bars.

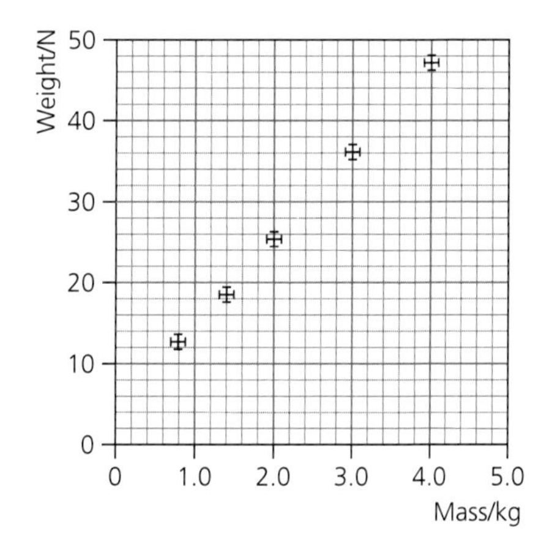

These experimental results suggest that:
A the measurements show a significant systematic error but small random error
B the measurements show a significant random error but small systematic error
C the measurements are precise but not accurate
D the weight of an object is proportional to its mass.

Q5 Which of the following is a fundamental SI unit?
A newton
B coulomb
C ampere
D joule

Q6 The distance travelled by a car in a certain time was measured with an uncertainty of 6%. If the uncertainty in the time was 2%, what would the uncertainty be in a calculation of the car's speed?
A 3%
B 4%
C 8%
D 12%

Q7 Which of the following quantities is a scalar?
A pressure
B acceleration
C gravitational field strength
D displacement

Q8 The current in a resistor is measured as 2.00 A ± 0.02 A. Which of the following correctly identifies the absolute uncertainty and the percentage uncertainty in the current?

	Absolute uncertainty	Percentage uncertainty
A	± 0.02 A	± 1%
B	± 0.01 A	± 0.5%
C	± 0.02 A	± 0.01%
D	± 0.01 A	± 0.005%

Q9 Which of the following is a reasonable estimate of the order of magnitude of the mass of a large aircraft?
A 10^3 kg
B 10^5 kg
C 10^7 kg
D 10^9 kg

Q10 Which of the following is equivalent to the SI unit of force (the newton)?
A kg m s^{-1}
B kg m^2 s^{-1}
C kg m s^{-2}
D kg m^2 s^2

2 Mechanics

ESSENTIAL IDEAS

- Motion may be described and analysed by the use of graphs and equations.
- Classical physics requires a force to change a state of motion, as suggested by Newton in his laws of motion.
- The fundamental concept of energy lays the basis on which much of science is built.
- Conservation of momentum is an example of a law that is never violated.

2.1 Motion – *Motion may be described and analysed by the use of graphs and equations*

Kinematics is the study of moving objects. The ideas of classical physics presented in this chapter can be applied to the movement of all masses, from the very small (freely moving atomic particles) to the very large (stars).

To completely describe the motion of an object at any one moment we need to say where it is, how fast it is moving and in what direction. For example, we might observe that a car is 20 m to the west of an observer, and moving northeast at a speed of 8 m s^{-1} (see Figure 2.1).

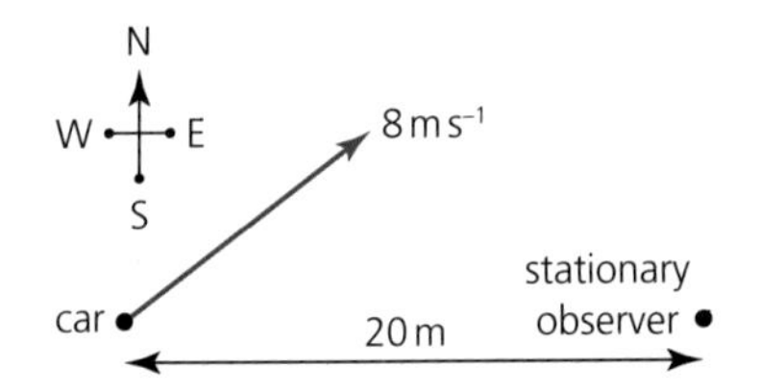

Figure 2.1 Describing the position and motion of a car

Of course, any or all of these quantities might be changing. In real life the movement of many moving objects can be complicated; they do not often move in straight lines and they might even rotate or have different parts moving in different directions.

In this chapter we will develop an understanding of the basic principles of kinematics by dealing with single objects moving in straight lines, and calculations will be confined to those objects that have a regular motion. We will consider the effects of air resistance later in this chapter.

Nature of Science

Everything is moving

The study of motion must be a cornerstone of science because *everything* moves. Stars and galaxies are moving apart from each other at enormous speeds, the Earth orbits the Sun and everything on Earth is rotating around the axis once every day. Atoms and molecules are in constant motion, as are the sub-atomic particles within them. Of course in everyday life many objects appear to be stationary, but only because we are only comparing them with their surroundings. If we were to imagine that an object was truly, absolutely, *not* in motion, we would have no way to prove it because all motion is relative to something else.

Distance and displacement

Displacement is defined as the distance in a given direction from a fixed reference point.

The displacement of an object is its position compared with a known reference point. For example, the displacement of the car in Figure 2.1 is 20 m to the west of the observer. To specify a displacement we need to state a distance *and* a direction from the reference point. The reference point is often omitted because it is obvious – for example, we might just say that an airport is 50 km to the north. Although a displacement can be anywhere in three dimensions, in this topic we will usually restrict our thinking to one or two dimensions.

Displacement and distance are both given the symbol s. This should not be confused with the symbol for speed (and velocity), which is v. The symbol h is also widely used for vertical distances (heights). The SI unit for distance is the metre, m, although other units, such as mm, cm and km, are in common use.

Because a direction is specified as well as a magnitude (size), displacement is a vector quantity. Distance is a scalar quantity because it has magnitude, but no direction.

Figure 2.2 shows the route of some people walking around a park. The total *distance* walked was 4 km, but the displacement from the reference point varied and is shown every few minutes by the vector arrows (**a**–**e**). The final *displacement* is zero because the walkers returned to their starting place.

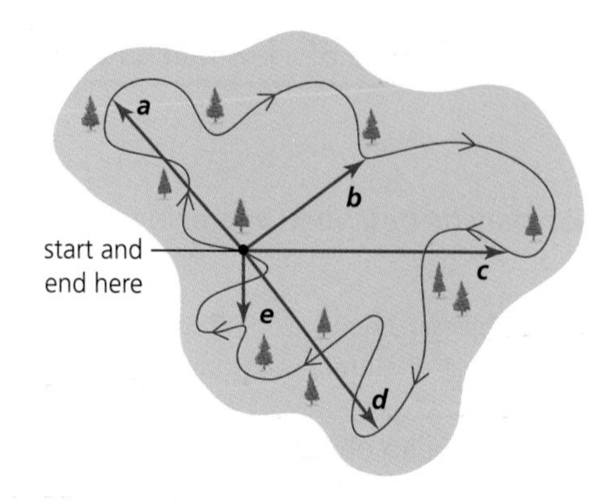

Figure 2.2 A walk in the park

The transport of various vehicles, goods and people around the world is big business, and is monitored and controlled by many countries and international companies. This requires accurate means of tracking the location and movement of a large number of vehicles (ships, aircraft etc.) and the rapid communication of this information between countries.

Figure 2.3 Tracking transportation

Speed and velocity

Speed is defined as the rate of change of distance with time.

Speed is a scalar quantity and it is given the symbol v. Its SI unit is metres per second, $\mathrm{m\,s^{-1}}$. Speed is calculated from:

$$\text{speed} = \frac{\text{distance travelled}}{\text{time taken}}$$

The delta symbol (Δ) is used wherever we want to represent a (small) change of something, so we can define speed in symbols, as follows:

$$v = \frac{\Delta s}{\Delta t}$$

If an object is moving with a constant speed, determining its value is a straightforward calculation. However, the speed of an object often changes during the time we are observing it, and the calculated value is then an average speed during that time. For example, if a car is driven a distance of 120 km in 1.5 h, its average speed is $80\,\text{km}\,\text{h}^{-1}$, but its actual speed will certainly have varied during the journey. At any one time we could look at the car's speedometer to find out the instantaneous speed – that is, the speed at that exact *instant* (moment). In kinematics we are usually more interested in instantaneous values of speed (and velocity and acceleration) than average values.

Average speeds are calculated over lengths of time that are long enough for the actual speeds to have changed. Instantaneous values have to be calculated from measurements made over very short time intervals (during which time we can assume that the speed was constant).

Speed is calculated using the distance travelled in the time being considered, regardless of the direction of motion. If the walkers in Figure 2.2 took 2 hours to walk around the park, their average speed would be $\Delta s/\Delta t$ (= 4/2) = $2\,\text{km}\,\text{h}^{-1}$.

Utilizations

Travel timetables

Figure 2.4 shows a timetable for the Ghan, a train that travels across Australia between Adelaide and Darwin, a distance of 2979 km along the track.

Adelaide – Alice Springs – Katherine – Darwin	**Operates all year**	**Additional services operate Jun to Aug**
	Sun	*Wed*
Depart Adelaide	12.20pm	12.20pm
	Mon	*Thurs*
Arrive Alice Springs	1.45pm	1.45pm
Depart Alice Springs	6.00pm	6.00pm
	Tues	*Fri*
Arrive Katherine	9.00am	9.00am
Depart Katherine	1.00pm	1.00pm
Arrive Darwin	5.30pm	5.30pm

■ **Figure 2.4** Ghan train timetable

1 a Calculate the journey time and hence the average speed.
  b Why is your answer to **a** misleading?

We are often concerned not only with how fast an object is moving, but also with the direction of movement. If speed *and* direction are stated then the quantity is called *velocity*.

Velocity is defined as the rate of change of displacement with time (speed in a given direction):

$$v = \frac{\Delta s}{\Delta t}$$

Note that Δs in this equation refers to displacement and not to the overall distance. (To avoid confusion, it is often better to define speed and velocity in words, not symbols.)

Velocity has the same symbol and unit as speed, but the direction should usually be stated as well, since velocity is a vector quantity. However, if the direction of motion does not change, it is not uncommon to refer to a speed, of say $4\,\text{m}\,\text{s}^{-1}$, as velocity because the direction is understood from the context.

Returning to the walkers in the park – at the end of their walk their average speed was $2\,\text{km}\,\text{h}^{-1}$, but their average velocity was zero because the final displacement was zero. This might not be a very useful piece of information; we are more likely to be interested in the instantaneous velocity at various times during the walk.

When the velocity (or speed) of an object changes during a certain time, the symbol u is used for the *initial velocity* and v is used for the *final velocity* during that time. These velocities are not necessarily at the beginning and end of the entire motion, just the velocities at the start and end of the period of time that is being considered.

The distance travelled in time t can be determined using the equation:

distance = average speed × time

For an object with *constant* acceleration:

average speed = $\frac{1}{2}$(initial speed + final speed)

For example, if a car accelerates uniformly from $12\,m\,s^{-1}$ to $16\,m\,s^{-1}$, then its average speed during that time was $14\,m\,s^{-1}$.

In symbols, this is shown as:

$$s = \left(\frac{v+u}{2}\right)t$$

This equation is given in the *Physics data booklet*.

Data logging in motion investigations

The use of motion sensors and data loggers (see Figure 2.5), light gates and electronic timers, and video recording have all made the investigation of various kinds of motion more interesting, much easier and more accurate.

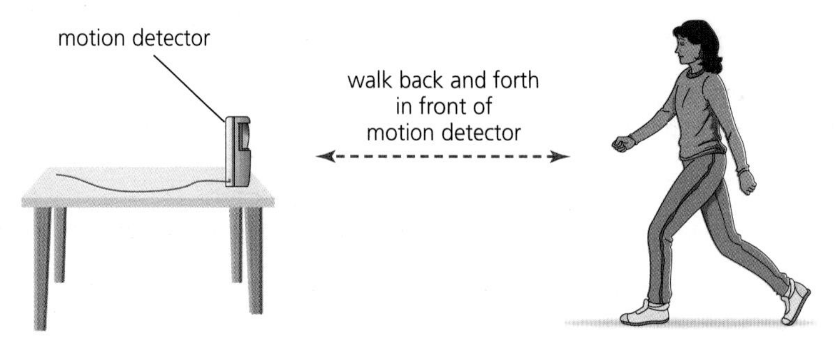

■ **Figure 2.5** Using a motion detector

■ Acceleration

Any variation from moving at a constant speed in a straight line is described as an *acceleration*. It is very important to realise that going faster, going slower and/or changing direction are all different kinds of acceleration (changing velocities).

Acceleration, a, is defined as the rate of change of velocity with time:

$$a = \frac{\Delta v}{\Delta t} = \frac{v-u}{t} \quad \text{(if the acceleration is constant over time } \Delta t\text{)}$$

The SI unit of acceleration is metres per second squared, $m\,s^{-2}$ (the same as the units of velocity/time, $m\,s^{-1}/s$). Acceleration is a vector quantity.

Acceleration can be:

- an increase in velocity (positive acceleration)
- a decrease in velocity (negative acceleration – sometimes called a *deceleration*)
- a change of direction.

Additional perspectives

Reaction times when timing motions

The delay between seeing something happen and responding with some kind of action is known as *reaction time*. A typical value is about 0.20 s, but it can vary considerably depending on the conditions involved. A simple way of measuring a person's reaction time is by measuring how far a metre rule falls before it can be caught between thumb and finger. The time can then be calculated using the equation $s = 5t^2$.

The measurement can be repeated with the person tested being blindfolded to see if the reaction time changes if the stimulus (to catch the ruler) is either sound or touch, rather than sight. Whatever tests are carried out, our reaction times are likely to be inconsistent. This means that whenever we use stopwatches operated by hand, the results will have an unavoidable uncertainty (see Chapter 1). It is sensible to make time measurements as long as possible to decrease the significance of this problem. (This reduces the percentage uncertainty.) Repeating measurements and calculating an average will also reduce the effect of random errors.

1 Use the method described above (or any other) to measure your reaction time when the stimulus is sight. Repeat the measurement 10 times.

  a What was the percentage variation between your average result and your quickest reaction time?

  b Did your reaction times improve with practice?

Graphs describing motion

Graphs can be drawn to represent any motion and they provide extra understanding and insight (at a glance) that very few people can get from written descriptions or equations. Furthermore, the gradients of graphs and the areas under graphs often provide additional valuable information.

Displacement–time graphs and distance–time graphs

Displacement–time graphs, similar to those shown in Figure 2.6, show how the displacements of objects from a reference point vary with time. All the examples shown in Figure 2.6 are straight lines and can be described as representing **linear relationships**.

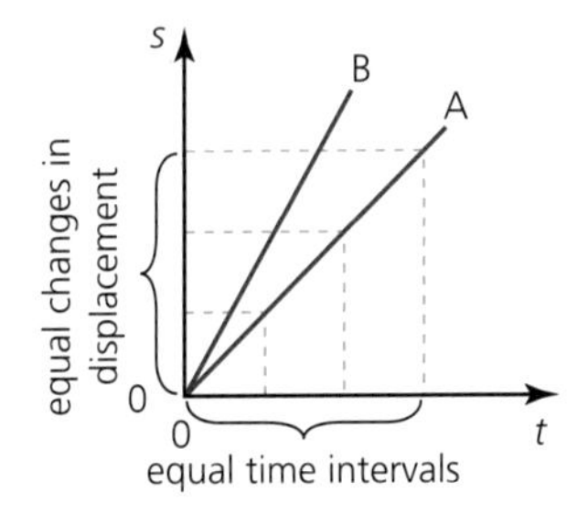

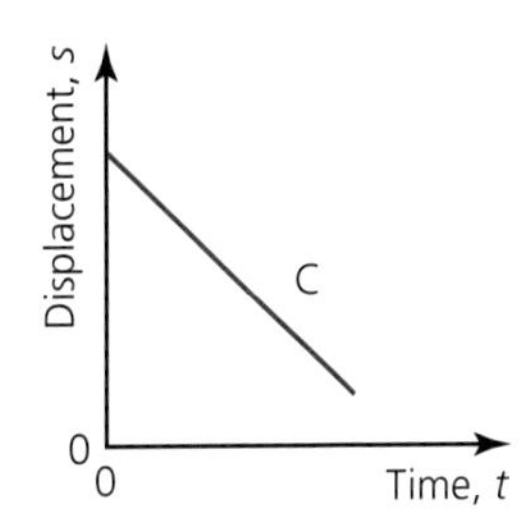

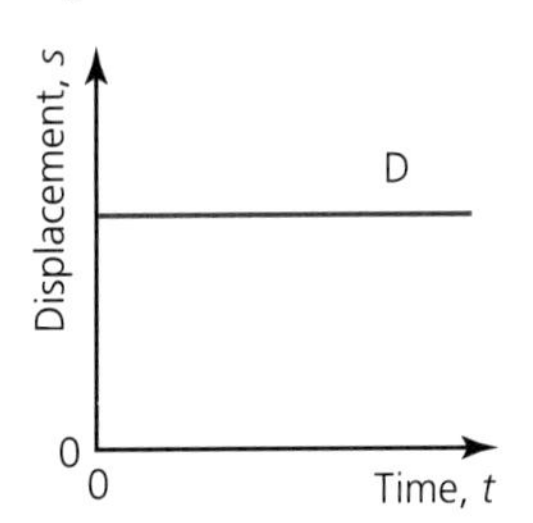

■ **Figure 2.6** Constant velocities on displacement–time graphs (*s*–*t* graphs)

- Line A represents an object moving away from a reference point such that equal displacements occur in equal times. That is, the object has a constant velocity. *Any* linear displacement–time graph represents a constant velocity (it does not need to start or end at the origin).
- Line B represents an object moving with a higher velocity than A.
- Line C represents an object that is moving closer to the reference point.
- Line D represents an object that is stationary (at rest). It has zero velocity and stays at the same distance from the reference point.

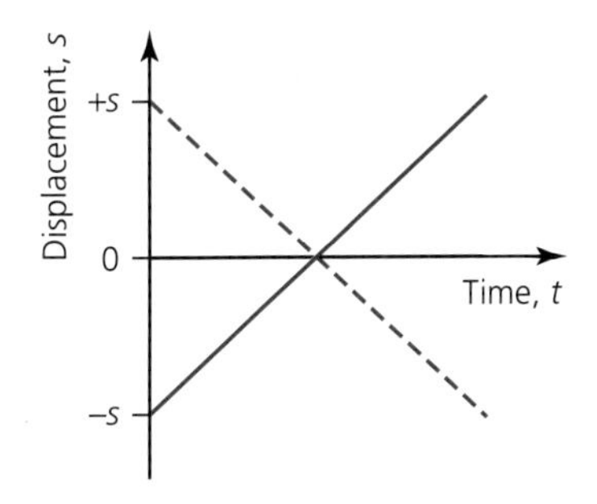

■ **Figure 2.7** Motion in opposite directions represented on a displacement–time graph

Displacement is a vector quantity, but displacement–time graphs like these are usually used in situations where the motion is in a known direction, so that the direction may not need to be stated again. Displacement in opposite directions is represented by the use of positive and negative values. This is shown in Figure 2.7, in which the solid line represents the motion of an object moving with a constant (positive) velocity. The object moves towards a reference point (when the displacement is zero), passes it, and then moves away in the opposite direction with the same velocity. The dotted line represents an identical speed in the opposite direction (or it could also represent the original motion if the directions chosen to be positive and negative were reversed).

Any curved (non-linear) line on a displacement–time graph represents a changing velocity, in other words, an acceleration (or deceleration). This is illustrated in Figure 2.8.

■ **Figure 2.8** Accelerations on displacement–time graphs

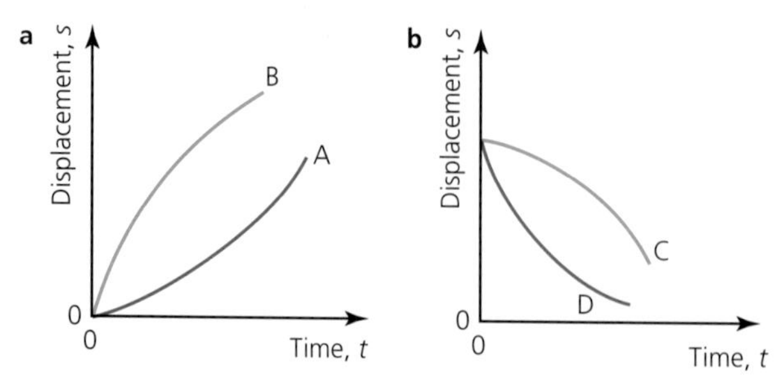

Figure 2.8a shows motion *away* from a reference point. Line A represents an object accelerating. Line B represents an object decelerating (negative acceleration).

Figure 2.8b shows motion *towards* a reference point. Line C represents an object accelerating. Line D represents an object decelerating (negative acceleration).

The values of the accelerations represented by these graphs may, or may not, be constant (this cannot be determined without a more detailed analysis).

In physics, we are usually more concerned with displacement–time graphs than distance–time graphs. In order to explain the difference, consider Figure 2.9. Figure 2.9a shows a displacement–time graph for an object thrown vertically upwards with an initial speed of $20\,\mathrm{m\,s^{-1}}$, without air resistance. It takes 2 s to reach a maximum height of 20 m. At that point it has an instantaneous velocity of zero, before returning to where it began after 4 s and regaining its initial speed. Figure 2.9b shows how the same motion would appear on an overall distance–time graph.

■ **Figure 2.9** Displacement–time and distance–time graphs for an object moving up and then down

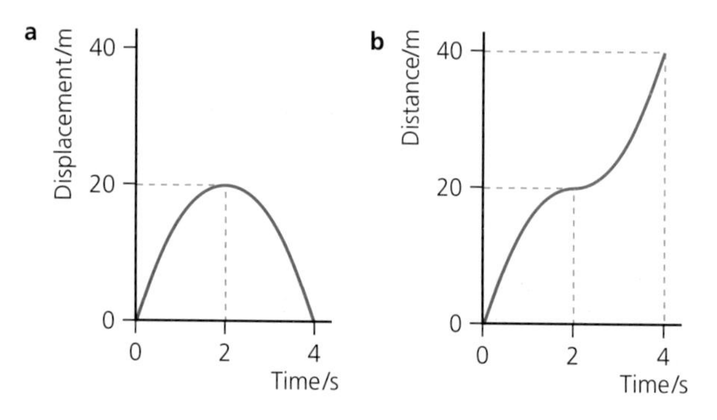

Gradients of displacement–time graphs

Consider the motion at constant velocity shown in Figure 2.10.

■ **Figure 2.10** Finding a constant velocity from a displacement–time graph

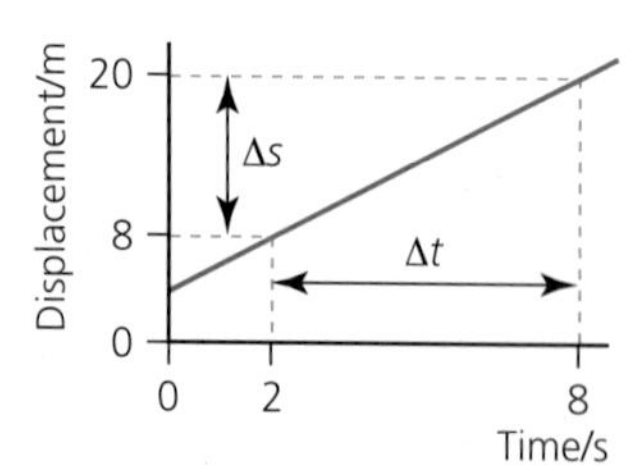

From the graph, the velocity v is given by:

$$v = \frac{\Delta s}{\Delta t} = \frac{20 - 8.0}{8.0 - 2.0} = 2.0\,\mathrm{m\,s^{-1}}$$

Note that the velocity is numerically equal to the gradient (slope) of the line. This is always true, whatever the shape of the line.

The instantaneous velocity of an object is equal to the gradient of the displacement–time graph at that instant.

Figure 2.11 shows an object moving with increasing velocity. The velocity at any time (for example t_1) can be determined by calculating the gradient of the tangent to the line at that instant.

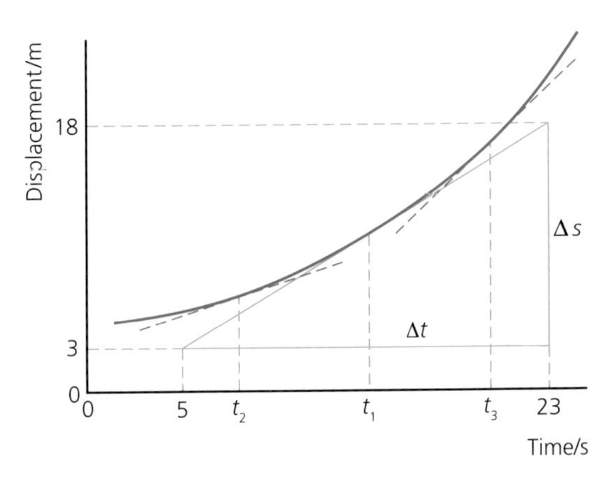

■ **Figure 2.11** Finding an instantaneous velocity from a curved displacement–time graph

The triangle used should be large, in order to make this process as accurate as possible. The tangent drawn at time t_2 has a smaller gradient because the velocity is smaller. At time t_3 the velocity is higher and the gradient steeper. So, in this example:

$$\text{velocity at } t_1 = \frac{18 - 3.0}{23 - 5.0} = \frac{15}{18} = 0.83\,\text{m}\,\text{s}^{-1}$$

1 Figure 2.12 represents the motion of a train on a straight track between two stations.
  a Describe the motion.
  b How far apart are the stations?
  c Calculate the maximum speed of the train.
  d What was the average speed of the train between the two stations?

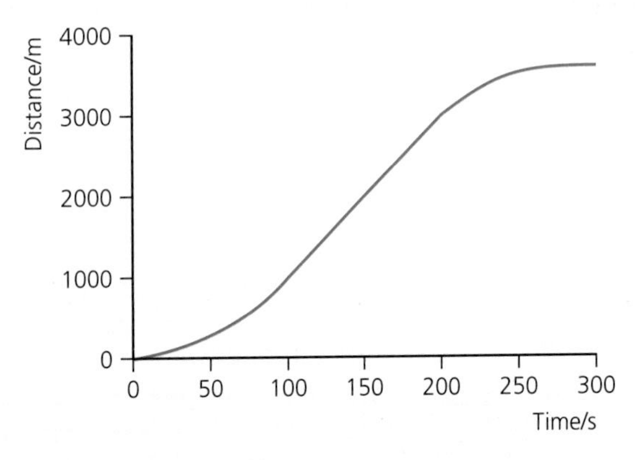

■ **Figure 2.12**

2 a Draw a displacement–time graph for a swimmer who swims 50 m at a constant speed of 1.0 m s^{-1} if the swimming pool is 25 m long and the swimmer takes 1 s to turn around half way through the race.
  b Find out the average speed of the world freestyle record holder when the 100 m record was last broken.
  c The world record for swimming 50 m in a pool of length 25 m is quicker than for swimming in a pool of length 50 m. Suggest why.

3 Draw a displacement–time graph for the following motion: a stationary car is 25 m away; 2 s later it starts to move further away in a straight line from you with a constant acceleration of 1.5 m s^{-2} for 4 seconds; then it continues with a constant velocity for another 8 s.

4 Describe the motion of the runner shown by the graph in Figure 2.13.

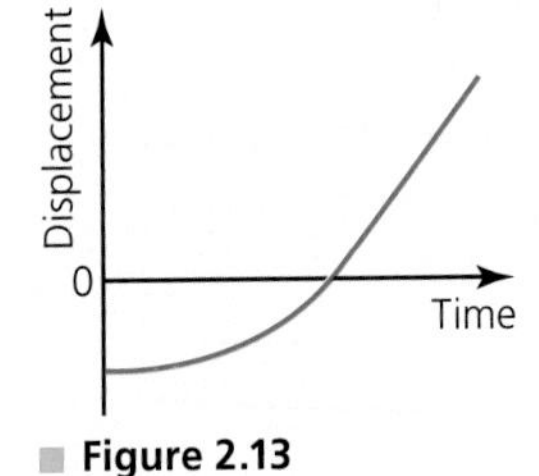

■ **Figure 2.13**

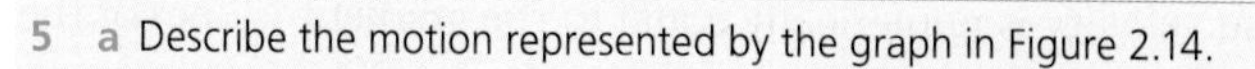

5 a Describe the motion represented by the graph in Figure 2.14.

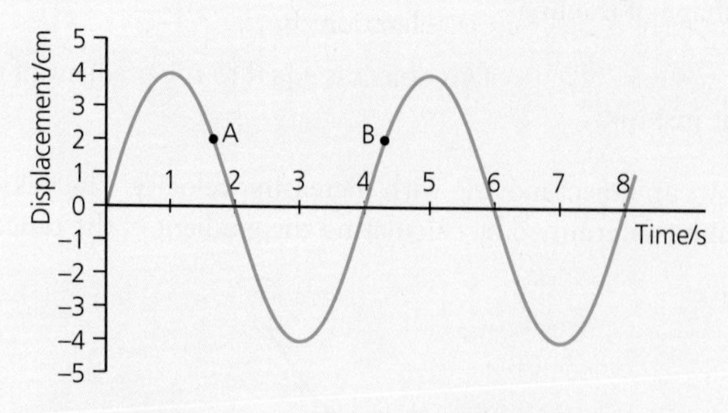

■ **Figure 2.14**

b Compare the velocities at points A and B.
c When is the object moving with its maximum and minimum velocities?
d Estimate values for the maximum and minimum velocities.
e Suggest what kind of object could move in this way.

Velocity–time graphs

Any velocity–time graph, like those shown in Figure 2.15, shows how the velocity of an object varies with time. *Any* straight (linear) line on any velocity–time graph shows that equal changes of velocity occur in equal times – that is, a constant acceleration.

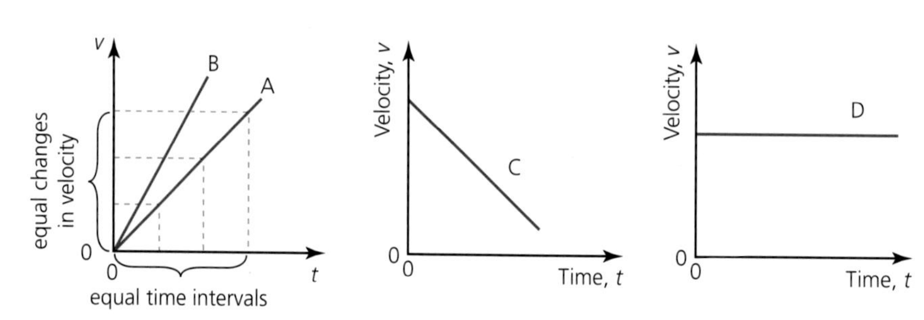

■ **Figure 2.15** Constant accelerations on velocity–time graphs

- Line A shows an object that has a constant positive acceleration.
- Line B represents an object moving with a higher positive acceleration than A.
- Line C represents an object that is decelerating (negative acceleration).
- Line D represents an object moving with a constant velocity – that is, it has zero acceleration.

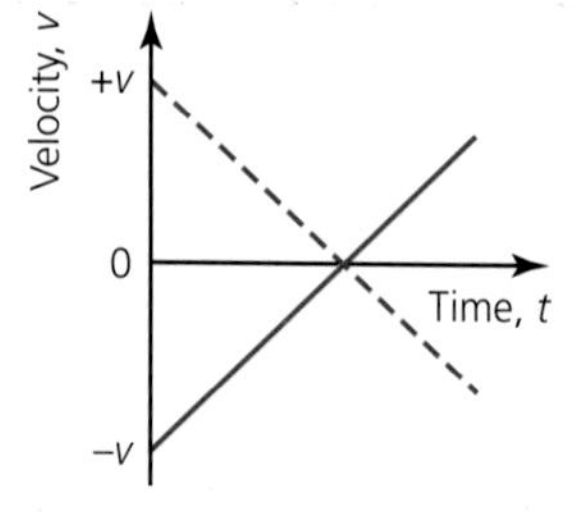

■ **Figure 2.16** Velocities in opposite directions

Curved lines on velocity–time graphs represent changing accelerations. Velocities in opposite directions are represented by positive and negative values. The solid line in Figure 2.16 represents an object that decelerates uniformly to zero velocity and then moves in the opposite direction with an acceleration of the same magnitude. This graph could represent the motion of a stone thrown in the air, reaching its maximum height and then falling down again. The acceleration remains the same throughout (9.81 $\mathrm{m\,s^{-2}}$ downwards). In this example velocity and acceleration upwards have been chosen to be negative, and velocity and acceleration downwards are positive. The dashed line would represent exactly the same motion if the directions chosen to be positive and negative were reversed.

Gradients of velocity–time graphs

Consider the motion at constant acceleration shown in Figure 2.17.

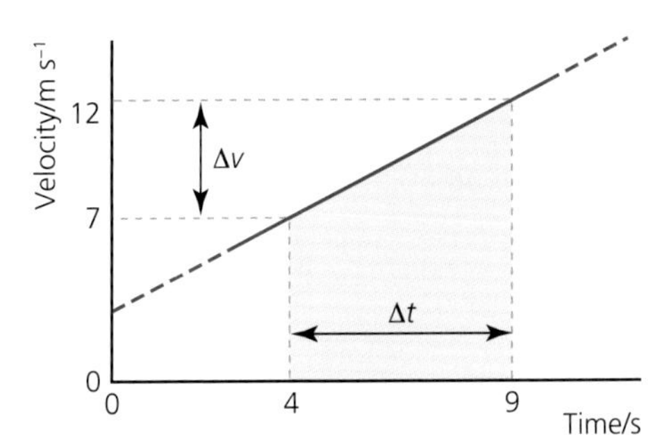

■ **Figure 2.17** Finding the gradient of a velocity–time graph

From the graph:

$$\text{acceleration, } a = \frac{\Delta v}{\Delta t} = \frac{12 - 7.0}{9.0 - 4.0} = 1.0\,\text{m s}^{-2}$$

Note that the acceleration is numerically equal to the gradient (slope) of the line. This is always true, whatever the shape of the line.

The instantaneous acceleration of an object is equal to the gradient of the velocity–time graph at that instant.

Worked example

1 The red line in Figure 2.18 shows an object decelerating (with a decreasing negative acceleration). Use the graph to find the instantaneous acceleration at 10 s.

A tangent drawn at the time of 10 s can be used to determine the value of the acceleration at that instant:

$$\text{acceleration, } a = \frac{\Delta v}{\Delta t} = \frac{0 - 12}{22 - 0} = -0.55\,\text{m s}^{-2}$$

In this example the large triangle used to determine the gradient accurately was drawn by extending the tangent to the axes for convenience.

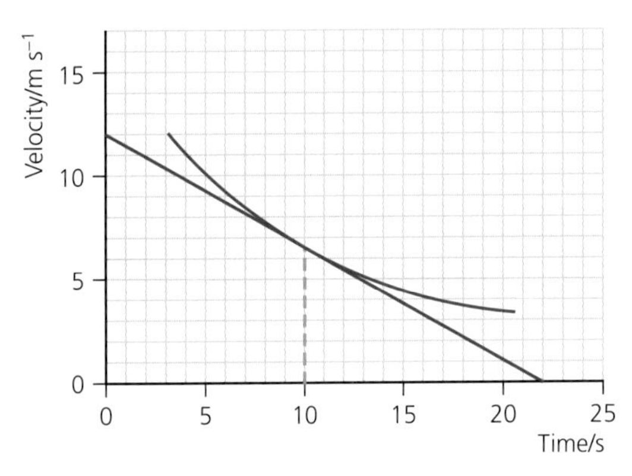

■ **Figure 2.18** Finding an instantaneous acceleration from a velocity–time graph

Areas under velocity–time graphs

Consider again the motion represented in Figure 2.17. The change of displacement, s, between the fourth and ninth second can be found from (average velocity) × time.

$$s = \frac{12 + 7.0}{2} \times (9.0 - 4.0) = 48\,\text{m}$$

This is numerically equal to the area under the line between $t = 4.0$ s and $t = 9.0$ s (as shaded in Figure 2.17). This is always true, whatever the shape of the line.

The area under a velocity–time graph is equal to the change of displacement in the chosen time.

Worked example

2 Figure 2.19a shows how the velocity of a car changed in the first 5 s after starting. Use the graph to estimate the distance travelled in this time.

In Figure 2.19b the blue line has been drawn so that the area under it and the area under the original line are the same (as judged by eye).

$$\text{distance} = \text{area under graph} = \frac{1}{2} \times 16 \times 5.0 = 40\,\text{m}$$

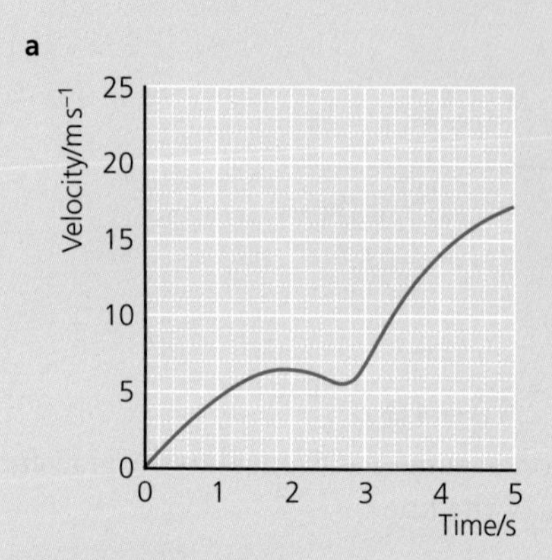

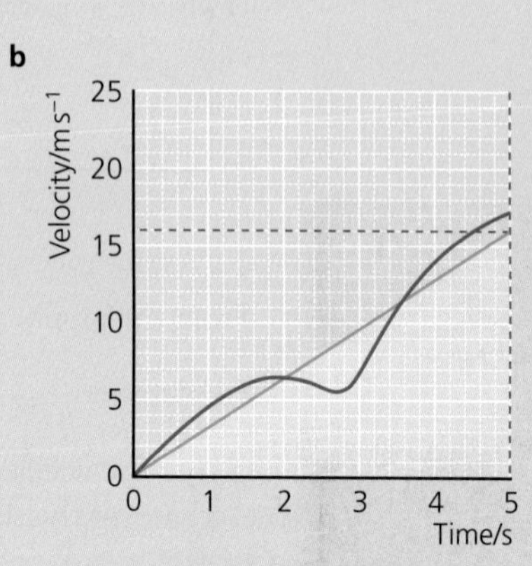

■ **Figure 2.19** Determining the displacement of a car during acceleration

6 **a** Describe the motion represented by the graph in Figure 2.20.
b Calculate accelerations for the three parts of the journey.
c What was the total distance travelled?
d What was the average speed?

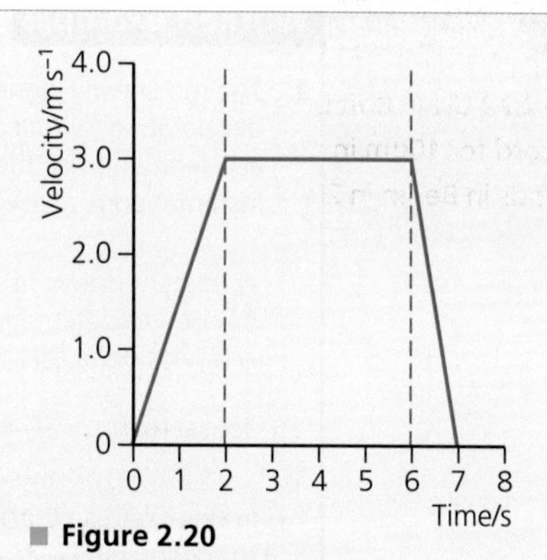

■ **Figure 2.20**

7 The velocity of a car was read from its speedometer at the moment it started and every 2 s afterwards.

The successive values (m s^{-1}) were: 0, 1.1, 2.4, 6.9, 12.2, 18.0, 19.9, 21.3 and 21.9. Plot a graph of these readings and use it to estimate the maximum acceleration and the distance covered in 16 s.

8 **a** Describe the motion of the object represented by the graph in Figure 2.21.
b Calculate the acceleration during the first 8 s.
c What was the total distance travelled in 12 s?
d What was the total displacement after 12 s?
e What was the average speed during the 12 s interval?

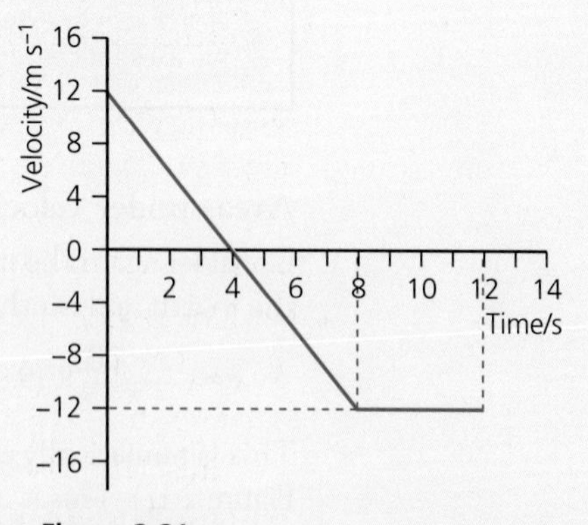

■ **Figure 2.21**

9 Sketch a velocity–time graph of the following motion: a car is 100 m away and travelling along a straight road towards you at a constant velocity of 25 m s^{-1}. Two seconds after passing you, the driver decelerates uniformly and the car stops 62.5 m away from you.

Utilizations

Biomechanics and 100 m sprinters

World-class sprinters can run 100 m in about 10 s (see Figure 2.22). The average velocity is easy to calculate: v = 100/10 = 10 m s^{-1}. Clearly they start from 0 m s^{-1}, so their highest instantaneous velocity must be greater than 10 m s^{-1}.

Trainers use the science of *biomechanics* to improve an athlete's techniques, and the latest computerized methods are used to analyse every moment of their races. The acceleration off the blocks at the start of the race is all important, so that the highest velocity is reached as soon as possible. For the rest of the race the athlete should be able to maintain the same speed, although there may be a slight decrease towards the end of the race. Figure 2.23 shows a typical velocity–time graph for a 100 m race completed in 10 s.

■ **Figure 2.22** Usain Bolt broke the world record for 100 m in a time of 9.58 seconds in Berlin in 2009

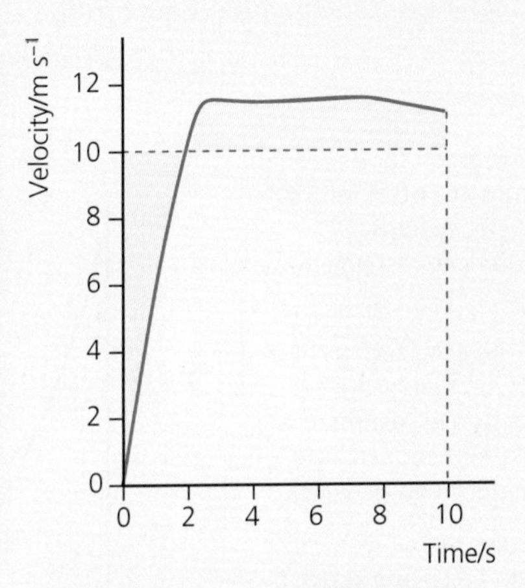

■ **Figure 2.23** Velocity–time graph for an athlete running 100 m

1 a Estimate the highest acceleration achieved during the race illustrated in Figure 2.23.

b When does the athlete reach their greatest velocity?

c Explain why the two shaded areas on the graph are equal.

d Using the internet to collect data, draw a graph showing how the world (or Olympic) record for the 100 m has changed over the last 100 years.

e Predict the 100 m record for the year 2040.

Acceleration–time graphs

An acceleration–time (a–t) graph, like those shown in Figure 2.24, shows how the acceleration of an object changes with time. In this chapter, we are mostly concerned with constant accelerations (it is less common to see motion graphs showing changing acceleration). The graphs in Figure 2.24 show five lines representing constant accelerations.

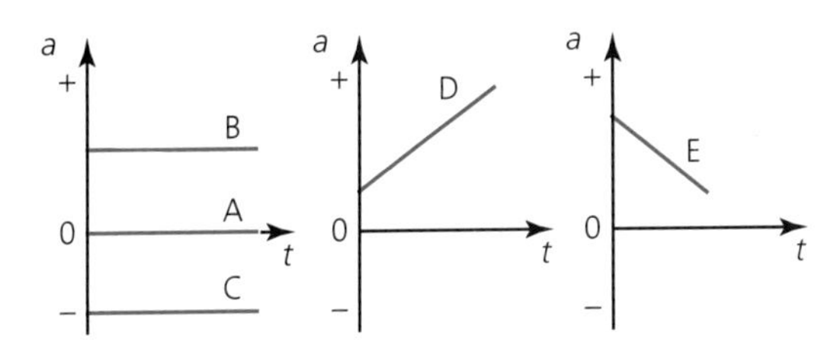

■ **Figure 2.24** Graphs of constant acceleration

- Line A shows zero acceleration, constant velocity.
- Line B shows a constant positive acceleration (uniformly increasing velocity).
- Line C shows the constant negative acceleration (deceleration) of an object that is slowing down at a constant rate.
- Line D shows a (linearly) increasing positive acceleration.
- Line E shows an object that is accelerating positively, but at a (linearly) decreasing rate.

Areas under acceleration–time graphs

Figure 2.25 shows the constant acceleration of a moving car. Using $a = \Delta v/\Delta t$, between the fifth and thirteenth seconds, the velocity of the car increases by:

$$\Delta v = a\Delta t = 1.5 \times 8.0 = 12\,\text{m s}^{-1}$$

The change in velocity is numerically equal to the area under the line between $t = 5$ s and $t = 13$ s (shaded in Figure 2.25). This is always true, whatever the shape of the line.

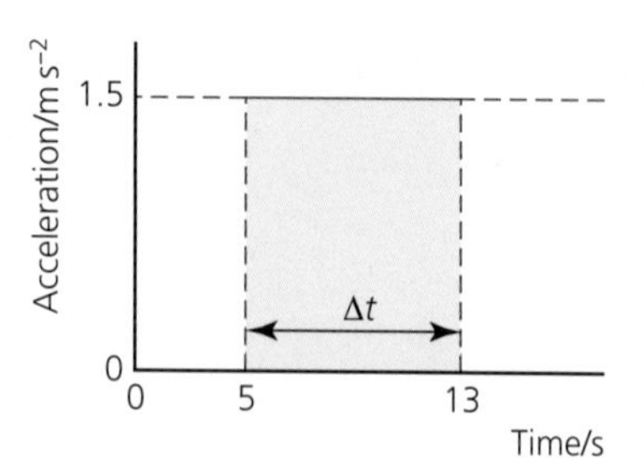

■ **Figure 2.25** Calculating change of velocity from an acceleration–time graph

> The area under an acceleration–time graph is equal to the change of velocity in the chosen time.

10 Draw an acceleration–time graph for a car that starts from rest, accelerates at $2\,\text{m s}^{-2}$ for 5 s, then travels at constant velocity for 8 s, before decelerating uniformly to rest again in a further 2 s.

11 Figure 2.26 shows how the acceleration of a car changed during a 6 s interval. If the car was travelling at $2\,\text{m s}^{-1}$ after 1 s, estimate a suitable area under the graph and use it to determine the approximate speed of the car after 5 s.

12 Figure 2.27 shows a tennis ball being struck by a racquet. Sketch a possible velocity–time graph and an acceleration–time graph from 1 s before impact to 1 s after the impact.

13 Sketch possible displacement–time and velocity–time graphs for a bouncing ball dropped from rest. Continue the sketches until the third time that the ball contacts the ground.

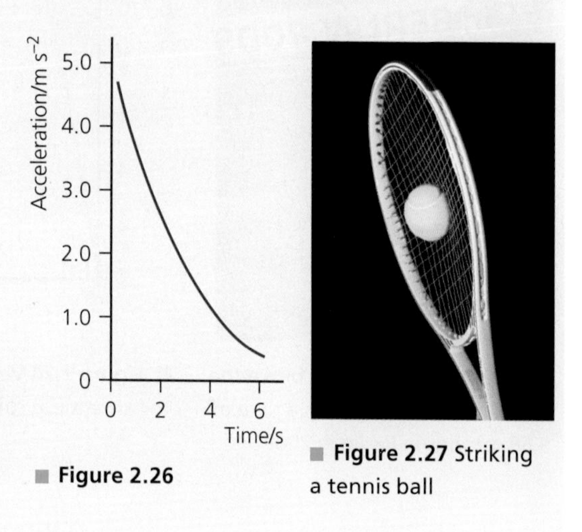

■ **Figure 2.26**

■ **Figure 2.27** Striking a tennis ball

Graphs of motion: summary

If any one graph of motion is plotted (s–t, v–t or a–t), then the motion is fully defined and the other two graphs can be drawn with information about gradients and/or areas taken only from the first graph. This is summarized in Figure 2.28.

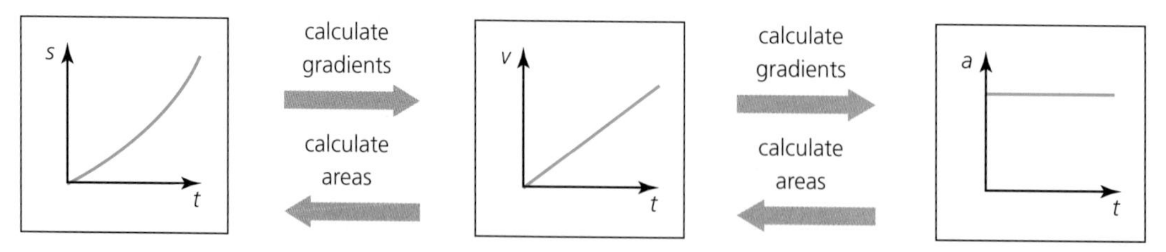

■ **Figure 2.28** The connections between the different graphs of motion

To reproduce one graph from another by hand is a long and repetitive process, because in order to produce accurate graphs a large number of similar measurements and calculations need to be made over short intervals of time. Of course, computers are ideal for this purpose.

In more mathematically advanced work, which is not part of this course, *calculus* can be used to perform these processes using differentiation and integration.

Utilizations

Kinematic equations: vehicle braking distances

Figure 2.29 represents how the velocities of two identical cars changed from the moment that their drivers saw danger in front of them and tried to stop their cars as quickly as possible. It has been assumed that both drivers have the same reaction time (0.7 s) and both cars decelerate at the same rate ($-5.0\,\text{m}\,\text{s}^{-2}$).

The distance travelled at constant velocity before the driver reacts and depresses the brake pedal is known as the 'thinking distance'. The distance travelled while decelerating is called the 'braking distance'. The total stopping distance is the sum of these two distances.

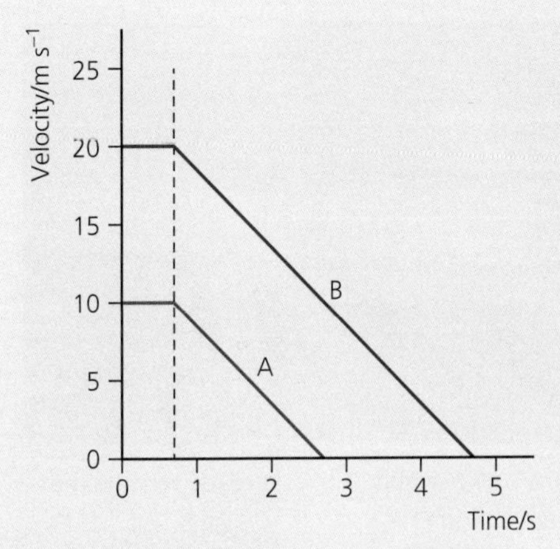

■ **Figure 2.29** Velocity–time graphs for two cars braking

Car B, travelling at twice the velocity of car A, has twice the thinking distance. That is, the thinking distance is proportional to the velocity of the car. The distance travelled when braking, however, is proportional to the velocity *squared*. This can be confirmed from the areas under the v–t graphs. The area under graph B is four times the area under graph A (during the deceleration). This has important implications for road safety and most countries make sure that people learning to drive must understand how stopping distances change with the vehicle's velocity. Some countries measure the reaction times of people before they are given a driving licence.

Set up a spreadsheet that will calculate the total stopping distance for cars travelling at initial speeds, u, between 0 and $40\,\text{m}\,\text{s}^{-1}$ with a deceleration of $-6.5\,\text{m}\,\text{s}^{-2}$. (Make calculations every $2\,\text{m}\,\text{s}^{-1}$.) The thinking distance can be calculated from $s_t = 0.7u$ (reaction time 0.7 s). In this example the braking time can be calculated from $t_b = u/6.5$ and the braking distance can be calculated from $s_b = (u/2)t_b$. Use the data produced to plot a computer-generated graph of stopping distance (y-axis) against initial speed (x-axis).

■ Equations of motion for uniform acceleration

The five quantities u, v, a, s and t are all that is needed to fully describe the motion of an object moving with uniform (constant) acceleration.

- u = velocity (speed) at the start of time t
- v = velocity (speed) at the end of time t
- a = acceleration (constant)
- s = distance travelled in time t
- t = time taken for velocity (speed) to change from u to v and to travel a distance s

If any three of the quantities are known, the other two can be calculated using the two equations below. If we know the initial velocity u and acceleration a of an object, and the acceleration is uniform, then we can determine its final velocity v after a time t by rearranging the equation used to define acceleration. This gives:

$$v = u + at$$

This equation is given in the *Physics data booklet*.

We have also seen that the distance travelled while accelerating uniformly from a velocity u to a velocity v in a time t can be calculated from:

$$s = \frac{(v + u)t}{2}$$

This equation is given in the *Physics data booklet*.

These two equations can be combined mathematically to give two further equations, shown below, which are also found in the *Physics data booklet*. These very useful equations do not involve any further physics theory; they just express the same physics principles in a different way.

$$v^2 = u^2 + 2as$$
$$s = ut + \frac{1}{2}at^2$$

Remember that, the four equations of motion can only be used if the acceleration is *uniform* during the time being considered.

The equations of motion are covered in the IB Mathematics course (and also treated in calculus form).

Worked examples

3 A Formula One racing car (see Figure 2.30) accelerates *from rest* (i.e. it was *stationary* to begin with) at $18\,m\,s^{-2}$.

a What is its speed after 3.0 s?

b How far does it travel in this time?

c If it continues to accelerate at the same rate, what will its velocity be after it has travelled 200 m from rest?

d Convert the final velocity to $km\,h^{-1}$.

■ **Figure 2.30** Formula One cars ready to start the Canadian Grand Prix

a $v = u + at$

$v = 0 + (18 \times 3.0)$

$v = 54\,m\,s^{-1}$

b $s = \frac{v+u}{2}t$

$s = \frac{0+54}{2} \times 3.0$

$s = 81\,m$

But note that the distance can be calculated directly, without first calculating the final velocity, as follows:

$s = ut + \frac{1}{2}at^2$

$s = (0 \times 3.0) + (0.5 \times 18 \times 3.0^2)$

$s = 81\,m$

c $v^2 = u^2 + 2as$

$v^2 = 0^2 + (2 \times 18 \times 200)$

$v^2 = 7200$

$v = 85\,m\,s^{-1}$

d $85\,m\,s^{-1} = 85 \times 3600 = 3.1 \times 10^5\,m\,h^{-1}$

$3.1 \times 10^5\,m\,h^{-1} = \frac{3.1 \times 10^5}{10^3} = 310\,km\,h^{-1}$

4 A train travelling at $50\,m\,s^{-1}$ ($180\,km\,h^{-1}$) needs to decelerate uniformly so that it stops at a station 2 kilometres away.

a What is the necessary deceleration?

b How long does it take to stop the train?

a $v^2 = u^2 + 2as$

$0^2 = 50^2 + (2 \times a \times 2000)$

$a = \frac{-50^2}{2 \times 2000}$

$a = -0.63\,m\,s^{-2}$

b $v = u + at$

$0 = 50 + (-0.625) \times t$

$t = \frac{50}{0.625} = 80\,s$

(Alternatively, $s = \frac{u+v}{2}t$ could have been used.)

Assume that all accelerations are constant.

14 A ball rolls down a slope with a constant acceleration. When it passes a point P its velocity is $1.2\,m\,s^{-1}$ and a short time later it passes point Q with a velocity of $2.6\,m\,s^{-1}$.
 a What was its average velocity between P and Q?
 b If it took 1.4 s to go from P to Q, what is the distance PQ?
 c What is the acceleration of the ball?

15 A plane accelerates from rest along a runway and takes off with a velocity of $86.0\,m\,s^{-1}$. Its acceleration during this time is $2.40\,m\,s^{-2}$.
 a What distance along the runway does the plane travel before take-off?
 b How long after starting its acceleration does the plane take off?

16 An ocean-going oil tanker can decelerate no quicker than $0.0032\,m\,s^{-2}$.
 a What is the minimum distance needed to stop if the ship is travelling at 10 knots? (1 knot = $0.514\,m\,s^{-1}$)
 b How much time does this deceleration require?

17 An advertisement for a new car states that it can travel 100 m from rest in 8.2 s.
 a What is the average acceleration?
 b What is the speed of the car after this time?

18 A car travelling at a constant velocity of $21\,m\,s^{-1}$ (faster than the speed limit of $50\,km\,h^{-1}$) passes a stationary police car. The police car accelerates after the other car at $4.0\,m\,s^{-2}$ for 8.0 s and then continues with the same velocity until it overtakes the other car.
 a When did the two cars have the same velocity?
 b Has the police car overtaken the other car after 10 s?
 c By equating two equations for the same distance at the same time, determine exactly when the police car overtakes the other car.

19 A car brakes suddenly and stops 2.4 s later, after travelling a distance of 38 m.
 a What was its deceleration?
 b What was the velocity of the car before braking?

20 A spacecraft travelling at $8.00\,km\,s^{-1}$ accelerates at $2.00 \times 10^{-3}\,m\,s^{-2}$ for 100 hours.
 a What is its final speed?
 b How far does it travel during this acceleration?

21 Combine the first two equations of motion (given on page 33) to derive the second two ($v^2 = u^2 + 2as$ and $s = ut + \frac{1}{2}at^2$).

Nature of Science

Observations

Scientific knowledge only really developed after the importance of experimental evidence was understood.

The equations of motion (and Newton's laws of motion) are a very important part of classical physics that all students should understand well. They were first proposed at an early stage in the historical development of physics, when experimental techniques were not as developed as they are today. However, these basic ideas about motion still remain just as important in the modern world.

Early scientists, like Galileo and Newton, were able to make careful observations and gather enough evidence to support their theories about idealized motion despite the fact that friction and air resistance always complicate the study of moving objects. This is especially impressive because some of their theories contradicted ideas that had been accepted for 2000 years.

Acceleration due to gravity

We are all familiar with the motion of objects falling towards Earth because of the force of gravity. Figure 2.31 shows an experiment to gather data on distances and times for a falling mass, so that a value for its acceleration can be calculated. The electronic timer starts when the electric current to the electromagnet is switched off and the steel ball starts to fall. When the ball hits the trapdoor at the bottom, a second electrical circuit is switched off and the timing stops. Alternatively, a position sensor could be used to track the fall of the ball.

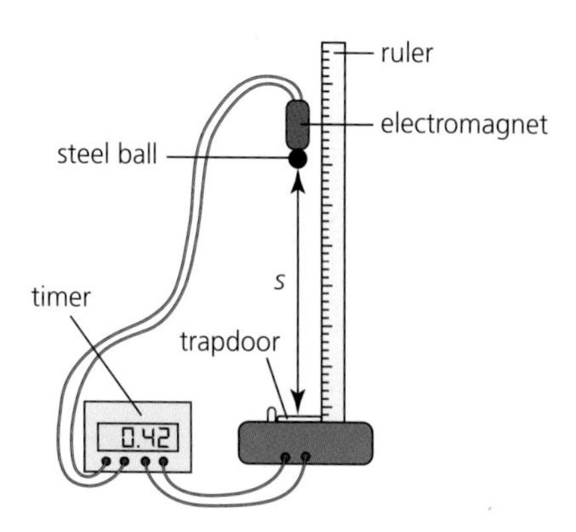

■ **Figure 2.31** An experiment to measure the acceleration due to gravity

Worked example

5 Suppose that when the mass fell 0.84 m the time was measured to be 0.42 s. Calculate its gravitational acceleration.

$s = ut + \frac{1}{2}at^2$

$0.84 = 0 + (0.5 \times a \times 0.42^2)$

$a = \frac{0.84}{0.088} = 9.5\,\text{m}\,\text{s}^{-2}$

City	g/m s^{-2}
Auckland	9.799
Bangkok	9.783
Buenos Aires	9.797
Cape Town	9.796
Chicago	9.803
Kuwait	9.793
London	9.812
La Paz	9.784
Mexico City	9.779
Tokyo	9.798

■ **Table 2.1** Values of g in some cities around the world

Of course, obtaining an accurate and reliable result will require further measurements. The measurement could be repeated for the same height, so that averages could be calculated. But it would be better to take measurements for different heights, so that an appropriate graph can be drawn, which will provide a better way of assessing random and systematic errors.

If accurate measurements are made in a vacuum (to be sure that there is no air resistance), the results are very similar (but not identical) at all locations on the Earth's surface. Some examples are shown in Table 2.1.

The **acceleration due to gravity** in a vacuum near the Earth's surface is given the symbol g. This is also called the **acceleration of free fall**. The accepted value of g is 9.81 m s^{-2}. This value should be used in calculations and is listed in the *Physics data booklet*. Anywhere on the Earth's surface (or in an airplane) can be considered as 'near to the Earth's surface'.

It is very important to remember that *all freely moving* objects close to the Earth's surface experience this same acceleration, g, downwards. This is true whether the object is large or very small, or whether it is moving upwards, downwards, sideways or in any other direction. 'Freely moving' means that the effects of air resistance can be ignored and that the object is not powered in any way. In reality, however, the effects of air resistance usually cannot be ignored, except for large, dense masses moving short distances from rest. But, as is often the case in science, we need to understand simplified examples first before we move onto more complicated situations.

Worked examples

6 A coin falls from rest out of an open window 16 m above the ground. Assuming that there is no air resistance:

a what is its velocity when it hits the ground?

b how long did it take to fall that distance?

a $v^2 = u^2 + 2as$

$v^2 = 0^2 + (2 \times 9.81 \times 16) = 314$

$v = 18\,\text{m s}^{-1}$

b $v = u + at$

$18 = 0 + 9.81t$

$t = \frac{18}{9.81} = 1.8\,\text{s}$

7 A ball is thrown vertically upwards and reaches a maximum height of 21.4 m.

a Calculate the speed with which the ball was released.

b What assumption did you make?

c Where will the ball be 3.05 s after it was released?

d What will its velocity be at this time?

a $v^2 = u^2 + 2as$

When the ball has travelled a distance $s = 21.4\,\text{m}$, its speed, v, at the highest point will be zero.

$0^2 = u^2 + (2 \times -9.81 \times 21.4)$

$u^2 = 419.9$

$u = 20.5\,\text{m s}^{-1}$

In this example, the vector quantities directed upwards (u, v, s) are considered positive and the quantity directed downwards (a) is negative. The same answer would be obtained by reversing all the signs. Using positive and negative signs to represent vectors (like displacement, velocity and acceleration) in opposite directions is common practice.

b It was assumed that there was no air resistance.

c $s = ut + \frac{1}{2}at^2$

$s = (20.5 \times 3.05) + \left(\frac{1}{2} \times -9.81 \times 3.05^2\right)$

$s = 16.9\,\text{m}$ above the ground

d $v = u + at$

$v = 20.5 + (-9.81 \times 3.05)$

$v = -9.42\,\text{m s}^{-1}$ (moving downwards)

In all of the following questions, ignore the possible effects of air resistance. Use $g = 9.81\,\text{m s}^{-2}$.

22 Suggest possible reasons why the acceleration due to gravity is not the same everywhere on the Earth's surface.

23 a How long does it take a stone dropped from rest from a height of 2.1 m to reach the ground?

b If the stone was thrown downwards with an initial velocity of $4.4\,\text{m s}^{-1}$, with what speed would it hit the ground?

c If the stone was thrown vertically upwards with an initial velocity of $4.4\,\text{m s}^{-1}$, with what speed would it hit the ground?

24 A small rock is thrown vertically upwards with an initial velocity of $22\,\text{m s}^{-1}$. When will its speed be $10\,\text{m s}^{-1}$? (There are two possible answers.)

25 A falling ball has a velocity of $12.7\,\text{m s}^{-1}$ as it passes a window 4.81 m above the ground. When will it hit the ground?

26 A ball is thrown vertically upwards with a speed of $18.5\,\text{m s}^{-1}$ from a window that is 12.5 m above the ground.

a When will it pass the same window moving down?

b With what speed will it hit the ground?

c How far above the ground was the ball after exactly 2 s?

27 Two balls are dropped from rest from the same height. If the second ball is released 0.750 s after the first, and assuming they do not hit the ground, how far apart are the two balls:

a 3.00 s after the second ball was dropped?

b 2.00 s later?

28 A stone is dropped from rest from a height of 34 m. Another stone is thrown downwards 0.5 s later. If they both hit the ground at the same time, what was the initial velocity of the second stone?

29 In Worked example 3 an acceleration of $18\,m\,s^{-2}$ was quoted for a Formula One racing car. The driver of that car could be said to experience a '*g*-force' of nearly 2*g*, and during the course of a typical race a driver may have to undergo *g*-forces of nearly 5*g*. Explain what you think is meant by a '*g*-force' of 2*g*.

30 Stone A is dropped from rest from a cliff. After it has fallen 5 m, stone B is dropped.
a How does the distance between the two stones change (if at all) as they fall?
b Explain your answer.

31 a A flea accelerates at the enormous average rate of $1500\,m\,s^{-2}$ during a vertical take-off that lasts only about 0.0012 s. What height will the flea reach?
b Measure how high you can jump vertically (standing in the same place), and use the result to calculate your take-off velocity.
c In order to jump up you had to bend your knees and reduce your height. Measure by how much your height was reduced just before jumping, then use the result to estimate your average acceleration during take-off.
d What was the duration of your take-off?
e Compare your performance with the flea's.

32 Use the internet to learn more about the GOCE project, which ended in 2013 (Figure 2.32).

Figure 2.32 The Gravity Field and Steady-State OceanCirculation Explorer (GOCE) satellite was launched by the European Space Agency in 2009

Figure 2.33 Burj Khalifa in Dubai

33 Figure 2.33 shows the tallest building in the world: Burj Khalifa in Dubai.
a How long would it take an object to reach the ground if it was dropped from 828 m (the height of Burj Khalifa)?
b With what speed would it hit the ground?

34 The times of fall for a ball dropped from different heights (Figure 2.31) were measured.
a Sketch the height–time graph you would expect to get from these results.
b By considering the equation $s = ut + \frac{1}{2}at^2$, what would be the best graph to draw to produce a straight best-fit line from which the acceleration due to gravity could be determined?

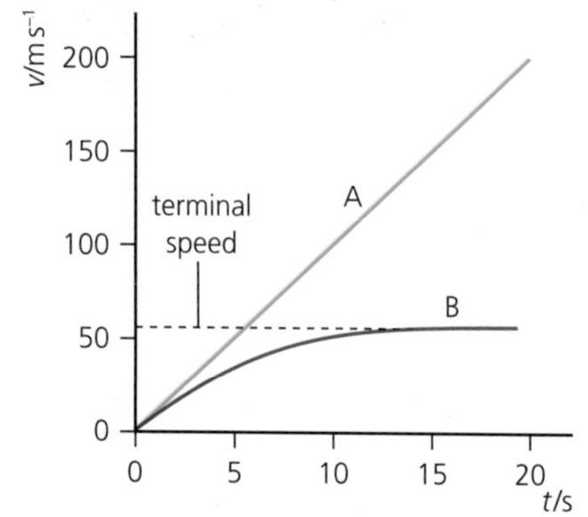

Figure 2.34 An example of a graph of velocity against time for an object falling under the effect of gravity, with and without air resistance

Fluid resistance and terminal speed

As any object moves through air, the air is forced to move out of the path of the object. This causes a force opposing the motion called **air resistance**, or **drag**.

Similar forces will oppose the motion of an object moving in any direction through *any* gas or liquid. (Gases and liquids are both described as **fluids** because they can flow.) Such forces opposing motion are generally described as **fluid resistance**.

Figure 2.34 represents the motion of an object falling towards Earth. Line A shows the motion without air resistance and line B shows the motion, more realistically, with air resistance.

When any object first starts to fall, there is no air resistance. The initial acceleration, g, is the same as if it was in a vacuum. As the object falls faster, the air resistance increases, so that the rate of increase in velocity becomes less. This is shown in the Figure 2.34 by the line B becoming less steep. Eventually the object reaches a constant, maximum speed known as the **terminal speed** or *terminal velocity* ('terminal' means final). The value of an object's terminal speed will depend on its cross-sectional area, shape and weight, as discussed in Section 2.2. The terminal speed of a skydiver is usually quoted at about $200\,km\,h^{-1}$ ($56\,m\,s^{-1}$) – Figure 2.35. Terminal speed also depends on the density of the air – in October 2012 Felix Baumgartner (Figure 2.36), an Austrian skydiver, reached a world record speed of $1358\,km\,h^{-1}$ by starting his jump from a height of about 39 km above the Earth's surface where there is very little air.

■ **Figure 2.35** Skydivers in free fall

■ **Figure 2.36** Felix Baumgartner about to jump from a height of 39 km

The design and motion of simple parachutes make interesting investigations, especially if they can be videoed falling near to vertical scales. The movement of an object falling vertically through a liquid (oil for example) is slower and can also be investigated in a school laboratory. It may also reach a terminal speed, and have a pattern of motion similar to that shown in Figure 2.34. Computer simulations are also useful for gaining a quick appreciation of the factors that affect terminal speeds. Air resistance is discussed in greater detail later in this chapter (page 52).

Additional Perspectives

Galileo

It is a matter of common observation that 'heavier objects fall to Earth quicker than lighter objects'. This is easily demonstrated by dropping, for example, a ball and a piece of paper side by side. The understandable belief that heavier objects fall faster was a fundamental principle in 'natural philosophy' (the name for early studies of what is now known as science) for more than 2000 years of civilization. In ancient Greece, Aristotle had closely linked the motion of falling objects to the belief that all processes have a purpose and that the Earth was the natural and rightful resting place for everything.

■ **Figure 2.37** Galileo Galilei

In the sixteenth century the Italian scientist Galileo (Figure 2.37) was among the first to suggest that the reason why various objects fall differently is only because of air resistance. He predicted that, if the experiment could be repeated in a vacuum (without air), all objects would have exactly the same pattern of downwards motion under the effects of gravity.

In one of the most famous stories in science, Galileo dropped different masses off a balcony on the Tower of Pisa in Italy to show to those watching on the ground below that falling objects are acted on equally by gravity. This story may or may not be true, but one of the reasons that Galileo is so respected as a great scientist is that he was one of the first to actually do experiments, rather than just think about them.

It was many years later, after the invention of the first vacuum pumps, that Isaac Newton and others were able to remove the effects of air resistance and demonstrate that a coin (a 'guinea') and a feather fall together.

In 1971 that famous experiment was repeated on the Moon (Figure 2.38) when astronaut David Scott dropped a hammer and feather side by side. Millions of people all over the world were watching while he reminded them of Galileo's achievements. The strength of gravity is less on the Moon than on the Earth because the Moon is smaller. Objects accelerate towards the Moon at about $\frac{1}{6}$ of the rate that they would on the Earth ($g = 1.6\,\text{m}\,\text{s}^{-2}$).

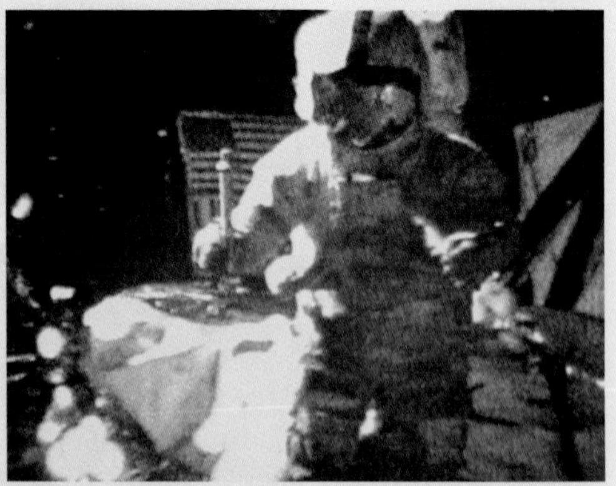

■ **Figure 2.38** Dropping objects on the Moon

1 Galileo's achievements were specifically mentioned when the experiment was repeated on the Moon, but do you think that there were other scientists who were equally deserving of credit for advancing understanding of motion and gravity? Give the names of two such pioneers of science and list their greatest achievements.

Nature of Science

What is science?

The Italian scientist Galileo Galilei (1564–1642) is famous for his pioneering work on kinematics and falling objects, and it has been acknowledged that he was one of the first *practical* scientists (in the modern meaning of the word). But what, exactly, is science and what makes science different from other human activities?

This is not an easy question to answer in a few words, although there are certainly important characteristics that most scientific activities share:

- Science attempts to see some underlying simplicity in the vast complexity around us.
- Science looks for the logical patterns and rules that control events.
- Science seeks to accumulate knowledge and, wherever possible, to build on existing knowledge to make an ever-expanding framework of understanding.

Most importantly, science is based on experimentation and evidence – that is, science relies on 'facts' that are, at the current time, accepted to be 'true'. No good scientist would ever claim that something must be absolutely 'true' for all time – one of the leading characteristics of science is the constant independent and widespread testing of existing theories by experiment. No fact or theory can ever be proven to be true for all times and all places, so science often advances through experiments that try to disprove new theories or existing knowledge.

The question 'what is science?' is often answered by explaining *how* scientists work, the so-called 'scientific method', which can be summarized as follows, although any particular scientific process can show variations from this generalized pattern:

- Choose a topic for investigation (for example, the design of golf balls – Figure 2.39).
- Research available information on the chosen topic (maybe use the internet to find out about the design of golf balls).
- Ask a suitable question for investigation (for example, would a larger golf ball travel further than a smaller golf ball, if struck in the same way?).
- Use theory to predict what you think will happen in the investigation (for example, you might think that a smaller ball has less air resistance and so will go further).

■ **Figure 2.39** Why are golf balls a certain size?

- Design and carry out an investigation to test your prediction.
- Process the results and evaluate their uncertainties.
- Draw conclusions, accepting or rejecting your predictions.
- If the conclusions are unsatisfactory, repeat them and/or redesign the investigation.
- If the conclusions are satisfactory and can be repeated, present your findings to other people.

35 A well-known old Chinese proverb says 'I hear and I forget, I see and I remember, I do and I understand'. Consider your own knowledge of physics. To what extent has doing experimental work improved your understanding? Do you think doing more experimental work (and less theoretical work) would improve:
a your interest?
b your examination results?

Explain your answers.

Projectile motion

In our discussion of objects moving through the air, we have so far only considered motion vertically up or down. Now we will extend that work to cover objects moving in *any* direction. A **projectile** is an object that has been projected through the air (for example, fired, launched, thrown, kicked or hit) and which then moves only under the action of the forces of gravity (and air resistance, if significant). A projectile has no ability to power or control its own motion.

Components of a projectile's velocity

The instantaneous velocity of any projectile at any time can conveniently be resolved into vertical and horizontal components, v_V and v_H, as shown in Figure 2.40.

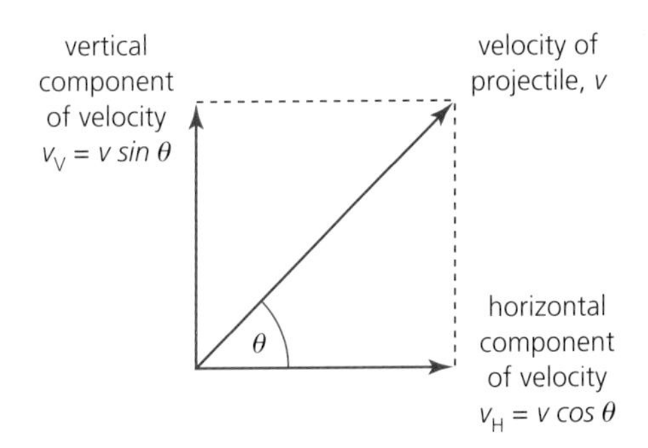

Figure 2.40 Vertical and horizontal components of velocity

Because these components are perpendicular to each other, they can be treated independently (separately) in calculations.

When there is no air resistance, all objects moving through the air in the uniform gravitational field close to the Earth's surface will accelerate vertically downwards at 9.81 $m\,s^{-2}$ because of the force of gravity. This is true for *all* masses and for *all* directions of motion (including moving upwards). In other words, any object that is projected at any angle will *always* accelerate vertically downwards at the same rate as an object dropped vertically (in the absence of air resistance).

Because of the acceleration due to gravity, the values of the vertical component and the resultant velocity of a projectile will change continuously during the motion, but it is important to realise that the horizontal component will remain the same, *if* the air resistance is negligible, because there are no horizontal forces acting on the projectile.

ToK Link

Intuition

The independence of horizontal and vertical motion in projectile motion seems to be counter-intuitive. How do scientists work around their intuitions? How do scientists make use of their intuitions?

Human **intuition** has played a significant part in many scientific discoveries and developments, but scientists also need the imagination to propose theories that may sometimes seem contrary to 'common sense'.

This is especially true in understanding the weird realm of quantum physics, where relying on everyday experiences for inspiration is of little or no use. But it is worth remembering that many of the well-established concepts and theories of classical physics that are taught now in schools would have seemed improbable to scientists at the time they were first proposed.

36 At one particular moment a tennis ball is moving upwards with velocity of 28.4 m s^{-1} at an angle of 15.7° to the horizontal. Calculate the vertical and horizontal components of this velocity.

37 An aircraft is descending with a constant velocity of 480 km h^{-1} at an angle of 2.0° to the horizontal.
  a What is the vertical component of the plane's velocity?
  b How long will it take to descend by 500 m on this flight path? (Give the answer to the nearest minute.)

38 A stone is projected upwards at an angle of 22° to the vertical. At that moment it has a vertical component of velocity of 38 m s^{-1}.
  a What is the horizontal component of velocity at this time?
  b After another second will:
    i the horizontal component
    ii the vertical component
  be greater, smaller or the same as before? (Ignore the effects of air resistance.)

Parabolic trajectory

Figure 2.41 shows a *stroboscopic photograph* of a bouncing ball. In a stroboscopic photograph the time intervals between the different positions of the ball are always the same.

The typical **trajectory** of a projectile is **parabolic** (shaped like a parabola or part of a parabola) when air resistance is negligible. For example, Figure 2.42 shows the trajectory of an object projected horizontally compared with that of an object dropped vertically at the same time. Note that both objects fall the same vertical distance in the same time.

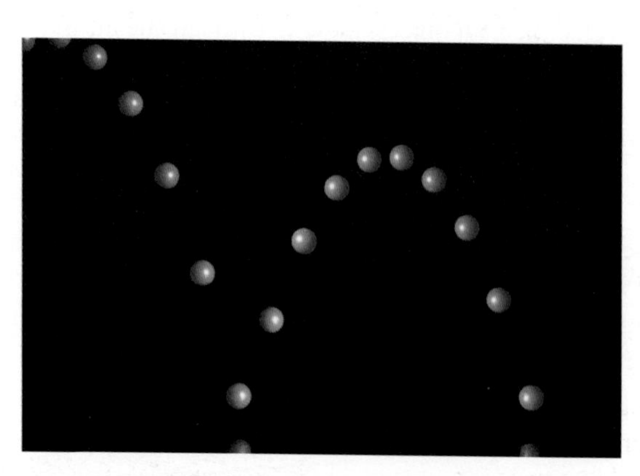

■ **Figure 2.41** Parabolic trajectory of a bouncing ball

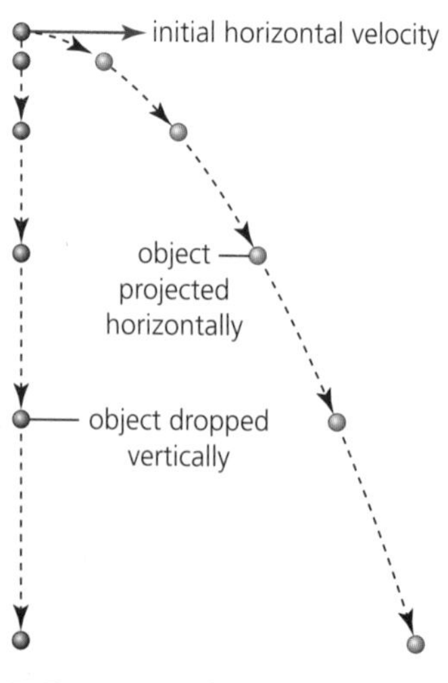

■ **Figure 2.42** The parabolic trajectory of an object projected horizontally compared with an object dropped vertically

The distance travelled and time of flight of various projectiles have always made popular and interesting physics investigations. Video recording and analysis make this much easier and more accurate. Many computer simulations are also available that enable students to quickly compare trajectories under different conditions.

Utilizations

Ballistics

The study of the use of projectiles is known as *ballistics*. Because of its close links to hunting and fighting, this is an area of science with a long history, going all the way back to spears, and bows and arrows. Figure 2.43 shows a common medieval misconception about the motion of cannon balls: they were thought to travel straight until they ran out of energy.

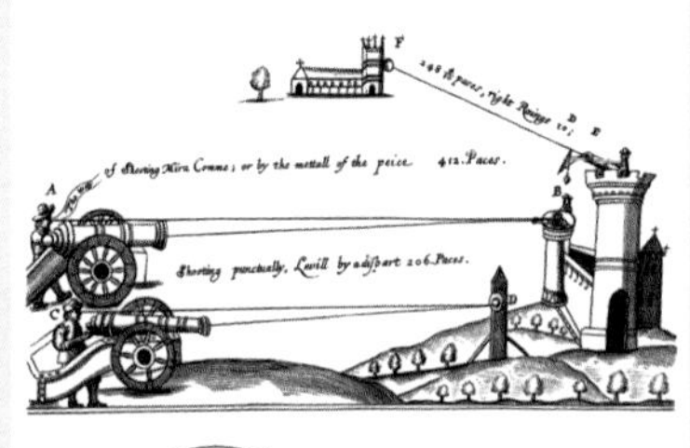

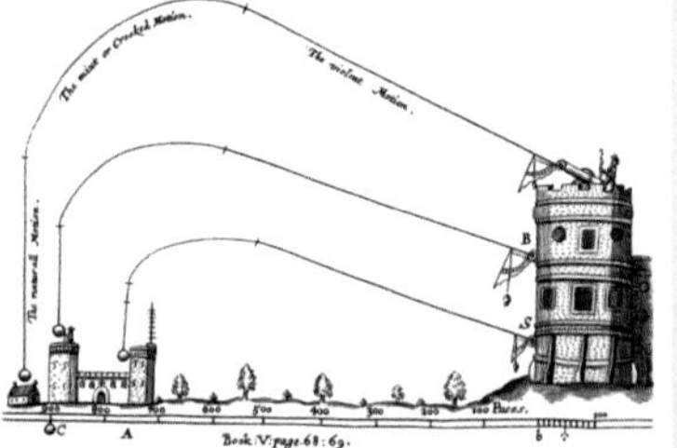

■ **Figure 2.43** Trajectories of cannon balls were commonly misunderstood

■ **Figure 2.44** A bullet 'frozen' by high-speed photography

Photographs taken in quick succession became useful in analysing many types of motion in the nineteenth century, but the trajectories of very rapid motion (like projectiles) was difficult to understand until they could be filmed, or illuminated by lights flashing very quickly (stroboscopes). The photograph of the bullet from a gun shown in Figure 2.44 required high-technology, such as a very high-speed flash and very sensitive image recorders, in order to 'freeze' the projectile (bullet) in its rapid motion (more than $500\,m\,s^{-1}$).

Use the internet to find out about the work of Eadweard Muybridge.

Effects of air resistance

In practice, ignoring the effects of air resistance can be unrealistic, especially for smaller and/or faster-moving objects. So it is important to understand, in general terms, how air resistance affects the motion of projectiles.

Air resistance (*drag*) provides a force that opposes motion. Without air resistance we assume that the horizontal component of a projectile's velocity is constant, but with air resistance it decreases. Without air resistance the vertical motion always has a downwards acceleration of $9.81\,m\,s^{-2}$, but with air resistance the acceleration will be reduced for falling objects and the deceleration increased for objects moving upwards.

Figure 2.45 shows typical trajectories with and without air resistance (for the same initial velocity). Note that with air resistance the path is no longer parabolic or symmetrical.

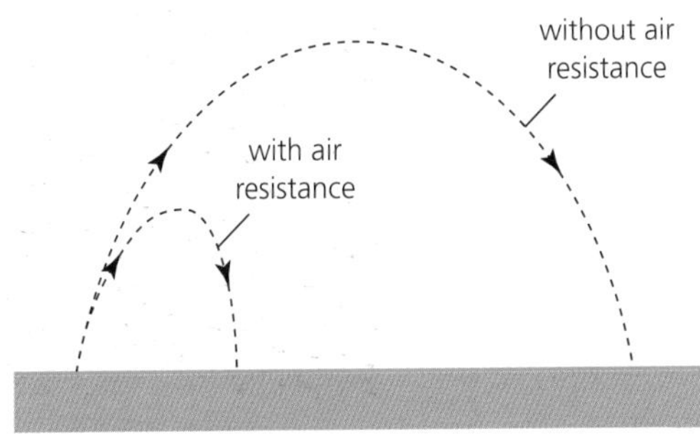

■ **Figure 2.45** Effect of air resistance on the trajectory of a projectile

Calculations on projectile motion

If the velocity (speed and direction) of any projectile moving through the air is known at any moment, then the equations of motion can be used to determine the object's velocity at *any* time during its trajectory. *To carry out any of these* *calculations, we must assume that there is no air resistance and that the downwards acceleration due to gravity is always 9.81 m s*$^{-2}$.

The transfer of gravitational potential energy (mgh) to kinetic energy ($\frac{1}{2}mv^2$) can sometimes provide an alternative solution to a problem – by equating these two energies we see that for a mass falling from rest through a vertical height close to the Earth's surface:

$$mgh = \frac{1}{2}mv^2$$

or

$$v = \sqrt{2gh}$$

Objects projected horizontally

Worked example

8 A bullet is fired horizontally with a speed of 524 m s^{-1} from a height of 22.0 m above the ground. Calculate where it will hit the ground.

First we need to calculate how long the bullet is in the air. We can do this by finding the time that the same bullet would have taken to fall to the ground if it had been dropped vertically from rest (so $u = 0$):

$$s = ut + at^2$$

$$22.0 = \frac{1}{2} \times 9.81 \times t^2$$

$$t = 2.12\,\text{s}$$

Without air resistance the bullet will continue to travel with the same horizontal component of velocity (524 m s^{-1}) until it hits the ground 2.12 s later. Therefore:

horizontal distance travelled = horizontal velocity × time

horizontal distance = 524 × 2.12 = 1.11 × 10^3 m

39 Make a copy of Figure 2.45 and add to it the trajectories of an object projected in the same direction with:
 a lower initial velocity
 b higher initial velocity.

40 a Use a spreadsheet to calculate the vertical and horizontal displacements (every 0.2 s) of a stone thrown horizontally off a cliff (from a height of 48 m) with an initial velocity of 25 m s^{-1}. Continue calculations until it hits the ground.
 b Plot a graph of the stone's trajectory.

41 A rifle is aimed horizontally and directly at the centre of a target that is 52.0 m away.
 a If the bullet had an initial velocity of 312 m s^{-1}, how long would it take to reach the target?
 b How far below the centre would the bullet hit the target?

Objects projected at other angles

The physics is the same for all projectiles at all angles: trajectories are still parabolic and the vertical and horizontal components remain independent of each other. But the mathematics is more complicated if the initial motion is not vertical or horizontal.

The most common problems involve finding the maximum height and the maximum horizontal distance (**range**) of the projectile.

If we know the velocity and position of a projectile, we can always use its *vertical component* of velocity to determine:

- the time taken before it reaches its maximum height, and the time before it hits the ground
- the maximum height reached (assuming its velocity has an upwards component).

The *horizontal component* can then be used to determine the range.

If the velocity at any time is needed, for example when the projectile hits the ground, then the vertical and horizontal components have to be combined to determine the resultant in magnitude and direction.

Worked example

9 A stone was thrown upwards from a height 1.60 m above the ground with a speed of 18.0 m s^{-1} at an angle of 52.0° to the horizontal. Assuming that air resistance is negligible, calculate:

a its maximum height
b the vertical component of velocity when it hits the ground
c the time taken to reach the ground
d the horizontal distance to the point where it hits the ground
e the velocity of impact.

First we need to know the two components of the initial velocity:

$$u_V = u\sin\theta = 18.0\sin 52.0° = 14.2\,\text{m s}^{-1}$$
$$u_H = u\cos\theta = 18.0\cos 52.0° = 11.1\,\text{m s}^{-1}$$

a Using $v^2 = u^2 + 2as$ for the upwards vertical motion (with directions upwards considered to be positive), and remembering that at the maximum height $v = 0$, we get:

$0 = 14.2^2 + [2 \times (-9.81) \times s]$

$s = +10.3$ m above the point from which it was released; a total height of 11.9 m.

(Using $\frac{1}{2}v^2 = gh$ is an alternative way of performing the same calculation.)

b Using $v^2 = u^2 + 2as$ for the complete motion gives:

$v^2 = 14.2^2 + [2 \times (-9.81) \times (-1.60)]$

$v = 15.3$ m s^{-1} downwards

c Using $v = u + at$ gives:

$-15.3 = 14.2 + (-9.81)t$

$t = 3.0$ s

d Using $s = vt$ with the horizontal component of velocity gives:

$s = 11.1 \times 3.0 = 33.3$ m

e Figure 2.46 illustrates the information we have determined so far and the unknown angle and velocity.

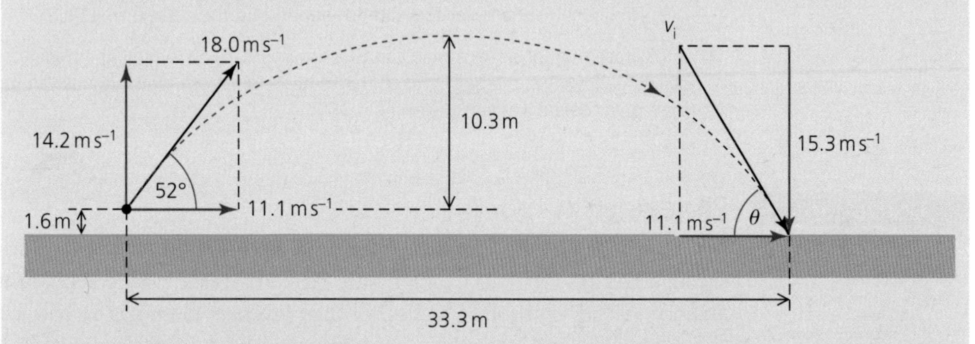

■ **Figure 2.46**

From looking at the diagram, we can use Pythagoras's theorem to calculate the velocity of impact:

(velocity of impact)^2 = (horizontal component)^2 + (vertical component)^2

$v_i^2 = 11.1^2 + 15.3^2$

$v_i = 18.9$ m s^{-1}

The angle of impact with the horizontal, θ, can be found using trigonometry:

$\tan\theta = \dfrac{15.3}{11.1}$

$\theta = 54.0°$

42 Repeat Worked example 9 for a stone thrown with a velocity of 26 m s^{-1} at an angle of 38° to the horizontal from a cliff top. The point of release was 33 m vertically above the sea.

43 The maximum theoretical range of a projectile occurs when it is projected at an angle of 45° to the ground (once again, ignoring the effects of air resistance). Calculate the maximum distance a golf ball will travel before hitting the ground if its initial velocity is 72 m s^{-1}. (Because you need to assume that there is no air resistance, your answer should be much higher than the actual ranges achieved by top-class golfers.)

44 A jet of water from a hose is aimed directly at the base of a flower, as shown in Figure 2.47. The water emerges from the hose with a speed of 3.8 m s^{-1}.

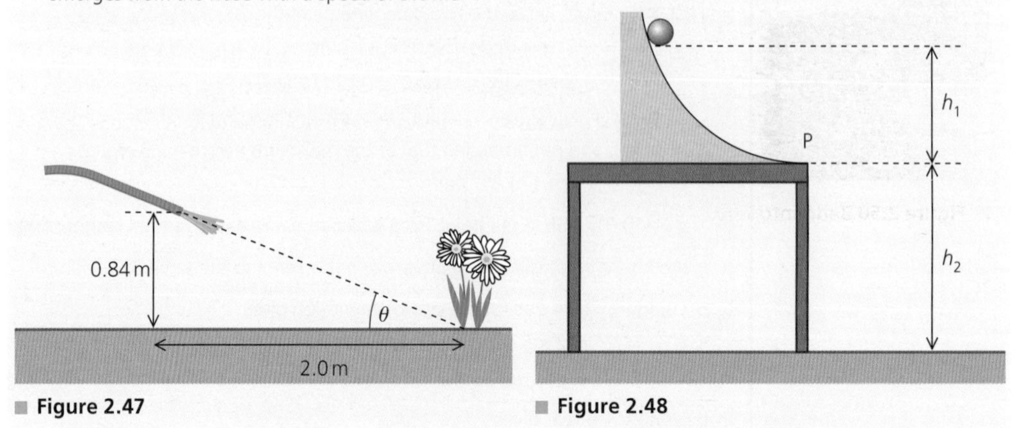

■ **Figure 2.47**

■ **Figure 2.48**

a Calculate the angle, θ, and the vertical and horizontal components of the initial velocity of the water.
b How far away from the base of the plant does the water hit the ground?

45 A ball rolls down the slope shown in Figure 2.48 and is then projected horizontally off the table top at point P.
a Show that the maximum range of the ball is given by $R = 2\sqrt{h_1h_2}$. (Ignore any effects due to the spinning of the ball.)
b What assumption(s) did you make?
c Explain why your answer to a did not depend on the mass of the ball.

46 If the maximum distance a man can throw a ball is 78 m, what is the minimum speed of release of the ball? (Assume that the ball lands at the same height from which it was thrown and that the greatest range for a given speed is when the angle is 45°.)

Utilizations

Projectiles in sport

Many sports and games involve some kind of object (often a ball) being thrown, kicked or hit through the air. Obvious examples are basketball (see Figure 2.49), tennis, football, badminton, archery, cricket and golf. The skill of the players is to make the ball, or other object, move with the right speed and trajectory, and often to also be able to judge correctly the trajectory of an object moving towards them.

■ **Figure 2.49** A basketball moves in approximately parabolic flight

The mass, shape, diameter and the nature of the surface of a ball will all affect the way in which it moves through the air after it has been 'projected'. Although in most sports it can be assumed that the ball will follow an *approximately* parabolic path through the air, if the ball always had a perfectly parabolic trajectory then the game would be predictable and less skilful. The effect of the air moving over the surface of the ball plays an important part in many sports and good players can use this to their advantage by putting spin on the ball. There is a difference in air pressure on opposite sides of a spinning ball, producing a force that affects the direction of motion.

■ **Figure 2.50** Badminton shuttlecock trajectories are not parabolic

Part of the fun of playing or watching sports is to see a ball being struck or thrown with such skill that it travels with great accuracy and speed, or goes a long distance. It is interesting to consider how the design of the balls in different sports has evolved.

Badminton is an unusual sport because the design of the shuttlecock deliberately produces non-parabolic trajectories (see Figure 2.50). A shuttlecock has a small mass for its cross-sectional area, which means that it can travel very fast when it is first hit; after that, air resistance has a significant effect, considerably reducing its range. Most of a shuttlecock's mass is concentrated in the 'cork' at the opposite end from the feathers, so that it always moves in flight such that the cork leads the motion.

1 In which sport can the struck ball travel the longest distance? Find out if there are any regulations in that game that try to limit how far the ball can travel.
2 If the balls from a variety of different sports were all dropped from the same height onto the same hard surface, which one would bounce up to the greatest height? Discuss possible reasons why that ball loses the smallest fraction of its energy when colliding with the surface and why that is important for the sport in which it is used.
3 Research an explanation of how spin can cause a ball to change direction.

2.2 Forces – *classical physics requires a force to change a state of motion, as suggested by Newton in his laws of motion*

At its simplest, a force is a push or a pull. A force acting on an object (a *body*) can make it start to move (Figure 2.51) or change its motion if it is already moving. In other words, a force can change the velocity of an object; accelerations are caused by forces.

Forces can also change the shape of an object. That is, a force can make an object become deformed in some way. For example, when we sit on a soft chair the deformation is easy to see. When we sit on a hard chair, or stand on the floor, there is still a deformation but it is usually too small to see.

■ **Figure 2.51** The upwards force on a rocket accelerates it into space

Clearly, the effect of a force will depend on the direction in which it acts. Force is a *vector* quantity. Like all vectors, a force can be represented by drawing a line of the correct length in the correct direction (shown with an arrow), to or from the correct *point of application*. The vector arrow should be clearly labelled with an accepted name or symbol. The length should be *proportional to* the magnitude of the force. For example, in Figure 2.52 vector arrows represent the different weights of two people.

When discussing the forces *acting* on an object, we may alternatively talk about *applying* a force to an object, or *exerting* a force on an object.

The symbol $\boldsymbol{F}$ is used for force and the SI unit of force is the newton, N. One newton is defined as the (resultant) force that makes a mass of 1 kg accelerate by $1\,\mathrm{m\,s^{-2}}$.

Objects as point particles

When a force is applied to an object in simple situations, the shape and size of the object are often not of much importance, and adding details to any drawing can lead to confusion. We might, for example, wonder if an extended object shown in a drawing will tip or rotate when acted on by forces. See Figure 2.61 for an example. For this reason, and for simplicity, objects are often represented as points – **point particles**.

Different types of force

Apart from obvious everyday pushes and pulls, we are surrounded by a number of different types of force. In the following section we will introduce and briefly discuss these types of force:

- weight
- tension and compression
- reaction forces
- friction and air resistance
- upthrust
- other non-contact forces (like weight)

Weight

The **weight**, W, of a mass is the **gravitational force** that pulls it towards the centre of the Earth. Weight is related to mass by the following equation:

$$W = mg$$

In this equation, W is the weight in newtons, m is the mass of the object in kilograms and g is the acceleration due to gravity in metres per second squared (m s^{-2}). The larger the mass of the object, the bigger its weight. (The symbol W is more commonly used for work.)

An alternative interpretation of g is as the ratio of weight to mass, $g = W/m$. Expressed in this way, it is known as the **gravitational field strength** with the unit of newtons per kilogram, N kg^{-1} ($1\,\text{N kg}^{-1} = 1\,\text{m s}^{-2}$). A certain mass would weigh less on the Moon because the Moon has a smaller gravitational field strength than the Earth.

The value of g on, or close to, the Earth's surface is assumed to be $9.81\,\text{m s}^{-2}$, although it does vary as we saw in Table 2.1. For quick approximations a value of $g = 10\,\text{m s}^{-2}$ is often used – which is only a 2% difference. The value of g decreases as the distance from the centre of the Earth increases. For example, at a height of 300 km above the Earth's surface the value of g has decreased slightly to $9.67\,\text{m s}^{-2}$. This means that objects at that height, such as a satellite or astronauts in orbit around the Earth, are *not* weightless (as is often believed), but weigh only a little less than on the Earth's surface.

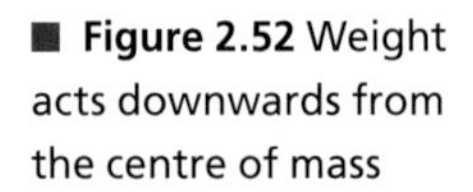
Figure 2.52 Weight acts downwards from the centre of mass

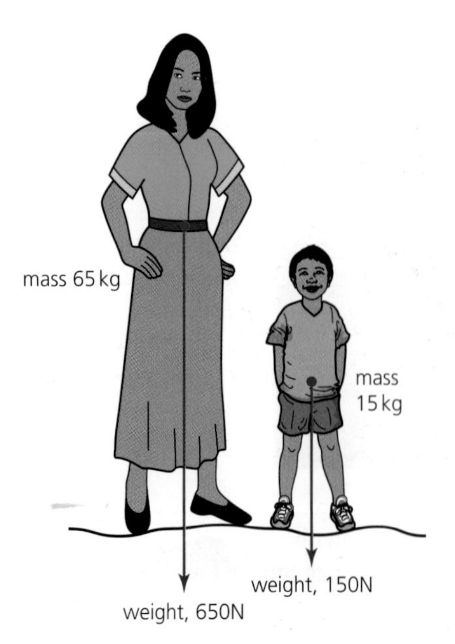

If we want to represent the weight of an object in a diagram, we use a vector arrow of an appropriate length drawn vertically downwards from the **centre of mass** of the object, as shown in Figure 2.52. The centre of mass of an object can be considered as the 'average' position of all of its mass. For symmetrical and uniform objects the centre of mass is at the geometrical centre.

The mass of an object stays the same wherever it is in the universe, but the gravitational force on an object (its weight) varies depending on its location. For example, the acceleration due to gravity (gravitational field strength) on the Moon is 1.6 $m s^{-2}$ and on Mars it is 3.7 $m s^{-2}$. The acceleration due to gravity is different on the Moon and Mars because they have different masses and sizes compared with the Earth. In deep space, a very long way from any star or planet, any object would be (almost) weightless.

Unfortunately, in everyday conversation the word 'weighing' is used for finding the mass of an object, in kg for example (not the weight in N), and the question 'what does that weigh?' would also usually be answered in kilograms, not newtons. This is a common confusion that every student and teacher of physics has to face.

Worked example

10 An astronaut has a mass of 62.2 kg. What would her weight be in the following locations?

a on the Earth's surface
b in a satellite 300 km above the Earth
c on the Moon
d on Mars
e a very, very long way from any planet or star

a $W = mg = 62.2 \times 9.81 = 610$ N
b $W = 62.2 \times 9.67 = 601$ N
c $W = 62.2 \times 1.6 = 100$ N
d $W = 62.2 \times 3.7 = 230$ N
e zero

47 Calculate the weight of the following objects on the surface of the Earth:
a a car of mass 1250 kg
b a newborn baby of mass 3240 g
c one pin in a pile of 500 pins that has a total mass of 124 g.

48 A girl has a mass of 45.9 kg. Use the data given in Table 2.1 to calculate the difference in her weight between Bangkok and London.

49 a It is said that 'an A380 plane has a maximum take-off weight of 570 tonnes' (Figure 2.53). A tonne is the same as a mass of 1000 kg. What is the maximum weight of the plane (in newtons) during take-off?
b The plane can carry a maximum of about 850 passengers. Estimate the total mass of all the passengers and crew. What percentage is this of the total mass of the plane on take-off?
c The maximum landing weight is '390 tonnes'. Suggest a reason why the plane needs to be less massive when landing than when taking off.
d Calculate the difference in mass and explain where the 'missing' mass has gone.

■ **Figure 2.53** The Airbus A380 is the largest passenger airplane in the world

50 The weight of an object decreases very slightly as its distance above the Earth's surface increases. Suggest why the weight of an object might not increase if it was taken down a mine shaft and closer to the centre of the Earth.

51 A mass of 50 kg would have a weight of 445 N on the planet Venus. What is the strength of the gravitational field there? Compare it with the value of g on Earth.

52 Consider two solid spheres made of the same metal. Sphere A has twice the radius of sphere B. Calculate the ratio of the two spheres' circumferences, surface areas, volumes, masses and weights.

Utilizations

Force meters and weighing

Forces are most easily measured by the changes in length they produce when they squash or stretch a spring (or something similar). Such instruments are called *force meters* (also called *newton meters* or *spring balances*) – see Figure 2.54. In this type of instrument the spring usually has a change of length proportional to the applied force. The length of the spring is shown on a linear scale, which can be calibrated (marked) in newtons. The spring goes back to its original shape after it has measured the force.

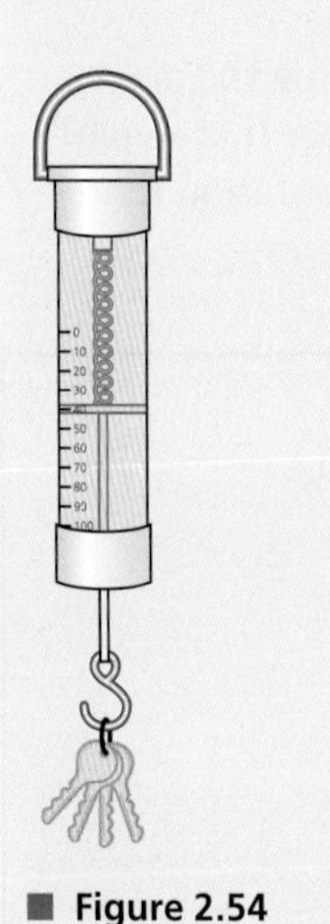

■ **Figure 2.54** Force meter

Such instruments can be used for measuring forces acting in any direction, but they are also widely used for the measurement of weight. The other common way of measuring weight is with some kind of 'balance' (scales). In an equal-arm balance, as shown in Figure 2.55, the beam will only balance if the two weights are equal. That is, the unknown weight equals the known weight.

In this type of balance, the pivot could be moved closer to the unknown weight if it is much heavier than the known weight(s). The balance then has to be calibrated using the *principle of moments*. This principle is not part of the course, but it may be familiar to students from earlier work.

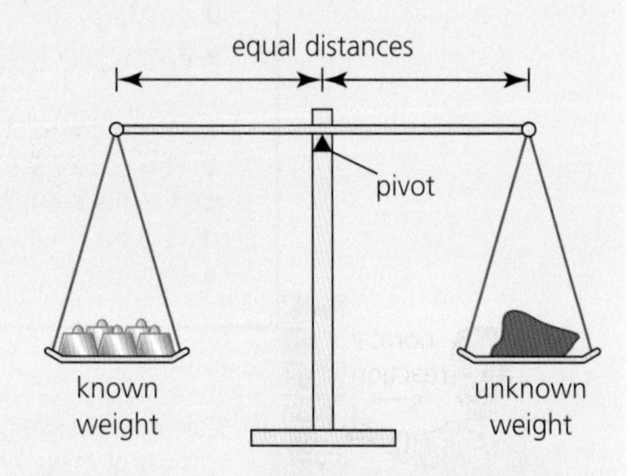

■ **Figure 2.55** An equal-arm balance

Either of these methods can be used to determine an unknown weight (N) and they rely on the force of gravity to do this, but such instruments are much more commonly calibrated to indicate mass (kg or g) rather than weight. This is because we are usually more concerned with the quantity of something, rather than the effects of gravity on it. We usually assume that mass (kg) = weight (N)/9.81 anywhere on Earth because any variations in the acceleration due to gravity, *g*, are insignificant for most, but not all, purposes.

1 If you were buying something small and expensive, like gold or diamonds (Figure 2.56), should the amount you are buying be measured as a mass or a weight? Explain your answer.

■ **Figure 2.56** An expensive ring

Tension and compression

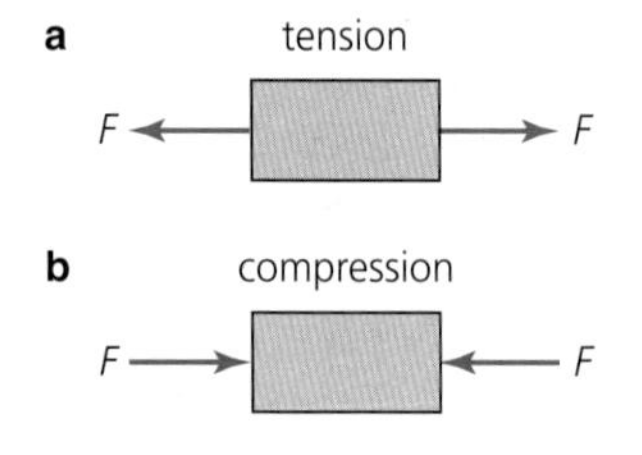

■ **Figure 2.57** Object under **a** tension and **b** compression

When an object is stretched by equal and opposite forces pulling it outwards, we describe it as being under **tension** (Figure 2.57a). Stretched strings or rubber bands are familiar examples of objects under tension, but *tensile* forces are also very common in more rigid objects, such as the horizontal tie in a stool or chair, where its purpose is to stop the legs from moving outwards.

When an object is squashed by equal and opposite forces pushing it inwards, we describe it as being under **compression** (Figure 2.57b).

All structures have parts that are under tension and parts under compression. The stone pillars at Stonehenge, in the UK (Figure 2.58) are strong under the compression caused by their own weight and the weight of the slabs resting on top.

■ **Figure 2.58** Stonehenge was built over 4000 years ago

The horizontal slabs on the top of stonehenge bend *very* slightly, so that the top surface is compressed, while the lower surface is under tension and this can result in destructive cracks spreading upwards. Similar principles apply to the construction of all modern buildings, bridges etc. As an example consider Figure 2.59, which shows a sketch of a suspension bridge, along with the parts that are under tension (T) and the parts under compression (C). Construction of model bridges of various designs, followed by observation of the testing of their strengths by adding increasing loads is a popular educational exercise for physics students.

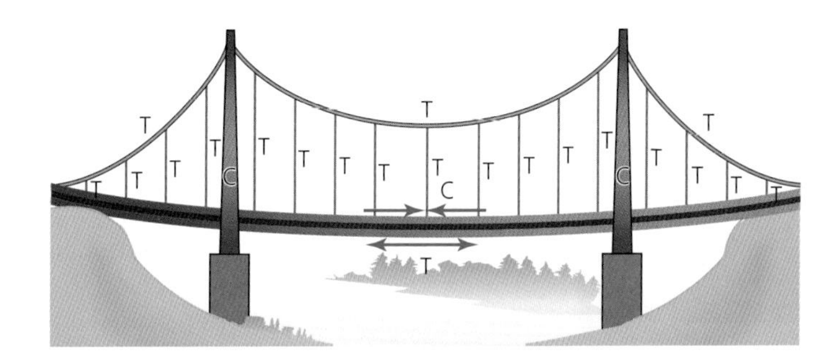

■ **Figure 2.59** Model suspension bridge

■ **Figure 2.60** Reaction forces

Reaction forces

If two objects are touching (in *contact* with) each other, then each must exert a force on the other. For example, if you push on a wall, then the wall must also push back on you; when you stand on the floor your weight presses down but the floor must also push up on you to support your weight. If this was not true you would fall through the floor or the wall.

In Figure 2.60 the boy's weight is pushing down on the ground and his hand is also pushing the wall. The force of the wall on the boy's hand and the force of the ground on his feet are examples of **contact forces** (also called *reaction forces*). These forces are always perpendicular to the surface and that is why they are often called **normal reaction forces** (the word 'normal' used in this way means perpendicular).

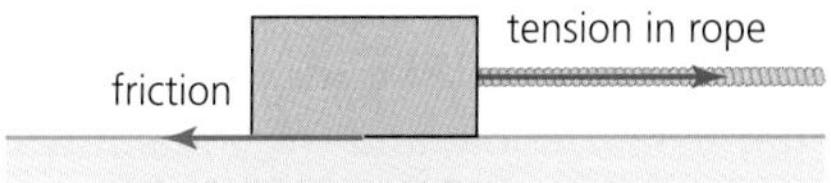

tension in rope

friction

■ **Figure 2.61** Friction opposing motion

Solid friction

When objects are moving (or trying to move) and they are in contact with other surfaces, forces between the surfaces act in such a way as to *oppose* (try to stop) the motion. This type of force is called **friction**.

There are many ways of trying to reduce the effects of friction in an attempt to make movement easier, but friction can never be completely overcome. Friction between two objects acts parallel to the surfaces of both, in the opposite direction from the motion (or intended motion). This is shown in Figure 2.61, in which a block is being accelerated by being pulled by a rope along the floor.

But without friction, movement would be very difficult. Consider how you walk across a room (Figure 2.62) – in order to take a step, the foot pushes backwards on the ground and, because of friction, the ground pushes forward on the foot. Without friction, walking and most methods of transportation would be impossible.

Friction is discussed in more detail later in this chapter.

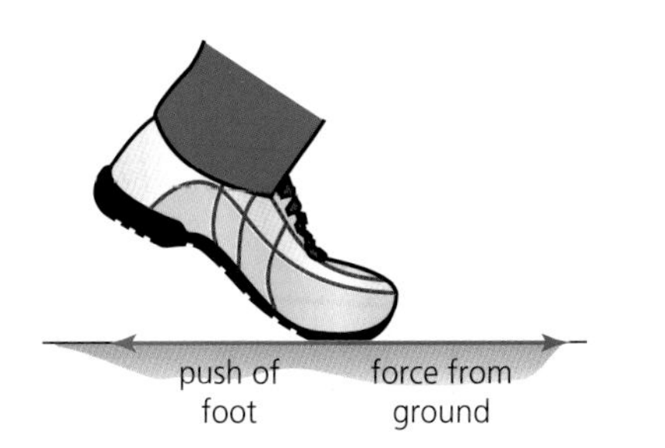

■ **Figure 2.62** We need friction to walk

Air resistance

Air resistance (sometimes called **drag**) is also a force that opposes motion. An object moving through air has to knock the air out of the way, and this produces a force in the opposite direction to motion.

■ **Figure 2.63** The LZR Racer swimsuit is built using NASA technology

■ **Figure 2.64** Shanghai maglev train

■ **Figure 2.65** Wind tunnel testing

The study of the factors affecting air resistance is of great importance when discussing falling objects, parachutes and all modes of transport (especially for vehicles moving fast) and it has many interesting sporting applications.

The amount of air resistance depends on the cross-sectional area of the moving object, but also on the way in which the air flows past the surfaces. Altering the shape of an object and the nature of its surfaces can have a considerable effect on the air resistance it experiences. Changing the shape and/or surface of an object, particularly at the front, in order to reduce air resistance is called *streamlining*.

Most importantly, the faster an object moves the higher the air resistance opposing its motion becomes. Typically, for a given object, it is often assumed that air resistance is proportional to the speed *squared*. This means that air resistance becomes *much* more important for objects moving very quickly. The air resistance opposing a 100 m sprinter moving at $10\,\text{m}\,\text{s}^{-1}$ could be 400 times bigger than on a casual walker moving at $0.5\,\text{m}\,\text{s}^{-1}$. The retarding effects of air resistance on a car driving a few blocks to the local shops at an average speed of $30\,\text{km}\,\text{h}^{-1}$ would be much less than the same car driving at $110\,\text{km}\,\text{h}^{-1}$ along a motorway (about 13 times greater).

Similar ideas apply to the drag effects when an object, person or animal moves through (or on) water (fluid resistance). For example, the amount of resistance experienced by a swimmer can be reduced by about 5% by wearing drag-reducing swimsuits, as shown in Figure 2.63. Of course it is very important that the suit does not affect the swimmer's movement in any way and that the extra weight of the swimsuit is insignificant.

Figure 2.64 shows one of the magnetic levitation ('maglev') trains that run between Shanghai and its main airport, which is about 30 km away. Magnetic forces lift the train above the surface of the track to eliminate friction, and the streamlined shape of the train is designed to reduce air resistance. The train completes the journey in about 7 minutes and reaches a top speed of about $430\,\text{km}\,\text{h}^{-1}$, although in test runs it exceeded $500\,\text{km}\,\text{h}^{-1}$.

The effect of air resistance on different objects is often tested in 'wind tunnels' (Figure 2.65). Instead of the object moving through stationary air, it is kept still while fast moving air is blown against it.

Utilizations

Air travel

Aircraft use a lot of fuel moving passengers and goods from place to place quickly, but we are all becoming more aware of the effects of planes on global warming and air pollution. Some people think that governments should put higher taxes on the use of planes to discourage people from using them too much. Improving railway systems, especially by operating trains at higher speeds, will also attract some passengers away from air travel. Of course, engineers try to make planes more efficient so that they use less fuel, but the laws of physics cannot be broken and jet engines, like all other heat engines, cannot be made much more efficient than they are already.

Planes will use a lower fuel if there is a lower air resistance acting on them. This can be achieved by designing planes with streamlined shapes, and also by flying at greater heights where the air is less dense. Flying more slowly (than their maximum speed) can also reduce the amount of fuel used for a particular trip, as it does with cars, but people generally want to spend as little time travelling as possible.

The pressure of the air outside an aircraft at its typical cruising height is far too low for the comfort and health of the passengers and crew, so the air pressure has to be increased inside the plane, but this is still much lower than the air pressure near the Earth's surface. The difference in air pressure between the inside and outside of the plane would cause problems if the plane had not been designed to withstand the extra forces.

Planes generally carry a large mass of fuel, and the weight of a plane decreases during a journey as the fuel is used up. The upwards force supporting the weight of a plane in flight comes from the air that it is flying through and will vary with the speed of the plane and the density of the air. When the plane is lighter towards the end of its journey it can travel higher, where it will experience less air resistance.

1 a Find out how much fuel is used on a long-haul flight of, say, 12 hours.

  b Compare your answer with the capacity of the fuel tank on an average sized car.

  c On a short-haul flight as much of 50% of a plane's fuel might be used for taxiing, taking off, climbing and landing, but on longer flights this can reduce to under 15%. Explain the difference.

2 Do you agree that the use of planes should be discouraged in some way? Does the rest of your group agree with you? Does the government have the right or obligation to try to change people's behaviour by the use of taxes? (Taxes on alcohol and cigarettes are similar examples.)

Upthrust

Upthrust is a force exerted vertically upwards on *any* object that is in a fluid (gas or liquid). This force arises because the pressure of the fluid on the object is greater at its bottom than at its top. Upthrust acts in the opposite direction to weight and its effect is to reduce the *apparent* weight of the object.

The upthrust from water is a familiar experience for swimmers and divers and is the same force that keeps a boat afloat. The weight of a floating object, or an object buoyant under water, is equal and opposite to the upthrust (see Figure 2.66). Upthrust forces also exist on objects in air, but they are less significant and are only normally noticeable on very light objects like balloons.

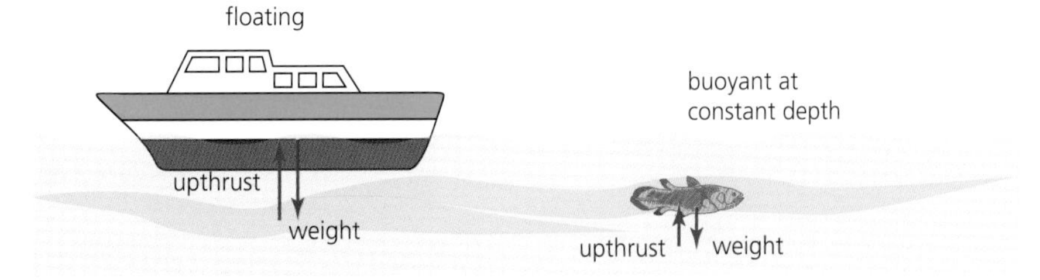

■ **Figure 2.66** Forces on objects that are floating or buoyant under water

Non-contact forces

Of all the forces discussed above, *gravitational force* (*weight*) is different from the others because it acts across space and there does not need to be any contact (between an object and the Earth). *Magnetic forces* and *electric forces* behave in a similar way, and non-contact *nuclear forces* also exist within atoms. Understanding these **fundamental forces** plays a very important part in physics. These forces are all covered in more detail in later chapters of this book.

In order to fully explain all the contact forces mentioned earlier in this chapter, it would be necessary to consider the electromagnetic forces acting between particles in the different objects/substances.

Free-body diagrams

When objects come into contact with each other they exert forces on each other. This means that even the simplest force diagrams can get confusing if all the forces are included. To avoid this confusion, we often draw only *one object* and show only the forces acting *on* that one object. Forces that act from the body onto something else are *not* included.

Drawings that show only one object and the forces acting on it are called **free-body diagrams**. Some simple examples are shown in Figure 2.67.

Figure 2.67 Free-body diagrams. The object has a solid outline and the forces are shown in red. The dotted lines might not be included

a A box on the ground

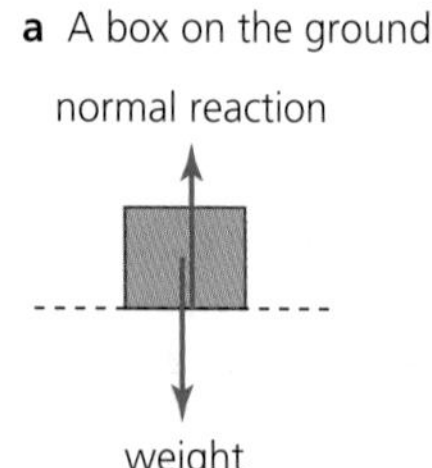

b The Moon orbiting the Earth

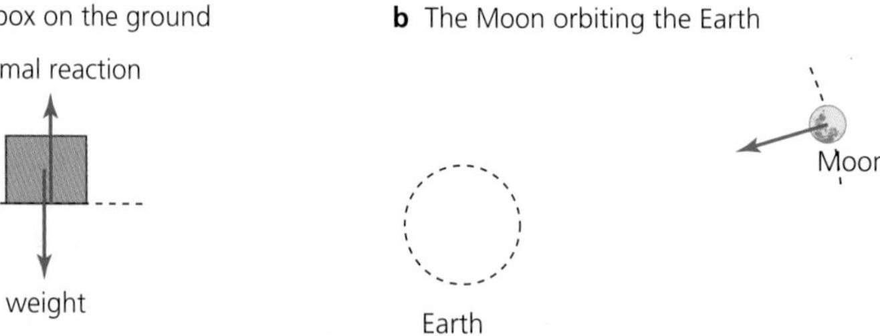

c A swinging pendulum

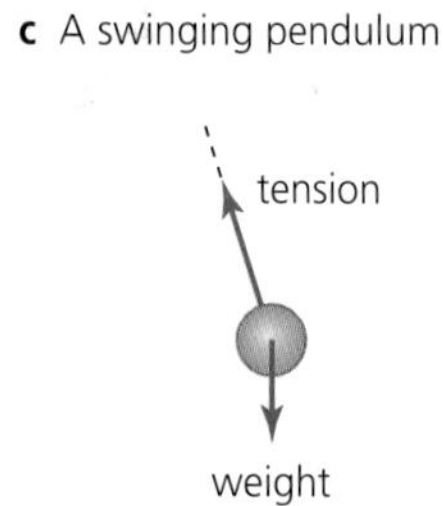

d A box pulled along the ground (at constant speed)

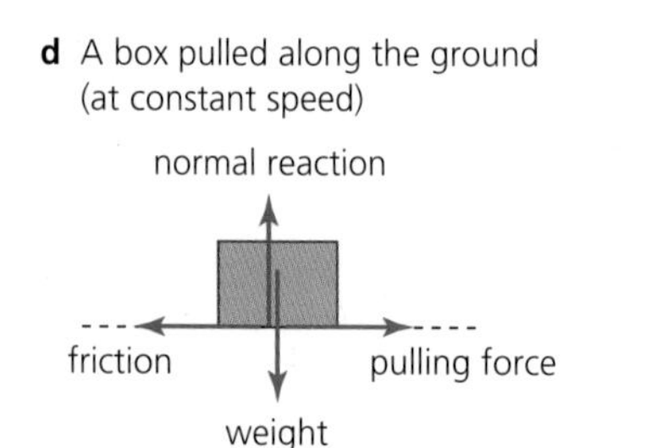

53 Figure 2.68 shows two unequal masses connected by string over a frictionless pulley. Draw free-body diagrams to show the forces acting on both masses.

54 Figure 2.69 represents a hot air balloon. The two ropes are stopping it from moving vertically away from the ground. Draw a free-body diagram for all the forces acting on the basket.

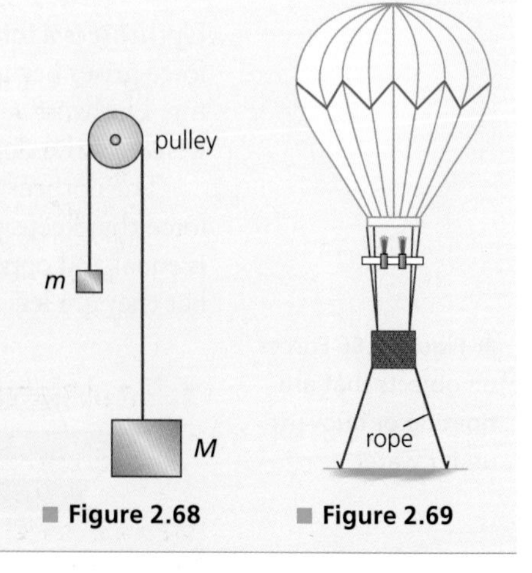

Figure 2.68 **Figure 2.69**

Resultant forces and component forces

Resultants

There is usually more than one force acting on an object. In order to determine the overall effect of two or more forces on an object, we must add up the forces, taking their directions into consideration. This gives the **resultant** (overall, net) force acting on the object.

Determining the sums and differences of vectors, such as forces, was covered in Chapter 1. It might be helpful to review that section before continuing.

55 Three separate forces of 1 N, 2 N and 3 N act on an object at the same time. If the forces are parallel to each other what are the possible values for the magnitude of the resultant force?

56 Use a scale drawing to determine the resultant of two forces of 8.5 N and 12.0 N acting at an angle of 120° to each other.

57 Calculate the resultant of forces of 7.7 N and 4.9 N acting perpendicularly to each other.

58 The resultant of two forces is 74 N to the west. If one force was 18 N to the south, what was the size and direction of the other force?

59 A 32 kg box is being pushed across a horizontal floor with a horizontal force of 276 N. The frictional force is equal to 76% of its weight.
a Draw a free-body diagram for the box.
b What is the resultant force on the box?

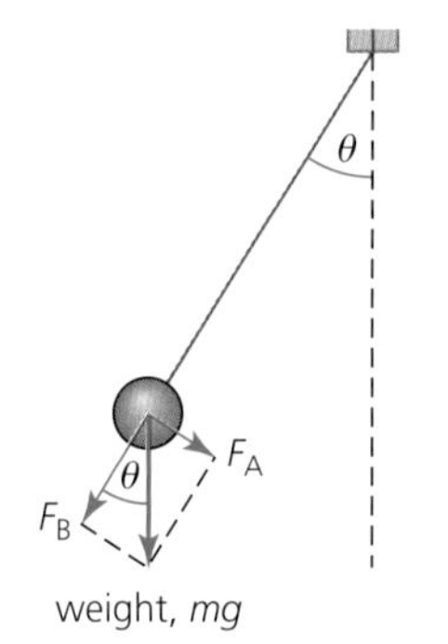

Figure 2.70 Resolving the weight of a pendulum into components:

$F_A = mg\sin\theta$
$F_B = mg\cos\theta$

Components

Two or more forces can be combined to give a resultant, but the 'opposite' process is just as important.

A single force can be considered as equivalent to two separate forces, which are usually chosen to be at right angles to each other ($F\cos\theta$ and $F\sin\theta$). This is called *resolving* a force into two components (Chapter 1).

Resolving forces into components is usually done when a single force is not acting in the direction of motion. As an example, consider the swinging pendulum shown in Figure 2.70. The single force of the weight, *mg*, can be resolved into a force, F_A, acting in the instantaneous direction of motion and a force, F_B, that is at right angles to F_A acting along the line of the string and equal and opposite to the tension in the string.

60 The mass of the pendulum shown in Figure 2.70 is 382 g and the angle θ is 27.4°.
a What is the tension in the string?
b What is the force acting in the direction of motion?

61 The mass shown in Figure 2.71 is stationary on the slope (*inclined plane*).
a Draw a free-body diagram showing the forces acting on the mass.
b Resolve the weight of the mass into two components that are parallel and perpendicular to the slope.

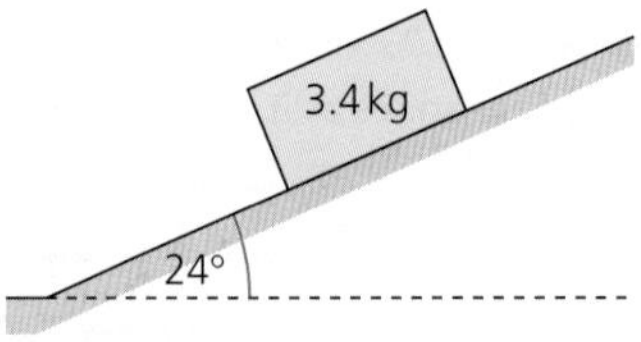

3.4 kg
24°

Figure 2.71

62 A resultant forward force of 8.42×10^4 N acting on a train of mass 3.90×10^5 kg accelerates it a rate of 0.216 m s^{-2} when it is travelling on a horizontal track.
a If the train starts to climb a slope of angle 1.00° to the horizontal, calculate the component of weight acting down the slope.
b What is the new resultant force acting on the train?
c Predict a possible acceleration of the train as it starts to climb the slope.
d Suggest why it is more difficult for trains to travel up steeper slopes than for cars.

Translational equilibrium

An object that has no resultant force acting on it is said to be in **translational equilibrium**.

The word 'translational' refers to movement from place to place. Being 'in equilibrium' means that the forces acting on an object are 'balanced', so that they have no overall effect and the object will therefore continue to move in exactly the same way (or remain stationary).

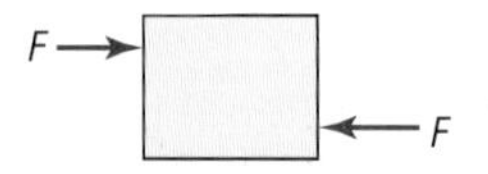

■ **Figure 2.72** The object is in translational equilibrium, but not in rotational equilibrium

It should be noted that it is possible for parallel, equal and opposite forces to act on an object *along different lines* and thereby cause it to rotate, as shown in Figure 2.72. The object will start to rotate under the turning effect of the two forces, but there will be no translational movement. It is not in *rotational* equilibrium.

Newton's first law of motion

Newton's first law of motion summarizes the conditions necessary for translational equilibrium:

> Newton's first law of motion states that an object will remain at rest, or continue to move in a straight line at a constant speed, unless a resultant force acts on it.

In other words, if there is a resultant force acting on an object, it will accelerate.

Examples of Newton's first law

It is not possible to have an object on Earth with *no* forces acting on it because gravity affects all masses. Therefore any object that is in equilibrium must have at least two forces acting on it, and quite possibly many more. All moving objects, or objects tending to move, will also have frictional forces acting on them. If an object is in translational equilibrium, then the forces acting on it (in any straight line) are balanced, so that the resultant force is zero. Consider the following examples.

- **Objects at rest with no sideways forces** – The box shown in Figure 2.67a is in equilibrium because its weight is equal to the normal reaction force pushing up on it.
- **Horizontal motion at constant velocity** – Consider Figure 2.67d, which also shows an object in translational equilibrium because the forces are balanced. It may be stationary or moving to the right with a constant velocity (we cannot tell from this diagram).
- **Vertical motion of falling objects** – Figure 2.73 shows a falling ball. In **a** the ball is just starting to move and there is no air resistance. In **b** the ball has accelerated and has some air resistance acting against its motion, but there is still a resultant force and acceleration downwards. In **c** the speed of the falling ball has increased to the point where the increasing air resistance has become equal and opposite to the weight. There is then no resultant force and the ball is in translational equilibrium, falling with a constant velocity called its terminal velocity or terminal speed.

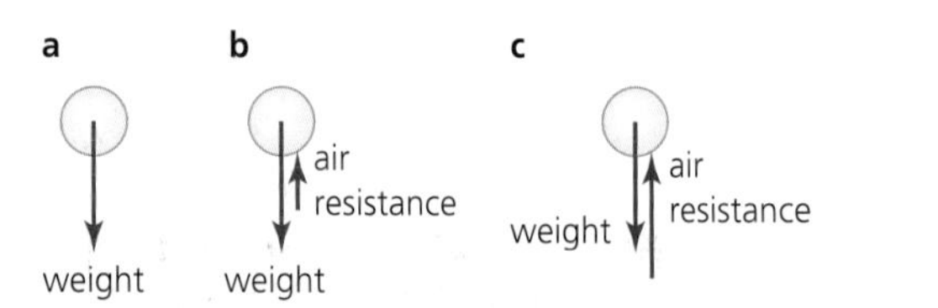

■ **Figure 2.73** The resultant force on a falling object changes as it gains speed

- **Horizontal acceleration** – Figure 2.74 shows the forces acting on a bicycle and rider. Because the force from the road is bigger than air resistance the cyclist will accelerate to the right. As the bicycle and rider move faster and faster, they will meet more and more air resistance. Eventually the air resistance becomes equal to the forward force (but opposite in direction) and a top speed is reached. This is similar to the ideas used to explain the terminal speed of a falling object and the same principles apply to the motion of all vehicles.

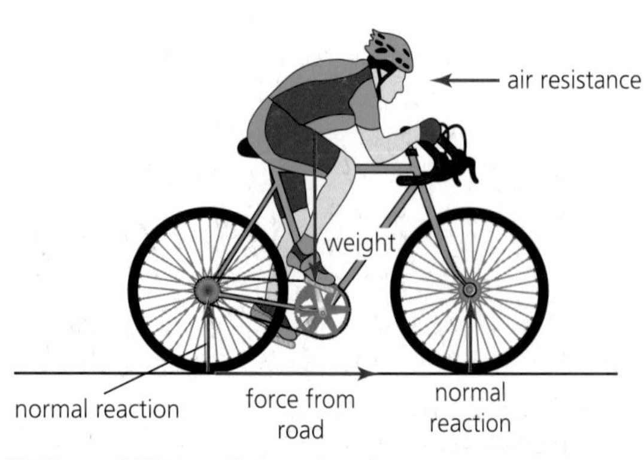

■ **Figure 2.74** A cyclist accelerating

We know that any object that is stationary for some time, like a book placed on a table, is in equilibrium, and an object moving with constant velocity is also in equilibrium. But it is important to realise that a moving object that is only at rest for a moment is *not* in equilibrium. For example, a stone thrown vertically in the air comes to rest for a moment at its highest point, but the resultant force on it is not zero and it is not in equilibrium. Similarly, at the instant that a race is started a sprinter is stationary, but that is the time of highest resultant force and acceleration.

Three forces in equilibrium

If two forces are acting on a mass such that it is *not* in equilibrium, then to produce equilibrium a third force can be added that is equal in magnitude to the resultant of the other two, but in the opposite direction. All three forces must act through the same point. For example, Figure 2.75 shows a free-body diagram of a ball on the end of a piece of string kept in equilibrium by a sideways pull that is equal in magnitude to the resultant of the weight and the tension in the string.

The equilibrium of three forces can be investigated simply by connecting three force meters together with string just above a horizontal surface, as shown in Figure 2.76. The three forces and the angles between them can be measured for a wide variety of different values, each of which maintains the system stationary.

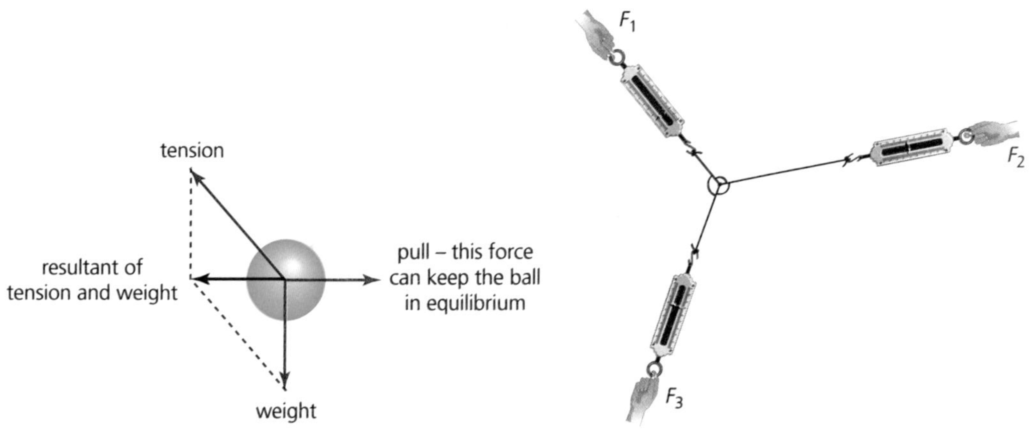

■ **Figure 2.75** Three forces keeping a suspended ball in equilibrium

■ **Figure 2.76** Investigating three forces in equilibrium

63 Draw fully labelled free-body diagrams for:
   a a car moving horizontally at constant velocity
   b an aircraft moving horizontally with a constant velocity
   c a boat decelerating after the engine has been switched off
   d a car accelerating up a hill.

64 Figure 2.77 shows the path of an object thrown through the air without air resistance. Make a copy of the diagram and add to it vector arrows to represent the forces acting on the object in each position.

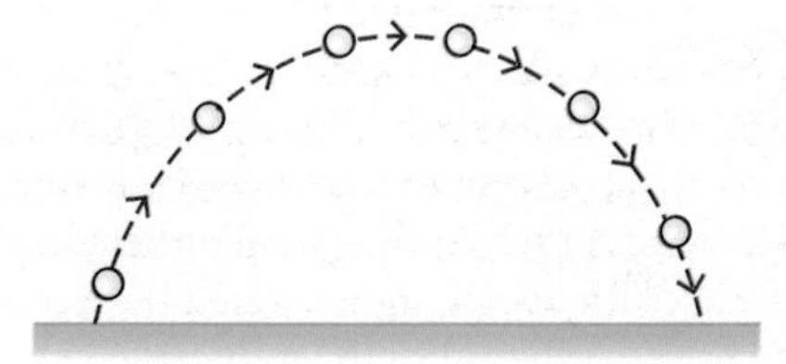

■ **Figure 2.77**

65 A heavy suitcase resting on the ground has a mass of 30.6 kg.
   a Draw a labelled free-body diagram to show the forces acting on the suitcase.
   b Re-draw the sketch to show all the forces acting if someone tries to lift up the case with a vertical force of 150 N.

66 Stand on some bathroom scales with a heavy book in your hand. Quickly move the book upwards while watching the reading on the scales. Repeat, but this time move the book quickly downwards. Explain your observations.

67 a If you are in an elevator (lift) with your eyes closed, is it possible to tell if you are stationary, or moving up or moving down? Explain.

b A person in an elevator (lift) experiences two forces: their weight downwards and the normal reaction force up from the floor. Sketch a free-body diagram to show the forces acting on a person in an elevator if:
  i they are moving at a constant velocity
  ii they are starting to move downwards
  iii they are starting to move upwards
  iv the elevator decelerates after it has been moving downwards
  v the elevator decelerates after it has been moving upwards.

68 A skydiver is falling with a terminal velocity of about 200 km h^{-1} when he opens his parachute.
  a Draw free-body diagrams showing the forces acting on the *skydiver*:
    i at the moment that the parachute opens
    ii just before the skydiver reaches the ground.
  b Sketch a fully labelled graph showing how the velocity of the skydiver changed from the moment he left the plane to the time he landed on the ground.

69 Refer back to Figure 2.71, which shows a box resting on a slope. It can only stay stationary because of the frictional force opposing its movement down the slope. What is the magnitude of the frictional force?

70 Figure 2.78 shows a climber using a rope to get up a mountain. Draw a free-body diagram to represent the forces acting on the climber.

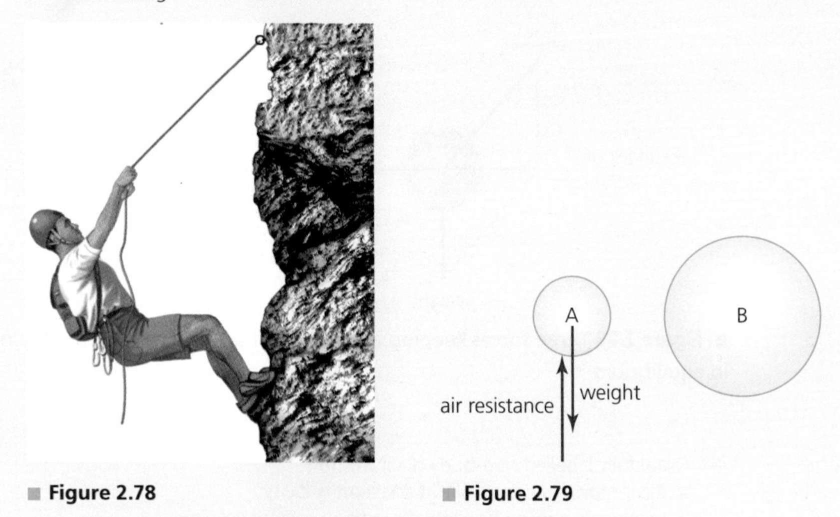

■ **Figure 2.78** ■ **Figure 2.79**

71 Figure 2.79 represents two raindrops falling side by side with the same instantaneous velocity. The forces acting on drop A are shown and it has a radius *r*.
  a Copy the diagram and show the forces acting on drop B, which has radius 2*r*.
  b Describe the immediate motion of the two drops and explain the difference.

Nature of Science

Aristotle and natural philosophy

Isaac Newton is widely quoted as writing modestly about his achievements: 'I do not know what I may appear to the world, but to myself I seem to have been only like a boy playing on the sea-shore, and diverting myself now and then finding a smoother pebble or a prettier shell than ordinary, whilst the great ocean of truth lay all undiscovered before me.'

Newton's apparent humility is also reflected in the following quotation: '…if I have seen a little further it is by standing on the shoulders of giants.' Although this is not an entirely original quotation, it is believed that Newton was giving credit to those scientists and philosophers who had preceded him. Among these was Aristotle.

Aristotle (384–322 BC) was a Greek philosopher and one of the most respected founding figures in the development of human thinking and philosophy. His work covered a very wide range of subjects, including his interpretation of the natural world and the beginnings of what we now call science, although it was called 'natural philosophy' and had a very different approach from modern scientific methods.

Although the 'science' of the time did not involve careful observation, measurements, mathematics or experiments (remember, this was more than 2300 years ago), Aristotle did appreciate the need for universal (all-embracing) explanations of natural events in the world around him.

■ **Figure 2.80** Aristotle

He believed that everything in the world was made of a combination of the four elements he called earth, fire, air and water. The Earth was the centre of everything and each of the four earthly elements had its natural place. When something was not in its natural place, then it would tend to return – in this way he explained why rain falls and why flames and bubbles rise, for example.

With our greatly improved knowledge in the modern world it can be easy to dismiss Aristotle's work and point out the inconsistencies. But his basic ideas on motion, for example, were so simple and powerful that they were widely believed for more than 1500 years, until the age of Galileo and Newton.

Solid friction

As discussed earlier, friction is a force that will try to stop any solid surface moving over another solid with which it is in contact. The origins of friction can be various and complicated but, in general terms, the magnitude of frictional forces will depend on:

- the nature of the two materials
- the 'roughness' of the surfaces of the two materials (roughness is not easily defined and although rougher surfaces often create more friction, it would be wrong to assume that rougher surfaces *always* increase frictional forces)
- the forces acting normally between the surfaces (pushing them together).

To reduce the friction between two given surfaces, something should be placed between them. This could be water, oil, air, graphite, small rollers or balls. Fluids used in this way are called *lubricants*.

The forces of friction can be investigated using simple apparatus such as that in Figure 2.81, which shows a wooden block being pulled on a horizontal table. The horizontal pulling force is increased until the block *just* starts to move.

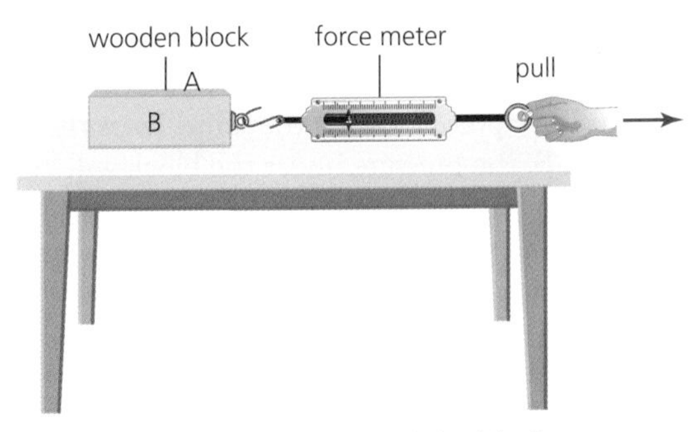

■ **Figure 2.81** A simple experiment to measure frictional forces

Applying Newton's first law, if the block is not moving then it must be in translational equilibrium, so that there is no resultant force and the frictional force must be equal and opposite to any pulling force, as indicated by the force meter. If the block is pulled with a bigger force but it still does not move then the frictional force must also have increased, remaining equal and opposite to the pulling force. But the frictional force will have an upper limit, and if the force is continually increased, at some time it will become greater than the maximum possible value of the frictional force. Then there will be a resultant force on the block and it will accelerate (see Figure 2.82).

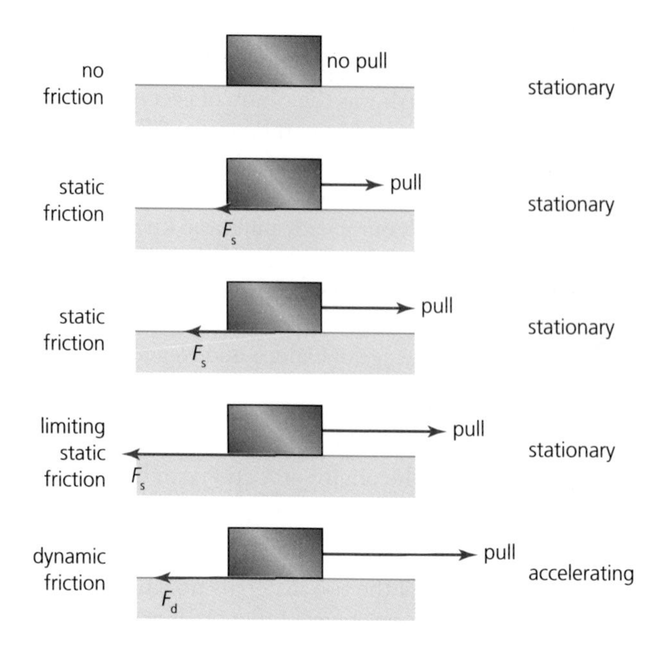

■ **Figure 2.82** How frictional forces change as the force applied increases

Usually the frictional force during motion (**dynamic friction**) is *less* than the maximum frictional force before movement has started (**static friction**). The force of dynamic friction can be considered to be approximately the same for all speeds. That is, for any given two surfaces, the force of dynamic friction has an approximately constant value, whereas the force of static friction can vary from zero up to a limiting value.

Figure 2.83 shows typical results showing how the maximum value of the static frictional force varies as the total weight pressing down is changed by loading masses on top of the block. Note that the total weight is equal to the normal reaction force between the surfaces, R.

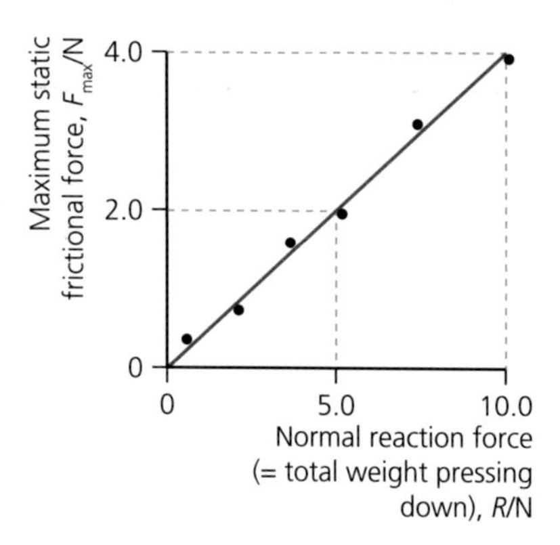

■ **Figure 2.83** Typical variation of maximum static frictional force with normal reaction force (a similar pattern of results will be obtained for dynamic friction)

If the experiment shown in Figure 2.81 is repeated with the same block, but with surface B resting on the table, the measurements of frictional forces will be (approximately) the same. Even though the area of A may be half that of B, the pressure under the block will be doubled ($p = F/A$ – see Chapter 3), pushing the surfaces closer together. In this simplified analysis, maximum values of frictional force depend *only* on the nature of the two surfaces and the normal reaction force between them, and not on the area involved.

Looking at Figure 2.83, we can see that maximum static frictional force, F_{max}, is proportional to normal reaction force, R, so that:

$$\frac{F_{max}}{R} = \mu$$

where μ is a constant for friction between these two materials, which can be determined from the gradient of the graph.

It is usual to quote two different values – for static friction μ_s and for dynamic friction μ_d. These constants are known as the **coefficients of friction**. (The term 'coefficient' simply means a constant used to multiply a variable, in this case a force.) Since these coefficients/constants are *ratios* of forces, they have no units.

Table 2.2 gives some examples of the coefficients of static friction. Values are quoted for two clean, dry, flat, smooth surfaces. Although such simplified situations are a very useful starting point in any analysis, it should be understood that in realistic situations frictional forces are often complex and unpredictable.

Table 2.2 Approximate values for coefficients of static friction

Materials		Approximate coefficients of static friction, μ_s
steel	ice	0.03
ski	dry snow	0.04
Teflon	steel	0.05
graphite	steel	0.1
wood	concrete	0.3
wood	metal	0.4
rubber tyre	grass	0.4
rubber tyre	road surface (wet)	0.5
glass	metal	0.6
rubber tyre	road surface (dry)	0.8
steel	steel	0.8
glass	glass	0.9
skin	metal	0.9

The following two equations for frictional forces can be found in the *Physics data booklet*.

$$F_f \leq \mu_s R$$
$$F_f \leq \mu_d R$$

Alternatively the maximum possible frictional forces are given by $F_{max} = \mu_s R$, or $F_{max} = \mu_d R$.

Worked example

11 a What is the coefficient of friction for the two surfaces represented in Figure 2.83?

b Assuming the results were obtained for apparatus like that shown in Figure 2.81, what minimum force would be needed to move a block of total mass

i 200 g

ii 2000 g?

iii Why is any answer to **ii** unreliable?

c Estimate a value for the dynamic frictional force for the same apparatus with a 200 g mass

i for movement at $1.0\,m\,s^{-1}$

ii for movement at $2.0\,m\,s^{-1}$

a $\mu_s = \frac{F_{max}}{R} = \frac{4.0}{10.0} = 0.40$ (This is equal to the gradient of the graph.)

b i $F_f = \mu_s R = \mu_s mg = 0.40 \times 0.200 \times 9.81 = 0.78\,N$

ii $0.40 \times 2.000 \times 9.81 = 7.8\,N$

iii Because the answer is extrapolated from well outside the range of experimental results shown on the graph.

c i We would expect the dynamic frictional force to be a little smaller than the maximum static frictional force, say about 0.6 N instead of 0.78 N.

ii The dynamic frictional force is usually assumed to be independent of speed, so the force would still be about 0.6 N at the higher speed.

Utilizations

Tyres and road safety

Much of road safety is dependent on the nature of road surfaces and the tyres on vehicles. Friction between the road and a vehicle provides the forces needed for any change of velocity – speeding up, slowing down, changing direction and going around corners. Smooth tyres will usually have the most friction in dry conditions, but when the roads are wet, ridges and grooves in the tyres are needed to disperse the water (Figure 2.84).

■ **Figure 2.84** Tread on a car tyre

To make sure that road surfaces produce enough friction, they cannot be allowed to become too smooth and they may need to be resurfaced every few years. This is especially important on sharp corners and hills. Anything that gets between the tyres and the road surface – for example oil, water, ice and snow – is likely to affect friction and may have a significant effect on road safety.

Increasing the area of tyres on a vehicle will change the pressure underneath them and this may alter the nature of the contact between the surfaces. For example, a farm tractor may have a problem about sinking into soft ground, and such a situation is more complicated than simple friction between two surfaces. Vehicles that travel over soft ground need tyres with large areas to help avoid this problem.

1 Use the internet to find out what materials are used in the construction of tyres and road surfaces to produce high coefficients of friction.

72 a Explain what a coefficient of friction equal to zero would mean.
 b Is it possible to have a coefficient of friction greater than 1? Explain your answer.

73 a A 30 kg wooden container rests on a dry concrete floor. Estimate the force needed to start to move it sideways.
 b Estimate the solid frictional force opposing the motion of a 55 kg girl skating across ice.

74 Consider question 61 and Figure 2.71. The mass can only be stationary on the slope because of friction.
 a Explain how an inclined plane can be used to determine a value for the coefficient of static friction between the surfaces.
 b If the block shown in the diagram just begins to slip down the slope when the angle is 45°, calculate a value for the coefficient of static friction.

75 The coefficient of friction between a moving car and the road surface on a dry day was 0.67.
 a If the car and driver has a total mass of 1400 kg, what frictional force acts between the road and the tyre?
 b If three passengers with a total mass of 200 kg get into the car, calculate a new value for the frictional force.
 c Discuss the possible effects on safety of having extra passengers in the car.

76 How can friction with roads be increased under icy conditions?

77 Suggest why Formula One racing drivers 'warm up' the tyres on their cars. Find out how this is this done. Also, suggest why F1 tyres are so large.

78 Suggest circumstances under which a 'rougher' surface might reduce (rather than increase) friction.

Nature of Science

Newton's *Principia*: an outstanding combination of using mathematics and intuition

Isaac Newton is generally accepted as having been one of the greatest scientists of all time. His three books collectively called *Philosophiæ Naturalis Principia Mathematica*, generally known as the '*Principia*', which include his laws of motion (and gravitation), are among the most influential books ever published (Figure 2.85).

Newton was English, but his books were written in Latin. In order to advance the understanding of mechanics, Newton is often credited with 'inventing' the mathematics of *calculus*, which now plays such an important role in the application of mathematics to many branches of advanced physics. But the 'invention' of calculus was a matter of considerable debate between Newton and the notable German mathematician Gottfried Leibniz, who also had reputable claims. (But was calculus invented or discovered?)

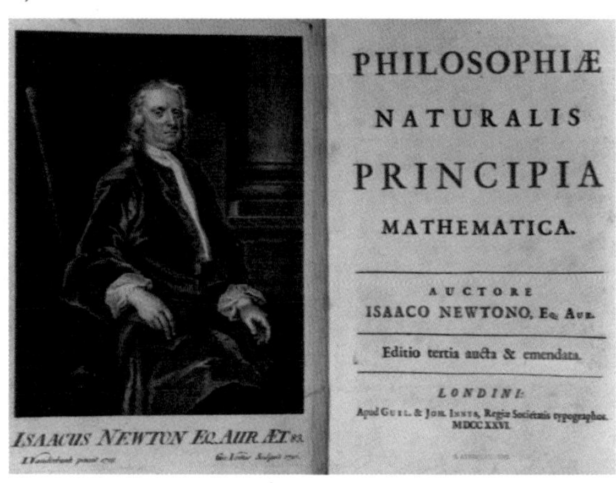

■ **Figure 2.85** Title page of Newton's *Principia*

One of the most famous stories in the history of science is that Newton was suddenly inspired to think of gravity in a new way when observing an apple falling from a tree. It makes a good story to think of this as a moment of *inspiration*, a flash of *intuition*, but we will never know if that was true or not. But certainly Newton himself did make reference to apples falling from trees while explaining his thought processes in devising his law of gravitation.

Sudden, unexpected moments of inspiration have always played an important part in the development of new ideas in all areas of human thinking. However, the American inventor Thomas Edison is famous for saying that 'genius is 1% inspiration and 99% perspiration', suggesting that hard work is also needed to convert great ideas into something useful.

■ Newton's second law of motion

Newton's first law establishes that there is a link between resultant force and acceleration. Newton's second law takes this further and states the mathematical connection – when a resultant force acts on a (constant) mass, the acceleration is proportional to the force:

$$a \propto F$$

Both force and acceleration are vector quantities and the acceleration is in the same direction as the force.

Investigating the effects of different forces and different masses on the accelerations produced is an important part of most physics courses, although reducing the effects of friction is essential for good results. (Air tracks or friction-compensated runways are useful in this respect.) Such experiments also show that if the same resultant force is applied to different masses, then the acceleration produced is inversely proportional to the mass, m – for example, doubling the mass results in half the acceleration.

Combining these results, we see that acceleration is proportional to F/m. Newton's second law can be written as:

$$F \propto ma$$

If we define the unit of force, the newton, to be the force that accelerates 1 kg by 1 m s^{-2}, then we can write:

force (N) = mass (kg) × acceleration (m s^{-2})

$$F = ma$$

This widely used equation is listed in the *Physics data booklet*.

This is just one version of Newton's second law and it is restricted to use with constant masses. A second version of the law will be discussed later in this section, after the concept of momentum has been introduced.

Note that for gravitational forces (weights, W), $F = ma$ can be written as $W = mg$, where g represents the acceleration due to gravity. ($W = mg$ was used earlier in this chapter.)

Newton's second law shows clearly that sudden changes of speed (large accelerations or decelerations) require large forces ($F \propto a$). This can be very helpful in explaining why, for example, a glass will break when dropped on the floor, but might not break if dropped onto a cushion (shorter time and larger deceleration on the hard floor, so that the forces acting on the glass are bigger).

Worked examples

12 a What resultant force is needed to accelerate 1.8 kg by 3.3 m s^{-2}?

b What acceleration would be produced if the same force were applied to a mass of 800 g?

a $F = ma$

$F = 1.8 \times 3.3$

$F = 5.9 N$

b $a = \frac{F}{m}$

$a = \frac{5.9}{0.800}$

$a = 7.4 \text{ m s}^{-2}$

13 A car engine produces a forward force of 4400 N, but the forces of air resistance and friction opposing the motion of the car are a total of 1900 N.

a What is the resultant force acting on the car?

b If the car accelerates under the action of this force by 1.8 m s^{-2}, what is its total mass?

a resultant force, $F = 4400 - 1900$

$F = 2500$ N forwards

b $m = \frac{F}{a}$

$m = \frac{2500}{1.8}$

$m = 1400$ kg

14 a What is the resultant force acting on a mass of 764 g falling towards Earth (assume there is no air resistance)?

b What is the common name of this force?

a $F = ma$

$F = 0.764 \times 9.81$

$F = 7.49$ N

b This force is usually called the weight of the object and, in this context, the equation $F = ma$ would usually be written as $W = mg$.

Newton's second law can also be applied to small masses, like atomic particles. For example, an electron has a mass of 9.1×10^{-31} kg and if a force of 1 N were to act on it, it would accelerate at a rate of 1.1×10^{30} m s^{-2}!

Forces acting for a limited time

The resultant forces that change the velocities of various objects do not continue to act forever. Often such forces act for a few seconds or less, and in the case of many **impacts** and **collisions** the duration of the forces involved may be only fractions of a second. (The word **interaction** is often used as a general term to include all possible situations.)

As an example, consider a falling object. If a mass is dropped onto a surface, the force exerted during the impact is *not* equal to its weight (as is often thought) but can be calculated from $F = ma$. For example, if a falling 2.0 kg mass lands in sand and its speed is reduced from 6.0 m s^{-1} to zero in 0.50 s, then the average deceleration is -12 m s^{-2} (0−6.0)/0.50), and the average retarding force is 24 N. If the same mass was dropped onto concrete at the same speed, the time for the impact would be much less and the force much bigger. For example, if the impact lasted 0.050 s, then the average retarding force would be 240 N.

The forces on the long jumper shown in Figure 2.86 are reduced because the time of impact is increased by landing on sand.

Figure 2.86 Impact in a sand-pit reduces force

Whenever any moving object is stopped, the longer the time of impact, the smaller the deceleration and, therefore, the smaller the forces involved. These ideas are very useful when using physics to explain, for example, how forces are reduced in accidents.

When the aim is to accelerate an object with a force, the longer the time for which the force is applied, the bigger the change of velocity. In many sporting activities a ball is struck with some kind of bat, club or racquet, and the longer the time of contact the better. This is one reason for the advice to 'follow through' with a strike.

79 What resultant force is needed to accelerate a train of mass 3.41×10^5 kg from rest to 15.0 m s^{-1} in exactly 20 s?

80 When a force of 5.6 N was applied to a 4.3 kg mass it accelerated by 0.74 m s^{-2}. Calculate the frictional force acting on the mass.

81 A small plane of mass 12 400 kg is accelerated from rest along a runway by a resultant force of 29 600 N.
 a What is the acceleration of the plane?
 b If the acceleration remains constant, what distance is needed before the plane reaches its take-off speed of 73.2 m s^{-1}?

82 A car of mass 1200 kg was travelling at 22 m s^{-1} when the brakes were applied. The car came to rest in a distance of 69 m.
 a What was the deceleration of the car?
 b What was the average resultant force acting on the car?

83 A big box of mass 150 kg is pulled by a horizontal, thin rope in an effort to move it sideways along the ground. The box just begins to move when the frictional force is 340 N.
 a Draw a free-body diagram to show all the forces acting on the box when it starts to move.
 b If the tension in the rope is 380 N, calculate the acceleration of the box.
 c Explain what might happen if an attempt was made to accelerate the box at 1 m s^{-2}.

84 A metal block of mass 650 g moves horizontally across a sheet of glass at a constant speed while a force of 2.8 N is applied.
 a What is the coefficient of dynamic friction for these surfaces?
 b If a 940 g block of the same metal is pulled across the glass with a force of 4.5 N, calculate the acceleration produced.

85 A small train carriage of mass 12 500 kg rests on a slope.
 a If the coefficient of static friction is 0.42, what is the biggest angle of a slope (to the horizontal) that will enable the carriage to remain stationary?
 b When the slope is at that angle, if the engine produces a force of 6.0×10^4 N parallel to the track, calculate the acceleration of the train up the slope.

86 Use Newton's second law to explain why it will hurt you more if you are struck by a hard ball than by a soft ball with the same mass and speed.

87 What force would be needed to accelerate a proton (mass 1.67×10^{-27} kg) by 1.0×10^{10} m s^{-2}?

88 A trolley containing sand is pulled across a frictionless horizontal surface with a small but constant resultant force. Describe and explain the motion of the trolley if sand can fall through a hole in the bottom of the trolley.

89 A man of mass 82.5 kg is standing still in an elevator (lift) that is accelerating upwards at 1.50 m s^{-2}.
a What is the resultant force acting on the man?
b What is the normal reaction force acting upwards on him from the floor?

90 Figure 2.87 shows two masses connected by a light string passing over a pulley.

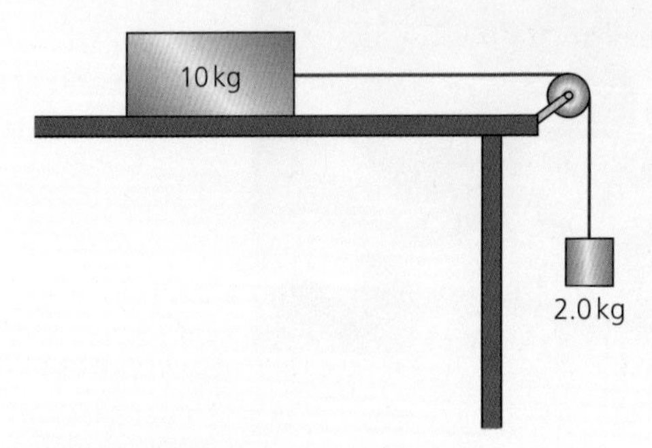

■ **Figure 2.87**

a Assuming there is no friction, calculate the acceleration of the two blocks.
b What resultant force is needed to accelerate the 2.0 kg mass by this amount?
c Draw fully labelled free-body diagrams for the two masses, showing the magnitude and direction of all forces.

91 A high-jumper (Figure 2.88) of mass 75.4 kg raised his centre of mass from 0.980 m to 2.05 m when clearing a high-jump bar set at 2.07 m.

■ **Figure 2.88**

a What vertical take-off speed was necessary in order for him to rise to this height?
b If the athlete took 0.22 s to project himself upwards, what was the average resultant force acting on him?
c What force did the ground exert upwards on him?
d What did the athlete do in order to make the ground push him upwards?
e Explain, using good physics, why it is sensible for him to land on foam when falling back to the ground.

92 a Explain why a person jumping down from even a small height should bend their knees as they land.
b Explain how air bags (and/or seat belts) reduce the injuries to passengers in car accidents.

93 A basketball of mass 0.62 kg was thrown downwards with an impact speed of 16 m s^{-1} onto some bathroom scales and the maximum force was estimated to be 280 N.
a If the average force exerted on the ball was about half the maximum force, what was the approximate deceleration of the basketball?
b Estimate the length of time that the basketball was in contact with the scales before it bounced away from the surface.

Newton's third law of motion

Newton's third law tells us about the forces acting *between* two bodies. Whenever *any* two objects come in contact with each other, or otherwise interact, they exert forces on each other (Figure 2.89). The two forces must *always* have the same magnitude; more precisely, the force that A exerts on B is equal *and opposite* to the force that B exerts on A.

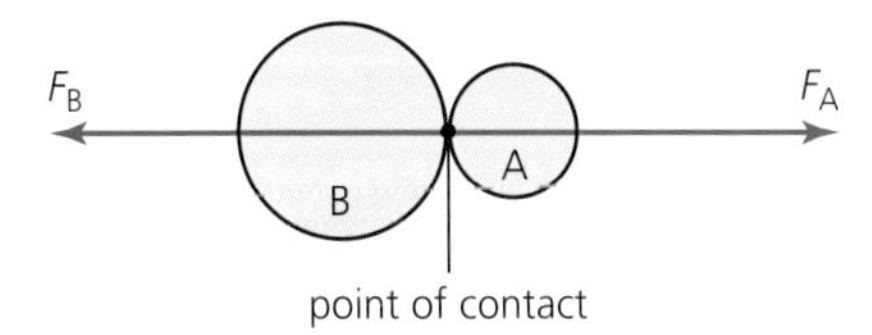

Figure 2.89 When two bodies interact $F_A = -F_B$

Newton's third law of motion states that whenever one body exerts a force on another body, the second body exerts exactly the same force on the first body, but in the opposite direction.

Essentially this law means that forces must *always* occur in equal pairs, although it is important to realize that the two forces must act on *different* bodies and in opposite directions. The two forces are always of the same type, for example gravity/gravity or friction/friction. Sometimes the law is quoted in the form used by Newton: 'to every action there is an equal and opposite reaction'. In everyday terms, it is simply not possible to push something that does not push back on you. Here are some examples:

- If you pull a rope, the rope pulls you.
- If the Earth pulls a person, the person pulls the Earth (Figure 2.90).
- If a fist hits a cheek, the cheek hits the fist (Figure 2.91).
- If you push on the ground, the ground pushes on you.
- If a boat pushes down on the water, the water pushes up on the boat (Figure 2.92).
- If the Sun attracts the Earth, the Earth attracts the Sun.
- If a plane pushes down on the air, the air pushes up on the plane.

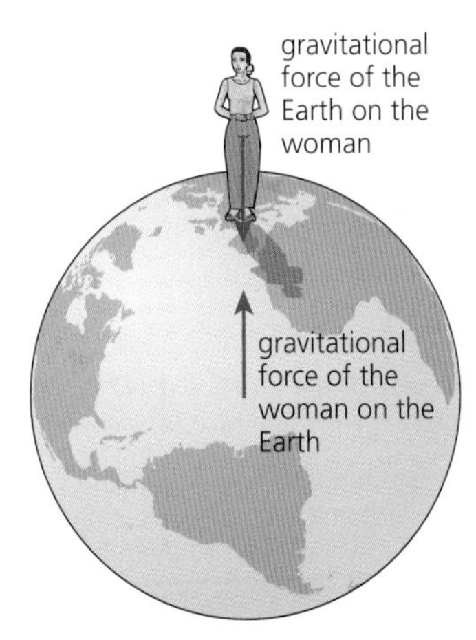

Figure 2.90 The force on the woman is equal and opposite to the force on the Earth

Figure 2.91 The force on the glove is equal and opposite to the force on the cheek

■ **Figure 2.92** The force pushing the man forward is equal to the force pushing the boat backwards

94 Figure 2.93 shows a suggestion to make a sailing boat move when there is no wind. Discuss how effective this method could be.

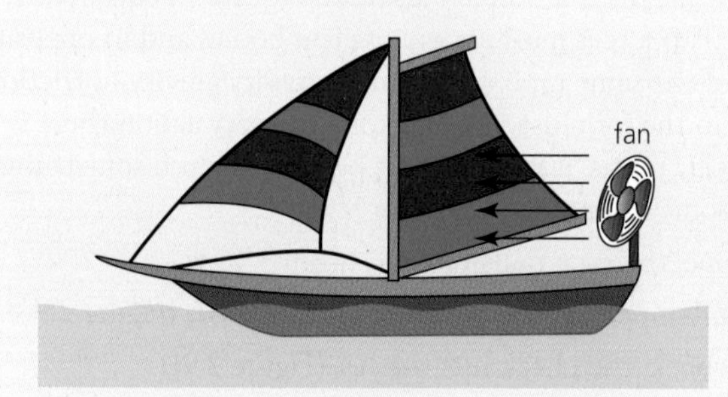

■ **Figure 2.93**

95 A book has a weight of 2 N and is at rest on a table. The table exerts a normal reaction force of 2 N upwards on the book. Explain why these two forces are *not* an example of Newton's third law.

96 Seven examples of pairs of Newton's third law forces are provided above. Give three more examples. Try to use different types of force.

97 A large cage with a small bird sitting on a perch is put on some weighing scales. What happens to the weight shown on the scales when the bird is flying around the cage (compared with when it is sitting still)? Explain your answer.

ToK Link

Predeterminism

Classical physics believed that the whole of the future of the universe could be predicted from knowledge of the present state. To what extent can knowledge of the present give us knowledge of the future?

All matter is made of particles and it has been suggested that if we could know everything about the present state of all the particles in a system (their positions, energies, movements etc.), then maybe we could use the laws of classical physics to predict what will happen to them in the future. *If* these ideas could be expanded to include *everything*, then the future of the universe would already be decided and **predetermined**, and the many apparently unpredictable events of everyday life and human behaviour (like you reading these words at this moment) would just be the laws of physics in disguise.

However, we now know that the laws of physics (as imagined by humans) are not always so precisely defined, nor as fully understood as physicists of earlier years may have believed. The principles of quantum physics in particular contrast with the laws of classical physics. Furthermore, in a practical sense it is totally inconceivable that we could ever know enough about the present state of everything in the universe in order to use those data to make detailed future predictions.

2.3 Work, energy and power – *the fundamental concept of energy lays the basis on which much of science is built*

Energy is probably the most widely used concept in the whole of science and the word is also in common use in everyday language. Despite that, the idea of energy is not easy to explain or define, although it is still a very useful concept.

■ **Figure 2.94** The toy dogs get their energy from batteries

When a battery is fitted in a child's toy dog (Figure 2.94), it moves, jumps up in the air and barks. After a certain length of time the toy stops working. In order to try to *explain* this we will almost certainly need to use the concept of energy. Chemical energy in the battery is transferred to electrical energy, which produces motion energy in a small electric motor. Some energy is also transferred from electricity to sound in a loudspeaker. Eventually all the useful energy in the battery is transferred to the surroundings. The concept of 'energy' makes this easier to explain.

We can talk about the energy 'in' the gasoline (petrol) we put in the tanks in our cars (for example) and go on to describe that energy being transferred to the movement of the car. But nothing has actually flowed out of the gasoline into the car, and all this is just a convenient way of expressing the idea that the controlled combustion of gasoline with oxygen in the air can do something that we consider to be useful.

Perhaps the easiest way to understand the concept of energy is this: energy is needed to make things happen. Whenever *anything* changes, energy is transferred from place to place or from one form to another. Most importantly, energy transfers can be *calculated* and this provides the basic 'accounting system' for science. Any event will require a certain amount of energy for it to happen and if there is not enough energy available, it cannot happen. For example, if you do not get enough energy (from your food) you will not be able to climb a 500 m hill; if you do not put enough gasoline in your car you will not get to where you want to go; if energy is not transferred quickly enough from an electrical heater it will not keep your room warm enough in cold weather.

Calculating energy transfers is a very important part of physics and we start with very common situations in which energy is transferred to objects to make them move.

■ Work done as energy transfer

One very common kind of energy transfer occurs when an object is moved (displaced) by using a force. This is called doing **work** (symbol W). The word 'work' has a very precise meaning in physics, which is different from its use in everyday language:

work done = force × displacement in the direction of the force

If the force is constant and in the same direction as movement, we can write:

$$W = Fs$$

However, in many examples either or both of these two assumptions may not be true.

The unit of work is the **joule**, J. 1 J is the work done when a force of 1 N displaces an object 1 m in the direction of the force. The same unit is used for measuring *all* forms of energy – kilojoule, kJ, and megajoule, MJ, are also in common use.

Work and energy are scalar quantities.

The weightlifter in Figure 2.95 is *not* doing any work on the weights over her head while they are being held still, because there is no movement. But energy *is* being transferred in her muscles and, no doubt, they hurt because of that.

If the movement is perpendicular to a force, no work is done by the force. For example, the force of gravity keeps the Moon in orbit around the Earth (Figure 2.96), but no work is being done.

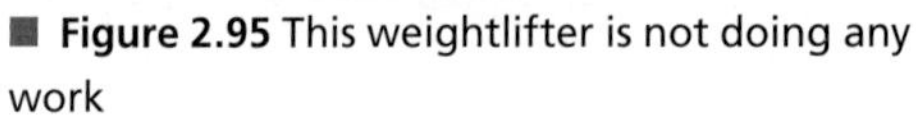

■ **Figure 2.95** This weightlifter is not doing any work

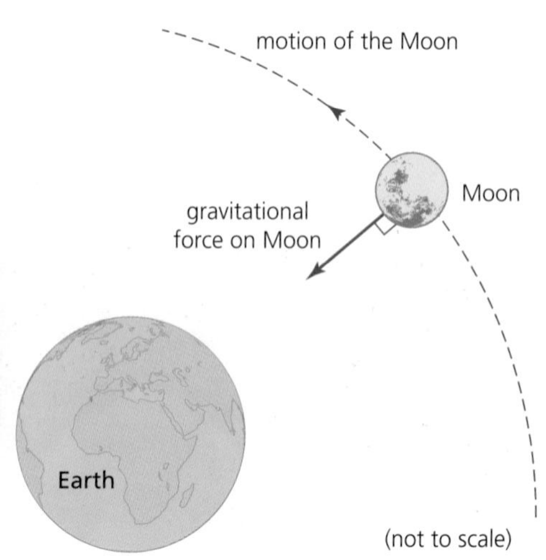

■ **Figure 2.96** No work is done as the Moon orbits the Earth.

Worked example

15 How much work is done when a 1.5 kg mass is raised 80 cm vertically upwards?

The force needed to raise an object (at constant velocity) is equal to its weight (mg) and the symbol h is widely used for vertical distances. (To avoid confusion, W will normally be used to represent work and not weight.)

$W = Fs$

$W = mg \times h$

$W = 1.5 \times 9.81 \times 0.80$

$W = 12\,\text{J}$

Calculating the work done if the force is not in the same direction as the displacement

Sometimes a displacement is produced that is not in the same direction as the force (Figure 2.97). For whatever reason, the object shown can only be displaced in the direction shown by the dashed red line, but the force is acting at an angle θ to that line.

■ **Figure 2.97** Movement that is not in the same direction as the force

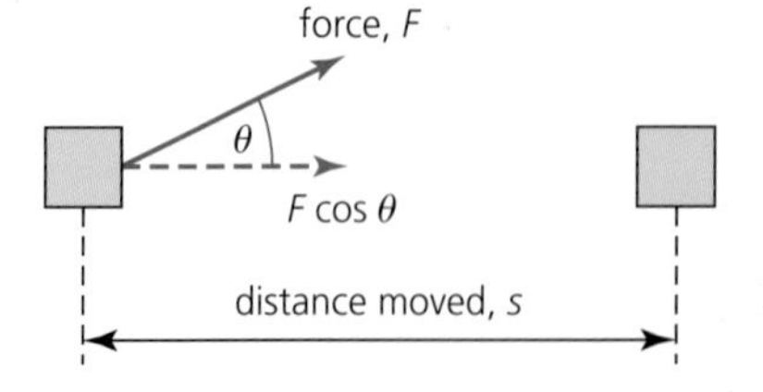

We know that work done = force × displacement in the direction of the force. In this case the force used to calculate the work done is the *component* of F in the direction of movement – that is $F\cos\theta$, so that:

$$W = Fs\cos\theta$$

This equation is listed in the *Physics data booklet*.

If force and movement are in the same direction, $\cos\theta$ equals 1 and the equation becomes $W = Fs$, as before. If the force is perpendicular to motion then $\cos\theta$ equals zero and no work is done.

Worked example

16 The 150 kg box in Figure 2.98 was pulled 2.27 m across horizontal ground by a force of 248 N, as shown.

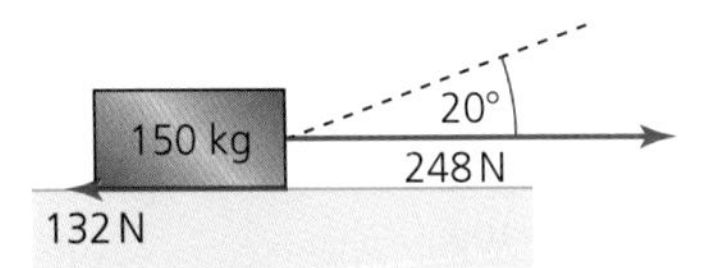

■ **Figure 2.98**

a How much work was done by the force?
b If the frictional force was 132 N, what was the acceleration of the box?
c Suggest why it may make it easier to move the box if it is pulled in the direction shown by the dashed line.
d When the box was pulled at an angle of 20.0° to the horizontal, the force used was 204 N. Calculate the work done by this force in moving the box horizontally the same distance.

a $W = Fs$

$W = 248 \times 2.27 = 563\,\text{J}$

b $a = \frac{F}{m}$

$a = \frac{(248 - 132)}{150}$

$a = 0.773\,\text{m s}^{-2}$

c When the box is pulled in this direction, the force has a vertical component that helps reduce the normal reaction force between the box and the ground. This will reduce the friction opposing horizontal movement.

d The force is not acting in the same direction as the movement. To calculate the work done we need to use the horizontal component of the 204 N force.

$W = F\cos 20.0° \times s$

$W = 204 \times 0.940 \times 2.27$

$W = 435\,\text{J}$

98 Calculate the work done when a 12 kg suitcase is:
a pushed 1.1 m across the floor with a force of 54 N
b lifted 1.1 m upwards.

99 The pram in Figure 2.99 is being pulled with a force that is not in the same direction as its movement. Calculate the work done when the pram is pulled 90 cm to the right with a force of 36 N at an angle of 28° to the horizontal.

■ **Figure 2.99**

100 In Figure 2.100 a gardener is pushing a lawnmower at a constant speed of $0.85\,m\,s^{-1}$ with a force, P, of 70 N at an angle of 40° to the ground.

a Calculate the magnitude of the frictional force, F.

b How much work is done in moving the lawnmower for 3 s?

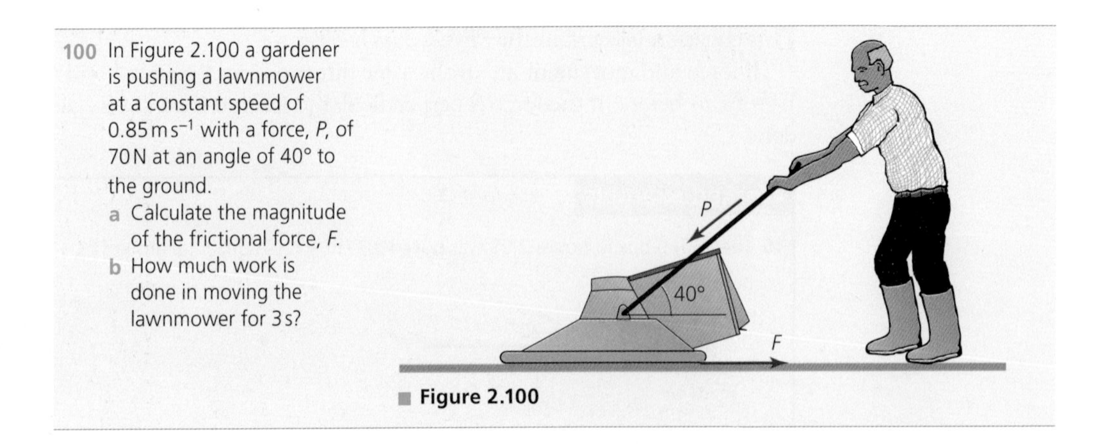

■ **Figure 2.100**

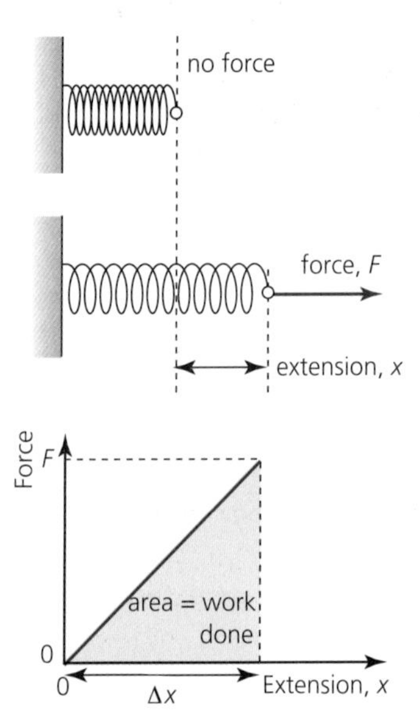

■ **Figure 2.101** Force and extension when stretching a spring

Calculating the work done by varying forces

When making calculations using the equation $W = Fs\cos\theta$ we must use a single, constant value for the force but, in reality, forces are rarely constant. In order to calculate the work done by a *varying* force we have to make an estimate of the *average* force involved. In some situations we can assume that the force varies in a predictable and linear way, so that the average force is halfway between the starting force and the final force. A simple example would be calculating the work done in stretching a spring, as shown in Figure 2.101. Provided that the spring is not overstretched, the force is proportional to the **extension** (displacement of the end of the spring), as shown in the force–extension graph. (This relationship is known as *Hooke's law*.)

Using $W = Fs$ in this situation involves using the *average* force during the extension of the spring. As the spring was stretched, the force increased linearly from zero to F and the average was $F/2$. The work done, W, was therefore $(F/2) \times \Delta x$. (The symbol x is commonly used to represent extensions, and Δx changes of extension.)

This calculation is identical to a calculation to determine the shaded area under the graph. This is *always* true – if the way a force varies with displacement is known in detail, a force–displacement graph can be drawn and then:

> The area under a force–displacement (distance) graph equals the work done.

The usefulness of determining the work done from the area under a graph becomes more obvious when we consider stretching a material like rubber, for which the displacement is *not* proportional to the force. Typically, rubber becomes stiffer the more it is stretched. Figure 2.102 shows a possible force–extension (displacement) graph for rubber.

The work done, W, in stretching the rubber by 40 cm (for example) is found from the area under the curved graph. This can be estimated from the shaded triangle, which is judged by eye to have a similar area:

$$W = 0.5 \times 50 \times 0.40 = 10\,J$$

■ **Figure 2.102** Force–extension (displacement) graph for rubber under tension

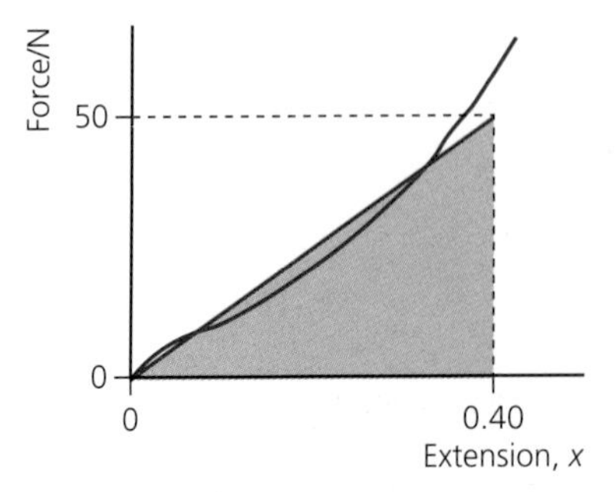

Worked example

17 a Calculate the work done in raising the centre of gravity of a trampolinist of mass 73 kg through a vertical height of 2.48 m (see Figure 2.103).

b When he lands on the trampoline he is brought to rest for a moment before being pushed up in the air again. If the maximum displacement of the trampoline is 0.8 m, sketch a force–displacement graph for the surface of the trampoline.

■ **Figure 2.103** The more the trampoline stretches, the higher the trampolinist can jump

a $W = Fs$ = weight × height = $mg \times h$

$W = 73 \times 9.81 \times 2.48 = 1800\,J$

b The shape of the graph is not known, so it has been drawn linearly for simplicity (Figure 2.104). The area under the graph must be approximately equal to 1800 J (ignoring any energy transferred to other forms). This means that the maximum force must be about 4500 N, so that 0.5 × 0.8 × 4500 = 1800 J.

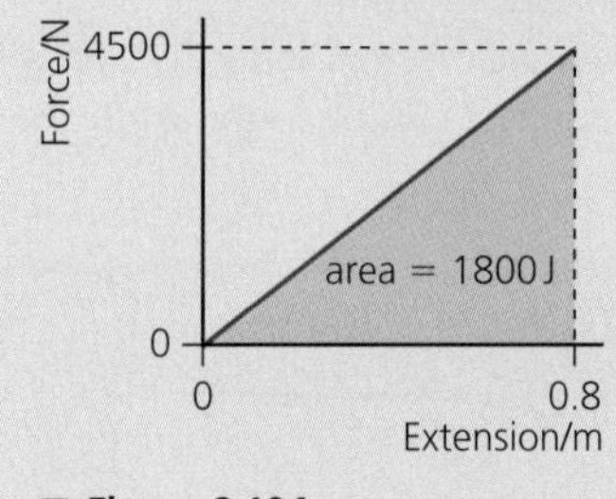

■ **Figure 2.104**

101 A spring was 48 mm long with no force applied. When a force of 8.3 N was applied its overall length extended to 74 mm. How much work was done on the spring (assuming that the spring obeyed Hooke's law)?

102 Some springs are designed to be compressed (squashed) as well as stretched. The suspension system of a car is a good example of this.
  a Estimate the maximum downwards force you can exert on the side of a car and how far down it will move because of your force.
  b Calculate an approximate value for the work done on the spring in this process.

103 A long thin strip of a plastic was stretched by hanging 100 g masses on the end of it. For masses up to 0.800 kg the extension was proportional to the force stretching it. The extension was 14 cm for a load of 0.800 kg. For heavier loads the plastic became more flexible and when it broke with a load of 1.2 kg, the extension was 30 cm and the plastic had permanently changed shape.
  a Sketch a force–distance graph to represent this behaviour.
  b Estimate the total work done in stretching the plastic to:
    i 14 cm
    ii 30 cm.
  c Suggest what happens to the energy transferred during the stretching of the plastic.

104 A steeper gradient on a force–extension (F–s) graph means that a spring requires a greater force to stretch it by a certain amount. In other words, the gradient of a force–extension graph represents the *stiffness* of the spring. $\Delta F/\Delta s$ is called the *force constant* of the spring, but a similar concept can be used when stretching materials other than springs.
  a Calculate the force constant for the spring represented in Figure 2.105.
  b How much work was done when the spring was extended by 8.0 cm?
  c What was the length of the spring when the force was 19 N if its original length was 27.0 cm?
  d Explain why it would be unwise to use information from this graph to predict the length of the spring if a mass of 10 kg was hung from its end.

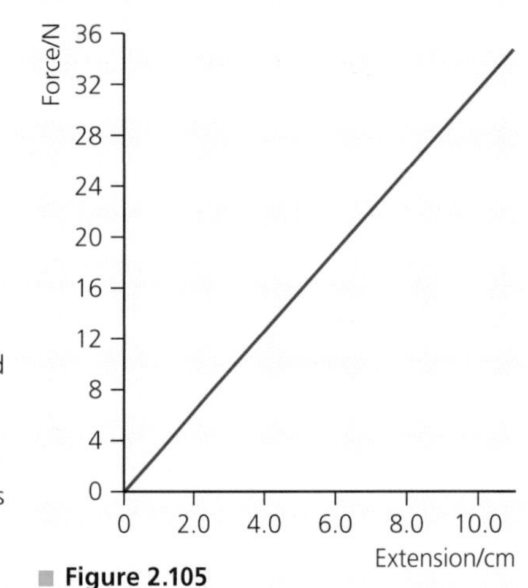

■ **Figure 2.105**

Additional Perspectives

Force and extension: stress and strain

Force–extension graphs can be drawn for specimens of different materials (often in the shape of wires). Such graphs provide important basic information about the *stiffness* and *strength* of the actual specimens tested, whether they remain out of shape and their ability to store energy when stretched. Of course, all this depends not only on what the specimen is made of but also on its original length and cross-sectional area. However, if we draw *stress–strain* graphs (rather than force–displacement graphs), they can be applied to specimens of *any* shape of the same material.

$$\text{stress} = \frac{\text{force}}{\text{area}} \qquad \text{strain} = \frac{\text{change in length}}{\text{original length}}$$

As an example consider Figure 2.106a, which shows the stress–strain relationship for a metal wire up to the point where it breaks.

From this graph we can make the following conclusions about this metal:

- For strains up to about 0.05 (5%), any extension produced will be proportional to the force applied (stress is proportional to strain).
- For bigger stresses, the metal becomes less stiff.
- The wire will be approximately twice its original length when it breaks.
- The work done in stretching the material can be found from the area under the graph if the dimensions of the wire are known.

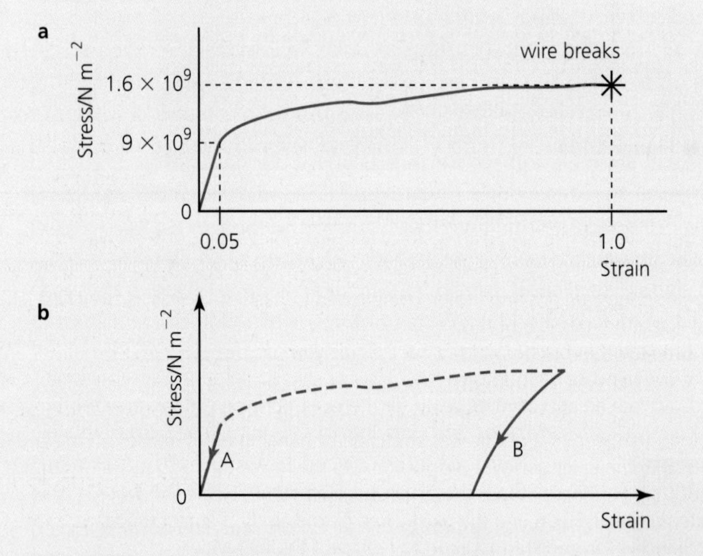

■ **Figure 2.106** Stress–strain graph

Figure 2.106b represents the same metal, but shows what happens as the force is *removed* in two different situations. For A the maximum force on the wire did not take it beyond its limit of proportionality and the wire returned to its original shape. This is called 'elastic' behaviour.

For B the wire stretched a lot more before the force was removed and the wire was permanently deformed from its original shape. This is known as 'plastic' behaviour.

1 a If the original length of the wire was 1.20 m and its cross-sectional area was $3.8 \times 10^{-8}\,\text{m}^2$, calculate the force required to break the wire and its length just before it broke.

 b Estimate the force needed to break another wire of the same metal, but of length 2.4 m.

2 Calculate the ratio stress/strain for the linear section of the graph. This is the stiffness of the metal. What are the units of stiffness?

3 Approximately how much work was done in stretching this wire to its breaking point?

Different forms of energy

When anything happens (i.e. something changes) it is because energy has been transferred from one form to another or from one place to another. It is important to have an overview of the most important forms of energy. In this section we will briefly introduce the following:

- gravitational (potential) energy
- nuclear (potential) energy
- chemical (potential) energy
- electric (potential) energy and energy transferred by an electric current
- elastic strain (potential) energy
- thermal energy (heat)
- radiant energy
- kinetic energy (including wind energy)
- internal energy
- mechanical wave energy (including sound and waves on water).

(Although it is not needed for this chapter, it may be of interest to note that *any* mass has an energy content that is equivalent to the magnitude of the mass. The two are related by Einstein's famous equation $E = mc^2$. This is covered in Chapter 7.)

Potential energies

When work is done to stretch a spring (see Figure 2.101), we say that energy has been transferred to it and that the energy is then in the form of elastic potential (strain) energy. (**Elastic** behaviour means that when the force is removed the material will return to its original shape. When we say that a material is **strained**, we mean that it has changed shape.) While the spring is stretched we can say that energy is 'stored' in it and that the energy is still available for later use, possibly to do something useful like running a toy or a wind-up watch. Elastic potential energy is just one of several more important forms of potential energy.

Potential energies are energies that are stored. In general, potential energies exist where there are forces between objects. The following is a list of the important potential energies:

- When a mass is lifted off the ground against the force of gravity, we say that it has gained **gravitational potential energy**. This energy is stored in the mass. If the mass is allowed to fall, it loses gravitational potential energy and this energy could be transferred to do something useful.
- **Nuclear (potential) energy** is associated with the forces between particles in the nuclei of all atoms.
- **Chemical (potential) energy** is associated with the forces between electrons and other particles in atoms and molecules.
- **Electric (potential) energy** is associated with the forces between electric charges.
- **Elastic strain (potential)** energy is associated with the forces needed to change the shape of objects.

Kinetic energy

Work has to be done on any object to make it move from rest or to make it move faster. This means that all moving objects have a form of energy. This is called **kinetic energy**. ('Kinetic' means related to motion.) Wind energy is the kinetic energy of moving air.

Energy being transferred

When energy is being transferred from place to place, we may refer to the following forms of energy:

- **electrical energy** carried by an electric current
- **thermal energy** (heat) transferred because of a difference in temperature

- **radiant energy** transferred by electromagnetic waves ('light energy' or 'solar energy' are specific examples of electromagnetic radiant energy)
- **mechanical waves** (like water and sound waves) involve oscillating masses. They combine kinetic energy and potential energy (see Chapter 4).

Internal energy

There is a large amount of energy *inside* substances because of the random motions of the particles they contain and the forces between them. The total random kinetic energy and electric potential energy of all the particles inside a substance is called its **internal energy**.

Whenever objects move in everyday situations, some or all of their kinetic energy will be transferred to internal energy because of frictional forces (and then thermal energy spreads out into the surroundings).

Examples of energy transfers

Everything that happens in our lives involves transfers of energy, so we do not need to look far for examples. When we switch on an electric fan, energy transferred from the electric current is changed into kinetic energy of the fan and then the kinetic energy of the moving air. After the fan is turned off, all of that energy will have spread out into the environment as thermal energy and then internal energy. But where did the current get its energy from? Maybe in a nearby power station the chemical energy in oil was transferred to internal energy and kinetic energy of steam, which was then converted to kinetic energy in the turbines that generated the electricity.

The Sun has provided most of the energy we use on this planet. Nuclear energy in hydrogen nuclei in the Sun is converted to internal energy and radiation. The radiation that reaches Earth is transferred to chemical energy in plants (and then animals). Oil comes from what remains of plants and small organisms that died millions of years ago and have then decayed in the absence of oxygen.

Worked example

18 Describe the energy transfers that occur when a mass hanging vertically on the end of a metal spring is displaced and allowed to move up and down (oscillate) freely until it stops moving.

Whenever the spring is stretched or squashed there will be changes in elastic potential energy. Whenever the mass moves up and down there will be transfers of gravitational potential energy. When the spring and mass lose potential energy, the mass will gain kinetic energy. The mass will lose kinetic energy when it is transferred back to potential energy. Some energy will be continually transferred to the surroundings in the form of thermal energy, so that eventually the motion will stop.

105 Make a single flow chart showing all the energy transfers listed in the two paragraphs above.

106 Name devices whose main uses are to perform the following energy transfers:
- a electricity to sound
- b chemical to electricity
- c nuclear to electricity
- d sound to electricity
- e chemical to electromagnetic radiation
- f chemical to kinetic
- g elastic strain to kinetic
- h kinetic to electricity
- i chemical to internal
- j electromagnetic radiation to electricity

Principle of conservation of energy

The study of the different forms of energy and the transfers between them is the central theme of physics. The main reason that the concept of energy is so important lies in the **principle of conservation of energy**. ('Conservation' means to keep the same.)

> The **principle of conservation of energy** states that energy cannot be created or destroyed; it can only be transferred from one form to another.

In other words, the total amount of energy in the universe is constant. This is perhaps the most important principle in the whole of science. If, in any particular situation, energy seems to have just appeared or disappeared, then we know that there must be an explanation for where it has come from or where it has gone to. It is worth repeating at this point that *all* everyday processes spread (**dissipate**) some energy into the surroundings and this dissipated energy is then of no further use to us, but it has not 'disappeared'.

In order to use the principle of conservation of energy, we must be able to measure and calculate energy transfers. Equations for calculating quantities of energy in its different forms (see the next section) are essential knowledge for all students of physics.

Nature of Science

The importance of the conservation of energy

Despite its considerable importance, the wide-ranging principle of conservation of energy cannot be directly verified by individual experiments. To prove that a physics principle (law/theory) is *always* true, for *any* process, at *any* place at *any* time is simply impossible. Added to which, all mechanical processes dissipate energy into the surroundings, making it difficult to quantify where energy is going. So it is perhaps not surprising that it took so many years for the principle to become fully established. But since then no experiment has ever produced results that have contradicted the principle, although the original concept of energy has had to be adapted and broadened over time to include heat and mass.

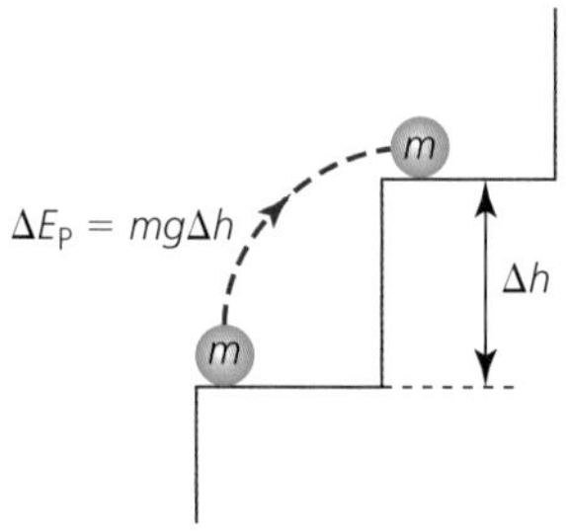

Calculating mechanical energies

Gravitational potential energy

When we lift a mass we transfer energy to it and, because we are using a force to move it, this kind of energy transfer can be described as an example of *doing work*. We have already seen (in Worked example 17) that we can calculate the work done from $W = mgh$. The difference in energy that a mass has, because of the work that was done to move it up or down to a new position, is called its *change in* gravitational potential energy (Figure 2.107).

Figure 2.107 A mass gaining gravitational potential energy

The symbol E is used to represent energy in general and E_p is used to represent gravitational potential energy in particular. When a mass m is raised a vertical height Δh, as shown in Figure 2.107, the change in gravitational potential energy, ΔE_p, can be calculated from:

$$\Delta E_p = mg\Delta h$$

This equation is included in the *Physics data booklet*.

Note that this equation enables us to calculate *changes* in gravitational potential energy between various positions. If we wanted to calculate the 'total' gravitational energy of a mass, we would have to answer the question, gravitational potential energy compared with where? This might be the floor, sea level or the centre of the Earth, for example. This is discussed further in Chapter 10.

Worked example

19 How much gravitational potential energy is gained in each of these situations?

a An 818 g book is raised 91 cm from the floor to a desk.

b A 56 kg boy walks up 18 14-cm steps from the first to the second floor of a building.

c The same boy walks down to the first floor from the second floor.

a $\Delta E_p = mg\Delta h$

$\Delta E_p = 0.818 \times 9.81 \times 0.91$

$\Delta E_p = 7.3\,\text{J}$

b $\Delta E_p = mg\Delta h$

$\Delta E_p = 56 \times 9.81 \times (18 \times 0.14)$

$\Delta E_p = 1400\,\text{J}$

c $-1400\,\text{J}$

The negative sign means that the gravitational potential energy is less than on the second floor.

107 a What is the difference in gravitational potential energy of a 74 kg mountain climber between the top and bottom of a 2800 m mountain?

b Is the value of the gravitational field strength the same at the top and bottom of the mountain? If not, does it affect your answer to a?

108 A cable car rises a vertical height of 700 m in a total distance travelled of 6 km (Figure 2.108).

■ **Figure 2.108** Ngong Ping cable car in Hong Kong

a How much gravitational potential energy must be transferred to a car of mass 1800 kg during the journey if it has eight passengers with an average mass of 47 kg?

b Suggest reasons why a lot more energy (than your answer to a) has to be transferred in making this journey.

109 a How much gravitational energy must be transferred *to* an elevator of total mass 2340 kg when it rises 72 floors, each of average height 3.1 m?

b The same amount of gravitational potential energy must be transferred *from* the elevator when it comes down again. Elevators use counterweights (see Figure 2.109). As the elevator comes down the counterweight goes up, and when the elevator goes up the counterweight comes down. Explain the advantage of this system. Funicular railways usually have two carriages: as one goes up, the other comes down.

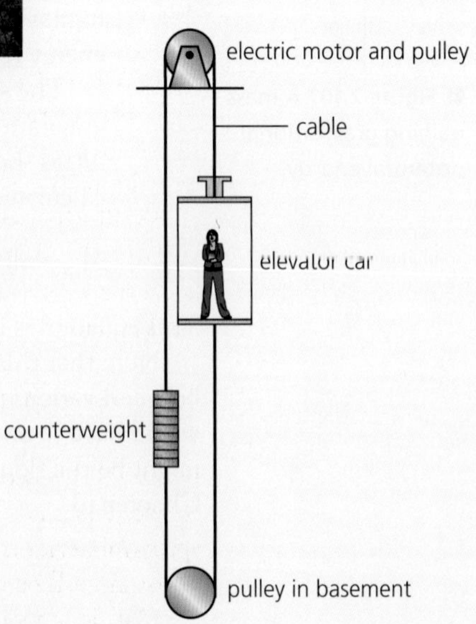

■ **Figure 2.109** An elevator and its counterweight

Kinetic energy

Consider a mass m accelerated uniformly from rest by a constant force, F, acting in the direction of motion over a distance s. The work done can be calculated from $W = Fs$. But $F = ma$ and $v^2 = u^2 + 2as$ (with $u = 0$), so the equation can be re-written as:

$$W = Fs = ma \times \frac{v^2}{2a} = \frac{1}{2}mv^2$$

That is, the kinetic energy of a moving mass, E_K, can be calculated from:

$$E_K = \frac{1}{2}mv^2$$

This equation is included in the *Physics data booklet*.

Worked example

20 Calculate the kinetic energy of a car of mass 1320 kg travelling at 27 m s^{-1}.

$$E_K = \frac{1}{2}mv^2$$

$$E_K = \frac{1}{2} \times 1320 \times 27^2$$

$$E_K = 4.8 \times 10^5 \text{ J}$$

110 a Calculate the kinetic energy of:
i a soccer ball (mass 440 g) kicked with a speed of 24 m s^{-1}
ii a 24 g bullet moving at 420 m s^{-1}.
b Estimate the kinetic energy of a woman sprinter in an Olympic 100 m race.

111 A mass of 2.4 kg in motion has kinetic energy of 278 J. What is its speed?

112 An electron has a mass of 9.1×10^{-31} kg. If it moves at 1% of the speed of light, what is its kinetic energy? (Speed of light = 3.00×10^8 m s^{-1})

113 A car travelling at 10 m s^{-1} has kinetic energy of 100 kJ. If its speed increases to 30 m s^{-1}, what is its new kinetic energy?

Elastic potential energy

Figures 2.101 and Figure 2.102 show force–extension graphs with the areas underneath representing the work done by the force (energy transferred) during the stretching of a spring or other material.

If the spring or material stretched elastically, it will recover its shape when the force is removed and the energy transferred again can be be used to do something useful. This energy stored is called *elastic potential energy* and, as discussed before, it is equal to the area under the graph. Elastic potential energy is a particularly helpful concept when explaining the behaviour and usefulness of springs and rubber.

If the stretching force, F, is proportional to the extension, Δx, as shown in Figure 2.101, then the gradient

$$\frac{F}{\Delta x} = \text{constant, } k$$

where k is known as the force constant of the spring or material. Sometimes the term '*spring constant*' is used. The units of k are N m^{-1} (or N cm^{-1}). A larger force constant means that a spring/material is *stiffer* – more force is required to produce the same extension. A larger force constant will be represented by a steeper graph.

Consider again Figure 2.101 in which the force is proportional to extension and a force F produces an extension Δx. An equation for the elastic potential energy, E_P, can be determined by considering the area of the triangle under the graph up to that point:

$$E_P = \frac{1}{2}F\Delta x$$

and because

$$F = k\Delta x$$

$$E_P = \frac{1}{2}k\Delta x^2$$

This equation is in the *Physics data booklet*.

If the force is *not* proportional to the extension then this equation cannot be used, but the energy transferred can still be determined from an estimate of the area. However, useful elastic potential energy is often associated with proportional stretching, such as in springs that are not overstretched. Non-linear force–extension graphs usually represent materials that do not return to their original shape when the force is moved, so that much of the energy that was transferred in stretching cannot be stored and then recovered. (The energy will tend to make the material get warmer.) Rubber is an important exception (see Figure 2.102).

Worked example

21 When a force of 38.2 N was applied to a steel spring its length increased linearly from 23.7 cm to 34.1 cm.

a Calculate the force constant of this spring.
b Calculate the length of the spring if the force is reduced to 22.4 N.
c Calculate a value for the extension produced by a force of 50.0 N. Why is your answer unreliable?
d How much elastic potential energy is stored in the spring:
  i when its extension is 5.9 cm?
  ii when the force was 7.6 N?
f Explain why doubling a stretching force can result in multiplying the stored elastic potential energy by four.

a $k = \frac{F}{\Delta x} = \frac{38.2}{(0.341 - 0.237)} = 367\,\text{N m}^{-1}$

b $\Delta x = \frac{F}{k} = \frac{22.4}{367} = 0.0610\,\text{m}$

Adding this to an original (unstretched) length of 0.237 m gives an overall length of 0.298 m.

c $\Delta x = \frac{F}{k} = \frac{50.0}{367} = 0.136\,\text{m}$

This value is unreliable because we do not know if the spring's extension is still proportional to this larger force (which is greater than the value quoted in the original question).

d $E_p = \frac{1}{2}k\Delta x^2$

$= \frac{1}{2} \times 367 \times 0.059^2 = 0.64\,\text{J}$

e $x = \frac{F}{k} = \frac{7.6}{367} = 0.021\,\text{m}$

$= \frac{1}{2} \times 367 \times 0.021^2 = 0.081\,\text{J}$

f If the extension is proportional to the force, doubling the force also doubles the extension.

114 Describe, including a suitable diagram, how you could perform an experiment to determine the force constant of a spring.

115 **a** If a spring has a force constant of 134 N m^{-1}, what extension is produced by a force 10 N? (Assume that the spring is not overstretched.)
b How much elastic potential energy is stored in the spring with this force?
c What force would be needed to store 1.51 J of elastic potential energy in the spring?

116 Estimate the amount of elastic potential energy stored in the suspension system of a car when a family of four people sit inside it.

Utilizations

Making work easier

When lifting a heavy object vertically, the amount of (gravitational potential) energy that we need to transfer to it is governed only by its weight and the vertical height (mgh). For example, to lift a 100 kg mass a height of 1 m requires about 1000 J of work to be done on it. Although 1000 J is not a lot of work that does not mean that we can do this job easily.

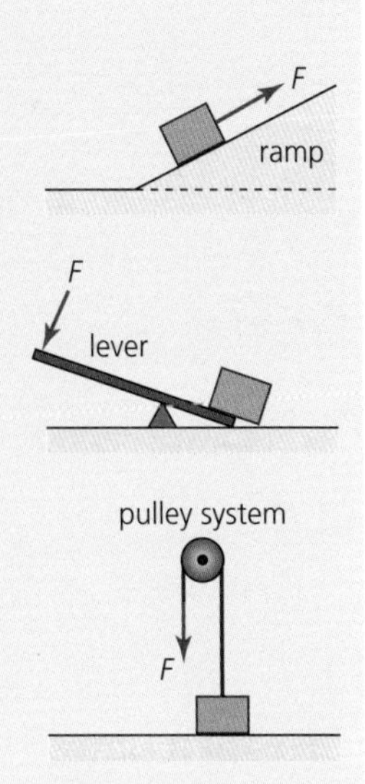

■ **Figure 2.110** Simple machines

There are two main reasons why this job could be difficult. Firstly, we may not be able to transfer that amount of energy in the time required to do the work. Another way of saying this is that we may not be *powerful* enough. (Power is discussed later in this chapter.) Secondly, we may not be *strong* enough because we are not able to provide the required upwards force of about 1000 N. Power and strength are often confused in everyday language.

Lifting (heavy) weights is a common human activity and many types of simple 'machine' were invented many years ago to make this type of work easier, by reducing the force needed. These include the ramp (*inclined plane*), the *lever* and the *pulley* (Figure 2.110).

In each of these simple machines the force needed to do the job is reduced, but the distance moved by the force is increased. If there were no energy losses (any energy losses would be mainly due to friction), the work done by the force (Fs) would equal the useful energy transferred to the object being raised (mgh). In practice, because of energy losses, we would transfer more energy using a machine than if we lifted the load directly without the machine. This is not a problem because we are usually much more concerned about how easy it is to do a job, rather than the total energy needed or the efficiency of the process.

1 Figure 2.111 shows a car *jack* being used to raise one side of a car.

a Estimate how much gravitational energy must be transferred to the car so that a tyre can be changed.

b By changing the design of the jack it is possible in theory to raise the car with almost any sized force that we choose. Estimate a convenient force that most people would be happy to use with a car jack.

■ **Figure 2.111** Changing a car tyre using a simple machine (car jack)

c Use your answers to parts **a** and **b** to estimate how far the force must move in order to raise the car.

d Explain how it can be possible for the force to move such a large distance.

e Discuss why it might be useful for there to be a lot of friction when using a car jack.

Mechanical transfers of energy

So far in this chapter we have introduced important equations for four different kinds of mechanical energy:

■ Work done (energy transferred) when a force moves an object:

$$W = Fs$$

■ Kinetic energy of a moving object:

$$E_K = \frac{1}{2}mv^2$$

- Changes of gravitational potential energy when an object moves up or down:

  $\Delta E_P = mg\Delta h$

- Elastic potential energy when something is deformed elastically:

  $E_P = \frac{1}{2}k\Delta x^2$

When objects change their motion, energy will be transferred between these forms, and we can equate the appropriate equations to each other in order to predict what will happen, as shown by the following examples. In order to do this we need to assume that no energy is transferred to other forms (such as thermal energy or internal energy). This usually means that fiction and/or air resistance have to be considered to be negligible.

1 Gravitational potential energy/kinetic energy

When an object falls towards the ground it transfers gravitational potential energy to kinetic energy. When we throw or project an object vertically upwards we have to give it kinetic energy, which is then transferred to gravitational potential energy. At its highest point the kinetic energy of the object will be zero and its gravitational potential energy will be highest.

For an object of mass m that falls from rest moving through a vertical height, Δh, without air resistance we can write:

$$\Delta E_P = E_K$$

$$mg\Delta h = \frac{1}{2}mv^2_{max}$$

It follows that:

$$v^2_{max} = 2g\Delta h$$

This useful equation is equivalent to the equation of motion: $v^2 = u^2 + 2as$, with $u = 0$, a written as g, and s written as Δh. The same equation can be used for objects moving upwards. The equation can also be used for objects moving up or down slopes (without rolling), but only if we assume that friction is negligible.

Experiments to measure the highest speed of an object that has fallen vertically from a known height will show that, in practice, the kinetic energy gained is less than the gravitational potential energy lost. The difference can provide some information about the forces of air resistance that were acting on the falling object. Figure 2.112 shows a related experiment in which a falling mass, m, pulls another mass, M, along a horizontal table. In this investigation air resistance might not be significant, but frictional forces probably are. Both masses will gain kinetic energy.

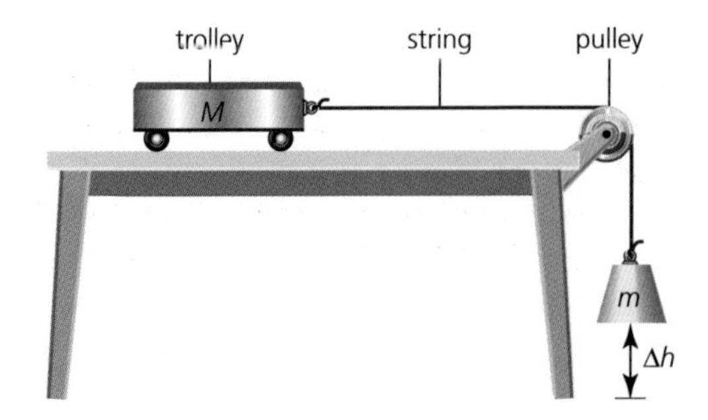

Figure 2.112 Transferring gravitational potential energy to kinetic energy

Worked examples

22 A ball is dropped from a height of 18.3 m. With what speed does it hit the ground? What assumption did you have to make in order to do this calculation?

$v^2_{max} = 2g\Delta h$

$v^2_{max} = 2 \times 9.81 \times 18.3$

$v_{max} = 18.9\,\text{m}\,\text{s}^{-1}$

This assumes that there was no energy dissipated because of air resistance. Note that this question could be written in a different way: with what speed would a ball have to be thrown upwards in order to reach a height of 18.3 m.

23 The ball shown in Figure 2.113 was released from rest in position A. It accelerated down the slope and had its highest speed at the lowest point. It then moved up the slope on the other side, reaching its highest point at B.

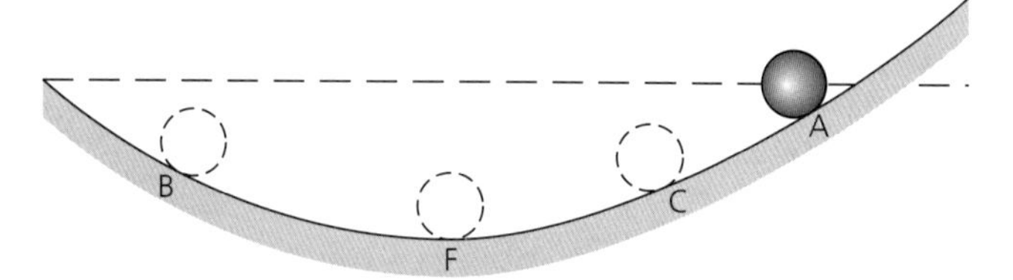

■ **Figure 2.113**

a Explain why B is lower than A.

b Describe the motion of the ball after leaving position B, explaining the energy transfers until it finally comes to rest at F.

a As the ball rolls backwards and forwards there is a continuous transfer of energy between gravitational potential energy and kinetic energy. The ball would only be able to reach the same height as A and regain all of its gravitational potential energy, if there was no friction. Because of friction some kinetic energy of the moving ball is transferred to internal energy in the ball and the slope, so it cannot reach the same height.

b The ball accelerates as it moves down the slope. When it reaches F, all of the extra gravitational potential it had because of its position at B has been transferred, and the ball has its highest speed. The ball decelerates as it moves up to C as kinetic energy is transferred back to gravitational potential energy. At all times that the ball is moving, some energy is being transferred to internal energy, so that the maximum height and maximum speed of the ball get less and less. Eventually the ball stops at F.

2 Elastic potential energy/kinetic energy

Sometimes we might use the energy stored in a spring, or a rubber band, to give an object kinetic energy and project it across the ground or through the air. Figure 2.114 shows an experiment in which an elastic cord is used to project a trolley across a horizontal surface. When the force is released the trolley will move to the right as the stretched elastic regains its shape. The effect on the initial speed of the trolley of varying the force, the extension or the mass of the trolley can be investigated.

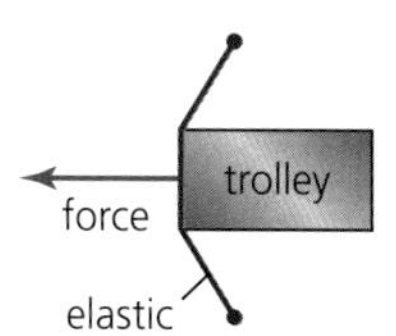

■ **Figure 2.114** Using stretched elastic to project a trolley

Firing an arrow from a bow uses the same kind of energy transfer from elastic potential energy. Assuming that the stretching force is proportional to the extension it produces, this kind of energy transfer can be expressed quantitatively as follows:

$$E_P = E_K$$

$$\frac{1}{2}k\Delta x^2 = \frac{1}{2}mv^2$$

Although it should be noted that much of the original energy can be transferred to other forms, rather than just the kinetic energy of the projected object. Consider the following example.

Worked example

24 a Assume that the force constant of a catapult is $480\,N\,m^{-1}$ and calculate the maximum possible speed of a 50 g stone fired from a catapult if the extension of the rubber on the catapult is 18 cm.

b Explain why in practice the actual speed will be much smaller.

a $\frac{1}{2}k\Delta x^2 = \frac{1}{2}mv^2$

$0.5 \times 480 \times 0.18^2 = 0.5 \times 0.05 \times v^2$

$v = 17.6\,m\,s^{-1}$

b The equation assumes that all of the elastic potential energy in the rubber is transferred to the kinetic energy of the stone, but in practice a lot of the stored energy will be transferred to kinetic energy of the rubber and to making the rubber slightly warmer.

117 From what height would you have to drop a mass for it to hit the ground with a speed of exactly $5\,m\,s^{-1}$? (Assume no air resistance.)

118 A stone is thrown vertically upwards with a speed of $14\,m\,s^{-1}$. What is the maximum possible height that it could rise to?

119 In order to estimate the height of a building, a ball is repeatedly thrown vertically upwards and its speed changed until it *just* reaches the top of the building (see Figure 2.115).

a If the ball takes 4.2 s from the moment it is released to the time it takes to hit the ground, estimate the height of the building.

b What are the main sources of error in this experiment?

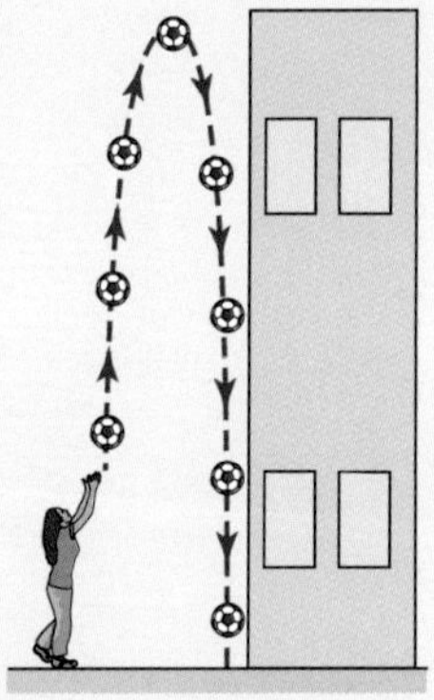

■ **Figure 2.115**

120 A ball is allowed to move down the slope shown in Figure 2.116a.

a If the surface is friction-free, what is its speed at point P? (Ignore any effects due to the ball rolling/spinning.)

b The same ball is then allowed to move down the friction-free slope shown in Figure 2.116b. What is its speed at Q?

c Explain why the ball reaches point P more quickly than point Q.

d But, of course, balls do *roll* down slopes. Suggest how this will affect your answers.

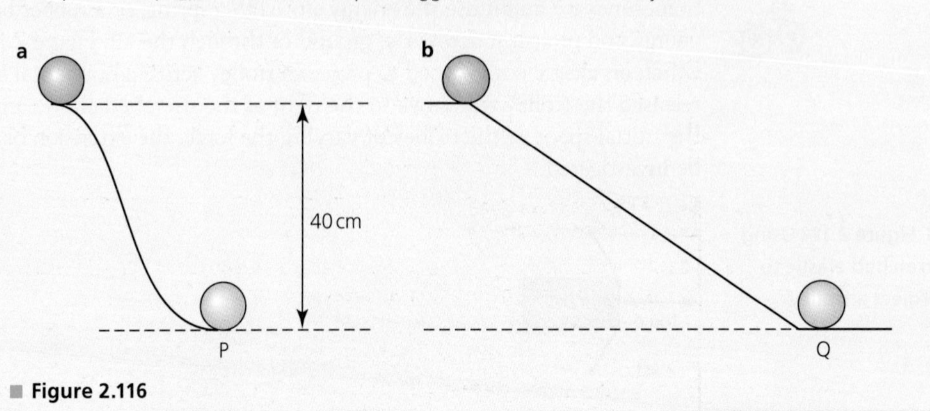

■ **Figure 2.116**

121 A rubber ball was dropped vertically onto the floor from a height of 1.0 m. After colliding with the floor it bounced up to a height of 60 cm. After the next bounce it only reached a height of 36 cm.
  a In what form is the energy when the ball is in contact with the ground?
  b List the energy transfers that take place from the release of the ball until the top of the first bounce.
  c Calculate the speed of the ball just before and just after it touches the floor during the first bounce.
  d Predict the height of the third bounce. What assumption did you make?

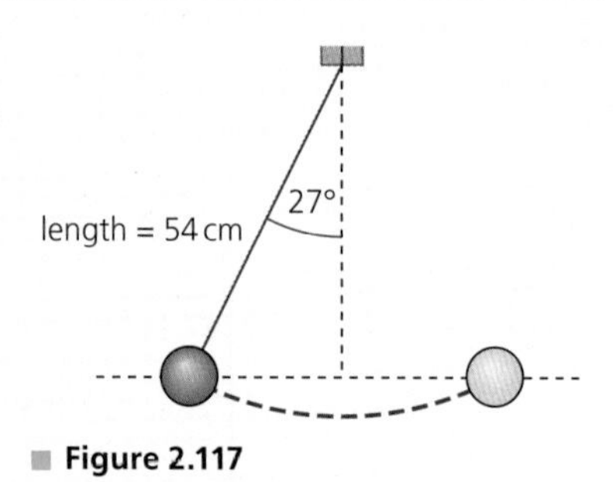

■ **Figure 2.117**

122 Figure 2.117 shows a pendulum. Calculate the maximum velocity of the pendulum when released from rest as shown.

123 A steel ball of mass 124 g, moving at 8.24 m s^{-1} on a horizontal frictionless surface, collides with an open-wound compression spring that has a force constant of 624 N m^{-1}. What is the maximum possible compression of the spring?

124 A rubber band of mass 4.7 g is stretched by 12 cm and then released so that it is projected vertically. The force and extension are proportional to each other and the force constant of the band is 220 N m^{-1} and only 10% of the elastic potential energy is transferred to kinetic energy. Calculate the height to which the rubber band will rise.

125 A gymnast of mass 39 kg falls from a height of 2.3 m onto the horizontal surface of trampoline. Estimate an effective combined force constant of the springs if the trampoline depresses by 43 cm before she comes to a momentary stop.

3 Doing work to change motion

Using forces to accelerate or decelerate objects is a very common human activity. The relationship work done equals change of kinetic energy therefore has many uses. The equation is:

$$Fs = \frac{1}{2}mv^2 - \frac{1}{2}mu^2$$

When using this equation, the force, F, must be assumed to be constant, or an average force can be used. Where the force does not vary linearly, the area under a force–distance graph can be used to determine the change in kinetic energy.

This equation can be used for objects that are accelerating or decelerating, but if the motion starts or ends with an object at rest (zero kinetic energy), we can write:

$$Fs = \frac{1}{2}mv^2$$

This equation shows us that when a force is used to stop a moving object, the smaller the force that is used, the longer the stopping distance of the object. (Or the greater the force used, the shorter the stopping distance.) Therefore, in order to reduce the forces in an impact, the stopping distance (and time) should be made as large as possible. For example, jumping into sand involves less force on our bodies than jumping onto solid ground.

If the speed of an object is decreasing because of work done by friction and/or air resistance, its kinetic energy will be transferred to internal energy, thermal energy and maybe some sound. We say that the energy has been **dissipated**, which means that it has spread out into the surroundings. This very common process is happening around us all the time.

Moving objects might also slow down because they collide with other things. As with friction, this will result in the dissipation of energy, but usually some energy is also used to *deform* the shape of the objects. The change of shape might be temporary, but it is often permanent.

Worked examples

25 A constant resultant force of 2420 N acts on a 980 kg car at rest. What is its speed after the car has moved 100 m? (Ignore resistive forces.)

$$Fs = \frac{1}{2}mv^2$$

$$2420 \times 100 = \frac{1}{2} \times 980 \times v^2$$

$$v = 22\ \text{m s}^{-1}$$

(Using $F = ma$ and the equations of motion will give the same answer.)

26 What average force is needed to stop a car of mass 1350 kg travelling at 28 m s^{-1} in a distance of 65 m?

$Fs = \frac{1}{2}mv^2$

$F \times 65 = \frac{1}{2} \times 1350 \times 28^2$

$F = 8100\,\text{N}$

27 In a car accident, the driver must be stopped from moving forward (relative to the car) and hitting the steering wheel in a distance of less than 25 cm.

a If he has a mass of 80 kg, what force is needed to stop him in this distance if he and the car were travelling at 10 m s^{-1} immediately before the impact?

b How is this force applied to him?

a $Fs = \frac{1}{2}mv^2$

$F \times 0.25 = \frac{1}{2} \times 80 \times 10^2$

$F = 16\,000\,\text{N}$

b The force is applied mainly by an inflated air bag (Figure 2.118) and/or from a seat belt. It is a very large force, but considerably less than the force that would act on him if he collided with the steering wheel. The effect of the force is also reduced because it is spread over a large area. (This reduces the pressure.)

■ **Figure 2.118** An air bag greatly reduces the forces in an accident

126 What average force is needed to reduce the speed of a 400 000 kg train from 40 m s^{-1} to 10 m s^{-1} in a distance of 1.0 km?

127 'Crumple zones' are a design feature of most vehicles (Figure 2.119). They are designed to compress and deform permanently if they are in a collision. Use the equation $Fs = \frac{1}{2}mv^2$ to help explain why a vehicle should not be too stiff and rigid.

■ **Figure 2.119** The front of the car is deformed but the passenger compartment is intact

128 A bungee jumper (Figure 2.120) of mass 61 kg is moving at 23 m s^{-1} when the rubber bungee cord begins to become stretched.

a Calculate her kinetic energy at that moment.

Figure 2.121 shows how the extension of the cord varies with the applied force.

b What quantity is represented by the area under this graph?

c Describe the relationship between force and extension shown by this graph.

d Use the graph to estimate how much the cord has extended by the time it has brought the jumper to a stop.

■ **Figure 2.120** Bungee jumping in Taupo, New Zealand

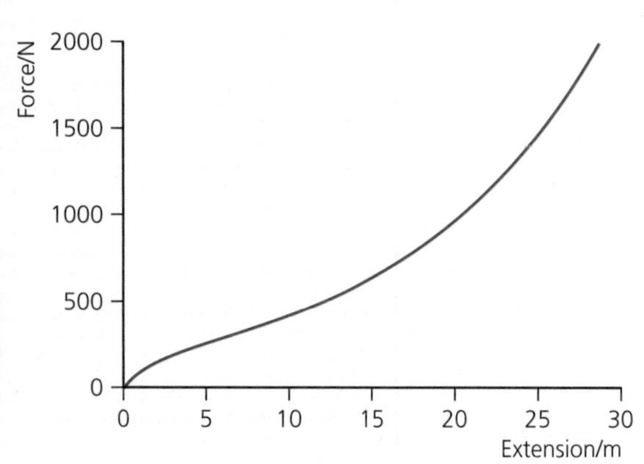

■ **Figure 2.121** Force–extension graph for a bungee cord

129 A pole-vaulter of mass 59.7 kg falls from a height of 4.76 m onto some foam.
  a Calculate the maximum kinetic energy on impact.
  b Will air resistance have had a significant effect in reducing the velocity of impact? Explain your answer.
  c If the foam deforms by 81 cm, calculate the average force exerted on the pole-vaulter.

130 Explain why you should always bend your legs when jumping down onto the ground.

131 a A nail in a piece of wood is struck once by a 1.24 kg hammer moving at 12.7 m s^{-1}. If the nail is pushed 15 mm into the wood what was the average force exerted on the nail?
  b What assumption did you make?

Utilizations

Regenerative braking systems in trains

The kinetic energy that must be given to a long, fast-moving train is considerable. Values of 10^8 J or more would not be unusual. When the train stops, all of that energy has to be transferred to other forms and, unless the energy can be recovered, the same amount of energy then has to be transferred to accelerate the train again. This is very wasteful, so the train and its operation should be designed to keep the energy wasted to a minimum. One way of doing this is to make sure that big, fast trains stop at as few stations as possible, perhaps only at their origin and final destination.

■ **Figure 2.122** The Light Rail Transit trains on the SBS network in Singapore have regenerative braking

Most ways of stopping moving vehicles involve braking systems in which the kinetic energy of the vehicle is transferred to internal energy because of friction. The internal energy is dissipated into the surroundings as thermal energy and cannot be recovered. A lot of research has gone into designing efficient 'regenerative braking systems' in recent years, usually involving the generation of an electric current, which can be used to transfer energy to chemical changes in batteries.

Small electric trains, which are often operated underground or on overhead tracks, are a feature of most large cities around the world (Figure 2.122). Such trains usually have stations every few kilometres or less, so regenerative braking systems and other energy-saving policies are very important. When designing a new urban train system, it has been suggested that energy could be saved by having a track shaped as shown in Figure 2.123.

■ **Figure 2.123** Possible track profile

1 Explain exactly what is meant by 'saving energy'.

2 Explain the reasoning behind the proposal shown in Figure 2.123.

3 Discuss other features of an urban train system that could 'save energy'.

4 Discuss which kind of engine/motor would be the best choice for such trains.

ToK Link

Fundamental concepts

To what extent is scientific knowledge based on fundamental concepts such as energy? What happens to scientific knowledge when our understanding of such fundamental concepts changes or evolves?

Any definition of physics will almost certainly include reference to energy, mass and force. These concepts are considered essential to an understanding of how the universe behaves. They are also described as *fundamental* because scientists currently believe that these concepts have no simpler explanations, so that an understanding of physics begins with them. But our understanding of these fundamental concepts has changed over the centuries and may continue to evolve in the future.

■ Power as a rate of energy transfer

When energy is transferred by people, animals or machines to do something useful, we are often concerned about how much time it takes for the change to take place. If the same amount of useful work is done by two people (or machines), the one that does it faster is said to be more *powerful*. (In everyday use the word 'power' is used more vaguely, often related to strength and without any connection to time.) If the same amount of useful work is done by two people (or machines), the one that does it by using the least overall energy is said to be more *efficient*.

Power is the rate of transferring energy. It is defined as:

$$\text{power} = \frac{\text{energy transferred}}{\text{time taken}}$$

The symbol for power is P and it has the SI unit the watt, W ($1\,\text{W} = 1\,\text{J}\,\text{s}^{-1}$). The units mW, kW, MW and GW are also in common use. The following are some examples of values of power in everyday life.

- A woman walking up stairs transfers chemical energy to gravitational energy at a rate of about 300 W.
- An 18 W light bulb transfers energy from electricity to light and thermal energy at a rate of 18 joules every second.
- A 0.0001 W calculator transfers energy at a rate of 0.0001 joules every second.

- A 2 kW water heater transfers energy from electricity to internal energy at a rate of 2000 joules every second.
- A typical family car might have a maximum output power of 100 kW.
- A 500 MW oil-fired power station transfers chemical energy to electrical energy at a rate of 500 000 000 joules every second.
- In many countries homes use electrical energy at an average rate of about 1 kW.

In this chapter, the energy transfer we are concerned with is that of doing work, *W*, so that we can write:

$$\text{power} = \frac{\text{work done}}{\text{time taken}}$$

$$P = \frac{\Delta W}{\Delta t}$$

This equation is *not* given in the *Physics data booklet*.

Worked examples

28 Calculate the average power of a 75 kg climber moving up a height of 30 m in 2 minutes.

$$P = \frac{\Delta W}{\Delta t}$$

$$P = \frac{mg\Delta h}{\Delta t}$$

$$P = \frac{75 \times 9.81 \times 30}{2 \times 60}$$

$$P = 180\,\text{W}$$

29 What average power is needed to accelerate a 1200 kg car from rest to 20 m s^{-1} in 8.0 s?

$$P = \frac{\Delta W}{\Delta t}$$

P = kinetic energy gained/time taken

$$P = \frac{\frac{1}{2}mv^2}{\Delta t}$$

$$P = \frac{0.5 \times 1200 \times 20^2}{8.0}$$

$$P = 3 \times 10^4\,\text{W} \ (= 30\,\text{kW})$$

Power transferred when travelling at constant velocity

The calculation for Worked example 29 ignored the large amount of work done in overcoming air resistance. But when any vehicle travels at a constant velocity, *all* of the work done is used to overcome resistance, rather than produce acceleration. If we replace ΔW with $F\Delta s$ in the equation for power (where F is the resistive force which, at constant velocity, is equal and opposite to the forward force), we get:

$$P = \frac{F\Delta s}{\Delta t}$$

and, because $\frac{\Delta s}{\Delta t} = v$, it follows that:

power to maintain constant velocity = resistive force × speed

$$P = Fv$$

This equation is given in the *Physics data booklet*.

Worked example

30 A small boat is powered by an outboard motor with a maximum output power of 40 kW. The greatest speed of the boat is 27 knots (1 knot = 1.85 km h^{-1}). What is the magnitude of the forward force provided by the motor at this speed?

$$1.85\,\text{km h}^{-1} = \frac{1.85 \times 1000}{3600} = 0.51\,\text{m s}^{-1}$$

At the maximum speed, resistive force, F = maximum force from engine.

$$P = Fv$$

$$40000 = F \times (27 \times 0.51)$$

$$F = 2900\,\text{N}$$

Efficiency

It is an ever-present theme of physics that, whatever we do, some of the energy transferred is 'wasted' because it is transferred to less 'useful' forms. In mechanics this is usually because friction or air resistance transfers kinetic energy to internal energy and thermal energy. The *useful* energy we get out of any energy transfer is *always* less than the *total* energy transferred.

When an electrical water heater is used, nearly all of the energy transferred makes the water hotter and it can therefore be described as 'useful', but when a mobile phone charger is used, for example, only some of the energy is transferred to the battery (most of the rest is transferred to thermal energy). Driving a car involves transferring chemical energy from the fuel and the useful energy is considered to be the kinetic energy of the vehicle, although at the end of the journey there is no kinetic energy remaining.

A process that results in a greater useful energy output (for a given energy input) is described as being more *efficient*. **Efficiency** is defined as follows:

$$\text{efficiency} = \frac{\text{useful energy output}}{\text{total energy input}} = \frac{\text{useful work out}}{\text{total work in}}$$

Because it is a ratio of two energies, efficiency has no units. It is often expressed as a percentage. It should be clear that, because of the principle of conservation of energy, efficiencies will *always* be less than 1 (or 100%).

An alternative definition of efficiency is obtained by dividing work by time to get power:

$$\text{efficiency} = \frac{\text{useful power out}}{\text{total power in}}$$

These two equations are given in the *Physics data booklet*

It is possible to discuss the efficiency of *any* energy transfer, such as the efficiency with which our bodies transfer the chemical energy in our food to other forms. However, the concept of efficiency is most commonly used when referring to electrical devices and engines of various kinds, especially those in which the input energy or power is easily calculated. Sometimes we need to make it clear exactly what we are talking about. For example, when discussing the efficiency of a car do we mean only the engine, or the whole car in motion along a road with all the energy dissipation due to resistive forces?

The efficiencies of machines and engines usually change with the operating conditions. For example, there will be a certain load at which an electric motor operates with maximum efficiency; if it is used to raise a very small or a very large mass it will probably be inefficient. Similarly, cars are designed to have their greatest efficiency at a certain speed, usually about 100 km h^{-1}. If a car is driven faster (or slower) then its efficiency decreases, which means that more fuel is used for every kilometre travelled.

Car engines, like all other engines that rely on burning fuels to transfer energy, are inefficient because of fundamental physics principles (Option B). There is nothing that we can do to change that, although better engine design and maintenance can make some improvements to efficiency.

In recent years we have all become very aware of the need to conserve the world's energy resources and limit the effects of burning fossil fuels in power stations and various modes of transport on global warming (see Chapter 8). Improving the efficiency of such 'heat engines' has an important role to play in this worldwide issue.

Many experiments can be designed to determine the efficiency of various machines or processes. Figure 2.124 shows how the efficiency of an electric motor can be found by comparing the electrical power input to the rate at which it can raise a known *load* (weight). The variation of efficiency with load and/or input power can be investigated.

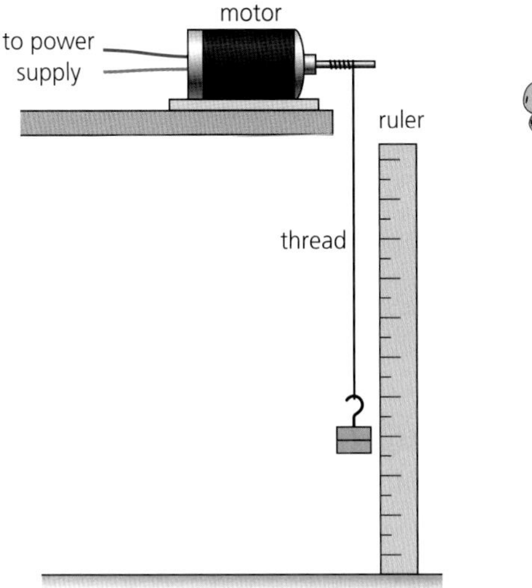

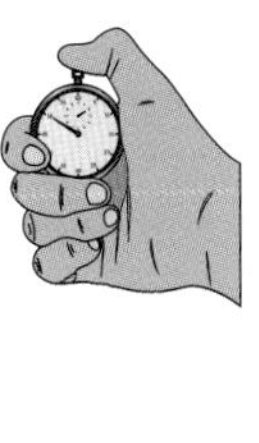

■ **Figure 2.124** Experiment to measure the efficiency of a motor

Worked example

31 a In an experiment, energy is provided to an electric motor at a rate of 0.80 W (Figure 2.124). If it raises a 20 g load by 80 cm in 1.3 s, what is its efficiency?

b What happens to all the energy that is not transferred usefully to the load?

a Power used to raise mass, $P = \frac{mg\Delta h}{\Delta t}$

$$P = \frac{0.02 \times 9.81 \times 0.80}{1.3}$$

$$P = 0.12\text{ W}$$

$$\text{efficiency} = \frac{\text{useful power output}}{\text{total power output}}$$

$$\text{efficiency} = \frac{0.12}{0.80} = 0.15 \text{ (or 15\%)}$$

b 85% of the energy transferred by electricity to the motor does not go usefully to the increased gravitational energy of the load. The wasted energy goes mostly to internal energy in the motor, which is then transferred as thermal energy to the surroundings. In addition, some energy will have been transferred to sound and some energy was used to stretch the string that connected the load to the motor.

132 a How much energy must be transferred to lift twelve 1.7 kg bottles from the ground to a shelf that is 1.2 m higher?

b If this task takes 18 s, what was the average useful power involved?

133 Estimate the output power of a motor that can raise an elevator of mass 800 kg and six passengers 38 floors in 52 s. (Assume there is no counterweight.)

134 How large are the resistive forces opposing the motion of a car that is operating with an output power of 23 kW at a constant speed of 17 m s^{-1}?

135 a What is the output power of a jet aircraft that has a forward thrust of 660 000 N when travelling at its top speed of 950 km h^{-1} (264 m s^{-1}) through still air?

b If homes use on average 1 kW of electrical power, how many homes could this same amount of power supply?
c Discuss whether the rapidly increasing use of aircraft for travel around the world should be discouraged.

136 The output from a power station is 325 MW. What is the power input if it is 36% efficient?

137 A man uses a ramp to push a 60.0 kg box onto the back of a truck, as shown in Figure 2.125. To lift the box directly would require a vertical force of 590 N, but by using the ramp this force is reduced.

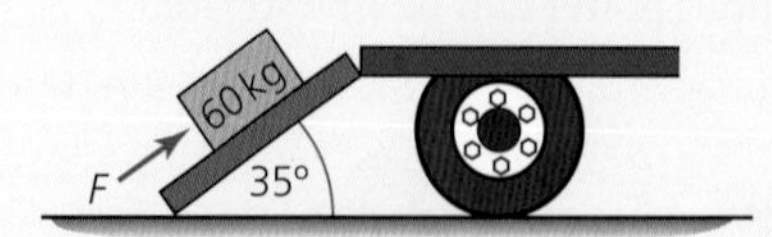

■ **Figure 2.125**

a Calculate the minimum force needed to push the box up the ramp if there is no friction.
b In practice the force needed was 392 N. What was the size of the frictional force? Explain why this force can be useful.
c The length of the ramp is 2.11 m. Calculate the work done in pushing the box along the ramp from the ground onto the truck.
d The useful energy transferred to the box is its gravitational potential energy. Calculate the efficiency of using the ramp.

2.4 Momentum and impulse – *conservation of momentum is an example of a law that is never violated*

■ Momentum

The product of mass and velocity is an important quantity in physics.

Linear momentum is defined as mass multiplied by velocity.

Momentum is given the symbol p and has the units kilogram metre per second, $kg\,m\,s^{-1}$.

$$p = mv$$

This equation is listed in the *Physics data booklet.*

Because kinetic energy, $E_K = \frac{1}{2}mv^2$, the connection between momentum and kinetic energy can be expressed as:

$$E_K = \frac{p^2}{2m}$$

This equation is included in the *Physics data booklet.*

In all calculations concerning momentum it is very important to remember that momentum is a vector quantity and the direction should always be given. In the core of this course we only deal with momentum in straight lines (*linear* momentum), but similar ideas can be applied elsewhere to the *angular* momentum associated with rotating objects or particles (Option B).

Worked example

32 A ball of mass 540 g moving vertically downwards hits the ground with a velocity of $8.0\,m\,s^{-1}$. After impact it bounces upwards with an initial velocity of $5.0\,m\,s^{-1}$.
a Calculate the momentum of the ball immediately before and after impact.
b What was the change of momentum during impact (Δp)?

a Initial momentum before impact, $p = mu = 0.54 \times 8.0 = 4.3\,kg\,m\,s^{-1}$ downwards. Final momentum after impact, $p = mv = 0.54 \times 5.0 = 2.7\,kg\,m\,s^{-1}$ upwards.

The opposite directions may be represented by positive and negative signs rather than written descriptions. That is, the momentum before impact was $+4.3\,kg\,m\,s^{-1}$ and the momentum afterwards was $-2.7\,kg\,m\,s^{-1}$. (The choice of signs is interchangeable.)

b $\Delta p = (-2.7) - (+4.3) = -7.0\,kg\,m\,s^{-1}$ (upwards)

Newton's second law expressed in terms of rate of change of momentum

Returning to Newton's second law of motion ($F = ma$), we can now rewrite it in terms of momentum by using the definition of acceleration, $a = (v - u)/t$:

$$F = \frac{m(v - u)}{t}$$

$$F = \frac{mv - mu}{t}$$

This equation can be expressed in words as: force equals change of momentum divided by time, or *force equals the rate of change of momentum*. This version of Newton's second law is commonly written in the form:

$$F = \frac{\Delta p}{\Delta t}$$

This equation is listed in the *Physics data booklet*.

For most situations the two versions of Newton's second law of motion are equivalent and the choice of which to use depends on the information provided. Because the version above does not require a constant mass, it is a more generalized statement of the law.

Figure 2.126 shows a karate chop being used to break some wooden boards. Consider $F = \Delta p/\Delta t$: to maximize the force acting between the hand and the boards, the mass striking the boards must be as large as possible and moving with a large speed. The time of impact needs to be as small as possible.

■ **Figure 2.126** Karate chop on wooden boards

Impulse and force–time graphs

Impulse

As discussed earlier in this chapter, many forces act on objects only for a limited time, and the longer the time that the force acts for, the greater the effect produced. For this reason the concept of *impulse* is introduced:

Impulse is defined as the product of force and the time for which the force acts:

$$\text{impulse} = F\Delta t$$

Rearranging the equation $F = \Delta p/\Delta t$ gives:

$$F\Delta t = \Delta p$$

This equation is given in the *Physics data booklet*. Expressed in words it is:

impulse = change of momentum

Impulse has the same units as momentum: kg m s^{-1} (or N s may be used). In this book we will not use any symbol to represent impulse.

Worked example

33 Calculate the average force exerted on the bouncing ball by the ground in Worked example 32 if the duration of impact was 0.18 s.

$$F = \frac{\Delta p}{\Delta t}$$

$$F = \frac{-7.0}{0.18} = -39\,\text{N}$$

(The negative sign represents a force upwards.)

Alternatively, the same answer could be obtained using $F = ma$:

$$a = \frac{(v - u)}{t} = \frac{(-5.0) - (+8.0)}{0.18} = -72\,\text{m s}^{-2}$$

Then,

$$F = ma = 0.54 \times (-72) = -39\,\text{N}$$

Force–time graphs

In many simple impulse calculations we can assume that the forces involved are constant, or that the average force is half of the maximum force. For more accurate work this is not good enough, and we need to know in detail how a force varies during an interaction. Such details are commonly represented by force–time graphs. The curved line in Figure 2.127 shows a typical example of a force varying over time Δt.

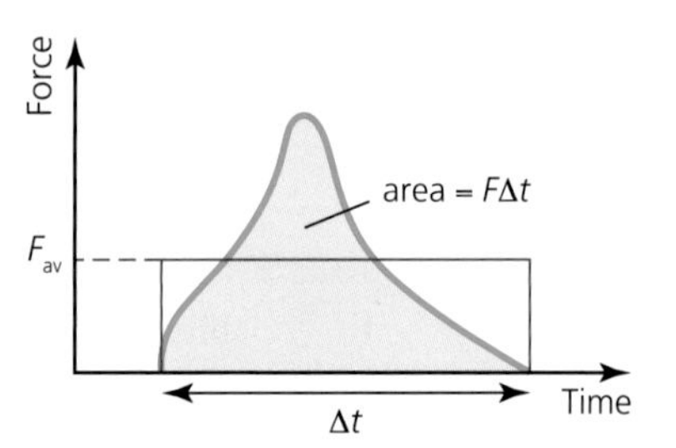

■ **Figure 2.127** Graph showing how a force varies with time

In general, the area under any force–time graph for an interaction equals force × time, which equals the impulse (or change of momentum). This is true whatever the shape of the graph. The area under the curve in Figure 2.127 can be estimated by drawing a rectangle of the same area (as judged by eye), as shown in red. F_{av} is then the average force during the interaction.

Force sensors that can measure the magnitude of forces over short intervals of time can be used with data loggers to gather data and draw force–time graphs for a variety of interactions, both inside and outside a laboratory. Stop-motion replay of video recordings of collisions can also be very interesting and instructive.

Force–time graphs can be helpful when analysing any interaction, but especially impacts involved in road accidents and sports.

Worked example

34 Figure 2.128 shows how the force on a 57 g tennis ball moving at 24 m s^{-1} varied when it was struck by a racquet moving in the opposite direction.

a Estimate the impulse given to the ball.

b What was the velocity of the ball after being struck by the racquet?

c The ball is struck with the same force with different racquets. Explain why a racquet with looser strings will return the ball with higher speed.

d What is the disadvantage of playing tennis with a racquet with loose strings?

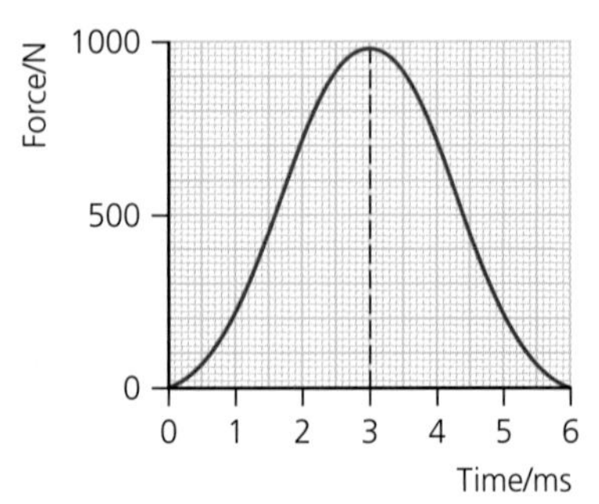

■ **Figure 2.128** Force–time graph for striking a tennis ball

a Approximate area under graph = $1000 \times (3.0 \times 10^{-3}) = 3.0\,\text{N s}$

b $m\Delta v = 3.0$

$$\Delta v = \frac{3.0}{0.057} = 53\,\text{m s}^{-1}$$

The initial velocity was 24 m s^{-1} towards the racquet. If the change of velocity was 53 m s^{-1}, then the ball must have moved away from the racquet with a velocity of (53 – 24) = 29 m s^{-1}.

c The time of contact with the ball, Δt, will be longer with looser strings, so that the same force will produce a bigger impulse (change of momentum).

d There is less control over the direction of the ball.

138 A train of mass 2.3 × 10^6 kg travelling east along a straight track at 14.3 m s^{-1} reduces its speed to 9.8 m s^{-1}. What is the change of momentum?

139 What average force is needed to reduce the speed of a 1200 kg car from 24 m s^{-1} to 11 m s^{-1} in 4.8 s?

140 A baseball bat hits a ball with an average force of 970 N that acts for 0.0088 s.
- a What impulse was given to the ball?
- b What was the change of momentum of the ball?
- c The ball was hit back in the same direction that it came from. If its speed before being hit was 32 m s^{-1}, what was its speed afterwards? (Mass of baseball was 145 g.)

141 Figure 2.129 shows how the force between two colliding cars changed with time. Both cars were driving in the same direction and after the collision they did not stick together.
- a Estimate the impulse.
- b Just before the collision the faster car (mass 1200 kg) was travelling at 18 m s^{-1}. Estimate its speed immediately after the collision.

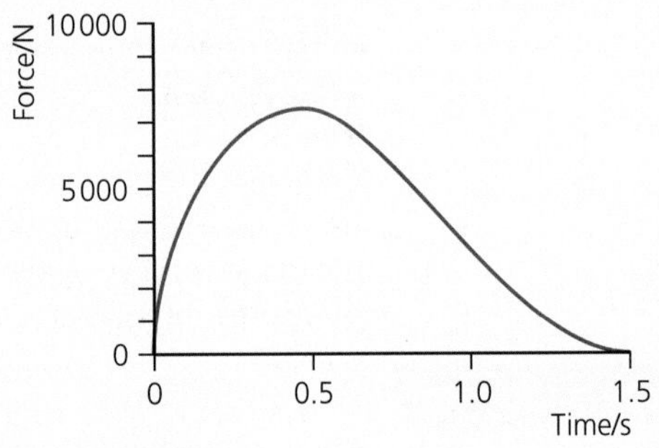

Force/N
10000
5000
0
0 0.5 1.0 1.5
Time/s

■ **Figure 2.129**

142 A soft ball (A) of mass 500 g is moving to the right with a speed of 3.0 m s^{-1} when it collides with another soft ball (B) moving to the left. The time of impact is 0.34 s, after which ball A rebounds with a speed of 2.0 m s^{-1}.
- a What was the change of velocity of ball A?
- b What was the change of momentum of ball A?
- c Calculate the average force exerted on ball A.
- d Sketch a force–time graph for the impact.
- e Add to your sketch a possible force–time graph for the collision of hard balls of similar masses and velocities.
- f Suggest how a force–time graph for ball B would be different (or the same) as for ball A.

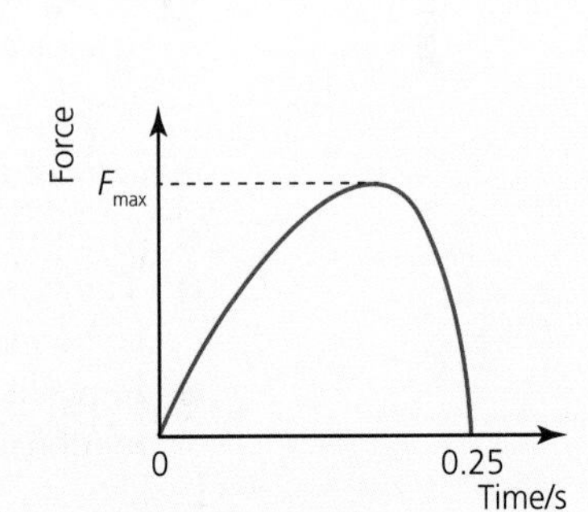

■ **Figure 2.130**

143 A car driver involved in an accident was prevented from hitting the steering wheel by the action of an airbag. Figure 2.130 shows how the force on the driver varied during the accident (until the car was stationary).
- a If the maximum force on the driver was 2800 N, estimate the average force acting during the 0.25 s.
- b What was the change of momentum of the driver?
- c If the mass of the driver was 73 kg, what was the speed in km h^{-1} of the car and driver before the accident?

144 Consider Figure 2.126.
- a Discuss in detail how the movement of the karate expert can maximize the force exerted on the boards.
- b Suggest what features of the boards will help to make this an impressive demonstration.

145
- a What is the momentum of a 1340 kg car that has 4.30 × 10^5 J of kinetic energy?
- b What is the kinetic energy of a 340 g mass that has a momentum of 8.3 kg m s^{-1}?
- c What is the mass of a sub-atomic particle that has a momentum of 2.50 × 10^{-23} kg m s^{-1} when its kinetic energy is 3.43 × 10^{-16} J?

Conservation of linear momentum

We know from Newton's third law that whenever *any* two objects interact (most commonly *collide*) with each other for time Δt, they must always exert equal and opposite forces, F, on each other. Consider again Figure 2.89.

$$F_A = -F_B$$

It follows that the impulses, $F\Delta t$, and changes of momentum, Δp, must also be equal and opposite:

$$F_A \Delta t = -F_B \Delta t$$
$$\Delta p_A = -\Delta p_B$$

In other words, if objects A and B exert forces on each other, any gain of momentum of A *must* be equal to the loss of momentum of B so that the overall momentum is unchanged. It should be stressed that this is true *only if no other, external, forces are acting.* This is expressed in the *law of conservation of momentum:*

> The total (linear) momentum of any system is constant, provided that no external forces are acting on it.

Although the principle of conservation of momentum contains essentially the same physics as Newton's law, it is usually of much greater use in everyday situations. This is because it relates to the masses and velocities before and after an interaction, rather than the varying forces *during* short time intervals.

Nature of Science

Why momentum is an important concept

There are no known exceptions to the principle of conservation of momentum (Newton's third law) and this is why momentum is such an important quantity in physics. Like energy, momentum is *always* conserved. For this reason these fundamental principles are extremely useful pieces of physics, which can be applied to help predict what happens in *any* interaction involving forces between any number of different objects, including everything from the astronomically large to the microscopically small (e.g. collisions of atomic particles).

The following points always need to be remembered when using the principle of conservation of momentum:

- Momentum is a vector quantity, so its direction must *always* be included in calculations.
- The system must be *isolated* – only the interacting objects can be considered and there can be no forces acting from outside of that system.
- Immediately *after* an interaction, external forces (like friction) will usually affect the motions of the objects.

It is easy to give examples in which the momentum of an object decreases to zero, which may appear to contradict the principle of conservation of momentum. This apparent loss of momentum is usually because the system is not isolated – it is acted on by external forces, such as friction. In other examples, some or all momentum may appear to be lost when something collides with an object that has a much greater mass. The motion after impact may be too small to observe or measure. A typical example could be a person jumping down onto the ground. The predicted motion of the person–Earth system after impact is insignificant.

The force of gravity commonly increases the momentum of falling objects, but the objects are not isolated systems – there are external forces acting on them. For example, a 3 kg rock experiences a gravitational force towards the Earth of approximately 30 N and therefore gains momentum as it accelerates down. The law of conservation of momentum predicts correctly that the Earth must gain an equal momentum upwards towards the rock. Because the mass of the Earth is so great, its gain of momentum is insignificant.

Figure 2.131 shows water guns. Before it is used a gun and the water inside it have no momentum, but when water is fired forward it gains momentum. The principle of conservation of momentum tells us that the gun must gain an equal momentum in the opposite direction.

■ **Figure 2.131** Water guns are popular during the Songkran festival in Thailand

Worked examples

35 Mass A (4.0 kg) is moving at $3.0\,\text{m s}^{-1}$ to the right when it collides with mass B (6.0 kg) moving in the opposite direction at $5.0\,\text{m s}^{-1}$.

a If they stick together after the collision, what is their velocity?

b What assumption did you make?

a The total momentum must be the same before and after collision. Choosing velocities and momenta (momentums) to the right to be positive and to the left to be negative:

$$\text{momentum of A} = m_A u_A = 4.0 \times (+3.0) = +12.0\,\text{kg m s}^{-1}$$

$$\text{momentum of B} = m_B u_B = 6.0 \times (-5.0) = -30\,\text{kg m s}^{-1}$$

Therefore, the total momentum (before) the collision is +12 +(−30) = $-18\,\text{kg m s}^{-1}$. The total momentum after the collision must be the same as that before, so:

$$m_{AB} v_{AB} = -18\,\text{kg m s}^{-1}$$

$$(4 + 6) v_{AB} = -18\,\text{kg m s}^{-1}$$

$$v_{AB} = \frac{-18}{10} = -1.8\,\text{m s}^{-1}$$

The negative sign means the velocity is to the left.

This explanation has been written out in detail to aid understanding. A more direct way of answering any such question involving two masses interacting is as follows:

momentum before interaction = momentum after interaction

$$m_A u_A + m_B u_B = m_A v_A + m_B v_B$$

In this example:

$$(4.0 \times 3.0) + (6.0 \times -5.0) = (4.0 + 6.0) \times v_{AB}$$

$$v_{AB} = -1.8\,\text{m s}^{-1}$$

b The assumption made is that there are no external forces acting on the system. If there are significant frictional forces involved, the calculated answer can be taken to be the instantaneous velocity immediately after collision, and the effects of friction can be considered afterwards.

36 A bus of mass 5800 kg travelling at a steady $24\,\text{m s}^{-1}$ runs into the back of a car of mass 1200 kg travelling at $18\,\text{m s}^{-1}$ in the same direction. If the car is pushed forwards with a velocity of $20\,\text{m s}^{-1}$, calculate the velocity of the bus immediately after the collision. (Assume this is an isolated system and ignore the actions of the drivers and engines.)

momentum before interaction = momentum after interaction

$$m_A u_A + m_B u_B = m_A v_A + m_B v_B$$

$$(5800 \times 24) + (1200 \times 18) = (5800 \times v_{bus}) + (1200 \times 20)$$

$v_{bus} = 23.6\,\text{m s}^{-1}$ in the original direction

37 A bullet of mass 12.0 g is fired at a speed of 550 m s^{-1} from a rifle of mass 1.40 kg (Figure 2.132). What happens to the rifle?

In this example the total momentum is zero. This means that the momentum of the bullet must be equal and opposite to the momentum of the rifle:

momentum before interaction = momentum after interaction

$$m_A u_A + m_B u_B = m_A v_A + m_B v_B$$

$$0 + 0 = (0.012 \times 550) + (1.40 \times v_{rifle})$$

$$v_{rifle} = \frac{-6.60}{1.40} = -4.71\ \text{m s}^{-1}$$

The negative sign shows that the rifle moves in the opposite direction from the bullet. (This is often called *recoil*.)

■ **Figure 2.132**

Additional Perspectives

The principle of conservation of momentum and road accidents

The principle of conservation of momentum is useful when determining the speeds of vehicles involved in accidents.

More than 1.2 million people are killed in road accidents every year in the world, and about 25 million are seriously injured. As a global average, people have about a 1 in 80 chance of dying in a road accident. However, there are very large differences in safety standards between countries, mostly because of the differences in the behaviour of drivers and other road users, as well as the standards of policing and the quality of the road systems.

■ **Figure 2.133** Momentum is conserved in this collision test

The design of new vehicles in most countries of the world now includes many features intended to protect the occupants in the event of an accident. These so called **passive safety standards** include seat belts, air bags, padding on interior surfaces and crumple zones, all of which reduce forces in impacts by increasing the times and distances involved. But worldwide the majority of people killed in road accidents are unaffected by these measures because they are not *inside* the vehicles: they are typically either pedestrians or on bikes of various kinds, and struck by faster-moving vehicles.

1 Find out from the internet where your country ranks in a list of the annual number of road accidents/fatalities. If possible, also determine the major causes of road accidents in your country and discuss how the situation could be improved.

2 It is widely accepted that vehicles travelling too fast (for the conditions) are a major cause of serious accidents. Remember that when a car's speed doubles, its kinetic energy increases by a factor of four. Reducing speed limits is an obvious suggestion to try to reduce the number of accidents, but many people are against this idea. Suggest why.

146 A body of mass 2.3 kg moving to the left at 82 cm s^{-1} collides with a stationary mass of 1.9 kg. If they stick together, what is their velocity after impact?

147 A 50.9 kg bag of cement is dropped from rest onto the ground from a height of 1.46 m.
a What is the maximum speed of the bag as it hits the ground?
b Will the actual speed be much different from the maximum theoretical speed? Explain your answer.
c Assuming that the bag does not bounce, predict the combined speed of the Earth and bag after the impact. (Mass of the Earth = 6.0×10^{24} kg.) Is it possible to measure this speed?

148 In an experiment to find the speed of a 2.4 g bullet, it was fired into a 650 g block of wood at rest on a friction-free surface. If the block (and bullet) moved off with an initial speed of 96 cm s^{-1}, what was the speed of the bullet?

149 A ball moving vertically upwards decelerates and its momentum decreases, although the law of conservation of momentum states that total momentum cannot change. Explain this observation.

150 An astronaut of mass 90 kg pushes a 2.3 kg hammer away from her body with a speed of 80 cm s^{-1}. What happens to the astronaut? How can she stop moving?

151 Two toy cars travel in straight lines towards each other on a friction-free track. Car A has a mass of 432 g and a speed of 83.2 cm s^{-1}. Car B has a mass of 287 g and speed of 68.2 cm s^{-1}. If they stick together after impact, what is their combined velocity?

152 A steel ball of mass 1.2 kg moving at 2.7 m s^{-1} collides head-on with another steel ball of mass 0.54 kg moving in the opposite direction at 3.9 m s^{-1}. The balls bounce off each other, each returning back in the direction it came from.
a If the smaller ball had a speed after the collision of 6.0 m s^{-1}, use the law of conservation of momentum to predict the speed of the larger ball.
b In fact, this result is not possible. Suggest a reason why not.

153 Figure 2.134 shows two trolleys on a friction-free surface joined together by a thin rubber cord under tension. When the trolleys are released, they accelerate towards each other and the cord quickly becomes loose. Where would you expect the two trolleys to collide with each other?

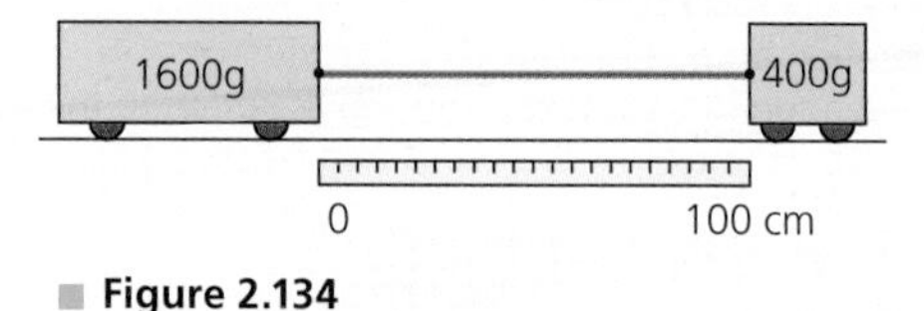

Figure 2.134

Vehicle propulsion

In order to accelerate a vehicle, or to keep it moving at a constant velocity overcoming friction and air resistance, a force is needed. This can be achieved by the vehicle pushing backwards on a road, so that the road pushes the vehicle forwards (Newton's third law). But for the motion of boats or vehicles travelling through the air or space there is better way of explaining the motion by considering the momentum given to something in the opposite direction from the intended vehicle motion.

When the propeller in Figure 2.135 spins, a force pushes the water backwards, and an equal and opposite force therefore pushes the boat forwards. The backwards momentum of the high-speed water would be equal and opposite to the forward momentum of the slower-moving boat if there were no other forces acting on the boat.

The use of oars and paddles in boats involves similar principles, while propellers on planes and helicopters make air move at high speed in the opposite direction from the vehicle's intended motion.

The jet engines on the plane shown in Figure 2.136 burn fuel combined with oxygen from the air that is taken in at the front of the engine. The resulting gases are ejected (thrown out) at the back of the engine with high speed. The gain in backwards momentum of the gases must be equal and opposite to the forward momentum of the engine and plane (if no other forces are acting).

Newton's third law offers an alternative and equivalent explanation – the force pushing the gases backwards is equal and opposite to the force pushing the plane forwards. Similar ideas can be applied to rockets, except that no air is taken in at the front.

■ **Figure 2.135** The propeller at the back of a boat exerts a force on the water and the water pushes back on the propeller

■ **Figure 2.136** A jet engine

154 A rocket engine on a spacecraft of total mass 10 000 kg ejected 1.4 kg of hot gases every second at an average discontinuous speed of $240\,m\,s^{-1}$.
a What was the force on the spacecraft?
b What was the acceleration of the spacecraft?
c If the engines were used for 30 s, what was the impulse given to the spacecraft?
d What was the change of velocity?

■ **Figure 2.137**

155 a Use Newton's third law to explain how the helicopter shown in Figure 2.137 can hover in the air.
b The helicopter blades give downwards momentum to the air. Apply the law of conservation of momentum to this situation.

ToK Link

Rules can restrict imagination

Do conservation laws restrict or enable further development in physics?

The 'simplicity' of conservation laws and the fact that they can be applied to anything and everything makes them very useful and conceptually appealing. However, it could be argued that their unquestioned acceptance among scientists actually inhibits original thinking.

Elastic collisions, inelastic collisions and explosions

When objects come into contact (or otherwise interact) with each other and exert forces on each other for relatively short periods of time, the interactions can be described as 'collisions'. We normally expect some or all of the kinetic energy of macroscopic objects to be dissipated in a collision, but it is important to define the *extreme* case:

A collision in which the total kinetic energy of the masses is the same before and after the collision is known as an elastic collision.

In our everyday, large-scale world *elastic collisions* are not possible because some energy is *always* dissipated into the surroundings.

Momentum is conserved in *all* collisions and in the theoretical extreme of an elastic collision, *kinetic* energy is also conserved. The two conservation equations representing this situation can be combined together simultaneously to predict exactly what would happen if an elastic collision occurred (but this is *not* required in this course). For example, if a mass collides elastically with another identical mass at rest, the only possibility is that the moving mass will stop and the other mass will move off with the same velocity as the first mass. The scientific purpose of a Newton's cradle (Figure 2.138) is to demonstrate this effect.

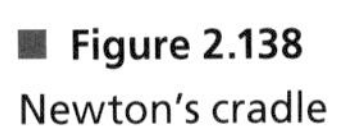

Figure 2.138 Newton's cradle

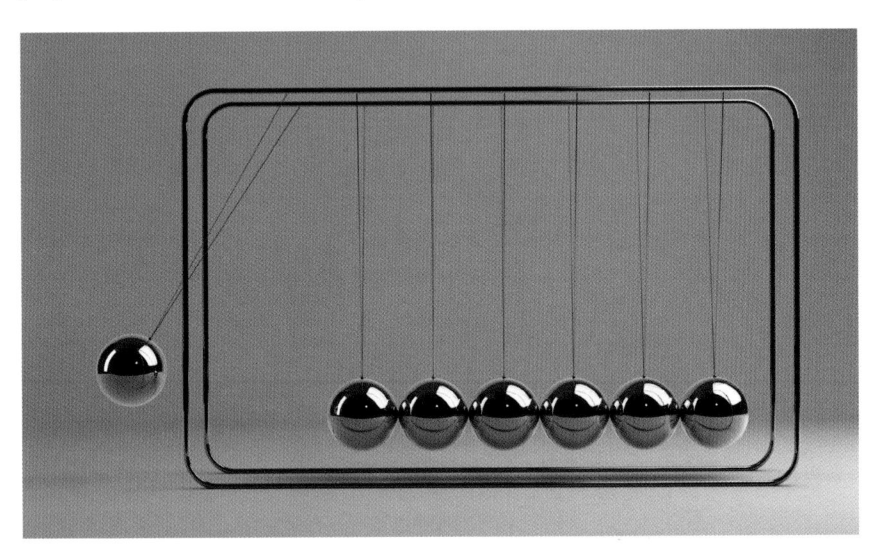

Collisions in which some or all kinetic energy is transferred to other forms of energy are called inelastic collisions. All collisions of everyday objects are inelastic.

A collision in which the objects stick together is called a totally inelastic collision.

In an 'explosion', masses that were originally at rest with respect to each other are propelled in different directions, so that there is higher kinetic energy after the explosion than before. By definition this type of interaction clearly cannot be described as *elastic* and can be considered to be similar to a totally *inelastic* collision in reverse.

The percentage of the total kinetic energy retained in collisions between masses moving together along a straight line can easily be investigated by measuring their masses and their speeds before and after the impact, although a low-friction surface is necessary for reliable results. It is instructive to investigate how the results change when the masses are varied, or the natures of the colliding surfaces are changed. Typically, objects made from elastic materials, like steel and rubber, retain the most kinetic energy. (Remember that a material described as elastic regains its shape after a force has been removed.) Conversely, inelastic materials deform permanently and much energy is dissipated as internal energy and thermal energy.

The concepts of internal energy, sound and deformation cannot be used sensibly to describe individual molecules. Therefore, on the microscopic scale, collisions between particles such as molecules in a gas are usually elastic and easily modelled by computer simulations.

156 A railway truck of mass 8340 kg travelling at 14.3 m s^{-1} collides with another truck of mass 6420 kg travelling at 8.78 m s^{-1} in the same direction.
a If after collision the two trucks become joined together, what is their initial speed?
b Calculate the percentage of kinetic energy retained in the collision.

157 A trolley of mass 2.0 kg is moving at a speed of 1.3 m s^{-1} directly towards another stationary trolley of mass 1.0 kg.
a If immediately after the collision the 1 kg trolley moves at a speed of 1.4 m s^{-1}, what is the speed of the other trolley?
b Calculate the amount of energy dissipated in this collision.

158 If in the previous question the speed of the 1 kg trolley after the collision was stated to be 4.1 m s^{-1} (instead of 1.4 m s^{-1}), explain why it would still be possible to calculate an answer for part a of the question, but not part b.

159 A cannon of mass 1100 kg fires a cannonball of mass 6.2 kg at a speed of 190 m s^{-1} (Figure 2.139).

■ **Figure 2.139** Firing a cannon

a Calculate the initial recoil speed of the cannon.
b The purpose of firing the cannon is to transfer chemical energy of the explosive into kinetic energy of the cannon ball, but the cannon is also given kinetic energy. Calculate the percentage of the total kinetic energy that is carried by the cannon ball.

Examination questions – a selection

Paper 1 IB questions and IB style questions

Q1 M is a small mass on the end of a lightweight string. It has been pulled to one side with a force that keeps it stationary.

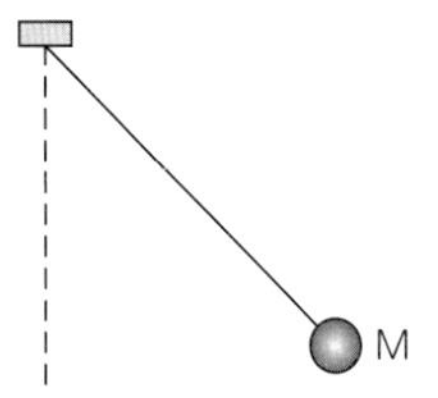

Which of the following four diagrams is the correct free-body representation of the forces acting on the mass?

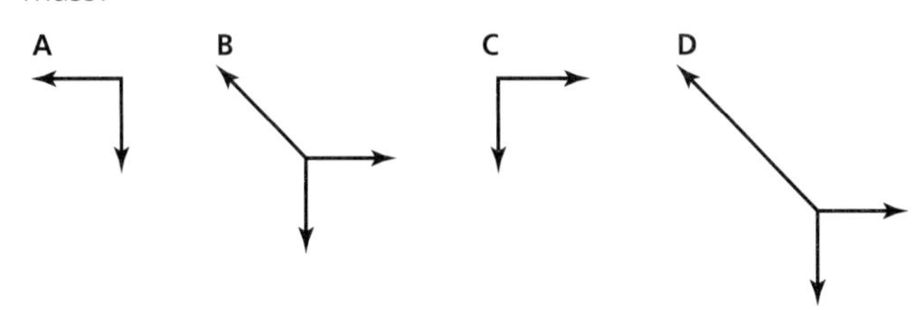

Q2 An object of weight W is slipping down a slope (inclined plane) that makes an angle of θ with the horizontal. If the object is moving with constant speed, what is the magnitude of the frictional force up the slope?

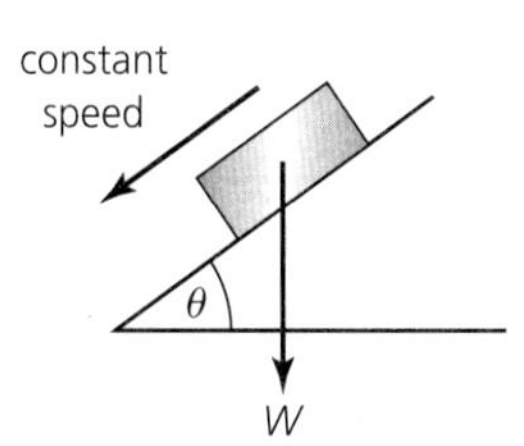

A W

B $W \sin \theta$

C $W \cos \theta$

D $\frac{W}{2}$

Q3 The work done when a constant force acts on a mass is always equal to:

A the magnitude of the force multiplied by the distance moved by the mass
B the magnitude of the force multiplied by the displacement perpendicular to the force
C the magnitude of the force multiplied by the displacement in the direction of the force
D the vector sum of the force and the distance moved by the mass.

Q4 A rocket is travelling across space when its engine ejects gases of total mass m in a time t. The speed of the gases relative to the rocket is v.
Which of the following is the correct expression for the force exerted by the gases on the rocket?

A mv

B $\frac{mv}{2}$

C mvt

D $\frac{mv}{t}$

Q5 Which of the following is a correct definition of the instantaneous velocity of a moving object at time *t*?

A $\frac{\text{distance moved}}{\text{time taken}}$

B $\frac{\text{displacement}}{\text{time taken}}$

C rate of change of displacement at time *t*

D rate of change of distance at time *t*

Q6 A big object is dropped from a large height. It hits the ground at time *T* after being dropped. Which of the following graphs best represents how the speed, *v*, of the object varies with time, *t*, until just before it hits the ground?

A

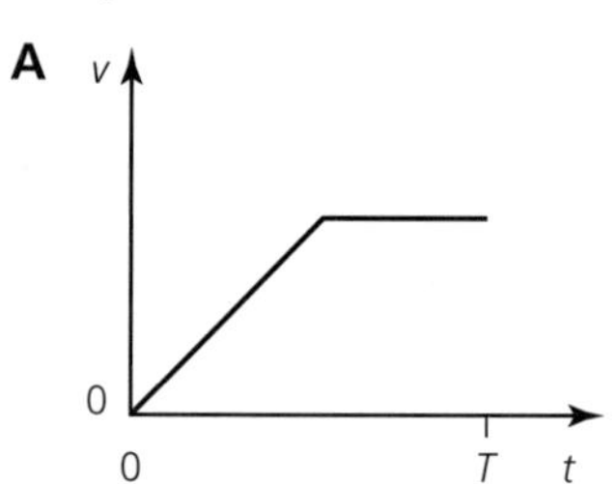

B

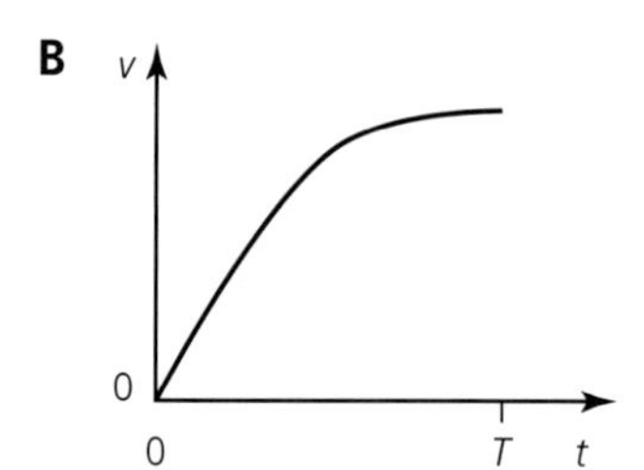

C

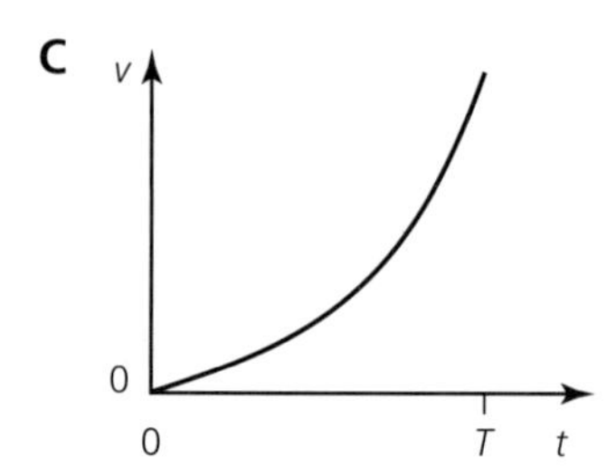

D

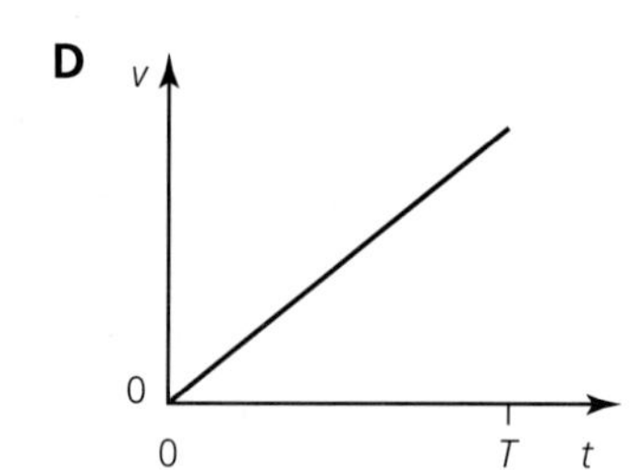

Q7 When the motion of two cars was compared, it was found that car A was more powerful than car B. Which of the following statements must be true?

A Car A produces more useful energy than car B.

B Car A produces a greater force than car B.

C In the same time, car A does more useful work than car B.

D In the same time, car A moves a greater distance than car B.

Q8 A mass of 5 kg is pulled up a slope at a constant speed of $2\,\mathrm{m\,s^{-1}}$. After rising a vertical height of 4 m, the total work done was 1200 J. The work done in overcoming friction was:

A 1000 J **B** 200 J **C** 2400 J **D** 1400 J

Q9 An increasing force acts on an object and its acceleration increases as shown in the graph.

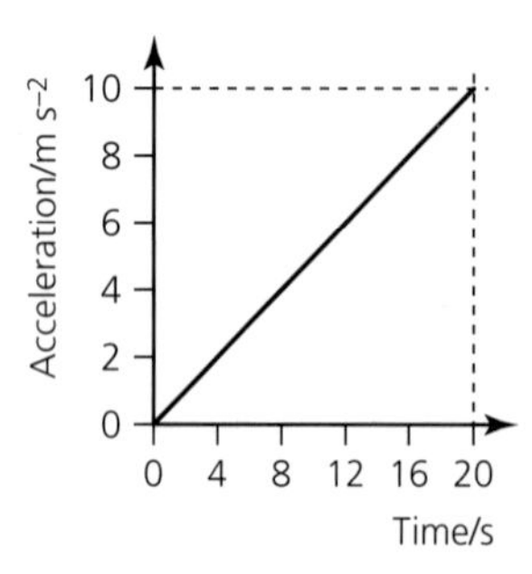

If the object was initially at rest, what is the speed of the object after 20 seconds?

A $0.5\,\mathrm{m\,s^{-1}}$ **B** $2.0\,\mathrm{m\,s^{-1}}$ **C** $100\,\mathrm{m\,s^{-1}}$ **D** $200\,\mathrm{m\,s^{-1}}$

Q10 If there is no resultant force acting on an object, which of the following quantities must also be zero?
A speed **B** velocity **C** acceleration **D** momentum

Q11 An electric motor raises a 2.5 kg mass a distance of 12 m in a time of 6 s. If the efficiency of the process was 20%, what was the input power to the motor?
A 10 W **B** 25 W **C** 250 W **D** 500 W

Q12 A mass is moving at a constant speed with a kinetic energy E_K. What is the kinetic energy of another mass that has twice the mass and half the speed of the first mass?
A $\frac{E_K}{2}$ **B** E_K **C** $2E_K$ **D** $4E_K$

Q13 A vehicle is driven up a hill at constant speed. Which of the following best describes the energy changes involved?
A Chemical energy is converted into gravitational potential energy.
B Chemical energy is converted into gravitational potential energy, sound and thermal energy.
C Gravitational potential energy is converted into chemical energy.
D Gravitational potential energy is converted into chemical energy, sound and thermal energy.

Q14 A weight *W* is suspended from the ceiling on the end of a length of string. According to Newton's third law of motion there must be another force equal and opposite to the weight. This second force is:
A the downwards force of the string on the ceiling
B the upwards force of the string on the weight
C the upwards force exerted by the weight on the Earth
D the tension in the string.

Q15 A stone is thrown up into the air at an angle to the horizontal. Assuming that air resistance is negligible, which of the following is *not* constant while the stone is moving through the air?
A horizontal component of velocity
B vertical component of velocity
C total energy of the stone
D acceleration of the stone

Q16 A steel ball is released from rest into a cylinder containing oil. Which of the following statements is incorrect?
A The force opposing motion is called drag.
B If the cylinder is large enough the ball will reach a terminal speed.
C The weight of the ball is reduced in the oil.
D A larger ball will experience a greater resistive force.

Q17 When an unstretched steel spring was extended by 10 cm the elastic potential energy stored in it was 0.20 J. What was the force constant of the spring?
A 4.0×10^{-3} N m^{-1}
B 4.0 N m^{-1}
C 10 N m^{-1}
D 40 N m^{-1}

Q18 When a ball was dropped onto a hard surface a student believed that the collision was elastic. For this to be true, the ball must:
A bounce up to the same height from which it was dropped
B stretch a lot
C be made of rubber
D get hotter.

Q19 A gas atom strikes a wall with speed v at an angle θ to the normal to the wall. The atom rebounds at the same speed v and angle θ.

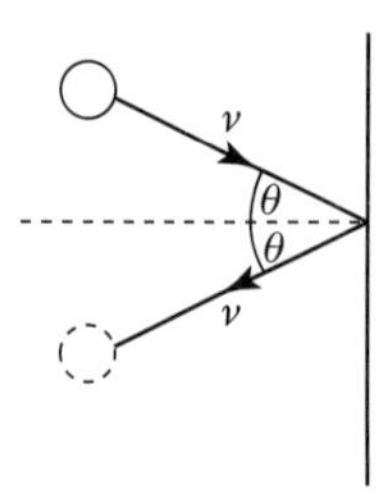

Which of the following gives the magnitude of the momentum change of the gas atom?

A zero
B $2mv\sin\theta$
C $2mv$
D $2mv\cos\theta$

Paper 2 IB questions and IB style questions

Q1 A bullet of mass 32 g is fired from a gun. The graph shows the variation of the force F on the bullet with time t as it travels along the barrel of the gun.

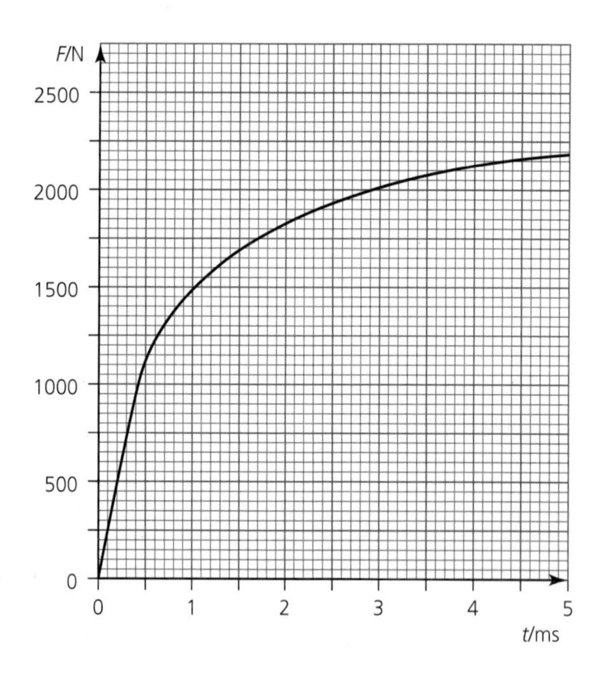

The bullet is fired at time $t = 0$ and the length of the barrel is 0.70 m.

a State and explain why it is inappropriate to use the equation $s = ut + \frac{1}{2}at^2$ to calculate the acceleration of the bullet. (2)

b Use the graph to:

i determine the average acceleration of the bullet during the final 2.0 ms of the graph (2)

ii show that the change in momentum of the bullet, as the bullet travels along the length of the barrel, is approximately 9 N s. (3)

c Use the answer in b ii to calculate:

i the speed of the bullet as it leaves the barrel (2)

ii the average power delivered to the bullet. (3)

d Use Newton's third law to explain why a gun will recoil when a bullet is fired. (3)

Q2 A clay block initially on the edge of a table is fired away from the table, as shown in the diagram.

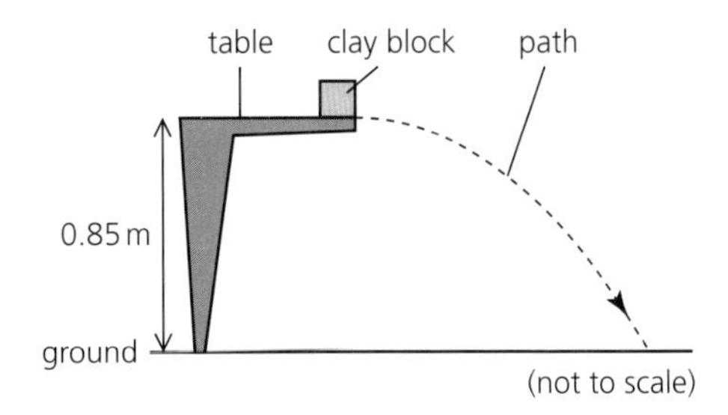

The initial speed of the clay block is $4.3\,\text{m}\,\text{s}^{-1}$ horizontally. The table surface is 0.85 m above the ground.

a Ignoring air resistance, calculate the horizontal distance travelled by the clay block before it strikes the ground. (4)

b The diagram shows the path of the clay block neglecting air resistance. Make a copy of the diagram and show on it the approximate shape of the path that the clay block will take assuming that air resistance acts on the clay block. (3)

© *IB Organization*

Q3 **a** A system consists of a bicycle and cyclist travelling at a constant velocity along a horizontal road.

i State the value of the net force acting on the cyclist. (1)

ii On a copy of the diagram, draw labelled arrows to represent the vertical forces acting on the bicycle. (2)

iii With reference to the horizontal forces acting on the system, explain why the system is travelling at constant velocity. (2)

b The total resistive force acting on the system is 40 N and its speed is $8.0\,\text{m}\,\text{s}^{-1}$. Calculate the useful power output of the cyclist. (1)

c The cyclist stops pedalling and the system comes to a rest. The total mass of the system is 70 kg.

i Calculate the magnitude of the initial acceleration of the system. (2)

ii Estimate the distance taken by the system to come to rest from the time the cyclist stops pedalling. (2)

iii State and explain one reason why your answer to c ii is only an estimate. (2)

© *IB Organization*

Q4 **a** Explain the difference between the coefficients of static and dynamic coefficients of friction. (2)

b The angle of a flat wooden slope to the horizontal is increased slowly until a metal cube *just* begins to slide down the surface. If the coefficient of static friction between the surfaces is 0.56, what is the greatest possible angle before the cube begins to slide? (2)

3 Thermal physics

ESSENTIAL IDEAS

- Thermal physics deftly demonstrates the links between the macroscopic measurements essential to many scientific models with the microscopic properties that underlie these models.
- The properties of ideal gases allow scientists to make predictions about the behaviour of real gases.

3.1 Thermal concepts – *thermal physics deftly demonstrates the links between the macroscopic measurements essential to many scientific models with the microscopic properties that underlie these models*

Molecular theory of solids, liquids and gases

An understanding of the particulate nature of matter is essential basic knowledge in physics, chemistry and most other branches of science.

- In solids the molecules (or atoms, or ions) are held close together by strong forces, usually in regular patterns. The molecules *vibrate* about their mean positions. See Figure 3.1.
- In liquids the molecules still vibrate, but the forces between some molecules are overcome, allowing them to move around a little. The molecules are still almost as close together as in solids, but there is little or no regularity in their arrangement, which is constantly changing.
- In gases the molecules are much further apart than in solids and liquids, and the forces between them are very, very small and usually negligible (except when they collide). This results in all molecules moving independently in random directions with a range of different (usually fast) speeds. These speeds are continually changing as a result of collisions.

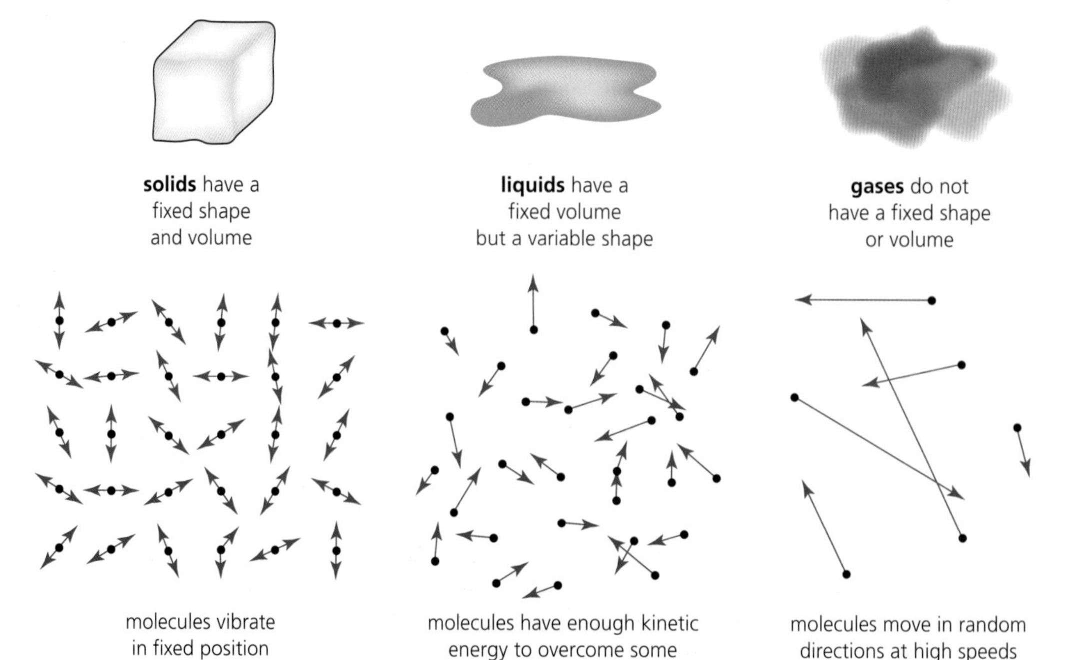

Figure 3.1 Differences between solids, liquids and gases; the vector arrows represent the velocities of the molecules

■ **Figure 3.2** This thermogram, taken using infrared radiation, uses colour to show different temperatures in a saucepan on a cooker. The scale runs from white (hottest), through red, yellow, green and blue to pink (coldest)

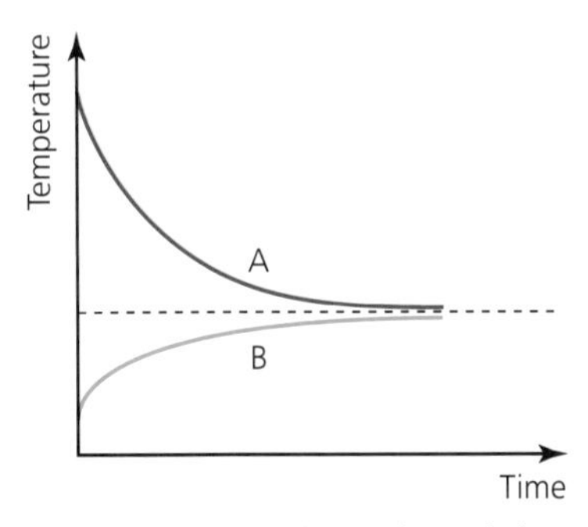

■ **Figure 3.3** Two objects (A and B) at different temperatures, insulated from their surroundings but not from each other, will reach thermal equilibrium

Thermal energy and temperature differences

Energy transferred from hotter to colder as a result of a temperature difference is called thermal energy. Thermal energy is often called 'heat', but this can be confusing because the word heat is also commonly, and *wrongly*, used for the energy *inside* substances. The energy inside substances is called internal energy.

Objects continuously exchange energy with their surroundings. An object that is hotter than its surroundings will give out (*emit*) more energy than it takes in, and a colder object will take in (*absorb*) more energy than it gives out.

Figure 3.2 shows the temperature differences in a saucepan and cooker, which will result in a flow of thermal energy. Thermal energy is *always* transferred from higher temperature to lower temperature.

> Temperature determines the direction of the net thermal energy transfer between two objects.

Consider the simple example of two objects (or substances) at different temperatures able to transfer thermal energy between themselves, but isolated from everything around them (their *surroundings*). The hotter object will transfer energy to the colder object and cool down, while at the same time the colder object warms up. As the temperature difference between the two objects gets smaller, so too does the *rate* of thermal energy transfer. This is represented in Figure 3.3, which shows how the temperature of two objects (A and B) might change when they are placed in thermal contact with each other. (Being in 'thermal contact' means that thermal energy can be transferred between them, by *any* means.) Eventually they will reach the same temperature.

If the temperatures have stopped changing and both objects are at the same temperature, the objects are said to be in thermal equilibrium and there will be no net flow of thermal energy between them.

In any realistic situation, it is not possible to completely isolate/insulate two objects from their surroundings, so the concept of thermal equilibrium may seem to be idealized. The concept of hotter objects getting colder and colder objects getting hotter suggests an important concept: eventually *everything* will end up at the same temperature.

Temperature and absolute temperature

The temperature scales that we use today were designed for simplicity and easy reproduction. On the Celsius scale (°C), sometimes called the centigrade scale, 0 °C is defined as the temperature at which pure water forms ice (at normal atmospheric pressure), and 100 °C is defined as the temperature at which pure water boils (at normal atmospheric pressure). It is important to realize that this temperature scale was devised for convenience – that is, these values were *chosen*, they were not discovered. In particular, 0 °C is definitely *not* a zero of temperature, nor a zero of energy. It has no significance other than being the melting point of ice. (For example, 10 °C cannot be considered to be 'twice as hot' as 5 °C.)

The Celsius temperature scale is used throughout the world and is a good example of how the effective communication of data between individuals and countries is dependent on an agreed system of units. For historical and cultural reasons, there a few countries (notably the USA) in which the Fahrenheit temperature scale is still in use.

After it was predicted that almost all molecular motion stops at −273 °C (see page 127), it made sense to make this the true zero of temperature. This temperature is commonly called **absolute zero**.

The **Kelvin (absolute) temperature scale** is an adaptation of the Celsius scale with its zero at −273 °C. (A more precise value is −273.15 °C.) On this scale the unit is **kelvin, K** (not °K). *Changes* in temperature of 1 °C and 1 K were chosen to be identical, which makes conversion from one scale to the other very straightforward:

$$T/\mathrm{K} = \theta/^\circ\mathrm{C} + 273$$

In the equation above, note that use of the symbol T for temperature implies the Kelvin scale and the symbol θ implies the Celsius scale. Table 3.1 compares some important temperatures on the two scales.

When making calculations involving temperature *changes*, either degrees celsius or kelvins may be used, but it is important to remember that when dealing with calculations involving just *one* temperature, kelvins must be used.

■ Table 3.1 A comparison of temperatures in degrees Celsius and in kelvin

Temperature	°C	K
absolute zero	−273	0
melting point of water	0	273
body temperature	37	310
boiling point of water	100	373

Worked example

1 The normal freezing point of mercury is −39 °C. What is this temperature in kelvin?

$T/K = \theta/^\circ C + 273$

$T = -39 + 273$

$T = 234\,\mathrm{K}$

1 a The world's highest and lowest recorded temperatures are reported to be 58 °C (in Libya) and −89 °C (in Antarctica). What are these temperatures on the Kelvin scale?
b The temperature of some water is raised from 17 °C to 55 °C. What is the temperature rise in kelvins?

2 a The volume of a gas is 37 cm^3 when its temperature is 23 °C. If the volume is proportional to the absolute temperature (K), at what temperature (°C) will the volume be 50 cm^3?
b What will the volume be when the gas is at −15 °C?

Internal energy

All substances contain moving particles. In the context of this chapter, the word 'particle' is a general term that might apply to a molecule, an atom or an ion. Although different substances may contain any or all of these, most substances are *molecular* and in the rest of this chapter the term 'molecule' will be used to describe the particles in any substance.

Moving molecules have *kinetic energy*. The molecules might be moving in different ways, which gives rise to three different forms of random molecular kinetic energy:

- Molecules might be vibrating about fixed positions (as in a solid) – this gives the molecules **vibrational kinetic energy**.
- Molecules might be moving from place to place (translational motion) – this gives the molecules **translational kinetic energy**.
- Molecules might also be spinning (rotating) – this gives the molecules **rotational kinetic energy**.

Molecules can have *potential energy* as well as kinetic energy. In solids and liquids, it is the *electrical forces* (between charged particles) that keep the molecules from moving apart or moving closer together. Wherever there are electrical forces there will be electrical potential energy in a system, in much the same way as gravitational potential energy is associated with gravitational force.

In gases, however, the forces between molecules are usually negligible because of the larger separation between molecules. This is why gas molecules can move freely and randomly. The molecules in a gas, therefore, usually have negligible electrical potential energy – all the energy is in the form of kinetic energy.

So to describe the total energy of the molecules in a substance, we need to take account of both the kinetic energies and the potential energies. This is called the *internal energy* of the substance and is defined as follows:

> The internal energy of a substance is the sum of the total random kinetic energies and total intermolecular potential energies of all the molecules inside it.

It is important not to call internal energy 'heat'. That is, we should not refer to the thermal energy (or heat) *in* anything.

In the definition of internal energy given above, the word 'random' means that the molecular movements are disordered and unpredictable. That is, they are not linked in any way to each other, or ordered – as their motions would be if they were all moving together, such as the molecules in a macroscopic motion of a moving car. The molecules in a moving car have both the ordered kinetic energy of macroscopic movement together and the random kinetic energy of internal energy.

Summarizing the differences between temperature, internal energy and thermal energy

Temperature, internal energy and thermal energy are widely used and are very important concepts throughout all of science, but they are commonly misunderstood and misused terms. To stress their importance, the meanings of these concepts are summarized below.

- *Internal energy* is the total energy (random kinetic and potential) of all the molecules inside a substance.
- If energy is transferred to a substance it gains internal energy and its molecules move faster. We say that it has become *hotter* and this is measured as an *increased temperature*. A more precise meaning of temperature is given on page 127.
- *Thermal energy* (heat) is energy flowing from a higher temperature to a lower temperature.

In any particular example, the object or substance we are considering is often called the **system**, and thermal energy flows between the system and the **surroundings**. Students may know that thermal energy is transferred by conduction, convection and radiation, but detailed knowledge of these processes is not needed in this chapter. They are discussed later, in Chapter 8.

In Figure 3.4 thermal energy is being conducted into one hand and out of the other.

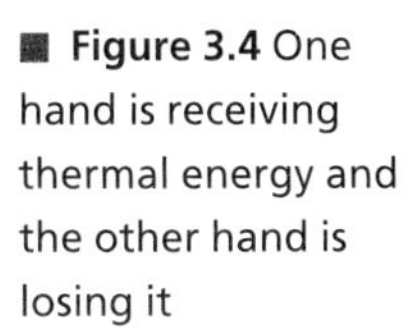

Figure 3.4 One hand is receiving thermal energy and the other hand is losing it

ToK Link

Sense perception

Observation through sense perception plays a key role in making measurements. Does sense perception play different roles in different areas of knowledge?

The scientific meaning of temperature is linked to microscopic molecular energies, but our appreciation of it is largely based on everyday **sense perceptions** of hot and cold (which are easily shown to be unreliable). Does this help or hinder our understanding of the temperature? Which of the other areas of knowledge are so dependent on information delivered directly to the body through the senses of touch, sight, hearing etc.?

3 Equal masses of two different solids at 20 °C are both heated to 40 °C.
 a Discuss whether they have the same amount of internal energy to begin with.
 b Discuss whether the same amount of thermal energy was transferred to both of them.

4 Figure 3.5 sketches how the resultant force between two molecules varies with separation; x_0 represents the average equilibrium separation between molecules in a solid. The molecules in a gas are typically ten times further apart than in a solid.

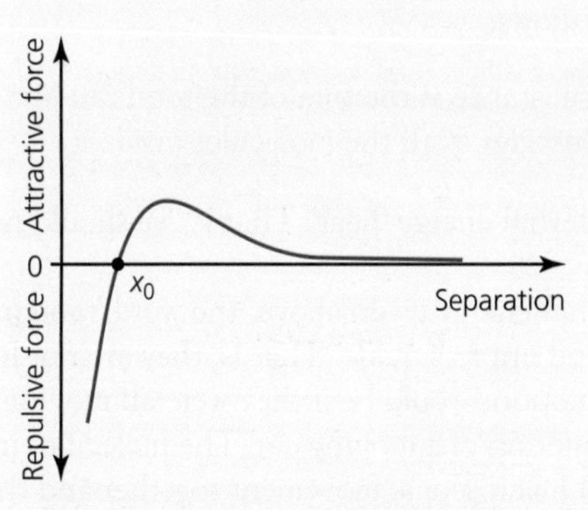

■ **Figure 3.5** How intermolecular force varies with molecular separation

 a Describe how the resultant force between molecules (in their equilibrium positions) changes if they move:
  i further apart
  ii slightly closer together.
 b What can you conclude from this graph about how the size of forces between molecules in a gas is different from forces in a solid?
 c Explain why you might expect the density of gases to be about 10^3 times smaller than the density of solids.

5 The 'sparkles' emitted from a sparkler (Figure 3.6) are very hot. Explain why they do not usually cause any harm when they land on people or their clothing. Use the terms 'temperature', 'internal energy' and 'thermal energy' in your explanation. (The hot sparkler itself will cause burns if touched.)

■ **Figure 3.6** Sparklers are usually not as dangerous as they may look!

Heating and working

Apart from supplying thermal energy to a system ('heating' it), there is another, very common and fundamentally different way to make something hotter: we can do *mechanical work* on it. A simple example would be the force of friction causing a temperature rise when surfaces rub together. Heating is a *non-mechanical* transfer of energy.

Figures 3.7 and 3.8 provide examples of these different ways of raising the temperature of an object. In Figure 3.7, the internal energy of the screw will rise as it gains thermal energy from the hand holding it. In Figure 3.8, its internal energy will rise because a force is twisting the screw into a piece of wood against resistive forces opposing it – doing mechanical work.

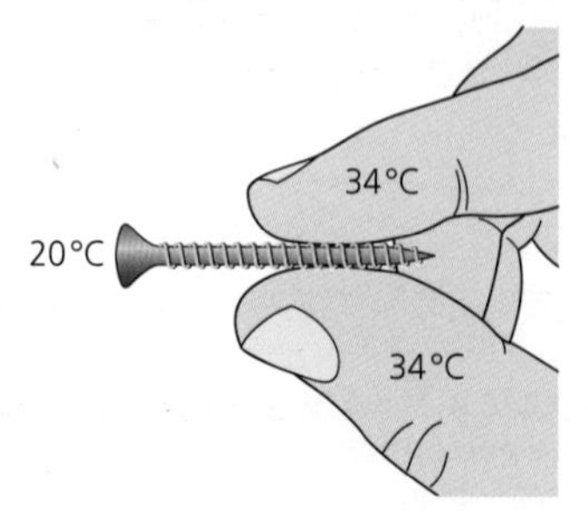

■ **Figure 3.7** Getting hotter by being in contact with something at a higher temperature

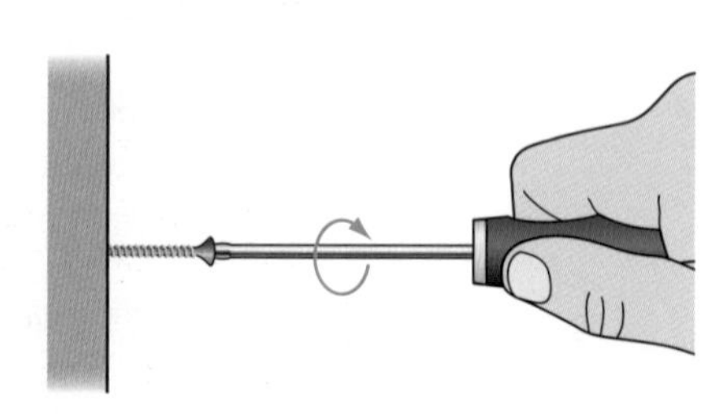

■ **Figure 3.8** Getting hotter because work is being done

Additional Perspectives

Understanding heat: James Prescott Joule

The SI unit of energy is named after the Englishman, James Prescott Joule (Figure 3.9), who was a nineteenth-century physicist and manager of a brewery. His experiences as a brewer may have contributed to his renowned skills in accurate measurement, especially those involving small temperature changes.

In the middle of the nineteenth century, 'heat' was believed to be an undetectable 'caloric fluid' that flowed from hotter to colder things. But Joule tried repeatedly to show that 'heat' was just another form of energy that could be transformed from an equal amount of other forms, such as kinetic energy or gravitational potential energy.

■ **Figure 3.9** James Prescott Joule

In particular, he is remembered for his 'mechanical equivalent of heat' experiments in which mechanical energy was used to raise the temperature of water. It is even claimed that he spent part of his honeymoon trying to measure a very small temperature difference between the top and bottom of waterfalls, which is not an easy thing to do!

His work united the separate topics of energy and 'heat', and was important in the later development of the law of conservation of energy and the first law of thermodynamics. Joule also worked with Lord Kelvin on thermometry and temperature scales.

1 The use of the word 'heat' can be confusing, especially if it is used to represent both energy inside objects as well as energy being transferred. Discuss the use of the word 'heat' in this Additional Perspective.

Nature of Science

Earlier ideas about heat were limited by a lack of knowledge about the particle nature of matter

A little more than 200 years ago, heat was explained in terms of a vague '**caloric fluid**' that flowed out of a hot object. This was an example of one of many serious scientific theories that were developed to explain observed phenomena, but which were never totally satisfactory because they could not explain *all* observations. The earlier '**phlogiston**' theory of combustion is another such theory related to heat.

Looking back from the twenty-first century, these theories may seem unsophisticated and inaccurate (but imaginative!). However, they should be judged in the context of their times, and at the time of these theories (seventeenth and eighteenth centuries) the molecular theory of matter had not been developed, so the current understanding of the flow of thermal energy was not possible then.

■ Heating and cooling graphs

The dotted blue straight line in Figure 3.10 shows how the temperature of an object heated at a constant rate changes with time under the idealized circumstances of no thermal energy (heat) losses. The temperature rises by equal amounts in equal times. However, thermal energy losses to the surroundings are unavoidable, so the curved red line represents a more realistic situation. The curve shows that the *rate* of temperature rise decreases as the object gets hotter. This is because thermal energy losses (from the object to the surroundings) are higher with larger temperature differences. If energy continues to be supplied, the object will eventually reach a constant temperature when the input power and rate of thermal energy loss to the surroundings are equal.

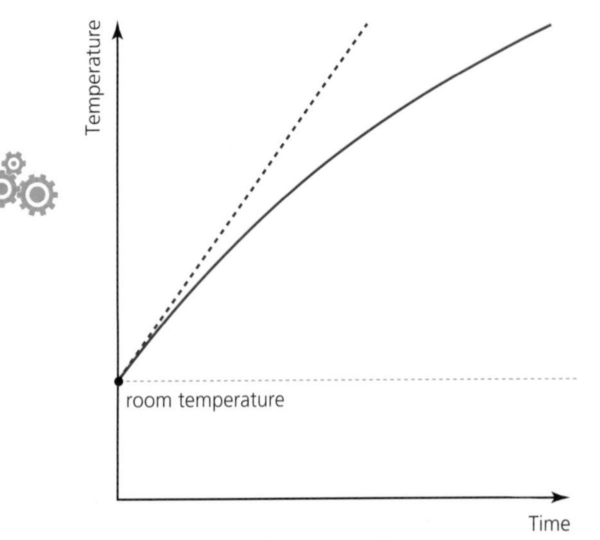

■ **Figure 3.10** A typical graph of temperature against time for heating at a constant rate

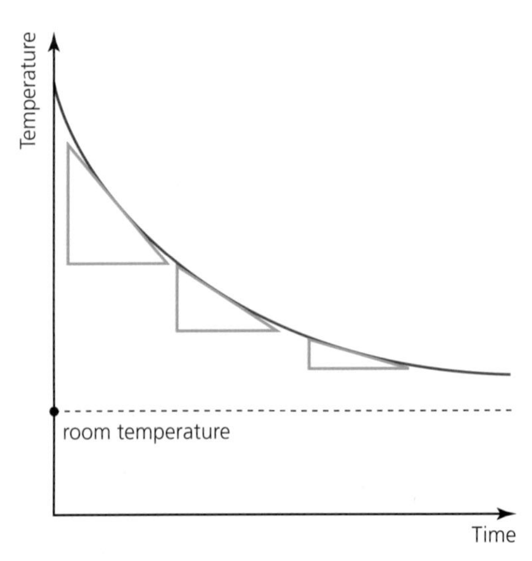

■ **Figure 3.11** A typical graph of temperature against time for an object cooling down to room temperature. Note how the gradient decreases with time

When something is left to cool naturally, the rate at which thermal energy is transferred away decreases with time because it also depends on the temperature difference between the object and its surroundings. This can be seen in Figure 3.11, in which the rate of cooling (as shown by the gradients at different times) gets less and less.

Specific heat capacity and thermal capacity

Specific heat capacity

To compare how different substances respond to heating, we need to know how much thermal energy will increase the temperature of the same mass (usually 1 kilogram) of each substance by the same amount (1 K or 1 °C). This is called the *specific heat capacity*, *c*, of the substance. (The word 'specific' is used here simply to mean that the heat capacity is related to a specified amount of the material, namely 1 kg.)

The specific heat capacity of a substance is the amount of energy needed to raise the temperature of 1 kg of the substance by 1 K. (Unit: $\text{J kg}^{-1}\text{K}^{-1}$, but $°\text{C}^{-1}$ can be used instead of K^{-1}.)

The values of specific heat capacity for some common materials are given in Table 3.2.

■ **Table 3.2** Some specific heat capacities of different materials

Material	Specific heat capacity/$\text{J kg}^{-1}\text{K}^{-1}$
copper	390
aluminium	910
water	4180
air	1000
dry earth	1250
glass (typical)	800
concrete (typical)	800

In simple terms, substances with high specific heat capacities heat up slowly compared with equal masses of substances with lower specific heat capacities (given the same power input). Similarly, substances with high specific heat capacities will cool down more slowly. It should be noted that water has an unusually large specific heat capacity. This is why it takes the transfer of a large amount of energy to change the temperature of water and the reason why water is used widely to transfer energy in heating and cooling systems.

If a quantity of thermal energy, Q, was supplied to a mass, m, and produced a temperature rise of ΔT, we could calculate the specific heat capacity from the equation:

$$c = \frac{Q}{m\Delta T}$$

(Remember that the delta sign, Δ, is used widely in science and mathematics to represent a small change in a quantity.)

This equation is more usually written as follows:

$$Q = mc\Delta T$$

This equation is included in the *Physics data booklet*.

When a substance cools, the thermal energy transferred away can be calculated using the same equation.

The simplest way of determining the specific heat capacity of a substance experimentally is to supply a known amount of energy from an electric heater placed inside (immersed in) the substance. Such a heater is called an **immersion heater**. This is easy enough with liquids, but for solids it is often necessary to drill a hole in the substance to allow the heater to be fitted inside, ensuring good thermal contact.

In the two experiments shown in Figures 3.12 and 3.13, a **joulemeter** is used to measure the energy transferred directly. (More commonly, the energy can be calculated using knowledge of electric circuits, from the equation: electrical energy = VIt, which is covered in Chapter 5.)

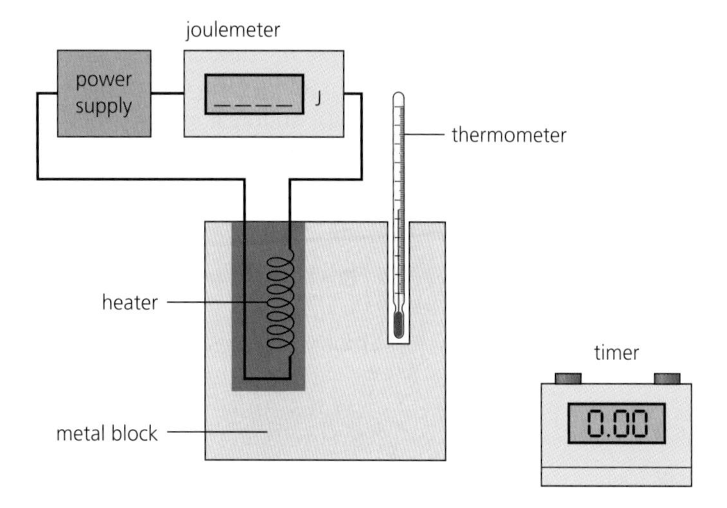

■ **Figure 3.12** Determining the specific heat capacity of a metal

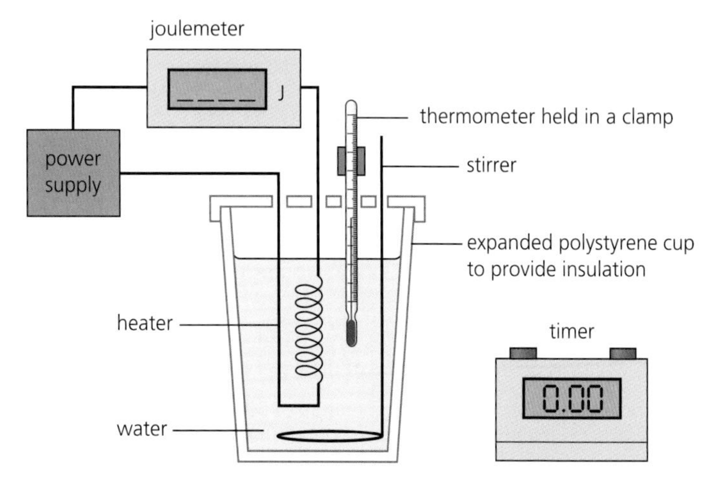

■ **Figure 3.13** Determining the specific heat capacity of water

Worked example

2 Suppose that the metal block shown in Figure 3.12 had a mass of 1500 g and was heated for 5 minutes with an 18 W heater. If the temperature of the block rose from 18.0 °C to 27.5 °C, calculate its specific heat capacity assuming that no energy was transferred to the surroundings.

$$c = \frac{Q}{m\Delta T}$$

and

$$Q = Pt$$

so,

$$c = \frac{18 \times (5 \times 60)}{1.5 \times (27.5 - 18.0)}$$

$$= 380\,\mathrm{J\,kg^{-1}\,°C^{-1}}$$

In making such calculations, it has to be assumed that all of the substance was at the same temperature and that the thermometer recorded that temperature accurately at the relevant times. In practice, both of these assumptions might lead to significant inaccuracies in the calculated value.

Furthermore, in any experiment involving thermal energy transfers and changes in temperature, there will be unavoidable losses (or gains) from the surroundings. If accurate results are required, it will be necessary to use **insulation** to limit these energy transfers, which in this example would have led to an overestimate of the substance's specific heat capacity (because some of the energy input went to the surroundings rather than into the substance). The process of insulating something usually involves surrounding it with a material that traps air (a poor conductor) and is often called **lagging**.

In the following questions, assume that no energy is transferred to or from the surroundings.

6 How much energy is needed to raise the temperature of a block of metal of mass 3.87 kg by 54 °C if the metal has a specific heat capacity of $456\,\mathrm{J\,kg^{-1}\,K^{-1}}$?

7 What is the specific heat capacity of a liquid that requires 3840 J to raise the temperature of a mass of 156 g by 18.0 K?

8 Air has a density of $1.3\,\mathrm{kg\,m^{-3}}$ and a specific heat capacity of $1000\,\mathrm{J\,kg^{-1}\,°C^{-1}}$. If 500 kJ was transferred to a room of volume $80\,\mathrm{m^3}$, what was the temperature rise?

9 If 1.0 MJ of energy is transferred to 15.0 kg of water at 18.0 °C, what will the final temperature be?

10 A drink of mass 500 g has been poured into a glass of mass 250 g (of specific heat capacity $850\,\mathrm{J\,kg^{-1}\,°C^{-1}}$) in a refrigerator. How much energy must be removed to cool the drink and the glass from 25 °C to 4 °C? (Assume the drink has the same specific heat capacity as water.)

11 A 20 W immersion heater is placed in a 2.0 kg iron block at 24 °C for 12 minutes. What is the final temperature? (Specific heat capacity of iron = $444\,\mathrm{J\,kg^{-1}\,°C^{-1}}$.)

12 How long will it take a 2.20 kW kettle to raise the temperature of 800 g of water from 16.0 °C to its boiling point?

13 An air conditioner has a cooling power of 1200 W and is left in a room containing 100 kg of air (specific heat capacity $1000\,\mathrm{J\,kg^{-1}\,°C^{-1}}$) at 30 °C. What will the temperature be after the air conditioner has been switched on for 10 minutes?

14 A water heater for a shower is rated at 9.0 kW. If water at 15 °C flows through it at a rate of 15 kg every 3 minutes, what will be the temperature of the water in the shower?

15 A burner on a gas cooker raises the temperature of 500 g of water from 24 °C to 80 °C in exactly 2 minutes. What is the effective average power of the burner?

Exchanges of thermal energy

Figure 3.3 shows the temperature–time graphs of two objects at different temperatures placed in good thermal contact so that thermal energy can be transferred relatively quickly, assuming that the system is insulated from its surroundings. Under these circumstances *the thermal energy given out by one object is equal to the thermal energy absorbed by the other object*. Exchanges of thermal energy can be used as an alternative means of determining a specific heat, as illustrated by question 16, or in the determination of the energy that can be transferred from a food or a fuel.

Calorimetry is the name used to describe experiments that try to accurately measure the temperature changes produced by various physical or chemical processes. Energy transfers can then be calculated if the masses and specific heat capacities are known. Calorimetric techniques may involve specially designed pieces of apparatus, called *calorimeters*, which are designed to limit thermal energy transfer to, or from, the surroundings.

16 A large metal bolt of mass 53.6 g was heated for a long time in an oven at 245 °C. The bolt was transferred as quickly as possible from the oven into a beaker containing 257.9 g of water initially at 23.1 °C (Figure 3.14). The water was stirred continuously and the temperature rose to a maximum of 26.5 °C.
   a Calculate the energy transferred to the water.
   b Why was the bolt kept in the oven for a long time?
   c Why was the transfer made quickly?
   d Calculate the specific heat capacity of the material of which the bolt is made.
   e Why was it necessary to stir the water?
   f Is the value for the specific heat capacity of the bolt likely to be an underestimate or an overestimate of its true value? Explain your answer.

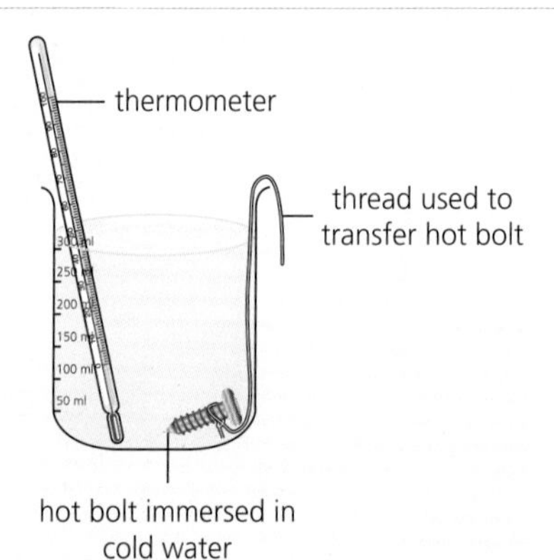

■ **Figure 3.14** A hot metal bolt placed in cold water

17 When 14.5 g of a certain fuel was burned, thermal energy was transferred to 63.9 g of water. The water temperature rose from 18.7 °C to 42.4 °C. Assuming that no thermal energy was transferred into the surroundings, calculate the maximum amount of energy that can be transferred from 1 kg of this fuel. (This is known as its *specific energy*.)

18 When running water into a bath tub, 84 kg of water at 54 °C was added to 62 kg of water at 17 °C.
   a What was the final temperature?
   b What assumption did you make?

19 Sand initially at 27.2 °C is added to an equal mass of water at 15.3 °C. If the specific heat capacity of sand is 822 J kg^{-1} °C^{-1} and the specific heat capacity of water is 4180 J kg^{-1} °C^{-1}, what will be the final temperature of the mixture? (Assume no energy is transferred to the surroundings.)

Thermal capacity

Many everyday objects are not made of only one substance, and referring to a specific amount (a kilogram) of such objects is not sensible. In such cases we refer to the thermal capacity of the whole object. For example, we might want to know the thermal capacity of a room and its contents when choosing a suitable heater or air conditioner.

The thermal capacity of an object is the amount of energy needed to raise its temperature by 1 K. (Unit: J K^{-1} or J °C^{-1}.)

$$\text{thermal capacity, } C = \frac{Q}{\Delta T}$$

Worked example

3 How much thermal energy is needed to increase the temperature of a kettle and the water inside it from 23 °C to 77 °C if its thermal capacity is 6500 J K^{-1}?

$Q = C\Delta T$

$Q = 6500 \times (77 - 23)$

$Q = 3.51 \times 10^5$ J

For the following questions, assume that no energy is transferred to or from the surroundings.

20 The thermal capacity of a saucepan and its contents is 7000 J K^{-1}. How long will it take a 1.5 kW electric cooker to raise the temperature of the saucepan and its contents from 22 °C to 90 °C?

21 If it took 32 minutes for a 2.5 kW heater to raise the temperature of a room from 8 °C to 22 °C, what was the room's thermal capacity?

22 If the contents of a refrigerator with a cooling power of 405 W have a thermal capacity of 23.9 kJ K^{-1}, how long will it take for the average temperature to be reduced from 19.50 °C to 5.20 °C?

Utilizations

Thermal capacity and building design

In countries with hot, dry climates, keeping cool in the daytime can be a major problem, especially if is not possible or desirable to use air conditioners. People in these countries have known for centuries about the advantages of building homes with large mass and, therefore, large thermal capacity (Figure 3.15). In some climates, it is also common for the air temperature to fall significantly at night because of the low humidity and the lack of clouds, which allows thermal energy to be radiated away.

■ **Figure 3.15** A thick-walled traditional African house

During the day, the radiated thermal energy from the Sun warms the building and using large amounts of materials like earth and stone (which have relatively high specific heat capacities) to make a home with a high thermal capacity ensures that the temperature rise is not too quick. At night, for the same reason, the temperature of the building will not fall suddenly and the people inside can keep warm. This effect produces a pleasant thermal 'lag', with the temperature inside the building 'cooler' in the morning and 'hotter' in the late afternoon or evening.

In essence, buildings with high thermal capacity 'average out' the temperature extremes that would otherwise be caused by the weather conditions and any significant changes in temperature between day and night. This is a useful property of nearly all buildings, whether they are located in hot or cold climates, and most buildings are best designed to have high thermal capacities. However, the economic costs of using large quantities of otherwise unnecessary building materials will usually limit the mass of buildings.

1 a Make estimates of the various quantities of building materials needed, and then calculate the approximate thermal capacity of a house similar to the one shown in the foreground of Figure 3.15.

b Assuming that on a hot day the radiated thermal energy from the Sun arrives perpendicularly down on the building's roof at a rate of 850 W m^{-2}, calculate the maximum temperature rise produced during 1 hour in the middle of the day. Assume that 5% of the thermal energy is absorbed in, and spread evenly throughout, the building. (It is likely that every effort will be made to ensure that the building does *not* absorb radiant energy falling on it – by using light colours to reflect back the energy.)

c The temperature rise you calculated in **b** was probably much higher than that which actually happens. Suggest why.

d Sketch a graph to show how you think that the air temperature in a hot, dry country may vary over a period of two cloudless days and nights. Then add to your graph a sketch showing how the inside temperature of a building such as that shown in Figure 3.15 varies over the same time.

e Suggest why air temperatures are always measured in the shade.

Nature of Science

The inevitable transfer of mechanical energy to internal energy

We know that, without a force pushing them forwards, all moving objects will tend to decelerate and eventually stop because of the forces of friction. Sometimes we need to provide extra forces to stop a moving object, such as a car. This too is usually done with the help of friction. When friction acts to slow motion, the macroscopic ordered kinetic energy of the moving molecules is transferred to the random, disordered kinetic energies of the molecules in both surfaces. There will be a rise in internal energy and temperature. Thermal energy will also be transferred to the surroundings.

The transfer of ordered energy to disordered energy cannot be stopped or reversed. If a moving object has been stopped by friction, it is simply not possible to transfer the increased internal energy (involving the random kinetic energies of molecules) back into the macroscopic overall, ordered kinetic energy of the moving object. Because of this, we may consider that useful energy has been 'wasted' by the frictional braking.

Consider the example of a car stopping under the action of its brakes. Friction between the tyres and the road, as well as air resistance on the surface of the vehicle, will contribute to the resistive forces, but for simplicity we can assume that all the kinetic energy of the car is transferred to raising the temperature of the brakes (Figure 3.16). An example of this kind of calculation can be found in question 23.

Falling objects transfer gravitational potential energy to kinetic energy. If they do not bounce when they hit the ground, we can assume in the short term (for ease of calculation) that all of the energy is transferred into the internal energy of the object. In other words, no thermal energy is transferred to the ground or surroundings. An interesting example is the expected small rise in temperature of water after it has fallen over a waterfall (Figure 3.17).

■ **Figure 3.16** The disc brakes on a car can get very hot

■ **Figure 3.17** The water temperature at the bottom of the waterfall will be a little higher than at the top

23 A car of total mass 1200 kg travelling at $17.3\,m\,s^{-1}$ has four disc brakes, each of mass 424 g and specific heat capacity $1580\,J\,kg^{-1}\,°C^{-1}$.
a Calculate the kinetic energy of the car.
b Estimate the maximum temperature rise of the brakes when they are used to decelerate the car to rest.
c The real temperature rise will be less than that calculated in part **b**, but why would a temperature rise be greater if the deceleration was greater?

24 A 12 g bullet travelling at $670\,m\,s^{-1}$ (twice the speed of sound) is fired into a wooden block of mass 650 g.
a Estimate the temperature rise if the combined thermal capacity of the bullet and wood was $1880\,J\,K^{-1}$.
b What assumption have you made?
c Without using a detailed calculation, what temperature rise would be produced by a similar bullet travelling at half the speed?

25 Using a steady force of 160 N, a paddle wheel was turned inside a tank containing 8.00 kg of water at 18.6 °C (Figure 3.18).
a If the rotating force was moved a total distance of 250 m, how much work was done?
b Estimate the final temperature of the water.
c Why is it difficult to demonstrate that mechanical energy can be directly transferred to an equivalent amount of internal energy?

26 A 1800 g piece of lead fell 14.8 m off the roof of a building.
a If the specific heat capacity of lead is $130\,J\,kg^{-1}\,°C^{-1}$ estimate the temperature rise of the lead after it hit the ground.
b What assumptions did you have to make in order to be able to do this calculation?
c Explain why a piece of lead of twice the mass would have approximately the same temperature rise if it fell the same distance.

27 Horseshoe Falls at Niagara has a vertical height of 53 m. What is the maximum possible temperature rise of water falling from the top to the bottom of this waterfall?

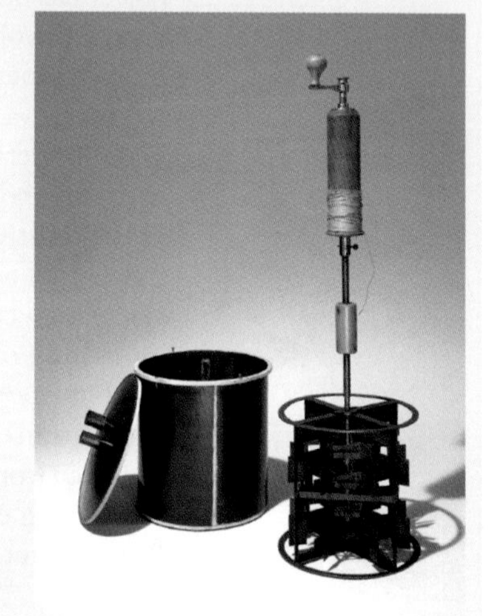

■ **Figure 3.18** Joule's apparatus for turning mechanical energy into internal energy.

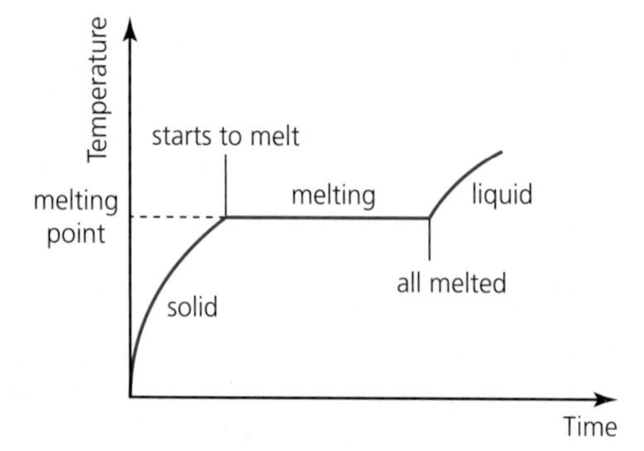

■ **Figure 3.19** Temperature changes as a solid is heated and melted (note that the lines are curved only because of energy transferred to the surroundings)

■ Phase change

A **phase** is a region of space in which all the physical and chemical properties of a substance are the same. A particular substance can exist in a solid phase, a liquid phase or a gaseous phase. These are sometimes called the three *states* of matter. For example, water can exist in three phases (states): liquid, ice (solid) and steam (gas). A bottle containing oil and water would have two phases, both in the liquid state.

When thermal energy is transferred to a solid it will usually get hotter. For many solid substances, once they reach a certain temperature they will begin to **melt** (change from a solid to a liquid), and while they are melting the temperature does not change (Figure 3.19). This temperature is called the **melting point** of the substance, and it has a fixed value at a particular air pressure (Table 3.3). Melting is an example of a **phase change**. Another word for melting is **fusion**.

Similarly, when a liquid cools, its temperature will be constant at its melting point while it changes phase from a liquid to a solid (Figure 3.20). This process is known as solidifying or **freezing**. Be careful – the word 'freezing' suggests that this happens at a low temperature but this is not necessarily true (unless we are referring to turning water into ice, for example). The phase changes of water are such common events in everyday life that we all tend to think of them as the obvious examples but, of course, many other substances can melt and freeze. For example, molten iron 'freezes' at 1538 °C.

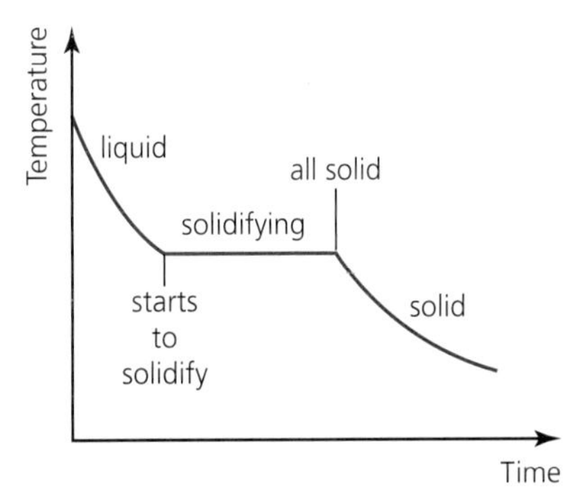

■ **Figure 3.20** Temperature changes when a liquid cools and freezes (solidifies)

(The graphs shown in Figures 3.19 and 3.20 have *time* represented on their horizontal axes, but the shape of the graph would be the same if *time* were replaced by *energy transferred* (assuming the power was constant).)

A change of phase also occurs when a liquid becomes a gas, or when a gas becomes a liquid. The change of phase from a gas to a liquid can be by boiling or evaporation, but generally this process can be called **vaporization**. Changing from a gas to a liquid is called **condensation**. The temperature at which boiling occurs is called the **boiling point** of the substance, and it has a fixed value for a particular air pressure (see Table 3.3). Boiling points can vary considerably with different surrounding air pressures.

A graph showing the temperature change of a liquid being heated to boiling will look very similar to that for a solid melting (Figure 3.19), while the graph for a gas being cooled will look very similar to that for a liquid freezing (Figure 3.20).

To melt a solid or boil a liquid, it is necessary to transfer thermal energy to them. However, as we have seen, melting and boiling occur at constant temperatures, so that the energy supplied must be used to overcome intermolecular forces and to increase molecular separations. In the case of melting, some forces are overcome, but in the case of boiling *all* the remaining forces are overcome.

When a liquid freezes (solidifies) the same amount of energy per kilogram is emitted as was needed to melt it (without a change in temperature). Similarly, boiling and condensing involve equal energy transfers.

Figure 3.21 represents the four principal phase changes. Table 3.3 lists the melting and boiling points of some common substances.

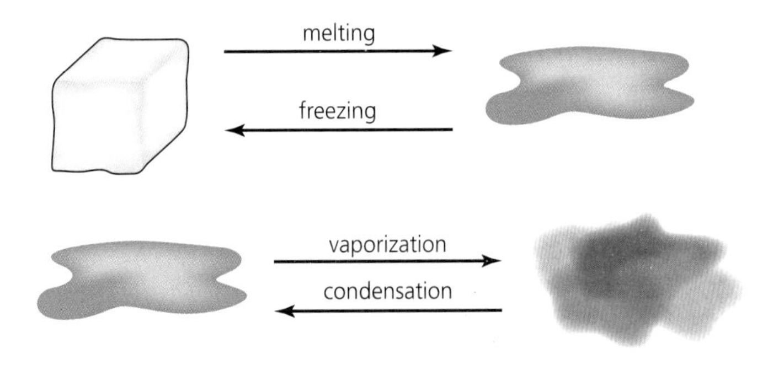

■ **Figure 3.21** Changes of phase

■ **Table 3.3** Melting points and boiling points of some substances (at normal atmospheric pressure)

Substance	Melting point °C	Melting point K	Boiling point °C	Boiling point K
water	0	273	100	373
mercury	−39	234	357	630
alcohol (ethanol)	−117	156	78	351
oxygen	−219	54	−183	90
copper	1083	1356	2580	2853
iron	1538	1811	2750	3023

Evaporation and boiling

Molecules in a liquid have a range of different kinetic energies that are continuously transferred between them. This means there will always be some molecules near the surface that have enough energy to overcome the attractive forces that hold the molecules together in the liquid.

Such molecules can escape from the surface and this macroscopic effect is called evaporation. The loss of the most energetic molecules means that the average kinetic energy of the molecules remaining in the liquid must decrease (until thermal energy flows in from the surroundings). This microscopic effect explains the macroscopic fall in temperature (cooling) that always accompanies evaporation from a liquid.

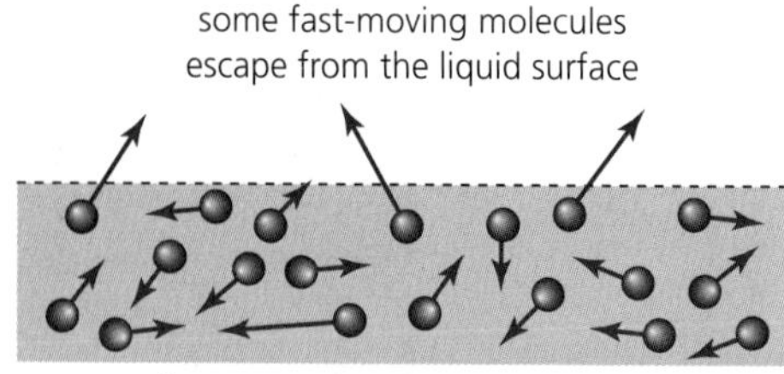

■ **Figure 3.22** Molecules leaving a surface during evaporation

Evaporation occurs only from the surface of a liquid and can occur at any temperature, although the rate of evaporation increases significantly with rising temperature (between the melting and boiling points). Boiling occurs at a precise temperature – the temperature at which the molecules have enough kinetic energy to form bubbles inside the liquid.

Utilizations

Evaporation, cooling and refrigeration

The cooling effect produced by water evaporating has been used for thousands of years to keep people and buildings cool. For example, in central Asia open towers in buildings encouraged air flow over open pools of water, increasing the rate of evaporation and the transfer of thermal energy up the tower by *convection currents*. The flow of air past people in buildings also encourages the human body's natural process of cooling by sweating.

Modern refrigerators and air conditioners also rely on the cooling produced when a liquid evaporates. The liquid or gas used is called the *refrigerant*. Ideally it should take a large amount of thermal energy to turn the refrigerant from a liquid into a dense gas at a little below the desired temperature.

In a refrigerator, for example, after the refrigerant has removed thermal energy from the food compartment, it will have become a gas and be hotter. In order to re-use it and turn it back into a cooler liquid again, it must be compressed and its temperature reduced. To help achieve this thermal energy is transferred from the hot, gaseous refrigerant to the outside of the refrigerator (Figure 3.23).

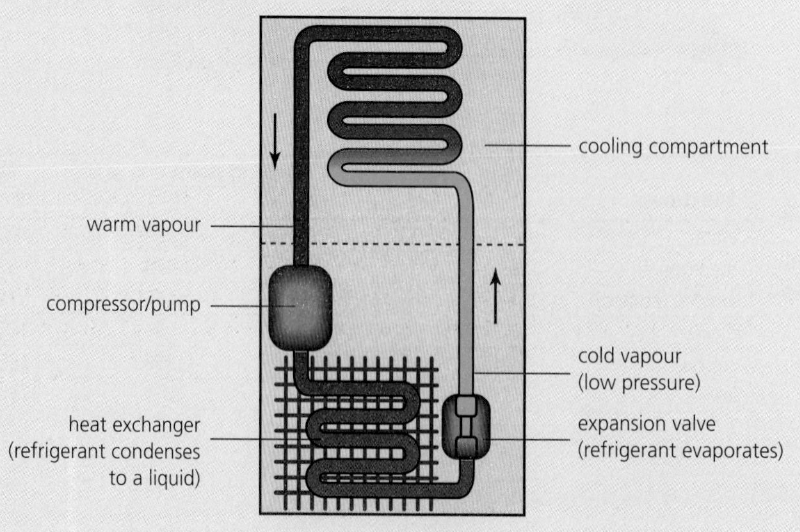

■ **Figure 3.23** Schematic diagram of a refrigerator

As well as having suitable thermal properties, a refrigerant must also be non-poisonous and must not be harmful to the environment. This is because, over a period of time, some of the refrigerant may leak out of the system into the environment. In the 1980s it was realized that the refrigerants being widely used (CFCs) were damaging the environment (the ozone layer in particular) and their use was discouraged and banned in many countries. More recently concern has centred on the effect of refrigerants on global warming. A lot of research has gone into producing a variety of synthetic refrigerants with the right combinations of physical and chemical properties.

1 Draw a simple sketch of an old central Asian building showing the tower, pool and convection currents.
2 Water is cheap, non-poisonous to us and to our environment, and needs a large amount of energy to make it evaporate. So why is it not used widely as a refrigerant in air conditioners?
3 Discuss whether or not a fountain in the living room of your home could help to keep you cool on a hot day.
4 As explained above, thermal energy has to be removed from the refrigerant during its cycle. Suggest how this can be done.

Specific latent heat

Consider the energy transfers involved with changes of phase. As an example, Figure 3.24 shows the changes in temperature that might occur when a quantity of crushed ice is heated *continuously* – first to become water, and then to become steam as the water boils. There are two flat sections on the graph. The first flat section, at 0 °C (the melting point of water), shows where the additional energy input from the heater is used to overcome some intermolecular forces in the ice. The temperature does not begin to rise until all the water is in the liquid phase. The second flat section occurs at 100 °C, the boiling point of water. Here the energy being supplied is used to free all the molecules from intermolecular forces.

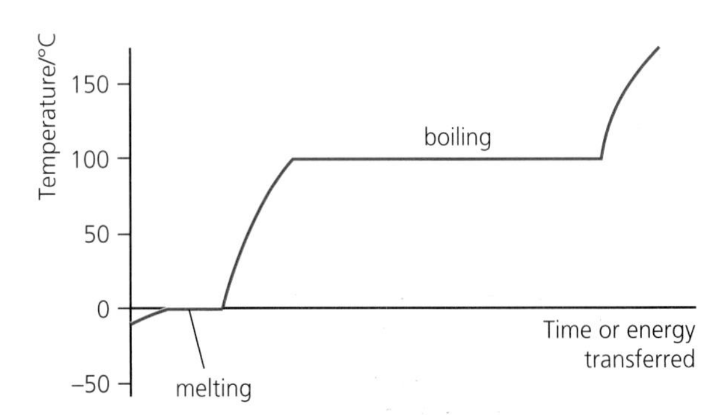

Figure 3.24 A graph of temperature against time for the heating of crushed ice

The thermal energy involved with changing potential energies during any phase change is called latent heat ('latent' means hidden). During melting or boiling, latent heat must be transferred *to* the substance. During condensing or freezing, latent heat is transferred *from* the substance.

The latent heat associated with melting or freezing is called latent heat of fusion, L_f. The latent heat associated with boiling or condensing is known as latent heat of vaporization, L_v.

The specific latent heat of a substance, L, is the amount of energy transferred when 1 kilogram of the substance changes phase *at a constant temperature*. (The units are $J\,kg^{-1}$.)

That is,

$$\text{specific latent heat, } L = \frac{\text{thermal energy transferred}}{\text{mass}} = \frac{Q}{m}$$

which is more commonly written as:

$$Q = mL$$

This equation is given in the *Physics data booklet*.

For example, the specific latent heat of fusion of lead is $2.45 \times 10^4\,J\,kg^{-1}$ and its melting point is 327 °C. This means that $2.45 \times 10^4\,J$ is needed to melt 1 kg of lead at a constant temperature of 327 °C.

Experiments to determine specific latent heats (water is often used as a convenient example) have many similarities with specific heat capacity experiments. Usually either an electric heater of known power is used to melt or boil a substance (or a warmer liquid is used to melt a cooler solid) – question 31 describes such an experiment.

Worked example

4 The latent heats of vaporization of water and ethanol are $2.27 \times 10^6\,J\,kg^{-1}$ and $8.55 \times 10^5\,J\,kg^{-1}$.

a Which one is 'easier' to turn into a gas/boil (at the same pressure)?

b How much thermal energy is needed to turn 50 g of ethanol into a gas at its boiling point of 78.3 °C?

a It is 'easier' to boil ethanol because much less energy is needed to turn each kilogram into a gas.

b $Q = mL$

$Q = 0.050 \times (8.55 \times 10^5)$

$Q = 4.3 \times 10^4\,J$

28 If the latent heat of fusion of a certain kind of chocolate is $160\,000\,J\,kg^{-1}$, how much thermal energy is removed from you when a 10 g bar of chocolate melts in your mouth?

29 Water is heated in a 2250 W kettle. When it reaches 100 °C it boils and in the next 180 s the mass of water reduces from 987 g to 829 g. Use these figures to estimate the latent heat of vaporization of water.

30 Why should you expect that the latent heats of vaporization of substances are usually larger than their latent heats of fusion?

31 0.53 g of steam at 100 °C condensed on an object and then the water rapidly cooled to 35 °C.
a How much thermal energy was transferred from the steam:
i when it condensed?
ii when the water cooled down?
b Suggest why a burn received from steam is much worse than from water at the same temperature (100 °C).

32 In the experiment shown in Figure 3.25, two identical 50 W immersion heaters were placed in some ice in two separate funnels. The heater above beaker A was switched on, but the heater above B was left off. After 5 minutes it was noted that the mass of melted ice in beaker A was 54.7 g, while the mass in beaker B was 16.8 g.
a What was the reason for having ice in *two* funnels?
b Use these figures to estimate the latent heat of fusion of ice.
c Suggest a reason why this experiment does not provide an accurate result.
d Describe one change to the experiment that would improve its accuracy.

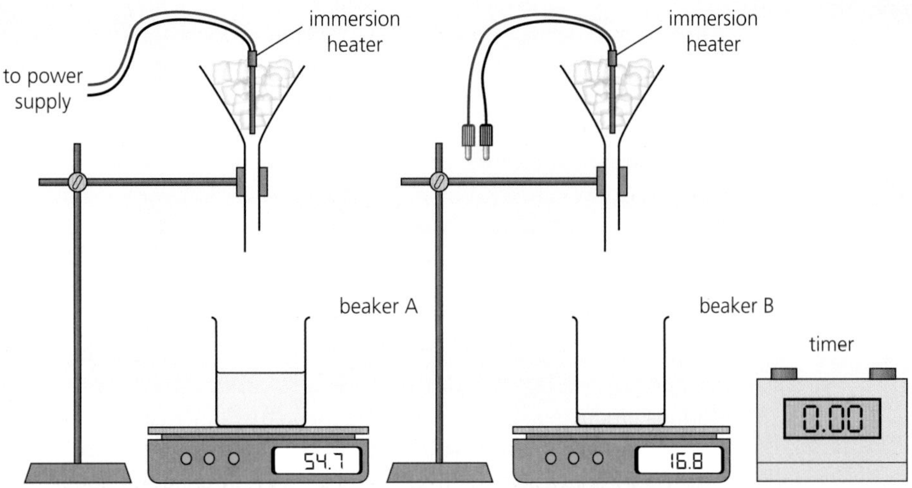

■ **Figure 3.25** An experiment to determine the latent heat of fusion of ice

33 120 g of water at 23.5 °C was poured into a plastic tray for making ice cubes. If the tray was already at 0 °C, calculate the thermal energy that has to be removed from the water to turn it to ice at 0 °C. (The latent heat of fusion of water is 3.35×10^5 J kg^{-1}.)

34 Clouds are condensed droplets of water and sometimes they freeze to become ice particles. Suppose a cloud had a mass of 240 000 kg, how much thermal energy would be released if it all turned to ice at 0 °C?

35 Some water and a glass container are both at a temperature of 23 °C and they have a combined thermal capacity of 1500 J K^{-1}. If a 48 g lump of ice at −8.5 °C is placed in the water and the mixture is stirred until all the ice has melted, what is the final temperature? (The specific heat capacity of ice is 2100 J kg^{-1} K^{-1}. The latent heat of fusion of water is 3.35×10^5 J kg^{-1}.)

3.2 Modelling a gas – *the properties of ideal gases allow scientists to make predictions about the behaviour of real gases*

Thermodynamics is a branch of physics that involves the study of transfers of thermal energy to do useful work. Usually this involves hot gases. In order to understand thermodynamics it is necessary to develop a good appreciation of some of the most basic scientific concepts, including molecular behaviours, energy, temperature and pressure. These concepts are used throughout *all* branches of science.

■ Kinetic model of an ideal gas

We will begin our introductory study of thermodynamics by considering a simplified model of how molecules move in a gas.

Of the three phases of matter, gases are the easiest to understand because we can usually assume that the motions of the molecules are random and independent. The random microscopic behaviour of countless billions of individual molecules results in a totally predictable macroscopic behaviour of gases that is much simpler to understand than the more complex molecular interactions in solids and liquids.

We can start by making some simplifying assumptions about the behaviour of gas molecules. This theory is known as the kinetic model of an ideal gas. It is called the 'kinetic' theory because it involves *moving* molecules. The model can then be used to help to explain pressure, temperature and the macroscopic behaviour of real gases. It should be noted at this early stage that this theoretical model is very good at describing the properties of real gases under most, but not all, circumstances.

Nature of Science

Using hot gases

The invention of devices that could continuously use the thermal energy (heat) transferred from a burning fuel to do useful mechanical work changed the world completely. No longer did people and animals have to do such hard work – they could get engines to do it for them, and much more quickly than they could do the same work themselves.

The idea that burning a fuel to heat water to make steam, which could then be used to make something move, had been understood for a very long time. Using this in a practical way was much more difficult and it was not until the early eighteenth century that the first commercial steam engines were produced. It was about 100 years afterwards that George Stephenson built the *Locomotion* for the first public steam railway, opened in Britain in 1825.

Nearly 200 years later, things are very different. We live in a world that is dominated by **heat engines** (devices that get useful mechanical work from the flow of thermal energy). We are surrounded by all sorts of different engines – in cars, boats, trains, planes, factories and power stations producing electricity (see Figure 3.26). All these need a transfer of thermal energy from fuels in order to work. It is difficult to overstate the importance of these devices in modern life, because without them our lives would be very different. Of course, we are now also very much aware of the problems associated with the use of heat engines: limited fossil fuel resources, inefficient devices, pollution and global warming.

■ **Figure 3.26** Using hot gases in heat engines

Assumptions of the kinetic model of an ideal gas

The microscopic model of a gas consists of a large number of molecules moving randomly. We can imagine these as the 'molecules in a box', as shown in Figure 3.27. Arrows of different lengths represent the random velocities. There are many computer simulations of gas behaviour to represent how the molecules move over a period of time and it is recommended to view one of these.

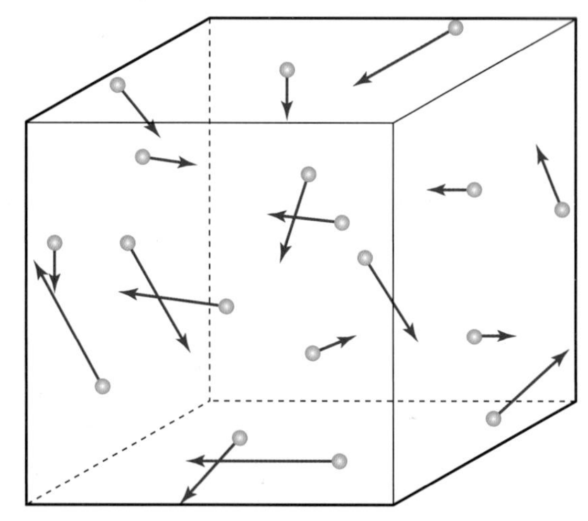

■ **Figure 3.27** Gas molecules moving around at random in a container

To simplify the theory, the following assumptions are made, and then the gas can be described as 'ideal':

- The gas contains a *very large number of identical molecules*.
- The *volume of the molecules is negligible* compared with the total volume occupied by the gas.
- The molecules are *moving in completely random directions*, at a wide variety of speeds.
- *There are no forces between the molecules, except when they collide*. Because there are no forces, the molecules have no (electrical) potential energy. This means that any changes of internal energy of an ideal gas are assumed to be only in the form of changes of random kinetic energy.
- *All collisions are elastic* – that is, the total kinetic energy of the molecules remains constant at the same temperature. This means that there is no energy transferred from a gas to its surroundings and the average random speed of its molecules will not decrease. If this were not true, all gases would cool down and their molecules would fall to the bottom of their containers and condense to liquids!

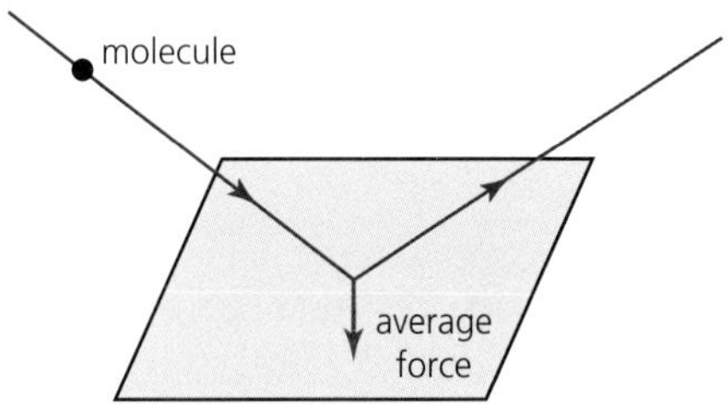

■ **Figure 3.28** Every molecular collision with a wall creates a tiny force on the wall

Using the kinetic theory of an ideal gas to explain gas pressure

When a molecule hits (collides with) a wall of a container, it exerts a force on the wall (Figure 3.28).

The size of the force could be calculated if we knew the mass and change in momentum of the molecule, together with the duration of the impact (see Chapter 2). Each and every collision can result in a different sized force, and it is not realistic to know all the forces that individual molecules exert on the walls. However, because the molecular motions are random and because of the incredibly large number of them, the total force caused by many molecular collisions on any unit area of container wall will be completely predictable, and will (usually) be constant at all places in the container. This is called the **pressure** of the gas. Both P and p are in widespread use as symbols for pressure.

Collisions between molecules (intermolecular collisions) are happening all the time, but they simply result in random changes to molecular velocities and have no overall effect on pressure or other macroscopic properties of the gas.

Pressure

The effect of a force often depends on the area on which it acts. For example, when the weight of a solid pushes down on a surface, the consequences usually depend on the area underneath it, as well as the magnitude of the weight.

Pressure is defined as force per unit area:

$$p = \frac{F}{A}$$

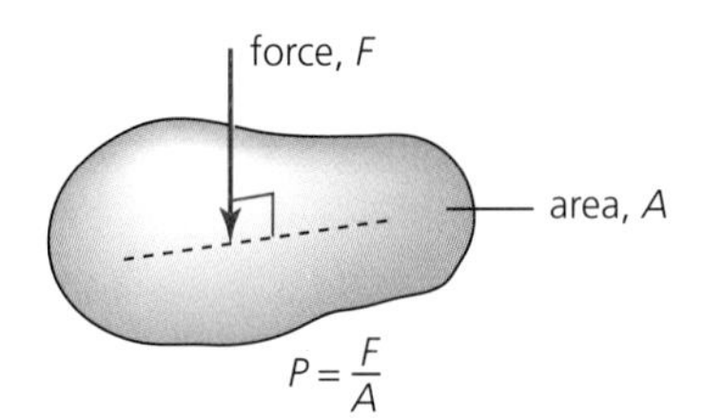

■ **Figure 3.29** Defining pressure

This equation is in the *Physics data booklet*. The SI unit of pressure is the pascal, Pa. 1 Pa = 1 N m^{-2}.

Note that the force used to calculate pressure is perpendicular to the surface, as shown in Figure 3.29. That is, it is a **normal** force.

The pressure caused in a gas, by random molecular collisions, acts equally in *all* directions and has a typical value in the order of magnitude of 10 N on every square centimetre. In SI units, the usual pressure of the air around us (**atmospheric pressure**) is 1.0×10^5 Pa at sea level. Atmospheric pressure acts upwards and sideways as well as downwards.

Using the kinetic theory of an ideal gas to explain gas temperature

Earlier in this chapter, temperature was explained as a way of determining the direction of thermal energy transfer. Now we can interpret temperature in a more profound way: in terms of molecular energies. When a gas is heated, the average speed of its molecules increases. The thermal energy transferred into the gas increases the random kinetic energy of the molecules.

When a substance is cooled its molecules move slower. Eventually almost all molecular motion stops and it is not possible to reach a lower temperature. Experiments with gases predict that this occurs at −273.15 °C which, as we saw earlier, is known as *absolute zero*.

The **absolute temperature** (T) of an ideal gas (in kelvin) is a measure of the mean random translational kinetic energy of its molecules ($\overline{E}_K$).

The bar above the E shows that an average value is being used.

This important relationship can be expressed mathematically as:

$$\overline{E}_K = \frac{3}{2} k_B T$$

This equation is in the *Physics data booklet*

k_B is a very important constant that links macroscopic temperature to microscopic energies. It is called the **Boltzmann constant** and has a numerical value of $1.38 \times 10^{-23}\,J\,K^{-1}$ (provided in the *Physics data booklet*).

Although we are discussing *ideal* gases, this equation and interpretation of temperature can also be applied to *real* gases. Note that there is nothing in the equation that applies only to a particular gas. In other words, at the same temperature, molecules of *all* gases have the same average *translational* kinetic energy. By considering KE = $\frac{1}{2}mv^2$, we can see that molecules of different gases must have different average speeds at the same temperature: more massive molecules move slower.

k_B can be interpreted in terms of the specific heat capacity of a gas: it is $\frac{2}{3}$ of the average amount of energy per molecule that must be supplied to raise the temperature of a *monatomic* ideal gas by 1 K. But note that more energy is needed to change the temperature of *molecular* gases because rotational and vibrational kinetic energies are also involved.

36 Normal air pressure is about 1×10^5 Pa.
  a What total force acts on an area of 1 mm^2?
  b If a student had a body with a total surface area of 1.5 m^3, what would the total force on their skin be?
  c Why does such a large force not have any noticeable effect?

37 Why does the pressure on underwater divers increase when they go deeper?

38 a The pressure under our feet acts down on the ground, but the pressure in a gas acts in all directions. Explain this.
  b In what direction(s) does pressure in a liquid act?

39 a Calculate the average translational kinetic energy of molecules in the air at 27 °C.
  b Calculate the temperature (°C) at which the average translational kinetic energy of the gas molecules is 1.0×10^{-20} J.
  c Estimate the average translational kinetic energy of the particles in the surface of the Sun. (Use the internet to find the data required.)

40 a Calculate an average speed for oxygen molecules at 0 °C – the mass of an oxygen molecule = 5.32×10^{-26} kg.
  b Calculate an average speed for nitrogen molecules at the same temperature – the mass of a nitrogen molecule = 4.65×10^{-26} kg.

It is important to realize that individual molecules of the same gas at the same temperature will be moving with a wide range of different speeds and kinetic energies. Any calculated value is just an average. Temperature is a concept that can only be applied to a collection of a very large number of molecules – not to individual molecules. That is, temperature is a macroscopic concept, although statistically it has a microscopic interpretation.

Additional Perspectives

Kinetic theory and the distribution of molecular speeds

Figure 3.30 shows the range and distribution of molecular speeds in a typical gas, and how it changes as the temperature increases. This is known as the *Maxwell–Boltzmann distribution*.

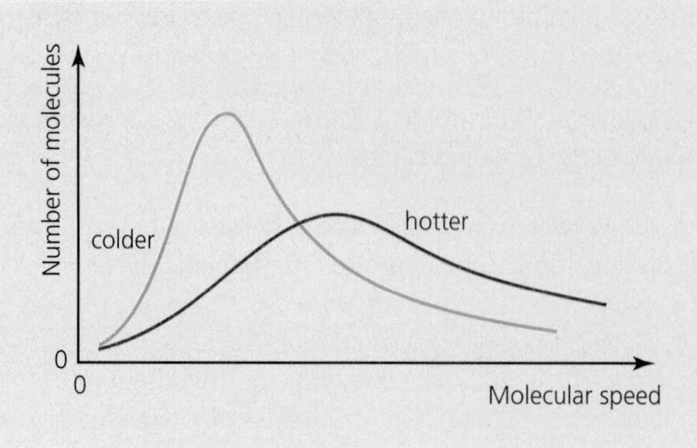

■ **Figure 3.30** Typical distributions of molecular speeds in a gas

Note that there are no molecules with zero speed and few with very high speeds. Molecular speeds and directions (that is, molecular velocities) are continually changing as the result of intermolecular collisions. As we have seen, higher temperature means higher kinetic energies and therefore higher molecular speeds. But the range of speeds also broadens and so the peak becomes lower, keeping the area under the graph, which represents the number of molecules, constant.

Figure 3.31 illustrates the ranges of molecular speeds for different gases at the same temperature. It shows that, as we have noted before, at the same temperature less massive molecules have higher speeds than more massive molecules.

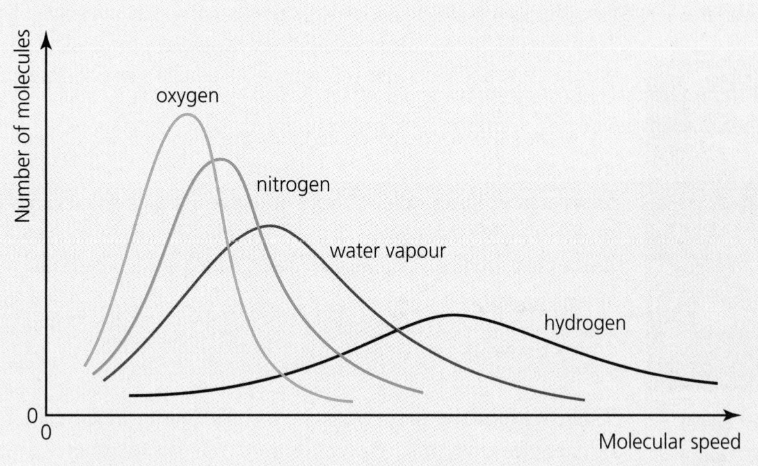

■ **Figure 3.31** Distribution of molecular speeds in different gases at the same temperature

1 a Explain why hydrogen gas *diffuses* more quickly than other gases at the same temperature.

b Suggest the name of a gas that diffuses very slowly.

c If a cylinder of nitrogen gas at 20 °C was placed in a plane at an airport and later travelled at $500\,\mathrm{m\,s^{-1}}$, what would happen to the average speed of the nitrogen molecules and the temperature? Explain your answer.

d Suggest why a person's voice sounds unusually high-pitched when they have just inhaled a small amount of helium gas from a balloon.

e Explain why it might be reasonable to assume that the volumes of all four gases represented in Figure 3.31 are approximately equal.

Mole, molar mass and the Avogadro constant

If we want to make calculations about the mass, speed and kinetic energy of gas molecules, then we need to understand the link between the macroscopic measurement of mass and the microscopic numbers of molecules.

The *amount* of a substance (symbol n) is a measure of the number of particles it contains and it is measured in moles:

One mole (mol) of a substance is the amount of the substance that contains as many (of its defining) particles as there are atoms in exactly 12 g of carbon-12.

The number of particles in a mole is called the Avogadro constant, N_A.

The value of the Avogadro constant is given in the *Physics data booklet*: $N_A = 6.02 \times 10^{23}\,\mathrm{mol^{-1}}$ (to three significant figures). So that for a sample containing N molecules, the number of moles is given by:

$$n = \frac{N}{N_A}$$

This equation is given in the *Physics data booklet*.

A value for the Avogadro constant can be determined experimentally from an understanding of how electric currents (see Chapter 5) are carried by moving ions during electrolysis, and from precise measurements of the changes of mass involved.

By definition, 1 mole of carbon-12 atoms has a mass of exactly 12 g and contains 6.02×10^{23} atoms. The same number of atoms of hydrogen would have a mass of only 1.0 g because each hydrogen atom has only $\frac{1}{12}$ the mass of a carbon-12 atom. Similarly, 63.5 g of copper would contain the same number of atoms because, on average, each copper atom has 63.5 times more mass than a hydrogen atom.

■ **Figure 3.32** One mole of water (in the form of ice), sugar, copper and aluminium

Most substances are molecular, and each molecule will consist of two or more atoms. One mole of a molecular substance contains Avogadro's number of *molecules*. For example, 1 mole of oxygen *atoms* has a mass of 16 g, but oxygen is a molecule with two atoms, O_2, so 1 mole of oxygen *molecules* (6.02×10^{23} molecules) has a mass of 32 g. The molar mass of oxygen is $32\,\text{g mol}^{-1}$.

The molar mass of a substance is defined as the mass that contains 1 mole of (its defining) particles (unit: g mol^{-1}) .

Table 3.4 lists the molar masses of some common substances, as well as the number of particles that this involves. Figure 3.32 shows what 1 mole of a number of different substances looks like.

■ **Table 3.4** Molar masses

Substance	Molar mass/g mol^{-1}	Particles
aluminium	27.0	6.02×10^{23} atoms of aluminium
copper	63.5	6.02×10^{23} atoms of copper
gold	197	6.02×10^{23} atoms of gold
hydrogen	2.02	12.04×10^{23} atoms of hydrogen combined to make 6.02×10^{23} molecules
oxygen	32.0	12.04×10^{23} atoms of oxygen combined to make 6.02×10^{23} molecules
water	18.0	6.02×10^{23} atoms of oxygen combined with 12.04×10^{23} atoms of hydrogen to make 6.02×10^{23} molecules of water
air (at normal temperature and pressure)	≈ 29	6.02×10^{23} molecules from a mixture of gases
sugar (sucrose)	342	6.02×10^{23} large molecules

Worked example

5 The molar mass of water is $18\,\text{g mol}^{-1}$. How many molecules are there in 1 kg of water?

number of moles (amount of water) in 1 kg $= \frac{1000}{18} = 55.6$ mol

number of molecules in 55.6 mol $= 55.6 \times (6.02 \times 10^{23}) = 3.34 \times 10^{25}$ molecules

You will need to use data from Table 3.4 to answer these questions.

41 a What mass of aluminium will contain exactly 4 mol of atoms?
b What mass of sucrose will contain 1.0×10^{22} molecules?
c How many moles are there in 2.00 kg of carbon?
d How many molecules are there in 1.0 kg of carbon dioxide (CO_2)?

42 A gold ring had a mass of 12.3 g.
  a How many moles of gold atoms did it contain?
  b When it was weighed 50 years later its mass had decreased to 12.2 g. How many atoms were 'lost' from the ring on average every second during this time?

43 a If the density of air in a classroom is 1.3 kg m^{-3}, what is the total mass of gas in a room with dimensions 2.5 m × 6.0 m × 10.0 m?
  b Approximately how many moles of air are in the room?
  c Approximately how many molecules are in the room?

44 The density of aluminium is 2.7 g cm^{-3}.
  a What is the volume of one mole of aluminium?
  b What is the average volume occupied by one atom of aluminium?
  c Approximately how far apart are atoms in aluminium?

45 What mass of oxygen contains 2.70 × 10^{24} molecules?

Equation of state for an ideal gas

We now turn our attention to the large-scale (*macroscopic*) physical properties of an ideal gas: amount, n, pressure, p, volume, V, and absolute temperature, T. Together, these four properties completely describe the physical characteristics of any amount of any gas.

Using the assumptions of the kinetic theory for ideal gases and the laws of mechanics (Chapter 2), it is possible to show that these four variables are linked by the equation:

$$\frac{pV}{nT} = \text{constant}$$

The value of the 'constant' in this equation is the same for *all* ideal gases. This is because, at the same temperature, the molecules of all ideal gases have the same average translational kinetic energy: more massive gas molecules travel slower than lighter molecules. The result is that equal amounts of all ideal gases at the same temperature exert the same pressure in the same volume. There is nothing in this equation that is used to represent the properties of any particular gas.

The 'constant' in this equation is called the universal (molar) gas constant. (This is often simply reduced to 'the *gas constant*'.) It is given the symbol R and has the value of 8.31 J K^{-1} mol^{-1}. This value is given in the *Physics data booklet*.

The equation can be re-written as follows:

$$pV = nRT$$

This equation is given in the *Physics data booklet*.

This equation is known as the equation of state for an ideal gas. This important equation defines the *macroscopic* behaviour of an ideal gas. For example the pressure, volume or temperature of a known amount, n, of an ideal gas can be calculated using the equation of state if the other two variables are known. The *microscopic* meaning of an ideal gas was discussed earlier in this chapter. The equation also defines the meaning of the universal gas constant ($R = pV/nT$).

By considering its units, it should be clear that the universal gas constant, R, can be interpreted as being related to the amount of energy needed to raise the temperature of 1 mole of an ideal gas by 1 K.

total random translational kinetic energy (= total internal energy) of all the molecules in *1 mole* of an ideal gas = $\frac{3}{2}RT$

This equation is *not* given in the *Physics data booklet*.

But we have already seen that the *average* translational kinetic energy of *one molecule* of an ideal gas = $\frac{3}{2}k_BT$ and because 1 mole contains N_A molecules, we can deduce that:

the Boltzmann constant, $k_B = \frac{R}{N_A}$

This equation is *not* given in the *Physics data booklet*.

So the equation for average translational kinetic energy of molecules:

$$\bar{E}_K = \frac{3}{2}k_B T$$

can also be written as:

$$\bar{E}_K = \frac{3}{2}\frac{R}{N_A}T$$

This equation is in the *Physics data booklet*.

Nature of Science

Effective collaboration is often needed for theories to develop

Unlike Newton's laws, for example, the kinetic model of an ideal gas is not usually attributed to any particular famous scientific figure. The kinetic model of an ideal gas was not a theory proposed by a single scientist after a discovery or moment of inspiration. It was the result of collaboration and the gradual combination of ideas from many scientists, some lesser, some greater, over many years.

Modern scientific research and development is usually characterized by collaborative teamwork, with members of a team usually having different specialities, different skills and different perspectives. The expectation is that this encourages a productive scientific environment in which new and original thinking can prosper. But it was not always this way. In the past, scientific research tended to be more individualized and sometimes secretive.

Worked examples

6 What is the pressure of 23 mol of a gas behaving ideally in a 0.25 m^3 container at 310 K?

$$p = \frac{nRT}{V}$$

$$p = \frac{23 \times 8.31 \times 310}{0.25}$$

$$p = 2.4 \times 10^5 \text{ Pa}$$

7 a A fixed mass of an ideal gas has a volume of 23.7 cm^3 at 301 K. If its temperature is increased to 365 K at the same pressure, what is its new volume, V_2?

b Explain why the volume has increased.

a $\frac{V}{T} = \frac{nR}{p}$ = constant

$$\frac{23.7}{301} = \frac{V_2}{365}$$

$$V_2 = 28.7 \text{ cm}^3$$

b The volume has increased because the pressure has not changed. The molecules are moving faster and would collide with the walls more often (and with more force) creating a higher pressure unless the volume increased.

8 Estimate the total translational kinetic energy of all the molecules in an average sized room.

approximate volume = $4 \times 4 \times 3 \approx 48 \text{ m}^3$

approximate mass of air = $48 \times$ density of air (1.3 kg m^{-3}) = 62 kg

approximate number of moles = $\frac{\text{mass of air}}{\text{average molar mass}} = \frac{62}{0.029} = 2100$ mol

approximate kinetic energy = number of moles × KE mol^{-1} = $2100 \times \frac{3}{2}RT$

At a temperature of 300 K (27 °C), this gives a value of:

$2100 \times 1.5 \times 8.31 \times 300 \approx 8 \times 10^6 \text{ J}$

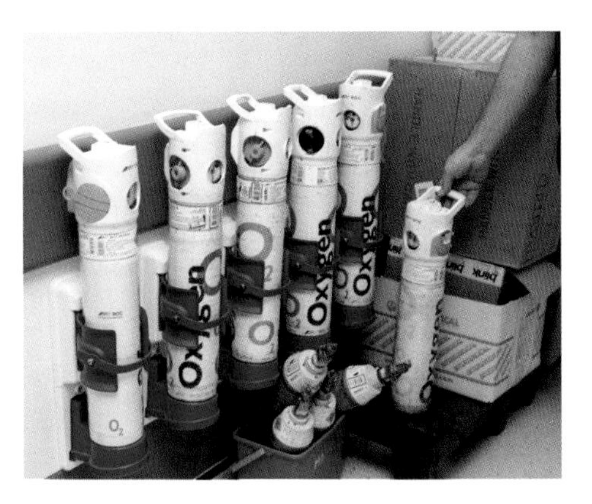

■ **Figure 3.33** Oxygen gas stored in strong metal cylinders

Differences between ideal gases and real gases

Of course, *ideal* gases are not exactly the same as *real* gases. However, the differences are usually, but not always, insignificant so that we often assume that the equation of state for an ideal gas can be used to make calculations for a sample of a real gas. But real gases will *not* behave like ideal gases (and obey the equation of state) if the intermolecular forces become significant because the molecules are closer together, or when molecular sizes can no longer be considered negligible compared with their separations. This can happen at high pressures and densities or at low temperatures.

When a gas is cooled down the molecules move slower and slower and the pressure decreases (assuming that the mass and volume do not change). As the temperature falls, real gases turn into liquids and then solids, if the temperature becomes so low that the molecules do not have enough kinetic energy to overcome the intermolecular forces. However, an ideal gas cannot be liquefied.

In order to store or transport gases it is usually necessary to reduce their volume by using high pressures and/or by turning them into liquids (Figure 3.33). The containers need to be very strong and treated carefully, and under these conditions we would not expect the gases to behave *ideally*.

46 What volume of gas (in cm^3) contains 0.780 mol of a gas at 264 K at atmospheric pressure (1.00×10^5 Pa)?

47 At what temperature (°C) is 0.46 mol of gas if the pressure exerted in an 8520 cm^3 container is 1.4×10^5 Pa?

48 Approximately what amount of air would exert a pressure of 2.60×10^7 Pa in a 12 000 cm^3 cylinder used by a scuba diver (Figure 3.34)?

■ **Figure 3.34** Becoming a scuba diver involves learning about variations of gas pressure under water (see question 48)

49 a What is the molar mass of oxygen?
b What volume will be occupied by 1.0 kg of oxygen at 25 °C and 1.3×10^5 Pa?

50 What pressure will be exerted by putting 2.49 g of helium gas in a 600 cm^3 container at −30.5 °C?

51 a A container has 4060 cm^3 of a gas at twice atmospheric pressure (2.00×10^5 Pa). If the volume is reduced to 3240 cm^3 at the same temperature, what is the new pressure?
b Explain why the pressure has increased.

52 a Some helium gas in a flask exerts a pressure of 2.12×10^6 Pa at −234 °C. If the temperature is increased by exactly 100 °C, calculate a value for the new pressure (assuming that the flask does not expand).
b Explain why the pressure increases.
c Explain why the use of the equation of state might lead to an inaccurate result in this example.

53 A gas at 287 K has a volume of 2.4 m^3 and a pressure of 1.2×10^5 Pa. When it is compressed the pressure rises to 1.9×10^5 Pa and the temperature rises by 36 K. What is the new volume?

54 Container C has 4 moles of gas. Container D is at the same temperature but has 3 moles of gas in twice the volume. What is the value of the ratio p_C/p_D?

55 The volume of a gas is to be doubled, but the pressure must be kept the same. What must the final temperature (Celsius) be if it was originally at 17 °C?

56 Explain how the cruising height of the hot-air balloon shown in Figure 3.35 can be controlled.

■ **Figure 3.35** A hot-air balloon

ToK Link

Models always have limitations

When does modelling of 'ideal' situations become 'good enough' to count as knowledge?

We cannot directly use our human perception (sight for example) to help understand the behaviour of molecules in a gas. That is one reason why we need a 'model'. But can we *ever* be sure that a model is a true representation of reality, if we can never observe events directly? Or will there always be some doubt? And, if the model is useful, does it really matter if the model is a 'true' representation of something that we cannot observe anyway?

Can a model of the solar system, for example, be considered to be knowledge of a higher level, because we can directly observe and record the motion of the planets?

Experimentally investigating the macroscopic behaviour of real gases

Given a closed container with a gas sealed inside it, there are four macroscopic, physical properties of the gas that we can measure relatively easily: mass, volume, temperature and pressure. *For a fixed mass of gas*, the other three properties will be interlinked – change one and at least one of the other two must change, or maybe all three will change.

Additional Perspectives

Measuring gas pressures

A small *difference* in gas pressures can be measured using a simple U-tube containing a liquid, as shown in Figure 3.36. The gas from a container on the left-hand side (not shown) is at a pressure higher than the pressure from the atmosphere on the right-hand side. This piece of apparatus is called a simple *manometer*.

The difference in pressure, Δp, causes a difference in liquid levels and the height Δh is a measure of that pressure difference. For example, it could be recorded as 14.0 cm of water. If mercury were used in the tube instead of water, Δh would be about 1 cm because mercury is approximately 14 times denser than water. To convert the pressure difference to pascals, the formula $\Delta p = \Delta h \rho g$ can be used, where ρ is the density of the liquid in the manometer. (This equation can be derived by considering $p = F/A$ with respect to the extra weight of the liquid on one side and the cross-sectional area of the tube.)

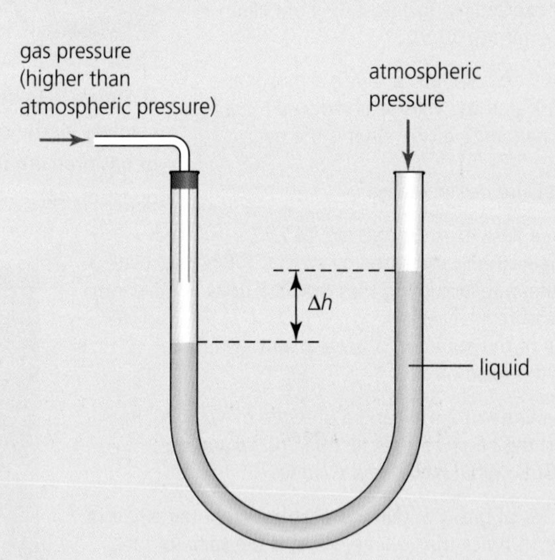

■ **Figure 3.36** Simple manometer

To directly measure the actual pressure of the gas in the container using a manometer (rather than a pressure difference) it would be necessary to completely remove the air from the right-hand side (create a *vacuum*). Δh would then be larger and it would require a very long tube, unless mercury was used.

An instrument designed for measuring atmospheric pressure is called a *barometer*. More generally, instruments for measuring gas pressure are often called *pressure gauges*. There are many different designs.

1 What is a pressure of 14 cm of water expressed in pascals?

2 a Suggest a suitable length of mercury-filled tubing for measuring atmospheric pressure (about 1×10^5 Pa). Explain your answer.

b Why would it be necessary for the tubing to be strong?

The gas laws

Three classic physics experiments showed that *all* real gases, under most circumstances, follow the same simple patterns of behaviour (these are often called the three **gas laws**). Students are also recommended to use computer models to investigate the physical properties of gases over a much wider range of circumstances than would normally be possible in a school laboratory.

Variation of gas volume with pressure: Boyle's law

Figure 3.37 shows two sets of apparatus that could be used to investigate how changing the pressure on a fixed mass of gas affects its volume. Using a force to change the volume of a gas will tend to change its temperature, which will complicate the results. To minimize this unwanted effect the changes should be made slowly.

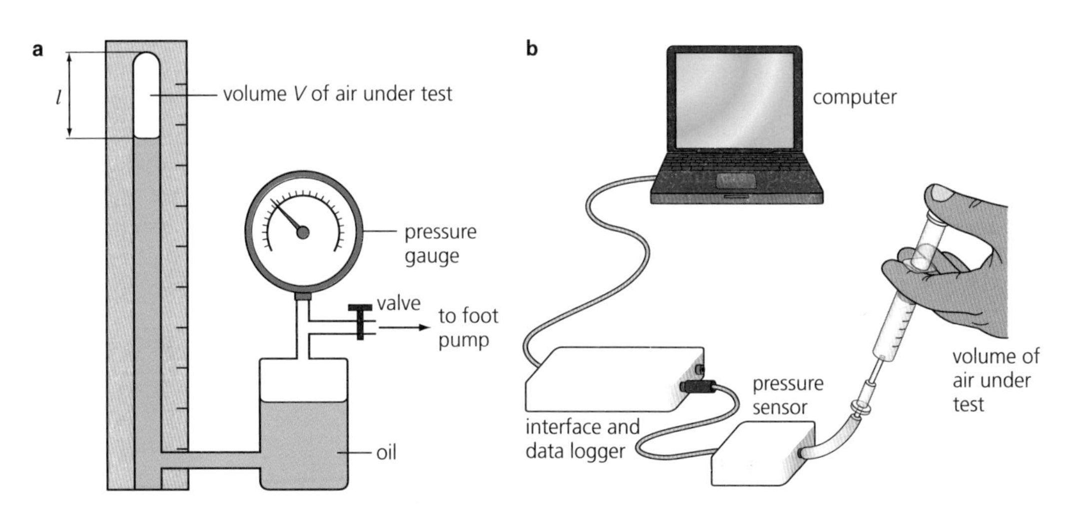

■ **Figure 3.37** Two methods for investigating Boyle's law

Typical experimental results are represented in Figure 3.38a. The line on this graph is called an isotherm ('iso' means 'the same' – all the data points on an isotherm are for the gas at the same temperature).

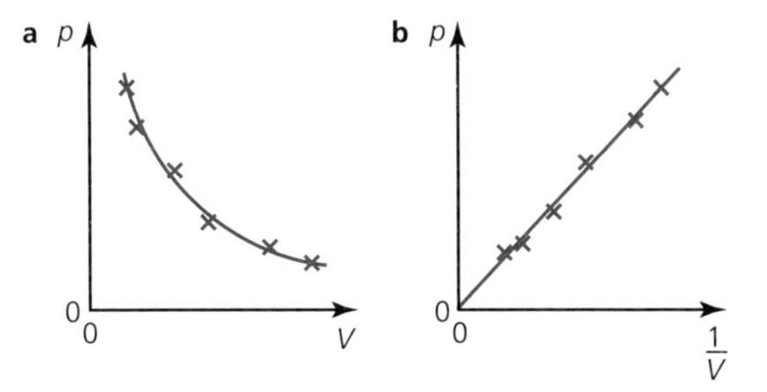

■ **Figure 3.38** Two graphs showing that gas pressure is inversely proportional to volume

Figure 3.38b represents the same data re-drawn to produce a linear graph through the origin, confirming that, for a fixed mass of gas at constant temperature, pressure is inversely proportional to volume (*Boyle's law*):

$$p \propto \frac{1}{V}$$

Variation of gas pressure with temperature: the pressure law

Figure 3.39 shows two sets of apparatus that could be used to investigate how changing the temperature of a fixed mass of gas in a constant volume affects its pressure.

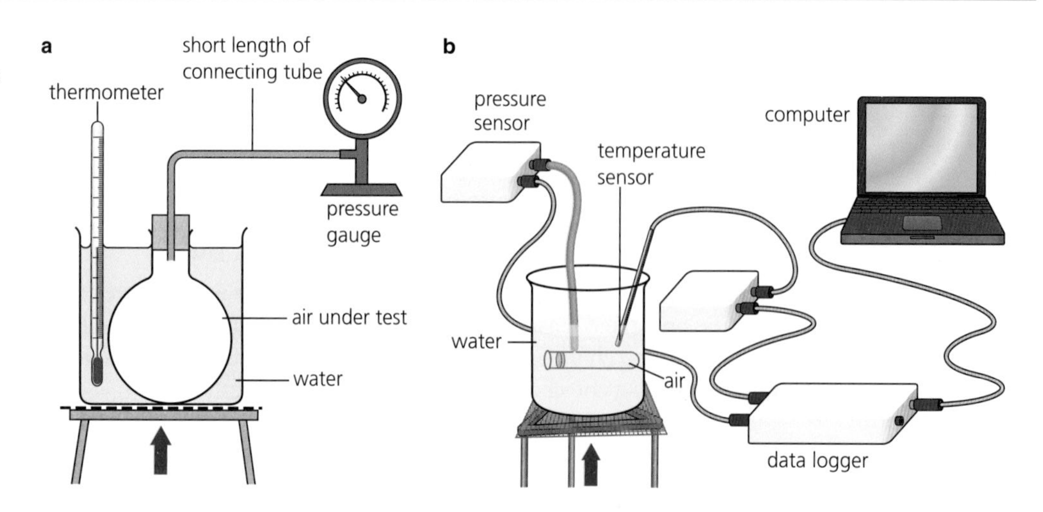

■ **Figure 3.39** Two sets of apparatus that can be used to investigate the pressure law

Typical experimental results are represented in Figure 3.40. Experimental results are usually taken within the range 0 °C to 100 °C (273 K to 373 K) and then the graph is extrapolated back to predict pressures at lower temperatures. The pressure is predicted to reduce to zero close to −273 °C (**absolute zero**, 0 K). The pressure is zero at this temperature because almost all molecular motion has stopped.

For a fixed mass of gas at constant volume, the pressure is proportional to the temperature (K) (the *pressure law*):

$$p \propto T$$

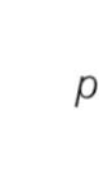
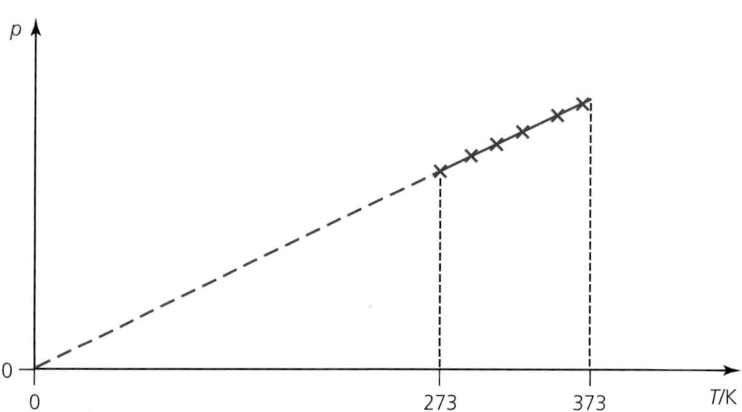

■ **Figure 3.40** Gas pressure is proportional to absolute temperature (K)

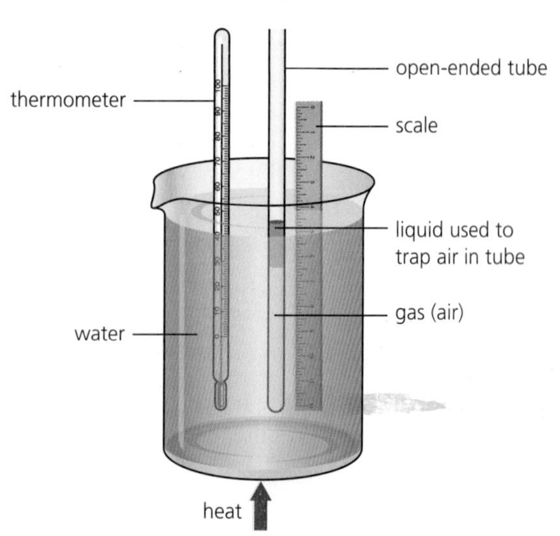

■ **Figure 3.41** Simple apparatus for investigating Charles's law

Variation of gas volume with temperature: Charles's law

Figure 3.41 shows simple apparatus that could be used to investigate how changing the temperature of a fixed mass of gas maintained at constant pressure affects its volume. As the water is heated, thermal energy flows into the trapped gas, which expands (the liquid moves) maintaining a constant pressure because the open end of the tube will remain at atmospheric pressure.

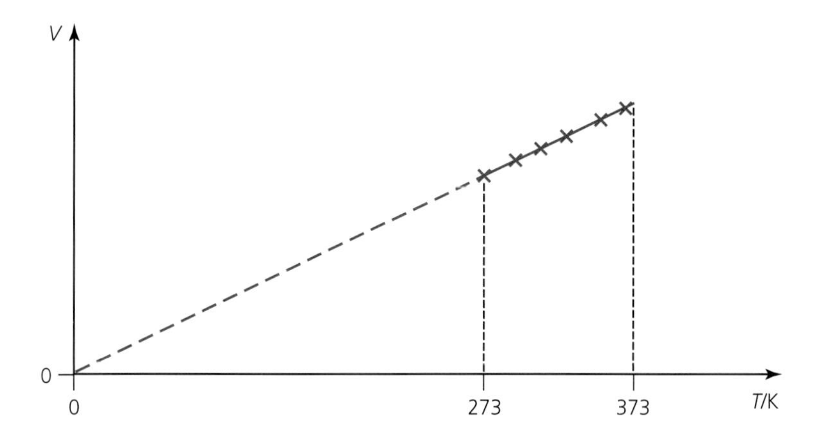

Figure 3.42 Gas volume is proportional to absolute temperature (K)

Typical experimental results are represented in Figure 3.42. The volume is predicted to reduce to zero close to −273 °C (absolute zero, 0 K).

For a fixed mass of gas at constant pressure, the volume is proportional to the temperature (K) (*Charles's law*):

$$V \propto T$$

Combined gas laws

The results of the three gas law experiments can be combined to give:

$$pV \propto T$$

Further experimental investigations confirm that, under most circumstances, the physical properties of real gases are described by the equation:

$$pV = nRT$$

This *empirical* equation (based on experimental results) is the same as the *theoretical equation of state* discussed on page 131. In other words, when the equation was first developed it confirmed earlier experimental discoveries.

Understanding changes in the *state* of a gas, and how they can be represented graphically, plays an important part in the study of heat engines (Option B).

Using kinetic theory to qualitatively explain the gas laws

- *Boyle's law* – As the volume of a gas is decreased, the molecules will hit the walls more frequently because they have less distance to travel between collisions. So, the smaller the volume, the higher the pressure.
- *Pressure law* – When the temperature of a gas increases, the average molecular speed increases. The molecules collide with the walls with more force and more frequently. (There are more collisions every second with the walls.) So, the higher the temperature, the higher the pressure.
- *Charles's law* – When the temperature of a gas increases, the average molecular speed increases, so the molecules collide with the walls more frequently and with more force. This results in a net outward force on the walls of the container, but if any of the walls are moveable the gas will expand, keeping the pressure on the inside the same as the pressure on the outside of that wall. So, the higher the temperature, the larger the volume (but only if the gas is able to expand freely).

57 a Explain why the forces between molecules in a solid or liquid are much larger than the intermolecular forces in a gas.
b Can intermolecular forces between molecules ever be truly zero? (Look back at Figure 3.5) Explain your answer.

58 a Explain in detail what happens to the molecules in a warm gas placed inside a colder container.
b What happens to the internal energy and temperature of the walls and the gas?

59 a Explain why the pressure in an ideal gas is predicted to become zero if it is cooled to −273 °C.
b If they are cooled enough, real gases will condense to liquids long before they get to −273 °C. Explain this observation with reference to molecular speeds and intermolecular forces.

60 Make a copy of Figures 3.38a and 3.38b and then add another isothermal line to each to represent the results that would be obtained with the same gas in the same apparatus, but at a higher temperature.

Additional Perspectives

Understanding randomness

Throughout this topic there have been frequent uses of the word 'random' with respect to energy or motion. But what exactly does 'random' mean? The word has various uses throughout science and more generally, often with slight differences in meaning. For example, we might say that the result of throwing a six-sided die is random because we cannot predict what will happen, although we probably appreciate that there is a one-in-six chance of any particular number ending up on top. In this case, all outcomes should be equally likely. Another similar example could be if we were asked to 'pick a card at random' from a pack of 52. Sometimes we use the word random to suggest that something is unplanned, for example a tourist might walk randomly around the streets of a town.

Unpredictability is a key feature of random events and that certainly is a large part of what we mean when we say a gas molecule moves randomly. All possible directions of motion may be equally likely, but the same cannot be said for speeds. Some speeds are definitely more likely than others. For example, at room temperature a molecular speed of $500\,\text{m}\,\text{s}^{-1}$ is much more likely than one of $50\,\text{m}\,\text{s}^{-1}$. Similarly when we refer to random kinetic energies of molecules (in any of the three phases of matter) we mean that we cannot know or predict the energy of individual molecules, although some values are more likely than others. But there is a further meaning: we are suggesting that individual molecules behave independently and that their energies are disordered.

Perhaps surprisingly, in the kinetic theory the random behaviour of a very large number of individual molecules on the macroscopic scale leads to complete predictability in our everyday macroscopic world. Similar ideas occur in other areas of physics, notably in radioactive decay (Chapter 7), where the behaviour of an individual atom is unknowable, but the total activity of a radioactive source is predictable. Of course, insurance companies and casinos can make good profits by understanding the statistics of probability.

1 If ten coins were tossed and three came down as heads and seven came down as tails, it would not be too surprising. But if 100 were tossed and only 30 were heads it would be amazing; and if a thousand tosses produced 300 heads and 700 tails it would be almost completely unbelievable. Explain these comments.

2 Explain why individual gas molecules colliding with the walls of their container produce a predictably constant pressure.

3 Choose five numbers between 1 and 10 at random. How did you make your choice? How could anyone tell by looking at the numbers you have chosen, that they were really picked at random?

Examination questions – a selection

Paper 1 IB questions and IB style questions

Q1 The temperature of an ideal gas is a measure of:

A the average momentum of the molecules
B the average speed of the molecules
C the average translational kinetic energy of the molecules
D the average potential energy of the molecules.

Q2 If the volume of a fixed mass of an ideal gas is decreased at a constant temperature, the pressure of the gas increases. This is because:

A the molecules collide more frequently with each other
B the molecules collide more frequently with the walls of the container
C the molecules are moving at a higher average speed
D the molecules exert greater average forces on the walls during collisions.

Q3 The specific heat capacity of a substance is defined as the amount of thermal energy needed to raise the temperature of:

A the mass of the substance by 1K
B the volume of the substance by 1K
C unit volume of the substance by 1K
D unit mass of the substance by 1K.

Q4 Which of the following is an important assumption of the kinetic theory of ideal gases?

A The forces between molecules are zero.
B All the molecules travel with the same speed.
C The molecular potential energies are constant.
D The molecules have zero momentum.

Q5 A system consists of an ice cube placed in a cup of water. The system is thermally insulated from its surroundings. The water is originally at 20°C. Which graph best shows the variation of total internal energy U of the system with time t?

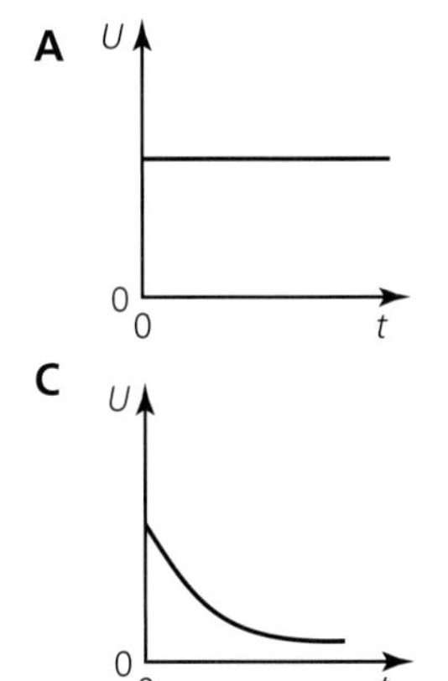

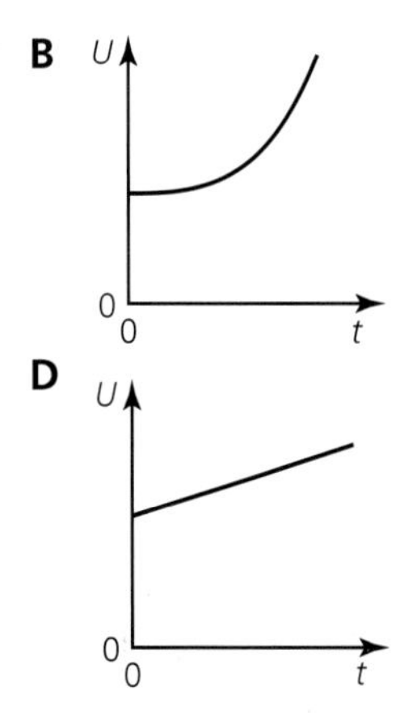

Q6 Which of the following is the correct conversion of a temperature of 100K to degrees Celsius?

A –373°C **B** –173°C **C** 173°C **D** 373°C

Q7 Which of the following is a correct statement about the energy of the molecules in an ideal gas?

A The molecules only have kinetic energy.
B The molecules only have thermal energy.
C The molecules only have potential energy.
D The molecules have kinetic and potential energy.

Q8 A copper block of mass M was heated with an immersion heater. The graph shows how the temperature of the block was affected by the thermal energy supplied to it. The gradient of the line is m.

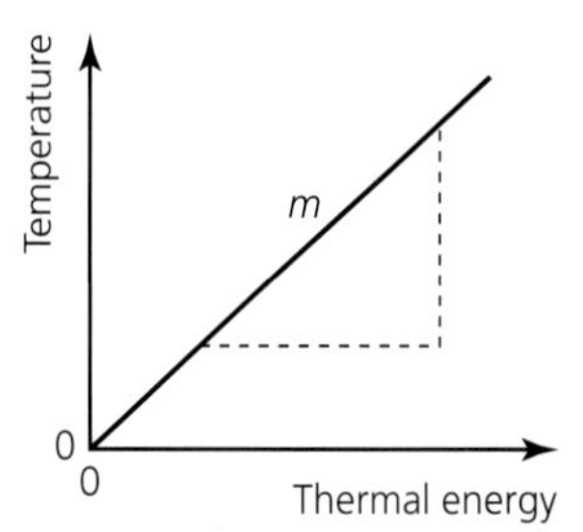

Which of the following expressions equals the specific heat capacity of the block?

A m **B** $\frac{1}{m}$ **C** mM **D** $\frac{1}{mM}$

Q9 The temperature of an ideal gas was increased from 50°C to 100°C. The average translational kinetic energy of its molecules increased by a factor of:

A 1.2 **B** 2.0 **C** 4.0 **D** 50

Q10 Two objects near each other are at the same temperature. Which of the following statements has to be true?

A The objects have the same internal energy.
B The objects have the same thermal capacity.
C No thermal energy is exchanged between the objects.
D The net thermal energy exchanged between the objects is zero.

Paper 2 IB questions and IB style questions

Q1 This question is about change of phase of a liquid and latent heat of vaporisation.

a A liquid in a calorimeter is heated at its boiling point for a measured period of time. The following data are available.
power rating of heater = 15 W
time for which liquid is heated at boiling point = 4.5×10^2 s
mass of liquid boiled away = 1.8×10^{-2} kg
Use the data to determine the specific latent heat of vaporisation of the liquid. (1)

b State and explain one reason why the calculation in part a will give a value of the specific latent heat of vaporisation of the liquid that is greater than the true value. (2)

Q2 **a** The internal energy of a piece of copper is increased by heating.

i Explain what is meant, in this context, by internal energy and heating. (3)
ii The piece of copper has mass 0.25 kg. The increase in internal energy of the copper is 1.2×10^3 J and its increase in temperature is 20 K. Estimate the specific heat capacity of copper. (2)

b An ideal gas is kept in a cylinder by a piston that is free to move. The gas is heated such that its internal energy increases and the pressure remains constant. Use the molecular model of ideal gases to explain:

i the increase in internal energy (1)
ii how the pressure remains constant. (3)

Q3 **a** Under what conditions can the equation $pV = nRT$ be applied to real gases? (2)

b A balloon containing air was kept in a freezer at a temperature of −8.80°C. When its volume was 210 cm^3 and its pressure was 1.60×10^5 Pa it was moved and placed outside where the temperature was 31°C. If the volume increased to 224 cm^3, what was the new pressure inside the balloon? (2)

c If the balloon contained a little water, explain what differences you would expect. (2)

4 Waves

ESSENTIAL IDEAS

- A study of oscillations underpins many areas of physics with simple harmonic motion (SHM), a fundamental oscillation that appears in various natural phenomena.
- There are many forms of waves available to be studied. A common characteristic of all travelling waves is that they carry energy, but generally the medium through which they travel will not be permanently disturbed.
- All waves can be described by the same sets of mathematical ideas. Detailed knowledge of one area leads to the possibility of prediction in another.
- Waves interact with media and each other in a number of ways that can be unexpected and useful.
- When travelling waves meet they can superpose to form standing waves in which energy may not be transferred.

4.1 Oscillations *– a study of oscillations underpins many areas of physics with simple harmonic motion (SHM), a fundamental oscillation that appears in various natural phenomena*

The various kinds of oscillation listed below show the extent to which oscillations are found all around us and suggest why the study of oscillations is such an important part of physics. Although there are many different objects that may oscillate, most of them can be better understood by using the same simple model, called simple harmonic motion.

■ **Figure 4.1** Oscillations of a humming bird's wings

■ Examples of oscillations

There are repeating motions of many diverse kinds, both in nature and in the manufactured world in which we live. Examples of these oscillations include:

- tides on the ocean
- the movement of our legs when we walk
- a heart beating
- clock mechanisms
- atoms vibrating
- machinery and engines
- a guitar string playing a musical note
- our eardrums (when we hear a sound)
- electronic circuits that produce radio waves and microwaves
- waves of various kinds, including light, which are transmitted by oscillations.

And there are many, many more examples to be found throughout mechanical and electrical engineering.

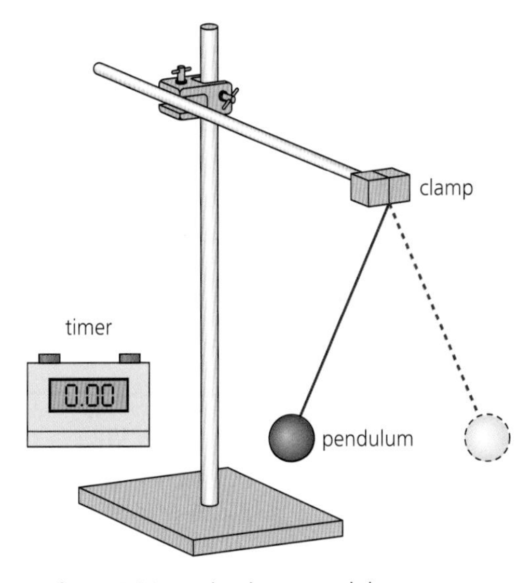

■ **Figure 4.2** Investigating a pendulum

Oscillations can be very rapid and difficult to observe. In a school laboratory we usually begin the study of oscillations with experiments on very simple kinds of oscillators that are easy to see and that oscillate at convenient rates, for example a pendulum or a mass on a spring (or between springs) (see Figures 4.2 and 4.3). These two simple oscillators behave in very similar ways to many other more complicated oscillators and we can use them as basic **models** that we can apply to other situations.

An air-track that is gently curved in the vertical plane can be very useful for detailed investigations of very slow oscillations. Other oscillations that can be investigated in a school laboratory include a mass vibrating on the end of a metre ruler, a liquid in a U-tube, a ball rolling on a curved surface and ice floating on water.

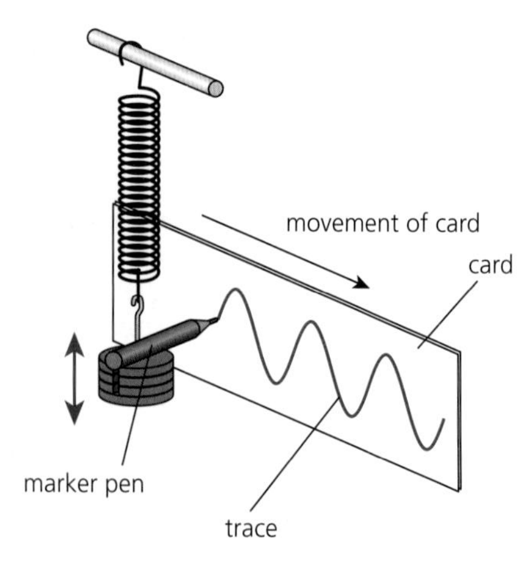

■ **Figure 4.3** Investigating a mass oscillating on a spring

■ Simple harmonic oscillations

Experiments such as those shown in Figure 4.2 and Figure 4.3 demonstrate an important property of many oscillators: for any particular oscillator, the time taken for each oscillation is the same, whatever the starting displacement. This means that when energy is dissipated, so that maximum speeds and displacements decrease, the time taken for each oscillation *still* remains the same.

The word **isochronous** is used to describe events that occupy equal times. Perfectly isochronous oscillations may not be possible in the macroscopic world, but many real oscillators (a swinging pendulum, for example) are almost isochronous. Oscillations like these are known as **simple harmonic oscillations**.

Utilizations

Oscillations and measurement of time

From the middle of the seventeenth century to less than a century ago, swinging pendulums were used in the world's most reliable and accurate clocks. Isochronous electronic oscillations are now used widely to measure time. For many reasons in this modern world (including travel, computing, communications and location finders), it is necessary for different countries to agree on time-keeping procedures and to synchronize clocks around the world. This is easily achieved using the internet.

Our impression of the flow of time is an interesting philosophical and scientific issue based largely on repeating events, such as the daily spin of the Earth on its axis that gives us night and day, and the yearly orbit of the Earth around the Sun that gives us seasons.

■ **Figure 4.4** A digital watch relies on electrical oscillations.

Planet Earth's time system of years, days, hours, minutes and seconds would seem strange to aliens from another planet. In fact, the Earth spins once on its axis not every 24 hours, but about every 23 hours and 56 minutes. Our 24-hour day is based on the time it takes for the Sun to return to its highest elevation as viewed from Earth and, apart from the rotation of the Earth itself, this is also affected, to a much lesser extent, by the Earth's daily movement in its orbit around the Sun.

The Earth's rotation is also decreasing at a rate of about 2 seconds every 100 000 years, so the length of the day is minutely but relentlessly increasing. This is mainly due to the gravitational interaction between the Earth and the Moon.

1 Before the invention of pendulum clocks, how did people measure short intervals of time?

2 Imagine you are watching a video of a pendulum swinging. Explain how you would know whether the video was being played forwards or backwards. Discuss whether this helps us understand the meaning of time.

Time period, frequency, amplitude, displacement and phase difference

The reason why an oscillation may be simple harmonic is discussed later in this chapter. First, we need to define the basic terminology that we use to describe oscillations.

Oscillations may occur whenever an object is displaced from its equilibrium position and then experiences a 'restoring' force pulling, or pushing, it back where it came from. The object then gains kinetic energy and passes through the equilibrium position, so that its displacement is now in the opposite direction to the original. The restoring force then also acts in the opposite direction so that the object decelerates, stops and then accelerates back towards the equilibrium position again. The process keeps repeating.

The equilibrium (or mean) position is the position at which there is no resultant force on the object. In other words, the equilibrium position is the place where the object would stay if it had not been disturbed and the place to which it tends to return after the oscillations have stopped.

The displacement, x, of an oscillator is defined as the distance in a specified direction from its equilibrium position.

(This is similar to the more general definition of displacement given on page 21.) The displacement varies continuously during an oscillation.

When describing an object that keeps repeating its motion, there are two very obvious questions to ask – how large are the oscillations and how quickly does it oscillate?

The amplitude of an oscillator is defined as its maximum displacement. The maximum displacement is the distance from the equilibrium position to the furthest point of travel.

We use the symbol x_0 (or sometimes A) for amplitude. A bigger amplitude means that there is more energy in the system.

The (time) period, T, of an oscillation is the time it takes for one complete oscillation. That is, the minimum time taken to return to the same position (moving in the same direction).

A complete oscillation is sometimes called a cycle.

Most oscillations are quick – there may be a great many oscillations every second. Therefore it is usually more convenient to refer to the number of oscillations every second (rather than the period).

The frequency, f, of an oscillator is defined as the number of oscillations in unit time (usually every second).

When something is disturbed and then left to oscillate without further interference, it is said to oscillate at its natural frequency.

Of course, the frequency and time period of an oscillation are essentially the same piece of information expressed in two different ways. They are connected by the simple formula:

$$T = \frac{1}{f}$$

This equation is given in the *Physics data booklet*.

The concept of frequency is so useful that it has its own unit. A frequency of one oscillation per second is known as one hertz, Hz. The units kHz (10^3 Hz), MHz (10^6 Hz) and GHz (10^9 Hz) are also often used.

Phase difference

We may want to compare two or more similar oscillations that have the same frequency. For example, children side by side on swings in a park, as shown in Figure 4.5.

Figure 4.5 Four oscillations out of phase

If two oscillators continue to do exactly the same thing at the same time, they are described as being **in phase**: they always pass through their equilibrium positions moving in the same directions at the same time. In this context the word 'phase' means a stage in a process (oscillation). It should not be confused with a phase of matter (Chapter 3).

If similar oscillations are not moving together in phase, then they are described as having a **phase difference**.

If one oscillation is always 'ahead' of the other by exactly half an oscillation (or 'behind' by half an oscillation), then they will be moving in opposite directions as they pass through the equilibrium position and are described as being **exactly out of phase**.

Oscillations may be out of phase by any amount between these two extremes, but the detailed measurement of phase difference is not needed in this chapter (see Chapter 9).

Worked example

1 A child on a swing went through exactly five complete oscillations in 10.4 s.

a What was the period?

b What was the frequency?

a $T = \frac{10.4}{5.00} = 2.08\,\text{s}$

b $f = \frac{1}{T} = \frac{1}{2.08} = 0.481\,\text{Hz}$

1 A pendulum was timed and found to have exactly 50 oscillations in 43.6 s.

a What was its period?

b What was its frequency?

c How many oscillations will it undergo in exactly 5 minutes?

d Explain why it is a good idea, in an experiment to find the period of a pendulum, to measure the total time for a large number of oscillations.

2 What is the period of a sound wave that has a frequency of 3.2 kHz?

3 Radio waves oscillate very quickly.
 a What is the frequency in Hz of a radio wave with a time period of 5×10^{-9} s?
 b Express the same frequency in MHz.

4 A pendulum has a frequency of 0.5 Hz and an amplitude (measured along the arc) of 2.4 cm. After release from its maximum displacement on the left-hand side,
 a what is its displacement after:
  i 0.5 s
  ii 3.0 s?
 b What distance has it travelled after 5.0 s?
 c A second, identical, pendulum is placed next to the first and is released from its maximum displacement on the right-hand side 3 s after the first. Describe the phase difference between the two pendulums.

Conditions for simple harmonic motion

The simplest kind of oscillation will occur when an object, such as a mass attached to a spring (Figure 4.3), is displaced from its equilibrium position and experiences a restoring force, F, which is proportional to the displacement and in the opposite direction. Under these circumstances, doubling the amplitude will result in twice the restoring force, so that the object will oscillate with a *constant time period*, even if the amplitude varies. In general, an isochronous oscillation may occur if:

$$F \propto -x$$

The negative sign in this equation indicates that the force is in the opposite direction to the displacement. In other words, the force opposes the motion. That is why it is called it is a *restoring* force.

This kind of oscillation is called **simple harmonic motion** (SHM). Although it is a theoretical model, many real-life oscillators approximate quite closely to this ideal model of SHM. SHM is defined in terms of accelerations, but because acceleration is proportional to force the relationship has the same form.

Simple harmonic motion is defined as an oscillation in which the acceleration, a, of a body is proportional to its displacement, x, from its equilibrium position and in the opposite direction:

$$a \propto -x$$

That is, the acceleration is always directed towards the equilibrium position (as shown by the negative sign) or, in other words, the object decelerates as it moves away from the equilibrium position (see Figure 4.6) in which the proportionality is confirmed by a straight line through the origin. A force–displacement graph will be similar, with a gradient equal in magnitude to the *force constant*.

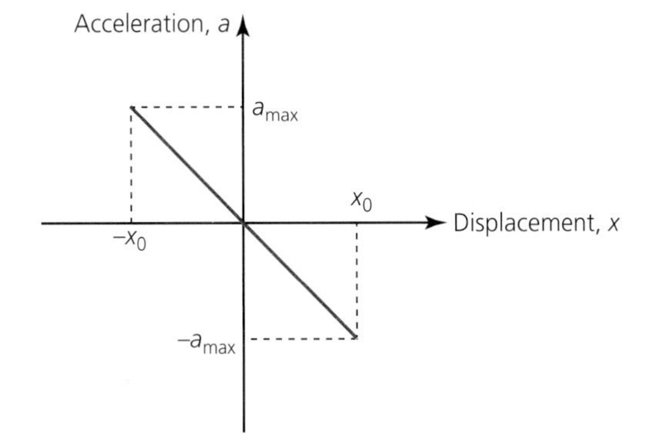

Figure 4.6 Acceleration–displacement graph for SHM

Nature of Science

The importance of the SHM model

Repetitive motions (oscillations) are to be found in many aspects of our lives, and we can begin to understand most of them by using the model of simple harmonic oscillation, the study of which plays a very important role in physics (and which is covered in more detail in Chapter 9 for Higher Level students). A number of simple oscillators that are described well by the basic theory can be investigated in the laboratory, but many others do not match the theory so conveniently. Advanced mathematical models and computer simulations of oscillators can be developed from basic SHM to include the dissipation of energy and thereby represent reality more realistically.

5 a Make a list of six things in your home or school that can oscillate.
  b For each oscillator, discuss whether the oscillations are likely to be isochronous or not.

6 a Do you think that a mass moving freely up and down vertically on the end of a rubber band will be undergoing simple harmonic motion?
  b Explain your answer.

7 Make a copy of Figure 4.6. Suppose that the graph you have drawn represents the SHM of a mass, m, oscillating vertically on the end of a spring of force constant k. Add two other lines to the graph to represent the oscillations of:
  a a mass of $0.5m$ on the same spring
  b a mass m on a spring with a force constant of $0.5k$.

Graphs of simple harmonic motion

Using data-gathering sensors connected to a computer, as shown in Figure 4.7, it is relatively easy to produce displacement–time graphs for a variety of oscillators.

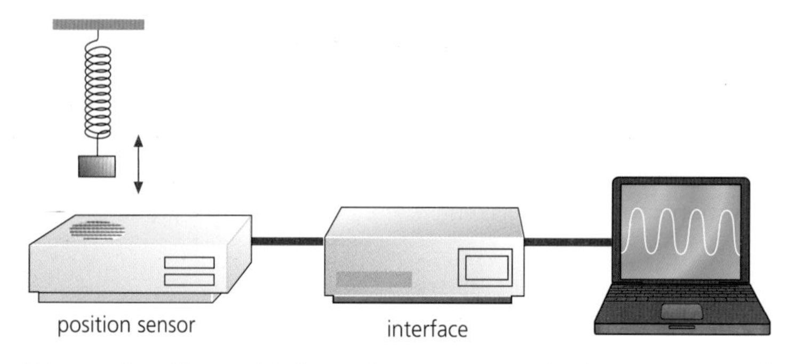

Figure 4.7 Collecting data on the oscillations of a mass on a spring using a sensor and data logger

The graph in Figure 4.8 shows the variation in displacement, x, with time, t, for our idealized model of a particle moving with simple harmonic motion. Here we have *chosen* that the particle has zero displacement at the start of the timing and has an amplitude of x_0. The graph has a simple **sinusoidal** shape (like a sine wave). Equally, we might have chosen the graph to start with the particle at maximum displacement. It would then appear as a cosine wave.

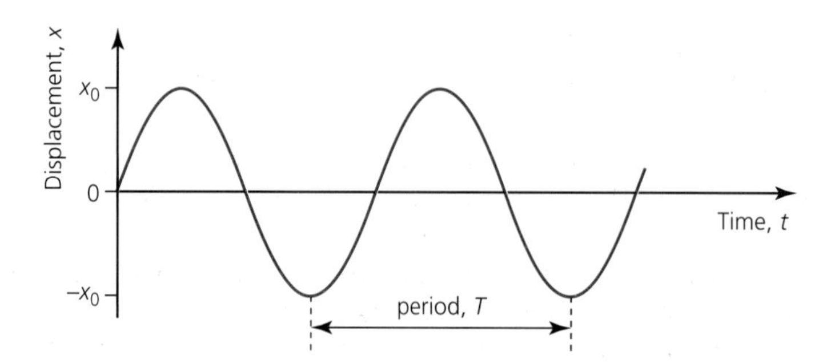

Figure 4.8 Displacement–time graph for simple harmonic motion represented by a sine wave – timing was started when the particle had zero displacement

The velocity at a particular time can be found from the gradient of the displacement–time graph at that point:

$$\text{velocity, } v = \frac{\text{change in displacement}}{\text{change in time}} = \frac{\Delta s}{\Delta t}$$

Similarly, the acceleration at any given time is the gradient of the velocity–time graph:

$$\text{acceleration, } a = \frac{\text{change in velocity}}{\text{change in time}} = \frac{\Delta v}{\Delta t}$$

Using this information, three separate, but interconnected, graphs of motion can be drawn and compared, as shown in Figure 4.9.

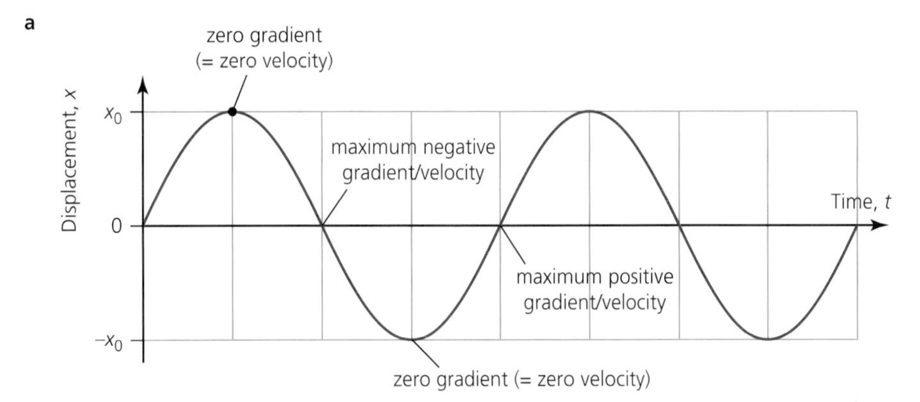

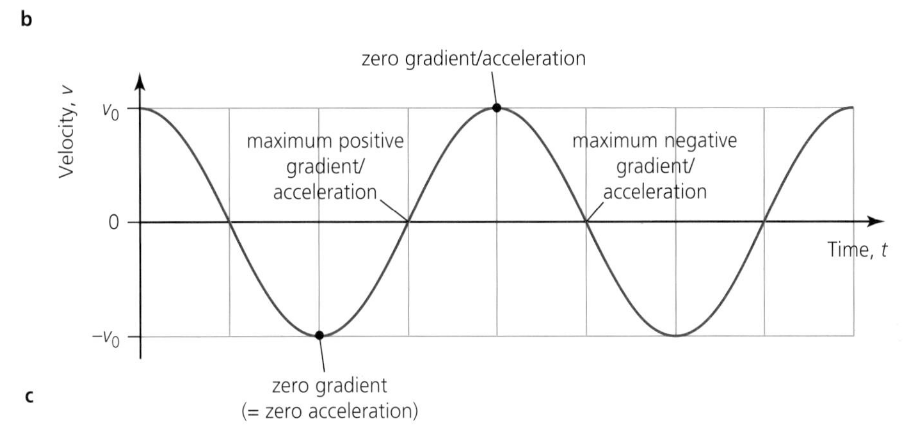

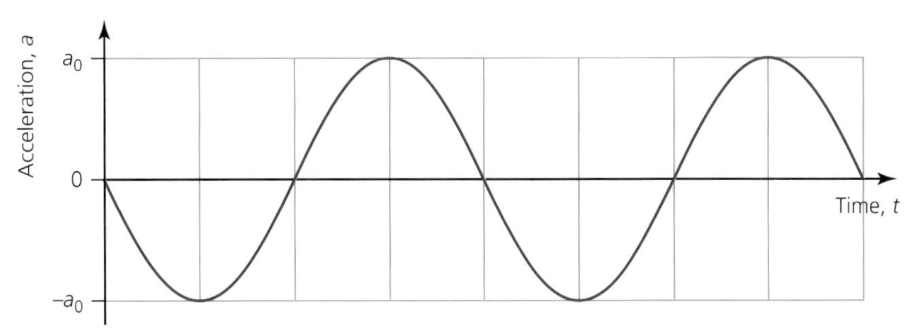

■ **Figure 4.9** Graphs for simple harmonic motion starting at displacement $x = 0$: **a** displacement–time; **b** velocity–time; **c** acceleration–time

We see that the velocity has its maximum value, v_0, when the displacement x is zero, and the velocity is zero when the displacement is at its maximum, x_0. In other words, the velocity graph is one quarter of an oscillation out of phase with the displacement graph.

We can see from Figure 4.9 that the acceleration has its maximum value when the velocity is zero and the displacement is greatest. This is to be expected because when the displacement is highest, the restoring force, acting in the opposite direction, is highest. In terms of phase difference, the acceleration graph is one quarter of an oscillation out of phase with the velocity graph and half an oscillation out of phase with the displacement graph.

Figure 4.10 shows all three graphs drawn on the same axes, so that they can be compared more easily. (Note that the amplitudes of the three graphs are arbitrary; they are not interconnected.)

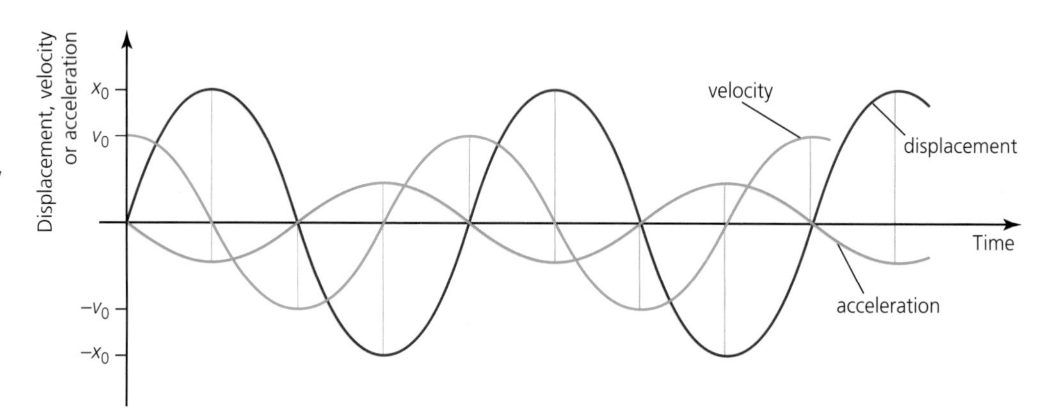

■ **Figure 4.10** Comparing displacement, velocity and acceleration for SHM, with timing starting at displacement $x = 0$

Students are recommended to make use of a computer simulation of an object oscillating with SHM, combined with the associated graphical representations.

8 Look at the graph in Figure 4.11, which shows the motion of a mass oscillating on a spring. Determine:
a the amplitude
b the period
c the displacement after 0.15 s
d the displacement after 1.4 s.

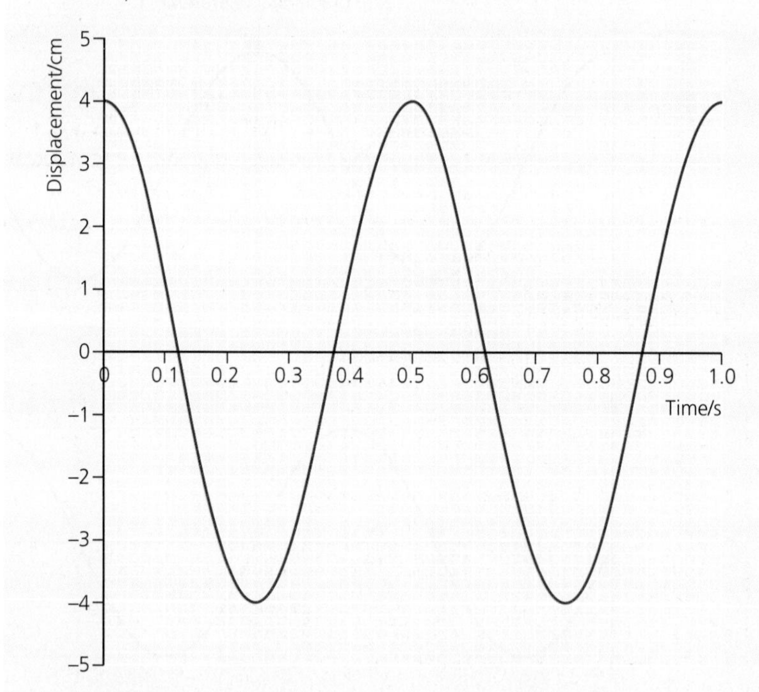

■ **Figure 4.11**

9 a Sketch a displacement–time graph showing two complete oscillations for a simple harmonic oscillator with a time period of 2.0 s and an amplitude of 5.0 cm.
b Add to the same axes the waveform of an oscillator that has twice the frequency and the same amplitude.
c Add to the same axes the waveform of an oscillator that has an amplitude of 2.5 cm and the same frequency, but which is one quarter of an oscillation out of phase with the first oscillator.

10 a Sketch a velocity–time graph showing two complete oscillations for a simple harmonic oscillator that has a frequency of 4 Hz and a maximum speed of $4.0\,cm\,s^{-1}$.
b On the same axes sketch waveforms to show the variation of:
i displacement
ii acceleration for the same oscillation.
c Describe the phase differences between these three graphs.

Energy changes during simple harmonic oscillations

When an object that can oscillate is pushed, or pulled, away from its equilibrium position against the action of a restoring force, work will be done and potential energy will be stored in the oscillator. For example, a spring will store elastic potential energy and a simple pendulum will store gravitational potential energy. When the object is released it will gain kinetic energy and lose potential energy as the force accelerates it back towards the equilibrium position. Its kinetic energy has a maximum value as it passes through the equilibrium position and, at the same time, its potential energy is zero. As it moves away from the equilibrium position, kinetic energy decreases as the restoring force opposes its motion and potential energy increases again.

As an example, consider Figure 4.12 in which a simple pendulum has been pulled away from its equilibrium position, C, to position A. While it is held in position A, it has zero kinetic energy and its change of gravitational potential energy (compared to position C) is highest. When it is released, gravity provides the restoring force and the pendulum exchanges potential energy for kinetic energy as it moves through position B to position C. At C it has maximum kinetic energy and the change in potential energy has reduced to zero. The pendulum then transfers its kinetic energy back to potential energy as it moves through position D to position E. At E, like A, it has zero kinetic energy and a maximum change of potential energy. The process then repeats every half oscillation.

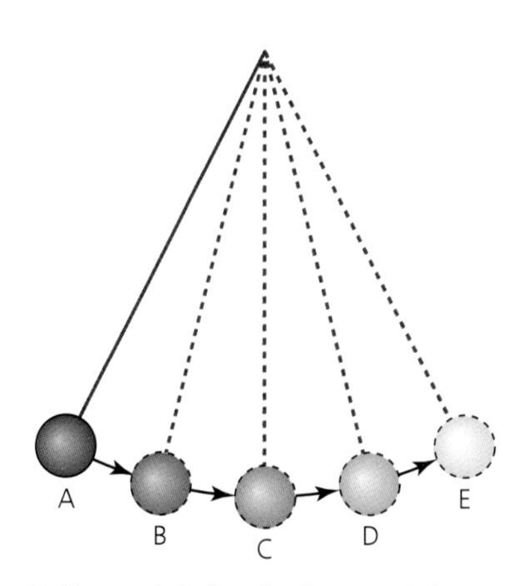

Figure 4.12 A swinging pendulum

If the pendulum was a perfect simple harmonic oscillator there would be no energy losses, so that the sum of the potential energy and the kinetic energy would be constant and the pendulum would continue to reach the same maximum vertical height and maximum speed every oscillation. In practice, frictional forces will result in energy dissipation and all the energies of the pendulum will progressively decrease.

Worked example

2 A pendulum, such as that shown in Figure 4.12, has a time period of 1.5 s.
a Sketch a graph to show how the kinetic energy of the pendulum might vary over a time of 3.0 s. (Start from position A and assume that no energy is transferred to the surroundings.)
b On the same axes, draw a second graph to represent the changes in gravitational potential energy of the pendulum.
c Add a line to represent the variation of total energy.

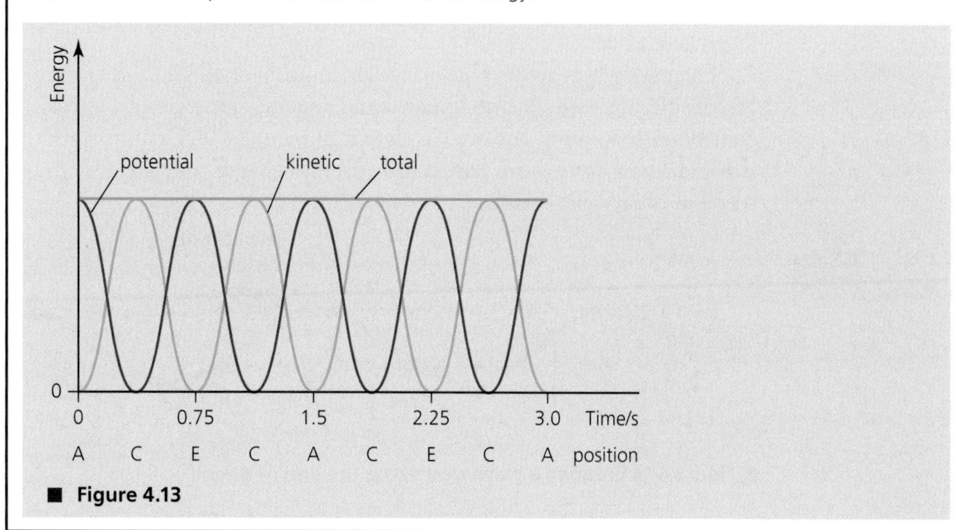

Figure 4.13

11 Describe the energy transfers that occur during one cycle of a mass oscillating on the end of a spring (see Figure 4.3).

12 Make a copy of the potential energy graph from Figure 4.13. Add a second line to show how the graph would change if there were significant energy losses to the surroundings.

13 Explain why a liquid oscillating in a U-tube will only complete one or two oscillations before stopping.

ToK Link

A primary intention of models is to simplify

The harmonic oscillator is a paradigm for modelling where a simple equation is used to describe a complex phenomenon. How do scientists know when a simple model is not detailed enough for their requirements?

Is it sufficient to say that scientists should be satisfied if a particular theory/model explains all available observations fully? If there are differences between the predictions of a theory/model and repeated observations, does that require the model to be improved or replaced? Or do we have to accept that some situations are so complex that no model can ever truly represent them?

4.2 Travelling waves – *there are many forms of waves available to be studied. A common characteristic of all travelling waves is that they carry energy, but generally the medium through which they travel will not be permanently disturbed*

We are all familiar with water waves but there are many other types of wave, all of which have similar characteristics. These include many important waves that we cannot see. **Mechanical waves**, including sound, involve oscillating masses and need a material through which to travel, but waves like light (**electromagnetic waves**) can also travel across empty space.

Clearly, the study of waves is a very important topic in physics and we can summarize the importance of travelling waves by saying that they can transfer energy from place to place without transferring matter. Furthermore, certain types of waves can be modified to transfer information – to communicate. Sound waves and radio waves are the most obvious examples.

Nature of Science

Patterns, trends and discrepancies

Looking for **patterns** and similarities is a major aspect of all branches of science. Without patterns and trends, a collection of a large quantity of disconnected data and observations would be impossible to analyse. We now appreciate that light, sound, disturbances on water, vibrating strings (for example, on musical instruments) and earthquakes are all examples of waves, but this was far from obvious. It took centuries of careful observation to reach this conclusion, which greatly simplifies our understanding of the natural world. If we understand basic wave theory, we can apply our knowledge to many different branches of science, from the motion of distant galaxies to the structure of the atom.

■ Travelling waves

Waves that transfer energy away from a source are known as continuous *travelling waves* or *progressive waves*.

The easiest way to develop an understanding of *all* kinds of travelling waves is to begin by considering a simple one-dimensional system – for example, a wave sent along a rope by continuously shaking one end, as shown in Figure 4.14. Demonstrations with waves on ropes, strings or long springs are particularly useful because they are easy to produce and the waves travel at observable speeds.

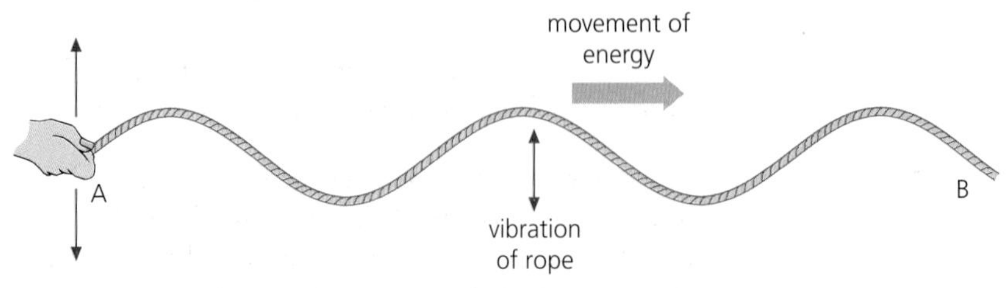

■ **Figure 4.14** Creating a wave by shaking the end of a rope

Consider the idealized example of the end of a rope being shaken (oscillated) up and down, or from side to side, with SHM. The oscillations will be passed along the rope, from each part to the next, with a delay. Each part of the rope will, in turn, do exactly the same thing (if there is no energy dissipation). That is, each part of the rope will oscillate with SHM parallel to the original disturbance. It is easy to imagine how the characteristically shaped sinusoidal wave (as shown in Figure 4.14) is produced. The oscillations that transfer many different kinds of waves can be considered to be simple harmonic.

■ **Figure 4.15** Ocean waves transferring a large amount of energy at Brighton, England – there is no continuous net movement of the water itself

Scientists describe the motion of a wave away from its source as propagation. Any substance through which it is passing is called the medium of propagation. (The plural of medium is *media*.)

It is very important to realize that when we observe a wave, despite our impression of movement, the medium itself does *not* move continuously in the direction of apparent wave motion.

Travelling waves transfer energy without transferring matter. For example, ocean waves may 'break' and 'crash' on to a shore or rocks (Figure 4.15), but there is no overall (net) continual movement of water from the sea because of the waves. A wooden log floating on water will just tend to oscillate up and down as waves pass (if there is no wind).

If a disturbance is not continuous but a single event (perhaps just one oscillation), then we may describe the spreading disturbance as a pulse rather than a wave.

■ Transverse and longitudinal waves

All waves are one of only two kinds: transverse waves or longitudinal waves. We will first consider transverse waves.

Transverse waves

It helps our understanding of waves passing through a continuous medium, like a rope, to imagine a model in which the medium is represented by separate (discrete) particles, as shown in Figure 4.16. The black line represents a transverse wave moving to the right; the arrows show which way the particles are moving at that moment. The second wave, shown in red, represents the position of the same wave a short time later.

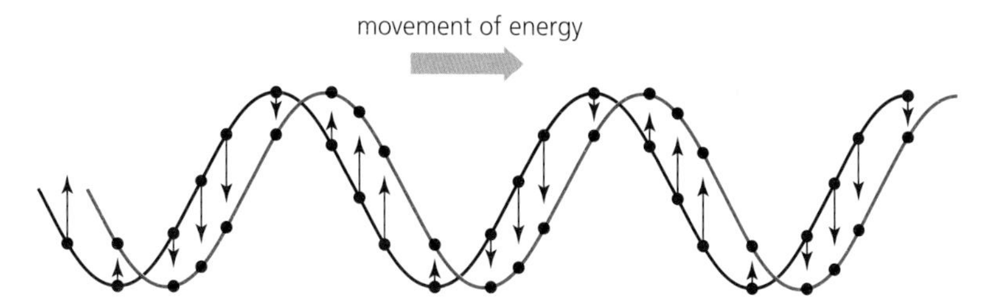

■ **Figure 4.16** Movement of particles as a transverse wave moves to the right

Each part of the medium is doing the same thing (oscillating with the same frequency and same amplitude). But the different parts of the medium have different displacements at any particular time. That is, the different parts of the medium are not all moving *in phase* with each other. The kind of wave shown in Figure 4.16 is described as transverse.

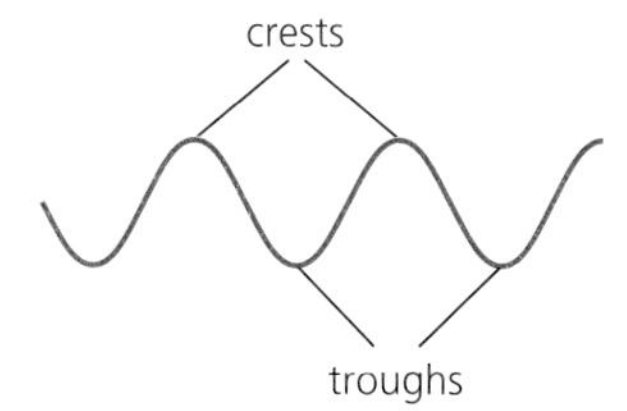

■ **Figure 4.17** Crests and troughs of a transverse wave

In a transverse wave, each part of the medium oscillates *perpendicularly* to the direction in which the wave is transferring energy.

The 'tops' of transverse waves (especially water waves) are called crests and the 'bottoms' of the waves are called troughs (Figure 4.17).

Examples of transverse waves include light (and all the other parts of the electromagnetic spectrum), water waves and waves on stretched strings and ropes.

Transverse mechanical waves cannot pass through gases because of the random way in which molecules move in gases.

Longitudinal waves

In a longitudinal wave, each part of the medium oscillates *parallel* to the direction in which the wave is transferring energy.

We can demonstrate a longitudinal wave using a spring, but we need to stretch it first so that the coils are not touching. (Demonstrations often use 'slinky' springs.) In order to make a longitudinal wave, the end of the spring is then oscillated 'backwards and forwards' (ideally with SHM) along the line of the spring.

A diagram of a longitudinal wave modelled by a slinky spring is shown in Figure 4.18. Characteristic **compressions** (where the spring is compressed) and **rarefactions** (where the spring is stretched) are marked. Longitudinal waves are sometimes called **compression waves**.

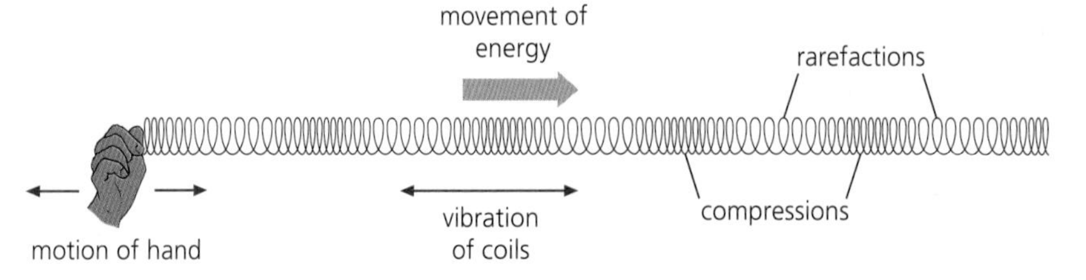

■ **Figure 4.18** Oscillations of a spring transferring a longitudinal wave

As with transverse waves, when a longitudinal wave transfers energy through a medium, the medium itself does not undergo net translational motion away from the source.

Examples of longitudinal waves include sound and compression waves in solids. Earthquakes create both longitudinal waves and transverse waves in the Earth's rocks.

■ Representing waves graphically

Two similar graphs can be drawn to represent waves (transverse or longitudinal) and they are easily confused because they look similar. These are displacement–distance graphs and displacement–time graphs. Compare Figures 4.19 and 4.20.

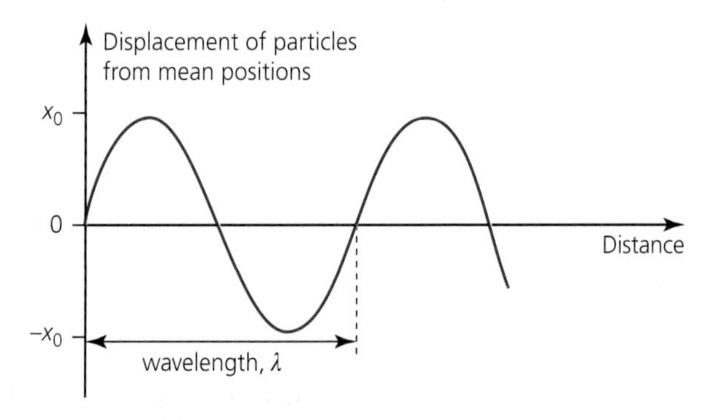

■ **Figure 4.19** Displacement–distance graph for a medium showing the meaning of wavelength, λ

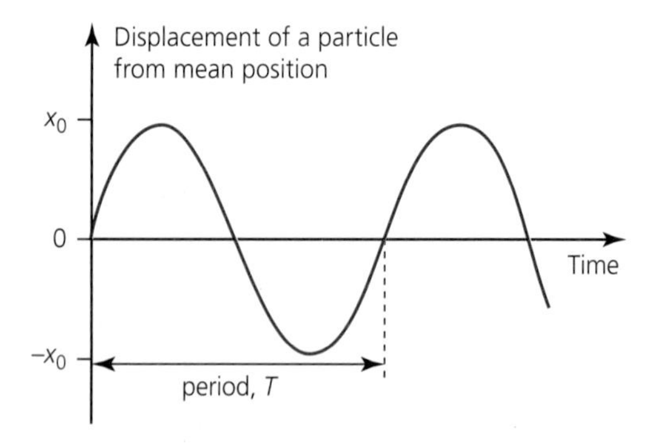

■ **Figure 4.20** Displacement–time graph for an individual particle showing the meaning of period, T

Figure 4.19 shows the position of *many* parts of a medium, each moving with amplitude x_0, at one particular instant in time (a 'snapshot'). Figure 4.20 shows the movement of an *individual* part of the medium as the wave passes through it. (This is similar to the SHM graphs shown earlier in this chapter.)

The shape of these graphs may suggest that they only represent transverse waves, but it is important to realize that similar graphs can also be used to represent the displacements in longitudinal waves (or even the variations in pressure when sound waves travel through a gas). Figure 4.21 shows how the random arrangement of molecules in air might change as a longitudinal sound wave passes through. The compressions correspond to regions of higher than average air pressure and the rarefactions are regions of lower than average air pressure.

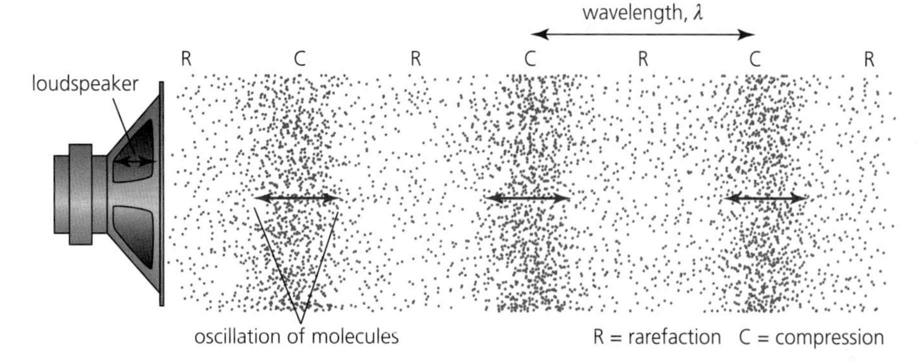

■ **Figure 4.21** Arrangement of molecules in air as sound passes through

■ Wavelength, frequency, period and wave speed

The **wavelength**, λ, of a wave is defined as the shortest distance between two points moving in phase.

Wavelength is easily shown on a displacement–position graph (Figure 4.19) and is usually measured in metres.

The (time) **period**, T, of a wave is defined as the time it takes for one complete wave to pass a given point (or the time for one complete oscillation of a point within the medium).

Period is easily represented on a displacement–time graph (Figure 4.20).

The **frequency**, f, of a wave is defined as the number of waves that pass a given point in unit time (or the number of oscillations at a point within the medium in unit time).

Wave speed, c, is defined as the distance travelled by a wave in unit time.

The symbols v and c are both commonly used for wave speed, but the speed of electromagnetic waves is always represented by c. All wave speeds are usually measured in m s^{-1}.

Wave equation

We know that, in general, speed equals distance moved divided by time taken. A wave moves a distance of one wavelength in one time period, so that $c = \lambda/T$.

Or, because $T = 1/f$:

$$c = f\lambda$$

wave speed = frequency × wavelength

This equation is in the *Physics data booklet*.

This is a simple, but very important and widely used, equation. It can be used for all types of wave.

It is important to realize that once a wave has been created, its frequency cannot change. So if there is a change of speed, for example after entering a different medium, there will be a corresponding change of wavelength (slower speed, shorter wavelength; higher speed, longer wavelength).

■ The nature of electromagnetic waves

Light travels as a transverse wave and the **spectrum** of visible 'white' light, from red to violet, is a familiar sight (Figure 4.22). The wavelengths of light waves are very small and the different colours of the spectrum have different wavelengths. Red light has the longest wavelength (approximately 7×10^{-7} m) and violet has the shortest wavelength in the visible spectrum (approximately 4×10^{-7} m).

■ **Figure 4.22** Spectrum of visible light

But visible light is only a very small part of a much larger group of waves, called the **electromagnetic spectrum**, the major sections of which are listed in Table 4.1. It is important to understand that the spectrum is *continuous* and there is often no definite boundary between one section and another; the sections may overlap. For example, gamma rays from a radioactive source may be identical to X-rays from an X-ray tube.

We are surrounded by many of these waves all the time and they come from a variety of very different sources. 'Light' is just the name we give to those electromagnetic waves that human beings can detect with their eyes. The only fundamental difference between the various kinds of electromagnetic waves is their wavelength, so it is not surprising that, under appropriate circumstances, all of these transverse waves exhibit the same basic wave properties.

All electromagnetic waves can travel across empty (free) space (a vacuum) with exactly the same speed, $c = 3.00 \times 10^8\,\text{m}\,\text{s}^{-1}$.

This constant is listed in the *Physics data booklet*.

Electromagnetic waves do not *need* a medium through which to travel, but when electromagnetic waves pass through other materials, their speeds are lower than in a vacuum (although their speed in air is almost identical to their speed in a vacuum).

Table 4.1 gives a very brief summary of the major parts of the spectrum. Waves of wavelength shorter than about 10^{-8} m can cause damage to the human body, even at low intensity.

■ **Table 4.1** The different sections of the electromagnetic spectrum

Name	A typical wavelength	Origins	Some common uses
radio waves	10^2 m	electronic circuits/aerials	communications, radio, TV
microwaves	10^{-2} m	electronic circuits/aerials	communications, mobile phones, ovens, radar
infrared (IR)	10^{-5} m	everything emits IR but hotter objects emit *much* more IR than cooler things	lasers, heating, cooking, medical treatments, remote controls
visible light	5×10^{-7} m	very hot objects, light bulbs, the Sun	vision, lighting, lasers
ultraviolet (UV)	10^{-8} m	the Sun, UV lamps	fluorescence
X-rays	10^{-11} m	X-ray tubes	medical diagnosis and treatment, investigating the structure of matter
gamma rays	10^{-13} m	radioactive materials	medical diagnosis and treatment, sterilization of medical equipment

The range of different wavelengths, from 10^{-14} m or shorter up to 10^4 m or longer, is enormous and it is important to remember an order of magnitude for the principal radiations.

The exact nature of these waves was a major puzzle in science for a long time because all other wave types need a medium in which to travel, while electromagnetic waves can transfer energy across free space. We now know that these transverse waves are not carried by oscillations of a medium but by linked oscillating electric and magnetic fields (which do not need a medium), as shown in Figure 4.23, hence their name *electromagnetic* waves. However, electromagnetic waves also have some properties that can only be explained by thinking of them as 'particles' (and not waves) called **photons**. Each photon transfers an individual amount of energy dependent on its frequency. These ideas are not needed now and they are discussed in more detail in Sections 7.3 and 12.1.

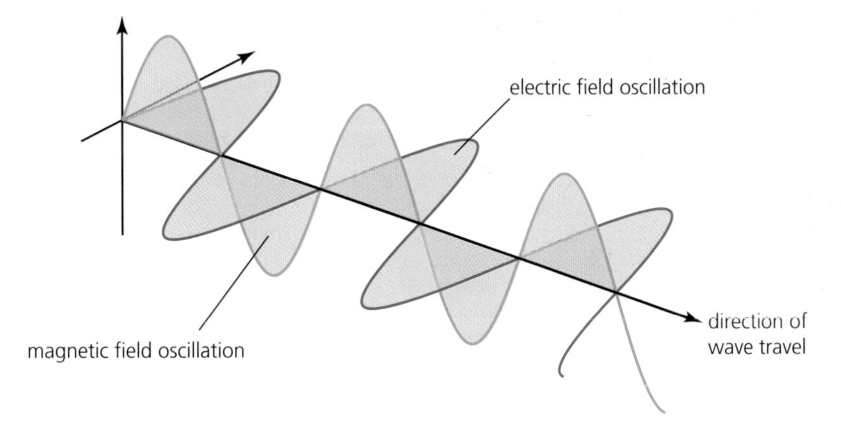

■ **Figure 4.23** Electromagnetic waves are combined electric and magnetic fields.

The means of detecting the various electromagnetic radiations have not been included in Table 4.1. Observing the origin, detection and absorption of the invisible electromagnetic radiations from different parts of the spectrum can make them seem more 'real'.

Two uses of electromagnetic waves deserve special mention:

- Microwaves and radio waves are generated easily and can be detected with oscillating currents in electric circuits. The properties of these waves can be modified to carry information (communicate) very quickly over large distances.

■ **Figure 4.24** Using electromagnetic waves for national and international communication

- The spectra emitted by energized gases can be analysed to provide information about the energy levels inside atoms.

Worked example

3 The crests of waves passing into a harbour are 2.1 m apart and have an amplitude of 60 cm. Ten waves pass an observer every minute.

a What is their frequency?

b What is their speed?

a $T = \frac{60}{10} = 6.0\,\text{s}$

So $f = \frac{1}{T} = \frac{1}{6.0} = 0.17\,\text{Hz}$

b $v = f\lambda$

$v = 0.17 \times 2.1 = 0.36\,\text{m}\,\text{s}^{-1}$

14 a Calculate the wavelength in air of a high-pitched sound of frequency 2.20 kHz. (Speed of sound in air = $340\,\text{m}\,\text{s}^{-1}$)
b When the same sound passed into a metal, its wavelength increased to 2.05 m. What was the speed of sound in the metal?
c Suggest a reason why sound travels faster in solids than in gases like air.

15 a What is the frequency of radio waves of wavelength 560 m? (Speed of electromagnetic waves = $3.0 \times 10^{8}\,\text{m}\,\text{s}^{-1}$)
b How long would it take a radio signal to travel from:
i Delhi to Mumbai (1407 km)
ii a satellite orbiting at a height of 33 000 km to the Earth's surface
iii Earth to Mars? (Use the internet to help determine the maximum and minimum separations of the two planets.)

16 a What is the frequency (in MHz) of a gamma ray with a wavelength of 4.4×10^{-12} m?
b Suggest a possible reason why electromagnetic radiation of higher frequencies is more dangerous to humans than that of lower frequencies.

17 a Calculate the wavelength of visible light of frequency of 5.5×10^{14} Hz.
b Suggest what colour the light may be.
c Some radiation from the Sun has a wavelength of 6.0×10^{-8} m. In what part of the electromagnetic spectrum is this radiation?

18 The total separation of five wave crests on a ripple tank is 5.6 cm. If the waves are made by a vibrator with a frequency of 12 Hz, how long will it take the waves to travel a distance of 60 cm?

19 As you are reading this, which electromagnetic radiations are in the room?

ToK Link

Explaining the invisible

Scientists often transfer their perception of tangible and visible concepts to explain similar non-visible concepts, such as in wave theory. How do scientists explain concepts that have no tangible or visible quality?

Invisible electromagnetic waves cannot be observed directly, but they are considered to be wave-like because in some ways they behave similarly to waves that we can see (like waves on strings and water waves). But that should not imply that scientists believe that describing light, for example, as waves is an adequate or complete description. There should be no expectation that things we cannot observe directly behave in the same ways as things that we can see. Some properties of electromagnetic radiation cannot be explained by wave theory and require a particle (photon) theory explanation.

We also cannot see sound, but does that mean that, as with light, describing sound as *waves* is just a provisional theory that is useful on some occasions? That is, can all the properties of sound be explained using wave theory?

Additional Perspectives

Heinrich Hertz

■ **Figure 4.25** Heinrich Hertz

The unit for frequency is named after the German physicist, Heinrich Hertz. His most famous achievement, in 1886, was to prove experimentally for the first time that electromagnetic waves (radio waves) could be produced, transmitted and detected elsewhere. Although the distance involved was very small, it was the start of modern wireless communication. It was left to others (such as Guglielmo Marconi) to develop techniques for the transmission over longer and longer distances and then to modify the amplitude, frequency or phase of the radio waves to transfer information (for example, speech).

Hertz is often quoted as saying that his discovery was 'of no use whatsoever' and he was not alone in expressing that opinion at the time. The history of science contains many such statements and predictions that later turn out to be incorrect. The importance of discoveries or inventions may only be realized many years later. This is one reason why most scientists think it would be foolish to limit research to those projects that have immediate and obviously useful applications.

Tragically, Hertz contracted a fatal disease and died at the age of 36, less than eight years after his discovery and a long time before the full implications of his work had become clear.

1 Modern scientific research can be very expensive, so would countries be wiser to spend their money on other things (for example, improving general medical care), unless the research is obviously leading towards a definite and useful aim?

The nature of sound waves

When the surface of a solid vibrates, it will disturb the air (or any other medium) that surrounds it and produce a series of compressions and rarefactions (variations in pressure) that travel away from the surface as a longitudinal wave. If the waves have a frequency that can be detected by our ears (heard), then they are described as sound. Figure 4.21 shows the production of sound waves in air by a loudspeaker.

The normal range of hearing for humans, called the **audible range**, is approximately 20 Hz to 20 kHz. Higher frequencies are called **ultrasonics**.

Sound is a mechanical wave transferred by oscillating molecules and as such it clearly needs a medium of propagation. Sound cannot travel through a vacuum. Understanding the nature of sound waves leads us to expect that they will travel better through materials in which the molecules are closer together and have larger forces between them. Sound waves travel better in liquids than gases/air and even better in solids.

Speed of sound

Because speed = distance/time taken, in order to determine the speed of sound it is necessary to measure the time it takes sound to travel a known distance. Because the speed of sound is large, in order to get good results it is often necessary to use large distances and measure short time intervals accurately and precisely. Hand-held stopwatches may be used, but they are unlikely to produce satisfactory results because of the problems associated with human reaction times. Sometimes large, isolated outdoor surfaces are used to reflect sound waves back to their source to make the experiment easier to do, but in order to hear an **echo** clearly (separate from the original sound), the surface normally needs to be at least 50 m away.

Figure 4.26 shows the principle of a smaller-scale experiment that can be done accurately inside a laboratory. A short, sharp sound is made at position A. When the sound wave reaches microphone 1, the signal from the microphone travels along the connecting lead to start the timer. When the wave reaches microphone 2, the signal from the microphone to the timer stops the timing.

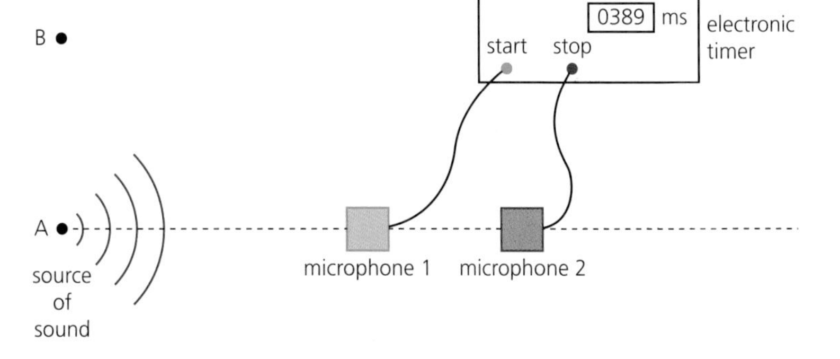

Figure 4.26 A laboratory experiment to determine the speed of sound

Worked example

4 Consider the experiment shown in Figure 4.26.

a i Suggest how a 'short, sharp' sound could be made.

ii Why is this kind of sound preferred for this experiment?

b What are the 'problems associated with human reaction times' when using hand-held stopwatches for measuring short intervals of time?

c When the microphones were separated by 1.35 m, the time shown on the timer was 389 ms. Calculate a value for the speed of sound in the laboratory.

d Is it necessary for the length of the connecting leads from the two microphones to the timer to be the same? Explain.

e What difference would it make to the results if the sound was produced at position B, rather than A?

f When the sound was made by hitting the laboratory bench, explain why the time shown on the timer was much less.

a **i** Hitting a retort stand with a small hammer.
ii A loud sound of short duration is needed to ensure that the microphones send clear signals to the timer.

b The times between events being observed and stopwatches being turned on, or off, is similar to the times being measured. This means that the percentage uncertainty is large. The reaction times are also unknown and variable.

c $c = \frac{\Delta s}{\Delta t}$

$c = \frac{1.35}{0.389} = 347\,\text{m}\,\text{s}^{-1}$

d No, because the speed of the electrical signals along the connecting leads (approximately the speed of electromagnetic waves in air) is *much* faster than the speed of sound being determined.

e A smaller time difference would be shown on the timer, resulting in an overestimate for the speed of sound (unless the change of geometry was taken into account in the calculation).

f A sound wave is now travelling through the bench at a speed higher than that of the sound wave in air. This can be used as a method for determining the speed of sound in some solids.

The speed of sound in different gases will depend on the average speed of the molecules within the gas. It varies with temperature because, on average, the molecules move faster at higher temperatures. At the same temperature, the speed will be higher in gases that have molecules of low mass and, therefore, high average molecular speeds (hydrogen for example). Apart from determining the speed of sound in different media, experiments can be designed to investigate how the speed of sound in a gas depends on the type of gas, the frequency of sound and the temperature.

Utilizations

Sound reflections in large rooms

Sound reflects well off hard surfaces like walls, whereas soft surfaces, like curtains, carpets, cushions and clothes, tend to absorb sound. A sound that reaches our ears may be quite different from the sound that was emitted from the source because of the many and various reflections it may have undergone. Because of this, singing in the shower will sound very different from singing outdoors or singing in a furnished room.

In a large room designed for listening to music (such as an auditorium, Figure 4.27), sounds travel a long way between reflections. Since it is the reflections that are responsible for most of the absorption of the sound waves, it will take a longer time for a particular sound to fall to a level that we cannot hear. This effect is called *reverberation*. The longer reverberation times of bigger rooms mean that a listener may still be able to hear reverberation from a previous sound at the same time as a new sound is received. That is, there will be some 'overlapping' of sounds.

■ **Figure 4.27** A large auditorium designed for effective transmission of sound.

Reflections of sounds off the walls, floor and ceiling are also an important factor when music is being produced in a recording studio, although some effects can be added or removed electronically after the original sound has been recorded.

1 Suggest why soft surfaces are good absorbers of sound.

2 Will the acoustic engineers, who are responsible for the sound quality in an auditorium, aim for a short or a long reverberation time? How can a reverberation time be changed?

3 Find out what anechoic chambers are, and what they are used for.

20 The speed of sound in air at a temperature of 20°C is $343\,m\,s^{-1}$.
  a Calculate how long it takes a sound to be reflected back to its source from a wall 50 m away.
  b Estimate the percentage uncertainty if this time was measured with a hand-held stopwatch.

21 a Describe an outdoor experiment using hand-held stopwatches to determine the speed of sound in air.
  b Give two reasons why the experiment cannot be carried out inside a building.

22 Explain why hydrogen molecules have a greater average speed than oxygen molecules at the same temperature.

23 Consider Worked example 4. Use the internet to find out the approximate temperature of the laboratory.

24 The speed of sound in sea water is $1540\,m\,s^{-1}$ and the speed of sound in pure water at the same temperature is $1490\,m\,s^{-1}$.
  a Suggest a reason for the difference in speeds.
  b Whales can communicate with each other over a distance of about 100 km. How long does it take sound to travel this distance?
  c It is believed that hundreds of years ago, whales could communicate over *much* greater distances. Suggest a reason for this.
  d A certain kind of whale 'sings' at a frequency of 50 Hz. What wavelength is this?

25 Ultrasound scans are widely used in hospitals to help to diagnose medical problems. Different wavelengths are needed to penetrate and examine different parts of the body and they produce images that vary in quality. If the speed of ultrasonic waves in muscle is $1580\,m\,s^{-1}$, what frequency is needed to produce a wavelength of 0.50 mm?

4.3 Wave characteristics *– all waves can be described by the same sets of mathematical ideas. Detailed knowledge of one area leads to the possibility of prediction in another*

■ Waves in two dimensions

We have discussed waves in a one-dimensional medium, like a rope, and must now develop our understanding further by considering progressive waves travelling in *two* dimensions. It will be helpful to think first of the waves on the surface of water, with which we are all familiar (Figure 4.28). Small waves (ripples) in tanks of shallow water are often used in laboratories to observe the behaviour of water waves. Figure 4.29 shows such a *ripple tank*.

To make waves on a flat water surface we need to disturb the water in some way, for example by repeatedly sticking a finger (or a stick) in and out of the water. Waves on the ocean are produced mainly by wind. Each part of the water affects the water around it and so the oscillation is passed on and away from the source, with a time delay. Waves on a ripple tank can also be made continuously by a motor that vibrates a small dipper or beam suspended on the water surface.

■ **Figure 4.28** Circular waves spreading out on a pond.

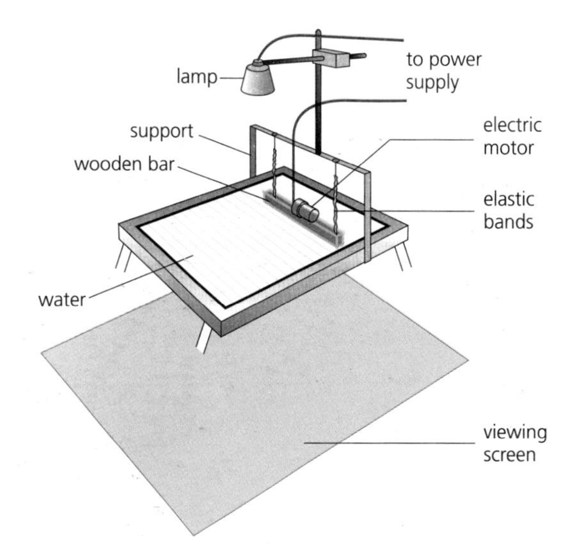

■ **Figure 4.29** A ripple tank is used to investigate wave behaviour.

Wavefronts and rays

When observing waves, we typically concentrate our attention on the crests of waves and, seen from above, the waves spread out on a smooth water surface in a circular manner with equal speeds in all directions.

Although a travelling wave is moving, we often want to represent the pattern of the waves (rather than their motion) on paper or a screen. If we draw a momentary position of the moving waves on paper, then the lines that we draw are called wavefronts. A wavefront is a line joining adjacent points moving in phase (for example, a line joining points where there are wave crests or where there are troughs). The distance between adjacent wavefronts is one *wavelength*.

Figure 4.30 shows circular wavefronts spreading from a point source. Lines pointing in the directions in which the wave energy is being transferred are called rays. A ray is always perpendicular to the wavefronts that it is representing. The movement of circular wavefronts is represented by *radial* rays spreading from a point source, as shown in Figure 4.30.

Figure 4.31 shows waves which could have been made by a straight beam oscillated on a water surface. Wavefronts like this, which are parallel to each other, are called plane wavefronts. The movement of plane wavefronts is represented by *parallel* rays.

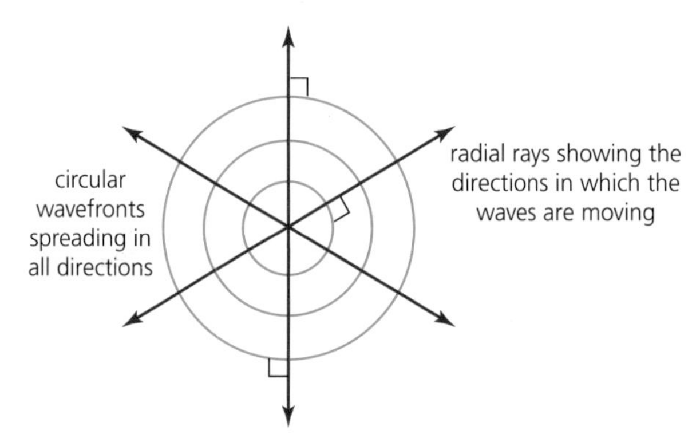

Figure 4.30 Circular wavefronts and radial rays spreading from a point source

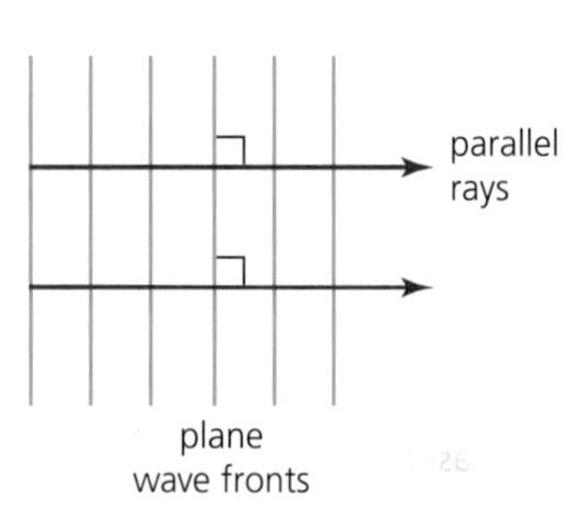

Figure 4.31 Plane wavefronts and parallel rays that are not spreading out

When circular wavefronts have travelled a long way from their source, they become very nearly parallel and will approximate to plane wavefronts. An important example of this is light waves coming from a long way away, which we usually consider to be plane (for example, wavefronts from the Sun).

We have described wavefronts in terms of two-dimensional waves on water surfaces, but similar ideas and terminology can be used to describe *all* waves in two or three dimensions. Computer simulations can be particularly useful for representing waves propagating in three dimensions.

ToK Link

Is physics 'simpler' than other areas of knowledge?

Wavefronts and rays are visualizations that help our understanding of reality, characteristic of modelling in the physical sciences. How does the methodology used in the natural sciences differ from the methodology used in the human sciences?

The study of all aspects of how people interact with each other and the world around them is called human science. It shares with the natural sciences the aims of objectivity and truth, but its **methodologies** usually need to be very different. Simple visual models (like wavefronts) and mathematical models (like the wave equation), which play such an important role in the study of physics, are not applied easily to the complexity of human existence. Furthermore, repeated, controlled experimentation is a cornerstone of much of **empirical** science (like physics), and this is often inappropriate in human science.

■ Amplitude and intensity

The amount of energy associated with waves is related to their amplitudes. Waves with higher amplitudes transfer more energy. When circular wavefronts travel away from a point source, their circumferences increase, so that the same amount of energy becomes spread over a longer length. As a result, the *amplitude* of the wave decreases (see Figures 4.28 and 4.32). This is also true for spherical wavefronts spreading in three dimensions, such as the spreading of light, sound or radio waves.

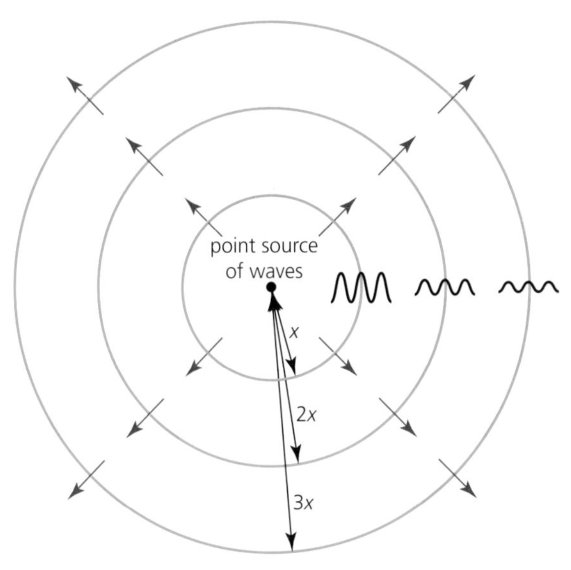

■ **Figure 4.32** Amplitude of circular wavefronts decreases with the distance from source.

At this point it becomes useful to introduce the concept of the **intensity** of waves:

Intensity, I, is defined as the wave power passing perpendicularly through a unit area.

$$\text{intensity, } I, = \frac{\text{power}}{\text{area}}$$

This equation is given in the *Physics data booklet* (for Topic 8).

The unit of intensity is W m^{-2}.

Clearly, wave intensity reduces with increasing distance from a point source and must be related to the amplitude of the waves.

In general the intensity of a wave is proportional to its amplitude squared:

$$I \propto A^2$$

This equation is given in the *Physics data booklet*. Note that here we are using A as a generalized symbol for amplitude (rather than x_0). It should not be confused with the use of A to represent area.

Worked examples

5 The wave intensity reaching a particular location on the Earth's surface on a sunny day might have an average of 600W m^{-2}.

a Calculate the average power incident perpendicularly on a solar panel, which has a surface area of 1.50m^2.

b What total energy would be incident on the panel in six hours?

c Discuss how the intensity incident on the panel might vary if it was moved up to a nearby mountain location.

a power, P, = intensity × area = 600 × 1.5 = 900 W

b energy = Pt = 900 × 6 × 3600 = 19.4 MJ

c The change of distance between the solar panel and the Sun is totally insignificant, but there will be less of the Earth's atmosphere above the solar panel and this will probably result in an increased incident intensity.

6 Ocean waves coming into a beach have an amplitude of 2.0 m and an average power of 5.2×10^4 W for every metre of their length.

a Estimate the power per metre which would be transferred if their amplitude reduced to 1.3 m.

b What amplitude would produce an overall power of 10 MW arriving at a beach of length 1.0 km?

c Where do the ocean waves get their energy from?

a $\frac{I}{A^2}$ = constant

$$\frac{(5.2 \times 10^4)}{2.0^2} = \frac{I}{1.3^2}$$

$I = 2.2 \times 10^4$ W

b 10 MW km^{-1} is equal to $\frac{1.0 \times 10^7}{10^3} = 1.0 \times 10^4$ W m^{-1}

$\frac{I}{A^2}$ = constant

$\frac{5.2 \times 10^4}{2.0^2} = \frac{1.0 \times 10^4}{A^2}$

$A = 0.88$ m

c From the wind, which in turn gets it from the Sun.

An enormous amount of energy is transferred by ocean waves. The technology to transfer this wave energy into useful electricity has proved to be difficult and expensive, although much research and development is still being carried out. Figure 4.33 shows a large buoy that floats on the water and generates electricity as waves pass. The energy is then transferred to the land using an underwater cable. This kind of technology will be used in the world's largest wave energy generator (62.5 MW), which is being planned for the ocean off the coast of Victoria, Australia.

■ **Figure 4.33** A wave energy generator

Inverse square law

For waves spreading out evenly in three dimensions from a point source without any loss of energy, their intensity, I, is inversely proportional to the distance from the source, x, squared:

$$I \propto \frac{1}{x^2}$$

This equation is given in the *Physics data booklet*.

Alternatively, Ix^2 = constant for waves spreading from a given source.

Inverse square law relationships play an important part in physics. As well as waves, a force field (for example, gravity) that spreads out evenly in all directions from a point obeys an inverse square law.

Consider light radiating from a point source without being absorbed (see Figure 4.34). The further from the source, the larger the area over which the light is spread and the fainter it becomes. The amount of light falling on each unit of area decreases with the square of the distance. This is because at twice the distance away from the source, the light has to spread to cover four times the area, and at three times the distance it has to spread to cover nine times the area.

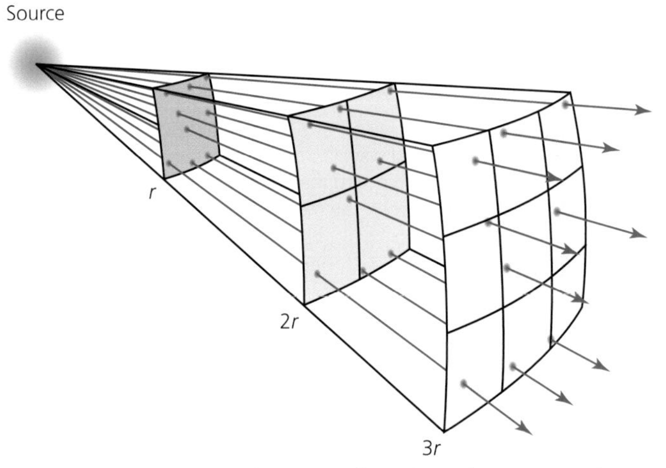

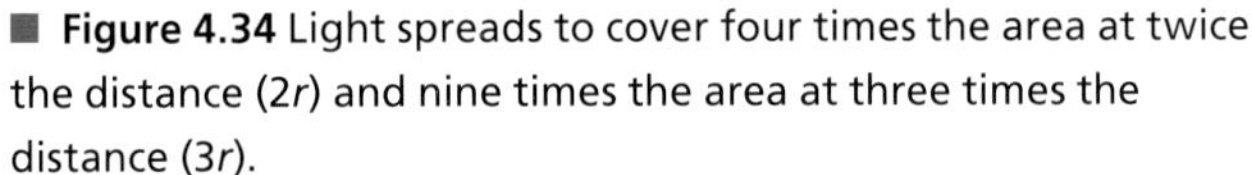

■ **Figure 4.34** Light spreads to cover four times the area at twice the distance (2*r*) and nine times the area at three times the distance (3*r*).

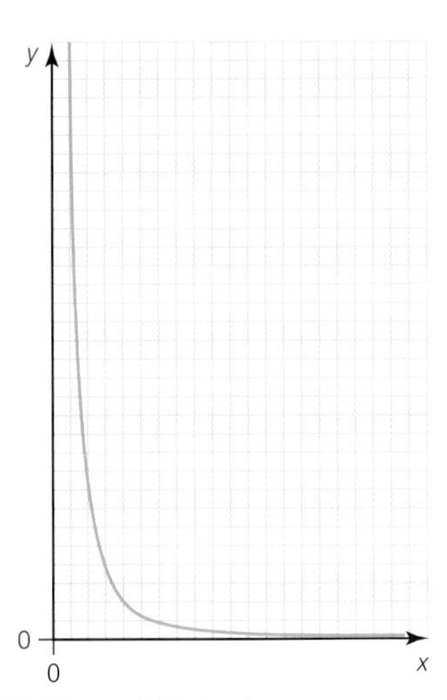

■ **Figure 4.35** An inverse square relationship

■ **Table 4.2** Are y and x^2 inversely proportional?

x	y
1.34	8.8
0.96	17
0.81	24
0.70	32
0.64	38
0.59	45
0.55	52
0.52	59
0.49	66

Consider the data for x and y shown in Table 4.2. If $y \propto 1/x^2$, then x^2y = constant.

It is important to know how an inverse square relationship appears on a graph. If a graph of y against x is drawn for the data in Table 4.2, its shape should be similar to Figure 4.35. Note that the line does not cross the axes. However, it may not be easy to be sure from a graph like this whether there is an inverse square relationship between two variables, so a more detailed check is needed.

Worked example

7 A family is worried because a telephone company is planning to build a transmitting aerial for mobile phones (a *base station*) 100 m from their home. They have read that it is recommended that the maximum intensity they should receive is 4.5 W m^{-2}.

a If the intensity 1 m from an aerial on the base station is 3.0 W m^{-2}, calculate an approximate intensity near their home.

b Why is this answer only an estimate?

c Compare your answer to **a** with the intensity received from a mobile phone held 5 cm from the head, assuming that the power 1 cm from the phone's transmitting aerial (inside the phone) is 50 mW.

a Ix^2 = constant

$3.0 \times 1^2 = I \times 100^2$

$I = 3.0 \times 10^{-4}$ W m^{-2}

b The aerials are designed to send the signals in certain directions, rather than equally in all directions (as is assumed when using the inverse square law). The actual intensity received will also depend on what is between the house and the aerial.

c $50 \times 1^2 = I \times 5^2$

I = 2mW, which is bigger than the answer to **a**.

26 A transverse wave is propagated along a rope, as shown in Figure 4.14.
  a If the hand moves with twice the amplitude (at the same frequency), what happens to the power given to the waves?
  b Explain why the amplitude of the wave decreases as it moves away from its source.

27 A 2 kW infrared grill radiates energy over an effective area of 500 cm^2.
  a What is the intensity of the infrared waves?
  b What total power is received by a piece of meat of area 92 cm^2 under the grill?
  c How many minutes will it take to cook the meat if it requires a total of 1.4×10^5 J?

28 A wave of amplitude 12 cm has an intensity of 54 W m^{-2}.
  a What is the intensity of the wave after it has spread out so that its amplitude has reduced to 7.0 cm?
  b What will the amplitude be when the intensity has decreased to 10 W m^{-2}?

29 It is considered a health risk to expose our ears to sounds of intensity 10 mW m^{-2} or greater for more than a few minutes.
  a What total power will be received by an eardrum of area 0.50 cm^2 at this intensity?
  b If the sound intensity 2.0 m in front of a loudspeaker at a rock concert was 12 W m^{-2}, how far away would you have to be to receive 10 mW m^{-2} or less?

30 A man is standing 1.0 m from a source of gamma radiation.
  a How far away would he have to move in order for the radiation intensity he received to be reduced by 99%?
  b What assumption did you make in answering a?

31 Do some research into the latest developments in the production of electricity from ocean waves.

32 A power station on the sea shore transfers the wave energy of water to electricity at a rate of 50 kW on a day when the average wave amplitude is 1.2 m.
  a Estimate the output power if the wave amplitude doubles.
  b What assumption did you make?
  c What amplitude of waves might be expected to produce an output of 150 kW?
  d What power could be transferred from waves of amplitude 1.0 m?

33 Use a spreadsheet to calculate values for x^2, $1/x^2$ and x^2y for the data in Table 4.2. Are x^2 and y inversely proportional?

34 Another way of checking data to see if there is a certain relationship is to draw a suitable graph to see if it produces a straight line. Use the values you calculated in question 33 to draw a graph of y against $1/x^2$. Does it produce a straight line through the origin (which would confirm inverse proportionality)?

■ Superposition

We turn our attention now to what happens when wavefronts from different places come together. In general, we can predict what will happen when waves meet by using the principle of **superposition**.

The principle of superposition states that at any moment the overall displacement at any point will be the vector sum of all the individual wave displacements.

This principle is illustrated by Figure 4.36. If wave A and wave B meet at a point, the resulting disturbance at any time is determined by adding the two individual displacements at that moment.

In this example waves A and B have different frequencies, but in the rest of this section we will only deal with the combination of two waves of the same frequency.

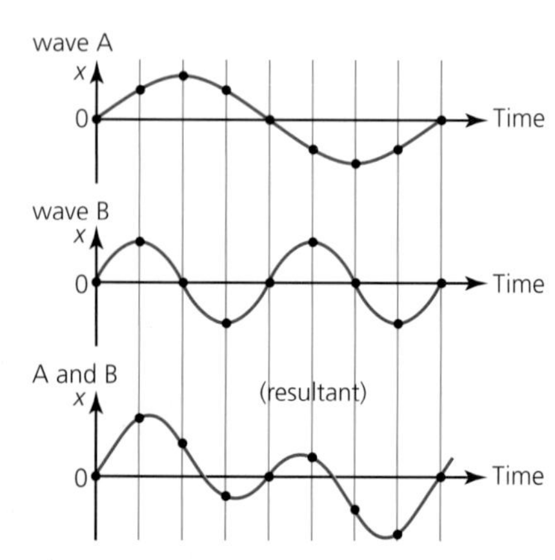

■ **Figure 4.36** Adding wave displacements using the principle of superposition.

Adding two (or more) waves together to find their resultant is an easy task for a computer. Computer modelling of superposition effects can be particularly instructive, because it is easy to adjust the amplitudes and frequencies of the waves concerned and then observe the consequences. Laboratory observation of the momentary superposition of waves (travelling in opposite directions) passing through each other is not so easy!

We will use the principle of superposition in Section 5 of this chapter to explain stationary interference patterns.

35 a Sketch a displacement–time graph for 1 s for a sinusoidal oscillation of amplitude 4 cm and frequency 2 Hz.
b On the same axes draw a graph representing an oscillation of amplitude 2 cm and frequency 4 Hz.
c Use the principle of superposition to draw a sketch of the resultant of these two waves.

36 Figure 4.37 shows two idealized square pulses moving towards each other. Draw the resultant waveform after
a 6 s
b 7 s
c 8.5 s
d 10 s.

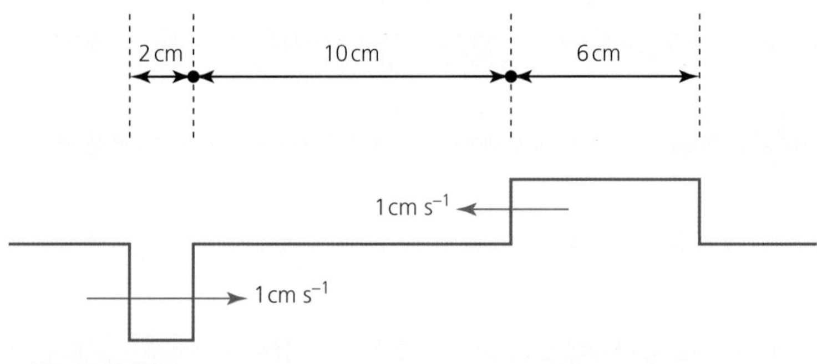

■ **Figure 4.37**

37 Two sinusoidal waves (A and B) from different sources have the same frequency and pass through a certain point, P, with the same amplitude at the same time.
a Sketch the two wave forms on a displacement–time graph assuming that they arrive in phase, and then draw the resultant waveform.
b Repeat for two waves that arrive at P exactly out of phase.

■ Polarization

Consider sending transverse waves along a rope. If your hand only oscillates vertically, the rope will only oscillate vertically. The wave can be described as **plane polarized** because it is only oscillating in one plane (vertical), as shown in Figure 4.38. If your hand only oscillates horizontally, it will produce a wave polarized in the horizontal plane.

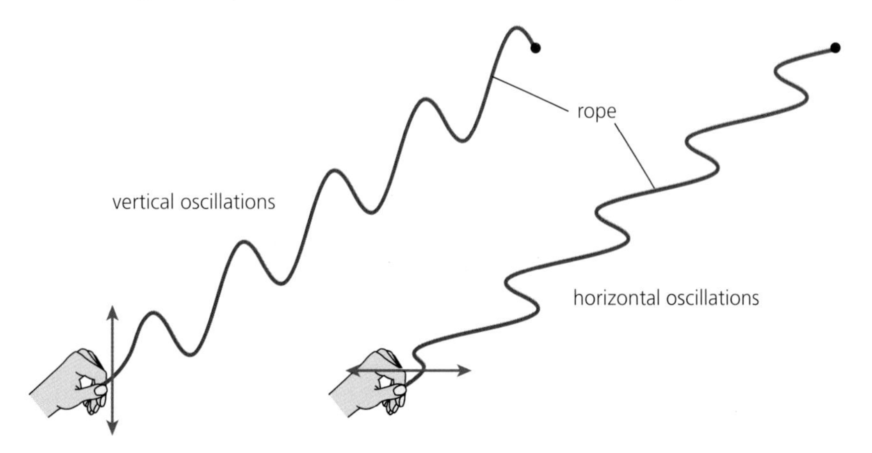

■ **Figure 4.38** Transverse waves on a rope

A transverse wave is (plane) polarized if all the oscillations transferring the wave's energy are in the same plane (called the plane of polarization).

The oscillations of a polarized wave must be perpendicular to the direction of wave travel, so it is impossible for longitudinal waves, like sound, to be polarized.

Polarized light and other electromagnetic waves

Consider again Figure 4.23 which shows the nature of electromagnetic waves. In that example, the waves are moving to the right, the electric field oscillations are in the horizontal plane and the magnetic field oscillations are in the vertical plane, but the oscillations of unpolarized waves could be in any plane perpendicular to the direction of wave travel (as long as the electric fields and magnetic fields are perpendicular to each other).

Electromagnetic waves, including light, are mostly emitted during random, unpredictable processes, so we would expect them to oscillate in random directions and not be polarized (see Figure 4.39a).

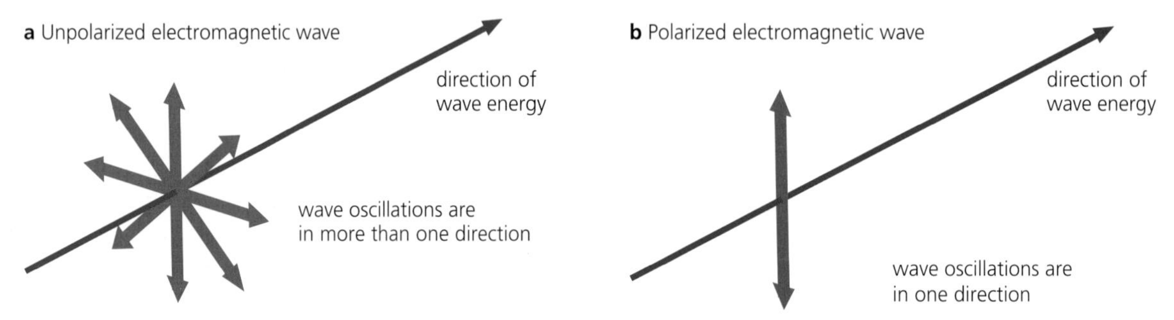

■ **Figure 4.39 a** Unpolarized electromagnetic radiation; **b** polarized electromagnetic radiation

Electromagnetic waves, such as light, are said to be (plane) polarized if all the electric field oscillations (or the magnetic field oscillations) are in one plane, as shown by Figure 4.39b.

Electromagnetic waves produced and transmitted by currents oscillating in aerials (radio waves and microwaves) will be polarized, with their electric field oscillations parallel to the transmitting aerial.

Figure 4.40 shows the transmission and reception of microwaves. In Figure 4.40a a strong signal is received because the transmitting and receiving aerials are aligned, but in Figure 4.40b no signal is received because the receiving aerial has been rotated through 90°.

■ **Figure 4.40** The receiver must be aligned in the same plane as the transmitter to detect microwaves

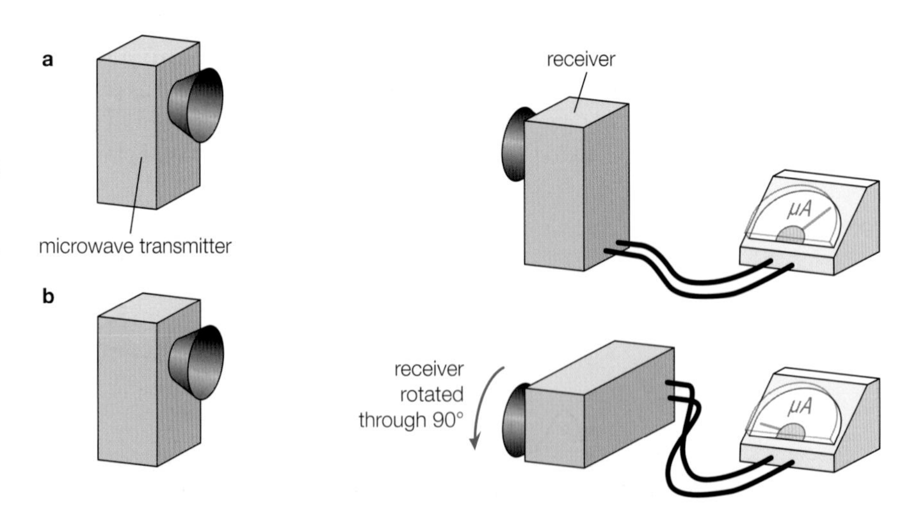

Polarization by absorption

Unpolarized light can be converted into polarized light by passing it through a special filter, called a polarizer, which absorbs oscillations in all planes except one.

Polarizing filters for light are made of long-chain molecules mostly aligned in one direction. Components of the electric field parallel to the long molecules are absorbed; components of the electric field perpendicular to the molecules are transmitted. Because of

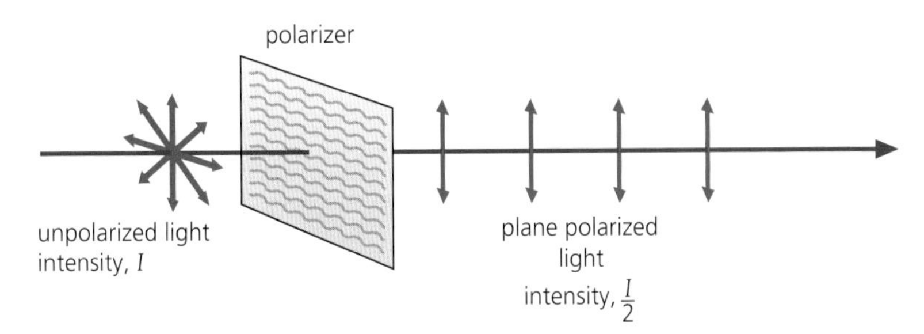

■ **Figure 4.41** Polarizing light using a polarizer (polarizing filter)

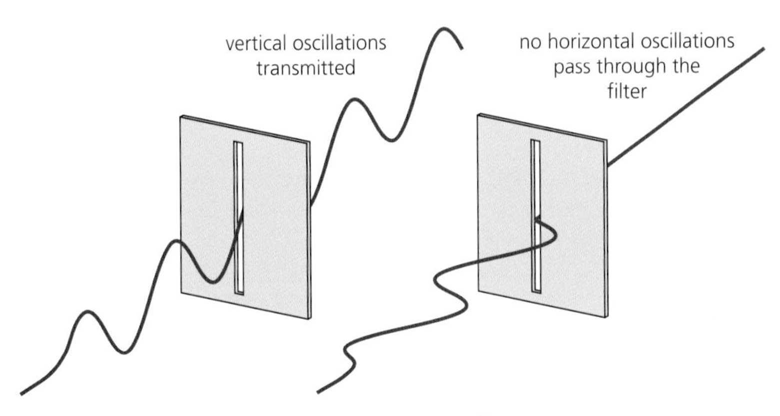

■ **Figure 4.42** A vertical slit acts as a polarizing filter.

this, we would expect the transmitted intensity to be about half the incident intensity, as shown in Figure 4.41.

In order to understand this, it may be helpful to consider how transverse waves made by oscillating a rope (as shown in Figure 4.38) would behave if they had to pass through a vertical slit (see Figure 4.42). Waves oscillating parallel to the slit would pass through, but others would be blocked. The slit acts as a polarizer.

What happens if polarized light is then incident on a second polarizing filter? This depends on how the filters are aligned. If the second filter (sometimes called an analyser) transmits waves in the same plane as the first (the polarizer), the waves will pass through unaffected, apart from a possible small decrease in intensity. The filters are said to be in parallel, as shown in Figure 4.43b. Figure 4.43a shows the situation in which the analyser only allows waves through in a plane that is at right angles to the plane transmitted by the polarizer. The filters are said to be 'crossed' and no light will be transmitted by the analyser. (See also Figure 4.44.)

■ **Figure 4.43** Polarizer and analyser: **a** crossed; **b** in parallel

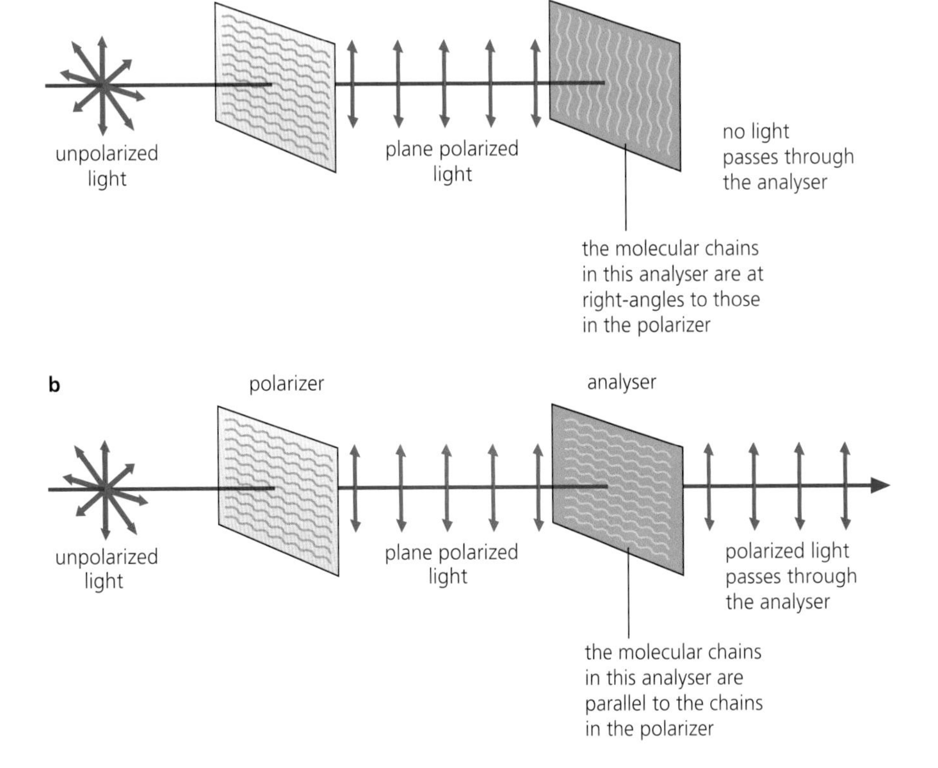

■ **Figure 4.44** Crossed polarising filters

The second filter is called an **analyser** because it can be rotated to analyse the light and determine if it is polarized and, if so, in which direction.

When you rotate a polarizing filter in front of your eye, (for example, as you look around at the light reflected from various objects) if the intensity changes then the light must be at least partly polarized. The most common type of transparent plastic used for polarizers and analysers is called **Polaroid®**.

Malus's law

Figure 4.45 represents a polarized light wave travelling through a polarizing filter (analyser) and perpendicularly out of the page, so that the electric field oscillations are in the plane of the paper. There is an angle, θ, between the oscillations and the plane in which the analyser will transmit all of the waves.

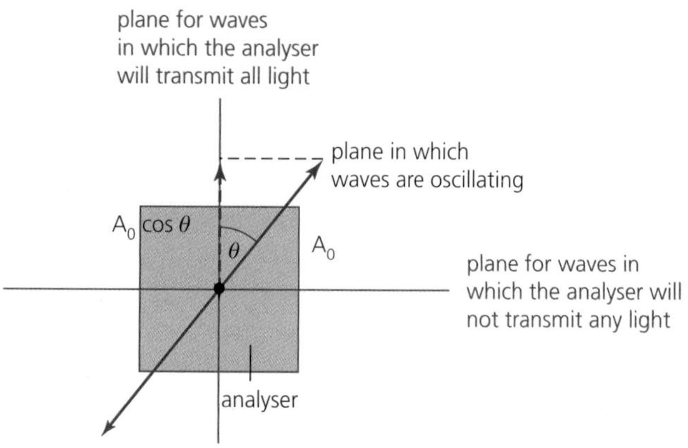

If the amplitude of the oscillations incident on the analyser is A_0, then the component in the direction in which waves can be transmitted equals $A_0 \cos\theta$.

We know, from earlier in this chapter, that the intensity of waves is proportional to amplitude squared, $I \propto A^2$, so the transmitted intensity is represented by the equation:

$$I = I_0 \cos^2\theta$$

This equation (known as Malus's law) is given in the *Physics data booklet*.

plane for waves in which the analyser will transmit all light
plane in which waves are oscillating
$A_0 \cos\theta$
θ
A_0
plane for waves in which the analyser will not transmit any light
analyser

■ **Figure 4.45** Angle between oscillations and the polarizer

Worked example

8 **a** If vertically polarized light falls on a polarizing filter (analyser) positioned so that its transmission direction (axis) is at 30° to the vertical, what percentage of light passes through the analyser?
b Repeat the calculation for an angle of 60°.
c What angle would allow 50% of the light to pass?
d Sketch a graph to show how the transmitted intensity would vary if the analyser was rotated through 360°.

a $I = I_0 \cos^2\theta$

$$\frac{I}{I_0} = \cos^2 30°$$

$$\frac{I}{I_0} = 0.75 \text{ or } 75\%$$

b $\frac{I}{I_0} = \cos^2 60° = 0.25$ or 25%

c $\frac{I}{I_0} = 0.50 = \cos^2 \theta$

$\cos \theta = \sqrt{0.50} = 0.71$

$\theta = 45°$

d See Figure 4.46. This graph shows the variation with angle of intensity transmitted through a polarizing filter.

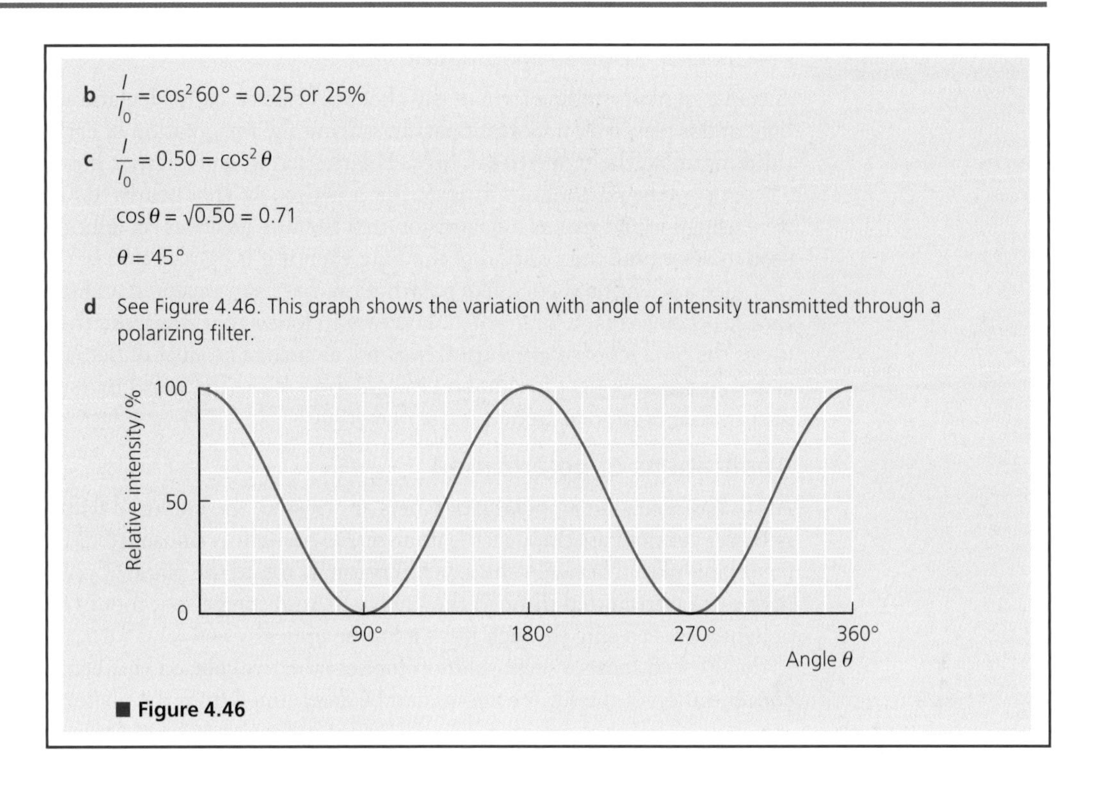

■ **Figure 4.46**

Polarization by reflection

When (unpolarized) light reflects off an insulator, the waves may become polarized and then the plane of polarization will be parallel to the reflecting surface.

The most common examples of polarization by reflection are the reflection of light from water and from glass. Such reflections are usually unwanted and the reflected light (sometimes called *glare*) entering the eyes can be reduced by wearing polarizing (*Polaroid®*) *sunglasses*, which also reduce the intensity of unpolarized light by half. Figure 4.47 shows an example. The fish under the water can be seen clearly when Polaroid® sunglasses are used. The sunglasses greatly reduce the amount of light reflected off the water's surface entering the eye. But the reduction is not the same for all viewing angles because the amount of polarization depends on the angle of incidence. Photographers may place a rotatable polarizing filter over the lens of their camera to reduce the intensity of reflected light.

■ **Figure 4.47** The same scene with and without Polaroid®

38 How could you check quickly whether sunglasses were made with Polaroid® or not?

39 Plane polarized light passes through an analyser that has its transmission axis at 75° to the plane of polarization. What percentage of the incident light emerges?

40 When unpolarized light was passed through two polarizing filters, only 20% of the incident light emerged. What was the angle between the transmission axes of the two filters?

41 Suggest why the blue light from the sky on a clear, sunny day is partly polarized.

42 Use the internet to find out about polarizing microscopes and their uses.

Nature of Science

Polarization inspired imagination

A transparent crystalline form of the chemical calcite, called Iceland spar, can partially polarize light and it played an important part in stirring the imagination of early physicists who were thinking about the properties of light. Polarization in Iceland spar was first described nearly 350 years ago by Bartholinus in 1669. But it is thought that nearly 1000 years earlier, the Vikings were using Iceland spar as a navigation tool because its effect on light rays from the Sun can be used to determine the position of the Sun, even if it is behind clouds.

Once a scientific theory, like polarization, has been accepted and used for a long period of time it becomes taken for granted. It is easy to forget that when the theory was first introduced, it was the product of original thinking and, as such, a product of human imagination, perhaps genius. It requires great insight and imagination to see the world in a new way and pioneering scientists deserve great credit for their creativity.

Applications of polarization

As we have seen, if two polarizing filters are crossed, the polarized light that passes through the polarizer cannot pass through the analyser, so no light is transmitted. However, if a transparent material is placed between the two filters it may rotate the plane of polarization, allowing some light to be transmitted through the analyser. A substance that rotates the plane of polarization of light waves passing through it is said to be **optically active**.

Figure 4.48 shows a sugar solution (optically active) placed between polarising filters. The concentration of the sugar solution can be determined from the magnitude of the rotation.

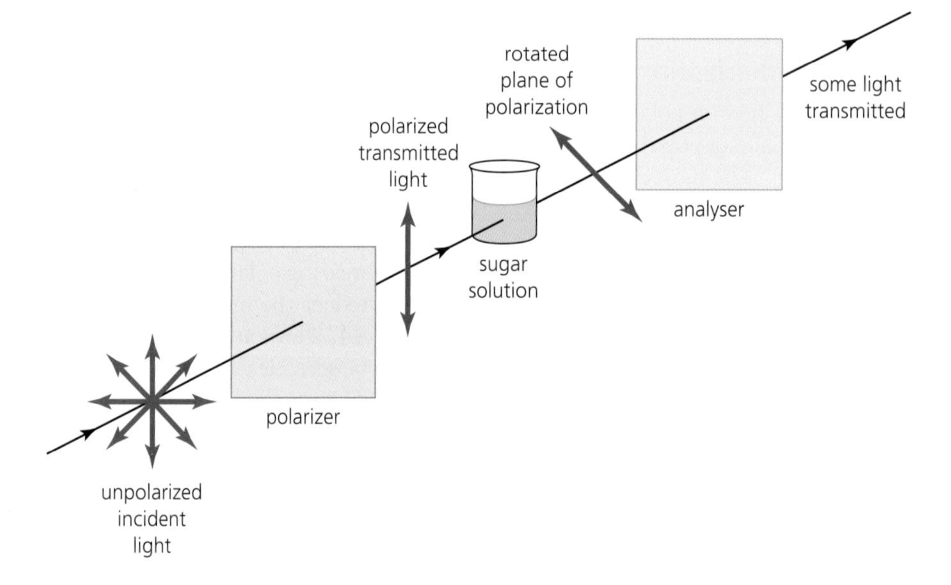

■ **Figure 4.48** Rotating the plane of polarization with sugar solution

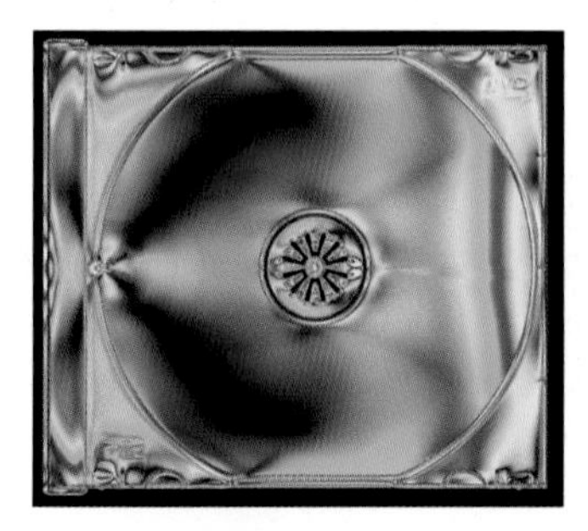

■ **Figure 4.49** Stress concentration in a DVD case seen with polarized light

Some plastics and glasses become optically active and rotate the plane of polarization of light when they are stressed (see Figure 4.49). This can be useful for engineers who can analyse possible concentrations of stress in a model structure before it is made.

Liquid-crystal displays (LCDs)

A **liquid crystal** is a state of matter that has physical properties between those of a liquid and a solid (crystal). Most interestingly, the ability of certain kinds of liquid crystal to rotate the plane of polarization of light can be changed by applying a small potential difference across them, so that their molecules twist in the electric field. Figure 4.50 represents a simplified arrangement. If there is no potential difference (p.d.) across the liquid crystal, no light is transmitted out of the analyser.

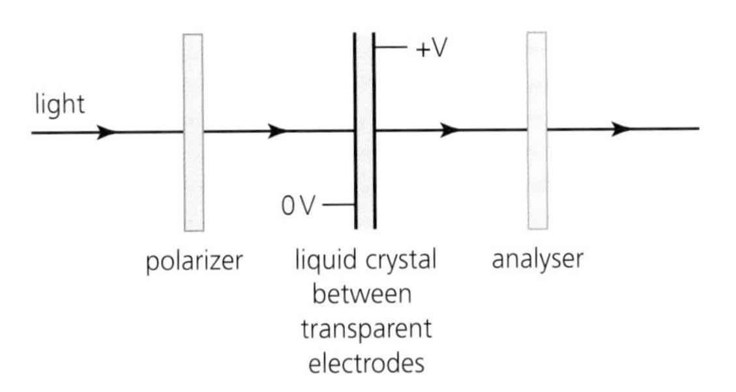

■ **Figure 4.50** Arrangement of parts in a liquid-crystal display

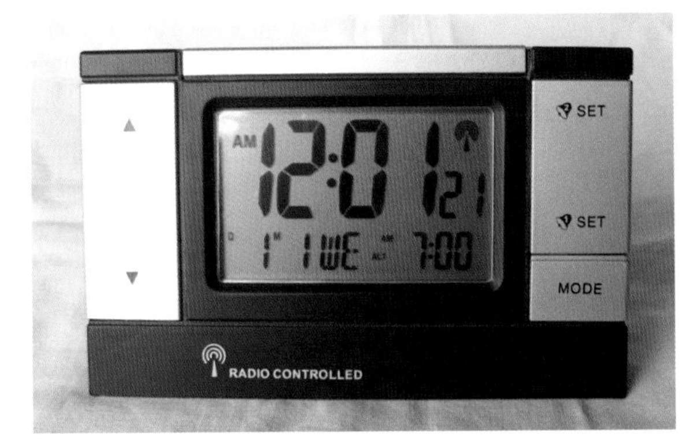

■ **Figure 4.51** A seven-segment liquid-crystal display

When a p.d. is applied to the liquid crystal, its molecules change orientation to align with the electric field and the plane of polarization rotates so that some, or all, of the light is now transmitted. The amount of rotation of the plane of polarization and the amount of light transmitted depend on the size of the p.d.

In simple displays (for example those used on many calculators, digital clocks and watches), light entering through the front of the display passes through the liquid crystals and is then reflected back to the viewer. Each segment of the display will then appear dark or light, depending on whether or not a p.d. has been applied to the liquid crystal (see Figure 4.51).

Using the same principles, large numbers of liquid crystals are used to make the tiny picture elements (pixels) in many computer and mobile phone displays, and televisions. Colours are created by using filters and the light is provided by a fluorescent lamp or LEDs behind the display.

43 Unpolarized light of intensity 48 mW is incident on two polarizing filters that have an angle of 20° between their transmission axes.
 a What is the intensity entering the second filter?
 b What intensity emerges from the second filter?
 c If a sugar solution placed between the filters rotates the plane of polarization by 28°, what intensity then emerges from the second filter?

44 Use the internet to research and compare the main advantages and disadvantages of using liquid crystal and LED displays.

Utilization

Polarization and 3-D cinema

Our eyes and brain see objects in three dimensions (3-D) because when our two eyes look at the same object, they each receive a slightly different image. This happens because our eyes are not located at exactly the same place. This is known as *stereoscopic vision*. Our brain merges the two images to give the impression of three dimensions or 'depth'. But when we look at a two-dimensional image in a book or on a screen, both eyes receive essentially the same image.

If we want to create a 3-D image from a flat screen, we need to provide a different image for each eye and the use of polarized light makes this possible. (Some earlier, and less effective, systems used different coloured filters.) In the simplest modern systems, one camera is used to take images that are projected in the cinema as vertically polarized waves. At the same time a second camera, located close by, takes images that are later projected as horizontally polarized waves. Sometimes a second image can be generated by a computer program (rather than by a second camera) to give a 3-D effect.

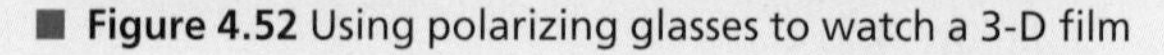

■ **Figure 4.52** Using polarizing glasses to watch a 3-D film

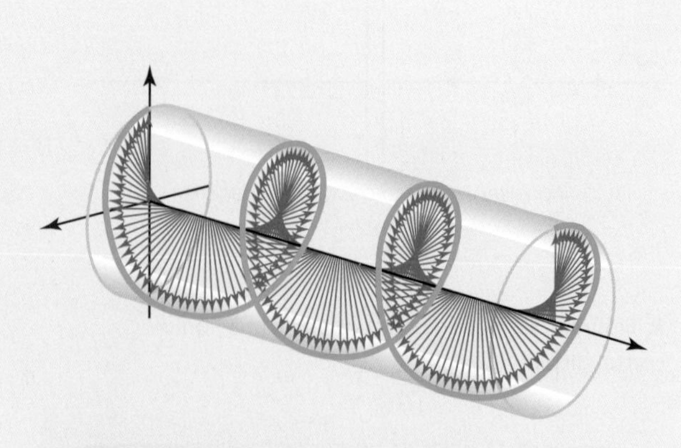

■ **Figure 4.53** Circularly polarized waves

To make sure that each eye receives a different set of images, the viewer wears polarizing glasses, allowing the vertically polarized light into one eye and the horizontally polarized light into the other. One problem with using plane polarized waves is that viewers need to keep their heads level, but this can be overcome by using circularly polarized waves in which the direction of the electric field oscillations continually rotates in circles, as in Figure 4.53. A single projector sends images alternating between clockwise and anti-clockwise polarization to the screen.

1 Use the internet to investigate the latest developments in 3-D television techniques.

4.4 Wave behaviour – *waves interact with media and each other in a number of ways that can be unexpected and useful*

Changes in the speed of waves, or obstacles placed in their path, will change their shape and/or direction of motion, with very important consequences. These effects are called **reflection**, **refraction** and **diffraction**. **Interference** happens when waves combine. We will discuss each of these four wave properties in this section.

■ Reflection and refraction

Reflection

When a wave meets a *boundary* between two different media, usually some or all of the waves will be reflected back. Under certain circumstances some waves may pass into or through the second medium (we would then say that there was some **transmission** of the waves). An obvious example would be light waves passing through **transparent** materials, for example liquids or various kinds of glass. Figure 4.54 shows both reflection and transmission at the same window.

■ **Figure 4.54** Light reflected off and passing through a window

We will refer again to our two wave models (waves on springs and waves on water) to develop our understanding of reflection.

Reflection in one dimension

Figure 4.55 shows a single pulse on a spring or string travelling towards a fixed boundary where it is totally reflected, with no loss of energy. Note that the reflected wave is inverted. We say that it has undergone a phase change of half a wavelength.

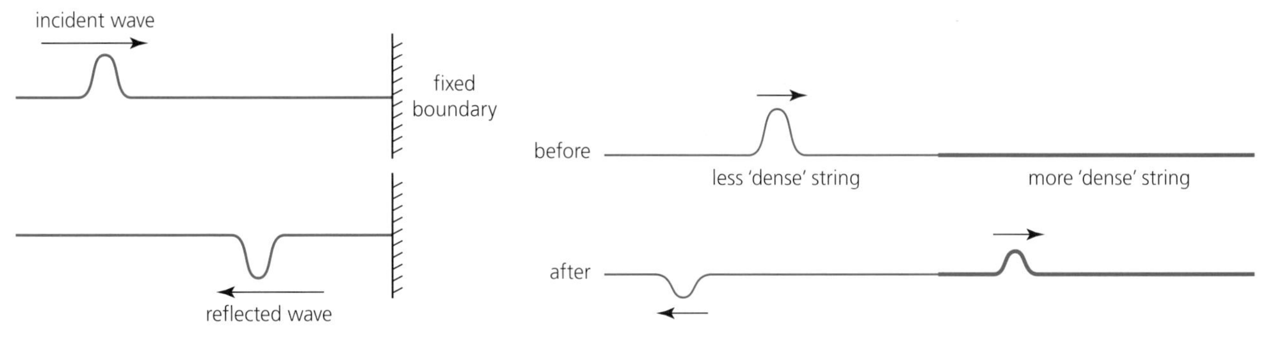

■ **Figure 4.55** Reflection of a pulse off a fixed boundary ■ **Figure 4.56** A pulse travelling into a 'denser' medium

Pulses will change speed if they cross a boundary between springs/strings of different mass per unit length. In Figure 4.56 a transverse pulse is transmitted from a less 'dense' string to a more 'dense' string (higher mass per unit length), where it travels more slowly.

Note that there are now two pulses and both have reduced amplitude because the energy has been split between the two. The transmitted pulse will now have a slower speed but its phase has not changed. The reflected pulse returns with the same speed but has undergone a phase change of half a wavelength.

In Figure 4.57 the transverse pulse is passing from a more 'dense' string to a less 'dense' string. This time there is no phase change for the reflected pulse.

■ **Figure 4.57** A pulse travelling into a less 'dense' medium

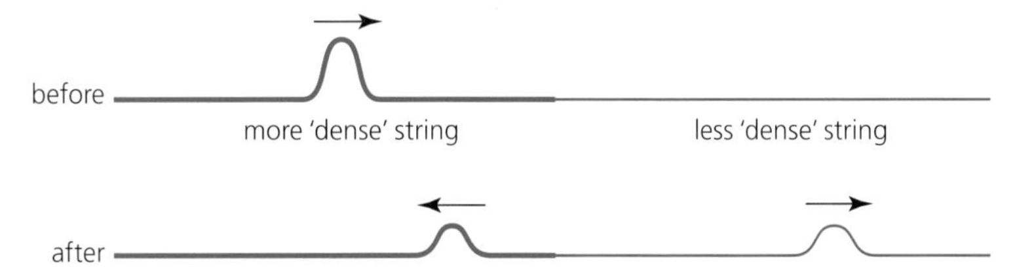

Longitudinal waves and pulses behave in a similar way to transverse waves.

Reflection in two dimensions

If plane waves reflect off a straight (plane) boundary between two media, they will reflect so that the reflected waves and the incoming (**incident**) waves make equal angles with the boundary (see Figure 4.58).

The **angle of incidence**, i, is equal to the **angle of reflection**, r.

■ **Figure 4.58** Reflection of plane waves from a straight boundary

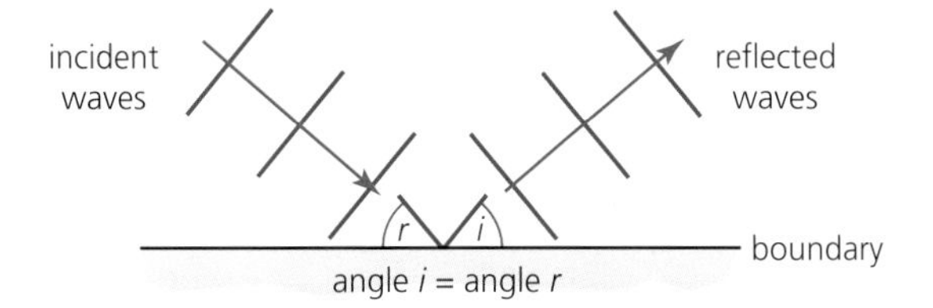

When discussing the reflection of light, it is more common to represent reflection by a **ray diagram**, such as shown in Figure 4.59.

■ **Figure 4.59** Reflection of rays from a reflecting surface

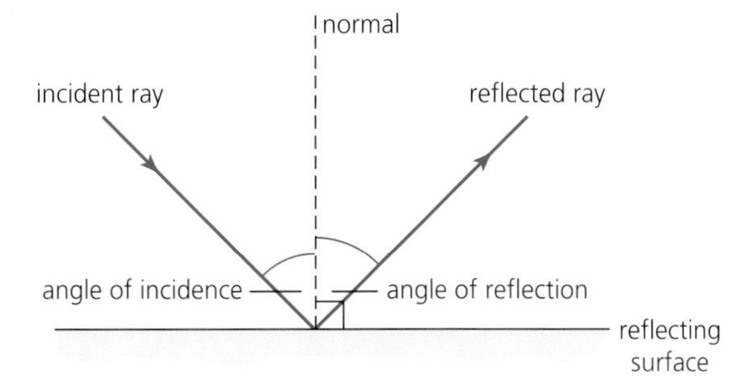

As before, the angle of incidence, i, is equal to the angle of reflection, r. But in this diagram the angles are measured between the rays and the '**normal**'. The normal is a construction line we draw on diagrams that is perpendicular to the reflecting surface.

In practice, as with the one-dimensional waves, some wave energy may also be transmitted as well as reflected. The transmitted waves may change direction. This effect is called *refraction* and it is discussed next.

Refraction

Typically, waves will change speed when they travel into a different medium. The speed of water waves decreases as they pass into shallower water. Such changes of speed may result in a change of direction of the wave.

Figure 4.60 shows plane wavefronts arriving at a medium in which they travel slower. The wavefronts are parallel to the boundary and the ray representing the wave motion is perpendicular to the boundary.

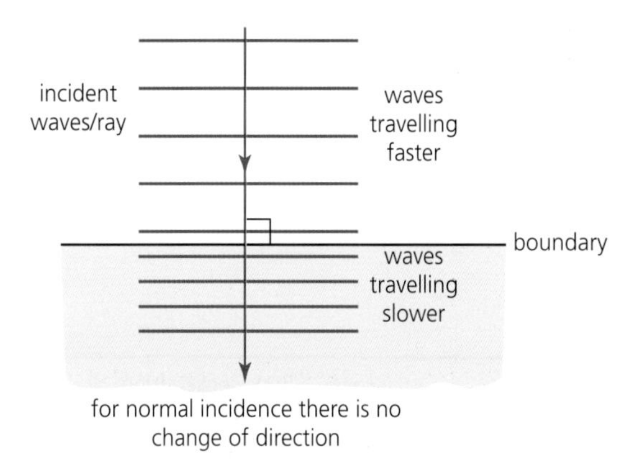

■ **Figure 4.60** Waves slowing down as they enter a different medium

In this case there is no change of direction but because the waves are travelling slower, their wavelength decreases, although their frequency is unchanged (consider $c = f\lambda$). Now consider what happens if the wavefronts are not parallel to the boundary, as in Figure 4.61.

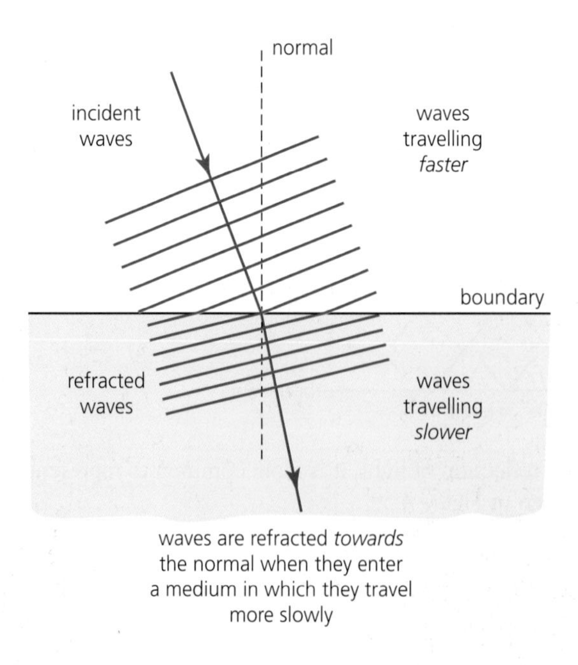

■ **Figure 4.61** Waves refracting as they enter a denser medium

Different parts of the same wavefront reach the boundary at different times and, consequently, they change speed at different times. There is a resulting change of direction that we call *refraction*. The bigger the change of speed, the greater the change of direction.

When waves enter a medium in which they travel more slowly, they are refracted *towards the normal*. Conversely, when waves enter a medium in which they travel faster, they are refracted *away from the normal*. This is shown in Figure 4.62, but note that this is similar to Figure 4.61 with the waves travelling in the opposite direction.

■ **Figure 4.62** Waves refracting as they enter a less dense medium

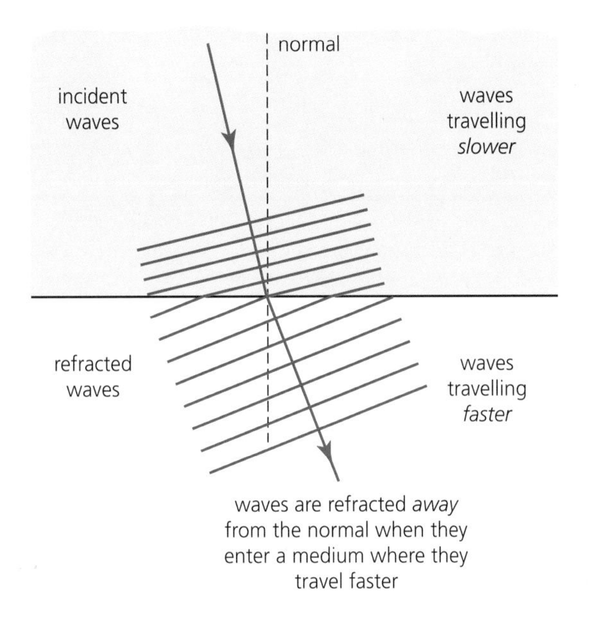

The refraction of light is a familiar topic in the study of physics, especially in optics work on lenses and prisms, but all waves tend to refract when their speed changes. Often this is a sudden change at a boundary between media, but it can also be a gradual change, for example if the density of a medium changes gradually.

The photograph in Figure 4.63 shows the gradual refraction of water waves approaching a beach. It is possible to learn about how the water depth is changing by observing the refraction of waves in shallow water. Waves travel slower in shallower water, so the wave crests get closer together (shorter wavelength) because the frequency does not change.

The focusing of a glass lens occurs because the light waves slow down and are refracted in a systematic way by the smooth curvature of the glass (see Figure 4.64). By using the refraction of light by more than one lens, we are able to extend the range of human sight to the very small (microscopes) and the very distant (telescopes). The *reflection* of light by curved mirrors is also able to focus light and is used in some telescopes and microscopes.

■ **Figure 4.63** Ocean waves refracting (and diffracting) as they approach a beach

■ **Figure 4.64** A lens uses refraction to focus light

In Figure 4.65 light waves are refracted in a more disorganized way by the irregular changes in the density of the hot air moving above the runway and behind the plane.

■ **Figure 4.65** Diffuse refraction of light by hot gases

Snell's law, critical angle and total internal reflection

Figure 4.66 shows a single ray of light representing the direction of waves being refracted when entering a medium where they travel slower (Figure 4.66a) and a medium where they travel faster (Figure 4.66b).

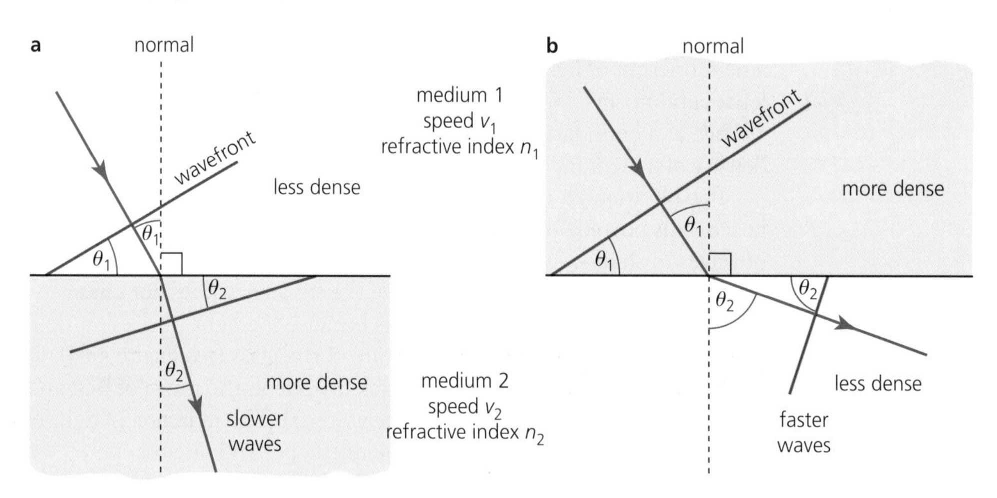

■ **Figure 4.66** Light rays being refracted **a** towards the normal and **b** away from the normal

For a boundary between any two given media, it was discovered experimentally that any angle of incidence, θ_1, was connected to the corresponding angle of refraction, θ_2, by the equation:

$$\frac{\sin \theta_1}{\sin \theta_2} = \text{constant, } n$$

Using trigonometry, it can be shown that this ratio is a constant because the ratio of the wave speeds in the two media (v_1/v_2) is constant.

$$\frac{\sin \theta_1}{\sin \theta_2} = \text{constant} = \frac{v_1}{v_2}$$

For light passing *from air* (or, more precisely, vacuum) into a particular medium, the constant is known as the **refractive index**, n, of that medium.

$$n_{\text{medium}} = \frac{\sin \theta_{\text{air}}}{\sin \theta_{\text{medium}}} = \frac{v_{\text{air}}}{v_{\text{medium}}}$$

Because it is a ratio, refractive index has no unit.

Because the speed of light in air is (almost) the same as the speed in a vacuum, the refractive index of air is 1.0.

For example, the speed of light in air (or vacuum) is $3.0 \times 10^8\,\mathrm{m\,s^{-1}}$ and in a certain kind of glass it might be $2.0 \times 10^8\,\mathrm{m\,s^{-1}}$, so that the refractive index of that kind of glass would be 1.5. With this information we can then calculate the angle of refraction for any given angle of incidence.

Consider again Figure 4.66a, or Figure 4.66b, the refractive index of medium 1, $n_1 = \frac{v_{air}}{v_1}$ and the refractive index of medium 2, $n_2 = \frac{v_{air}}{v_2}$, so that:

$$\frac{n_1}{n_2} = \frac{\sin\theta_2}{\sin\theta_1} = \frac{v_2}{v_1}$$

This is known as **Snell's law** and it is given in the *Physics data booklet.*

The refractive index of a transparent solid can be determined experimentally by locating the paths of light rays passing through a parallel-sided block. Figure 4.67 shows the path of a single ray entering and leaving a block. It could be travelling in either direction. Once these two rays have been located, the path of the ray within the block can be drawn in, and the angles of incidence and refraction measured so that a value for the refractive index can be calculated. Further values can be determined for other angles of incidence and an average value of refractive index calculated. Alternatively, a graph of sin θ_1 against sin θ_2 should be a straight line with a gradient equal to the refractive index.

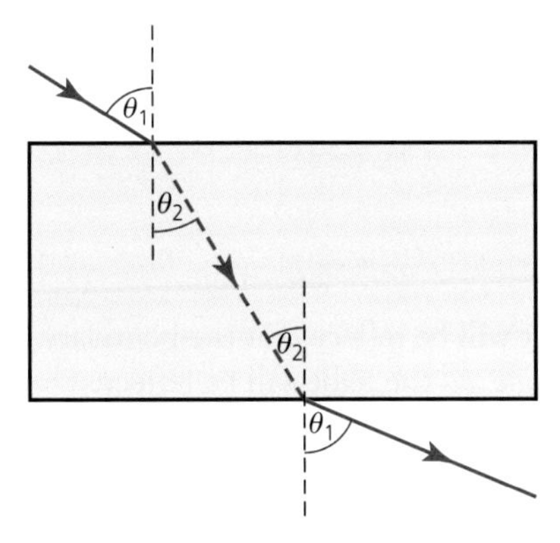

■ **Figure 4.67** Light rays passing through a parallel-sided transparent block

Worked example

9 **a** If light waves represented by a ray making an angle of incidence of 60° enter into glass of refractive index 1.52, calculate the angle of refraction.

b What is the speed of light in this glass? (Speed of light in air = $3.0 \times 10^8\,\mathrm{m\,s^{-1}}$)

a $n_{glass} = \frac{\sin\theta_{air}}{\sin\theta_{glass}}$

$1.52 = \frac{\sin 60}{\sin\theta_{glass}}$

angle of refraction = $\theta_{glass} = 35°$

b $n_{glass} = \frac{v_{air}}{v_{glass}}$

$1.52 = \frac{(3.0 \times 10^8)}{v_{glass}}$

$v_{glass} = 2.0 \times 10^8\,\mathrm{m\,s^{-1}}$

45 Light rays in air enter a liquid at an angle of 38°. If the refractive index of the liquid is 1.4, what is the angle of refraction?

46 Plane water waves travelling at $48\,\mathrm{cm\,s^{-1}}$ enter a region of shallower water with the wavefronts at an angle of 34° to the boundary. If the waves travel at a speed of $39\,\mathrm{cm\,s^{-1}}$in the shallower water, predict the direction in which they will move.

47 Light rays travel at $2.23 \times 10^8 \, m s^{-1}$ in a liquid and at $3.00 \times 10^8 \, m s^{-1}$ in air.
  a What is the refractive index of the liquid?
  b Light rays coming out of the liquid into air meet the surface at an angle of incidence of 25°. What is the angle of the emerging ray to the normal in air?

48 A certain kind of glass has a refractive index of 1.55. If light passes into the glass from water (refractive index = 1.33) and makes an angle of refraction of 42°, what was the angle of incidence?

49 a Use trigonometry to show that the refractive index between two media is equal to the ratio of wave speeds $\left(\frac{v_1}{v_2}\right)$ in the media.
  b Show that the refractive index for waves going from medium 1 into medium 2 is given by ${}_1n_2 = \frac{n_2}{n_1}$.

50 Explain why it is impossible for any medium to have a refractive index of less than one.

Total internal reflection

Consider again Figure 4.66b, which shows a wave/ray entering an optically less dense medium (a medium in which light travels faster). If the angle of incidence, θ_1, is gradually increased, the refracted ray will get closer and closer to the boundary between the two media. At a certain angle, the refracted ray will be refracted at an angle of exactly 90° along the boundary (see Figure 4.68). This angle is called the **critical angle**, θ_c, shown in red in the diagram.

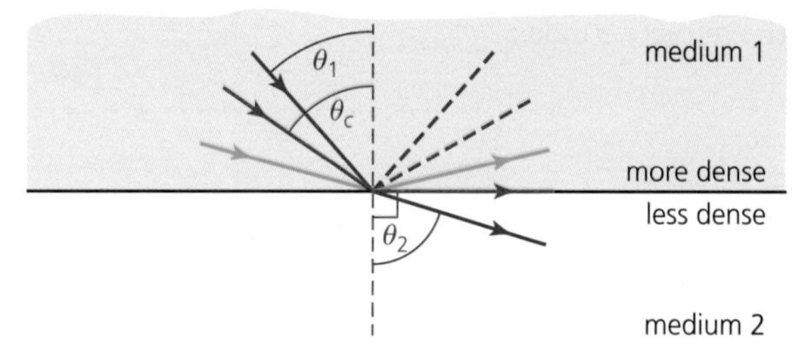

■ **Figure 4.68** Total internal reflection occurs if the angle of incidence is greater than the critical angle, θ_c

For any angle of incidence some light will be reflected at the boundary, but for angles of incidence bigger than the critical angle, *all* the light will be reflected back and remain in the denser medium. This is called **total internal reflection**.

We know that $\frac{n_1}{n_2} = \frac{\sin \theta_2}{\sin \theta_1}$, but at the critical angle, $\theta_1 = \theta_c$ and $\theta_2 = 90°$, so that $\sin \theta_2 = 1$, and then:

$$\frac{n_1}{n_2} = \frac{1}{\sin \theta_c}$$

Most commonly, the light will be trying to pass from an optically denser material (medium 1) like glass, plastic or water, *into air* (medium 2), so that $n_2 = n_{air} = 1$, and so:

$$n_{\text{denser medium}} = \frac{1}{\sin \theta_c}$$

This equation is given in the *Physics data booklet* for Option C, but *not* for Chapter 4.

In order to investigate critical angles experimentally, light rays have to be traced through a material that is *not* parallel sided. This is most conveniently done with a semi-circular block of glass or plastic. The critical angle of a liquid could be determined if a low-voltage, low-power lamp could be supported safely inside a container of the liquid.

A material with a higher refractive index will have a smaller critical angle, which means that light is more likely to be internally reflected. The high refractive indices of certain kinds of glass and precious stones (diamond, for example) are responsible for their 'sparkly' appearance.

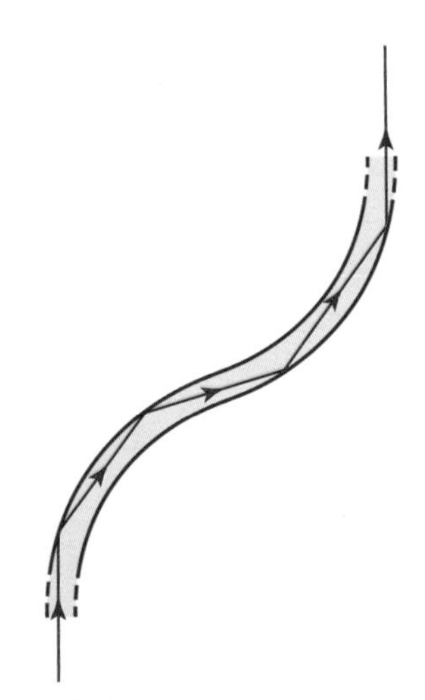

■ **Figure 4.69** Total internal reflection along a glass fibre

One important application of total internal reflection is in communication. Light passing into a glass fibre can be trapped within the fibre because of multiple internal reflections and it will then be able to travel long distances, following the shape of the fibre (see Figure 4.69). The light can be modified to transmit digital information very efficiently.

Worked example

10 A certain kind of glass has a refractive index of 1.54 and water has a refractive index of 1.33.

a In which medium does light travel faster?

b Under what circumstances might light be internally reflected when meeting a boundary between these two substances?

c Calculate the critical angle for light passing between these two media.

a Water (because it has a lower refractive index).

b When travelling from glass into water at an angle bigger than the critical angle.

c $\sin\theta_c = {}_{glass}n_{water} = \frac{v_{glass}}{v_{water}}$

but $n_{glass} = \frac{v_{air}}{v_{glass}}$ and $n_{water} = \frac{v_{air}}{v_{water}}$

so that $\frac{v_{glass}}{v_{water}} = \frac{n_{water}}{n_{glass}}$

hence, $\sin\theta_c = \frac{n_{water}}{n_{glass}} = \frac{1.33}{1.54} = 0.863$

$\theta_c = 59.7°$

51 a If the speed of light in sea water is $2.21 \times 10^8\,\text{m s}^{-1}$, calculate its refractive index.
b Calculate the critical angle for light passing between sea water and air.

52 a Outline how you could carry out an experiment to measure the critical angle of light in glass.
b Suggest a method for measuring the critical angle of light in water.

53 Use the internet to find out about the *cladding* of optical fibres and *acceptance angles*. Write a short summary of your findings.

54 Find out how light travelling along optical fibres can be used to transmit data.

Utilization

Total internal reflection and endoscopy

The total internal reflection of light along flexible optical fibres is used in *endoscopes* for carrying out medical examinations. Light from a source outside is sent along fibres to illuminate the inside of the body. Other optical fibres, with lenses at each end, are used to bring a focused image outside for viewing directly or via a camera and monitor (see Figure 4.70).

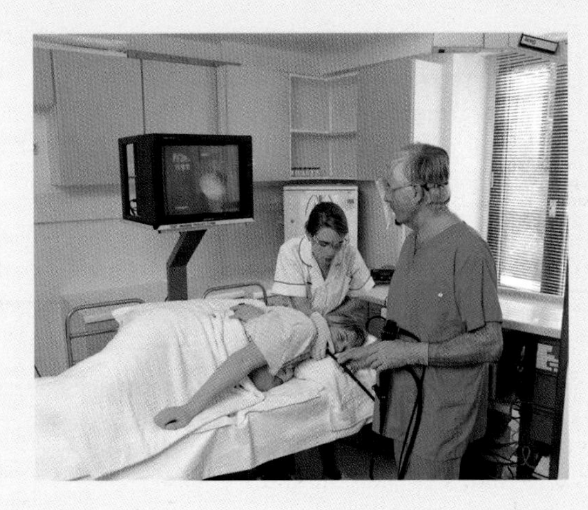

■ **Figure 4.70** An endoscope can be used to inspect a patient's stomach

Endoscopes are used widely for diagnosing medical problems and, increasingly, they are being used in treatments such as small-scale surgery or taking a biopsy. Endoscopes can be inserted into any natural openings in the body, but they are sometimes also inserted through incisions made by a doctor.

1 There are many different types of endoscope used for different parts of the body. Use the internet to learn more about one particular kind of endoscopy and write a short report for presentation to your fellow students.

Dispersion of light

The speeds of different colours (wavelengths) of light in a particular medium (glass, for example) are not exactly the same. Red light travels the fastest and violet is the slowest. This means that different colours travelling in the same direction from the same source will not travel along exactly the same paths when they are refracted. When light goes through parallel-sided glass (like a window), the effect is not usually significant. However, when white light passes into and out of other shapes of glass (like **prisms** and **lenses**), or water droplets, it can be **dispersed** (separated into different colours). A triangular prism, as shown in Figure 4.71, is commonly used to disperse white light into a spectrum (Figure 4.22).

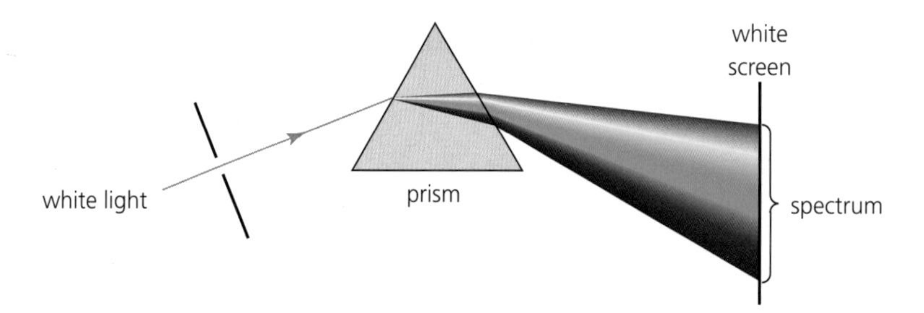

■ **Figure 4.71** Using a prism to produce a spectrum of white light

Additional Perspectives

Ibn al-Haytham

Until recently the important role played by Islamic scientists has been somewhat neglected by other cultures. An eleventh-century scientist, Ibn al-Haytham (Figure 4.72), also known as Alhazen, is arguably one of the greatest physicists of all time. He was a pioneer of modern 'scientific method', with its emphasis on experimentation and mathematical modelling, but he lived hundreds of years before Galileo and the others who are widely credited with similar innovations.

■ **Figure 4.72** Ibn al-Haytham

His work was wide ranging, but included experimental and quantitative investigation of the refraction of light (similar to the work carried out by Snell centuries later). Incidentally, he is also credited by many with being the first to realize that the 'twinkling' of stars is due to refraction of light passing through the Earth's atmosphere.

1 Find out the names of some other scientists and mathematicians from around the eleventh century or a few hundred years before. Where did they live and what were their major achievements?

2 Are scientific developments sometimes achieved in isolation by lone individuals, or is collaboration, rather than secrecy, an important aspect in most research?

Diffraction through a single slit and around objects

When waves pass through gaps (*apertures*) or pass around obstacles in their path, they will tend to 'spread' or 'bend' around them. This important effect is called **diffraction** (a term that should not be confused with *refraction*). Waves often encounter objects in their path and the study of diffraction is vital in appreciating how waves travel from place to place. This has become especially important in this age of wireless communication.

All waves diffract under suitable conditions and the fact that something diffracts is clear evidence of its wave nature. Sometimes the effects of diffraction are very noticeable, as they usually are with water waves and sound waves, but diffraction may also be difficult to observe, as it is with light waves. This is because the amount of diffraction is dependent on how the size of the wavelength compares to the size of the obstacle or gap.

Diffraction is most important when the wavelength and the gap, or obstacle, are about the same size.

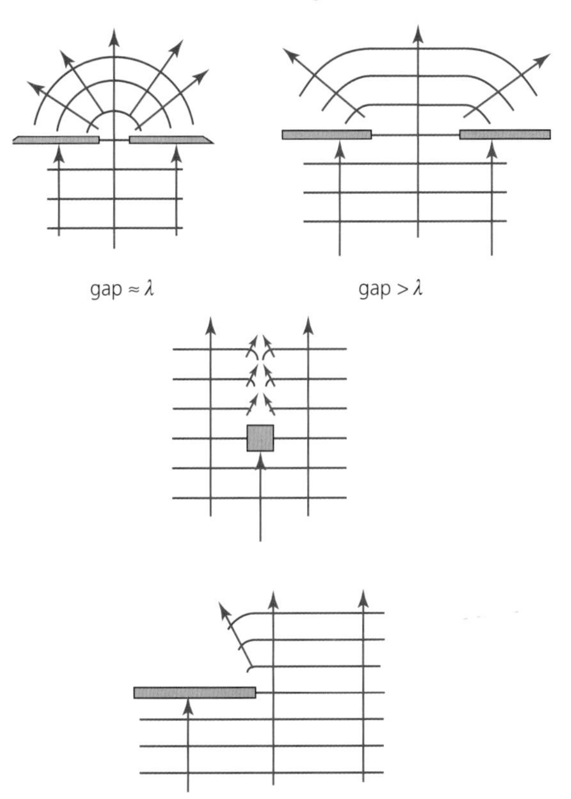

■ **Figure 4.73** Diffraction of plane waves through gaps and around obstacles (reflected waves not shown)

Figure 4.73 represents the two-dimensional diffraction of plane waves through apertures and around obstacles. The diagrams can be applied to the diffraction of any waves, including those in three dimensions, although it is important to realize that they are simplified. These kinds of patterns are easily observed with water waves using a ripple tank (Figure 4.29).

Examples of diffraction

Sound

Wavelengths typically vary from about 2 cm to 20 m. This means that sound is easily diffracted around corners, buildings, doors and furniture, for example. As a result, we can hear sounds even when we cannot see where they are coming from.

Lower-pitched sounds have longer wavelengths and therefore diffract better around larger objects like buildings, so we tend to hear them from further away. Low-pitched sounds also spread away better from larger loudspeakers (often called 'woofers'), while high-pitched sounds benefit from smaller speakers ('tweeters').

■ **Figure 4.74** This large loudspeaker is good at emitting longer wavelengths at high volume

Light

All the various colours of light have wavelengths of less than 10^{-6} m (10^{-3} mm). This means that the diffraction of light tends to go unnoticed because only very small gaps diffract light significantly. However, the diffraction of light entering our eyes does limit our ability to see (resolve) details and it also limits the resolution of telescopes and microscopes. You can see some effects of diffraction by looking at a white surface through a narrow gap made between your fingers: dark lines are seen, which are parallel to the length of the gap.

In order to observe the diffraction of light on a screen in a darkened laboratory it is easiest to use monochromatic light. Monochromatic light means light of a single colour or, more precisely, light with a single wavelength (or a very narrow range of wavelengths). Lasers are an excellent source of monochromatic light for observing diffraction. Figure 9.6 on page 388 shows a typical experimental arrangement.

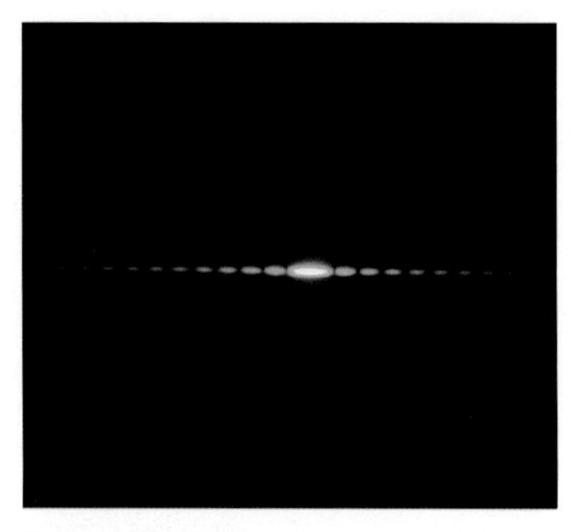

■ **Figure 4.75** Diffraction pattern of monochromatic light passing through a narrow slit

When light passes through a very narrow vertical slit and then strikes a screen some distance away, a diffraction pattern, such as that shown in Figure 4.75, will be seen. A series of light and dark bands is observed, with the central band being brighter and wider than the others. This pattern is discussed further in Chapter 9, but it is important to realize that the pattern can only be explained by a *wave* theory of light.

Radio waves

Radio waves (including microwaves) have a very wide range of wavelengths, from a few centimetres up to a kilometre or more. When engineers design radio communication systems for radio, TV, satellite broadcasts and mobile phones, for example, they have to choose a suitable wavelength to use. This involves considering how far they want the waves to travel between transmitter and receiver, and whether or not there are obstacles such as buildings or hills in between. Ideally the size of the transmitting and receiving aerials will be comparable to the size of the wavelength used, although cost and convenience may reduce aerial sizes. For example, the wavelengths used for mobile phones are typically a few centimetres.

■ **Figure 4.76** Microwaves diffract when they are emitted from transmitting aerials

X-rays

Because diffraction effects (for a given wavelength) depend on the size of the diffracting object, it is possible to learn something about an object by observing and measuring how it diffracts a wave of known wavelength. This has many important applications. X-rays, for example, have wavelengths comparable to the size of atoms and the diffraction of X-rays has been very important for scientists learning about the spacing of atoms and how they are arranged in crystalline solids.

Additional Perspectives

Tsunamis

The consequences of the tsunamis following the massive earthquakes off the Indonesian island of Sumatra on 26 December 2004 and the north-eastern Japanese coastline on 11 March 2011 were tragic and overwhelming. Sudden and massive motion of the Earth's crust along a fault line passed energy to the ocean above, resulting in the movement of an enormous volume of water.

Any tsunami waves resulting from an earthquake travel at high speed (maybe for thousands of kilometres) with little loss of energy, and hence possibly devastating consequences when they reach land.

But why are some places affected much worse than others? Of course the height of the land near the shore is a major factor, as is the distribution of homes and people. A full explanation must also take into account the refraction, reflection and diffraction of the incident waves as they approach a coast. Changes in depth of water and the orientation of such changes (relative to the coast) will affect the height and shape of the waves, and the direction in which they are moving. The shape of the coastline can result in reflections and diffractions that have a focusing effect.

Similar explanations can be used to show why some beaches are much better for surfing than others.

■ **Figure 4.77** The tsunami of December 2004 had devastating effects on low-lying areas

1 What causes waves on the oceans and why do they always seem to come into shore (rather than travelling outwards)?

Utilizations

Satellite 'footprints'

Figure 4.78 represents the intensity of signal arriving at the surface of the Earth from a TV broadcasting satellite positioned somewhere above the equator. The different coloured rings show the diameter of aerial ('dish') required to receive a signal of sufficient power. For example, homes in the outer ring need an aerial with a diameter 120/50 = 2.4 times wider than homes in the centre. This suggests that the received intensity is about six times higher in the centre than in the outer ring.

The transmitting aerial on the satellite does not send the TV signals equally in all directions, but directs the waves to the required locations on the Earth's surface (see Figure 4.79). As the waves emerge from the aerial and reflectors they are diffracted, and this diffraction is responsible for the size and shape of the 'footprint'.

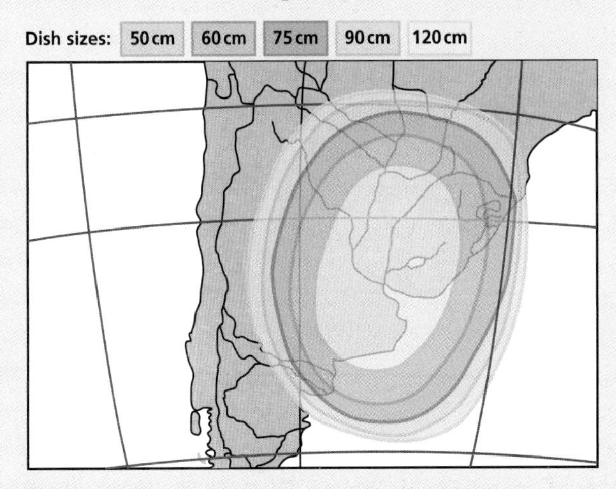

■ **Figure 4.78** Satellite footprint and dish sizes

■ **Figure 4.79** A satellite broadcasting TV signals

1 Explain why the information in Figure 4.78 suggests that the intensity of the signal is about six times higher in the centre than in the outer ring.

2 Find out the typical wavelengths of electromagnetic waves used in satellite TV transmissions and compare this to the size of the transmitting and receiving aerials.

Interference patterns

We are surrounded by many kinds of waves and, of course, the paths of these waves cross each other all the time. When different waves cross, or 'meet', they usually pass through each other without any significant effect, but if the waves are similar to each other (in amplitude and wavelength), then the results can be important. This *superposition* effect is known as the **interference** of waves.

Constructive and destructive interference

Consider Figure 4.80. Suppose that waves of the same frequency are emitted from sources A and B in phase (or with a constant phase difference). Such sources are described as being **coherent**.

If these waves travel equal distances and at the same speed to meet at a point such as P_0, which is the same distance from both sources, they will arrive *in phase*. Using the principle of superposition, we know that if they have the same amplitude, the result will be an oscillation at P_0 that has twice the amplitude of the individual oscillations. This is an example of **constructive interference**, as shown in Figure 4.81a.

■ **Figure 4.80** Interference and path difference

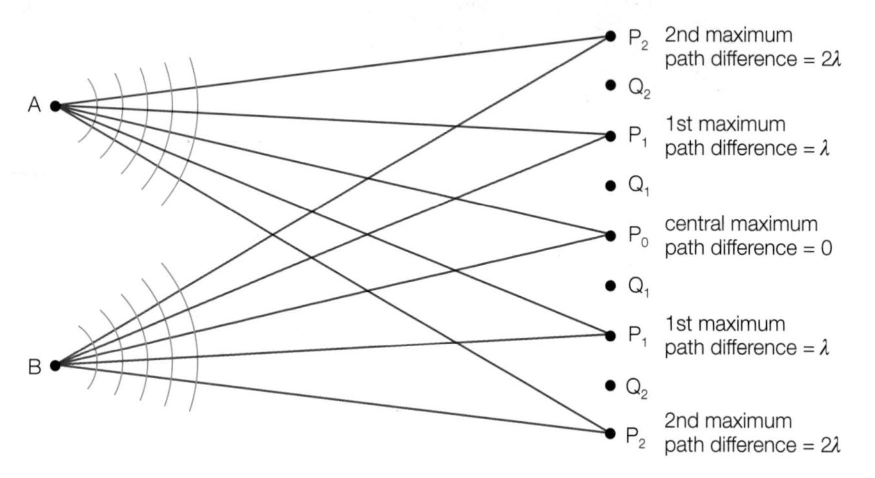

Similarly, there will be other places, such as P_1 and P_2, where the waves are *in phase* and interfere constructively because one wave has travelled one wavelength further than the other, or two wavelengths further, or three wavelengths etc.

■ **Figure 4.81** Constructive and destructive interference

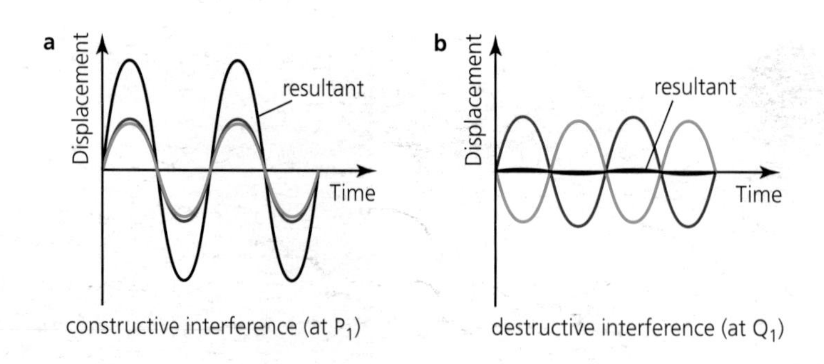

Path difference

In general, we can say that under these conditions, constructive interference will occur anywhere where there is a path difference equal to a whole number of wavelengths. Path difference is the difference in the distance travelled by waves from two separate sources that arrive at the same point.

In other places, such as at points Q_1 and Q_2, the waves will arrive *exactly out of phase* because one wave will have travelled half a wavelength further than the other, or one and a half wavelengths, or two and a half wavelengths and so on. In these places the result will be a minimal oscillation, an effect called destructive interference (as shown in Figure 4.81b). The resultant will probably not be zero because one wave has a greater amplitude than the other, since they have travelled different distances. The overall pattern will appear as shown in Figure 4.82.

The fact that there are places where waves can come together to produce no waves is especially important because *only* waves can show this behaviour. For example, when it was discovered that light can interfere, there was only one possible conclusion – light must travel as a wave.

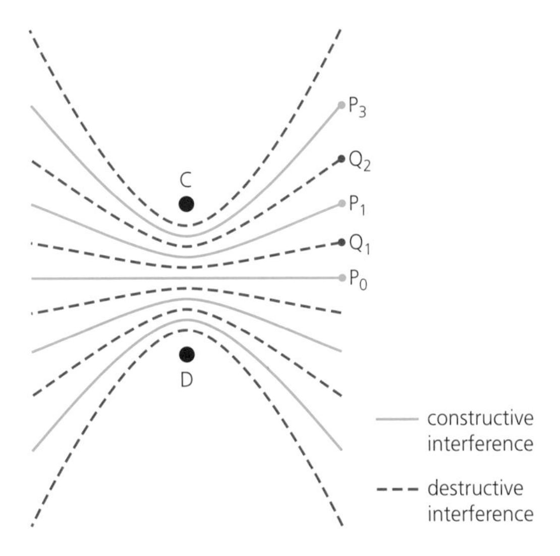

■ **Figure 4.82** The interference pattern produced by coherent waves from two sources, C and D

Two waves combining to give no waves at particular places may seem to contradict the principle of conservation of energy, but the 'missing' energy appears at other places in the interference pattern where there is constructive interference, giving twice the amplitude. (Remember that doubling the amplitude of an oscillation implies four times the energy.)

The interference of waves from two sources is conveniently demonstrated using a ripple tank (Figure 4.29). Figure 4.83 shows a typical interference pattern. It should be compared to the right-hand side of Figure 4.82.

■ **Figure 4.83** The interference of water waves on a ripple tank

Summary of conditions needed for interference

The condition for constructive interference is that coherent waves arrive at a point in phase. This will occur when the path difference is equal to a whole number of wavelengths.

For constructive interference, path difference = $n\lambda$ (where n is an integer, 1, 2, 3 etc.)

The condition for destructive interference is that coherent waves arrive at a point exactly out of phase. This will occur when the path difference equals an odd number of half wavelengths.

For destructive interference, path difference = $(n + \frac{1}{2})\lambda$

These two conditions are given in the *Physics data booklet*.

At most places in an interference pattern there is neither perfectly constructive nor perfectly destructive interference, but something in between these two extremes.

Examples of interference

Different sources of waves (light waves, for example) are not normally coherent because the waves are not produced in a co-ordinated way. So, although in principle *all* waves can interfere, in practice examples are limited to those waves that can be made to be coherent. This may involve using a single source of waves and splitting the wavefronts into two.

Interference of microwaves

In Figure 4.84, a single microwave source is being used, but the wavefronts are split into two by the slits between the aluminium sheets and, if the receiver is moved around, it will detect the interference pattern produced when the microwaves diffract through the two slits and then overlap.

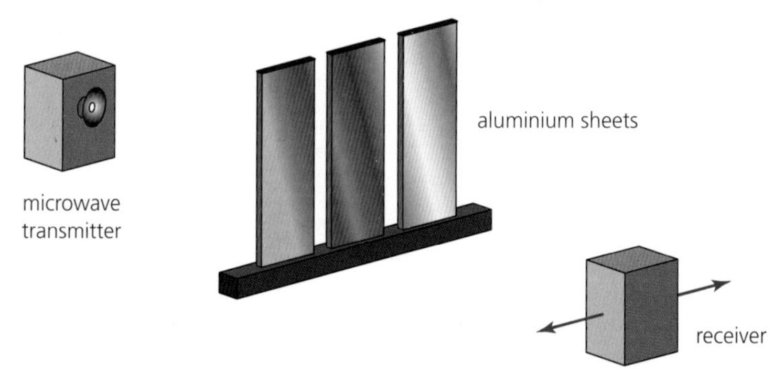

■ **Figure 4.84** Interference of microwaves

Interference of sound waves

Identical waves can be produced using two sources driven by the same electronic signal, such as with radio or microwave transmitters, or with sound loudspeakers. In Figure 4.85 the listener hears the sound intensity changing as she walks past the speakers. (Unwanted reflections from walls may make this difficult to hear clearly inside a room.)

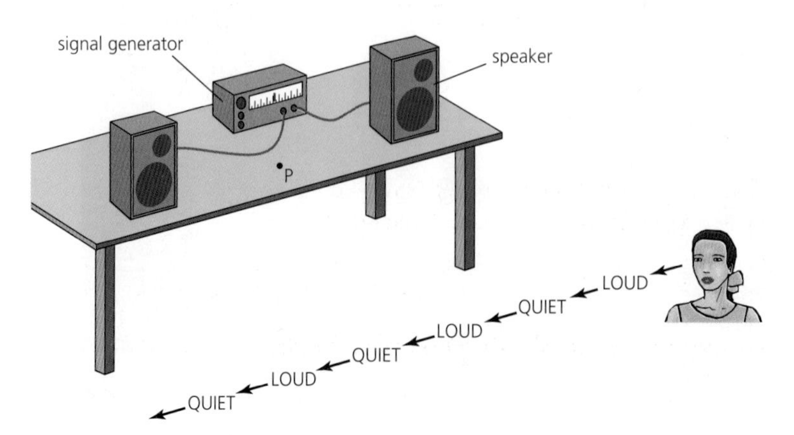

■ **Figure 4.85** Interference of sound waves

Double-slit interference

The interference of light waves can be demonstrated in a darkened room by passing monochromatic laser light through two narrow slits that are very close together. The resulting interference pattern can be seen on a distant screen (see Figure 4.86). The interference of light can also be observed with white light (with or without filters), but the pattern is more difficult to observe.

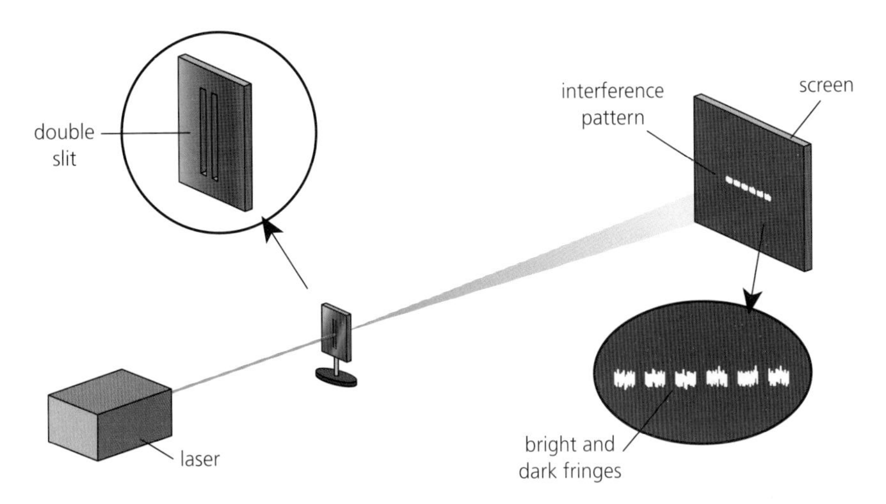

■ **Figure 4.86** Interference of light waves

A series of light and dark 'fringes' is seen on the screen. The pattern is similar to the diffraction pattern seen when light passes through a single slit, although in the interference pattern all the fringes are about the same width. The closer the two slits, the wider the spacing of the interference pattern.

This experiment was first performed by Thomas Young and it is of great historical importance because the observation of the interference of light confirmed its wave-like nature (because only waves can interfere or diffract). Measurements of the geometry of the experiment can lead to a determination of the wavelength of the light used (see Figure 4.87).

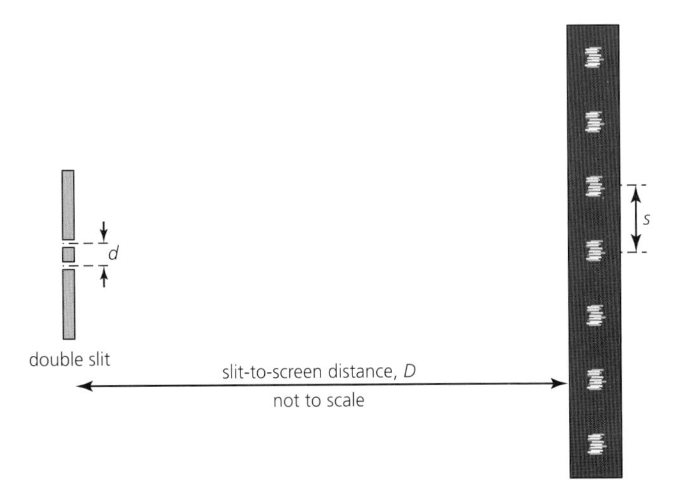

■ **Figure 4.87** Geometry of the double-slit experiment

The wavelength of the light used is related to the geometry of the experiment by the following equation (which is explained in Chapter 9):

$$s = \frac{\lambda D}{d}$$

This equation is given in the *Physics data booklet*.

Worked example

11 Monochromatic laser light was directed at some double slits and an interference pattern was observed on a screen placed 4.78 m from the slits. The distance between the centre of a bright fringe and the centre of another, eight fringes away, was 5.8 cm. If the slit separation was 0.33 mm, calculate the wavelength of the radiation.

$$s = \frac{\lambda D}{d}$$

$$\frac{(5.8 \times 10^{-2})}{8} = \frac{4.78\lambda}{(0.33 \times 10^{-3})}$$

$$\lambda = 5.0 \times 10^{-7}\,\text{m}$$

55 Why are microwaves often used in school laboratories to demonstrate the interference of electromagnetic waves?

56 The girl shown in Figure 4.85 found that when she moved in the direction shown by the arrows, there was approximately 50 cm between successive positions where the sound was loud. The speakers were 120 cm apart and her closest distance to point P was 80 cm. Estimate the approximate wavelength and frequency of the sources.

57 Figure 4.88 shows two sources of waves on a ripple tank. The diagram is one quarter of full size. Each source produces waves of wavelength 2.8 cm. Take measurements from the diagram to determine what kind of interference will occur at P.

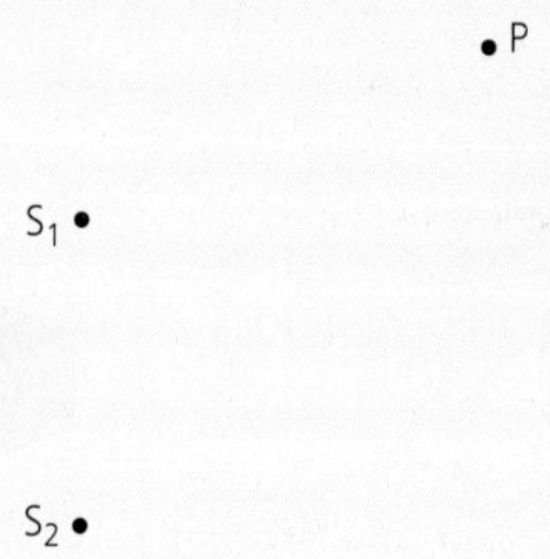

• P

S_1 •

S_2 •

Figure 4.88

58 An observer stands halfway between two speakers facing each other and he hears a loud sound of frequency 240 Hz.
a Explain why the intensity of the sound decreases if he walks slowly in any direction.
b What is the shortest distance he would have to move to hear the sound rise to a maximum again? (The speed of sound in air = 340 m s^{-1}.)

59 Explain why no interference pattern is seen when the light beams from two car headlights cross over each other.

60 In Figure 4.84 the centres of the slits were 6 cm apart.
a Suggest why the width of the two slits was chosen to be about the same size as the wavelength.
b The receiver detected a maximum signal when it was 45 cm from one slit and 57 cm from the other. Suggest possible values for the wavelength of the microwaves.
c How could you determine the actual wavelength?

61 A helium laser with a wavelength of 633 nm (1 nm = 1 × 10^{-9} m) was used by a teacher to demonstrate double-slit interference with a pair of slits of spacing 0.50 mm. If the teacher wanted the fringe separation on the screen to be 1.0 cm, how far away from the slits would he have to put the screen?

Figure 4.89 How are supernumerary rainbows explained?

62 a Find out what causes the colours seen in rainbows (Figure 4.89).
b If you look closely, you can see light and dark bands inside the rainbow. These *supernumerary* rainbows are a diffraction and interference effect. Research into how they are formed.

Utilizations

Using interference to store digital data

Digital data is stored and transmitted in *binary* form as a very long series of 0s and 1s (offs and ons). This means that each tiny location in the medium that stores the data must be able to distinguish between (only) two states. An optical storage medium (like CD, DVD or Blu-ray) uses the constructive and destructive interference of laser light to create these two states.

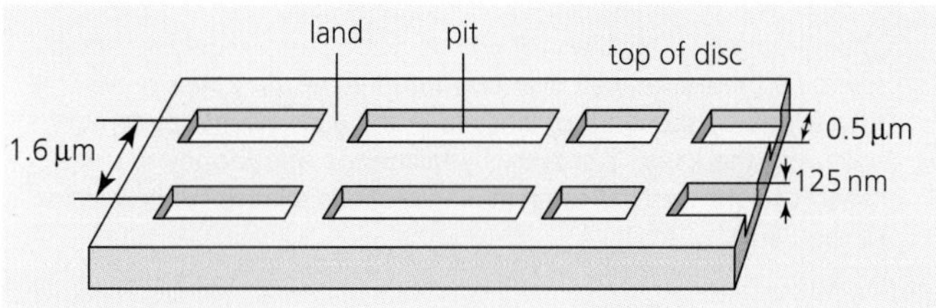

■ **Figure 4.90** The three-dimensional arrangement of pits and lands in a track on a CD (not to scale). The length of the pits ranges from 830 to 3560 nm

The data on a CD or DVD is stored in a track of microscopic 'pits' and 'lands', which are moulded into a thin layer of transparent plastic (Figure 4.90). These indentations are then covered with a thin layer of reflective aluminium. The pits and lands are arranged one after another in a spiral track, starting from the centre of the disc.

To recover the data, a laser beam is focused on the CD surface and is reflected back onto a detector. How it is reflected depends on whether it falls on a pit or a land. If the laser beam is entirely incident on a pit (or on a land) then all the waves in the reflected beam are in phase with each other. Constructive interference takes place and a strong signal (a binary one) is detected.

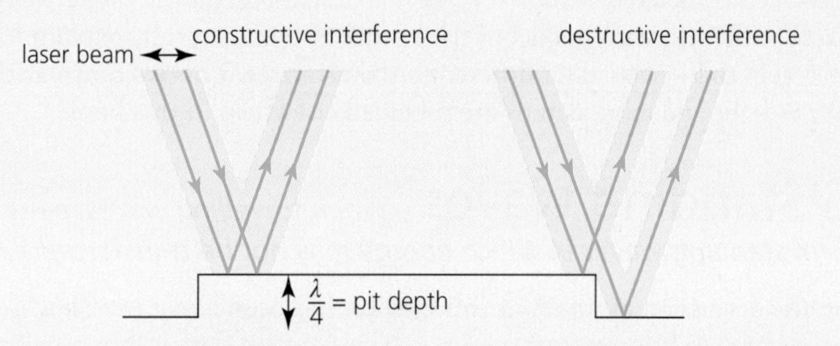

■ **Figure 4.91** The reflection of the laser beam from a pit and from a land – here the light is shown falling at an angle, but in a CD player it is *almost* normal to the disc surface

If part of the light beam is incident on a pit and part on a land, then there is a path difference between the two parts of the laser beam. The pit depth is such that, for a laser beam of wavelength λ, the path difference is $\lambda/2$, which means that destructive interference occurs and there is no signal produced (a binary zero). Because the path difference is twice the pit depth, to obtain a path difference of $\lambda/2$, pit depth = $\lambda/4$.

The laser beam reflects off the rotating spiral track and the signal received by the detector changes as the beam travels from pit to land to pit. This produces digital signals of 1s and 0s, varying according to whether the interference is constructive or destructive and also according to the lengths of the pits and lands.

It is estimated that the world's ability to store data doubles every two to three years, reflecting people's desire to keep pictures and videos, and the desire of organizations to keep all the records that they can. Data are mostly stored electronically, magnetically or optically, although small solid-state systems (with no moving parts) are preferable if they can have a large enough data storage capacity.

While storing information may be easy, removing it when it is no longer of use can get overlooked and may not be as simple as the touch of a button. Information that may be untrue, outdated or just embarrassing often remains stored, and is therefore accessible to large numbers of people, indefinitely. This is especially true for any use of the internet, where our 'digital footprint' (the data we leave behind us) may be much larger than we think.

1 Compare the advantages of storing data on an optical disk, flash drive, external hard disk (or solid-state drive) or internet 'cloud'.

ToK Link

Competing theories

Huygens and Newton proposed two competing theories of the behaviour of light. How does the scientific community decide between competing theories?

Christiaan Huygens was a notable Dutch scientist and mathematician who believed that the known properties of light at that time (in the 1670s) could be explained if light consisted of *waves*. This was in conflict with the theory of the nature of light generally known as Newton's 'corpuscular theory'. Newton's theory could be used to explain the properties of light (like reflection and refraction) by assuming it to consist of tiny *particles*, but Huygens' wave theory was also able to explain the diffraction and polarization of light.

Particle theory could not be used to explain the interference of light, as demonstrated by Young more than one hundred years later in the early nineteenth century. If light interfered it *must* have wave-like properties, but all the other known waves needed a medium to travel through. For example, sound can travel through air, but not through a vacuum because there are no oscillating molecules to transfer the waves.

About 150 years before, Descartes is credited with developing the concept of the 'ether' – a mysterious, undetectable substance that was everywhere, filling all of space. Scientists adopted this idea to help explain how light can travel across space. (Descartes had proposed the ether to help explain the idea that forces (magnetic, gravitational, electrical) can act 'at a distance' between objects with nothing in the intermediate space.)

The ether was a less than totally convincing, but widely accepted, scientific theory for well over two hundred years. It was finally discredited by the work of Einstein in the early twentieth century following the earlier discovery by Michelson and Morley that the speed of light was the same in all directions relative to the motion of the Earth.

Nature of Science

Wave–particle duality

The nature of light has been a central issue and a matter for great debate in the development of physics for centuries. Many famous physicists have proposed useful theories in the past, but none seemed completely satisfactory, or able to explain all the properties of light. Scientists now accept that no *single* model of the nature of light can fully explain its behaviour. Different models seem to be needed for different circumstances. This is known as the **wave–particle duality** of light and more details are included elsewhere in this book.

4.5 Standing waves – *when travelling waves meet they can superpose to form standing waves in which energy may not be transferred*

So far, the discussion of waves in this chapter has been about *travelling waves*, which transfer energy progressively away from a source. Now we turn our attention to waves that remain in the same position.

■ The nature of standing waves

Consider two travelling waves of the same shape, frequency, wavelength and amplitude moving in opposite directions, such as shown in Figure 4.92, which could represent transverse waves on a string or rope.

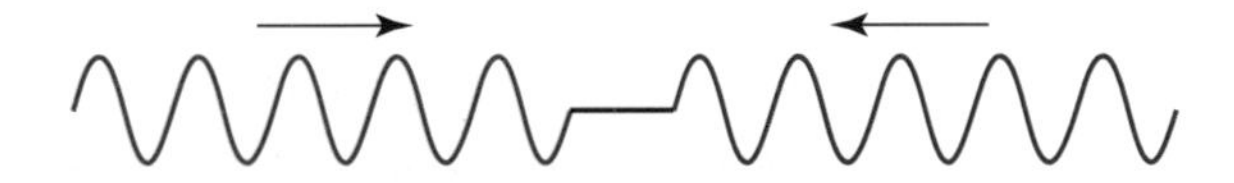

■ **Figure 4.92** Two sinusoidal waves travelling towards each other

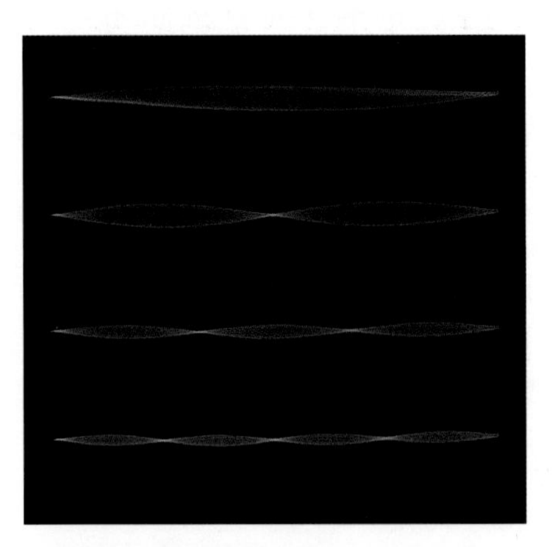

■ **Figure 4.93** Standing waves on a stretched string

As these waves pass through each other, they may combine to produce an oscillating wave pattern that does not change its position. Such patterns are called **standing waves** (or sometimes stationary waves). Standing waves usually occur in an enclosed system in which waves are reflected back on each other repeatedly. Typical examples of standing-wave patterns are shown in Figure 4.93. Note that a camera used for this produces an image over a short period of time (not an instantaneous image) and that is why the fast-moving string appears blurred. This is equally true when we view such a string with our eyes.

Simple standing-wave patterns can be produced easily by shaking one end of a rope, or a long stretched spring, at a suitable frequency, while someone holds the other end fixed. Patterns like those shown in Figure 4.93 require higher frequencies (because a string is much less massive than a rope), but can be produced by vibrating a stretched string with a mechanical vibrator controlled by the variable electrical oscillations from a **signal generator**. This apparatus can be used to investigate the places at which the string appears to be stationary and the frequencies at which they occur.

Nodes and antinodes

There are points in a standing wave where the displacement is *always* zero. These points are called **nodes**. At positions between the nodes, the oscillations of all parts of the medium are *in phase*, but the amplitude of the oscillations will vary. Midway between nodes, the amplitude is at its maximum. These positions are called **antinodes**. Figure 4.94 represents the third wave from the photograph in Figure 4.93 diagrammatically. Note that the distance between alternate nodes (or antinodes) is one wavelength.

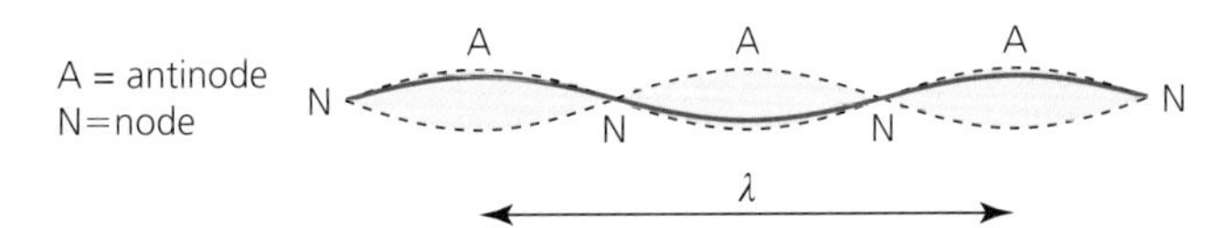

■ **Figure 4.94** Nodes and antinodes in a standing wave on a stretched string – the solid line represents a possible position of the string at one moment.

There is energy associated with a standing wave and, without dissipative forces, the oscillations would continue forever. But energy is not *transferred* by a standing wave out of the system.

We can explain the formation of a standing wave pattern by determining the resultant at any place and time. We can do this by using the *principle of superposition*. The overall displacement is the sum of the two individual displacements at that moment. Nodes occur at places where the two waves are *always* out of phase. At other places, the displacements will oscillate between zero and a maximum value that depends on the phase difference. At the antinodes the two waves are always perfectly in phase. (Students are recommended to use a computer simulation to illustrate this time-changing concept.)

Standing waves are possible with any kind of wave moving in one, two or three dimensions. For simplicity, discussion here has been confined to one-dimensional waves, such as transverse waves on a stretched string.

Boundary conditions

Standing waves occur most commonly when waves are reflected repeatedly back from boundaries in a confined space, like waves on a string or air in a pipe. The frequencies of the standing waves will depend on the nature of the end of the string or the end of the pipe. These are called **boundary conditions**. For example, the ends of a string, or rope, may be fixed in one position or be free to move; the ends of a pipe could be closed or open. When the ends are free to move we would expect the standing wave to have antinodes and there will be nodes when the ends are fixed.

■ Modes of vibration of transverse waves on strings

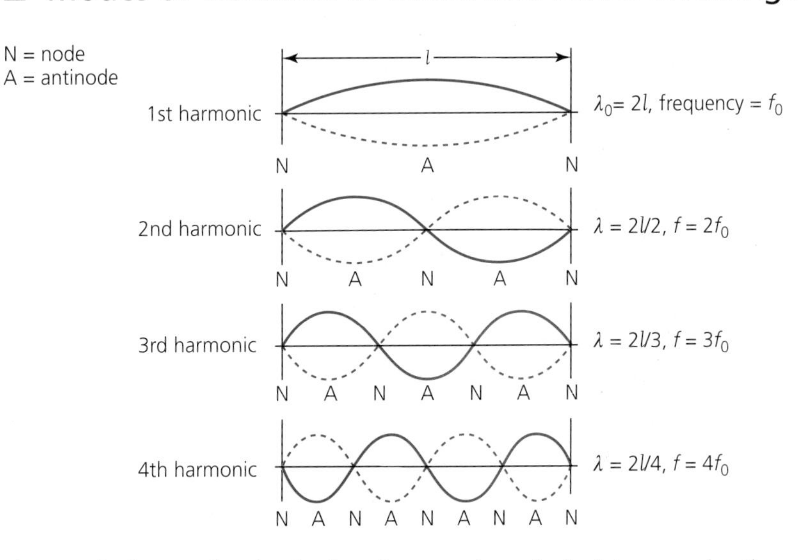

■ **Figure 4.95** Modes of vibration of a stretched string fixed at both ends

If a stretched string fixed at both ends is struck or plucked, it can only vibrate as a standing wave with nodes at both ends. The simplest way in which it can vibrate is shown at the top of Figure 4.95. This is known as the **first harmonic**. (Also sometimes called the *fundamental mode* of vibration.) It is usually the most important mode of vibration, but a series of other harmonics is possible and can occur at the same time as the first harmonic. Some of these harmonics are shown in Figure 4.95. The wavelength, λ_0, of the first harmonic is $2l$, where l is the length of the string. The speed of the wave, v, along the string depends on the tension and the mass per unit length. The frequency of the first harmonic, f_0, can be calculated from $v = f\lambda$:

$$f_0 = \frac{v}{\lambda_0} = \frac{v}{2l}$$

This shows us that, for a given type of string under constant tension, the frequency of the first harmonic is inversely proportional to the length. The first harmonic of a longer string will have a lower frequency.

The wavelengths of the harmonics are, starting with the first (longest), $2l$, $\frac{2l}{2}$, $\frac{2l}{3}$ $\frac{2l}{4}$ and so on.

The corresponding frequencies, starting with the lowest, are f_0, $2f_0$, $3f_0$, $4f_0$ and so on.

Worked example

12 A string has a length of 1.2 m and the speed of transverse waves on it is $8.0\,\text{m s}^{-1}$.

a What is the wavelength of the first harmonic?

b Draw sketches of the first four harmonics.

c What is the frequency of the third harmonic?

a $\lambda_0 = 2l = 2 \times 1.2 = 2.4\,\text{m}$

b See Figure 4.95.

c $\lambda = \frac{2.4}{3} = 0.8\,\text{m}$

$f = \frac{v}{\lambda} = \frac{8.8}{0.8} = 10\,\text{Hz}$

Standing waves on strings are usually between fixed ends, but it is possible that one, or both, boundaries could be free. If there are antinodes at each boundary, the frequency of the first harmonic will be the same as for fixed boundaries, with nodes at each end. If there is a node at one end and an antinode at the other, the frequency of the first harmonic will be lower. An example of this situation would be a standing wave produced on a chain hanging vertically.

Musical instruments

An amazing variety of musical instruments have been used all around the world for thousands of years (see Figure 4.96). Most involve the creation of standing wave patterns (of different frequencies) on strings, wires, surfaces or in tubes of some kind. The vibrations disturb the air around them and thereby send out sound waves (music).

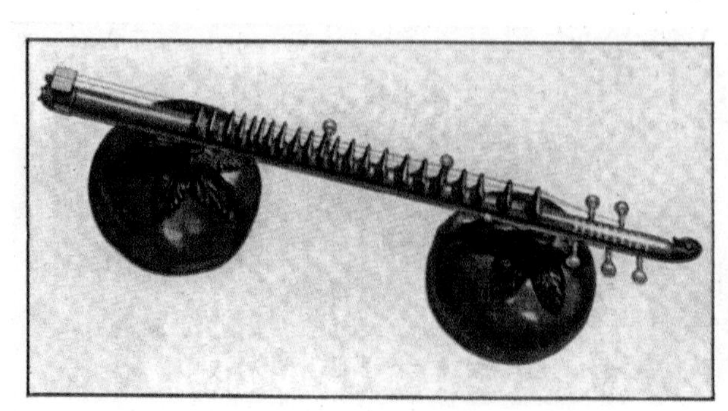

■ **Figure 4.96** A Vina, an ancient Indian musical instrument. Composed of bamboo and two gourds

When musical notes are played on stringed instruments, like guitars, cellos and pianos, the strings vibrate mainly in their first harmonic modes, but various other harmonics will also be present. This is one reason why each instrument has its own, unique sound. Figure 4.97a shows a range of frequencies that might be obtained from a guitar string vibrating with a first harmonic of 100 Hz. The factors affecting the frequency of the first harmonic are the length of the string, the tension and the mass per unit length. For example, middle C has a frequency of 261.6 Hz. The standing transverse waves of the vibrating strings are used to make the rest of the musical instrument oscillate at the same frequency. When the vibrating surfaces strike the surrounding air, travelling longitudinal sound waves propagate away from the instrument to our ears.

It can stimulate discussion if students bring different musical instruments into the laboratory to compare how the sounds are made and how the frequencies are controlled. A frequency analysis can be particularly interesting (as Figure 4.97a). Conversely, there are computer programs that enable the synthesis of sounds by adding together (superposing) basic waveforms.

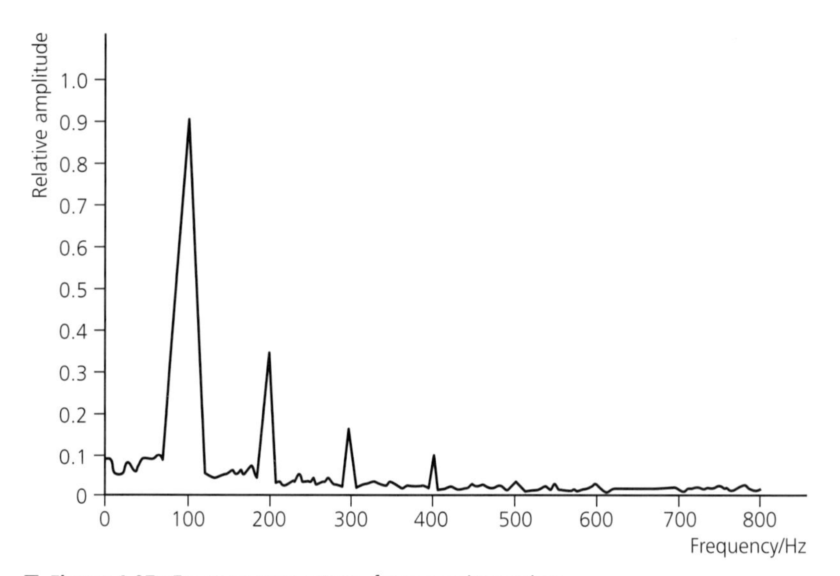

■ **Figure 4.97a** Frequency spectrum from a guitar string

■ **Figure 4.97b** Creating standing waves on a cello

ToK Link

Understanding things we cannot see requires imagination

There are close links between standing waves in strings and Schrodinger's theory for the probability amplitude of electrons in the atom. Application to superstring theory requires standing-wave patterns in 11 dimensions. What is the role of reason and imagination in enabling scientists to visualize scenarios that are beyond our physical capabilities?

Standing waves on strings can only have one of a series of well-defined and mathematically related frequencies. After it was discovered that an electron can only exist in hydrogen in one of a series of mathematically related energy levels, it was realized that standing-wave theory could be applied to atoms. However, that should not imply that there are physical similarities.

Nature of Science

Common reasoning processes

The connections between simple musical notes and mathematics have been known for more than 2000 years. Pythagoras is often credited with being the first to appreciate this relationship, in about the year 600 BCE. Whether this is true or not, the 'Music of the Spheres' was a long-established and foresighted philosophy that sought to find number patterns and harmonies, not only in music but in the world in general and also in astronomical observations.

Ideas related to standing waves can be identified over a period of more than 2500 years, from ancient musical instruments all the way through to the way electrons fit into atoms and the very latest *superstring theories of everything*. This is a common theme of the various areas of knowledge within science: a shared terminology and **common reasoning process** develops and evolves over time. When faced with a new discovery, or innovation, our natural instinct is to interpret it in terms of existing knowledge.

■ Standing sound waves in pipes

Air can be made to vibrate and produce standing longitudinal sound waves in various containers and tubes. The sound produced by blowing across the top of an empty bottle is an everyday example of this. Many musical instruments, such as a flute or a clarinet, use the same idea. For simplicity, we will only consider standing waves in *pipes* of uniform shape (sometimes called air *columns*).

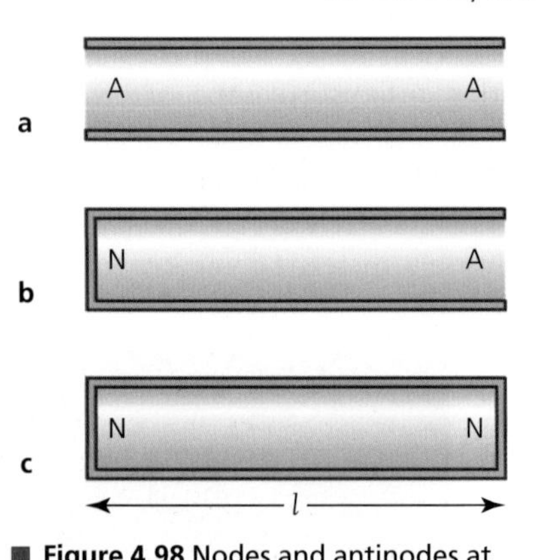

■ **Figure 4.98** Nodes and antinodes at the ends of open and closed pipes

As with strings, in order to understand what wavelengths and frequencies can be produced, we need to consider the length of the pipe and the boundary conditions.

This is illustrated in Figure 4.98. In diagram **a**, the pipe is open at both ends, so it must have antinodes, A, at the ends, and at least one node in between. In diagram **b**, the pipe is open at one end (antinode) and closed at the other end (node, N). In diagram **c**, the pipe is closed at both ends, so it must have nodes at the ends and at least one antinode in between.

Figure 4.99 shows the first three harmonics for a pipe open at both ends. The wavelength of the first harmonic (twice the distance between adjacent nodes or antinodes) is $2l$. (Note that drawings of standing longitudinal waves can be confusing: the curved lines in the diagram are an indication of the amplitude of vibration. They should not be mistaken for transverse waves.)

A pipe closed at both ends also has a first harmonic with a wavelength of $2l$.

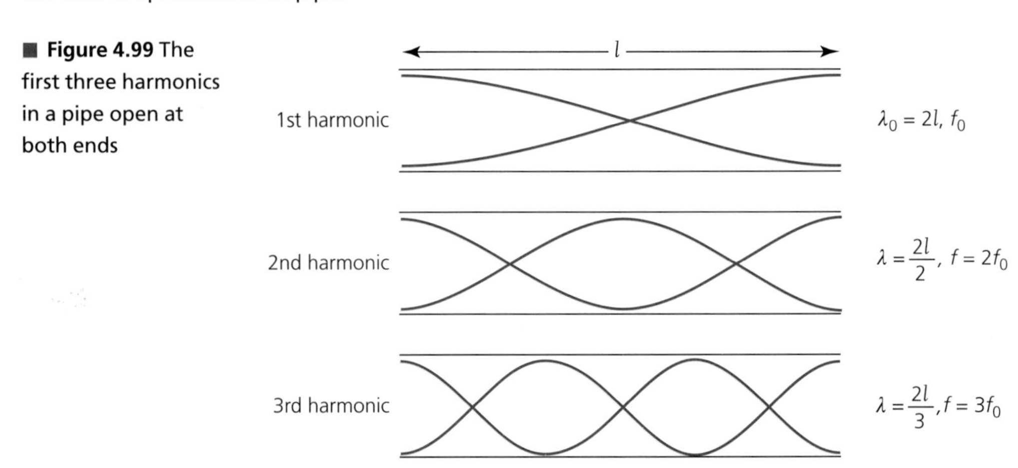

■ **Figure 4.99** The first three harmonics in a pipe open at both ends

Figure 4.100 shows the first three possible harmonics for a pipe open at one end and closed at the other. Only odd harmonics are possible under these circumstances. The first harmonic has a wavelength of $4l$.

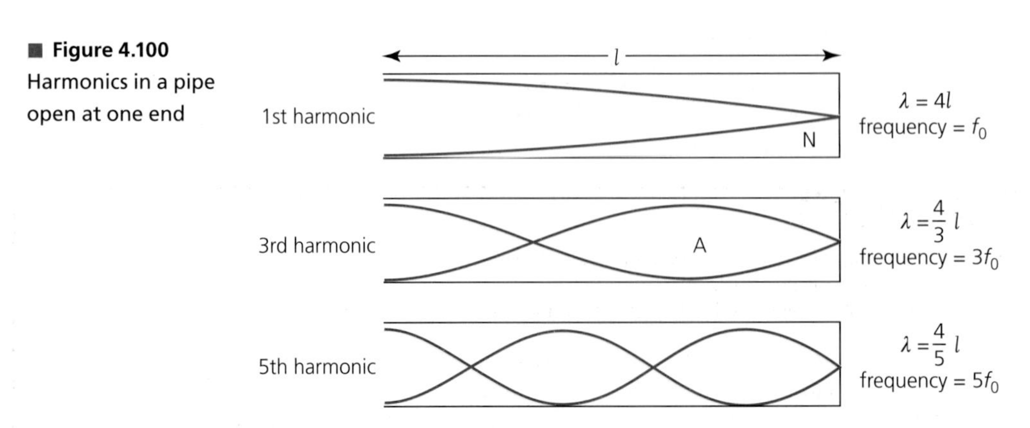

■ **Figure 4.100** Harmonics in a pipe open at one end

Figure 4.101 shows a way of demonstrating standing waves with sound. A speaker is placed close to the open end of a long transparent pipe that is closed at the other end. Some powder is scattered all along the pipe and when the loudspeaker is turned on and the frequency carefully adjusted, the powder is seen to move into separate piles. This is because the powder tends to move from places where the vibrations are large (antinodes) to the nodes, where there are no vibrations.

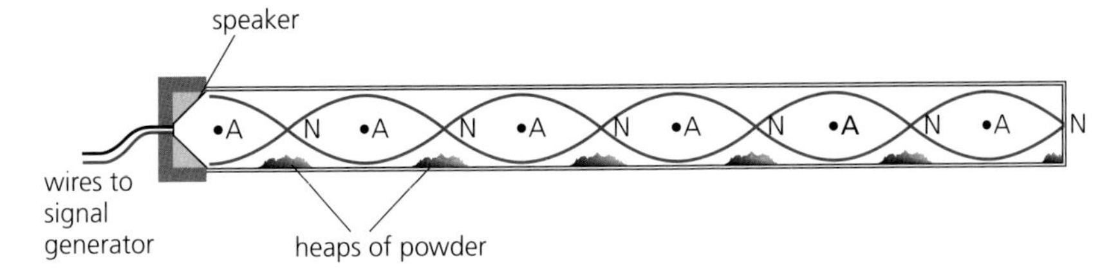

■ **Figure 4.101** Demonstrating a standing wave with a loudspeaker

Another way of demonstrating standing waves of air in a pipe is by using a **tuning fork**, as shown in Figure 4.102. A vibrating tuning fork is held above the open end of a pipe. The pipe is open at the top and closed at the bottom by the level of water. The height of the pipe above the water is slowly increased until a louder sound is heard, which occurs when the tuning fork's frequency is the same as the frequency of the first harmonic of the air column. If the height of the pipe above water is increased again, then further harmonics may be heard.

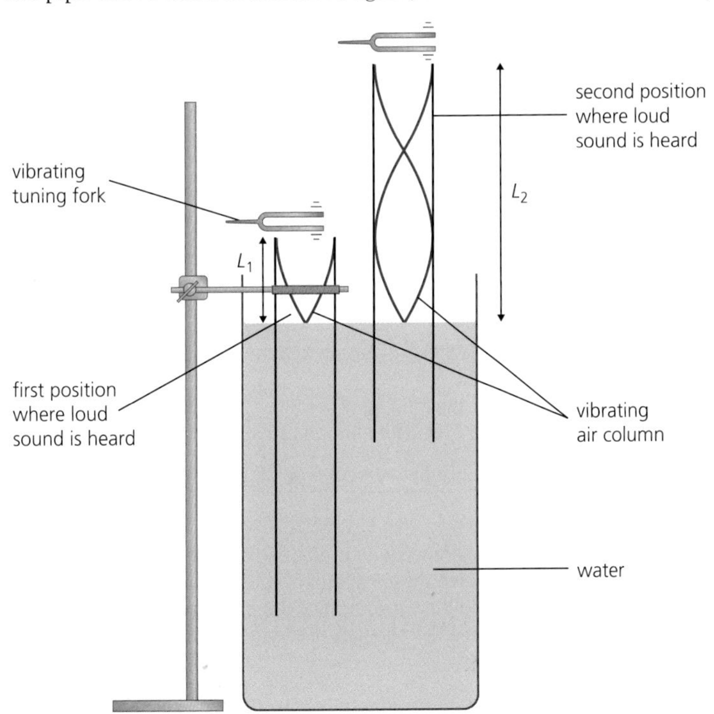

■ **Figure 4.102** Demonstrating standing waves with a tuning fork

Measurements can be made during this demonstration that will enable a value for the speed of sound to be determined. Alternatively, if the speed of sound in air is known, the experiment can be used to determine the frequency of a tuning fork.

Worked example

13 a If the tuning fork in Figure 4.102 had a frequency of 659 Hz, calculate the length L_1. (Assume the speed of sound in air is $340\,\text{m}\,\text{s}^{-1}$.)

b How far will the pipe need to be raised to obtain the next position when a loud sound is heard?

a

$$\lambda = \frac{v}{f} = \frac{340}{659} = 0.516\,\text{m}$$

This wavelength will be four times the length of the tube (see Figure 4.100).

So,

$$L_1 = \frac{0.516}{4} = 0.129\,m$$

In reality this is only an approximate answer because antinodes do not occur exactly at the open ends of tubes. (If a more accurate answer is needed in a calculation, it is possible to use an 'end correction', which is related to the diameter of the tube.)

b Refer to Figure 4.102. In the first position, the air column contains one-quarter of a wavelength. In the second position, the air column contains three-quarters of a wavelength. Therefore, it must have been raised half a wavelength, or 0.258 m.

Summary of the differences between standing waves and travelling waves

	Standing waves	**Travelling waves**
Wave pattern	standing (stationary)	travelling (progressive)
Energy transfer	no energy is transferred	energy is transferred through the medium
Amplitude (assuming no energy dissipation)	amplitude at any one place is constant but it varies with position between nodes; maximum amplitude at antinodes, zero amplitude at nodes	all oscillations have the same amplitude
Phase	all oscillations between adjacent nodes are in phase	oscillations one wavelength apart are in phase; oscillations between are not in phase
Frequency	all oscillations have the same frequency	all oscillations have the same frequency
Wavelength	twice the distance between adjacent nodes	shortest distance between points in phase

Table 4.3 Comparison of standing waves and travelling waves

63 A first harmonic is seen on a string of length 123 cm at a frequency of 23.8 Hz.
  a What is the wave speed?
  b What is the frequency of the third harmonic?
  c What is the wavelength of the fifth harmonic?

64 a What is the phase difference between two points on a standing wave which are:
    i one wavelength apart
    ii half a wavelength apart?
  b The distance between adjacent nodes of the third harmonic on a stretched string is 18.0 cm, with a frequency of 76.4 Hz. Sketch the waveform of this harmonic.
  c On the same drawing add the waveforms of the first harmonic and the fourth harmonic.
  d Calculate the wavelength and frequency of the fifth harmonic.
  e What was the wave speed?

65 a The top of a vertical chain of length 2.1 m is shaken from side to side in order to set up a standing wave. Assuming that the wave speed along the chain is a constant 5.6 m s^{-1}, what frequency will be needed to produce the second harmonic?
  b In fact, the wave speed along the chain will *not* be constant and this will result in a standing wave of varying wavelength. Suggest a reason for the variable speed.

66 A certain guitar string has a first harmonic with a frequency of 262 Hz.
  a If the tension in the string is increased, suggest what happens to the speed of waves along it.
  b If the string is adjusted so that the speed increases, explain what will happen to the frequency of the first harmonic.
  c Suggest why a wave will travel more slowly along a thicker string of the same material, under the same tension.
  d Explain why thicker strings of the same material, same length and same tension produce notes of lower frequency.

67 Sketch the first three harmonics for a pipe that is closed at both ends.

68 A and B are two similar pipes of the same length. A is closed at one end, but B is open at both ends. If the frequency of the first harmonic of A is 180 Hz, what is the frequency of the first harmonic of B?

69 If the frequency used in the demonstration shown in Figure 4.101 was 6.75 kHz and the piles of powder were 2.5 cm apart, what was the speed of the sound waves?

70 What length must an organ pipe (open at one end, see Figure 4.103) have if it is to produce a first harmonic of frequency 90 Hz? (Use speed of sound = 340 m s^{-1})

71 An organ pipe open at both ends has a second harmonic of frequency 228 Hz.
a What is its length?
b What is the frequency of the third harmonic?
c What is the wavelength of the fourth harmonic?
d Suggest one advantage of using organ pipes that are closed at one end, rather than open at both ends.

72 a In an experiment such as that shown in Figure 4.102 using a tuning fork of frequency 480 Hz, the first harmonic was heard when the length of the tube was 18.0 cm. Use these figures to determine a value for the speed of sound in air.
b The experiment was repeated with a different tuning fork of unknown frequency. If the third harmonic was heard when the length of the air column was 101.4 cm, what was the frequency of the tuning fork?

Figure 4.103 On this church organ the wooden pipes are open at one end and the metal pipes behind are open at both ends

Utilizations

Designing microwave ovens

The frequency of the microwaves that are used to cook food is chosen so that they are absorbed by water and other polar molecules in the food. Such molecules are positively charged at one end and negative at the other end. They respond to the oscillating electromagnetic field of the microwave by gaining kinetic energy because of their increased vibrations, which means that the food gets hotter. Most microwave ovens operate at a frequency of 2.45 GHz. This frequency of radiation is penetrating, which means that the food is not just heated on the outside.

To ensure that the microwaves do not pass out of the oven into the surroundings, the walls, floor and ceiling are metallic. The door may have a metallic mesh (with holes) that will reflect microwaves, but allow light to pass through (because of its much smaller wavelength), so that the contents of the oven can be seen from outside.

The microwaves reflect off the oven walls, which means that the walls do not absorb energy, as they do in other types of oven. The 'trapping' of the microwaves is the main reason why microwave ovens cook food quickly and efficiently. However, reflected microwaves can combine to produce various kinds of standing wave in the oven and this can result in the food being cooked unevenly. To reduce this effect the microwaves can be 'stirred' by a rotating deflector as they enter the cooking space, or the food can be rotated on a turntable or stirred by hand.

1 Calculate the wavelength of microwaves used for cooking.

2 Design an experiment that would investigate if there were significant nodes and antinodes in a microwave oven. How far apart would you expect any nodes to be?

Examination questions – a selection

Paper 1 IB questions and IB style questions

Q1 A transverse wave is travelling along a stretched rope. Consider two points on the rope that are exactly half a wavelength apart. The velocities of the rope at these two points are always:

A the same as each other
B constant
C in opposite directions
D in a direction that is parallel to the direction in which the wave is travelling.

Q2 The graph represents a sound wave moving through air. The wave is moving in the direction shown by the arrow. Consider point A.

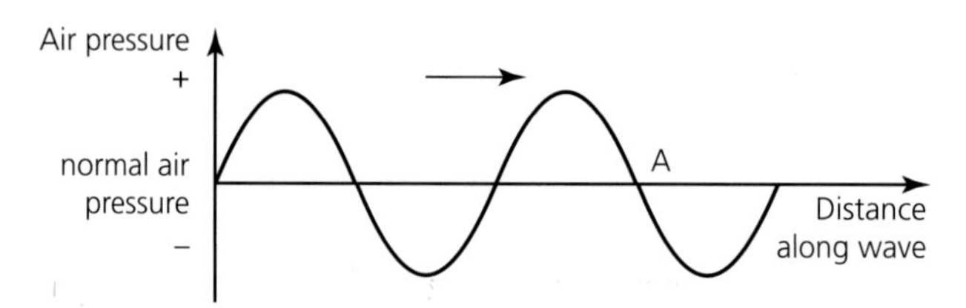

The pressure of the air at A is:

A zero
B always constant
C about to increase
D about to decrease.

Q3 A transverse progressive wave is being transmitted through a medium. The wave speed is best described as:

A the speed at which energy is being transferred
B the average speed of the oscillating particles in the medium
C the speed of the medium
D the speed of the source of the waves.

Q4 Light waves of wavelength λ, frequency f and travelling at speed c enter a transparent material of refractive index 1.6. Which of the following describes the wave properties of the transmitted light waves?

A wavelength 1.6λ; speed c
B frequency $1.6f$; speed $\frac{c}{1.6}$
C wavelength $\frac{\lambda}{1.6}$; frequency f
D wavelength λ; speed $\frac{c}{1.6}$

Q5 When plane wavefronts meet an obstacle they may be diffracted. Which of the following diagrams is the best representation of this effect?

A
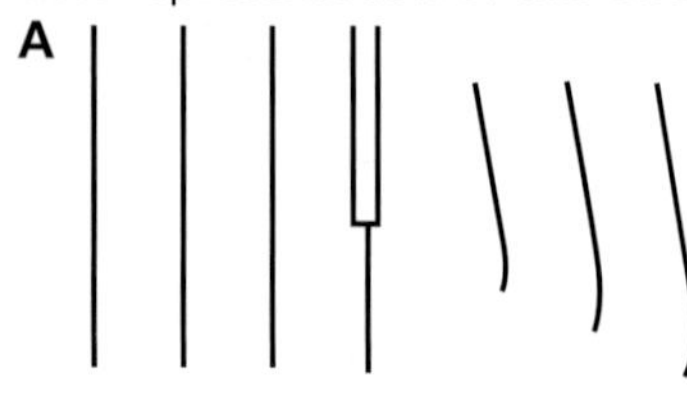

B
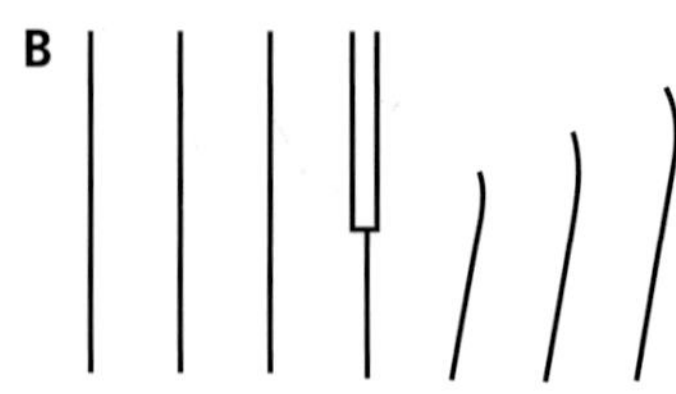

C
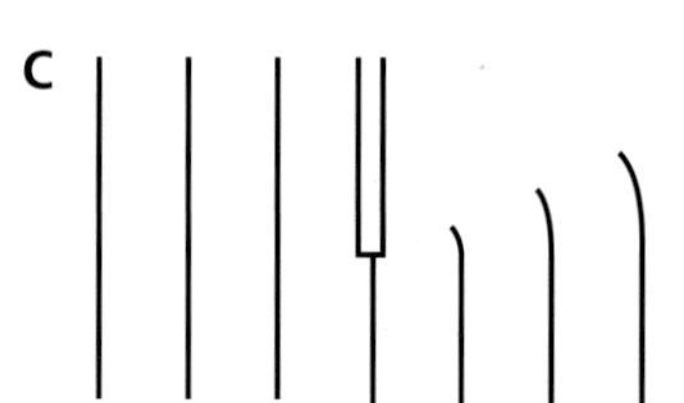

D
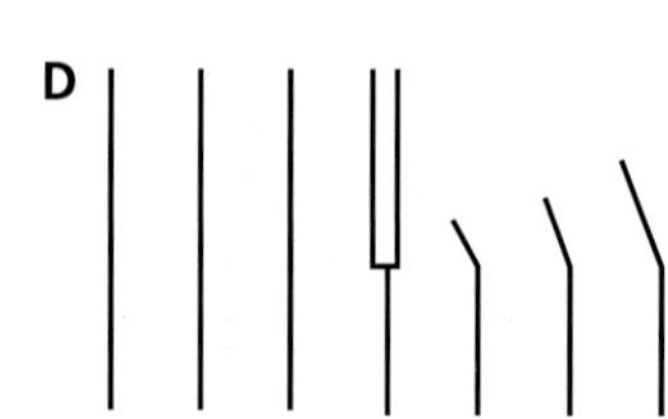

Q6 Which of the following is a possible definition of the wavelength of a progressive (travelling) transverse wave?

A the distance between any two crests of the wave
B the distance between a crest and an adjacent trough
C the amplitude of a particle in the wave during one oscillation of the source
D the distance moved by a wavefront during one oscillation of the source

Q7 The diagram shows a pulse travelling along a rope from X to Y. The end Y of the rope is tied to a fixed support.

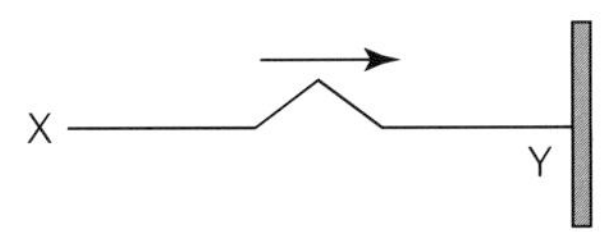

When the pulse reaches end Y it will:

A disappear
B cause the end of the rope at Y to oscillate up and down
C be reflected and be inverted
D be reflected and not be inverted.

Q8 Unpolarized light of intensity I_0 is incident on a polarizer. The transmitted light is then incident on a second polarizer. The axis of the second polarizer makes an angle of 60° to the axis of the first polarizer.

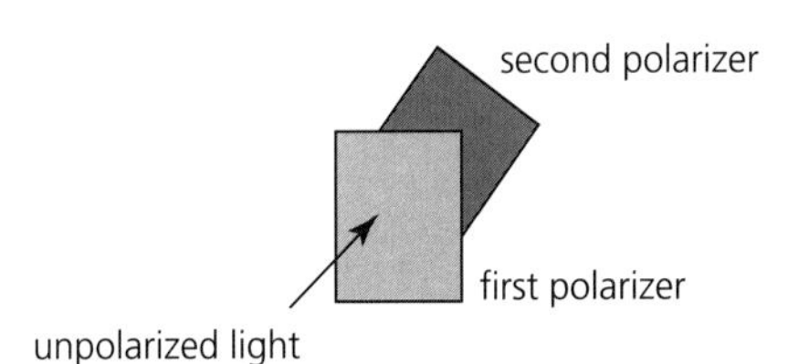

The cosine of 60° is $\frac{1}{2}$. The intensity of the light transmitted through the second polarizer is:

A I_0 **B** $\frac{I_0}{2}$ **C** $\frac{I_0}{4}$ **D** $\frac{I_0}{8}$

Q9 Which of the following describes standing waves on stretched strings?

A All points on the waves move with the same amplitude.
B All points on the waves move with the same phase.
C All points on the waves move with the same frequency.
D The wavelength is the distance between adjacent antinodes.

Q10 A source of sound of a single frequency is positioned above the top of a tube that is full of water. When the water is allowed to flow out of the tube a loud sound is first heard when the water is near the bottom, as shown in all the diagrams. Which diagram best represents the standing wave pattern that produced that loud sound?

A

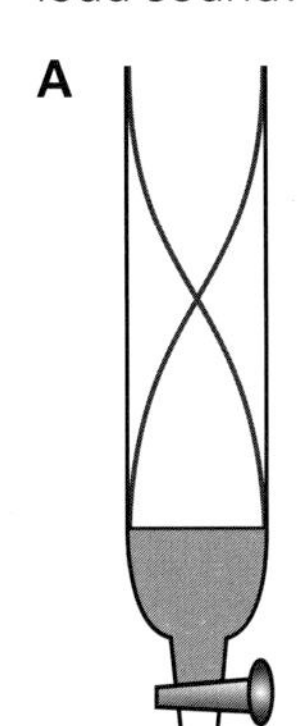

B

C

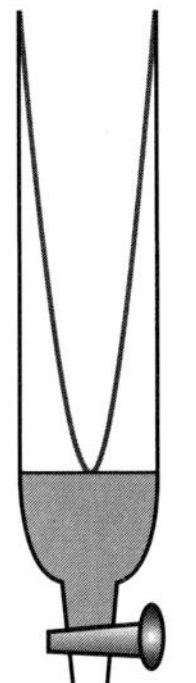

D

Q11 The intensity of a sound at a distance of 2 m from a point source is $1.8 \times 10^{-4}\,\mathrm{W\,m^{-2}}$. Which of the following is the best estimate of the intensity ($/10^{-4}\,\mathrm{W\,m^{-2}}$) 6 m from the source? Assume that there is no absorption of the sound in the air.

A 0.9 **B** 0.6 **C** 0.3 **D** 0.2

Q12 Light may be partially polarized when it is reflected off a surface. Which of the following statements about this effect is correct?

A The material must be a metallic conductor.
B The material must be transparent.
C The amount of polarisation depends on the angle of incidence.
D The effect only occurs for light over a threshold frequency.

Q13 Under certain conditions light travelling in a medium surrounded by air may be totally internally reflected. This effect can only occur if:

A the medium in which the light is travelling has a refractive index of less than 1
B the light strikes the surface of the medium with an angle of incidence greater than the critical angle
C the speed of the light in air is less than the speed of the light in the medium
D the light is monochromatic.

Paper 2 IB questions and IB style questions

Q1 **a** Graph 1 below shows the variation of time t with the displacement d of a travelling (progressive) wave. Graph 2 shows the variation with distance x along the same wave of its displacement d.

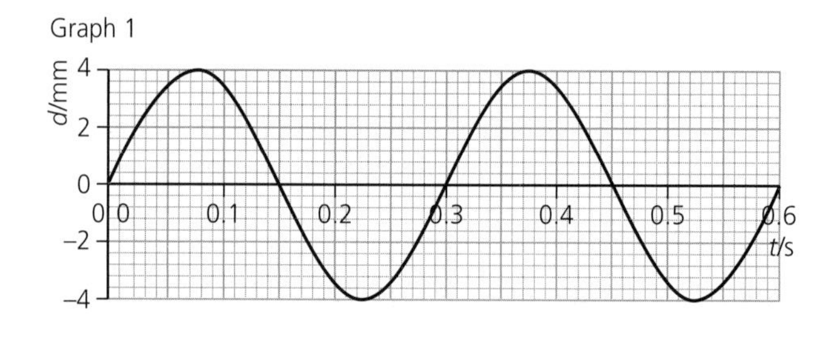

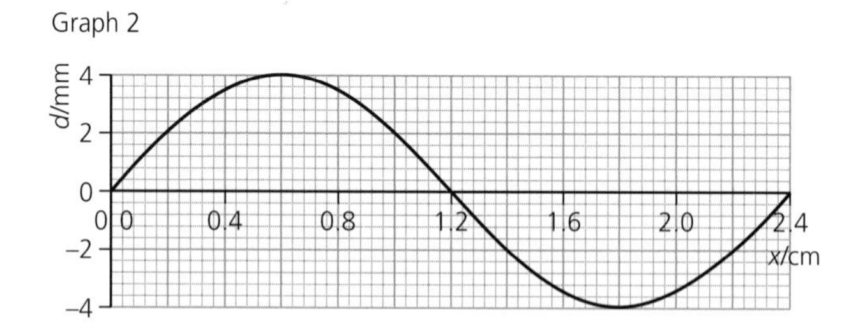

i State what is meant by a travelling wave. (1)
ii Use the graphs to determine the amplitude, wavelength, frequency and speed of the wave. (4)

b The diagram shows plane wavefronts incident on a boundary between two media A and B.

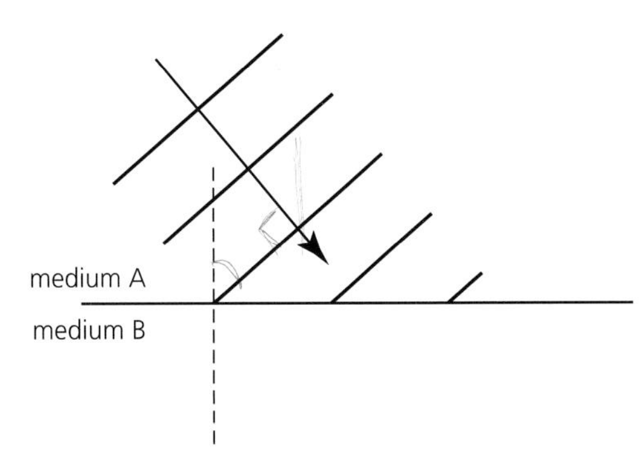

The ratio refractive index of medium B/refractive index of medium A is 1.4.
The angle between an incident wavefront and the normal to the boundary is 50°.

i Calculate the angle between a refracted wavefront and the normal to the boundary. (3)

ii Copy the diagram and construct three wavefronts to show the refraction of the wave at the boundary. (3)

© IB Organization

Q2 An object is vibrating in air. The variation with displacement x of the acceleration a of the object is shown below.

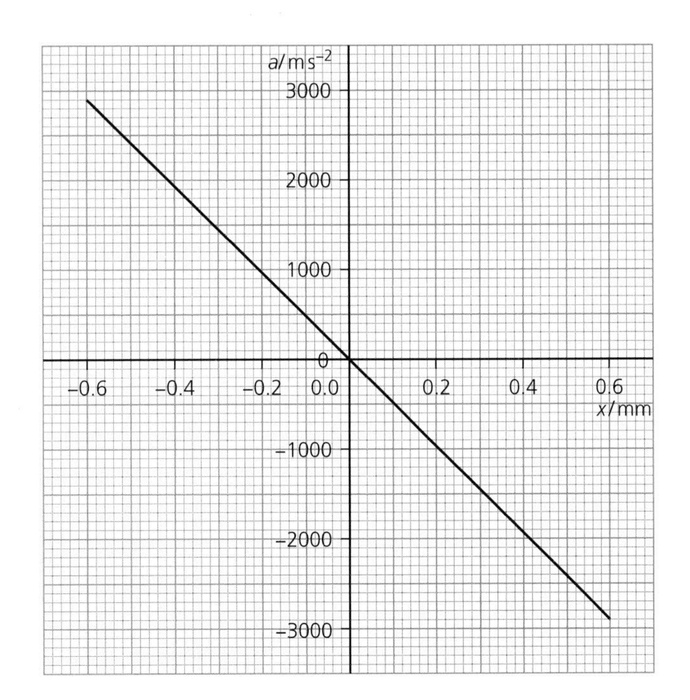

a State and explain two reasons why the graph indicates that the object is executing simple harmonic motion. (4)

b Use data from the graph to show that the frequency of oscillation is 350 Hz. (4)

c The motion of the object gives rise to a longitudinal progressive (travelling) sound wave.

i State what is meant by a longitudinal progressive wave. (2)

ii The speed of the wave is 330 m s^{-1}. Using the answer in b, calculate the wavelength of the wave. (2)

© IB Organization

Q3 A standing wave of sound is set up in an air column of length 38 cm that is open at both ends.

a Write down two ways in which a standing wave is different from a travelling wave. (2)

b If the speed of sound is 340 m s^{-1}, what is the frequency of the standing wave? (2)

5 Electricity and magnetism

ESSENTIAL IDEAS

- When charges move, an electric current is created.
- One of the earliest uses for electricity was to produce light and heat; this technology continues to have a major impact on the lives of people around the world.
- Electric cells allow us to store energy in a chemical form.
- The effect scientists call magnetism arises when one charge moves in the vicinity of another moving charge.

5.1 Electric fields – *when charges move, an electric current is created*

Charge

Electric charge is a fundamental property of some sub-atomic particles, responsible for the forces between them. Because there are two kinds of force (attractive and repulsive), we need two kinds of charge to explain them. We call these two kinds of charge **positive charge** and **negative charge**. The description of charges as 'positive' or 'negative' has no particular significance, other than to suggest that they are two different types of the same thing.

Charge is measured in **coulombs**, C. In the macroscopic world, one coulomb is a large amount of charge for an isolated 'charged' object and therefore we often use *microcoulombs* ($1\,\mu C = 10^{-6}\,C$) and *nanocoulombs* ($1\,nC = 10^{-9}\,C$).

All **protons** have a positive charge of $+1.60 \times 10^{-19}\,C$ and all **electrons** have a negative charge of $-1.60 \times 10^{-19}\,C$. This value for the basic quantity of charge is given in the *Physics data booklet*. **Neutrons** do not have any charge; they are **neutral**.

Charge is generally given the symbol Q, but q is also commonly used to represent the charge on an individual particle. The charge on an electron is represented by the letter e. One coulomb of negative charge is the total charge of 6.25×10^{18} electrons ($1/(1.60 \times 10^{-19})$).

When electrons are added to or removed from a neutral atom or molecule, it is then called an **ion**. This process is called **ionization**.

Law of conservation of charge

When more and more charged particles come together, the overall charge is found by simply adding up the individual charges, taking their signs into account. For example, a certain chlorine atom containing 17 protons ($17 \times (+1.6 \times 10^{-19}\,C)$), 18 neutrons (0 C) and 17 electrons ($17 \times (-1.6 \times 10^{-19}\,C)$) has no overall charge. The atom is neutral. If an electron were to be removed, however, it would become a positively charged ion of charge $+1.6 \times 10^{-19}\,C$.

The charge on electrons and protons ($\pm 1.6 \times 10^{-19}\,C$) is the basic quantity of charge. It is not possible to have a smaller quantity of charge than this in a free particle. All larger quantities of charge consist of various multiples of this fundamental charge. For example, it is possible for the charge on a particle to be $1.6 \times 10^{-19}\,C$, $3.2 \times 10^{-19}\,C$, $4.8 \times 10^{-19}\,C$, $6.4 \times 10^{-19}\,C$ and so on, but intermediate values (for example, $2.7 \times 10^{-19}\,C$) cannot exist. For this reason, charge is said to be **quantized**.

> The law of conservation of charge states that the total charge in any isolated system remains constant.

It is thought that the universe contains (almost) equal amounts of positive and negative charge, so that the total charge is zero. Under certain circumstances it may be possible to create or destroy individual charged particles, but only if the total charge of the system involved remains unchanged.

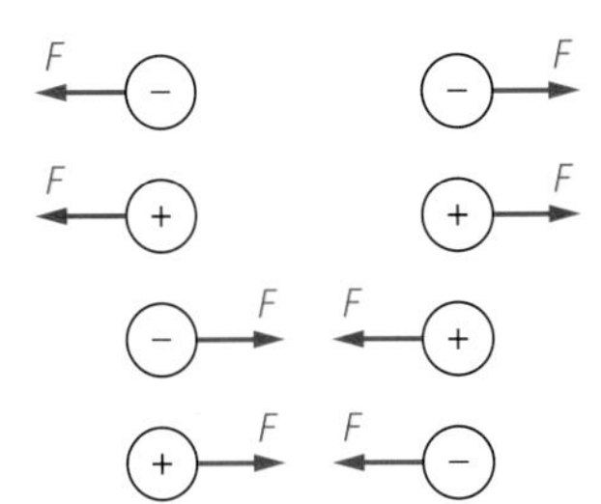

■ **Figure 5.1** Electric forces between similar and opposite charges

Conservation laws like this are very useful tools for physicists. For example, if there are two isolated, identical spheres and one is charged with 4.6×10^{-10} C and the other is neutral, when they touch the charge will be shared so that they both have 2.3×10^{-10} C. The total charge remains the same (assuming that the spheres remain isolated). If the spheres are not identical, the charge will still be shared, but not equally.

Fundamental forces between charges

If a particle has a positive charge it will *attract* a particle that has a negative charge, whereas similarly charged particles (for example, two positive charges) will *repel* each other. These fundamental forces of nature are called electric forces and they are illustrated in Figure 5.1.

Opposite charges attract; like charges repel.

Charging and discharging

Everyday objects contain an enormous number of charged particles (protons and electrons), with approximately equal numbers of positive and negative charges. When we talk about a (previously neutral) object becoming *charged*, it will be because a very large number of electrons have been added to it or removed from it. One common way for this to happen is by friction, such as when brushing dry hair with a plastic brush, as shown in Figure 5.2. If electrons are transferred in the process, one object (such as the hair brush) gains electrons, becoming negatively charged, while removing electrons from the other object (the hair), leaving it with a positive charge. The two objects then attract each other. In this example, individual hairs with similar charge are repelled from each other. Protons, unlike electrons, are located in the nuclei of atoms and cannot be separated and moved from their positions, so they are not involved in producing electrostatic effects.

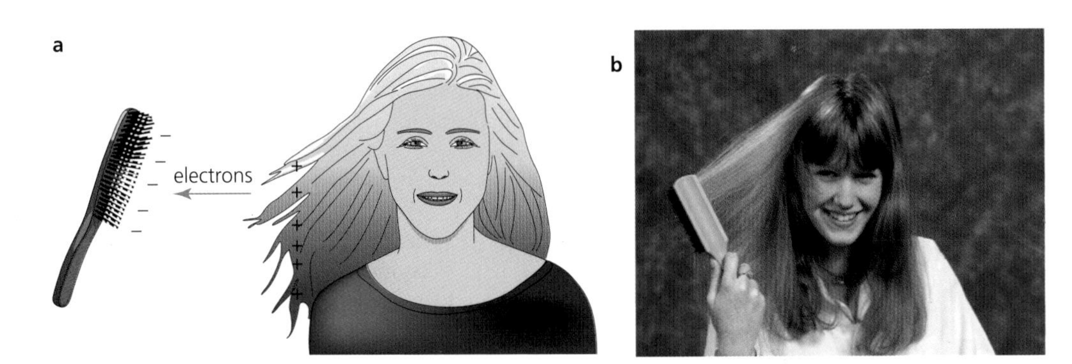

■ **Figure 5.2** When you brush your hair, individual hairs may move apart because of the repulsion between similar charges

Electrostatic effects tend to be noticeable under dry conditions and on insulators, rather than conductors. (Conductors and insulators will be discussed in greater depth later in this chapter see page 209.) This is because electrons move quickly through a metal conductor to discharge a charged object. If we want to be sure that there is no charge on an object, we make a good contact between the object and the ground. This is called earthing.

Large-scale electrostatic effects can be unwanted and even dangerous. For example, cars and planes can become charged as they move along the ground or through the air, and this can be a problem when they stop for refuelling – any sparks from a charged vehicle might cause an explosion of the fuel and air. This risk can be prevented by making sure that the vehicle and the fuel supply are well earthed.

Electric field

We have seen that any charge will exert a fundamental electric force on any other charge in the space around it (without the need for any physical contact). It should be clear that the space around a charge is not the same as a space where there is no charge. We say that an **electric field** exists in the space.

An electric field is a region of space in which a charge experiences an electric force.

In a similar way, a gravitational field exists in a region of space in which a mass experiences a gravitational force (see Chapter 6). We are all aware of the gravitational field around the Earth, but we also live in electric and magnetic fields.

Figure 5.3 shows the results of a demonstration of an electric field pattern around two charged points. The pattern was produced by using semolina seeds on oil. Readers may be familiar with magnetic fields of similar shape, produced with iron filings.

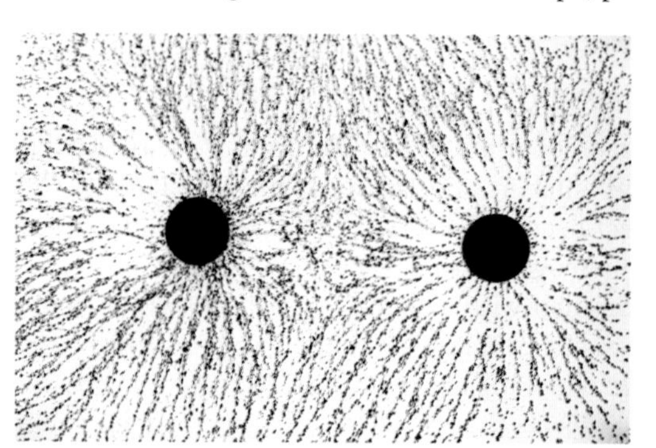

Figure 5.3 Demonstration of an electric field pattern

Electric field strength is defined as the force per unit charge that would be experienced by a small *positive* test charge placed at that point.

Reference is made here to a 'small positive test charge' because a large test charge (compared to the charge creating the original field) would disturb the field that it is measuring. In fact, even a small test charge would probably affect accurate measurement, so this definition may be viewed as theoretical rather than a practical means of measuring electric field strength.

Electric field strength is given the symbol E and has the unit **newtons per coulomb**, N C^{-1}.

$$\text{electric field strength} = \frac{\text{electric force}}{\text{charge}}$$

$$E = \frac{F}{q}$$

This equation is given in the *Physics data booklet*.

Electric field strength can be compared with gravitational field strength, g, which was briefly mentioned in Chapter 2 (g = gravitational force/mass). It is an easy matter to measure the Earth's gravitational field strength: simply determine the weight of a known mass. Similarly, in principle it should be easy to estimate the strength of an electric field by directly measuring the force on a small test charge placed in the electric field. In practical situations, however, this is not easy because the force will be very small and the charge will probably not remain constant.

Field strength is a *vector* quantity; it has a direction, the direction of force. For a gravitational field the force is always attractive and the direction is obvious. However, the direction of force in an electric field will depend on the sign of the test charge. By convention, the direction of the field is chosen to be that of the force acting on a *positive* charge placed in the field (as in the definition above).

Worked example

1 What force will be experienced by a charge of +6.3 μC placed at a point where the electric field strength is $410\,N\,C^{-1}$?

$$E = \frac{F}{q} \text{ so } F = Eq$$

$$F = 410 \times (6.3 \times 10^{-6})$$

$$F = 2.6 \times 10^{-3}\,N$$

Combined electrical fields

At any point on a straight line that passes through two point charges, we can calculate the magnitude of the *combined field* simply by adding the individual fields, taking their directions into account. The resultant is always directed along the line between the point charges. Consider Figure 5.4a. It shows two positive charges, A and B, and a test charge $+q$ in various positions. The direction of the field is the direction of the force experienced by the positive test charge. Along the line to the left of A, both fields are acting in the same leftwards direction. The field due to A will probably be stronger, unless B is a much bigger charge than A. Similarly, to the right of B, the fields are both acting in the same direction. To the left of A and to the right of B the combined field will be stronger than either individual field.

Between A and B the two fields act in opposite directions and there must be a position where the resultant electric field is zero. If the two charges are equal in magnitude as well as sign, this position will be the midpoint between them. If one charge is larger than the other, then the zero electric field will occur closer to the smaller charge.

a Similar charges

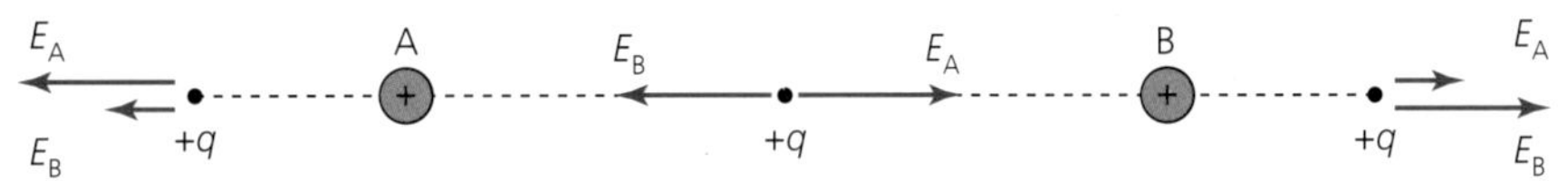

b Opposite charges

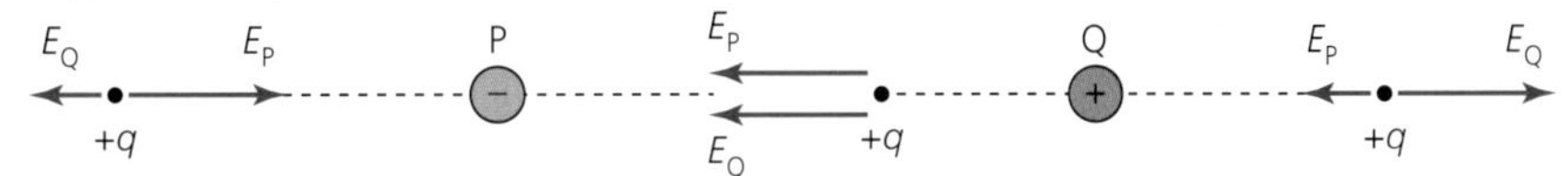

■ **Figure 5.4** The arrows represent the electric fields acting along a line between the two point charges: **a** A and B, **b** P and Q

In Figure 5.4b the two charges have opposite signs and, at any point on the line between them, the two fields will combine to give a stronger electric field. To the left of P and to the right of Q, the individual fields act in opposite directions and the combined electric field will be reduced. If the two charges have equal magnitudes, the combined field will never reduce to zero, but the field may be reduced to zero if one charge is greater than the other.

Experimental investigation of electrical fields in school laboratories is difficult. The use of computer simulations is recommended as an alternative because the combination of two or more electric fields can be investigated quickly and easily.

1 Describe and explain any one electrostatic event that you have experienced in your everyday life.

2 a What is the magnitude of an electric field at a point where a test charge of 47 nC experienced a force of 6.7×10^{-5} N?
  b If a charge of 0.28 μC was placed at the same point, what force would be exerted on it?

3 The electric field close to the Earth's surface is widely quoted to have a value of about $150\,N\,C^{-1}$ downwards.
  a Calculate the force on an electron in a field of this strength.
  b Compare your answer to the downwards force on the electron due to the gravitational field (that is, the electron's weight).

4 Suggest reasons why experimental investigations of electrical fields in school laboratories can be difficult.

5 The electric field strength at a certain point is $2.6 \times 10^6\,\mathrm{N\,C^{-1}}$ acting to the north.
a If this field is combined with another field of strength $4.3 \times 10^6\,\mathrm{N\,C^{-1}}$ acting towards the south, what is the combined field strength?
b Determine the magnitude of the resultant electric field if a third field of strength $3.5 \times 10^6\,\mathrm{N\,C^{-1}}$ to the east is also applied at the same point.

Coulomb's law

The greater the distance between two charged objects (or particles), the smaller the electrical force between them.

Figure 5.5 shows how the relationship between force and separation might be tested for two charged spheres. Two conducting spheres are charged and an electric top-pan balance is used to measure the change in force as the spheres are moved closer together or further apart.

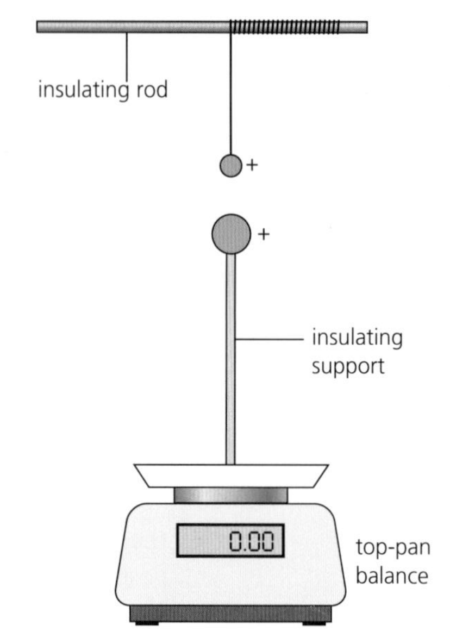

Figure 5.5 Testing Coulomb's law using a top-pan balance to measure the force between known charges

Experiments confirm that the relationship between force and distance follows an **inverse square law** (as discussed in Chapter 4). This is because the electric fields around charged points spread out equally in all directions. This is represented by vector arrows in Figure 5.6.

More detailed experiments show that:

$$F = \text{constant} \times \frac{q_1 q_2}{r^2}$$

where q_1 and q_2 are the two charges and r is the distance between them.

This relationship applies only for charges concentrated at a point. However, charged spherical objects behave as if all their charge were concentrated at a point (the centre of the sphere).

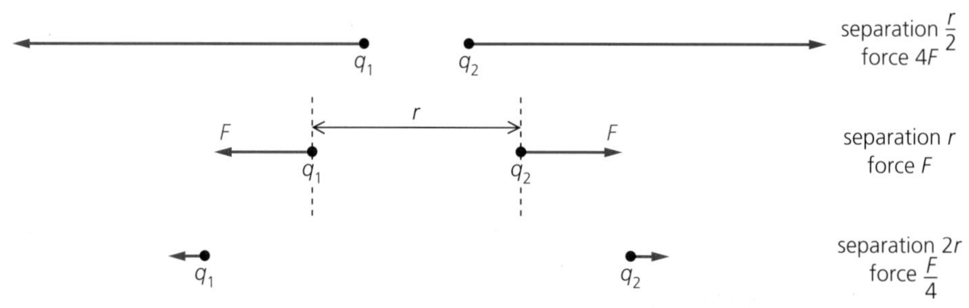

Figure 5.6 The repulsive force varies with distance between similar charges – the vector arrows are not drawn to scale

Putting in the symbol k for the constant (the Coulomb constant), we get:

$$F = k\frac{q_1 q_2}{r^2}$$

This is called Coulomb's law and it is given in the *Physics data booklet*, along with the value of k ($= 8.99 \times 10^9\,\mathrm{N\,m^2\,C^{-2}}$).

The law was first published by Charles Augustin de Coulomb (Figure 5.7) in 1783.

Figure 5.7 French physicist Charles Augustin de Coulomb (1736–1806)

Worked examples

2 What is the electric force on a negatively charged sphere, which has a charge of 0.82 μC, placed with its centre 21 cm away in air from the centre of a sphere charged with +0.47 μC?

$$F = k\frac{q_1 q_2}{r^2}$$

$$F = \frac{(8.99 \times 10^9)(8.2 \times 10^{-7})(4.7 \times 10^{-7})}{0.21^2}$$

$F = 0.079\,\text{N}$

The same force acts on both spheres, but in opposite directions.

3 Figure 5.8 shows two equally charged balls of the same mass suspended from the same point. They both hang at the same angle to the vertical.

Draw diagrams to show what would happen if:

a ball 1 was replaced with a heavier ball with the same charge

b the charge on the original ball 1 was doubled.

a Ball 1 would hang at a smaller angle. Ball 2 would be unchanged.

b The force on both balls would be bigger, so they would both hang at the same, larger angle to the vertical.

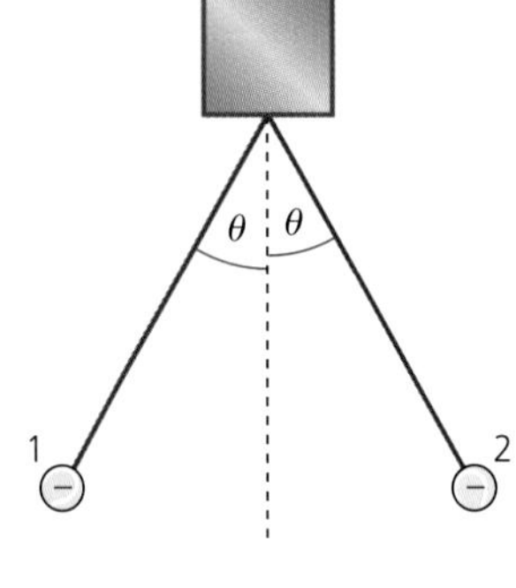

Figure 5.8

Permittivity

The Coulomb constant is sometimes expanded to the form given below. This equation is given in the *Physics data booklet*.

$$k = \frac{1}{4\pi\varepsilon_0}$$

The $\frac{1}{4\pi}$ indicates the radial nature of the force and ε_0 represents the electric properties of free space (a vacuum). ε_0 is called the electrical permittivity of 'free space' and has a value of $8.85 \times 10^{-12}\,\text{C}^2\,\text{N}^{-1}\,\text{m}^{-2}$(as given in the *Physics data booklet*).

Electric forces and fields can pass through a vacuum and the permittivity of free space, ε_0, is a fundamental constant that represents the ability of a vacuum to transfer an electric force and field.

The permittivities of other substances are all greater than ε_0, although dry air has similar electrical properties to free space. This means that the force between two charges in air would be reduced if the air was replaced by another medium.

The permittivity of a medium, ε, is divided by the permittivity of free space to give the relative permittivity, ε_r, of the medium. Some examples are shown in Table 5.1. (Relative permittivity is sometimes known as the *dielectric constant* of the medium.)

relative permittivity = permittivity of medium/permittivity of free space

$$\varepsilon_r = \frac{\varepsilon}{\varepsilon_0}$$

Because it is a ratio, relative permittivity does not have a unit. For example, if the permittivity of a certain kind of rubber was $5.83 \times 10^{-11}\,\text{C}^2\,\text{N}^{-1}\,\text{m}^{-2}$, its relative permittivity would be:

$$\frac{5.83 \times 10^{-11}}{8.85 \times 10^{-12}} = 6.59$$

Using relative permittivity, the general equation for the force between two charges in any medium becomes:

$$F = \frac{q_1 q_2}{4\pi\varepsilon_0\varepsilon_r r^2}$$

Note that the last two equations are *not* given in the *Physics data booklet*.

Free space (a vacuum)	1 (by definition)
dry air	1.0005
polythene	2
paper	4
concrete	4
rubber	6
water	80

■ **Table 5.1** The approximate relative permittivities of some common insulators

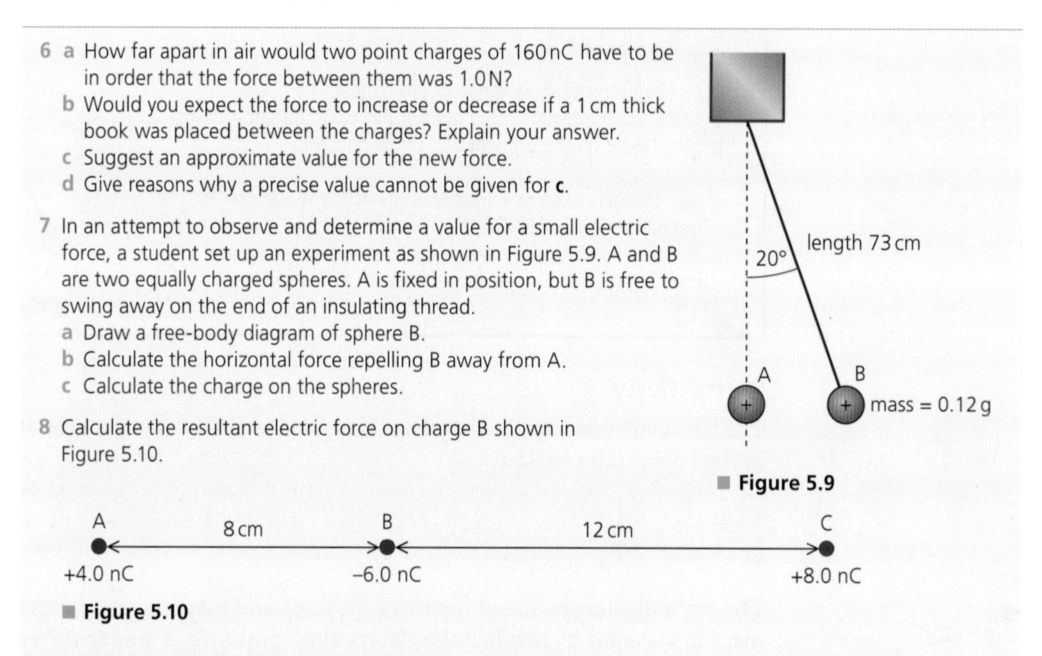

6 a How far apart in air would two point charges of 160 nC have to be in order that the force between them was 1.0 N?
  b Would you expect the force to increase or decrease if a 1 cm thick book was placed between the charges? Explain your answer.
  c Suggest an approximate value for the new force.
  d Give reasons why a precise value cannot be given for **c**.

7 In an attempt to observe and determine a value for a small electric force, a student set up an experiment as shown in Figure 5.9. A and B are two equally charged spheres. A is fixed in position, but B is free to swing away on the end of an insulating thread.
  a Draw a free-body diagram of sphere B.
  b Calculate the horizontal force repelling B away from A.
  c Calculate the charge on the spheres.

■ **Figure 5.9**

8 Calculate the resultant electric force on charge B shown in Figure 5.10.

■ **Figure 5.10**

9 Four equal charges of 0.14 μC are fixed in position at the corners of a square of sides 25 cm. Calculate the resultant electric force on any one charge.

10 A proton has a charge of $+1.6 \times 10^{-19}$ C and the charge on an electron is -1.6×10^{-19} C. Estimate the electric force between these two particles in a hydrogen atom, assuming that they are 5.3×10^{-11} m apart.

11 Outline how you would use the apparatus shown in Figure 5.8 to determine the relationship between force and distance between two charged spheres.

12 The gas freon has a relative permittivity of 2.4. Calculate its permittivity in $C^2 N^{-1} m^{-2}$.

13 Explain why Table 5.1 only lists the relative permittivities of insulators.

■ Electric current

If charged particles can be made to move (flow) from one place to another, they can transfer useful energy. That electrical energy can then be converted to other useful forms in devices like motors, heaters and computers. A flow of charge is called an **electric current**. Electric cables, like those shown in Figure 5.11, transfer large amounts of energy around countries.

■ **Figure 5.11** An electricity pylon of interesting design carrying electrical cables

The ability of electricity to be transmitted and transferred to other forms and to improve human lives cannot be underestimated, but it is thought that about 20% of the world's population still lives without electricity delivered through cables to their homes (see Figure 5.12).

The burning of oil and gas to generate electricity has very important and long-lasting implications (including climate change). This urgent issue, which affects everyone, is discussed at length in Chapter 8.

■ **Figure 5.12** Around 20% of people in the world live without mains electricity

Conductors and insulators

Currents can be made to flow around complete electric circuits because the wires and components are good conductors of electricity. An electrical insulator is a material through which currents cannot flow easily.

For a solid to be a good conductor it needs to contain a large number of electrons that can move around freely. Metals are good conductors because they contain large numbers of free electrons, which are not bound to any particular atom. Such electrons are sometimes described as being delocalized. Insulators have relatively few free electrons. Semi-conductors, such as silicon and germanium, have properties between these extremes and, importantly, their ability to conduct is easily altered by adding small amounts of impurities or by changing external properties such as temperature or light level.

When most solids gets hotter, the increasing vibrations of the atoms, ions or molecules get in the way of any electrons that are moving through the material, tending to make it a worse conductor. But in insulators and semi-conductors, the extra internal energy frees some electrons from their atoms and consequently the material may become a better conductor.

In solids, the atoms, ions or molecules cannot move around freely, so a flow of positive charges is not possible, but the structure of metals means that some electrons are free to move. Figure 5.13 shows free electrons moving through a metal. The electrons represent an electric current flowing through a conductor. Even when a current is not flowing, the free electrons are still moving around randomly at very high speeds, like molecules in a gas.

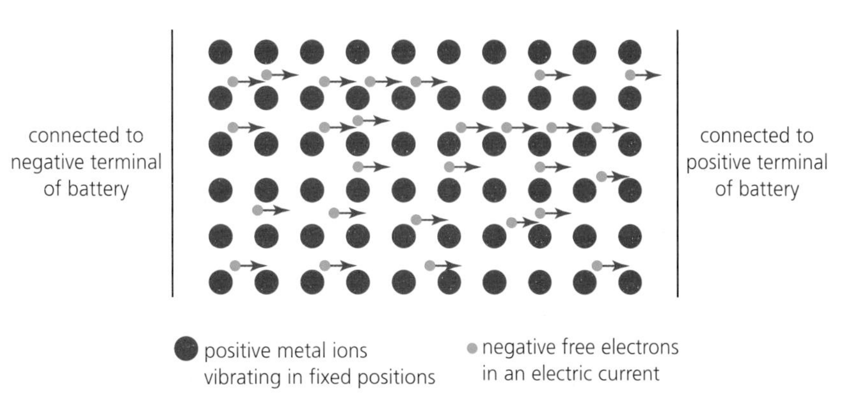

■ **Figure 5.13** Electric current flowing through a metal

The transfer of energy in electric circuits involves the movement of electrons. The energy is transferred from the energy source (such as a battery) to one or more components in the circuit. These components transfer the electrical energy into another form of energy, which

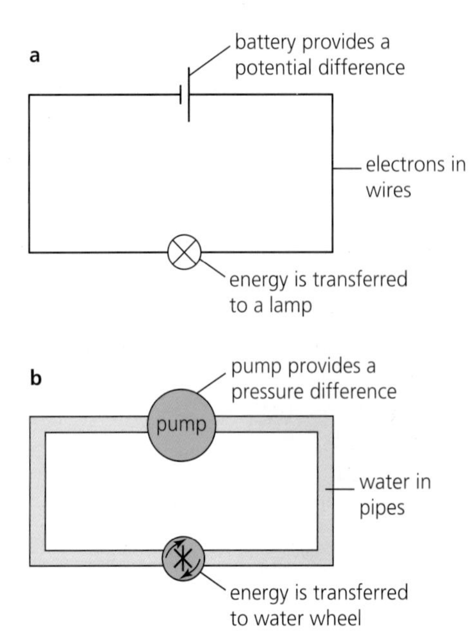

■ **Figure 5.14**
An electric circuit, **a**, can be compared to a water circuit, **b**.

can be put to good use. For example, a battery in a torch transfers chemical energy to electrical energy, and that energy is then transferred to light energy and internal energy in the torch bulb. Remember that electrons are present all around circuits all the time; they do not originate in the battery.

It may be useful to compare a battery in an electric circuit to a pump forcing water around pipes in a water circuit, as shown in Figure 5.14. The pump provides a *pressure* difference that moves the water. The water is then able to do useful work (maybe turning a water wheel) as it moves around the circuit. In an electric circuit, the battery is considered to provide a voltage (also called a *potential* difference). This idea will be discussed further later in this chapter.

Nature of Science

Modelling in things we cannot see

We cannot see a flow of electricity, unlike a flow of water, which can make electricity a difficult subject to study. But, more than that, an understanding of electricity requires an acceptance of a theory (model) about the fundamental particulate structure of matter and the nature of atoms.

Trying to explain everyday (macroscopic) observations by developing theories about the behaviour of microscopic particles that we cannot even see is a common activity among scientists. Understanding the nature of an electric current is just one example of this, but there are many other examples in which students of physics need to accept models of reality that they cannot see with their own eyes; the nature of electromagnetic waves for example, or the kinetic theory of gases. The scientists who have developed these models deserve great respect because of their considerable insight and imagination.

In the absence of a better model, it is sensible to accept the current model for conduction in metals, most especially because the model has been verified by countless experiments over many years, all of which have failed to disprove it.

Defining electric current

Electric current (given the symbol I) is the rate of flow of charge, defined by the equation

$$\text{current} = \frac{\text{charge flowing past a point}}{\text{time}}$$

$$I = \frac{\Delta q}{\Delta t}$$

This equation is given in the *Physics data booklet*.

The unit of current is the **ampere**, A, which is usually shortened to *amp*. If one coulomb of charge passes a point in a circuit in one second, we say that the current is one amp (1 A = $1\,\text{C}\,\text{s}^{-1}$). The units mA and μA are also in common use. A current of 1 A in a metallic conductor means that 6.25×10^{18} electrons are passing a point every second.

The ampere is one of the fundamental units of the SI system. It is defined in terms of the force per unit length between current-carrying conductors. The coulomb is a derived unit defined in terms of the amp: one coulomb is the charge flowing per second if the current is one amp.

Direct current (dc)

A current that always flows in the same direction around a circuit is called a **direct current** (dc). Batteries supply dc. If a current repeatedly changes direction, it is called an **alternating current** (ac). Usually dc is more useful than ac, but most electrical energy is transferred around countries and to our homes using alternating currents and high voltages because less energy is wasted to thermal energy (see Chapter 11).

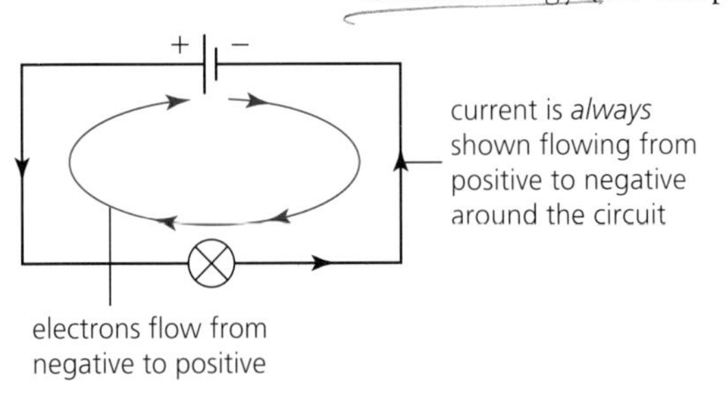

■ **Figure 5.15** Conventional current flow is from positive to negative around the circuit

When drawing circuits, it is common to draw an arrow to show the direction of flow of a direct current, but this cannot be done for alternating currents. For reasons of consistency, direct electric current is *always* shown flowing around the circuit from the positive terminal of the battery, or other energy source, to the negative terminal. This is the direction in which mobile positive charges would move (if they could). Because electric currents in solids are flows of negatively charged electrons, the arrows are in the *opposite direction from the movement of electrons* (as shown in Figure 5.15). The meaning of the positive and negative signs on batteries will be discussed later in this chapter (see Section 5.3).

ToK Link

Names, labels and conventions should not be confused with understanding

Early scientists identified positive charges as the charge carriers in metals, however the discovery of the electron led to the introduction of 'conventional' current direction. Was this a suitable solution to a major shift in thinking? What role do paradigm shifts play in the progression of scientific knowledge?

Atoms consist of a central nucleus containing positively charged protons, with negatively charged electrons around the outside. This is our understanding of atoms – and indeed what we teach in schools – but is this a scientific *fact*? If, instead, scientists said that electrons were positively charged and protons were negatively charged, it would not make any difference to our understanding of the universe. (In fact, negative protons and positive electrons *can* exist – they are called antimatter.)

The use of the terms *positive* and *negative* is just a way of describing the way in which they interact (such as positive attracts negative); reversing the labels does not change the meaning. But without using terms like these we would not be able to transfer our knowledge of fundamental physics to other people easily.

By agreement, the 'conventional' direction of electric current was chosen to be from positive to negative more than 200 years ago, a long time before the nature of an electric current was understood. After it was discovered that the currents in metallic conductors are actually a flow of negatively charged electrons (from negative to positive), there was no need to change the conventional direction that, under most circumstances, had been in use for a long time. As a result, the same convention remains in use today, despite the fact that it is not strictly 'true'. (Remember, however, that some electric currents *are* a flow of positive ions.)

Measuring electric currents with ammeters

Ammeters are the devices used to measure electrical currents. If we want to measure the current in a connecting wire in the circuit shown in Figure 5.15, an ammeter must be connected as shown in Figure 5.16. Because the battery produces a direct current, the ammeter must be connected the correct way around and this is indicated by the labelling, or colours, on the ammeter terminals. A red (+) terminal indicates that that side of the ammeter should be connected to the positive side of the battery. A typical laboratory has a selection of ammeters with different ranges and sensitivities and it is important to choose the correct ammeter for any particular circuit. Alternating currents cannot be measured using dc meters.

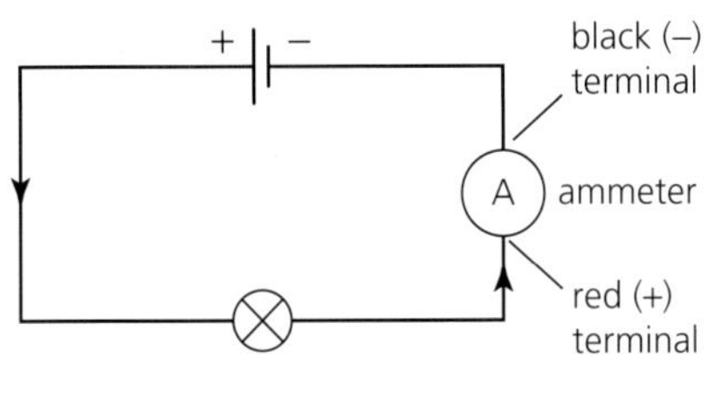

■ **Figure 5.16** Measuring current with an ammeter

When an ammeter is added to a circuit, it should not affect the current that it is trying to measure. This means that an ammeter needs to be a very good conductor of electricity.

14 a How much charge flows through a lamp in one second if the current is 0.25 A?
b How many electrons will have flowed through the lamp in this time?

15 a How much charge has passed through a battery in one hour if the current is 120 mA?
b How long will it take for a total charge of 100 C to flow through the battery?

16 Find out how it is possible for an electric current to flow through a fluorescent lamp.

17 Many people think that electricity pylons are ugly. Figure 5.11 shows a more interesting design but one which, no doubt, is much more expensive than usual.
a Suggest why this is an expensive design.
b Make a sketch of a pylon design that you hope could be practicable, good to look at and affordable.

Nature of Science

The observer effect

When taking any scientific measurement there is always the possibility that the act of taking the measurement will change what is being measured. In this section, the use of electric meters is discussed but there are many other examples, including the use of thermometers and pressure gauges. When measuring the pressure in a car tyre, as in Figure 5.17, some of the air in the tyre must flow into the pressure gauge. This will result in a reduction of pressure, although admittedly it will probably be a very small change. Doctors are well aware that a patient's blood pressure may well rise when it is being measured because of psychological effects.

■ **Figure 5.17** Measuring the air pressure in a car tyre

18 Explain how the use of a thermometer may affect the temperature that it is being used to measure.

19 Give another example of the 'observer effect' (other than those already mentioned).

How fast does electricity flow?

This interesting question has more than one answer. In our modern world, we have all become dependent on the fact that electric circuits can provide 'instant' power and communications. It is worth considering how this is possible.

A current in a metal wire is a flow of free electrons, so one way to answer this question is to consider the speed of the electrons. Even in a metal without a current, the free electrons move randomly like the molecules of a gas, only faster. A typical random speed might be $10^6\,\text{m s}^{-1}$. When a current is flowing, the electrons also move along the wire. Their net speed along the wire is called their **drift speed**.

In order to link the drift speed along the wire to the size of the current, we need to consider how many free electrons are available and the width of the wire. Figure 5.18 shows a current, I, flowing through part of a wire of cross-sectional area A.

During time Δt an electron travelling at a drift speed v along the wire will move a distance of $v\Delta t$. The number of electrons flowing past point P in time Δt is equal to the volume $v\Delta tA$ (shown by the shaded section in Figure 5.18) multiplied by the number of free electrons in unit volume, n. (Every material has a specific value for n, which is sometimes called **charge density**.)

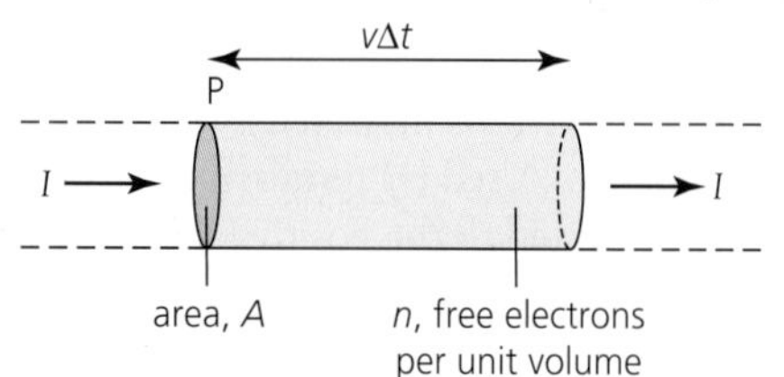

■ **Figure 5.18** Deriving $I = nAvq$

Because current = charge/time, the current in the wire equals the number of electrons passing a point in unit time multiplied by the charge on each electron, q. (Although we are considering electrons in this analysis, we are using q here, rather than e, to show that the equation could be applied to other types of charge.)

$$I = \frac{v\Delta tA \times n \times q}{\Delta t}$$

$$I = nAvq$$

This equation is given in the *Physics data booklet*.

20 Calculate the net speed (in mm s^{-1}) of free electrons in a wire of cross-sectional area 1.0 mm^2 if the current is 1.0 A and the metal has 1.0×10^{28} free electrons per cubic metre.

21 Consider a circuit that contains conductors of different thickness and different materials.
  a What must happen to the speed of free electrons as they move from a thicker wire to a thinner wire of the same material? (Assume that the current is the same.) Explain your answer.
  b Modern electronics makes much use of semi-conductors. Suppose that a particular semi-conductor has one million times fewer charges per m^3 than the metal in **a**. Compare the probable speeds of charges passing through these materials for the same current. What assumption(s) did you have to make?

22 a The density of copper is 8960 kg m^{-3} and its molar mass is 63.5 g. If each atom of copper provides one free electron, determine the number of free electrons in one cubic metre of copper.
  b Use the internet to find out which well-known metals have more free electrons per cubic metre than copper.

Calculations like that in question 20 show that the drift speed of electrons carrying currents in metal wires is usually less than a millimetre per second. This is a tiny fraction of their random speed. The low speed is surprising because we are used to electric devices working immediately when we turn them on. But remember that we should not think of the current as starting at the battery or the switch. It is better to think of all the free electrons, wherever they are in the circuit, starting and stopping at the same time. However, in reality, this cannot be perfectly true. When a switch is turned on, the battery or power supply sets up an electric field which moves around the circuit at a speed close to the speed of light (3×10^8 m s^{-1}). As the electric field reaches individual electrons they experience a force that starts their net motion.

Energy transfer in electric fields

To move a charge in an electric field usually involves a transfer of energy. Work is done because a force moves through a distance (unless the movement is perpendicular to the electric field and force). As a generalized example, consider Figure 5.19 which shows a small (test) charge, $+q$, in the electric field created by the larger charge, $+Q$. In this example, the charges are both positive, so there is a repulsive force, F_R, between them.

If q is moved in any direction other than along the circumference CC′, work has to be done because there is a component of the repulsive force acting on q. An external force will be needed to move q closer to Q, such as to position A, but if q moves away from Q, to B for example, then the field will do the work on q.

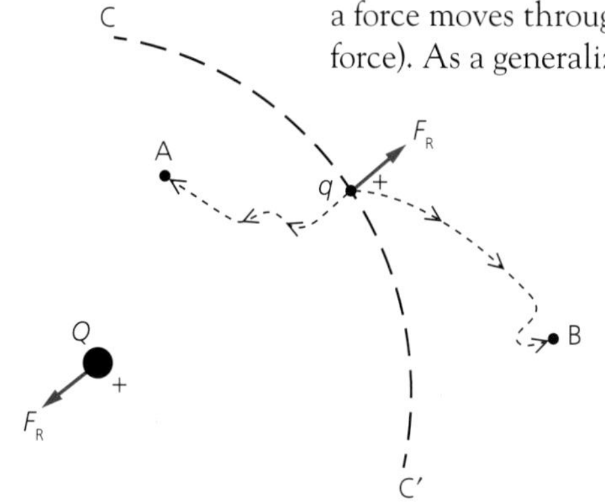

■ **Figure 5.19**

The work done (energy transferred) when moving *unit charge* between two points is a very important quantity in physics. It is called (electric) **potential difference**, commonly abbreviated to **p.d.**, and given the symbol *V*. The p.d. between two points does not depend on the path taken. (It is not necessary for Standard Level students to understand the concept of *potential*, which is covered in Chapter 10 for Higher Level students.)

Potential difference

The potential difference, *V*, between two points is defined as the work, *W*, that would have to be done *on* unit positive charge (+1 C) to move it between those points.

$$\text{potential difference} = \frac{\text{work done}}{\text{charge}}$$

$$V = \frac{W}{q}$$

This equation is given in the *Physics data booklet*.

(If a positive charge is attracted by a negative p.d., the work done will be *by* the electric field because the charge accelerates, so that W is negative. Similarly, the work done will be negative if a negative charge is attracted by, and is accelerated towards, a positive p.d.)

The unit of potential difference is the **volt**, V. (Unusually for physics, the quantity and the unit of potential difference have the same symbol, although quantities are shown in italics.) Potential difference is commonly referred to as **voltage**. A p.d. of 1 volt between two points means that 1 joule of energy is transferred when 1 coulomb of charge moves between the two points.

$$1\,\text{V} = 1\,\frac{\text{J}}{\text{C}}$$

The concept of potential difference is useful wherever and whenever charges may move, whether that is between points in space or between points along wires in an electrical circuit. Perhaps its most common everyday use is in the description of batteries and power supplies.

Batteries, and other electrical energy sources, provide potential differences (voltages) across the circuits in which they are connected because they transfer energy to the charge flowing through them. The positive terminal of a battery attracts electrons and the negative side repels them, so that a current can be sent around a circuit.

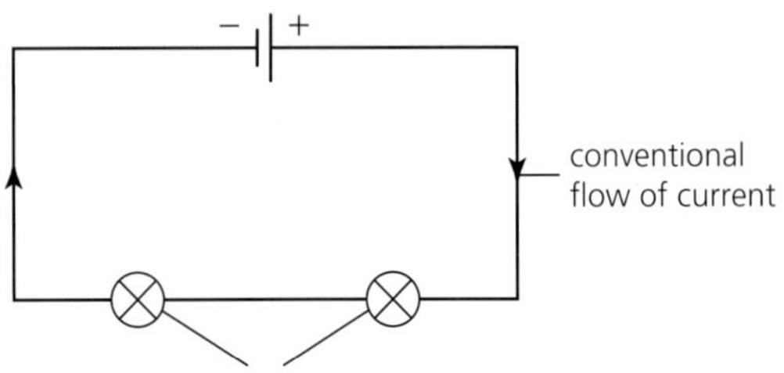

a 1.5 V cell transfers 1.5 J to every coulomb of charge passing through it

conventional flow of current

two lamps each transfering energy from electricity to light and thermal energy at a rate of 0.75 J per coulomb

■ **Figure 5.20** Energy transfers in a simple circuit

In Figure 5.20, a 1.5 V cell (battery) is transferring 1.5 J of energy *to* each coulomb of charge that flows through it. If the battery is connected to two identical lamps, each will have a p.d. of 0.75 V across it and energy will be transferred from electricity to light and thermal energy at a rate of 0.75 J *from* every coulomb flowing through them.

Measuring potential differences with voltmeters

Because p.d. = energy transferred/charge, there will be a p.d. *across* any component of any circuit that is transferring energy. ('Across' means between the two terminals.) **Voltmeters** are the devices used to measure potential difference (voltages) in circuits.

Consider Figure 5.20 which shows a circuit consisting of two lamps, a battery and three connecting leads. The lamps and the battery will have potential differences across them, but the three leads should be very good conductors so that no energy is transferred in them and there will be no p.d. across them. Figure 5.21 shows how voltmeters can be connected across components to measure potential differences.

Because the battery produces a direct current, the voltmeters must be connected the correct way around. This is indicated by the labelling, or colours, on their terminals. A typical laboratory has a selection of voltmeters with different ranges and sensitivities and it is important to choose the correct voltmeter for any particular circuit. Alternating voltages cannot be measured with dc meters.

It is important to realize that when a voltmeter is added to a circuit, it should not affect the p.d. that it is trying to measure. This means that a voltmeter needs to be a very poor conductor of electricity, so that very little current flows through it.

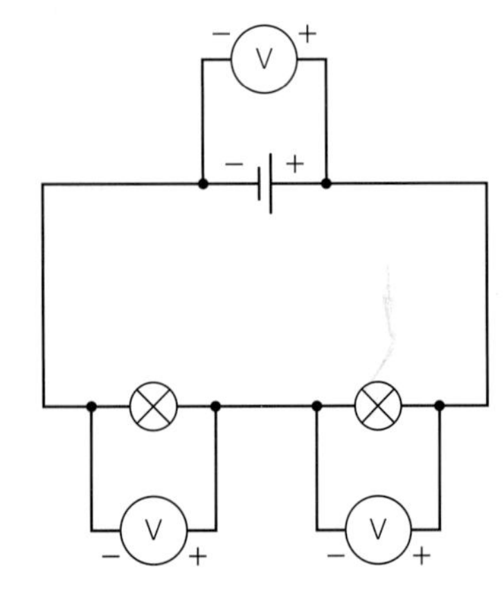

■ **Figure 5.21** Measuring potential differences

Worked examples

4 **a** How much work must be done to move a charge of +2.4 nC through a potential difference of +9.6 V?
b Is the work done on the charge by an external force or by the electric field?

a $W = Vq = 9.6 \times (2.4 \times 10^{-9}) = 2.3 \times 10^{-8}$ J

b The positive charge will experience a repulsive force in the field, so an external force must do work on the charge.

5 A torch bulb has a p.d. of 2.9 V across it when a current of 0.53 A is flowing through it.
a How much energy is transferred by each coulomb of charge flowing through the bulb?
b Describe the energy transfers in the bulb.
c How many electrons flow through the bulb every second?
d How much energy is transferred by each electron?

a 2.9 J

b Electrical energy is transferred to light and thermal energy.

c $\dfrac{0.53}{(1.6 \times 10^{-19})} = 3.3 \times 10^{18}$

d $W = Vq = 2.9 \times (1.6 \times 10^{-19}) = 4.6 \times 10^{-19}$ J

23 a What is the p.d. between two points if $+5.7 \times 10^{-6}$ J of work is needed (from an external force) to move a charge of $+1.3 \times 10^{-7}$ C between them?
b How would your answer change if the charge involved was -1.3×10^{-7} C?

24 a How much energy is transferred when 4.7 C of charge pass through a power supply rated at 110 V?
b How much charge must flow through the supply in order to transfer 1.0 MJ of energy?
c How many electrons are needed to transfer 1.0 MJ of energy?

25 a What is the p.d. across a water heater if 44 000 J of energy is transferred to internal energy when 200 C of charge flows through it?
b What charge has passed through a 1.5 V cell if 60 J of energy were transferred?

26 Suggest what would happen in the circuit shown in Figure 5.16 if the ammeter was removed and replaced by a voltmeter.

The electronvolt as a unit of energy in atomic physics

When an electron moves through a p.d. of 1.0 V, the work done = charge × p.d.:

$$W = (1.6 \times 10^{-19}) \times 1.0 = 1.6 \times 10^{-19}\ \text{J}$$

Physicists often avoid this kind of calculation (and the small numbers involved) by simply saying that energy of one **electronvolt**, eV, has been transferred to, or from, the electron when it moves through a p.d. of 1 V.

The electronvolt, eV, is defined as the work done when a charge of 1.6×10^{-19} C is moved through a p.d. of 1 V.

Clearly, 1 eV = 1.6×10^{-19} J

$$W(\text{J}) = W(\text{eV}) \times (1.6 \times 10^{-19})$$

This equation is *not* given in the *Physics data booklet.*

In fact, the electronvolt is a widely used unit of energy when referring to the energy of *any* atomic particle. The units keV and MeV are also in common use. When *any* singly charged particle (with charge of 1.6×10^{-19} C) moves through a p.d. of 1 V, the work done is 1 eV. The work done when a *doubly* charged ion moves through a p.d. of 1 V is 2 eV etc.

The electronvolt is also very widely used as a unit of kinetic energy for atomic particles (whether or not they move through a p.d.) and for the energy transferred by photons of electromagnetic energy.

Worked examples

6 How much energy is gained by a proton accelerated towards a negatively charged metal plate by a p.d. of 3500 V? Give your answer in
a electronvolts
b joules.

a 3500 eV
b $3500 \times (1.6 \times 10^{-19}) = 5.6 \times 10^{-16}$ J

7 How many keV of energy are transferred to, or from, an ion of charge 3.2×10^{-19} C when it moves through a p.d. of 4000 V?

$$\left(\frac{3.2 \times 10^{-19}}{1.6 \times 10^{-19}}\right) \times 4.0 = 8.0\,\text{keV}$$

8 An electron of mass 9.1×10^{-31} kg is accelerated by a p.d. of 3000 V.
a Calculate its final speed.
b What assumption(s) did you make?

a Energy transferred to electron = kinetic energy

$$Ve = \frac{1}{2}mv^2$$

$$3000 \times (1.6 \times 10^{-19}) = \frac{1}{2} \times (9.1 \times 10^{-31}) \times v^2$$

$$v = 3.2 \times 10^7\,\text{m s}^{-1}$$

b This calculation assumed that the electron started with negligible kinetic energy and that no kinetic energy was transferred away from it during the acceleration (because it was in a vacuum).

27 An electron is moving directly towards a negatively charged plate with a speed of $1.30 \times 10^7\,\text{m s}^{-1}$.
a What is the kinetic energy of the electron in:
i joules
ii electronvolts?
b If the voltage on the plate is −500 V, will the electron reach the plate? Explain.

28 a A particle emitted during radioactive decay had an energy of 2.2 MeV. What was its energy (in joules)?
b If its mass was 6.8×10^{-27} kg, what was its speed?

29 Calculate the gain in kinetic energy (in joules) of a proton that is accelerated across a vacuum by a p.d. of 5000 V.

30 A photon of light transfers about 2.0 eV of energy.
a Convert this energy to joules.
b How many photons would be emitted every second from a light bulb emitting 4W of light energy?

Nature of Science

The useful effects of electricity

Scientists have known for nearly 200 years that when an electric current flows through a substance, it can have (only) three basic effects:

- a *heating effect*, because the moving charges 'collide' with atoms and ions in the substance
- a *chemical effect* in some liquids and gases, because charges moving in opposite directions cause chemical decomposition
- a *magnetic effect* around the current.

Compared to this short list, any list of useful electric and electronic devices found in the modern world is both surprisingly long and extremely varied. Most of us have become very dependent on the use of these devices and if we are deprived of electrical power for our homes or mobile devices, we can get annoyed and unhappy very quickly.

The rate of scientific and technological advance in the use of electricity has been enormous. In the course of about 250 years we have gone from the first discovery of electric currents to, for example, smart phones, brain scanners, electric cars, espresso coffee machines and air conditioners. The world now, for most people and societies, is incredibly different from the way people lived 200 years ago, and most would say that the changes (delivered by electric currents) are generally for the better.

31 Find out which scientists are well known for their early research into electric currents, and when and where they carried out their investigations.

32 Write down the names of five electrical devices in your home that use:
  a the heating effect of a current
  b the magnetic effect of a current. (The magnetic effect of electricity is used to produce motion.)
  c Do you have any electric devices in your home that involve chemical reactions?

33 Some 'reality' TV shows have featured people trying to live without some or all modern 'conveniences'. The Amish people of Pennsylvania, USA famously live in a way which rejects modern technology (see Figure 5.22). What are the possible attractions of a life without electricity?

■ **Figure 5.22** An Amish buggy

5.2 Heating effect of electric currents – *one of the earliest uses for electricity was to produce light and heat; this technology continues to have a major impact on the lives of people around the world*

The heating effect of an electric current is used extensively throughout the world to heat water for cleaning and cooking, and to heat air (directly or indirectly) to keep homes and working environments warm. This heating effect is also employed in devices like irons and toasters, as well as filament lamps which need to get very hot before they will emit light.

ToK Link

There may be risks in exploring the unknown

Sense perception *in early electric investigations was key to classifying the effect of various power sources – however this is fraught with possible irreversible consequences for the scientists involved. Can we still ethically and safely use sense perception in science research?*

The dangers of electric currents passing through a human body are well understood now, but in early investigations, the risks were not fully controlled or understood and this resulted in several fatalities. Figure 5.23 illustrates a very famous and dangerous experiment carried out by Benjamin Franklin.

■ **Figure 5.23** Benjamin Franklin famously flew a kite in a lightning storm as part of his investigations into electricity

■ Circuit diagrams

Circuit diagrams are used extensively to show how electric components are connected together to perform useful tasks. The connecting wires (leads) should be drawn straight and parallel to the sides of the paper.

All circuits need at least one source of energy. A **cell** is a single component that transfers chemical energy to electrical energy. When cells are connected together they are called a **battery**, although in everyday language a cell is also commonly called a battery. The longer line represents the positive terminal of a cell.

Nature of Science

The use of common symbols

The communication of scientific information and ideas between different countries and cultures can be affected by language problems, but this is greatly helped by the use of standard symbols for physical quantities (and units) and for electrical components. Increasingly, English is being used as the international language of science but, naturally, there are many individuals, organizations and countries who prefer to use their own language. Imagine the confusion and risks that could be caused by countries using totally different symbols and languages to represent the circuitry on, for example, a modern international aircraft.

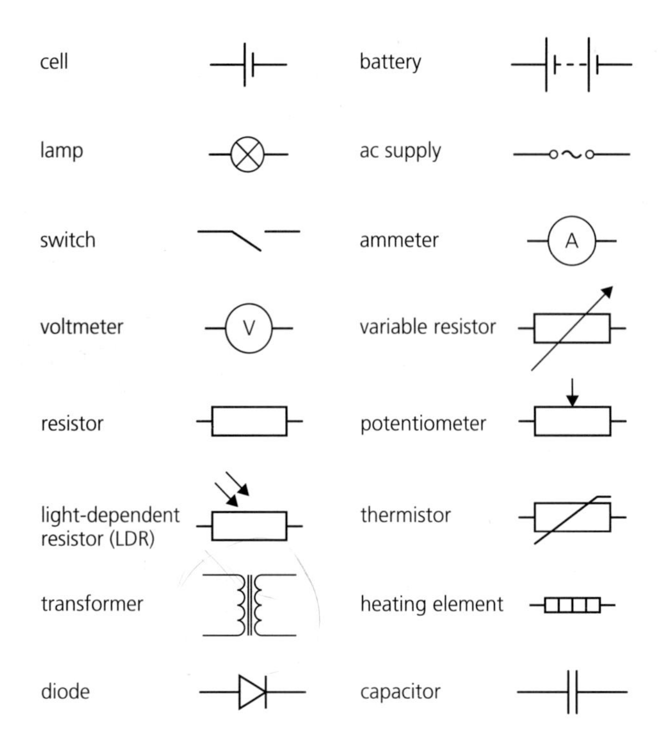

■ **Figure 5.24** Summary of the circuit symbols used in this course

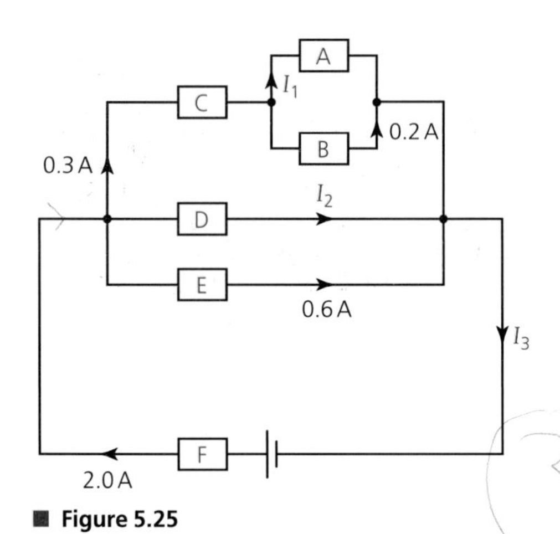

■ **Figure 5.25**

Figure 5.24 shows all the internationally agreed electrical circuit symbols that will be used in this course. You will find them in the *Physics data booklet.*

Components can be connected in series or in parallel. Connecting **in series** means that the components are connected one after another, with no alternative paths for an electric current. The same current flows through all the components if they are in series. If components are connected **in parallel** it means that the current divides because it can take two or more different paths between the same two points in the circuit. Any point in a parallel connection where wires join together and the current can take more than one path should be shown with a dot in the circuit diagram at that junction.

It is common to have components in series and components in parallel within the same circuit. Consider Figure 5.25: A and B are in parallel with each other, but together they are in series with C. The combination of A, B and C is in parallel with D and E. All of these are in series with F and the cell.

■ Kirchhoff's circuit laws

These two laws are used to help predict currents and voltages in circuits. They apply the laws of conservation of charge and conservation of energy to electric circuits.

Kirchhoff's first law

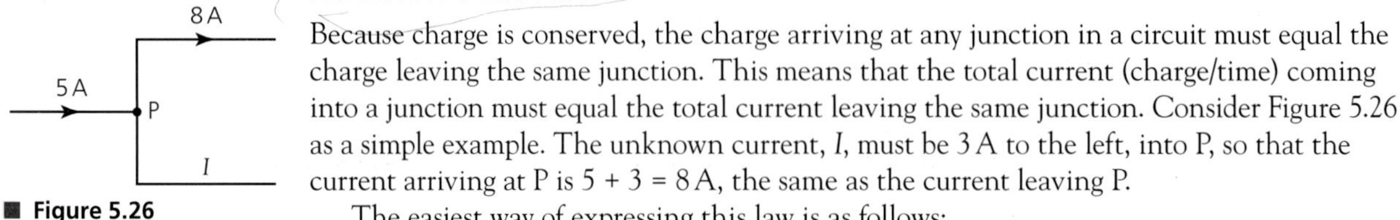

■ **Figure 5.26** Currents arriving at a point

Because charge is conserved, the charge arriving at any junction in a circuit must equal the charge leaving the same junction. This means that the total current (charge/time) coming into a junction must equal the total current leaving the same junction. Consider Figure 5.26 as a simple example. The unknown current, I, must be 3 A to the left, into P, so that the current arriving at P is 5 + 3 = 8 A, the same as the current leaving P.

The easiest way of expressing this law is as follows:

$$\Sigma I = 0 \text{ (junction)}$$

This equation is given in the *Physics data booklet.*

The Greek letter capital sigma, Σ, is used here to represent 'the sum of'. So, this equation means that the sum of all the currents at a junction is zero. Remember that currents arriving and currents leaving must be given opposite signs. In general, currents arriving are positive and currents leaving are negative, but they may be the other way round. In this way, the example shown in Figure 5.26 would be written as

$5 + I - 8 = 0$

$I = 8 - 5 = 3\,A$

Since '3' is positive, this shows that the current I is *into* P.

Kirchhoff's second law

Since p.d. = energy transferred/charge, any component in an electric circuit that is transferring energy as charge flows through it must have a p.d. across it.

From the law of conservation of energy, we know that the energy supplied to the moving charges in a circuit, or in any closed loop of a circuit, must be equal to the energy transferred to other forms by the components of that circuit loop.

It follows that the sum of the p.d.s supplying energy to any circuit loop must equal the sum of the p.d.s across components used for transferring energy to other forms (in the same circuit loop). Consider Figure 5.27, in which the two batteries are in opposition to each other and provide a combined p.d. of (24 − 6) = 18 V. If the p.d. across component 1 is 11 V, then the p.d. across component 2 must be (18 − 11) = 7 V.

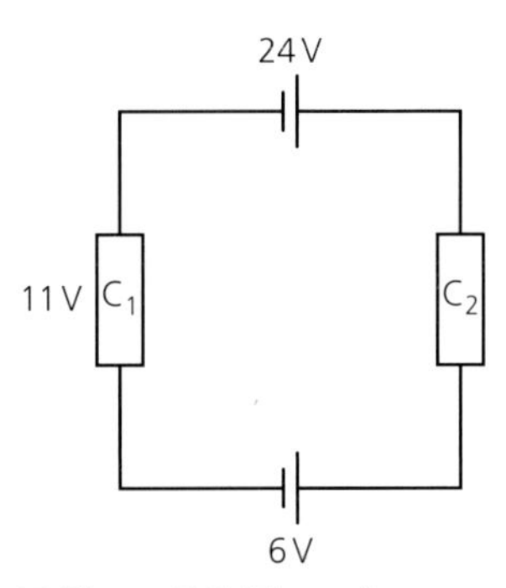

■ **Figure 5.27** The p.d.s around a circuit loop

The easiest way of expressing this law is:

$\Sigma V = 0$ (loop)

This equation is given in the *Physics data booklet*.

Applying Kirchhoff's second law to more complicated circuits can cause confusion, and we will return to this subject in Worked example 14.

34 Draw a circuit diagram to represent the following arrangement: two lamps, A and B, are connected to a 12 V battery with a switch such that it can control lamp A only (lamp B is always on). An ammeter is connected so that it can measure the total current in both lamps and a voltmeter measures the p.d. across the battery.

35 Determine the current at P in Figure 5.28.

36 Determine the values of the currents I_1, I_2 and I_3 in Figure 5.25.

37 Figure 5.29 shows the p.d.s across the components of an electrical circuit. Determine the magnitudes of V_1 and V_2.

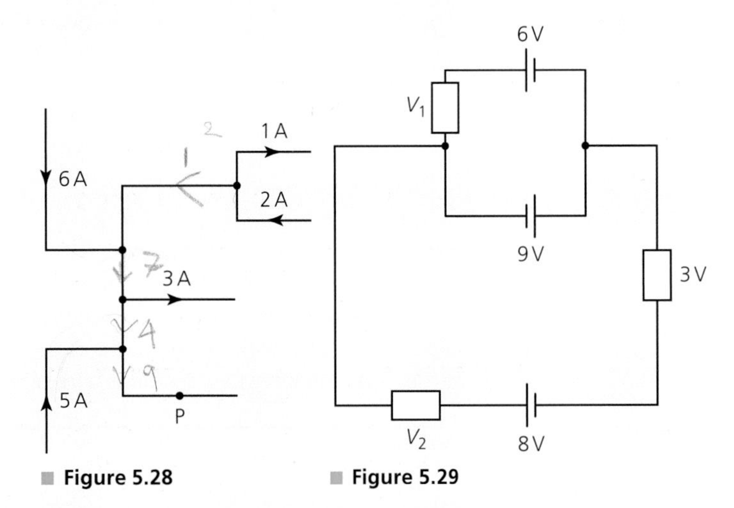

■ **Figure 5.28** ■ **Figure 5.29**

38 The use of an agreed common language and symbols for scientific documents has obvious advantages, but in other respects this can be seen as a serious threat to diversity and a country's culture and history.

a Write a short summary (about 100 words) of your own opinion on this issue.

b If we are to use a common language, what language should that be and why?

Heating effect of current and its consequences

When a current passes through a conductor, the free electrons transfer some energy to the ions or atoms with which they 'collide'. Any increase in the amplitude of vibrations of the ions or atoms is effectively an increase in the internal energy and temperature of the conductor.

In electric heaters this effect is very useful but in most other devices, and in all connecting wires and other connections (including plugs and sockets), any heating effect is unwanted, a waste of useful energy and a possible fire hazard. Connecting wires should be made of a very good metallic conductor (usually copper) and thick enough that there is no significant heating effect in normal use. (In household wiring, fuses or *circuit breakers* are used to disconnect currents if they become dangerously high.)

The heating effect of a current depends on the size of the current (obviously) and the extent to which the material through which it is flowing opposes/resists the flow of current. We refer to this as the **resistance** of the conductor. The following section shows how resistance can be calculated.

Resistance expressed as $R = \frac{V}{I}$

We have seen that a potential difference from an energy source is needed to make a current flow around a circuit. The obvious question to ask is 'how does the size of a current depend on the size of the p.d.?' The simple answer is that the current for a particular p.d. depends on the resistance of the component or the wire.

Resistance is defined as the ratio of the potential difference across a conductor to the current through it and is given the symbol *R*.

$$\text{resistance} = \frac{\text{potential difference}}{\text{current}}$$

$$R = \frac{V}{I}$$

The unit of resistance is the **ohm**, Ω. If a p.d. of 1 V produces a current of 1 A, then the resistance is 1 Ω. The units kΩ and MΩ are also in common use.

The resistance of a component, or wire, may change depending on circumstances. In particular, significant variations in temperature will often affect the value of a resistance.

A component that has been made to have a certain resistance is called a **resistor**. Resistors may have a fixed value or they can be variable. (See Figure 5.24 for the correct circuit symbols for the different types of resistor.) Some examples of fixed resistors are shown in Figure 5.30.

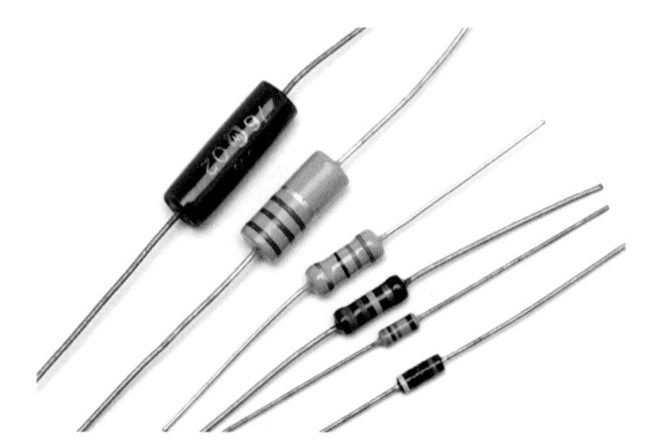

Figure 5.30 Fixed resistors

To determine the resistance of a component or wire experimentally, it is necessary to measure the current, *I*, through it for a known potential difference, *V*, across it. These figures can then be put into the equation, $R = V/I$, although a single pair of readings may not produce a reliable result. Figure 5.31 shows a simple circuit for determining the value of a resistance. (Most digital electric meters have a resistance setting by which a resistance can be determined without the need for other experimentation. In effect the meter is using its battery to send a current through the resistance, and then using the data to calculate its value.)

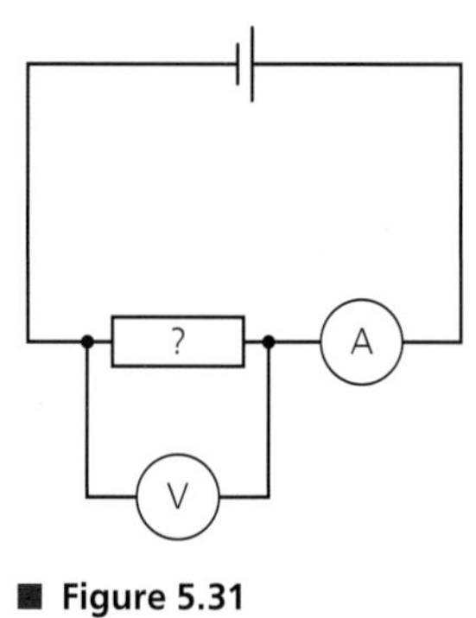

Figure 5.31 Determining the value of an unknown resistance

A calculation made using alternating currents and voltages will produce an answer with the same resistance as a calculation made with direct currents.

Worked example

9 **a** Calculate the resistance of a component if a p.d. of 8.7 V across it produces a current of 0.15 A through it.

b If the p.d. is increased to 12.4 V, the current rises to 0.17 A. What is the new resistance?

a $R = \frac{V}{I}$

$R = \frac{8.7}{0.15}$

$R = 58\,\Omega$

b $R = \frac{V}{I}$

$R = \frac{12.4}{0.17}$

$R = 73\,\Omega$

39 What voltage is needed to make a current of 560 mA pass through a 670 Ω resistor?

40 What is the resistance of a 230 V domestic water heater if the current through it is 8.4 A?

41 What current flows through a 37 kΩ resistor when there is a p.d. of 4.5 V across it?

42 What is the p.d. across a 68.0 Ω resistor if 120 C of charge flows through it in 60.0 s?

Ideal and non-ideal ammeters and voltmeters

Consider again Figure 5.31. It is intended that the ammeter should measure the current through the resistor only, but because of the way it is connected, the ammeter will measure the current through the resistor *plus* the current through the voltmeter. This problem could be solved by moving the voltmeter connection to the other side of the ammeter, but that introduces a different problem: the voltmeter would then be measuring the p.d. across the resistor *plus* the p.d. across the ammeter.

These problems can be overcome by using 'ideal' meters: an **ideal voltmeter** will have an infinitely high resistance and an **ideal ammeter** will have zero resistance. Ammeters with very low resistance, suitable for most experiments, are commonplace in laboratories. Digital voltmeters typically have resistances over a million ohms and are very versatile laboratory instruments. Voltmeters with moving coils will have much lower resistances and their use in circuits with high resistances may be inappropriate.

Figure 5.32 shows the principle of an early type of ammeter (certainly not an ideal ammeter!) that played an important role in the development of circuit measurements. As a current passes through the wire, it gets hot (because of its resistance), increases slightly in length (it expands) and the hanging weight causes the wire to sag. This sag can be arranged to produce a small deflection of a pointer and can make an interesting investigation.

In a modern school laboratory we may wish to measure currents of very different magnitudes: typically from a few amps all the way down to nano-amps (or lower). Possible voltage measurements also cover a very wide range. Digital multimeters can be very useful, but the measurement of very low currents and voltages will require specialized and sensitive instrumentation.

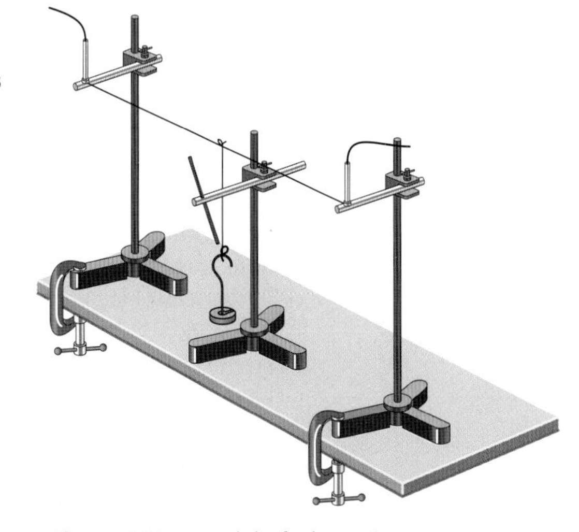

■ **Figure 5.32** A model of a hot-wire ammeter

Utilizations

Electric circuits: lighting our homes

The first inventions that produced light from electricity used the heating effect of a current in a metal wire to make it so hot that it emitted visible light. This principle is still used today; the wire is called a *filament* and the lamps are described as being *incandescent*. Incandescent lamps, as shown in Figure 5.33a, can be designed for use with a wide range of different voltages and powers. They can be made big enough to light a room or small enough for use in a pocket torch.

Incandescent lamps are easy to manufacture and inexpensive, but less than 10% of the electrical energy transferred to them becomes visible light (the rest becomes thermal energy). When the total number of filament lamps in use around the world is considered, this is obviously an enormous waste of energy and the generation of all that wasted power has a significant environmental effect on the planet.

An electric current passing through a gas at low pressure can also produce visible light, but a significant amount of ultraviolet radiation may also be emitted. If the inside of the glass container is given a suitable coating then most of the ultraviolet radiation can be transferred into visible light. The coating and the lamp are described as **fluorescent**. Fluorescent lamps are typically much more efficient at producing light than incandescent lamps, but they have the disadvantages of being bigger and more expensive, and they cannot be used with low voltages. Large fluorescent tubes (Figure 5.33b) have long been considered the best choice for lighting shops, offices, schools and advertising displays. In recent years, however, smaller fluorescent lamps have become cheaper and more widely available for home use, and the widespread use of incandescent lamps for general household lighting is now generally discouraged in most countries. It is likely that more and more countries will limit the use of incandescent lamps in the future.

Apart from being much more efficient than incandescent lamps, *compact fluorescent lamps* (CFLs) (Figure 5.33c) usually have a longer 'lifetime' before they need to be replaced, and this is important when considering the financial costs and the effects on the environment of their manufacture and disposal.

Light-emitting diodes (LEDs) provide another alternative for lighting our homes. They have been used for a long time as indicator lamps to show when electrical devices are on or off. They are useful for this purpose because of their small size, low power, low cost, reliability and long lifetime. Early designs of LEDs emitted green or red light and it is only in recent years that LEDs have begun to be used for more general lighting purposes. This is because of the invention in the 1990s of LEDs that emit blue light and then white light (Figure 5.33d). LEDs that emit light of different colours can be combined to create a vast range of different coloured effects.

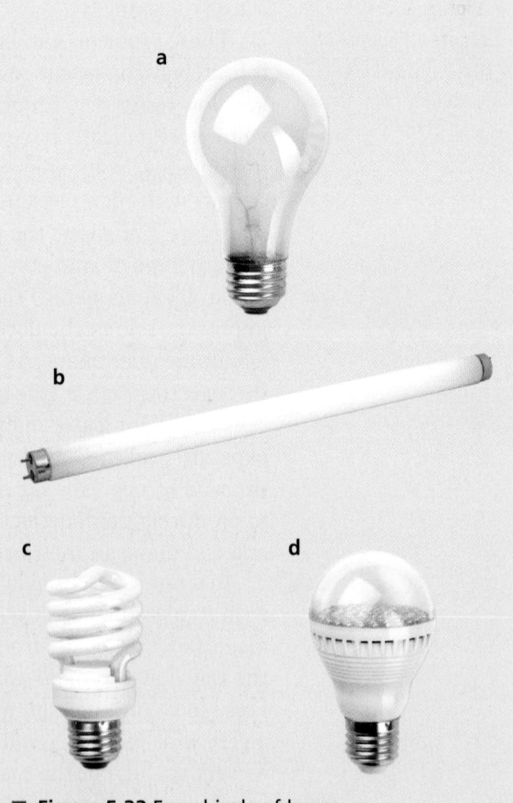

■ **Figure 5.33** Four kinds of lamp: **a** incandescent, **b** fluorescent tube, **c** CFL, **d** LED

A lot of research is still going on to improve the design of LEDs, particularly when it comes to increasing their overall efficiency in the conversion of electrical energy from a mains supply into visible light. Currently, the efficiency of using LEDs for home lighting is a little better than CFLs, but they are more expensive to buy. LEDs offer much greater possibilities for interesting and innovative lighting design and they are more reliable and longer lasting.

1 Make a checklist of the properties of a lamp which make it a suitable choice for lighting the homes of people around the world.

2 Use the internet to research the latest information on incandescent lamps, CFLs and LEDs. Use your checklist from question **1** to compare the advantages and disadvantages of the three types of lamp.

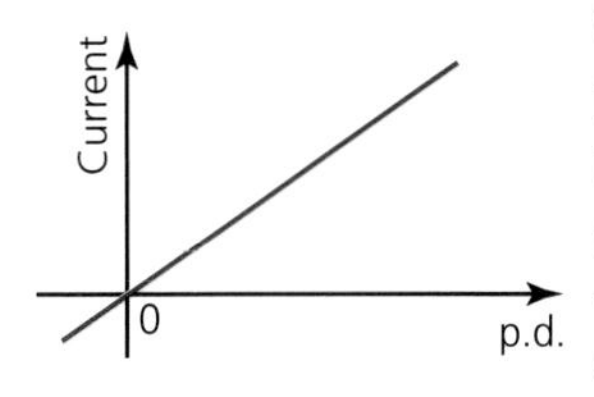

■ **Figure 5.34** Ohm's law for an ohmic resistor

Ohm's law

If the temperature of a metallic conductor is kept constant, its resistance will not change because any increase in the potential difference across it will produce a proportional increase in current, so that V/I (= R) is constant, as shown in Figure 5.34. This relationship was discovered by the German physicist Georg Ohm in 1827. (Note that the graph has been extended some way into negative values as an indication that the same behaviour is seen if the p.d. and the current both reverse directions.)

Ohm's law states that the current through a metallic conductor is proportional to the potential difference across it, if the temperature is constant.

$I \propto V$ at constant temperature

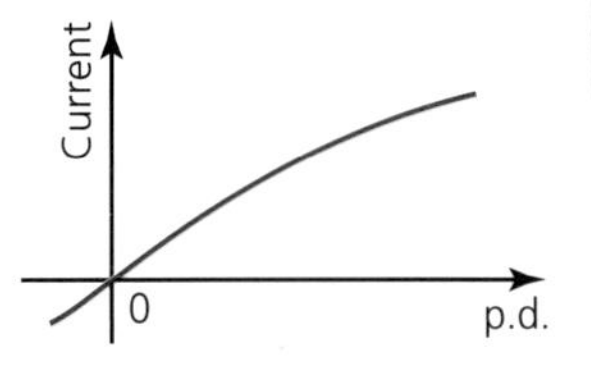

■ **Figure 5.35** A current–p.d. graph for a filament lamp

If the temperature of a metallic conductor does not change very much, it can usually be assumed that the resistance is constant and Ohm's law can be used. But if there are large changes in temperature, Ohm's law is not useful. For example, the operating temperature of a filament lamp may be 2500 °C and its resistance at that temperature will be significantly higher than its resistance at room temperature. Figure 5.35 shows how the increasing temperature and resistance of a filament lamp might affect its I–V graph.

Ohm's law is a starting point for understanding electrical conduction. It works equally well for dc and ac circuits, but there are many circuit components to which it cannot be applied. In order to investigate the electrical behaviour of a specific component (such as a filament lamp), either of the circuits shown in Figure 5.36 could be used. In circuit **a**, a variable electrical supply is used to obtain different voltages across the component. Circuit **b** uses a battery of fixed voltage and a **variable resistor** (a **rheostat**) to change the overall resistance of the circuit, therefore changing the current and voltage across the component. However, as we will see later in this chapter, using a battery and a *potential dividing circuit* is the best choice for this investigation.

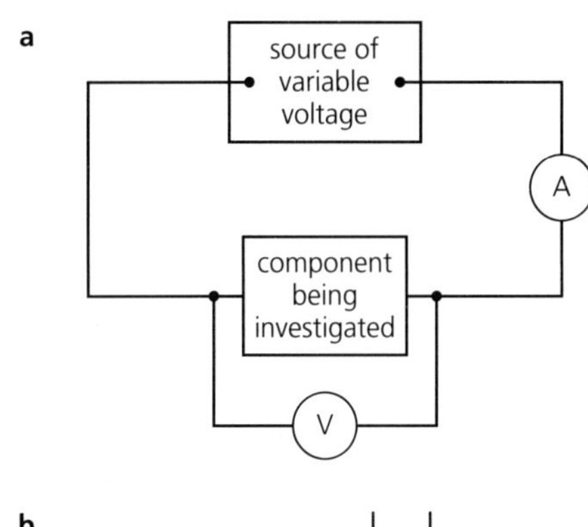

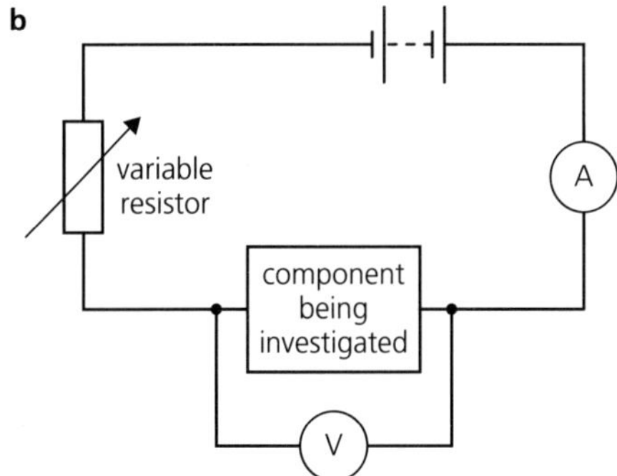

■ **Figure 5.36** Two methods to investigate the I–V characteristic of an electrical component

If the current through an electrical component is proportional to the p.d. across it, then it is described as being **ohmic**, because it 'obeys' Ohm's law and the *I–V* graph is a straight line through the origin (Figure 5.34). A metal wire at constant temperature is an ohmic device. All other components (including those not being used at approximately constant temperature) are described as **non-ohmic** (as in Figure 5.35). A filament lamp is a non-ohmic device because its temperature changes considerably when different currents pass through it.

The temperature dependence of a resistance wire can be investigated by taping it next to the bulb of a large range thermometer while measuring its resistance.

43 Suggest reasons why hot-wire ammeters are not used in modern laboratories.

44 Sketch *I–V* characteristic graphs for:
 a a device that has a resistance which decreases as the current increases
 b a non-ohmic device in which equal increases in voltage produce equal increases in current
 c a device that only allows current to flow through it in one direction (a diode).

45 Give an example of a non-electrical measuring device which may affect the quantity it is being used to measure. Explain why the results may not be accurate.

Nature of Science

Peer review or competition between scientists?

Georg Ohm's law was published in Germany in 1827 in the form that the current in a wire, *I*, is proportional to $(A/L)V$ where A is the cross-sectional area of a uniform metal wire of length *L*. Two years earlier, in England, Peter Barlow had incorrectly proposed 'Barlow's law' in the form *I* was proportional to $\sqrt{(A/L)}$, but with no reference to the key concept of voltage. It is not unusual for two or more different scientists, or groups of scientists, to be investigating similar areas of science at the same time, often in different countries.

In the worldwide, modern scientific community, with its quick and easy mass communication, new experimental results and theories are subjected to very close scrutiny and checking soon after they are published. In this way new ideas are reviewed carefully by other scientists and experts working in the same field, in a process commonly called **peer review**. But 200 years ago when Ohm was carrying out his research, things were very different. At that time, social factors and the reputation, power and influence of the scientist were sometimes as important in judging new ideas as the value of the work itself. The story of Barlow and Ohm is particularly interesting because in the early stages, the incorrect theory proposed by Barlow was more widely believed.

■ **Figure 5.37** Georg Ohm

Combining resistors in series and parallel

Resistors can be joined together to make new values of resistance. They may be connected in series, in parallel or in other combinations. The overall resistance of the combinations can be determined experimentally as described before. Computer programs also provide a valuable aid here and throughout the study of electric circuits because they enable students to construct and investigate circuits much quicker than in the laboratory.

Figure 5.38 shows three different resistors connected in series. Because of the law of conservation of charge, the charge per second (current) flowing into each resistor must be the same as the current, *I*, flowing out of it and into the next resistor.

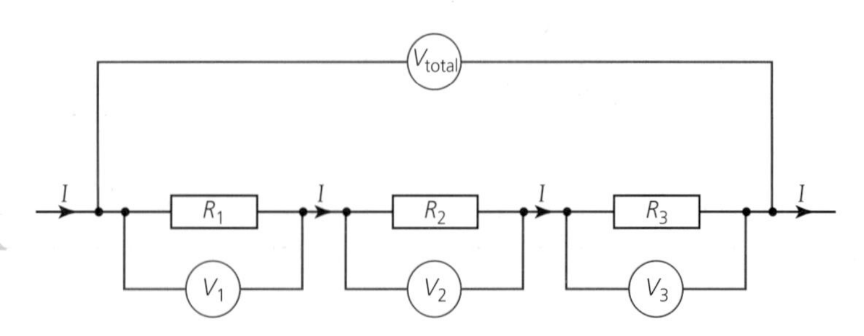

■ **Figure 5.38** Three resistors in series

The sum of the separate potential differences must equal the potential difference across them all, V_{total}, so that

$$V_{total} = V_1 + V_2 + V_3$$

Using $V = IR$ for the individual resistors, we get $IR_{total} = IR_1 + IR_2 + IR_3$, so that we can derive an equation for the single resistor, R, which has the same resistance as the combination:

$$R_{total} = R_1 + R_2 + \ldots$$

This equation is listed in the *Physics data booklet*.

Figure 5.39 shows three resistors connected in parallel with each other.

Because they are all connected between the same two points, they all have the same potential difference, V, across them.

The law of conservation of charge means that:

$$I_{total} = I_1 + I_2 + I_3$$

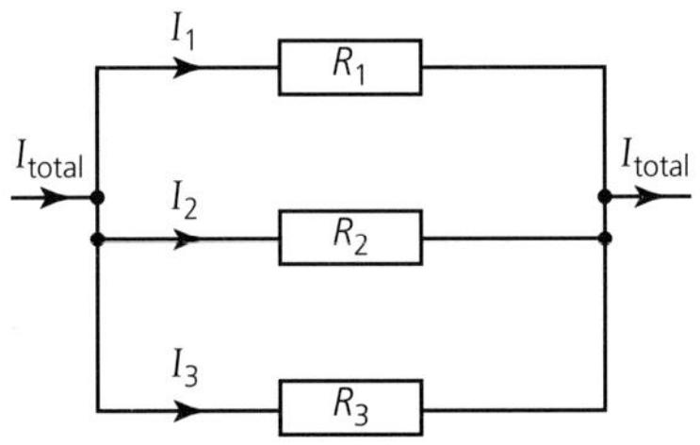

■ **Figure 5.39** Three resistors in parallel

Applying $V = IR$ throughout gives:

$$\frac{V}{R_{total}} = \frac{V}{R_1} + \frac{V}{R_2} + \frac{V}{R_3}$$

Cancelling the Vs gives us an equation for the single resistor, R_{total}, which has the same resistance as the combination:

$$\frac{1}{R_{total}} = \frac{1}{R_1} + \frac{1}{R_2} + \ldots$$

This equation is given in the *Physics data booklet*.

All the electrical equipment in our homes is wired in parallel because in that way, each device is connected to the full supply voltage and can be controlled with a separate switch.

Worked example

10 A 5000 Ω resistor and 8000 Ω resistor are connected in series.

a What is their combined resistance?

b What is the current through each of them if they are connected to a 4.5 V battery?

c What is the potential difference across the 5000 Ω resistor?

d Repeat these three calculations for the same resistors in parallel with each other.

a $5000 + 8000 = 13\,000\,\Omega$

b $I = \frac{V}{R} = \frac{4.5}{13\,000} = 3.5 \times 10^{-4}$ A through both resistors

c $V = IR = 3.5 \times 10^{-4} \times 5000 = 1.7$ V

d $\frac{1}{R} = \frac{1}{5000} + \frac{1}{8000} = \frac{13}{40\,000}$

$R = \frac{40\,000}{13} = 3080\,\Omega$

Both resistors have a p.d. of 4.5 V across them.

current through 5000 Ω, $I = \frac{V}{R} = \frac{4.5}{5000} = 9.0 \times 10^{-4}$ A

current through 8000 Ω, $I = \frac{V}{R} = \frac{4.5}{8000} = 5.6 \times 10^{-4}$ A

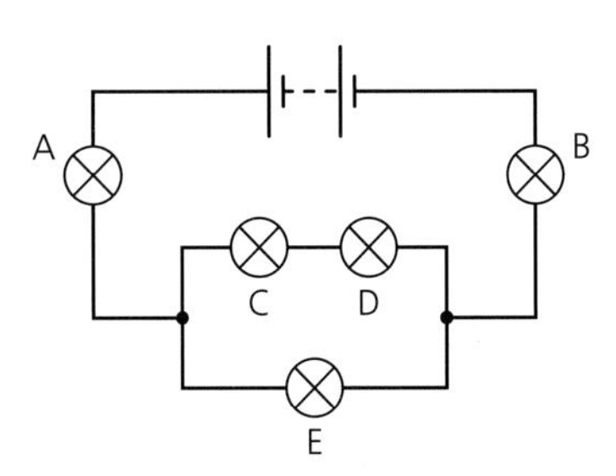

■ **Figure 5.40**

11 The lamps shown in Figure 5.40 are all the same.

a Compare the brightness of all the lamps.

b If all the lamps have the same constant resistance of 2 Ω, what is the total resistance of the circuit?

a Lamps A and B will have the same brightness because the same current flows through them both. That same current will be split between lamp E and lamps C and D, so that these three must all be dimmer than lamps A and B. Lamps C and D will have the same brightness because they are in series with each other. Lamp E will be brighter than lamps C or D because a higher current will flow through it.

b C and D together will have a resistance of 2 + 2 = 4 Ω.

E in parallel with C/D will have a resistance of 1.3 Ω ($\frac{1}{R} = \frac{1}{2} + \frac{1}{4}$).

Total resistance = 2 + 2 + 1.3 = 5.3 Ω

Connecting meters in series and parallel

If it is likely that the meters used in an experiment cannot be considered to be 'ideal', the effect of connecting an ammeter or voltmeter in a circuit can be determined by treating it the same as any other resistance, as shown in these worked examples.

Worked examples

12 a What current flows through a 12.0 Ω resistor connected to a p.d. of 9.10 V?

b An ammeter of resistance 0.31 Ω was used to measure the current. What value did it display?

c What was the percentage error when measuring this current?

a $I = \frac{V}{R} = \frac{9.10}{12.0} = 0.758\text{ A}$

b $I = \frac{V}{R} = \frac{9.1}{12.31} = 0.739\text{ A}$

c $\frac{0.739}{0.758} \times 100 = 97.5\%$. Percentage error was −2.5%.

13 Consider the circuit shown in Figure 5.41.

a What is the current in the circuit before the voltmeter is connected?

b What is the voltage across the 2000 Ω resistor?

c What voltages will be measured if voltmeters with the following resistances are connected in turn across the 2000 Ω resistor:

i 5000 Ω

ii 50 000 Ω?

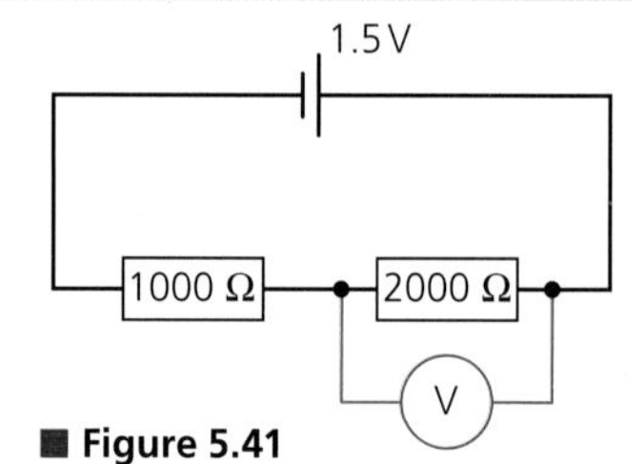

■ **Figure 5.41**

a $I = \frac{V}{R}$

$I = \frac{1.5}{(1000 + 2000)}$

$I = 5 \times 10^{-4}\text{ A}$ (0.5 mA)

b $V = IR$

$V = (5 \times 10^{-4}) \times 2000$

$V = 1.0\text{ V}$

c i First calculate the combined resistance of the 2000 Ω resistor and the 5000 Ω voltmeter in parallel:

$$\frac{I}{R} = \frac{I}{R_1} + \frac{I}{R_2}$$

$$\frac{I}{R} = \frac{1}{2000} + \frac{1}{5000}$$

$R = 1430\ \Omega$

Then find the voltage across a resistance of 1430 Ω connected in series with a 1000 Ω resistance:

$$V = 1.5 \times \frac{1430}{1430 + 1000}$$

$V = 0.88\text{ V}$ (instead of the 1.0 V predicted in **b**)

ii Repeating the calculation with a 50 000 Ω voltmeter gives $V = 0.99\text{ V}$, which is better because it is much closer to the value predicted without the voltmeter.

14 Use Kirchhoff's laws to determine the three unknown currents in the circuit shown in Figure 5.42.

In general when answering this kind of question, it will be necessary to write down as many different equations as there are unknowns (in this example there are three). The directions of the currents may be unknown, so they have been guessed, but this does not matter because if the wrong direction is chosen, the value calculated for the current will have the correct magnitude, but will be negative (as will be seen in this example).

1 Using Kirchhoff's first law: $I_2 = I_1 + I_3$

2 Using Kirchhoff's second law for the top circuit loop:
$12 = 10I_1 + 20(-I_3)$

The current I_3 has been given a negative sign because it is shown flowing towards the positive side of the battery, rather than in the conventional direction.

3 Using Kirchhoff's second law for the lower circuit loop: $9 = 20(-I_3)$

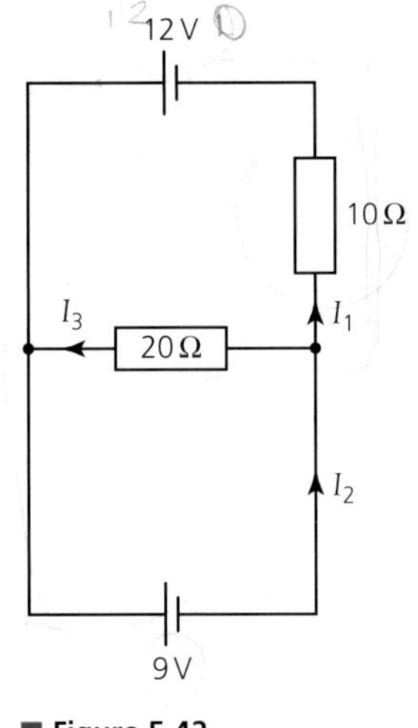

■ **Figure 5.42**

4 Kirchhoff's second law can also be used for the outer complete loop, but it will provide no additional information than a combination of equations 2 and 3: $12 - 9 = 10I_1$

From equation 4 we can see that $I_1 = 0.3\,\text{A}$

From equation 3 we can see that $I_3 = -0.45\,\text{A}$

So that $I_2 = I_1 + I_3 = 0.3 + (-0.45) = -0.15\,\text{A}$

The two negative signs show us that the directions chosen for I_3 and I_2 were wrong.

46 The two lamps shown in Figure 5.43 both have a resistance of 4.0 Ω when cold and 6.0 Ω when working normally with a p.d. of 6.0 V across them.

a What must be the value of the resistor R for the lamps to operate normally?

b What is the p.d. across the lamps when the switch is first turned on?

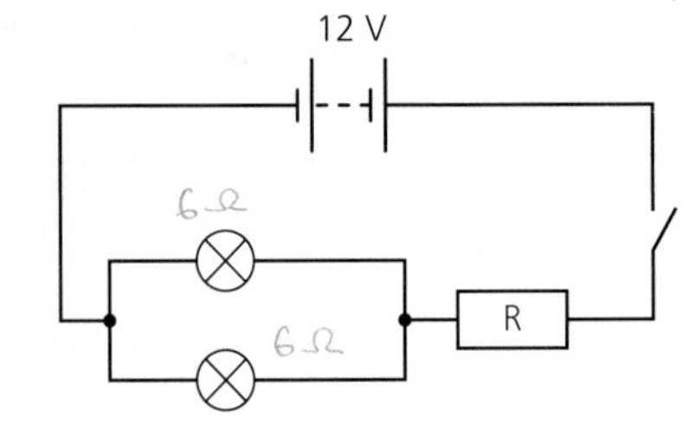

■ **Figure 5.43**

47 A fan heater, which is operated from the electricity mains, contains a heating element, a fan and a lamp to indicate when the heating element is on. The fan must always be on when the heater is on, but the fan can also be switched on if the heater is off. Draw a circuit with a *two-way switch* to show how they are all connected.

48 Consider the circuit shown in Figure 5.44. In which resistor will power be dissipated at the highest rate? Explain your answer.

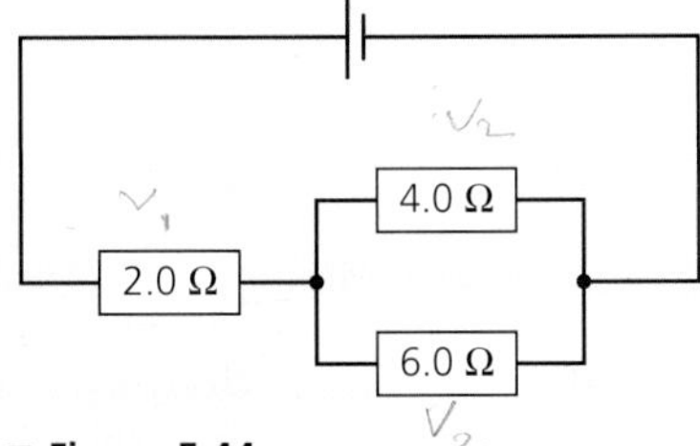

■ **Figure 5.44**

49 A 220 V mains electric heater has three settings on its switch: high, medium and low. Inside the heater there are two 40.0 Ω heating elements.

a Explain, with diagrams, how the use of just two heating elements can produce three different power outputs.

b Calculate the power output on each of the three settings.

50 Figure 5.45 shows a simple circuit in which the ammeter and voltmeter have been connected in the wrong positions.

a What (approximate) readings would you expect to see on the meters? Explain your answer.

b When the positions of the meters were swapped to their correct positions, what readings would you expect to see on the meters?

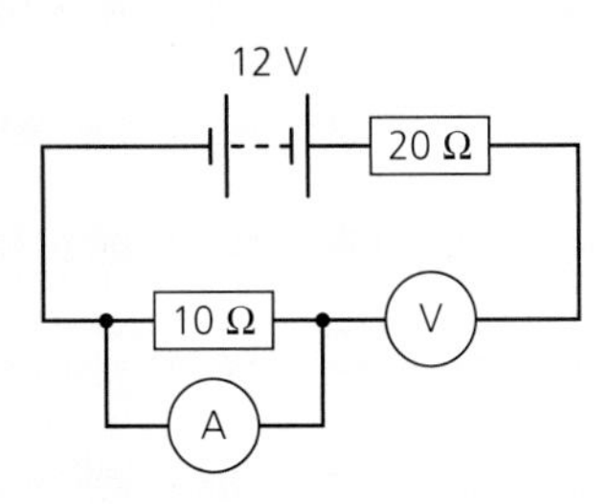

■ **Figure 5.45**

c The actual reading on the voltmeter was 3.9 V. Suggest a reason why it was lower than expected.
d Calculate the resistance of the voltmeter.
e What assumption(s) did you make about the ammeter?

51 In the circuit shown in Figure 5.46 a current of 830 mA flows through the lamp when the switch is closed.
a What is the p.d. across the lamp?
b Calculate the resistance and power of the lamp under these conditions.
c If the switch is opened, explain what happens to the brightness of the lamp.
d Calculate the new p.d. across the lamp.

52 The two ends of a 1 m length of resistance wire are joined together and the wire spread out into a circle. A student carries out an investigation to determine how the resistance between any two points on the circle varies with the length of wire between them. Sketch a graph to represent the results you would expect from this investigation.

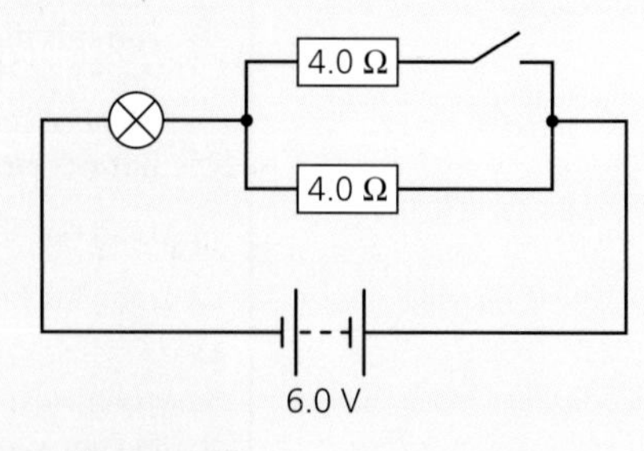

■ Figure 5.46

53 Determine the magnitudes and directions of the currents in the circuit shown in Figure 5.47

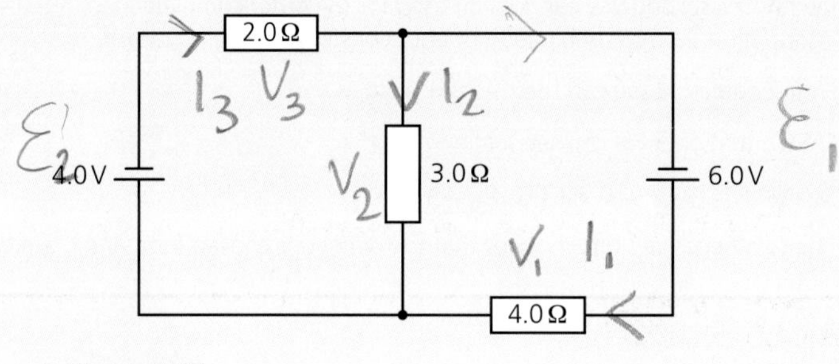

■ Figure 5.47

■ Resistivity

A material has electrical resistance because any electrons flowing through it have to pass atoms or ions vibrating in their path (see Figure 5.13). The longer the conductor, the more 'collisions' that are likely to occur and so the higher the resistance. The wider the conductor, the easier it is for electrons to flow, and so the lower the resistance.

Experiments to determine the resistance of wires of different lengths and diameters show that the resistance of wires (of the same metal or alloy) are proportional to their lengths, L, and inversely proportional to their cross-sectional areas, A at constant temperature:

$$R \propto \frac{L}{A}$$

But resistance also depends on the particular material being used, as is easily demonstrated by measuring the resistance of wires of different metals or alloys, all with the same dimensions. The more free electrons (per unit volume), n, in a material, the better it will be at conducting an electric current. We cannot simply look up the resistance of a material (like copper, for example) in a table of data, because resistance depends on shape as well as the material itself. Instead, we look up a material's **resistivity**, which is the resistance of a specimen of length 1 m and cross-sectional area 1 m^2 (as defined by the equation below).

Not surprisingly (for a theoretical wire of cross-sectional area 1 m^2), the resistivity of a good conductor has a very low numerical value. For example, the resistivity of copper is $1.7 \times 10^{-8}\,\Omega\,\text{m}$ at 20 °C.

Resistivity is given the symbol ρ and it has the unit ohm metre, Ω m (not ohms *per* metre):

$$\rho = \frac{RA}{L}$$

This equation is listed in the *Physics data booklet*.

The resistivity of different materials can be determined and compared by measuring the resistances of wires, or graphite pencils, of known dimensions. If the resistivity of graphite is known, it can be used with the equation above to estimate the thickness of a pencil line drawn on paper.

Resistance and resistivity usually change with temperature, sometimes significantly, and we should be careful about describing a material as an insulator or a conductor without also stating the temperature. Materials can vary enormously in their resistivities, as is shown in Table 5.2. The semi-conducting elements silicon and germanium are in the middle of the range.

Material	Resistivity/Ω m
silver	1.6×10^{-8}
copper	1.7×10^{-8}
aluminium	2.8×10^{-8}
iron	1.0×10^{-7}
nichrome (used for electric heaters)	1.1×10^{-6}
carbon (graphite)	3.5×10^{-5}
germanium	4.6×10^{-1}
sea water	$\approx 2 \times 10^{-1}$
silicon	6.4×10^{2}
glass	$\approx 10^{12}$
quartz	$\approx 10^{17}$
teflon (PTFE)	$\approx 10^{23}$

■ **Table 5.2** Resistivities of various substances at 20 °C

Worked examples

15 a Calculate the resistance of a 1.80 m length of iron wire of cross-sectional area 2.43 mm².

b A current of 2.4 A flowed through an 83 cm length of metal alloy wire of area 0.54 mm² when a p.d. of 220 V was applied across its ends. What was the resistivity of the alloy?

a $R = \frac{\rho L}{A}$

$R = \frac{(1.0 \times 10^{-7}) \times 1.80}{(2.43 \times 10^{-6})}$

$R = 7.4 \times 10^{-2}\,\Omega$

b $R = \frac{V}{I} = \frac{220}{2.4} = 91.7\,\Omega$

$\rho = \frac{RA}{L}$

$\rho = \frac{91.7 \times (5.4 \times 10^{-7})}{0.83}$

$\rho = 6.0 \times 10^{-5}\,\Omega\,\mathrm{m}$

16 Suggest possible reasons why aluminium is usually used for the cables that carry large electric currents around the world (rather than copper).

The resistivities of these two metals are in the ratio of 2.8/1.7 = 1.6. This means that an aluminium cable conducts equally as well as a copper wire if its cross-sectional area is 1.6 times larger. Aluminium is a better choice because it is much cheaper than copper and aluminium cables (of equal resistance) are lighter in weight.

Copper has other properties that make it a better choice for household wiring, such as its flexibility. Some more expensive metals like silver, gold and platinum are better conductors than copper and they are sometimes used where a very low resistance wire or connection is needed (for example, in some sound reproduction systems).

54 a What length of nichrome wire of cross-sectional area 0.0855 mm^2 is needed to make a 15.0 Ω resistor?
b What value resistor would be made with nichrome wire of double the length and half the diameter?

55 Suggest the possible relationship between the resistivity of a metal and the number of free electrons it has per cubic millimetre.

56 The nichrome heating element in an electric kettle has an overall length of 5.32 m. If its resistance is 25 Ω, what is the diameter of the wire?

57 Calculate the resistivity of a metal wire if a voltage of 2.5 V is needed to make a current of 26 mA pass through a wire of diameter 0.452 mm and length 745 cm.

Utilizations

Touchscreen technology

Touchscreens (Figure 5.48) have become very popular in devices such as mobile phones and tablets because they enable the user to control the display directly without the need for a keypad or mouse.

■ **Figure 5.48** A touchscreen tablet

There are several different technologies used in touchscreens; one of these uses the resistive properties of the display screen. When any object, such a finger, touches the screen the resistance at that point changes and the location is calculated by the central controller. If the finger moves, new locations are detected – the speed and direction of the movement is calculated and used to control the device. Multipoint screens can respond to two or more touches at the same time. One problem with resistive touchscreens is the fact that the screen reduces the quality of the image.

Another common touchscreen technology responds to changes in capacitance (capacity for storing charge) at the point where the screen is touched. These screens need to be touched with a conductor, like a finger.

1 Choose a particular touchscreen phone, tablet or other device and find out which technology is used for the screen.

2 Can you think of any disadvantages of using touchscreen devices?

Power dissipation

If the current through a resistor is, for example, 3 A then 3 C of charge is passing through it every second. If there is a potential difference across the resistor of 6 V, then 6 J of energy is being transferred by every coulomb of charge (to internal energy). The rate of transfer of energy is $3 \times 6 = 12$ joules every second (watts).

More generally, we can derive an expression for the power dissipated to internal energy in a resistor by considering the definitions of p.d. and current, as follows:

$$\frac{\text{energy transferred}}{\text{time}} = \frac{\text{energy transferred in resistor}}{\text{charge flowing through resistor}} \times \frac{\text{charge flowing through resistor}}{\text{time}}$$

$$\frac{W}{t} = \frac{W}{q} \times \frac{q}{t}$$

or:

power = potential difference × current

$$P = VI$$

Because $V = IR$, this can be rewritten as $P = (IR)I$, or:

$$P = I^2R$$

Alternatively, $P = V(V/R)$, or:

$$P = \frac{V^2}{R}$$

These three forms of the same equation are all given in the *Physics data booklet*.

To calculate the total energy transferred in a given time, we know that energy = power × time, so that:

electrical energy = VIt

This equation is *not* given in the *Physics data booklet*.

Worked examples

17 An electric iron is labelled as 220 V, 1200 W.

a Explain what the label means.

b What is the resistance of the heating element of the iron?

c Explain what would happen if the iron was used in a country where the mains voltage was 110 V.

a The label means that the iron is designed to be used with 220 V and, when correctly connected, it will transfer energy at a rate of 1200 joules every second.

b $P = \frac{V^2}{R}$

$1200 = \frac{220^2}{R}$

$R = 40.3\,\Omega$

c $P = \frac{V^2}{R}$

So:

$P = \frac{110^2}{40.3}$

$P = 300\,\text{W}$

The iron would transfer energy at $\frac{1}{4}$ of the intended rate and would not get hot enough to work properly. ($P = VI$, and both the p.d. and the current will be halved, assuming the resistance is constant.)

If an iron designed to work with 110 V was plugged into 220 V it would begin to transfer energy at four times the rate it was designed for; it would overheat and be permanently damaged.

58 A 12 V potential difference is applied across a 240 Ω resistor.
 a Calculate:
  i the current
  ii the power
  iii the total energy transferred in 2 minutes.
 b What value resistor would have twice the power with the same voltage?

59 A 2.00 kW household water heater has a resistance of 24.3 Ω.
 a What current flows through it?
 b What is the mains voltage?

60 a What would be the rate of production of thermal energy if a current of 100 A flowed through an overhead cable of length 20 km and resistance 0.001 ohm per metre?
 b Comment on your answer.

61 a What power heater will raise the temperature of a metal block of mass 2.3 kg from 23°C to 47°C in 4 minutes (specific heat capacity = 670 J kg^{-1} °C^{-1})?
 b Draw a circuit diagram to show how the heater should be connected to a 12 V supply and suitable electrical meters so that the power can be checked.

62 a An electric motor is used to raise a 50 kg mass to a height of 2.5 m in 74 s. The voltage supplied to the motor was 240 V but it was only 8% efficient. What was the current in the motor?
 b Suggest two reasons why the motor has a low efficiency.

63 a What value resistance would be needed to make a 1.25 kW water heater in a country where the mains voltage is 110 V?
 b What current flows through the heater during normal use?
 c Suggest why a larger current flows when it is first turned on.

Another unit for electrical energy

When we buy a battery or pay for the 'mains' electricity connected to our homes, we are really buying energy. In most countries mains electrical energy is sold by the **kilowatt hour**: 1 kW h is the amount of energy transferred by a 1 kW device in one hour, that is, the equivalent of 1000 J s^{-1} for 3600 s, or 3.6 MJ. This conversion rate is listed in the *Physics data booklet*.

64 a Use energy = *VIt* to estimate how much useful energy can be transferred from a single AA-sized battery.
 b What is the approximate cost of that energy per MJ?

65 a How much do householders have to pay for 1 kW h in your country?
 b Use the internet to compare your prices with those in some other countries.
 c What is the cost of 1 MJ of mains electrical energy in your country? By comparison, it should be clear that buying disposable batteries is a very expensive way of paying for energy and convenience. The disposal of the batteries is also a pollution problem.

66 In Chapter 8 we will discuss the effect of the generation and use of electrical energy on the environment. There are enormous differences in how much electrical energy is used in different countries. One way of discouraging the excessive use of electricity in developed countries is for governments to make the cost of a kW h greater for those homes that use more energy.
Discuss the advantages and disadvantages of the pricing system for electricity mentioned above.

■ Potential divider circuits

When two (or more) resistors are connected in series they share the total potential difference across them in the same ratio as the magnitudes of the resistances (as shown in Worked example 10). When this kind of arrangement of resistors is used deliberately for the control of p.ds around circuits, it is called a **potential divider**.

Typically, one of the two resistors will be variable, as shown in Worked example 18.

Worked example

18 The value of the variable resistor in Figure 5.49 can be changed continuously from 1 kΩ to 10 kΩ. What is the maximum and minimum potential difference, V_{out}, that can be obtained across R?

If the variable resistance is set to 1 kΩ,

$$V_{out} = \frac{6 \times 5}{5 + 1} = 5\,V$$

If the variable resistance is set to 10 kΩ,

$$V_{out} = \frac{6 \times 5}{5 + 10} = 2\,V$$

V_{out} may be a voltage that controls another part of the circuit; perhaps some kind of electronic switch that might be on if the voltage was, for example, higher than 4 V and off if it was lower than 3 V. In this way, another circuit can be switched on or off by changing the value of the variable resistor in this *potential divider*.

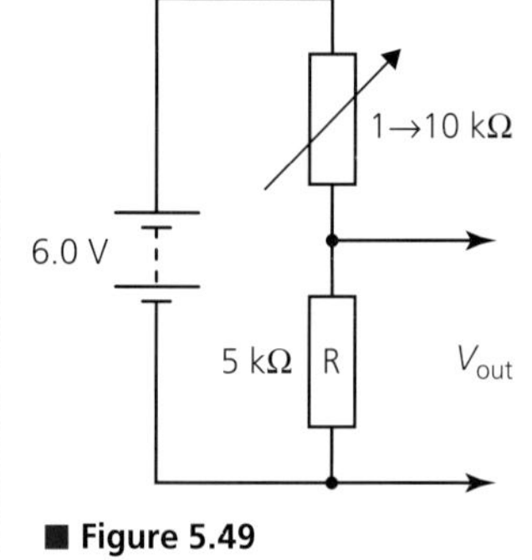

■ **Figure 5.49**

Sensors

Potential dividers are often used in automatic control systems in which a sensor is used as a variable resistor. A sensor is an electrical component that responds to a change in a physical property (temperature, for example) with a corresponding change in an electrical property (usually resistance).

Sensors are commonly available for the detection of most of the physical properties discussed in this book, but we shall limit our discussion to three sensors in particular.

- Light-dependent resistors (LDRs) (Figure 5.50): the resistance of an LDR decreases when increased light intensity provides the energy needed to release more free electrons in its semi-conducting materials.
- Temperature-dependent resistors (thermistors) (Figure 5.51): the resistance of a thermistor varies with changes of temperature. The resistance may increase or decrease, depending on the materials used in the particular thermistor.
- Strain gauges (Figure 5.52): the resistance of a strain gauge varies when its shape changes (remember $R = \rho L/A$).

■ **Figure 5.50** An LDR

■ **Figure 5.51** A thermistor

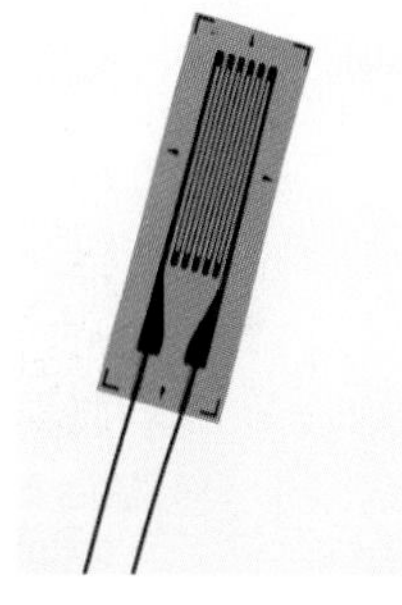
■ **Figure 5.52** A strain gauge

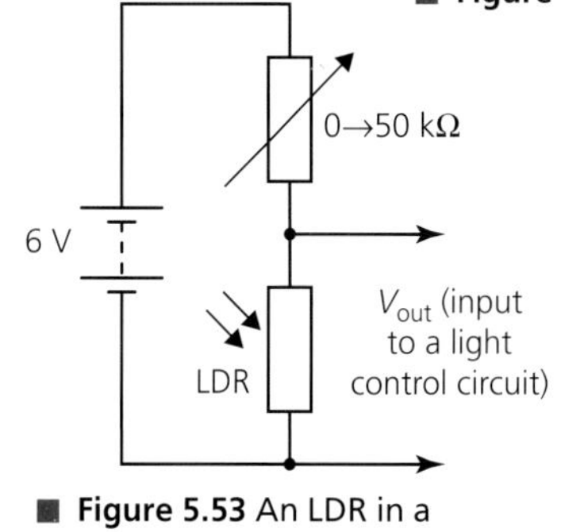

■ **Figure 5.53** An LDR in a potential divider circuit

Figure 5.53 shows an LDR connected as part of a potential divider. As the light intensity increases, the resistance of the LDR decreases and, therefore, the voltage across it, V_{out}, also decreases. V_{out} could be used to control an electronic switch that turns off lights when the light intensity rises to a certain level. The light level at which the lights are turned on or off can be changed by adjusting the value of the variable resistor.

A thermistor connected in a similar circuit could be used to turn a heater on when the temperature falls below a certain value, or turn a heater off when it gets too hot. Used in this way to control temperature, the thermistor would be part of a thermostat.

67 a What is the potential difference across points A and B in Figure 5.54?

b A student wants to connect a lamp which is rated at 15 W, 6 V. What is the working resistance of the lamp?

c The student thinks that the lamp will work normally if he connects it between A and B, in parallel with X. Explain why the lamp will not work as he hopes.

d Another student thinks that the lamp will work if resistor X is removed (while the lamp is still connected between A and B). Calculate the voltage across the lamp. Will the lamp work normally now?

e Suggest how the lamp could work normally using a 12 V battery.

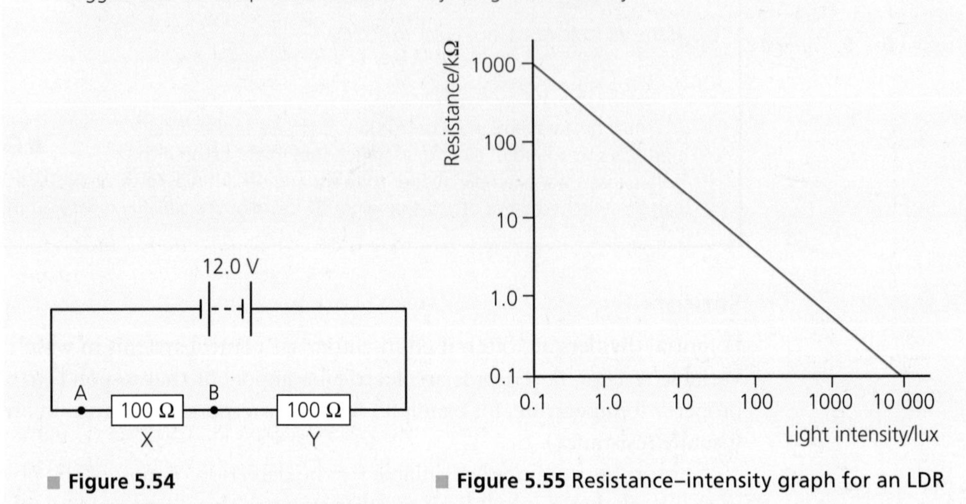

■ **Figure 5.54**

■ **Figure 5.55** Resistance–intensity graph for an LDR

68 a Draw a potential-dividing circuit that could be used to control the temperature in a refrigerator.

b Make a list of electrical devices that have thermostats inside them.

69 Figure 5.55 shows how a particular semi-conducting LDR responds to changes in light intensity. The LDR has a resistance of about 1300 Ω in normal room lighting (400 lux).

a Explain why the resistance decreases as the light intensity increases.

b The scales are logarithmic. Explain why this type of scale is used.

c Write an equation for the line.

d Calculate the resistance of the LDR when the light intensity is 200 lux.

e If the LDR was connected as shown in Figure 5.53, what value of the variable resistance would produce a p.d. of 1.2 V across the LDR under normal room lighting?

70 a If the length of a certain wire was increased by 1.00%, what percentage change would you expect in its resistance? (Assume that its volume remains constant.)

b Figure 5.56 shows a strain gauge connected in a potential dividing circuit. Calculate the reading on the voltmeter. What assumption(s) did you have to make about the voltmeter?

c If the resistance of the strain gauge increases by 0.570%, what will be the new reading on the voltmeter?

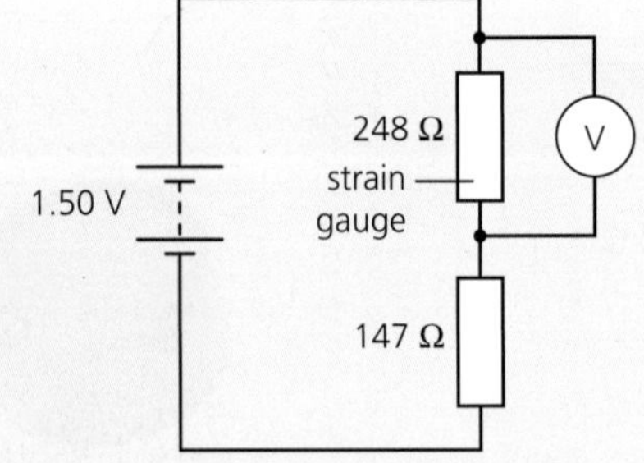

■ **Figure 5.56**

Using a variable resistor as a potential divider

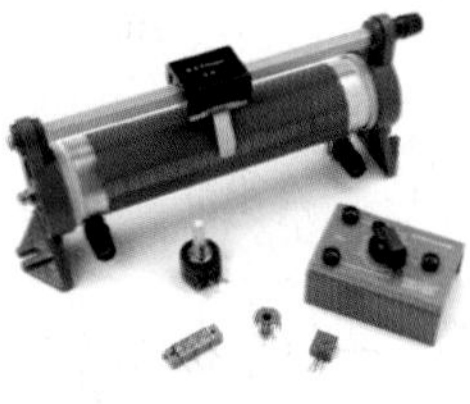

Variable resistors (Figure 5.57) are made in many different shapes and sizes. Most have three terminals, one at each end of the resistor and one sliding contact that can be moved along the resistor anywhere between the other two contacts.

The brightness of the bulb in Figure 5.58 can be adjusted using the variable resistance. Until now, we have only considered using two terminals. Used in this way, it may be best to think of a variable resistor as a way of changing the current, by changing the resistance in a series circuit. The p.d. is divided between the lamp and the variable resistor, which means that the circuit shown in Figure 5.58 is a simple example of a potential-dividing circuit. The range of the variable resistance must be chosen carefully for its intended use.

■ **Figure 5.57** A selection of variable resistors

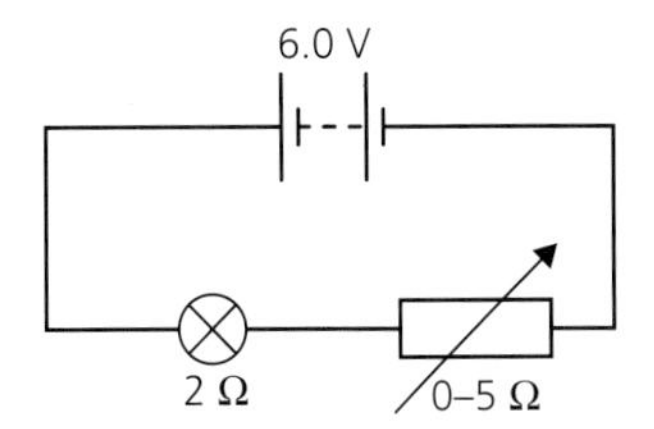

■ **Figure 5.58** Circuit using a variable resistance to change current

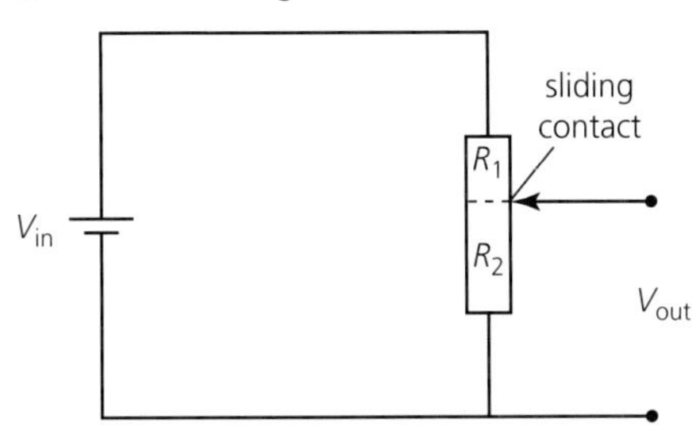

■ **Figure 5.59** A variable resistor used as a potentiometer

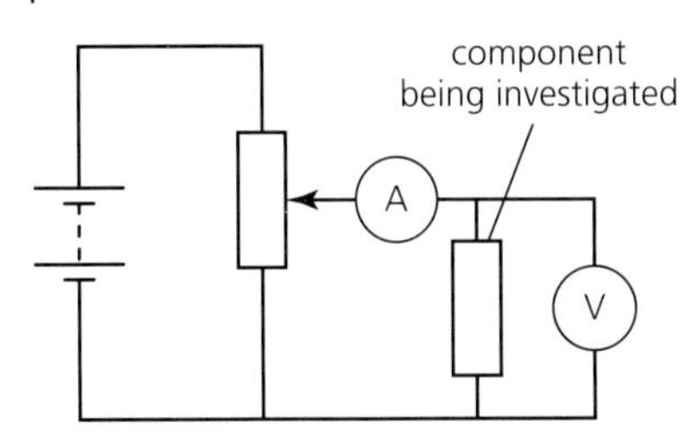

■ **Figure 5.60** A circuit for investigating *I–V* characteristics of electrical components

As an example, consider a 0–5 Ω variable resistor used to control a lamp of fixed resistance 2.0 Ω designed for normal use when there is a p.d. of 6 V across it. When the resistance is set to 0 Ω, the lamp will get the full 6 V; and when the resistance is set to its maximum of 5 Ω, the lamp will have a p.d. of (2.0/7.0) × 6.0 = 1.7 V across it. It is not possible to reduce the p.d. across the lamp to zero.

Lower p.d.s would be possible with a variable resistance that had a much wider range, but then making adjustments for higher voltages would become more difficult.

Alternatively, a variable resistor can be connected across a battery using all three terminals, as shown in Figure 5.59. Used in this way it can provide a potential difference, V_{out}, to another part of the circuit, which varies continuously from zero to the full p.d. of the battery, V_{in}. The maximum voltage will be obtained with the sliding contact at the top of the variable resistor, and the voltage will be zero with the contact at the bottom. A variable resistor used in this way is called a **potentiometer**.

It is essentially just a potential divider with the sliding contact on the variable resistor dividing it into two resistors of variable sizes (as shown by R_1 and R_2, where $R_1 + R_2$ is constant).

When a potentiometer is connected as the input into a circuit, the value of V_{out} cannot be calculated without considering the effect of the resistance of the rest of that circuit. Generally, the resistance of the circuit should be much higher than the resistance of the potentiometer.

A potentiometer provides the best way of varying the voltage to a component in order to investigate its *I–V* characteristics (see Figure 5.60).

71 a Draw a circuit that uses a potentiometer and a 12 V battery to provide an input that varies from 0 to 12 V for a small lamp rated at 12 V, 1.5 W.
b Label the lamp with its resistance (assume it is constant).
c Suggest a suitable resistance range for the potentiometer.
d If a 0–200 Ω potentiometer was used, estimate the p.d. across the lamp when the potentiometer was set to the middle of its range.
e Explain why the answer to d is not 6 V.

5.3 Electric cells – *electric cells allow us to store energy in a chemical form*

■ Cells

In the modern world, with so many mobile electronic devices, we are becoming increasingly dependent on compact, portable sources of electrical energy. An **electric cell** (sometimes called an **electrochemical cell**) is a component that uses chemical reactions to transfer chemical (potential) energy to the energy transferred by an electric current. Electric cells are also sometimes called *voltaic cells* (named after Alessandro Volta, an Italian scientist who is credited with the design of the first useful electric cells). The basic principle can be demonstrated by putting two different metals into a solution that conducts electricity (an **electrolyte**) as shown in Figure 5.61.

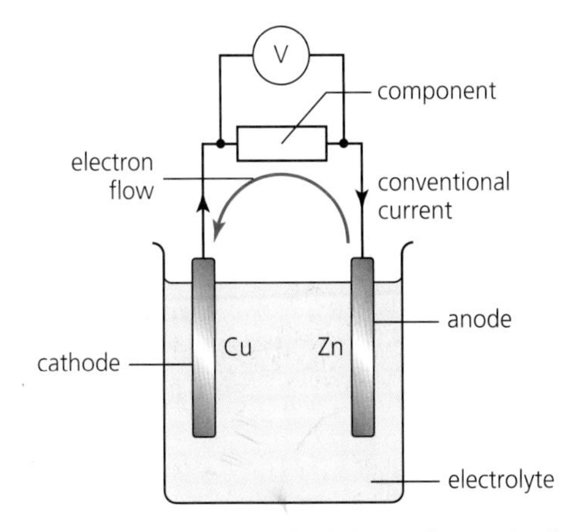

■ **Figure 5.61** A simplified electrochemical cell

The two conductors making contact with the electrolyte (copper and zinc in this example) are called the electrodes. In Figure 5.61, electrons flow around the external circuit from the zinc electrode to the copper electrode and in this process they can transfer useful energy to the circuit. Remember that conventional current is shown flowing in the opposite direction to the flow of electrons. The electrode from which electrons leave to travel around the circuit is called the anode and the electrode to which the electrons return is called the cathode. Details of the design of cells or chemical reactions are *not* needed for this course, but are included in the IB Chemistry course.

A very wide range of practical investigations into simple cells is possible: improving the design with a *salt bridge*, using different conductors, different temperatures, different electrode geometry or different electrolytes (or concentrations). Characteristics such as the voltage, current and power delivered to a circuit can be determined and how these properties vary with time. With such a wide selection of possible variables, the investigation of electrochemical cells is a topic well-suited to computer simulations.

Two or more cells connected together in a circuit are known as a *battery*, although the term 'battery' is also widely used for single cells.

Secondary cells

Simple cells that can only be used until the chemical reactions have stopped are called primary cells. In secondary cells, the chemical reactions can be reversed, so that the cells can be used again. This process is known as recharging a cell (battery). The batteries in mobile phones and notebooks are secondary cells and can be recharged many times, although there is a limit to how many times this can be done. Recharging batteries can take several hours or more. The lifetime of secondary cells and the times needed for recharging them are areas of continual technological research. The use of primary cells has decreased significantly in recent years.

In order to recharge a secondary cell (battery), a current must be made to pass through it in the opposite direction to the currents that flow when the battery is in normal use. The recharging voltage should usually be the same as the voltage produced by the cell in normal use (see Figure 5.62).

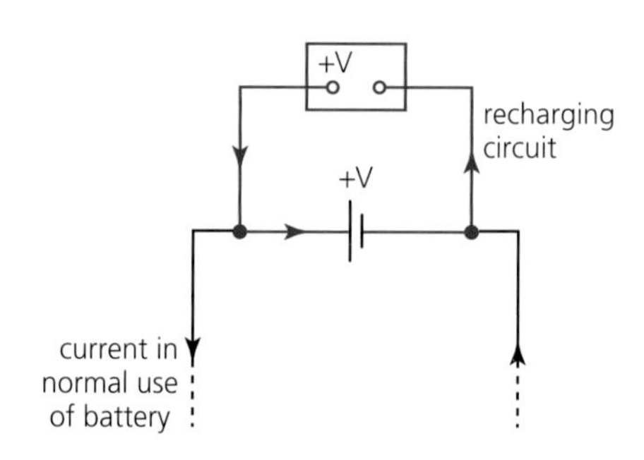

Figure 5.62 Recharging a battery

Connecting batteries into circuits

Electromotive force (emf)

Electrochemical cells use chemical reactions to provide potential differences and energy to circuits, but there are other devices that can provide p.d.s from different processes. For example, *photovoltaic cells* transfer light energy to electrical energy, our homes are provided with a *mains p.d.* and energy by *generators* at electric power stations, and the *dynamo* on a bicycle can transfer kinetic energy to electrical energy for the lamp.

The electromotive force (emf) of a battery, or any other source of electrical energy, is defined as the *total* energy transferred in the source per unit charge passing through it.

Most of this energy should be supplied to the circuit, but some energy will be transferred to internal energy within the battery itself.

Electromotive force is a potential difference (energy transferred/charge). It is given the symbol ε and its unit is the volt, V. The name *electromotive force* can cause confusion because it is not a force. For this reason it is commonly just called 'emf'.

Terminal potential difference

It is important to realize that when a current is flowing in a circuit, the potential difference across the terminals of the battery, or other power supply, is *not* equal to its emf. The p.d. across the battery is the same as the p.d. supplied to the circuit and is known as the **terminal potential difference**. When a current is flowing, the terminal p.d. is lower than the emf because of the *internal resistance* of the power source.

Internal resistance

Cells, batteries and other sources of electrical energy are not perfect conductors of electricity. The materials from which they are made all have resistance, called the **internal resistance** of the cell, which is given the symbol r. The internal resistance of a new battery bought for a torch might be about 1 ohm.

If the internal resistance of a battery (Figure 5.63) is much less than the resistance of the rest of the circuit, its effect can usually be ignored and, as a result, many examination questions refer to batteries or cells of 'negligible internal resistance'. But in other examples, the internal resistance of an energy source can have a significant effect on the circuit. The value of the internal resistance of a battery may vary when different currents flow through it, but we usually assume that it is constant.

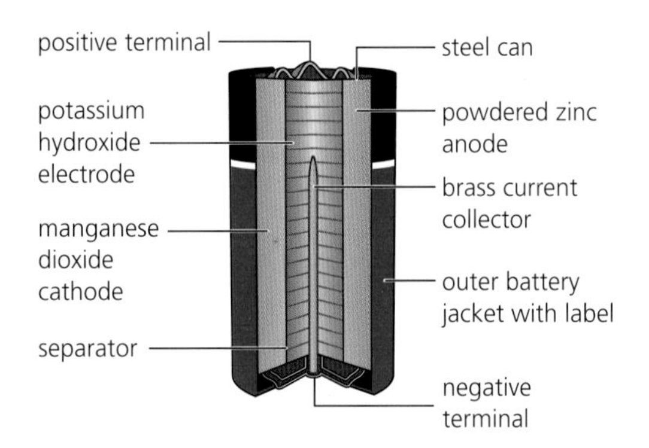

■ **Figure 5.63** The chemicals inside a battery provide internal resistance

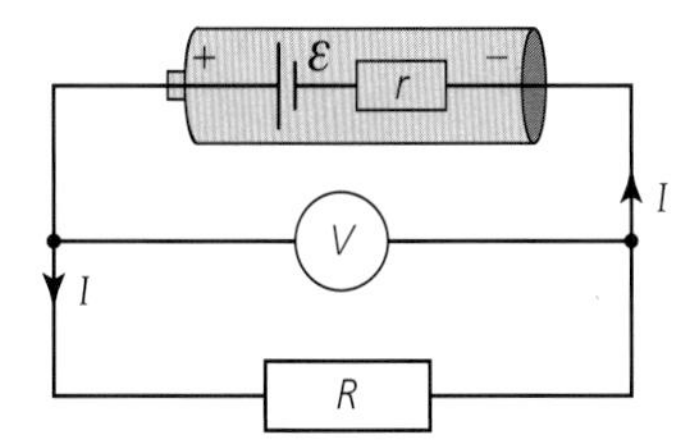

■ **Figure 5.64** A cell in a simple circuit

Consider Figure 5.64. As current flows through the cell, some energy will be transferred to internal energy because of its internal resistance, so that:

total energy transferred by the cell (per coulomb) = energy transferred to the circuit (per coulomb) + energy transferred inside the cell (per coulomb)

Emf of cell, ε = terminal p.d. across circuit, V, + 'lost volts' due to internal resistance of battery

The same current flows through both resistances and using $V = IR$ gives:

$$\varepsilon = IR + Ir = I(R + r)$$

This equation is given in the *Physics data booklet*.

Remember that the voltmeter shown in Figure 5.64 is *not* measuring the emf of the cell; it is measuring the terminal p.d. of the cell, which is the same as the useful p.d. supplied to the circuit.

If the current, I, in a circuit is zero, then the 'lost volts' (Ir) are also zero, so that the emf equals the terminal p.d. This means that the emf of a cell can be effectively measured by connecting a high-resistance voltmeter across it when it is not connected to anything else. (Because the voltmeter has a very high resistance, the current through it and the battery may be considered to be negligible.)

Worked example

19 a If the cell shown in Figure 5.64 has an emf of 1.5 V and an internal resistance of 0.90 Ω, calculate the current in the circuit if the resistor has a value of 5.0 Ω.

b What is the p.d. across the circuit?

a $\varepsilon = I(R + r)$

$1.5 = I(5.0 + 0.90)$

$I = \frac{1.5}{5.90} = 0.25\,\text{A}$

b $V = IR$

$V = 0.25 \times 5.0 = 1.3\,\text{V}$

The voltmeter will read 1.3 V, but the cell has an emf of 1.5 V so 0.2 V has been 'lost' because of the internal resistance. The lost volts cannot be measured directly with a voltmeter, but can be calculated from Ir. This shows us that the value of the lost volts will increase when a larger current, I, is flowing through the cell. If the power source has significant internal resistance, the useful p.d. supplied to the circuit will decrease significantly if a larger current flows.

To determine internal resistance experimentally, the circuit shown in Figure 5.64 can be used with an ammeter added to measure the current, I, through a known resistor, R. Then the equation $\varepsilon = IR + Ir = I(R + r)$ can be used to calculate r if the emf, ε, has been measured in a separate experiment. A better method is to use a variable resistor instead of R and measure the currents for different values of voltage. The gradient of a V–I graph will have an intercept equal to the emf and a gradient of magnitude equal to the internal resistance.

Similar cells may be connected in series or parallel to make a battery. If they are connected in series, the p.d.s and internal resistances simply add up. If they are connected in parallel, the p.d. provided will be the same as for a single cell, but the overall internal resistance can be calculated using the formula for resistances in parallel (page 225).

72 a When a battery of emf 4.5 V and internal resistance 1.1 Ω was connected to a resistor, the current was 0.68 A. What was the value of the resistor?
b If the resistor was replaced with another of twice the value, what would the new current be?
c What assumption(s) did you make?

73 When a cell of internal resistance 0.24 Ω was connected to a lamp, the current was 0.72 A and the p.d. across the lamp was 2.8 V.
a Calculate the resistance of the lamp.
b What was the emf of the cell?
c Calculate the rate of energy transfer (power) in:
i the lamp
ii the cell.

74 A high-resistance voltmeter shows a voltage of 12.5 V when it is connected across the terminals of a battery that is not supplying a current to a circuit. When the battery is connected to a lamp, a current of 2.5 A flows and the reading on the voltmeter falls to 11.8 V.
a What is the emf of the battery?
b Calculate the internal resistance of the battery.
c What is the resistance of the lamp?

75 The engine of a car is started at night with the car's headlights on. Starting a car requires a high current from a 12 V battery and the headlights become momentarily dim.
a Why is a high current needed?
b Suggest why the headlights become dim.

76 If a connecting wire is connected by mistake across a battery or power supply, it is an example of a *'short circuit'*.
a Calculate the current that flows through a battery of emf 12.0 V and internal resistance 0.25 Ω if it is accidentally 'shorted'.
b What assumptions did you make?
c Calculate the power transferred in the battery.
d Suggest what will happen to the battery.

77 Explain why it is not possible to measure the internal resistance of a battery in the same way as the resistance of a resistor in a circuit.

78 Three 1.5 V cells, each with an internal resistance of 1.0 Ω, are to be connected together to make a battery. Determine the combined emf and internal resistance if they are connected:
a in series
b in parallel.

79 a Draw a circuit for obtaining a set of potential difference–current readings for a battery connected to a variable resistance.
b Explain why the gradient of a *V–I* graph equals the internal resistance and why the intercept equals the emf.
c Explain why using a graph should provide a more reliable determination of internal resistance than a calculation based on a single pair of measurements.

80 Suggest reasons why in recent years primary cells have become used much less.

Utilizations

Electric cells: choosing a battery

When choosing a battery for a particular purpose, there is a lot more to consider than just its emf. It is also important to consider:

- internal resistance
- terminal p.d. in normal use (and how it changes with time)
- size
- total energy stored
- whether it is a primary cell or secondary cell
- recharging characteristics (if it is a secondary cell)
- whether its manufacture and disposal cause significant pollution

When we replace a battery in electrical or electronic equipment, we usually have no choice about which kind to use; that decision was made during the design process. Question 81 (page 241) shows why the internal resistance of a cell (and how it compares with the rest of the circuit) is important.

The dramatic increase in the use of high-technology batteries worldwide has important implications for both their supply and their disposal. The distribution, availability and cost of the materials used in battery manufacture may be an issue of increasing importance in the future. The chemicals used in batteries can become a major pollution concern if they are not disposed of properly.

Scientists are continually trying to improve the amount of energy stored per gram in batteries and this requires close collaboration with chemists. This is a major technological research area with far-reaching applications for mobile communications, emergency power supplies, battery-driven transportation and many other uses.

■ **Figure 5.65** An electric bicycle in China

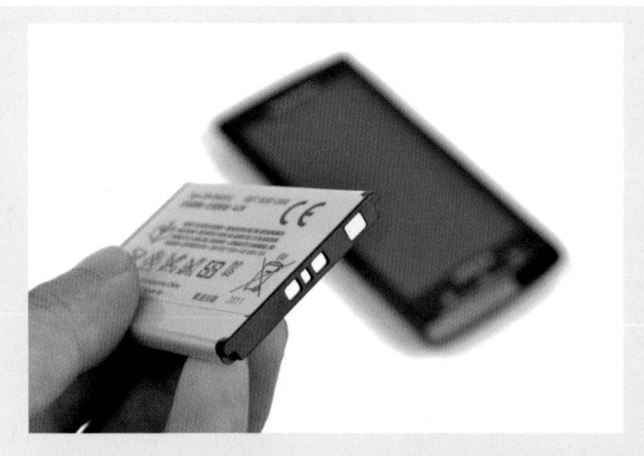

■ **Figure 5.66** A battery used in a mobile phone

The amount of energy stored in a battery is often given the unit *amp hour*, A h (or watt hour, W h), rather than joules. For example, a battery rated at 10 A h should be able to provide 1 A for 10 h or 0.01 A for 1000 h. If it is a 12 V battery then the nominal total energy stored will be $12 \times 10 \times 3600 = 4.3 \times 10^5$ J.

A typical mobile phone battery (Figure 5.66) is *lithium ion* in design and might be rated at 3.8 V, 8 W h. Approximately 1 kJ can be stored in every gram of chemicals in the battery.

1 Would doubling the size of the battery used in a mobile phone mean doubling the energy stored, and the time between recharges? If so, discuss why phone manufacturers don't do this.

Nature of Science

Scientific 'advances' can sometimes have unexpected long-term risks

Research into improved battery design seems to be of benefit to all of us, but as is often the case, there are negative aspects to consider as well: pollution and depletion of resources, for example. The disadvantages of a technological change may be well known *now*, but years ago the full consequences of improved battery design (especially the rapid increase in use of batteries) was not foreseen by many people. Perhaps it should have been, but even with such foresight would battery research have been any different?

Scientific and technological advances have produced really dramatic changes in lifestyles and living conditions (including health and sanitation) for nearly everyone on Earth within living memory. There seems little doubt that this astonishing rate of 'progress' will continue. However, for every development there are likely to be other consequences that may be predictable, or could be totally unexpected.

■ Discharge characteristics of batteries

We have seen that the terminal p.d. provided by a battery to a circuit will depend not only on its emf, but also on the resistance of the circuit in which it is placed and on the battery's internal resistance. However, another very important characteristic of a battery is how the terminal p.d. (and the current supplied) to the same circuit changes over time: its **discharge characteristic**. This can be investigated with voltage and current sensors set up with a data logger to take measurements at regular intervals over several hours or days. The *charging characteristic* of a secondary cell could be investigated in a similar way.

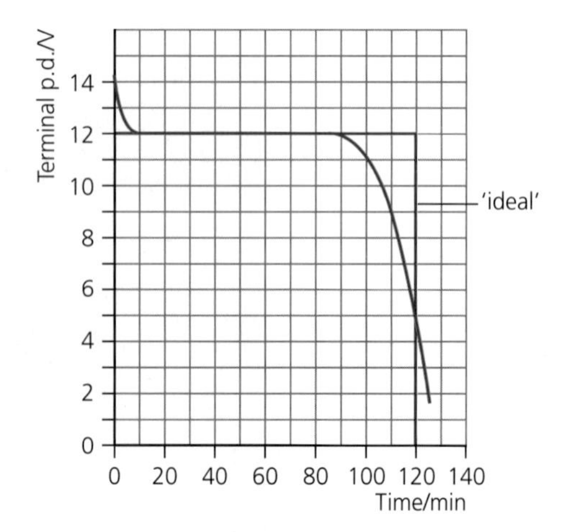

■ **Figure 5.67** Discharge characteristic of a battery

Ideally the terminal voltage provided by a battery should stay at the same value for as long as possible and then, when the stored energy has been transferred, the p.d. should fall quickly to zero. In this way, as little energy as possible would be provided at a reduced voltage, which would probably be too low for the correct functioning of the device that it is powering. Figure 5.67 shows a typical discharge characteristic. The initial p.d. drops quickly to the normal operating voltage and, in this example, it is hoped that the terminal p.d. will remain at 12 V for two hours.

81 The battery in Figure 5.68 has an emf of 6.0 V and an internal resistance of 0.45 Ω. The variable resistor, R, can be changed to any value between 0 Ω and 3 Ω.
 a Set up a spreadsheet to calculate the current in the circuit for a range of suitable values of R.
 b Add columns to calculate the p.d. across R and the power generated in the resistor.
 c Draw a graph showing how the power varies with the value of R.
 d Explain why the power is low for both high and low values of R.
 e Under what conditions can maximum power be obtained from the battery?

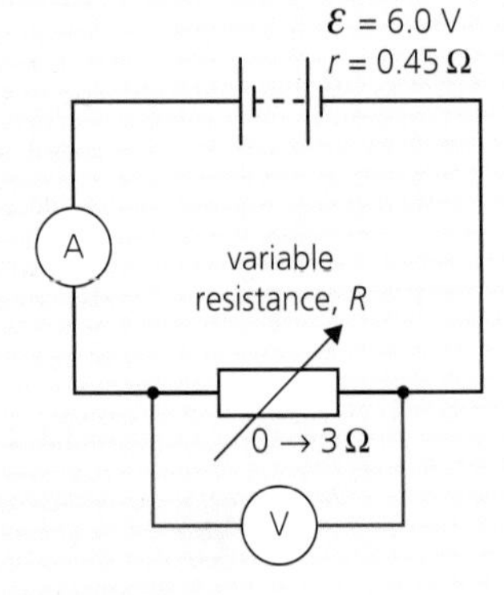

■ **Figure 5.68**

82 A major attraction of electric cars is that they are non-polluting, but the energy stored in their batteries had to be transferred from somewhere else. Much research is being carried out into the development of better batteries for electric cars so that they can supply the energy needed for longer distances and recharge much more quickly. Use the internet to find out the latest developments.

83 a Which major pollutants of land and water are present in used batteries?
 b Can the components of a used battery be recycled?
 c How does your community dispose of used batteries?

84 Carry out some research into the origin of the materials used in the latest high-technology batteries.

85 a Estimate the total energy stored in a typical torch battery by considering the power of the lamp and the number of minutes it can be used for.
 b Calculate the approximate amount of energy stored in each gram of the chemicals.
 c Compare your answer to the amount of energy stored in:
  i 1 g of chocolate
  ii 1 g of petrol (use the internet).

86 Consider Figure 5.67.
 a For how long could a device requiring 10–14 V be operated by this battery?
 b If the battery delivers a steady current of 0.46 A during this time, estimate the total energy transferred from the battery in this time.
 c Estimate the amount of energy still stored in the battery after the p.d. falls below 10 V.

87 Sketch a voltage–time graph to suggest how the terminal p.d. of a secondary cell might change during a recharging process.

ToK Link

Scientific responsibility

Battery storage is seen as useful to society despite the potential environmental issues surrounding their disposal. Should scientists be held morally responsible for the long-term consequences of their inventions and discoveries?

Most, if not all, scientific and technological developments have some unwanted, and/or unexpected, side-effects. Most commonly these may involve pollution, the threat of the misuse of new technologies or the implications for an overcrowded world. Or maybe a new technology will result in dramatic changes to how societies and cultures function; changes that will have both benefits and disadvantages, many of which will be a matter of opinion.

Should more effort be made to anticipate the possible negative aspects of scientific research and development? Perhaps that is unrealistic, because predicting the future of anything, especially the consequences of as-yet-unfinished research, is rarely successful. Of course, there are some extreme areas of research that most people would agree should *never* be allowed; nuclear weapons, for example. It is important to appreciate that a key feature of much scientific research is that it involves the investigation of the *unknown*.

If 'society' decides that it wishes to control some area of scientific and technological research because the possible negative consequences are considered to be greater than the possible benefits, who makes those decisions and who monitors and controls the research (especially in this modern international world)? Is it reasonable to expect scientists to be responsible for their own discoveries and inventions?

5.4 Magnetic effects of electric currents – *the effect scientists call magnetism arises when one charge moves in the vicinity of another moving charge*

■ Magnetic fields

Gravitational fields exist around masses and electric fields exist around charges, but what causes a **magnetic field**? The answer 'magnets' may be true, but the kind of '**permanent' magnets** with which we are all familiar (for example, that hold notes onto a refrigerator door) are only one small and fairly unimportant example of magnetic effects. *All* electric currents produce magnetic fields around them so it is difficult to separate the topics of electricity and magnetism. *Electromagnetic effects* dominate our lives and are essential for the operation of such things as power stations, trains, cars, planes, televisions, hairdryers and computers.

The Earth's magnetic field

Magnetic materials have been known for thousands of years, long before they were actually understood. Early civilizations recognized that a piece of magnetic material that is free to move will always twist until it is pointing approximately north–south. The end that points to the north was called the **north** (-seeking) **pole** and the other end was called the **south** (-seeking) **pole**. A magnet used in this way is called a **compass** and, for many centuries, compasses have been helping people to find their way around. In recent years, *magnetometers* (which detect the strength of a magnetic field) have become very popular, especially because they can be made very small and incorporated in devices like mobile phones.

The simplest bar magnets have one pole at each end and are called **dipole** magnets. It is not possible to have a magnet with only one pole and if a dipole magnet is cut in half, the result will be two smaller and weaker dipole magnets. Magnets can be designed to have complex shapes and multiple poles. When dipole magnets are brought close to each other, it quickly becomes obvious that opposite poles attract each other and similar poles repel each other. If at least one of a pair of such magnets is free to move, they will align with each other. This then helps us to explain the action of a compass: the Earth itself behaves like a very large bar magnet and the small, freely moving magnet of the compass is just twisting to line up with the Earth's magnetic field.

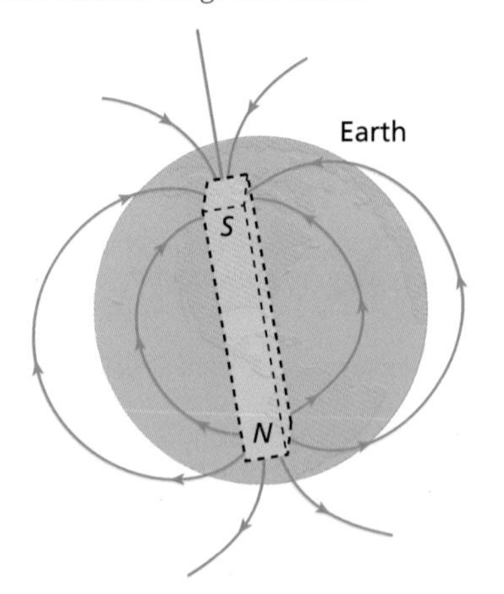

■ **Figure 5.69** The Earth's magnetic field

The terms *north pole* and *south pole* are still used for describing the ends of magnets, although this can very easily cause confusion with the Earth's geographic poles. By definition, the end we call the north pole of a magnet is attracted to the geographic north of the Earth. This means that the Earth's magnetic field has a *magnetic* south pole at the geographic north and a *magnetic* north pole at the geographic south. Figure 5.69 represents the Earth's magnetic field by field lines, which are discussed in greater depth in the next section.

The Earth's geographic poles are located where the axis of rotation reaches the surface, but the poles of the Earth's magnetic field are not in the same place, nor on the surface, and they slowly rotate. At the moment the geographic and magnetic poles at the north of the Earth are about 800 km apart. If this is not understood by a traveller, it could cause some problems with navigation over large distances using a magnetic compass, especially when close to the poles. Alternatively, a gyrocompass can be used (as on many large ships), which locates true north by responding to the Earth's rotation (with the added advantage of not being influenced by the steel of the ship). Or GPS can be used for navigation.

The Earth's magnetic field is caused by electric currents in its liquid outer core and is believed to reverse its polarity (north becomes south and south becomes north) on average about every 300 000 years, for reasons that are still not fully understood. The last such reversal was about 800 000 years ago.

88 The Earth's magnetic field extends a long way above the Earth's surface. Carry out some research into the Earth's magnetosphere and find out how it protects the Earth from the 'solar wind'.

89 It is often reported that some birds, fish, sea mammals and even bacteria can navigate by their natural ability to detect (in some way) the Earth's magnetic field. Use the internet to find out if this is scientific fact, theory or myth.

90 Do modern aircraft use magnetic compasses, gyrocompasses, magnetometers, radar, radio beacons, GPS or some other method of navigation? Carry out some research into the kinds of navigation system used on modern planes.

Magnetic field patterns

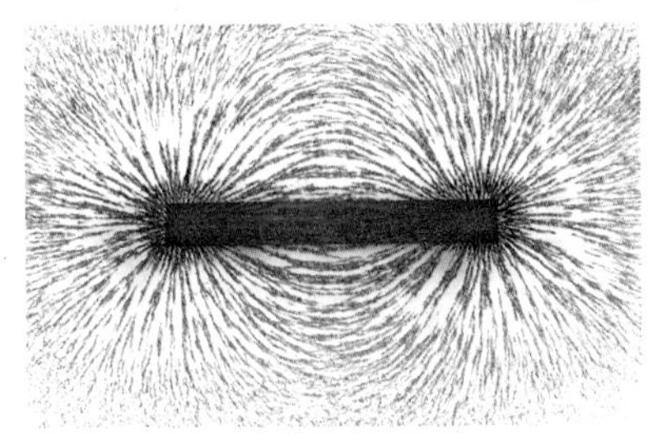

■ **Figure 5.70** Iron filings can be used to demonstrate the magnetic field around a bar magnet

Earlier in this chapter we saw that any charge (which may be moving or stationary) has an electric field around it. Magnetic fields are produced around *moving* charges.

Because electric currents are a flow of moving charge, *all* electric currents have magnetic fields around them.

The field around a permanent magnet (see Figure 5.70) has its origin in the movement of electrons in certain atoms (like iron, which is described as **ferrous**). In most non-ferrous materials these small-scale magnetic effects cancel each other out.

Magnetic fields can be represented on paper or screen by using **magnetic field lines**. (Gravitational and electric fields can also be represented by field lines.) By convention the direction of a magnetic field is the same as the direction in which a compass points: from *magnetic* north to *magnetic* south.

Fields are strongest where the lines are closest together, but field lines can never cross each other, because that would mean that the field pointed in two different directions at the same place.

Magnetic field patterns around permanent magnets or currents can be investigated experimentally in *two* dimensions by using small 'plotting' compasses or iron filings (such as shown in Figure 5.70), which individually twist until they are parallel with the field at that point. Of course, magnetic fields spread into *three*-dimensional space and computer simulations can help students to visualize the fields more realistically.

In this course we will discuss the magnetic fields around currents in long straight wires, around currents in solenoids and around dipole bar magnets, as well as the strong and useful uniform magnetic fields produced between opposite magnetic poles close together.

Field due to a current in a long straight wire

Figure 5.71a shows the field produced by a steady current flowing in a long straight wire. In the diagram we know that the current is flowing perpendicularly into the page by the use of the cross at the centre. The field is circular around the wire and gets weaker as the distance from the wire increases, shown by the increasing distance between the field lines. (The field strength is inversely proportional to the distance from the wire.) The direction of the field lines is shown and can be predicted using the **right-hand grip rule**: if the thumb points in the conventional direction of current flow, then the fingers indicate the direction of the magnetic field.

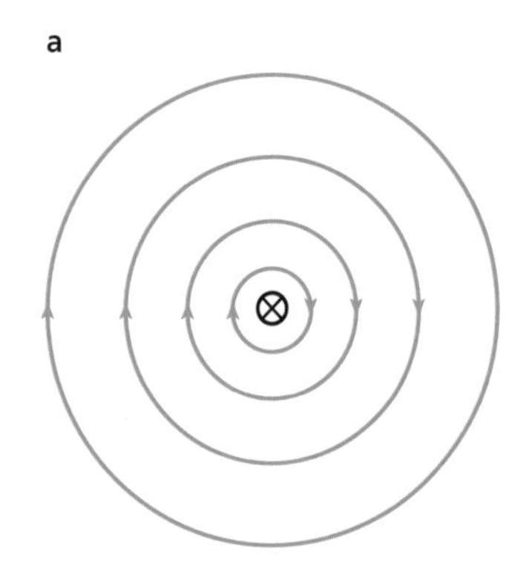

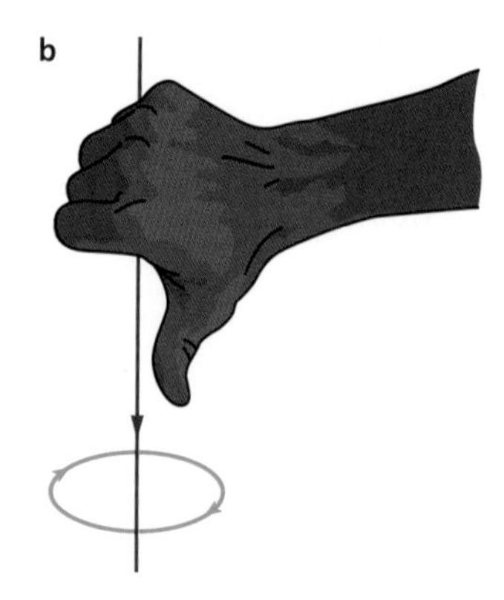

■ **Figure 5.71 a** The magnetic field due to a current flowing in a long straight wire into the plane of the paper; **b** using the right-hand grip rule to predict the direction of the field

Field due to a current in a solenoid

When a coil of insulated wire is wrapped regularly so that the turns do not overlap and it is significantly longer than it is wide, it is usually called a **solenoid**. Solenoids are useful for the strong, uniform magnetic fields produced inside them. Figure 5.72 represents a solenoid showing the parallel lines of a uniform magnetic field inside. The overall field pattern is similar in shape to that produced by a simple permanent bar magnet. One end of the solenoid acts like a north pole and the other like a south pole (confirmed by using the right-hand grip rule). Reversing the direction of the current changes the south pole to a north pole, and the north pole to a south pole. This is called reversing the **polarity** of the magnetic field. The more turns of wire in a given length, the stronger the magnetic field, for the same current.

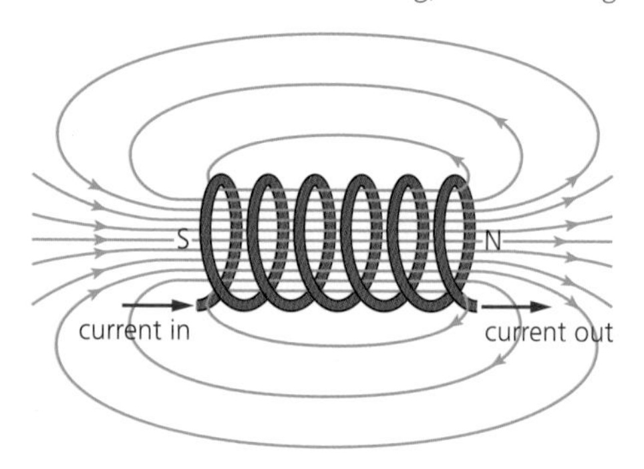

■ **Figure 5.72** Magnetic field due to the current in a solenoid

Different materials have different magnetic properties. Magnetic **permeability** is a measure of a medium's ability to transfer a magnetic field and it can be compared to electric **permittivity** for electric fields. The magnetic permeability of free space is given the symbol μ_0 and its value is given in the *Physics data booklet*.

Other materials have higher permeabilities. If a solenoid or coil of wire is wound around a material with high permeability (e.g. iron), the field produced can be very much stronger than it would be without the core. In this way very strong, but adjustable, electromagnets can be made (see Figure 5.73). The magnetic material used as the core must gain and lose its magnetic properties quickly and, for this reason, 'soft' iron is used as the core of most electromagnetic devices, such as transformers and motors.

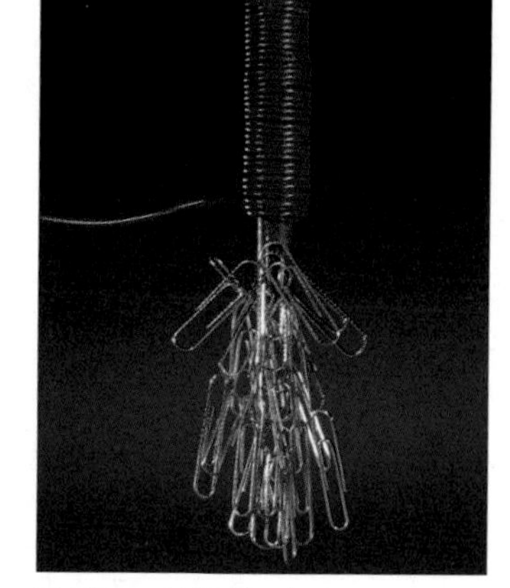

■ **Figure 5.73** Testing an electromagnet in a laboratory

The properties of electromagnets are an interesting topic for laboratory investigation not only because there are so many variables, but also because an accurate measurement of field strength needs to be devised.

Fields due to permanent magnets

Figure 5.74 shows the magnetic field around a simple dipole magnet, with a plotting compass indicating the direction of the field at one point. It should be compared to Figure 5.70.

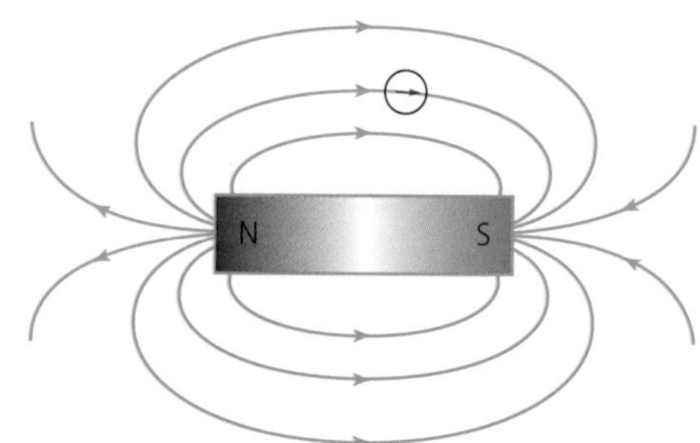

■ **Figure 5.74** The magnetic field around a simple dipole magnet

Figure 5.75 indicates the shape and direction of the fields between the poles of two magnets placed close together.

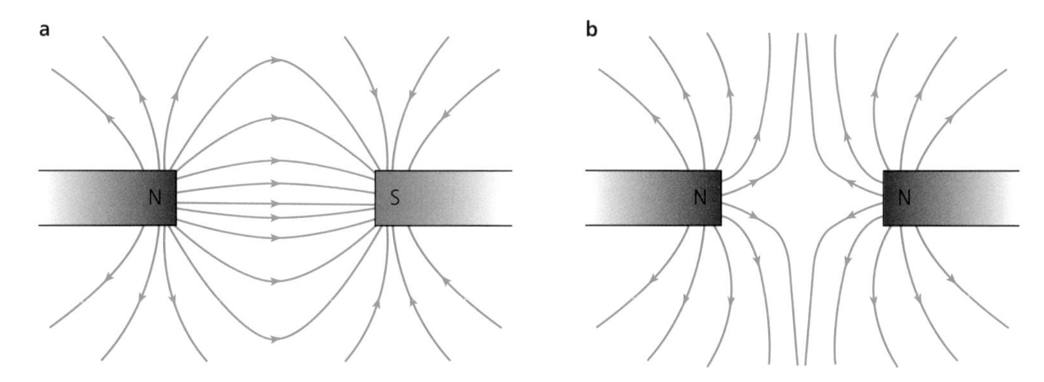

■ **Figure 5.75** Magnet fields between magnets: **a** between opposite poles and **b** between like poles

It can also be convenient to make a strong uniform field with an electromagnet as shown in Figure 5.76. Strong uniform magnetic fields, such as those formed inside solenoids, have many uses in science and technology, including medical scanners (see Figure 5.77).

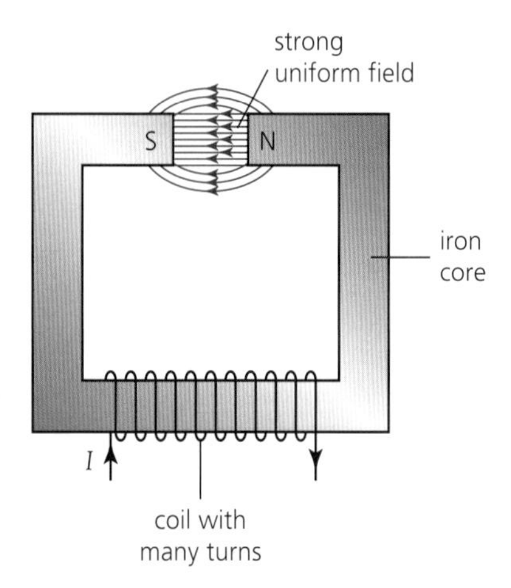

■ **Figure 5.76** Using an electromagnet to create a strong, uniform field

Nature of Science

Visual models can help understanding

Much of physics requires knowledge of things that cannot be seen. To help with this understanding, scientists develop *models* of various kinds to represent, in some way, the ideas they wish to communicate. Such models may be simplified written descriptions, two-dimensional drawings, three-dimensional structures or mathematical equations. Magnetic, electric and gravitational fields occupy the space around us, but they cannot be seen. This makes them especially challenging to understand, but the use of field lines to draw representations of them is a great aid to learning and to the communication of ideas. It should be clear that the field lines are not a physical reality, but the vizualization of fields in some way is essential for human minds to understand them.

ToK Link

Mapping is a common means of simplifying patterns of information

Field patterns provide a vizualization of a complex phenomenon, essential to an understanding of this topic. Why might it be useful to regard knowledge in a similar way, using the metaphor of knowledge as a map – a simplified representation of reality?

The use of lines and patterns to represent fields is accepted fully by the scientific community as possibly the only way of presenting these difficult ideas simply to the human mind, although no one thinks that they are 'real'. It might be argued that such a simplification in some ways restricts our understanding, or imagination, about the subject because it channels our thoughts in certain prescribed directions. The **mapping** (representing by drawing) of any knowledge is a simplification to aid understanding and one which has obvious appeal but, like all simplifications, has limitations.

Utilizations

Magnetic fields used in medical scanners

Magnetic resonance imaging (MRI) is used in modern hospitals to create images of organs within the human body. (This topic is covered in detail in Option C.) Although they are expensive to buy and to operate, such scanners are so useful for medical diagnoses that they are being introduced in hospitals worldwide.

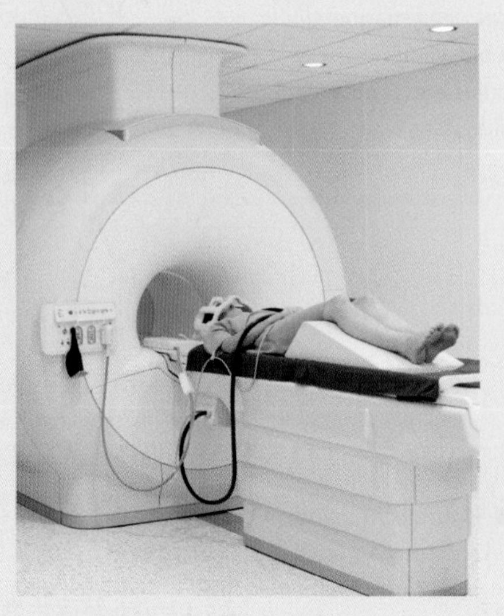

■ **Figure 5.77** An MRI scanner

Scanners, such as that shown in Figure 5.77, require very strong magnetic fields that are typically 10 000 times stronger than the magnetic field of the Earth. Very large currents are needed to produce such strong fields and this is only possible if the resistance of the wires can be reduced to very low values by operating at extremely low temperatures. Liquid helium at a temperature of about 3 K is used to coil the electromagnets so that the wires become superconducting (their resistance is almost zero).

1 Many different techniques are used for medical imaging. Find out the medical conditions that are best diagnosed by MRI scans.

■ Magnetic force

When a current is passing through a flexible, lightweight conductor in a strong magnetic field, as in Figure 5.78, the conductor may be seen to move. In this example, the aluminium starts to move upwards.

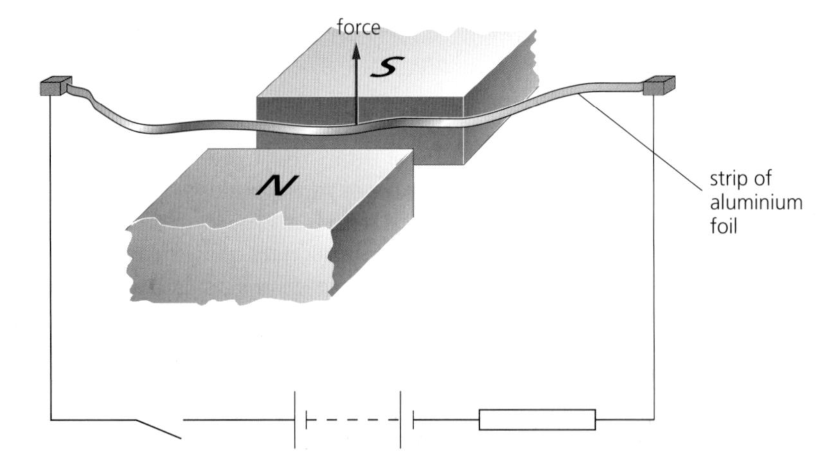

■ **Figure 5.78** Demonstrating the force on a current flowing in a piece of aluminium

In order to describe and explain this very important phenomenon, and many other electromagnetic effects, we need to represent and understand these situations in three dimensions. Figure 5.79a shows a situation similar to that in Figure 5.78 – a wire carrying an electric current across a magnetic field. The current is perpendicular to the magnetic field from the permanent magnets. In Figure 5.79b the same situation is drawn in two-dimensional cross-section, with the wire represented by the point P and the magnetic fields from the magnets (shown in green) and from the current (shown in blue) included.

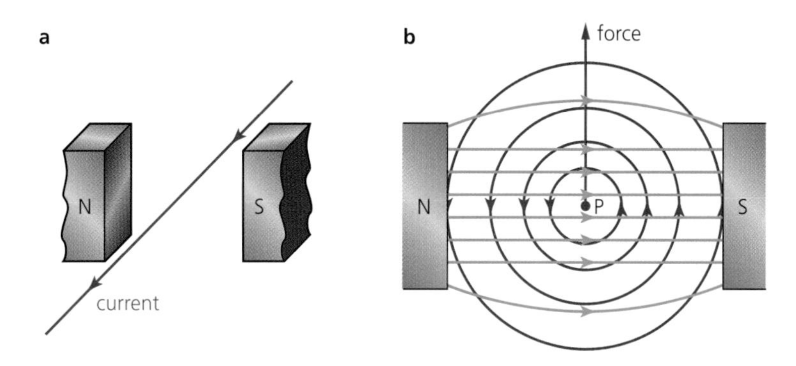

■ **Figure 5.79** Comparing the directions of current, field and force

The two fields are in the same plane, so it is easy to consider the combined field that they produce. Above the wire the fields act in opposite directions and they combine to produce a weaker field. Below the wire the fields combine to give a stronger field. This difference in magnetic field strength on either side of the wire produces an upwards force on the wire, which can make the wire move (if it is not fixed in position). This important and useful phenomenon is often called the **motor effect** and it provides the basic principle behind the operation of electric motors.

In order for there to be force on a current-carrying conductor in a magnetic field, the current must be flowing *across* the field. If the current is flowing in the same direction as the magnetic field lines from the permanent magnets, there will be no interaction between the fields and, therefore, no force produced. Figure 5.79 shows the situation in which the force is maximized because the field and current are perpendicular, but as the angle between field and current reduces, so too does the force. Consider Figure 5.80, which shows three different wires carrying the same current in different directions in the same uniform magnetic field.

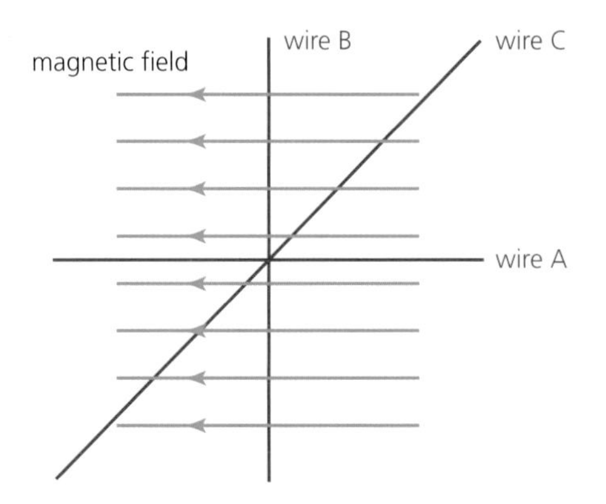

■ **Figure 5.80** How force varies with the angle of the current to the magnetic field: there will be no force on wire A and the biggest force per unit length will be on wire B. Wire C will experience a force, but the force per unit length of wire will be smaller than for wire B

Direction of the magnetic force on a current-carrying conductor in a magnetic field

The direction of the magnetic force is always perpendicular to both the direction of the current and the direction of the magnetic field. In Figure 5.80, the forces on B and C will be in or out of the plane of the paper, depending on the direction of the current. We can use **Fleming's left-hand rule** to predict the direction of force if the current and field are perpendicular to each other. This is shown in Figure 5.81. Remember that magnetic fields always point from magnetic north to magnetic south and the (conventional) direction of current is always from positive to negative.

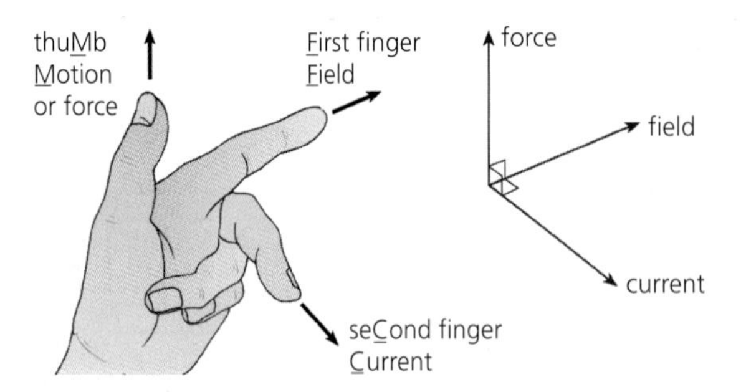

■ **Figure 5.81** Fleming's left-hand rule predicts the direction of the force

Magnetic field strength

We know that gravitational field strength, g, equals gravitational force divided by mass, and electric field strength, E, equals electric force divided by charge. So it would be consistent to suggest that **magnetic field strength** equals magnetic force divided by moving charge (current). However, the size of the magnetic force depends not only on the current, I, but also on the length of the conductor in the field, L, and its direction relative to the field, θ (see Figure 5.82).

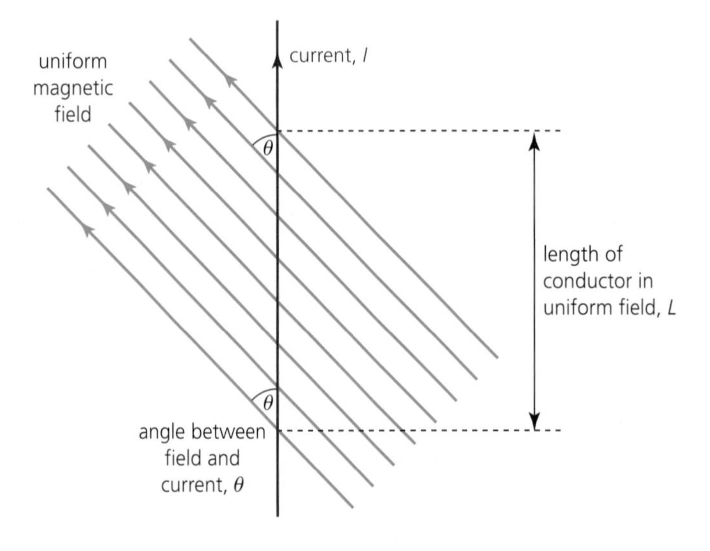

■ **Figure 5.82** Current I flowing at an angle θ across a magnetic field

We define magnetic field strength (given the symbol B) as follows:

$$B = \frac{F}{IL \sin \theta}$$

Magnetic field strength is also commonly called **magnetic flux density** (the reason for this will be given in Chapter 10 for Higher Level students). The units for magnetic field strength are newtons per amp metre, $\mathrm{N\,A^{-1}\,m^{-1}}$. Unlike gravitational and electric fields, the unit for magnetic field strength is also given another name: the **tesla**, T. (This unit is named after the eccentric physicist Nikola Tesla, who was born in Croatia to Serbian parents. He did a lot of important work on electromagnetism in Europe and, later, in the United States.)

The equation above is more usually written in the form:

$$F = BIL \sin \theta$$

This equation, which is given in the *Physics data booklet*, shows that the force on a current-carrying conductor in a magnetic field equals the magnitude of the current, multiplied by the length of the conductor, multiplied by the component of the field perpendicular to the current.

Worked example

20 In Figure 5.83, a measured current is flowing across a small, uniform magnetic field.

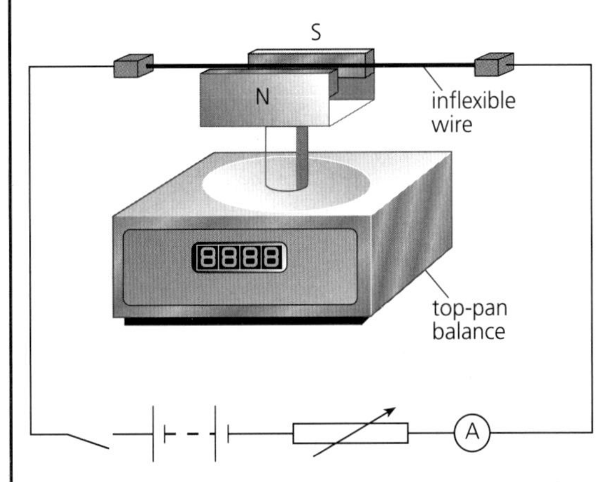

Figure 5.83 Investigating magnetic field strength

a In which direction is the force on the wire?
b In which direction is the force on the balance?
c When the current is flowing, the balance indicates that there is an extra mass of 4.20×10^{-2} g on the balance. What extra downwards force is this?
d If the current is 1.64 A and the length of the field is 8.13 cm, what is the strength of the magnetic field?

a Using the left-hand rule, the force is upwards.

b Using Newton's third law, the force is downwards.

c $W = mg = (4.2 \times 10^{-2} \times 10^{-3}) \times 9.81$

$W = 4.12 \times 10^{-4}$ N

d Using $F = BIL \sin\theta$, with $\sin\theta = 1$, gives:

$4.12 \times 10^{-4} = B \times 1.64 \times 0.0813$

$B = 3.09 \times 10^{-3}$ T

91 At point P in Figure 5.84 a large current is flowing vertically upwards, out of the plane of the paper.
a In what direction is the magnetic field (due to this current) at points Q and R?
b How does the strength of the field at Q compare to the strength at R?

92 Figure 5.85 shows a compass placed near to a solenoid carrying a small electric current I.
a Make a copy of the diagram showing where the compass might point if the current in the solenoid is doubled.
b Where would the same compass point if the direction of the current was reversed?
c Draw the compass in a position where it could point to geographic south (with the original current).

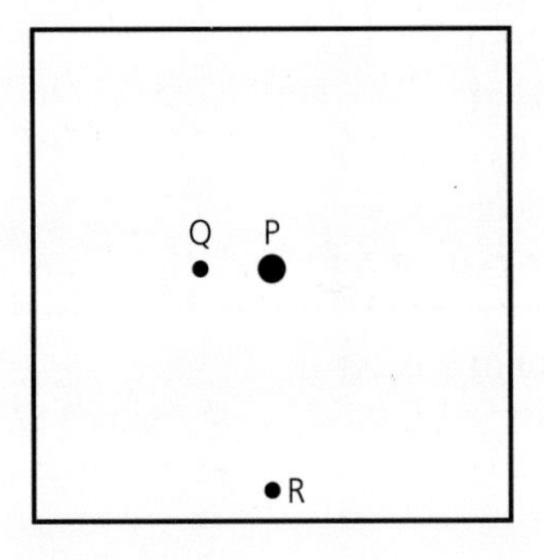

Figure 5.84

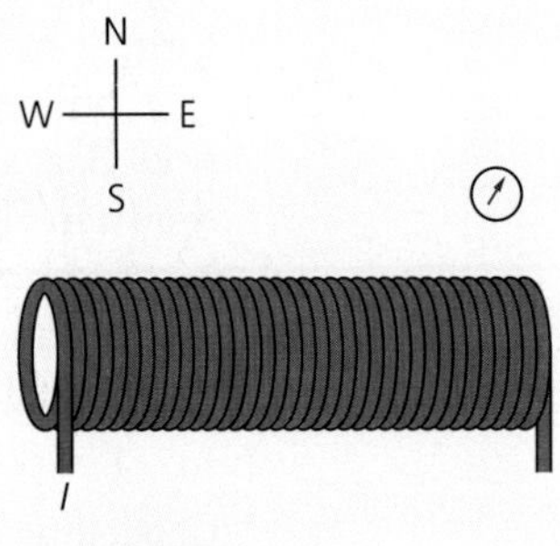

Figure 5.85

93 Calculate the magnetic force per metre on a wire carrying a current of 1.2 A through a magnetic field of 7.2 mT if the angle between the wire and the field is:
a 30°
b 60°
c 90°
d 0°.

94 a The Earth's magnetic field strength at a particular location has a horizontal component of 24 μT. What is the maximum force per metre that a horizontal cable carrying a current of 100 A could experience?
b In which direction would the current need to be flowing for this force to be vertically upwards?

95 A current is flowing in a horizontal wire perpendicularly across a magnetic field of strength 0.36 T. It experiences a force of 0.18 N, also horizontally.
a Draw a diagram to show the relative directions of the force, field and current.
b If the length of wire in the field is 16 cm, calculate the magnitude of the current.

96 a A current of 3.8 A in a long wire experiences a force of 5.7×10^{-3} N when it flows through a magnetic field of strength 25 mT. If the length of wire in the field is 10 cm, what is the angle between the field and the current?
b If the direction of the wire is changed so that it is perpendicular to the field, what would be the new force on the current?

97 Use the internet to learn about the suspension system used in Maglev trains.

Forces between currents in parallel wires

Consider the two parallel wires shown in Figure 5.86. If both wires are carrying a current, then each current is in the magnetic field of the other. Both wires will experience a force and, using the left-hand rule, the forces will be attractive between the wires if the currents are in the same direction. The forces are equal and opposite (remember Newton's third law). If the currents are in opposite directions, the wires will repel.

This arrangement is used to define the SI unit of current, the ampere. One ampere, 1 A, is the current flowing in two infinitely long, straight, parallel wires that produces a force of $2 \times 10^{-7}\,\mathrm{N\,m^{-1}}$ between the wires if they are 1 m apart in a vacuum.

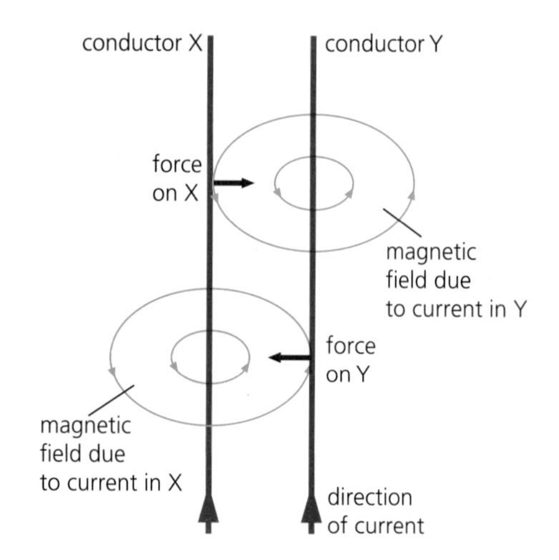

■ **Figure 5.86** Forces between parallel currents

■ Magnetic forces on individual charges moving across magnetic fields

It should be clear that the forces we have been discussing are not acting on the wires themselves, but on the *currents* in the wires. In fact, the force on a current is just the sum of the forces on the individual moving charges. *Any* moving charge crossing a magnetic field will experience a force, whether it is in a wire or moving freely (through a vacuum).

Consider a charge, q, moving across a magnetic field, B, at an angle, θ, and with a velocity, v, as shown in Figure 5.87. In time t, q moves from X to Y, a distance L.

We have seen that $F = BIL\sin\theta$. But in this case, $v = L/t$ and $I = q/t$, so the equation can be written as:

$$F = B(q/t)(vt)\sin\theta$$

Cancelling t, we get:

$$F = qvB\sin\theta$$

This equation is given in the *Physics data booklet*.

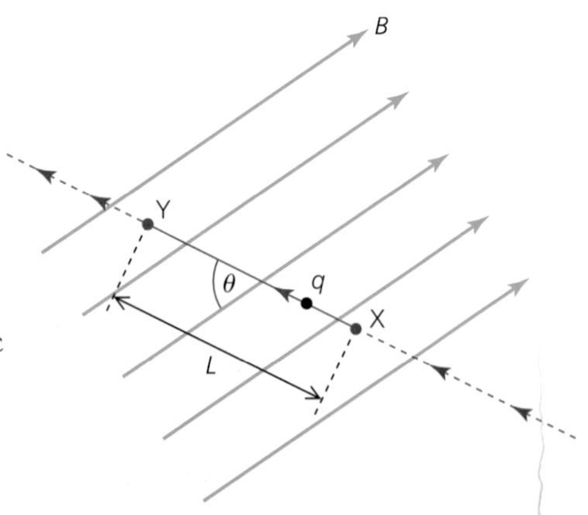

■ **Figure 5.87** An individual charge moving across a magnetic field

With this equation we can calculate the magnetic force acting on any charged particle moving in a uniform magnetic field. Note the important and interesting fact that the force is bigger if the particle moves faster. There is no force on a charge that is not moving (i.e. when $v = 0$).

It is not possible to track the progress of individual charged particles in school laboratories, but this equation is very useful when investigating the properties of charged particles that have been formed into particle *beams*. This includes beams of electrons, protons, ions and certain kinds of nuclear radiations (alpha particles and beta particles). The deflection of particle beams as they pass through magnetic fields (and/or electric fields) is a very useful way of determining the charge and/or mass of the individual particles, as well as their speeds and energies.

Direction of magnetic forces on charged particles moving across magnetic fields

Any charged particle moving perpendicularly across a magnetic field will experience a force that is always perpendicular to its instantaneous velocity. The direction of the force can be predicted by the left-hand rule, remembering that the conventional direction for current is that of the movement of positive charges. (This means that the direction of current for moving electrons will be opposite to their velocity.)

A force perpendicular to motion is the necessary condition for circular motion. (This will be explained further in Chapter 6.) Therefore, any charged particle moving perpendicular to a magnetic field will move along the arc of a circle.

Figure 5.88 shows examples of charges moving perpendicularly across strong magnetic fields. In each example, the speeds and kinetic energies of the particles do not change as a result of the perpendicular magnetic force. Note how the direction of a magnetic field perpendicular to the paper is represented:

- a ⊙ shows that the field is pointing out of the paper
- a × shows that the field is pointing into the paper.

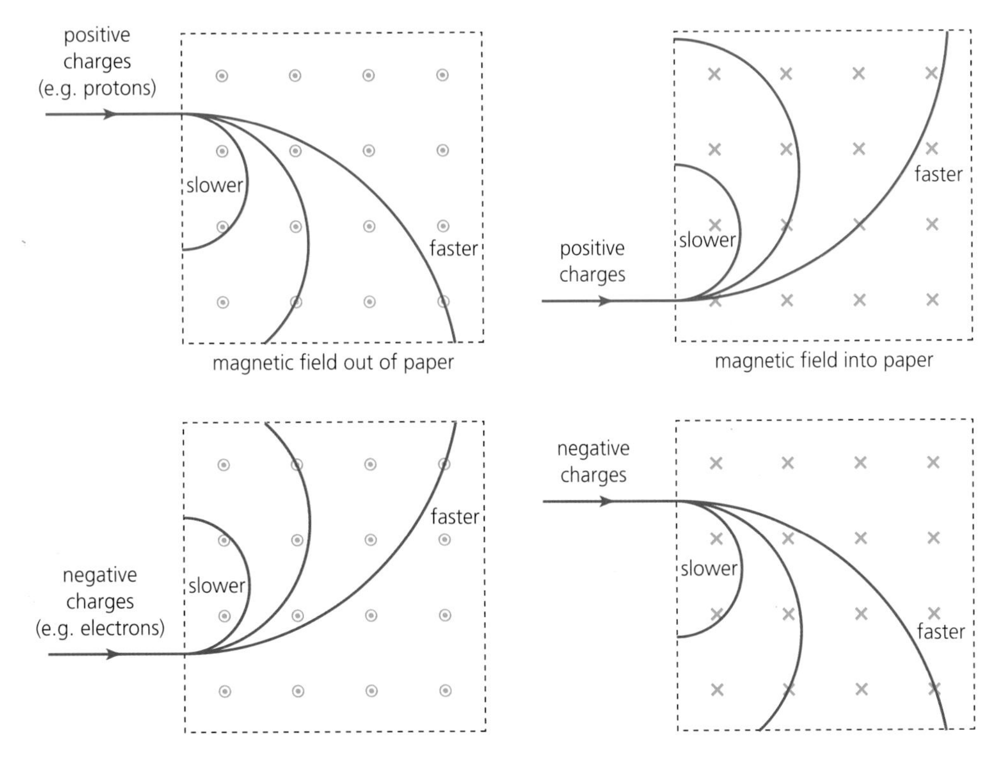

Figure 5.88 Circular paths of charges moving perpendicular to magnetic fields

Figure 5.88 also shows that the faster a charge is moving, the greater is the radius of its circular path, even though it is experiencing a bigger magnetic force. This can be explained by equating the equation for magnetic force to the equation for a force producing circular motion, $F = \frac{mv^2}{r}$ (see Chapter 6). For a charge of mass, m, moving in a circular path of radius r:

$$F = qvB \sin \theta = \frac{mv^2}{r}$$

Remembering that for a movement perpendicular to the field, sin θ = 1, we get:

$$qB = \frac{mv}{r}$$

$$r = \frac{mv}{qB}$$

This equation, which is *not* in the *Physics data booklet* and need not be remembered , shows us that for a given mass, charge and magnetic field, the radius is proportional to the velocity of the charge. A charge that is losing kinetic energy and velocity (because of collisions with other particles) will spiral inwards as its radius decreases.

We can also see that if the velocity and charge of particles are kept constant, their radius in a given magnetic field will depend only on their mass. Information about the mass and speed of sub-atomic particles can be determined by analysing their paths as they move across known magnetic fields (see Figure 5.89).

Figure 5.89 Curved paths of individual particles in a nuclear physics bubble chamber

If the motion of the charged particle is not perpendicular or parallel to the field, then it will move in a spiral-like path, called a helix.

Worked examples

21 What is the magnetic force acting on a proton (charge = +1.6 × 10^{-19} C) moving at an angle of 32° across a magnetic field of 5.3 × 10^{-3} T at a speed of 3.4 × 10^{5} m s^{-1}?

$F = qvB\sin\theta$

$F = (1.6 \times 10^{-19}) \times (3.4 \times 10^{5}) \times (5.3 \times 10^{-3}) \sin 32°$

$F = 1.5 \times 10^{-16}$ N

22 An electron of mass 9.1 × 10^{-31} kg and charge −1.6 × 10^{-19} C is moving at a speed of 1.6 × 10^{7} m s^{-1} perpendicularly to a magnetic field of 1.4 × 10^{-4} T. Calculate the radius of its path.

$$r = \frac{mv}{qB}$$

$$r = \frac{(9.1 \times 10^{-31}) \times (1.6 \times 10^{7})}{(1.6 \times 10^{-19}) \times (1.4 \times 10^{-4})}$$

$r = 0.65$ m

Utilizations

CERN

The world's largest nuclear physics laboratory is located near Geneva on the border between France and Switzerland (see Figure 5.90). Its work is done on behalf of 20 different European countries.

■ **Figure 5.90** The Large Hadron Collider at CERN is underground and has a radius of about 4 km

The letters of CERN originally represented the Conseil Européen pour la Recherche Nucléaire, but the organization is now called the European Organization for Nuclear Research.

The main activity at CERN is the use of particle accelerators to produce the extremely high energies needed to investigate the fundamental forces and particles of nature. This is achieved by the use of extremely large magnetic fields (using superconductors) to force charged particles to keep moving faster and faster in circular paths.

CERN provides an informative website, at http://home.web.cern.ch/.

1 Find out what research the Large Hadron Collider at CERN is used for.

---

98 a Explain how it is possible for a charged particle to move through a magnetic field without experiencing a force.
  b Explain whether it is possible for the same particle to move through electric and gravitational fields without experiencing forces.

99 An alpha particle has a charge of $+3.2 \times 10^{-19}$ C and a mass of 6.7×10^{-27} kg. It moves across a magnetic field of strength 0.28 T with a speed of $1.4 \times 10^{7}\,\text{m s}^{-1}$ at an angle of 33°.
  a What is the force on the particle?
  b Describe the path of the alpha particle across the field.

100 a What magnetic field strength is needed to provide a force of 1.0×10^{-12} N on each particle in a beam of singly charged ions ($q = 1.6 \times 10^{-19}$ C) moving perpendicularly across the field with a speed of $5.0 \times 10^{6}\,\text{m s}^{-1}$?
  b The ions move in the arcs of circular paths of radius 2.78 m. What is their mass?
  c If the beam was replaced with doubly charged ions of the same mass, but moving with half the speed, what would be the radius of their path?

101 a Electrons are accelerated into a beam by a voltage of 7450 V. What is the kinetic energy of the electrons in:
    i electronvolts
    ii joules?
  b Calculate the final speed of the electrons.
  c What strength of magnetic field is needed to make these electrons move in a circle of radius 14.8 cm?
  d If the accelerating voltage is halved, what would the radius of the electrons' path be in the same field?

102 A charge of $+4.8 \times 10^{-19}$ C moving perpendicularly across a magnetic field of 1.9×10^{-2} T experiences a force of 9.5×10^{-14} N.

a What was the speed of the particle?

b What electric field would be needed to produce the same force on this charge?

c Draw the path of the particle moving across the fields in a direction such that these two forces could be equal and opposite to each other (so that the resultant force on the particle was zero).

103 Find out how the Aurora Borealis is formed (see Figure 5.91).

Figure 5.91 The Aurora Borealis

Examination questions – a selection

Paper 1 IB questions and IB style questions

Q1 What are the ideal electrical properties of ammeters and voltmeters?
A ammeter zero resistance; voltmeter zero resistance
B ammeter zero resistance; voltmeter infinite resistance
C ammeter infinite resistance; voltmeter zero resistance
D ammeter infinite resistance; voltmeter infinite resistance

Q2 A voltage of 3 kV is used to accelerate an electron from rest in a vacuum. What is the maximum kinetic energy of the electron?

A 3 eV **B** 3 keV **C** 4.8 eV **D** 4.8 keV

Q3 The graph shows the I–V characteristics of a certain electrical component.

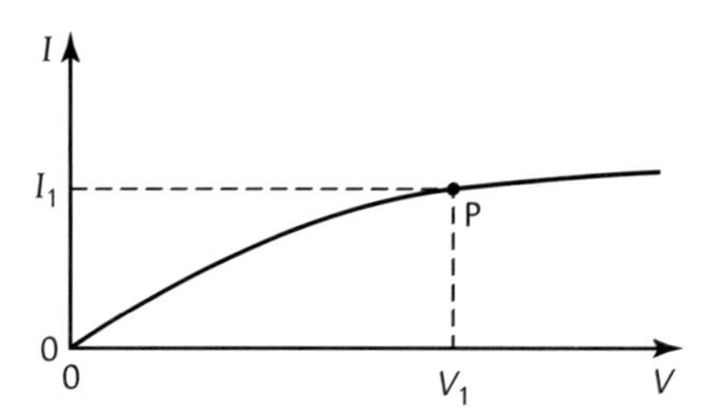

Which of the following shows how the resistance of the component at point P may be calculated?

A $\frac{V_1}{I_1}$ **B** $\frac{I_1}{V_1}$

C the gradient of the line at P **D** 1/gradient at P

Q4 The SI unit of current is based on:
A the charge passing a point in 1 second
B the power transformed by a p.d. of 1 volt
C the force on a conductor in a permanent magnetic field
D the force between parallel current-carrying conductors.

Q5 The diagram shows a potential divider circuit.

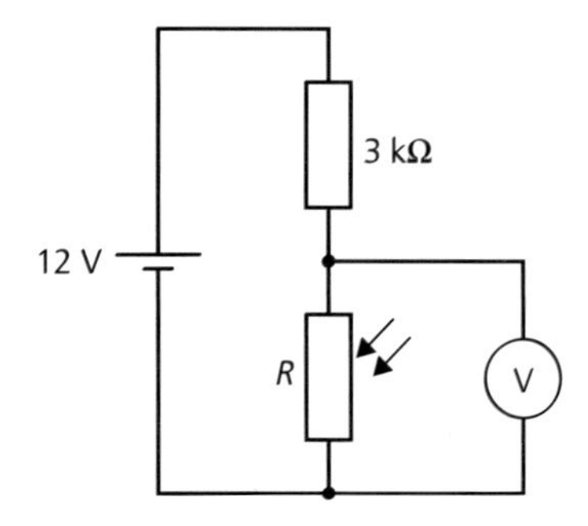

In order to increase the reading on the voltmeter:
A the temperature of R should be increased
B the temperature of R should be decreased
C the light intensity on R should be increased
D the light intensity on R should be decreased.

Q6 An electric heater has two elements, both connected in parallel to the same voltage supply. The first element has a resistance of R and a power output of P. The second element has a power output of $4P$. What is its resistance?

A $2R$ **B** $4R$ **C** $\frac{R}{2}$ **D** $\frac{R}{4}$

Q7 A cell of emf ε and internal resistance r delivers current to a small electric motor.

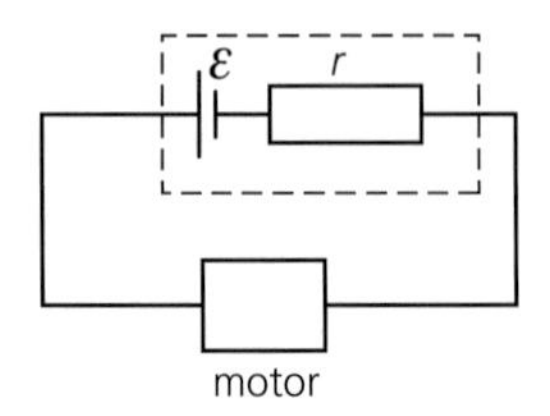

450 C of charge flows through the motor and 9000 J of energy are converted in the motor. 1800 J are dissipated in the cell. The emf of the cell is:

A 4.0 V **B** 16 V **C** 20 V **D** 24 V

© IB Organization

Q8 The diagram shows a circuit containing three equal resistances, a battery of emf ε and negligible internal resistance, and an ideal voltmeter.

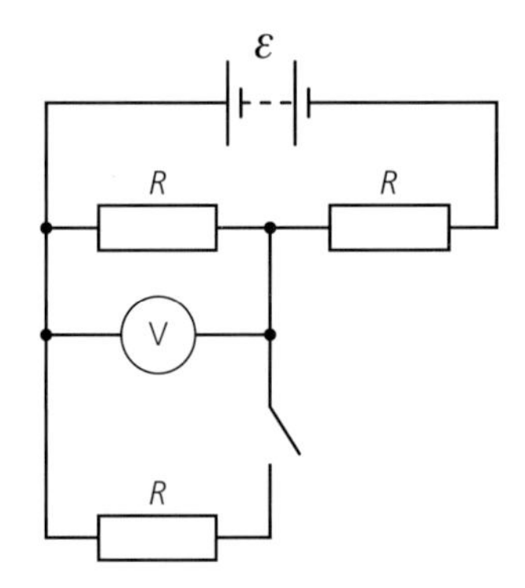

When the switch is closed the reading on the voltmeter:

A rises from zero to $\frac{\varepsilon}{2}$

B rises from zero to $\frac{\varepsilon}{3}$

C falls from $\frac{\varepsilon}{2}$ to $\frac{\varepsilon}{3}$

D falls from $\frac{\varepsilon}{2}$ to $\frac{\varepsilon}{4}$

Q9 A wire is made of a metal of resistivity ρ. The wire has length L, radius r and resistance R. What will be the resistance of another wire of resistivity 2ρ, length $L/2$ and radius $2r$?

A R **B** $\frac{R}{2}$ **C** $2R$ **D** $\frac{R}{4}$

Q10 Which of the following statements about the resistance of a conductor is a correct interpretation of Ohm's law?

A Resistance is proportional to temperature.
B Resistance is constant if temperature is constant.
C Resistance is proportional to potential difference if temperature is constant.
D Resistance is proportional to current if temperature is constant.

b If the force is horizontal, it cannot have a vertical component with which to support the weight of the ball (see Figure 6.14).

c As the speed of the ball is increased, a bigger centripetal force is needed for the same radius. If this force is greater than can be provided by the string, the string will break.

d The ball will continue its instantaneous velocity in a straight line after the string breaks. It will move at a tangent to the circle, but gravity will also affect its motion.

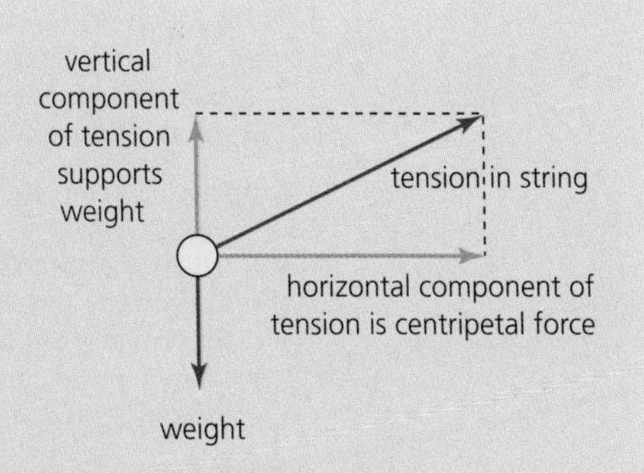

■ **Figure 6.14** Free-body diagram for a ball whirled in a circle

5 Young children enjoy playground rides and people from all countries seem to enjoy the thrills of amusement park rides (see Figure 6.15). Explain how it is possible for the passengers to be upside-down without falling out of the carriage.

The passengers do not fall out because their speed is sufficiently fast that the centripetal force required to keep them moving around the curved path is greater than their weight.

At the top of the circle the downward forces acting on a passenger are their weight (constant) and the reaction force from the seat (variable) – *if* they are in contact with it. The resultant force is found by adding these two forces. If the centripetal force required is greater than their weight they will stay in contact with their seat and the extra force needed is provided by the seat. If their weight is bigger than the centripetal force needed (because the speed is too slow), then they will fall out of the carriage.

■ **Figure 6.15** Upside-down on a fairground ride

6 A car of mass 1240 kg moves around a bend of radius 63 m on a horizontal road at a speed of $18\,\text{m s}^{-1}$. If the car was to be driven any faster there would not be enough friction and it would begin to skid.

a Determine a value for the coefficient of friction between the road and the tyres.
b Is this a coefficient of static friction or dynamic friction?
c Would a heavier car be able to drive faster around this bend?

a Centripetal force = frictional force

$$\frac{mv^2}{r} = \mu mg$$

$$\frac{1240 \times 18^2}{63} = \mu \times 1240 \times 9.81$$

$$\mu = 0.52$$

b Static, because before the skid the car is not moving in the direction of the frictional, centripetal force.
c No. Using the equation above, the mass cancels out. The extra centripetal force needed is provided by the extra friction.

10 A toy train of mass 432 g is moving around a circular track of radius 67 cm at a steady speed of $25\,\text{cm s}^{-1}$.
a Calculate the centripetal acceleration of the train.
b What is the resultant force acting on the train (in magnitude and direction)?
c What is the time taken for the toy train to go around the track once?

11 The hammer being thrown in Figure 6.16 completed two full circles of radius 2.60 m at a constant speed in 1.38 s just before it was released. Assuming that the motion was horizontal:
a What was its centripetal acceleration?
b What force did the thrower need to exert on the hammer if its mass was 4.0 kg?
c The thrower will aim to release the hammer when it is moving at an angle of 45° to the horizontal. Explain why.

12 a What angular velocity is produced when a centripetal force of 84 N is acting on a mass of 580 g moving in a circle of radius 1.34 m?
b How long will it take the mass to complete ten revolutions?

■ **Figure 6.16**

13 The Moon's distance from the Earth varies but averages about 380 000 km. The Moon orbits the Earth in an approximately circular path every 27.3 days.
   a Calculate the Moon's orbital speed in $m\,s^{-1}$.
   b What is the centripetal acceleration of the Moon towards the Earth?

14 a Calculate the centripetal force needed to keep a 1200 kg car moving at $15\,m\,s^{-1}$ around a curved road of radius 60 m.
   b What provides this force?
   c Explain why it might be dangerous for the diver to try to travel at twice this speed around the same bend.
   d If the road is wet or icy, why should the driver go even slower?
   e Discuss the possible effects on safety if the car was carrying passengers and luggage that increased the mass by 500 kg.

15 A bucket of water can be whirled in a vertical circle over your head without the water coming out if the centripetal force required is greater than the weight.
   a Calculate the minimum speed of the bucket of water if the radius of the circle is 90 cm.
   b How many revolutions is that every second?

16 A boy of mass 65 kg standing on the equator is spinning with the Earth at a speed of approximately $460\,m\,s^{-1}$.
   a What resultant centripetal force is required to keep him moving in a circle? (The Earth's radius is 6.4×10^6 m.)
   b What is the weight of the boy?
   c Draw a fully labelled free-body diagram of the boy.

17 Figure 6.17 shows a pendulum of mass 120 g being swung in a horizontal circle.
   a Draw a free-body diagram of the mass, *m*.
   b Calculate the centripetal force acting on the mass.
   c If the radius of the circle is 28.5 cm, what is the speed of the pendulum and how long does it take to complete one circle?

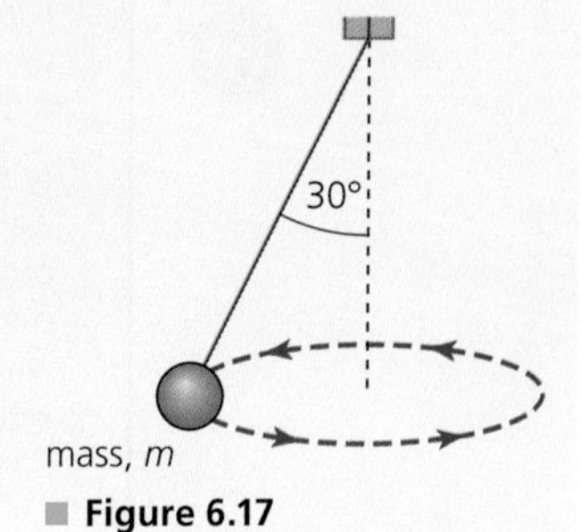

■ **Figure 6.17**

18 What is the fastest speed a car can drive around a curve of radius 49 m on a horizontal road if the coefficient of friction is 0.76?

6.2 Newton's law of gravitation – *the Newtonian idea of gravitational force acting between two spherical bodies and the laws of mechanics create a model that can be used to calculate the motion of planets*

■ Universal gravitation and the inverse square law

Isaac Newton was the first to realize that if the force of gravity makes objects (like apples) fall to the Earth and also keeps the Moon in orbit around the Earth, then it is reasonable to assume that the force of gravity acts between *all* masses. This is why he called it *universal* gravitation. Newton believed that the size of the gravitational force between two masses increases with the sizes of the masses, and decreases with increasing distance between them – following an *inverse square relationship* (see page 162).

The distance between the Earth and the Moon is equal to 60 Earth radii, and Newton was able to prove that the centripetal acceleration of the Moon towards the Earth (from v^2/r) was equal to $g/60^2$ (see Figure 6.18).

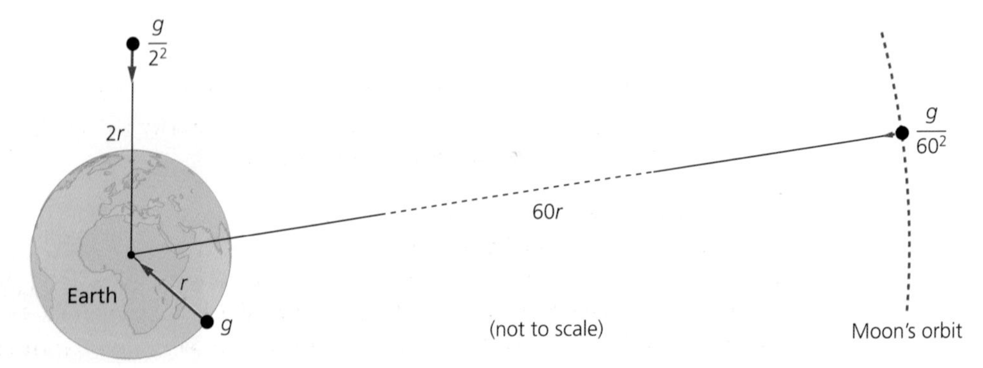

■ **Figure 6.18** How the acceleration due to gravity varies with distance from the Earth

Nature of Science

Diverse observations can have a common explanation

The preceding list offers a wide range of examples of circular motion, or motions in arcs of circles, or simply motions deviating from straight lines. Physicists now believe that *all* such examples require similar explanations in terms of forces acting radially inwards.

This is not at all obvious. It took centuries of observations and analysis to reach such a 'simple' conclusion. Even today, many people confuse *centripetal* forces with *centrifugal* forces (which are not mentioned in this course!).

Seeking simple, unifying explanations of diverse observations is a common theme of science, and one that requires the fundamental assumption that the universe may not be as complicated as it seems and is, therefore, capable of such analysis.

ToK Link

Should we believe in things that we cannot detect with our own senses?

Foucault's pendulum gives a simple, observable proof of the rotation of the Earth, which is largely unobservable. How can we have knowledge of things that are unobservable?

Figure 6.9 Foucault's pendulum

Another example of circular motion is that of the Earth, and the people on it, spinning in a circle once every day. Gravity provides the necessary centripetal force. Thousands of years ago most people would not have believed that the Earth is rotating. Today, most people accept this scientific 'fact', although they may not have seen any *direct* evidence. Are we sensible to believe what we are told by scientists and teachers, rather than to trust our own senses?

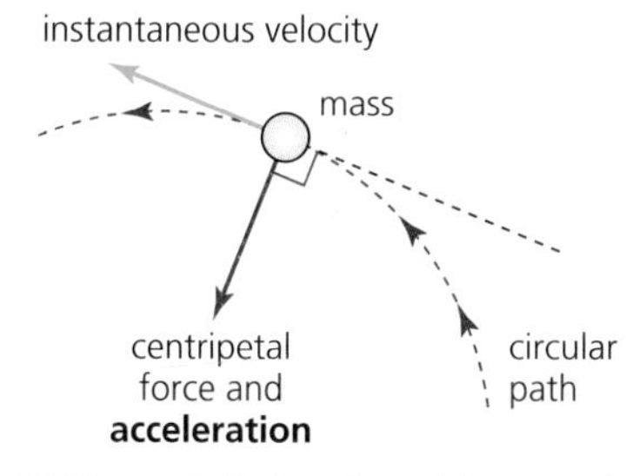

Figure 6.10 Centripetal force and acceleration

Centripetal acceleration

We know that a resultant force causes an acceleration, a. Therefore, a centripetal force towards the centre must result in a centripetal acceleration, also towards the centre of the circle. This is shown more clearly in Figure 6.10. Although there is an acceleration directed towards the centre, there is no movement in that direction or change in speed of the mass. Instead, the action of the force continually changes the *direction* of the motion of the mass. Remember that acceleration means a change of velocity, and the velocity of a mass can change by going faster, going slower or *changing direction*.

Equations for centripetal acceleration and centripetal force

When an object moving in a circle changes to have a faster speed, v, or smaller radius, r, the centripetal acceleration, and force, need to be larger. This is represented in the equation for calculating the magnitude of centripetal acceleration:

$$a = \frac{v^2}{r}$$

This equation is included in the *Physics data booklet*.

- Gravity provides the centripetal force that keeps planets, moons and satellites in orbit.
- Clothes in a washing machine or drier are kept moving in circles by the sides of their container pushing inwards. Water can pass out through holes where there is no centripetal force.
- A car can only move around a corner because of the force of friction between the road and the tyre (Figure 6.5).
- A person walking around a corner needs the force of friction between their feet and the ground to change direction.
- A passenger in a car can move in a circular path because of the force between them and their seat. If the car moves faster, the side of the car may also need to push inwards on them, although the passenger will feel that they are pushing outwards.
- An aircraft can change direction with the help of the force provided on the wings from the air when the plane tilts.
- A train on a track can go around bends because of the normal reaction force of the track on the wheel (Figure 6.6).
- When an object is moving in a curved path on a sloping surface (see Figure 6.7), the normal reaction force, R, will have a horizontal component, R_H, which can help provide more centripetal force if friction is not enough at high speeds. Such a surface, or track, is described as being banked (Figure 6.8).
- An electron may be considered to be in a circular orbit around a nucleus due to the electrical force between them.
- Charged particles (ions, electrons, protons) can be made to move in circular paths when they pass perpendicularly across magnetic fields in a vacuum (see Section 5.4).

■ **Figure 6.5**

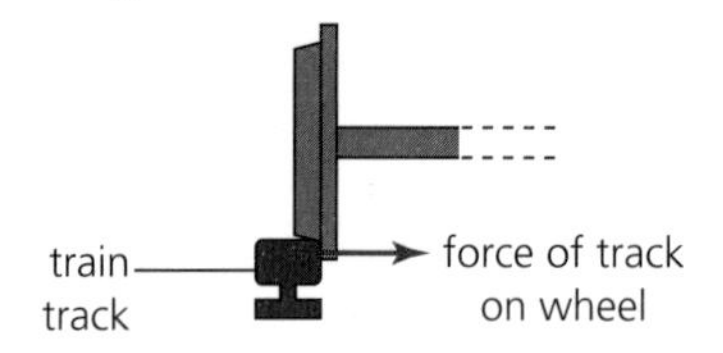

■ **Figure 6.6** The train track pushes inwards on the wheel of a train moving in a circular path

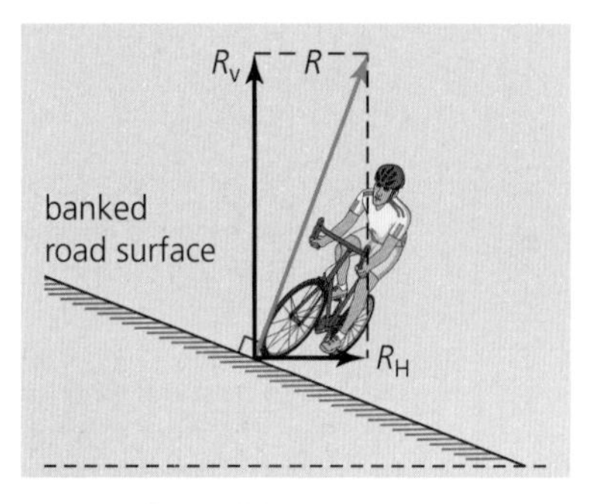

■ **Figure 6.7** Forces on a cyclist on a banked surface

■ **Figure 6.8** Cycling at the velodrome at the 2012 Olympic Games in London

6 Draw a free-body diagram of an airplane 'banking' around a corner.

7 Explain why a cyclist or runner can be seen to 'lean into' a bend when travelling quickly.

8 What provides the centripetal force for the rotating wheels of a car or bicycle?

9 Explain why a satellite orbiting the Earth in a circular path does not need an engine to keep it moving.

Worked example

3 A point mass completes 14.7 rotations in exactly one minute.
  a What is its angular velocity?
  b If its distance from the centre of rotation is 58 cm, what is its linear speed?

a $\omega = \dfrac{\Delta\theta}{\Delta t}$

$\omega = 14.7 \times \dfrac{2\pi}{60} = 1.54\,\text{rad}\,\text{s}^{-1}$

b $v = \omega r$

$v = 1.54 \times 0.58 = 0.89\,\text{m}\,\text{s}^{-1}$

1 Convert the following angles to radians:
  a 180°
  b 90°
  c 45°
  d 112°

2 If an object has an angular displacement of $3\pi/2$, what distance has it travelled along the circumference of a circle with a radius of 50 cm?

3 a Convert an angle of 5.00° to radians.
  b What is the sine of 5.00°?
  c What percentage error would be introduced by assuming that the sine of 5.00° was the same as an angle of 5.00° expressed in radians?

4 A bicycle is moving with linear speed of $8.2\,\text{m}\,\text{s}^{-1}$.
  a If the wheels each have a radius of 31 cm, what is their angular velocity?
  b Calculate the frequency of the wheels' rotation.

5 A small mass is moving in a circle of radius 1.32 m with a time period of 2.77 s.
  a What is its frequency?
  b What is its angular velocity?
  c What is its linear speed?

Centripetal force

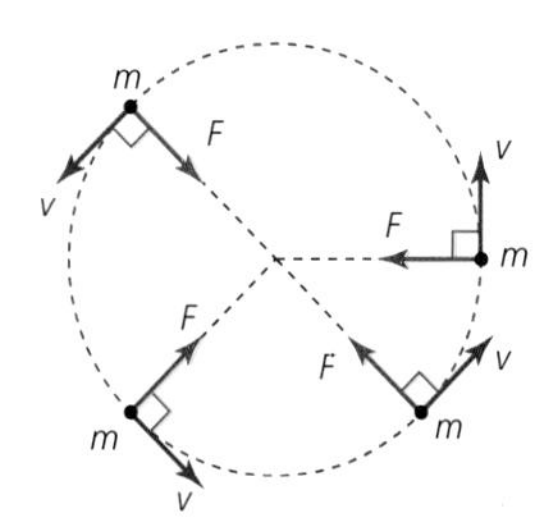

Figure 6.4 Velocity and centripetal force vectors during circular motion

Figure 6.4 shows four random positions of the same mass, m, moving in a circle with a constant speed. The blue arrows represent the instantaneous velocity, v, of the mass at various places around the circle. The velocities are always directed along *tangents* to the circle. Any force, F, causing circular motion is called a **centripetal force** and it always acts perpendicularly to the velocity, inwards towards the centre of the circle, as shown by the red arrows in Figure 6.4. Because the force is perpendicular to motion, it does not cause a change of speed, but it does continually cause a change of direction. This resultant centripetal force acting on the mass always has the same magnitude but it is continuously changing direction.

Because the force is always perpendicular to the motion, no work is done (energy transferred) by any centripetal force that makes an object move in a circle.

Examples of forces producing circular motion

The following list gives some examples of circular motion and the origins of their centripetal forces. It is important to understand that centripetal force is not a type of force in itself (like gravity or friction, for example), but it is the name used to describe any force that happens to be causing motion in a circle.

- If an object is whirled around on the end of a string in a (nearly) horizontal circle, the centripetal force is provided by the tension in the string (see Figure 6.13).

Angular velocity

A key quantity in the description of translational motion is velocity (change in linear displacement/change in time). Similarly, to describe circular motion, we use 'change in angular displacement/change in time'. This is called **angular velocity**. (Angular velocity is sometimes called *angular speed* or *angular frequency*.) Angular velocity has the symbol ω and the unit rad s^{-1}.

$$\text{angular velocity, } \omega = \frac{\Delta\text{angular displacement}}{\Delta\text{time}}$$

$$\omega = \frac{\Delta\theta}{\Delta t}$$

Because in one oscillation 2π radians are completed in period, T, for regular continuous rotation:

$$\omega = \frac{2\pi}{T}$$

Also, because $f = \frac{1}{T}$:

$$\omega = 2\pi f$$

These three widely used equations for ω are *not* given in the *Physics data booklet*.

Time period, frequency and angular velocity are three concepts that represent exactly the same information for uniform motion in a circle. For example, if an oscillator has a time period, T, of 0.20 s, we know that it has a frequency of 5.0 Hz (because $f = 1/T$) and an angular velocity of 31 rad s^{-1} (because $\omega = 2\pi/T$).

When information is presented to us, for example in a question, we are most likely to be given the period or the frequency of a rotation; for calculations, however, the angular velocity, ω, is usually needed.

Worked example

2 If an object completes exactly 20 rotations in 34.6 s, calculate:
a its frequency
b its angular velocity.

a $f = \frac{1}{T} = \frac{1}{(34.6/20.0)} = 0.578\text{ Hz}$
b $\omega = 2\pi f = 2\pi\,(0.578) = 3.63\text{ rad s}^{-1}$

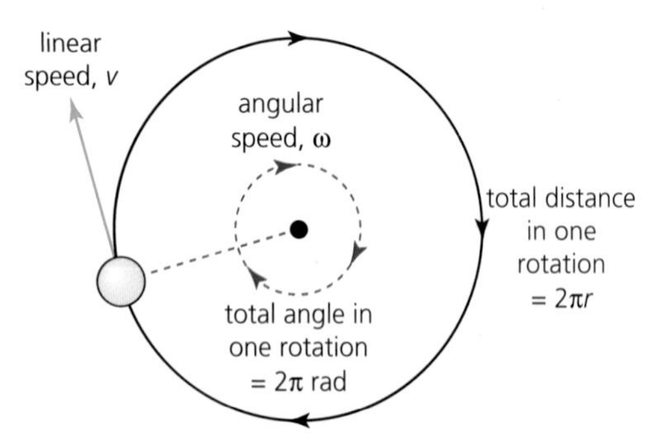

■ **Figure 6.3** Relating linear speed to angular velocity

Relationship between the speed of a mass undergoing circular motion, v, and its angular velocity, ω

If a (point) mass is undergoing circular motion with a constant linear speed, as shown in Figure 6.3, it also has a constant angular velocity and there must be a simple relationship between the two.

Because linear speed, $v = 2\pi r/T$ and angular velocity, $\omega = 2\pi/T$ (if we measure angles in radians):

$$v = \omega r$$

This equation is given in the *Physics data booklet*.

Period, frequency, angular displacement and angular velocity

Before we look more closely at the theory of circular motion, we will define some of the terms commonly used in this branch of physics.

The **period**, T, of a circular motion is defined as the time taken for the object to complete one rotation (360°).

The **frequency**, f, of circular motion is defined as the number of rotations in unit time (usually every second).

These definitions are similar to those presented previously in Chapter 4 (when discussing oscillations and waves). As before, period and frequency are connected by the simple equation:

$$f = \frac{1}{T}$$

Angular displacement

In Chapter 2, linear displacement was presented as a measure of the distance from a fixed reference point. For circular motion, displacement is measured as an angle.

Angular displacement, θ, is defined as the angle through which a rigid object has rotated from a fixed reference position.

For example, a body may be rotated through 198° in an anticlockwise direction, or 457° in a clockwise direction. A rotation of more than 360°, of course, implies more than one complete revolution.

At this point we will introduce an alternative and more convenient unit for the measurement of angles: the radian. In Figure 6.2 the angle θ (in radians) is equal to the distance, s, along the arc of the circle divided by the radius, r.

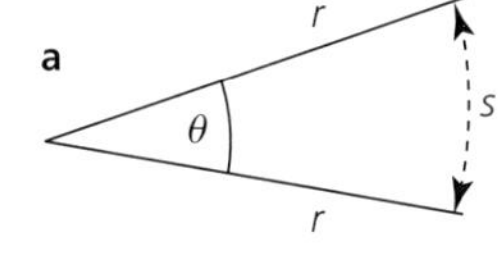

$$\theta \text{ (in radians)} = \frac{s}{r}$$

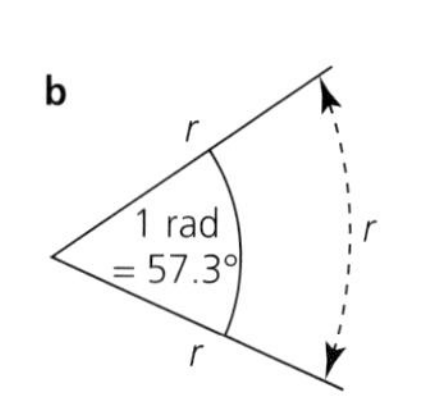

■ **Figure 6.2 a** Angle θ (in radians) is equal to s/r **b** one radian

Note that an angle expressed in radians is the ratio of two lengths and, as such, the unit (radian) is not based on an arbitrary/historical decision (like the choice of 360 degrees for one circle).

One radian is the angle subtended by an arc of length equal to one radius ($\theta = r/r = 1$), as shown in Figure 6.2b; 1 radian = 57.3°.

One rotation through an angle of 360° = $2\pi r/r = 2\pi$ radians; 180° is π radians and 90° is $\pi/2$ radians.

For small angles (less than about 0.1 rad, or 6°) the shape shown in Figure 6.2a can be considered as similar to a right-angled triangle. This enables a useful approximation to be made:

for small angles, θ (in radians) $\approx \sin\theta \approx \tan\theta$

Worked example

1 **a** Convert an angle of 137° to radians.
b How many **i** degrees, **ii** radians does a rotating object pass through in the process of five revolutions?
c If two circular motions are seen to be completely out of phase, what is the phase difference between them **i** in degrees, **ii** in radians?
d If two oscillations are $\pi/2$ out of phase, what fraction of an oscillation is between them?

a $\frac{137 \times 2\pi}{360} = 2.39\,\text{rad}$
b i $5 \times 360 = 1800°$
ii $5 \times 2\pi = 31.4\,\text{rad}$
c i 180°
ii π rad
d $\pi/2$ = ¼ of an oscillation

6 Circular motion and gravitation

ESSENTIAL IDEAS

- A force applied perpendicular to its displacement can result in circular motion.
- The Newtonian idea of gravitational force acting between two spherical bodies and the laws of mechanics create a model that can be used to calculate the motion of planets.

6.1 Circular motion – *a force applied perpendicular to its displacement can result in circular motion*

An object moving along a circular path with a constant speed has a continuously changing velocity because its direction of motion is changing all the time. From Newton's first law, we know that any object that is not moving in a straight line must be accelerating and, therefore, it must have a resultant force acting on it, even if it is moving with a constant speed.

Perfectly uniform motion in complete circles may not be a common everyday observation but the theory for circular motion can also be used with objects such as people or vehicles moving along arcs of circles and around curves and bends. Circular motion theory is also very useful when discussing the orbits of planets, moons and satellites. It is also needed to explain the motion of sub-atomic particles in magnetic fields, as discussed in Chapter 5.

Imagine yourself to be standing in a train on a slippery floor, holding on to a post (Figure 6.1). While you and the train are travelling in a straight line with a constant speed (constant velocity) there is no resultant force acting on you and you do not need to hold on to the post, but as soon as the train changes its motion (accelerates in some way) there needs to be a resultant force on you to keep you in the same place in the train. If there is little or no friction with the floor, the post is the only thing that can exert a force on you to change your motion. The directions of these forces (from the post) are shown in the diagram for different types of acceleration. If the post pushes or pulls on you, then by Newton's third law you must be pushing or pulling on the post, and that is the force you would be most aware of.

Figure 6.1 Forces that make a passenger accelerate in a train

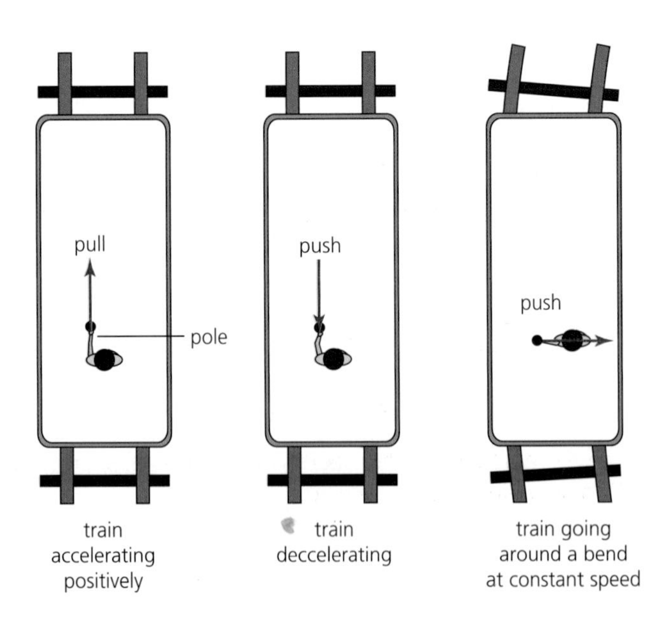

In particular, note that the direction of the force needed to produce a curved, circular path is perpendicular to the instantaneous motion.

b If the force is horizontal, it cannot have a vertical component with which to support the weight of the ball (see Figure 6.14).

c As the speed of the ball is increased, a bigger centripetal force is needed for the same radius. If this force is greater than can be provided by the string, the string will break.

d The ball will continue its instantaneous velocity in a straight line after the string breaks. It will move at a tangent to the circle, but gravity will also affect its motion.

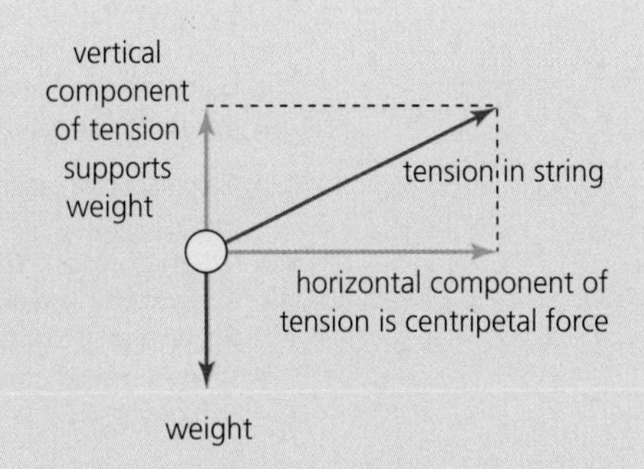

■ **Figure 6.14** Free-body diagram for a ball whirled in a circle

5 Young children enjoy playground rides and people from all countries seem to enjoy the thrills of amusement park rides (see Figure 6.15). Explain how it is possible for the passengers to be upside-down without falling out of the carriage.

The passengers do not fall out because their speed is sufficiently fast that the centripetal force required to keep them moving around the curved path is greater than their weight.

At the top of the circle the downward forces acting on a passenger are their weight (constant) and the reaction force from the seat (variable) – *if* they are in contact with it. The resultant force is found by adding these two forces. If the centripetal force required is greater than their weight they will stay in contact with their seat and the extra force needed is provided by the seat. If their weight is bigger than the centripetal force needed (because the speed is too slow), then they will fall out of the carriage.

■ **Figure 6.15** Upside-down on a fairground ride

6 A car of mass 1240 kg moves around a bend of radius 63 m on a horizontal road at a speed of $18\,\text{m s}^{-1}$. If the car was to be driven any faster there would not be enough friction and it would begin to skid.

a Determine a value for the coefficient of friction between the road and the tyres.
b Is this a coefficient of static friction or dynamic friction?
c Would a heavier car be able to drive faster around this bend?

a Centripetal force = frictional force

$$\frac{mv^2}{r} = \mu mg$$

$$\frac{1240 \times 18^2}{63} = \mu \times 1240 \times 9.81$$

$$\mu = 0.52$$

b Static, because before the skid the car is not moving in the direction of the frictional, centripetal force.
c No. Using the equation above, the mass cancels out. The extra centripetal force needed is provided by the extra friction.

10 A toy train of mass 432 g is moving around a circular track of radius 67 cm at a steady speed of $25\,\text{cm s}^{-1}$.
a Calculate the centripetal acceleration of the train.
b What is the resultant force acting on the train (in magnitude and direction)?
c What is the time taken for the toy train to go around the track once?

11 The hammer being thrown in Figure 6.16 completed two full circles of radius 2.60 m at a constant speed in 1.38 s just before it was released. Assuming that the motion was horizontal:
a What was its centripetal acceleration?
b What force did the thrower need to exert on the hammer if its mass was 4.0 kg?
c The thrower will aim to release the hammer when it is moving at an angle of 45° to the horizontal. Explain why.

■ **Figure 6.16**

12 a What angular velocity is produced when a centripetal force of 84 N is acting on a mass of 580 g moving in a circle of radius 1.34 m?
b How long will it take the mass to complete ten revolutions?

13 The Moon's distance from the Earth varies but averages about 380 000 km. The Moon orbits the Earth in an approximately circular path every 27.3 days.
a Calculate the Moon's orbital speed in $m s^{-1}$.
b What is the centripetal acceleration of the Moon towards the Earth?

14 a Calculate the centripetal force needed to keep a 1200 kg car moving at $15 m s^{-1}$ around a curved road of radius 60 m.
b What provides this force?
c Explain why it might be dangerous for the diver to try to travel at twice this speed around the same bend.
d If the road is wet or icy, why should the driver go even slower?
e Discuss the possible effects on safety if the car was carrying passengers and luggage that increased the mass by 500 kg.

15 A bucket of water can be whirled in a vertical circle over your head without the water coming out if the centripetal force required is greater than the weight.
a Calculate the minimum speed of the bucket of water if the radius of the circle is 90 cm.
b How many revolutions is that every second?

16 A boy of mass 65 kg standing on the equator is spinning with the Earth at a speed of approximately $460 m s^{-1}$.
a What resultant centripetal force is required to keep him moving in a circle? (The Earth's radius is 6.4×10^6 m.)
b What is the weight of the boy?
c Draw a fully labelled free-body diagram of the boy.

17 Figure 6.17 shows a pendulum of mass 120 g being swung in a horizontal circle.
a Draw a free-body diagram of the mass, *m*.
b Calculate the centripetal force acting on the mass.
c If the radius of the circle is 28.5 cm, what is the speed of the pendulum and how long does it take to complete one circle?

18 What is the fastest speed a car can drive around a curve of radius 49 m on a horizontal road if the coefficient of friction is 0.76?

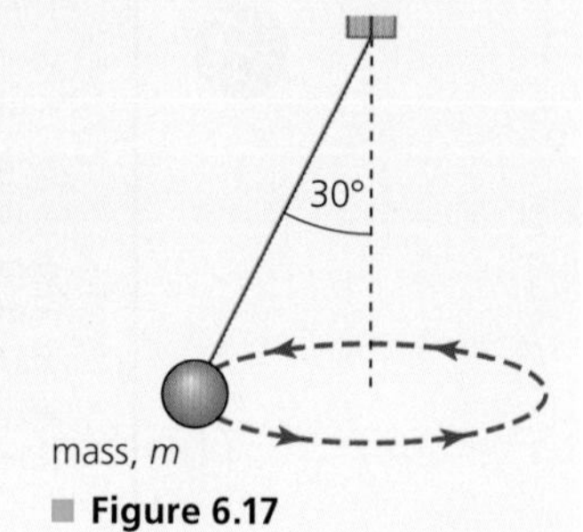

■ **Figure 6.17**

6.2 Newton's law of gravitation – *the Newtonian idea of gravitational force acting between two spherical bodies and the laws of mechanics create a model that can be used to calculate the motion of planets*

■ Universal gravitation and the inverse square law

Isaac Newton was the first to realize that if the force of gravity makes objects (like apples) fall to the Earth and also keeps the Moon in orbit around the Earth, then it is reasonable to assume that the force of gravity acts between *all* masses. This is why he called it *universal* gravitation. Newton believed that the size of the gravitational force between two masses increases with the sizes of the masses, and decreases with increasing distance between them – following an *inverse square relationship* (see page 162).

The distance between the Earth and the Moon is equal to 60 Earth radii, and Newton was able to prove that the centripetal acceleration of the Moon towards the Earth (from v^2/r) was equal to $g/60^2$ (see Figure 6.18).

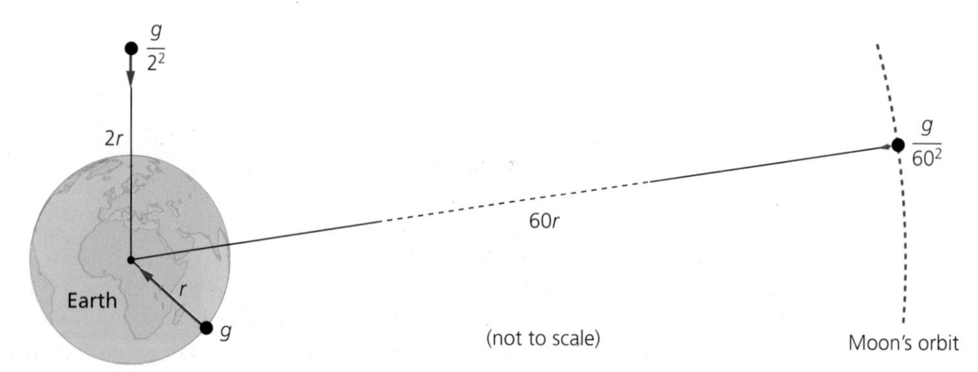

■ **Figure 6.18** How the acceleration due to gravity varies with distance from the Earth

Q11 Which of the following correctly describes the process of conduction?

A There are no electrons in insulating materials.

B A current in a solid conductor is carried by positive charges moving towards the negative terminal of the battery.

C Metals conduct well because they have plenty of free electrons.

D A current in a solid conductor consists of electrons moving in the direction of the electric field.

Q12 In the diagram, a long, current-carrying wire is normal to the plane of the paper. The current in the wire is directed into the plane of the paper.

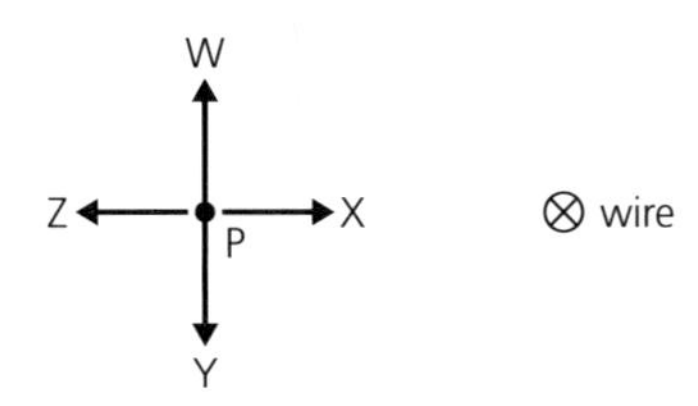

Which of the arrows gives the direction of the magnetic field at point P?

A W **B** X **C** Y **D** Z

Q13 A current of 2 A consists of electrons flowing through a metal wire of cross-sectional area 1 mm^2. If the metal contains 4×10^{28} free electrons per cubic metre, the best estimate for the drift speed of the electrons is:

A $1 \times 10^{-4}\,m\,s^{-1}$ **B** $3 \times 10^{-4}\,m\,s^{-1}$ **C** $1 \times 10^{-10}\,m\,s^{-1}$ **D** $3 \times 10^{-10}\,m\,s^{-1}$.

Q14 Which of the following is an *incorrect* statement about Kirchhoff's circuit laws?

A They can only be used if the temperature is constant.

B $\Sigma I = 0$ (junction) shows that charge is conserved.

C $\Sigma V = 0$ (loop) shows that energy is conserved.

D The laws can be applied to any electrical circuit.

Q15 The equation:

$$F = k\frac{q_1 q_2}{r^2}$$

describes the force between two electric charges. Which of the following is a correct statement about this equation?

A k is known as Boltzmann's constant.

B r is the radius of the spheres on which the two charges are placed.

C The equation applies to point charges.

D The equation can only be applied to uniform electric fields.

Paper 2 IB questions and IB style questions

Q1 The components shown below are to be connected in a circuit to investigate how the current I in a tungsten filament lamp varies with the potential difference V across it.

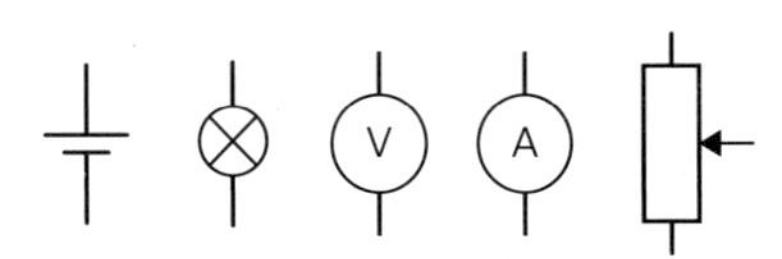

a Construct a circuit diagram to show how these components should be connected together in order to obtain as large a range as possible for values of potential difference across the lamp. (4)

b Copy the axes below and use them to sketch a graph of I against V for a filament lamp in the range $V = 0$ to its normal working voltage. (2)

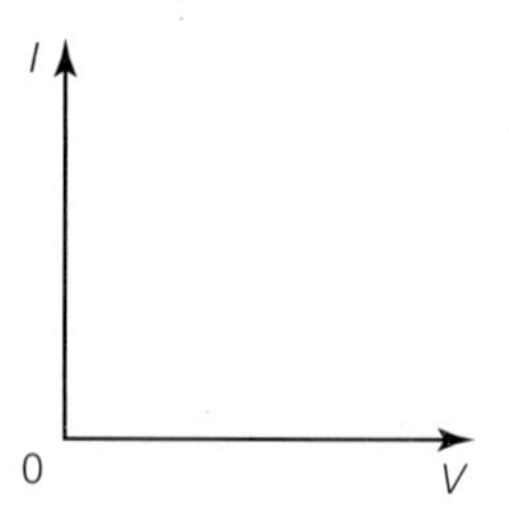

Q2 a A heating coil is to be made of wire of diameter 3.5×10^{-4} m. The heater is to dissipate 980 W when connected to a 230 V dc supply. The material of the wire has resistivity $1.3 \times 10^{-6}\ \Omega$ m at the working temperature of the heater.

i Define electrical resistance. (1)

ii Calculate the resistance of the heating coil at its normal working temperature. (2)

iii Show that the length of wire needed to make the heating coil is approximately 4 m. (2)

b Three identical electrical heaters each provide power P when connected separately to a supply S that has zero internal resistance. Copy the diagram and complete the circuit by drawing two switches so that the power provided by the heater may be either P or $2P$ or $3P$. (2)

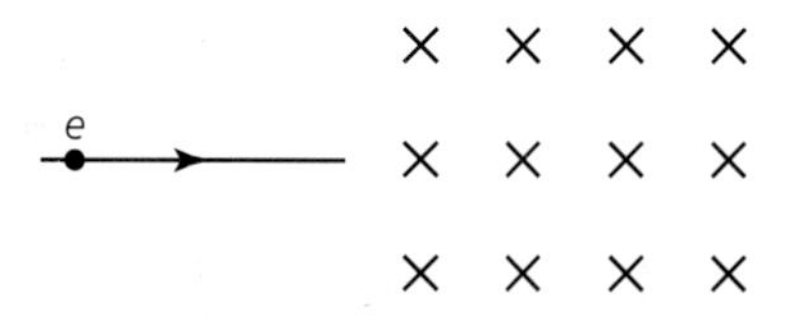

Q3 A student wishes to determine how much energy can be obtained from a simple electric cell made from two metal electrodes inserted into a potato.

a Draw a circuit diagram showing how the student should connect the measuring instruments. (2)

b List the measurements that the students would need to take. (3)

Q4 The diagram shows an electron travelling at a velocity of $6.9 \times 10^{6}\ \text{m s}^{-1}$ entering a uniform magnetic field of strength 2.3×10^{-2} T.

e

× × × ×
× × × ×
× × × ×

a Calculate the magnitude of the force acting on the electron while it is passing through the field. (2)

b Describe and explain the path of the electron in the field. (3)

Because the speed in a circle equals the circumference divided by the time taken to complete one circle (the period T), speed $v = \frac{2\pi r}{T}$.

The equation for centripetal acceleration can then be rewritten as:

$$a = \frac{v^2}{r} = \frac{(2\pi r/T)^2}{r}$$

so that:

$$a = \frac{4\pi^2 r}{T^2}$$

This equation is included in the *Physics data booklet*.

Because $F = ma$, the expression for the centripetal force needed to make a mass, m, move in a circle of radius, r, at a constant speed, v, is:

$$F = \frac{mv^2}{r}$$

This equation is included in the *Physics data booklet*.

Or, because $v = \omega r$, the equation for centripetal force can be rewritten as:

$$F = m\omega^2 r$$

This equation is included in the *Physics data booklet*.

In Figure 6.11, although they both have the same angular velocity, the bigger child needs a much bigger centripetal force because she has more mass and is travelling with a faster speed.

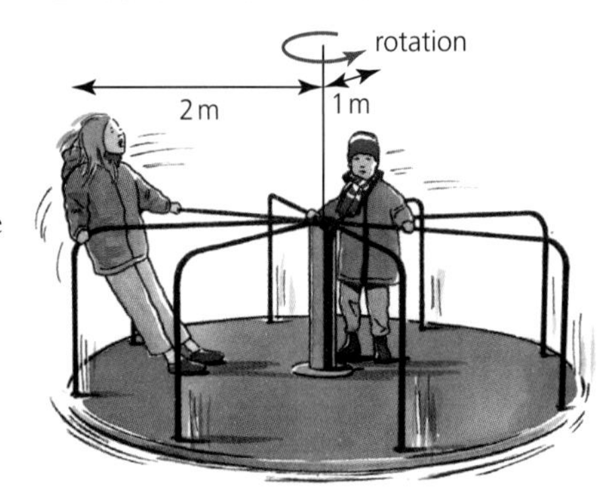

■ **Figure 6.11** Children on a playground ride

Additional Perspectives

Deriving the expression $a = \frac{v^2}{r}$

Consider a mass moving in a circular path of radius, r, as shown in Figure 6.12a. It moves through an angle, θ, and a distance, Δs, along the circumference as it moves from A to B, while its velocity changes from v_A to v_B.

To calculate acceleration we need to know the change of velocity, Δv. This is done using the vector diagram shown in Figure 6.12b. Note that the direction of the change of velocity (and therefore the acceleration) is towards the centre of motion.

The two triangles are similar and, if the angle is small enough that Δs can be approximated to a straight line, we can write:

$$\theta = \frac{\Delta v}{v} = \frac{\Delta s}{r}$$

(The magnitudes of v_A and v_B are equal and represented by the speed, v.)

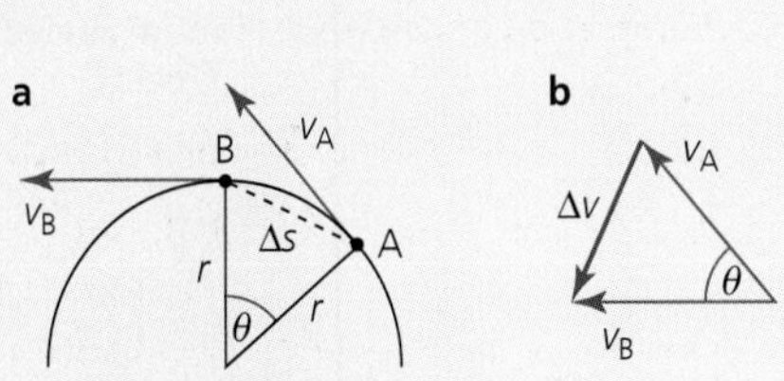

■ **Figure 6.12** Deriving an equation for centripetal acceleration

Dividing both sides of the equation by Δt we get:

$$\frac{\Delta v}{\Delta t \times v} = \frac{\Delta s}{\Delta t \times r}$$

Then, because $a = \Delta v/\Delta t$ and $\Delta s/\Delta t = v$:

$$a = \frac{v^2}{r}$$

1 The orbit of the Earth around the Sun is approximately circular, with an average radius of about 150 million kilometres. Calculate the centripetal acceleration of the Earth towards the Sun.

Experimental investigations of circular motion

- A straightforward investigation of circular motion can be carried out by whirling a small mass in circles on the end of a thin piece of string. See Figure 6.13. (The mass should be well secured and the string thick enough that it will not break.) It will be necessary to devise some way of measuring the force exerted on the string.
- Fairground rides offer many interesting examples of motion in arcs of circles, including those that offer passengers the opportunity to travel upside down ('looping the loop'). See Figure 6.15 and Worked example 5. Measurements of times and estimates of distances can lead to approximations of the magnitudes of the forces involved on the passengers.
- A coin (or other mass) on a rotating turntable can only stay in its position if there is enough friction to provide the necessary centripetal force. Observations can be made as the angular velocity is slowly increased (on different masses and/or different positions and/or different surfaces).
- A bucket of water can be safely whirled by hand in a vertical circle if it moves fast enough.
- Because of the rapid speed of many rotations, video analysis of circular motion can make observations and measurements much easier and more accurate. Using data loggers for measuring centripetal forces can also improve experiments that involve rapid movement.

Worked examples

4 Consider a ball of mass 72 g whirled with a constant speed of $3.4\,\text{m}\,\text{s}^{-1}$ around in a (nearly) horizontal circle of radius 65 cm on the end of a thin piece of string, as shown in Figure 6.13.

a Calculate the centripetal acceleration and force.

b Explain why the force provided by the string cannot act horizontally.

c Explain a probable reason why the string breaks when the speed is increased to $5.0\,\text{m}\,\text{s}^{-1}$.

d In what direction does the ball move immediately after the string breaks?

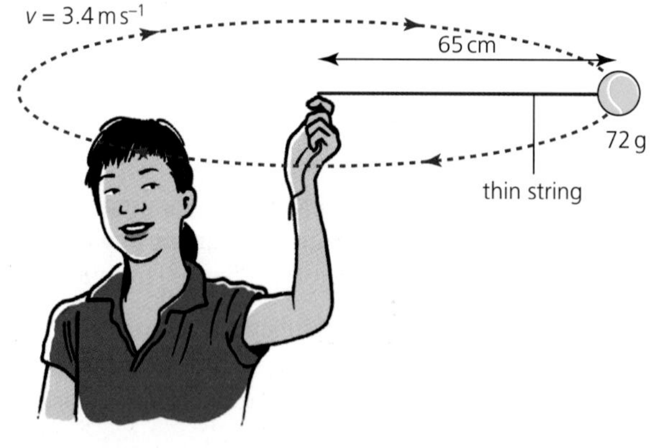

■ Figure 6.13

a $a = \frac{v^2}{r}$

$a = \frac{3.4^2}{0.65}$

$a = 18\,\text{m}\,\text{s}^{-2}$

$F = ma$

$F = 0.072 \times 18.0 = 1.3\,\text{N}$

Worked example

7 The average distance between the Earth and the Moon is 384000 km and the Moon takes 27.3 days to orbit the Earth.
a Calculate the average orbital speed of the Moon.
b What is the centripetal acceleration of the Moon towards the Earth?
c Compare your answer for **b** to $g/60^2$.

a $v = \frac{2\pi r}{T} = \frac{(2 \times \pi \times 3.84 \times 10^8)}{(27.3 \times 24 \times 3600)} = 1.02 \times 10^3\,\text{m s}^{-1}$

b $a = \frac{v^2}{r} = \frac{(1.02 \times 10^3)^2}{3.84} \times 10^8 = 2.71 \times 10^{-3}\,\text{m s}^{-2}$

c $9.81/60^2 = 2.73 \times 10^{-3}\,\text{m s}^{-2}$. The two answers are within one per cent of each other. This is very good evidence that gravitational accelerations (and forces) are represented by inverse square laws.

Newton's law of gravitation

The forces acting between two point masses (M and m) are proportional to the product of the masses and inversely proportional to their separation (r) squared.

$$F \propto Mm$$

$$F \propto \frac{1}{r^2}$$

Putting a constant of proportionality into the relationship, we get Newton's universal law of gravitation:

$$F = G\frac{Mm}{r^2}$$

This equation is given in the *Physics data booklet*.

G is known as the universal gravitation constant. Its value of $6.67 \times 10^{-11}\,\text{N m}^2\,\text{kg}^{-2}$ is given in the *Physics data booklet*. Its small value reflects the fact that gravitational forces are small unless one (or both) of the masses is very large. G is a *fundamental* constant which, as far as we know, always has exactly the same value everywhere in the universe. It should not be confused with g, the acceleration due to gravity, which varies with location. The relationship between g and G is covered later in this chapter.

The relationship between force and distance is illustrated in Figure 6.19. Note that exactly the same force always acts on *both* masses (but in opposite directions), even if one mass is larger than the other. This is an example of Newton's third law of motion.

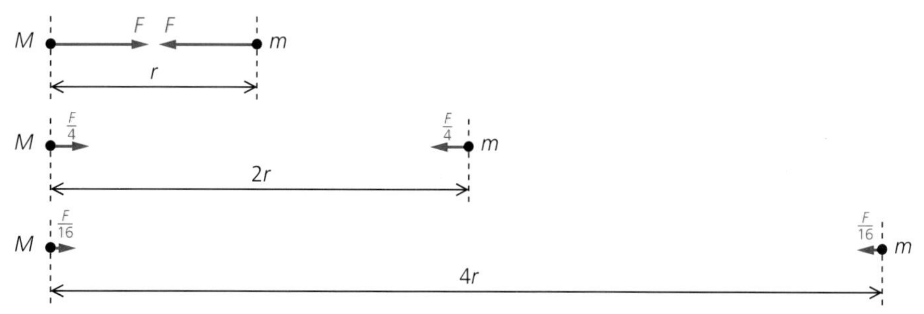

Figure 6.19 The gravitational force between point masses M and m decreases with increasing separation (the vectors are not drawn to scale)

Of course, the mass of an object is not all located at one point, but this does not mean that Newton's equation cannot be used for real masses. The forces between two spherical masses of uniform density located a long way apart are the same as if the spheres had all of their masses concentrated at their centre points. The gravitational field around a (spherical) planet is effectively the same as would be produced by a similar mass concentrated at the centre of the planet.

Worked example

8 Calculate the gravitational force acting between the Earth and a 1.0 kg book on the Earth's surface. (The Earth's mass is 6.0×10^{24} kg and its radius is 6.4×10^6 m.)

$$F = G\frac{Mm}{r^2}$$

$$F = (6.67 \times 10^{-11}) \times \frac{1.0 \times (6.0 \times 10^{24})}{(6.4 \times 10^6)^2}$$

$$F = 9.8\,\text{N}$$

This is the *weight* of a 1.0 kg mass on the Earth's surface. The book attracts the Earth up towards it with an equally sized force which, of course, has a negligible effect on the Earth. This is an example of Newton's third law.

19 Estimate the gravitational force between you and your pen when you are 1 m apart.

20 What is the gravitational force between two steel spheres each of radius 45 cm and separated by 10 cm? (The density of steel is 7900 kg m^{-3}.)

21 Calculate the average gravitational force between the Earth and the Sun. (You will need to research the relevant data.)

22 A proton has a mass of 1.7×10^{-27} kg, and the mass of an electron is 9.1×10^{-31} kg. Estimate the gravitational force between these two particles in a hydrogen atom, assuming that they are 5.3×10^{-11} m apart.

Similarities and differences between electric forces and gravitational forces

If Newton's law of gravitation is compared to Coulomb's law (page 206), it will be noted that there are striking similarities. Both laws describe forces that act across space between two points: one gravitational forces between masses, the other electric forces between charges.

Newton's and Coulomb's laws both apply to forces spreading radially into the space around them, so it is not surprising that the two equations have similar forms (inverse square laws). But, of course, there are also differences between electrical and gravitational forces:

- The electrical force between two isolated point charges is much, much larger than the gravitational force between two isolated point masses with the same separation. A comparison of the size of the two constants in the equations demonstrates this difference.
- As far as we know there is only one kind of mass, but there are two kinds of charge; it seems that there is no such thing as a repulsive gravitational force between masses.
- As masses get larger and larger, so do the gravitational forces involved. But when objects get larger and larger there will normally still be (approximately) equal numbers of positively and negatively charged particles, so electrical forces do not tend to increase with physical size.
- On the microscopic scale of atoms, ions, molecules and other particles, electrical forces dominate and gravitational forces are negligible. However, on the large scale, the only significant forces between planets and stars are gravitational.
- The value of the gravitational force between two masses in a certain arrangement will always be the same. The value of the electrical force between two charges depends on the electrical permittivity of the medium in which they are located, although we often assume that the medium is air or a vacuum.
- Electric forces are closely linked to magnetic forces, so much so that they are known together as the electromagnetic force. The electromagnetic force and gravitational force are two of the four fundamental forces of nature (see page 290).

Gravitational fields

We have seen in Chapter 5 that an electric field exists where a charge experiences a force, and a magnetic field exists where a moving charge (electric current) experiences a force. Gravitational fields are described similarly.

A region (around a mass) in which another mass would experience a gravitational force is called a gravitational field. We all live in the gravitational field of the Earth, while the Earth moves in the gravitational field of the Sun. Gravitational forces can be very small if the masses are small or far apart, in which case the fields may be totally insignificant but, in theory, they never reduce to zero completely.

We often want to represent a gravitational field on paper or on a screen, and this can be done with gravitational field lines as shown in Figure 6.20 and previously discussed for magnetic fields. The lines and arrows show the direction of the gravitational force that would be experienced by a mass placed at a particular place in the field. Figure 6.20a represents the spreading radial gravitational field lines around the Earth.

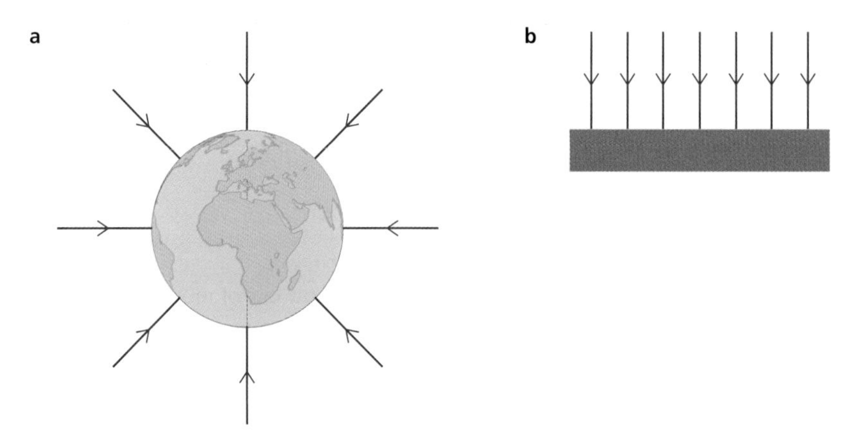

■ **Figure 6.20** Field lines are used to represent gravitational fields on paper or on screen: **a** radial field; **b** uniform field

The lines are closer together nearer to the Earth, which shows that the field is stronger. Field lines never cross each other; this would mean that gravitational force is acting in two different directions at the same place. The parallel lines in Figure 6.20b represent a uniform gravitational field, such as in a small region of the Earth's surface where variations in field are negligible.

ToK Link

Fields

The fact that one mass can affect another mass without any contact or other matter between them is very difficult to understand and explain: how can a mass 'know' that there is another mass affecting it when there is no connection between them? (This is sometimes called 'action at a distance'.) Similar comments can be made about the forces between electric charges. Adding to the puzzle, there seems to be no time delay between the movement of a mass and the gravitational effect of that movement somewhere else.

By giving a force a name (for example, 'gravitational force') and calling the space in which the force can be detected a 'field', we may feel that we understand it better, but do we really?

As always, the real usefulness of physics is in making calculations and predictions, and Newton's law and Coulomb's law are extremely useful in this respect. These laws predict that the fields which they describe extend indefinitely. In practice, of course, forces become immeasurably small if the distances involved get very large.

Einstein's theory of general relativity has provided a different interpretation of gravitational force and field in terms of the curvature of space–time, but that does not reduce the importance of Newton's universal law of gravitation any more than it reduces the usefulness of any of Newton's laws of motion. Quantum physics removes the need for the concept of electric field, instead using the idea of virtual particles to cause the forces.

There are certain phenomena in physics for which there seems to be no simpler explanation and we may be tempted to think 'that is just the way the universe is', not delving any deeper. However, the search for fundamental 'truths' is one of the defining features of physics.

The study of gravitational, electromagnetic and nuclear forces, and the fields we use to describe them, is a central part of physics because these forces have produced the world and the universe that we see around us. Trying to see similarities between these fundamental forces and fields has long been an issue for physicists trying to develop the concept of a single unified force.

Gravitational field strength

We may want to ask the question 'if a mass was put in a particular place, what would be the gravitational force on it?' The answer, of course, depends on the magnitude of the mass, so it is more helpful to generalize and ask 'what is the force on a unit mass (1 kg)?' If we know this, then we can easily calculate the gravitational force on any other mass.

Gravitational field strength is defined as the force per unit mass that would be experienced by a small test mass placed at that point.

Reference is made to a 'small **test mass**' because a large mass (compared to the mass creating the original field) would have a significant gravitational field of its own.

Gravitational field strength is given the symbol g and has the unit newtons per kilogram, N kg^{-1}. Field strength is a vector quantity and its direction is shown by the arrows on field lines.

$$\text{gravitational field strength} = \frac{\text{gravitational force}}{\text{mass}}$$

$$g = \frac{F}{m}$$

This equation is given in the *Physics data booklet*.

In general we know, from Newton's second law of motion, that $a = F/m$, so that gravitational field strength ($g = F/m$) in N kg^{-1} is numerically equal to the acceleration due to gravity in m s^{-2}.

Imagine you were on an unknown planet and wanted to find experimentally the gravitational field strength. This can be done easily by hanging a 'small' test mass of 1 kg on a forcemeter. The reading will be the strength of the gravitational field (in N kg^{-1}) and the direction of the field will be the same as the direction of the string– 'downwards' towards the centre of the planet.

The equation $g = F/m$ should be compared to $E = F/q$ and $B = F/IL\sin\theta$, each of which defines the strength of fundamental force fields.

Gravitational field strength around a planet

Since the gravitational force $F = GMm/r^2$, the gravitational field strength g around a point mass M can be found in terms of G by substituting for F in the equation $g = F/m$. If m is a test mass placed at a distance of r from a mass M, then:

$$g = \frac{GMm/r^2}{m}$$

$$g = G\frac{M}{r^2}$$

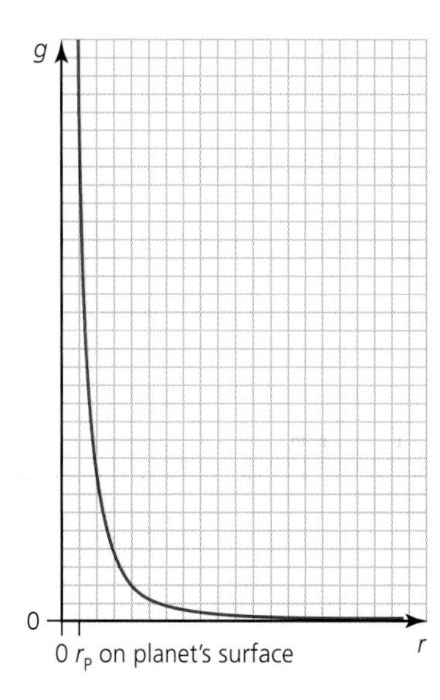

■ **Figure 6.21** A planet's gravitational field strength, g, decreases with distance from the centre of the planet, r, showing an inverse square relationship

This equation is given in the *Physics data booklet*. Although derived for a point mass, it can also be used to determine the gravitational field on the surface, or outside of, a spherical mass, such as a planet or moon. To determine the field on the surface of a planet, we replace r with r_p, the radius of the planet. We can use an average density to calculate the mass, but we must assume that the mass is concentrated at the centre of the sphere. Like gravitational force, gravitational field strength, g, follows an inverse square law with distance. This is sketched in Figure 6.21.

To determine how the gravitational field strength depends on a planet's radius we need to use these facts:

- mass, M, is equal to density, ρ, multiplied by volume, V
- the volume of a sphere equals $\frac{4}{3}\pi r^3$

So, we can write:

$$M = \rho\frac{4}{3}\pi r_P^{\ 3}$$

The density of a planet is not uniform, so the value used here is an average. Putting this equation for M back into the equation for g we get:

$$g = G\rho\frac{4}{3}\pi\frac{(r_P^{\ 3})}{(r_P^{\ 2})}$$

So that:

$$g = \frac{4}{3}G\pi\rho r_P$$

This equation predicts that the gravitational field strength at the surface of a planet is proportional to its radius. It is *not* in the *Physics data booklet.*

From the equation we would expect bigger planets to have stronger fields, but that is only true if they have equal average densities. (The Earth is the densest planet in our solar system, with an average density of 5510 kg m^{-3}. Venus and Mercury have similar densities to Earth but the density of Mars is significantly lower. The outer planets are gaseous and have lower densities. Saturn has the lowest density at 687 kg m^{-3}.)

Worked example

9 **a** Calculate the gravitational field strength on the surface of the Moon. The mass of the Moon is 7.35×10^{22} kg and its radius is 1740 km.

b Calculate the gravitational field strength at a point on the Earth's surface due to the Moon (not the Earth) assuming that the distance between the centre of the Moon and the Earth's surface is 3.8×10^8 m.

c Calculate the gravitational field strength on the surface of the planet Venus (radius = 6050 km, average density = 5.2×10^3 kg m^{-3}).

a $g = \frac{GM}{r^2}$

$g = \frac{(6.67 \times 10^{-11}) \times (7.35 \times 10^{22})}{(1.74 \times 10^6)^2}$

$g = 1.62$ N kg^{-1} (This is about one-sixth of the Earth's gravitational field strength.)

b $g = \frac{GM}{r^2}$

$g = \frac{(6.67 \times 10^{-11}) \times (7.35 \times 10^{22})}{(3.8 \times 10^8)^2}$

$g = 3.4 \times 10^{-5}$ N kg^{-1}

This shows us that on the Earth's surface, the gravitational field due to the Moon is about 300 000 times weaker than that due to the Earth. Although it is much weaker it still has some effects on the Earth, tides for example.

c $g = \frac{4}{3}G\pi\rho r_P$

$g = \frac{4}{3} \times (6.67 \times 10^{-11}) \times \pi \times (5.2 \times 10^3) \times (6.05 \times 10^6)$

$g = 8.8$ N kg^{-1}

23 The radius of the planet Mercury is 2440 km. Its mass is 3.3×10^{23} kg. Calculate the gravitational field strength on its surface.

24 a What is the gravitational field strength at a height of 300 km above the Earth's surface? (The radius of the Earth is 6.4×10^6 m.) Many satellites orbit at about this height.
b What percentage is this of the accepted value for the gravitational field strength on the Earth's surface?

25 Calculate the gravitational field strength on the surface of a planet that has a radius of 8560 km and an average density of 4320 kg m^{-3}.

26 What would the gravitational field strength be on a planet that had twice the radius of Earth and half the density?

27 Use a spreadsheet to calculate data and to draw a graph showing how the Earth's gravitational field strength varies from its surface up to a height of 50 000 km.

28 a Find out the name and details of the largest moon in our solar system and the planet around which it orbits.
b Calculate the gravitational field strength on its surface.

Additional Perspectives

Weighing the Earth

At the time Newton proposed his law of universal gravitation it was not possible to determine an accurate value for the gravitational constant, G. The only gravitational forces that could be measured were those of the weights of given masses on the Earth's surface. The radius of the Earth was known, but that still left two unknowns in the equation $F = GMm/r^2$; the gravitational constant and the mass of the Earth. If either of these could be found, then the other could be calculated using Newton's law of gravitation. That is why the determination of an accurate value for G was known as 'weighing the Earth'.

Certainly it was possible in the seventeenth century to get an approximate value for the mass of the Earth from its volume and estimated average density (using $m = \rho V$). But density estimates would have been little more than educated guesses. We know now that the Earth's crust has a much lower average density (about 3000 kg m^{-3}) than most of the rest of the Earth. However, it was possible to use an estimate of the Earth's mass to calculate an approximate value for the gravitation constant. The first accurate measurement was made more than 100 years later by Cavendish in an experiment that is famous for its precision and accuracy.

To calculate a value for G without needing to know the mass of the Earth (or the Moon, or another planet) required the direct measurement of the force between two known masses. Cavendish used lead spheres (see Figure 6.22) because of their high density (11.3 g cm^{-3}). The forces involved are very difficult to measure because they are so small, but also because similar-sized forces can arise from various environmental factors. (In fact, Cavendish's main aim was to get a value for the density of the Earth rather than to measure G.)

■ **Figure 6.22** A modern version Cavendish's apparatus

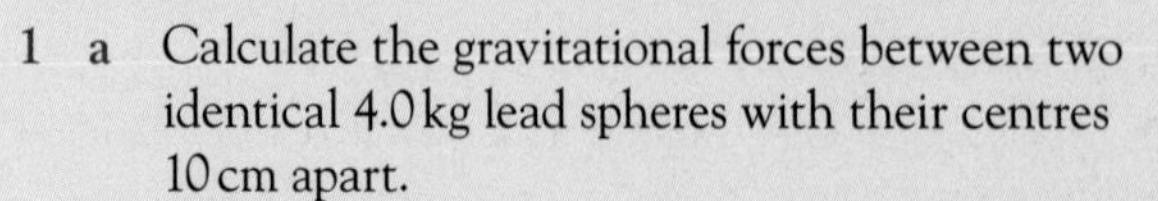

1 a Calculate the gravitational forces between two identical 4.0 kg lead spheres with their centres 10 cm apart.

b What is the distance between the surfaces of the spheres?

c Estimate the weight of a grain of salt and compare your answer to the gravitational force calculated in **a**.

2 In an early attempt to estimate the gravitational constant and calculate a value for the mass of the Earth, pendulums were suspended near mountains (see an exaggerated representation in Figure 6.23).

a What measurements should have been taken?

b Suggest why such experiments were unlikely to be very accurate.

c Find out what you can about Nevil Maskelyne and a Scottish mountain.

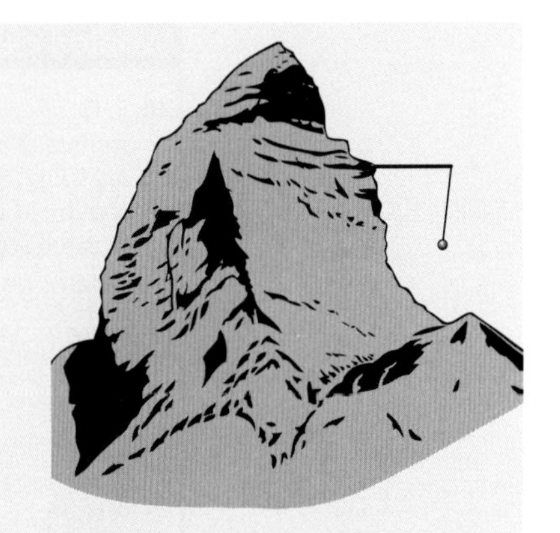

■ **Figure 6.23** A pendulum and a mountain attract each other

Calculating combined gravitational field strengths of planets and/or moons

It is possible that a mass may be in two or more separate gravitational fields. For example, we are in the fields of both the Earth and the Moon. However, the values calculated in Worked example 9 showed that on the Earth's surface the two fields are in the ratio $9.81/(3.4 \times 10^{-5})$, or about 300 000:1. In other words on the Earth's surface the Moon's gravitational field is almost negligible compared to the Earth's field. But if a spacecraft is travelling from the Earth to the Moon, the gravitational field due to the Earth will get weaker as the Moon's field gets stronger. There will be a point at which the two fields will be equal in strength but opposite in direction (shown as P, in Figure 6.24).

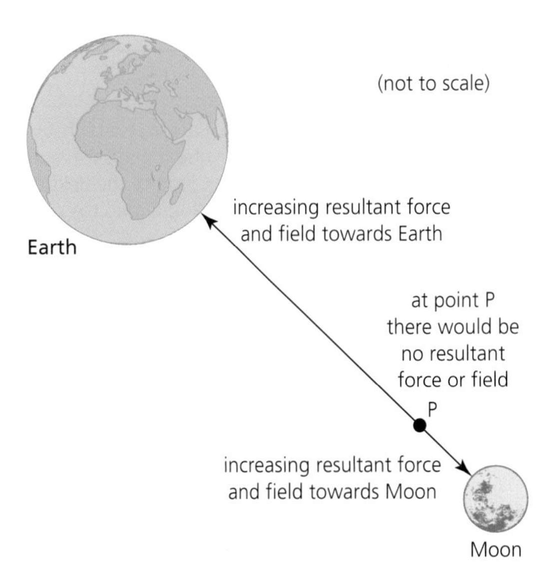

■ **Figure 6.24** Opposing fields cancel at a precise point, P, between the Earth and the Moon

At P the total gravitational field strength is zero and there will be no resultant force on the spacecraft because the pulls of the Moon and the Earth are equal and opposite. As the spacecraft travels from the Earth to P there is a resultant force pulling it back to Earth, but this is reducing in size. After the spacecraft passes P there will be an increasing resultant force pulling the spacecraft towards the Moon.

In general, if two or more masses are creating gravitational fields at a certain point, then the total field is determined by adding the individual fields, remembering that they are *vector quantities*.

In this chapter we will only be concerned with locations somewhere on the line passing through the two masses – then the vector addition of the two fields is straightforward, as shown in Worked example 10.

Worked example

10 In Figure 6.25 (which is not drawn to scale), P is a point midway between the centres of planets A and B. At P the gravitational field strength due to A is 4.0 N kg^{-1} and that due to B is 3.0 N kg^{-1}.

a What is the total gravitational field strength at P?

b What is the combined gravitational field strength at a point Q, which is the same distance from A as P?

a Taking the field towards the bottom of the diagram to be positive:

$$-4.0 + 3.0 = -1.0$$

The gravitational field strength is 1.0 N kg^{-1} towards A.

b The size of the field due to A is the same at Q as it is at P, but it is in the opposite direction. The strength of the field due to B at Q is 3^2 times smaller than at P because it is three times further away, but it is in the same direction.

$$4.0 + \left(\frac{3.0}{9}\right) = 4.3\,\text{N kg}^{-1} \text{ towards A and B}$$

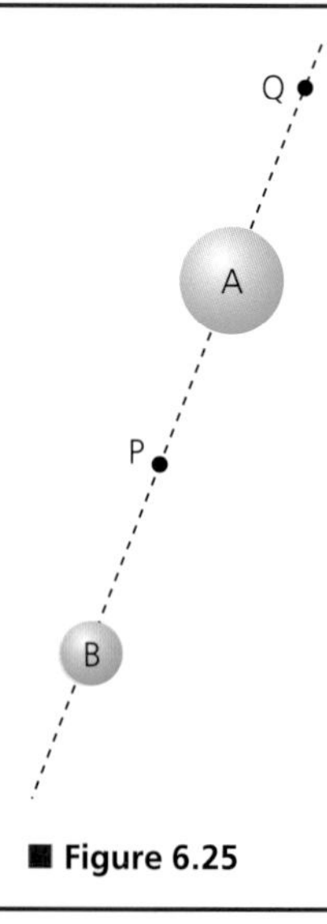

■ **Figure 6.25**

29 a Research data that will allow you to set up a spreadsheet to calculate the gravitational field strengths due to the Earth and the Moon at points along a straight line joining their surfaces.
b Combine the fields to determine the resultant field and draw a graph of the results.
c Where does the resultant gravitational field equal zero?

30 The gravitational fields of the Sun and the Moon cause the tides of the world's oceans. The highest tides occur when the resultant field is largest (at times of a 'new moon'). Draw a sketch to show the relative positions of the Earth, Sun and Moon when the resultant field on the Earth's surface is:
a strongest
b weakest (at the time of a 'full moon').

31 Consider Figure 6.25. If planet A has a gravitational field of 15 N kg^{-1} at Q, but the combined field at the same point is 16 N kg^{-1}, calculate the combined field at point P.

■ Orbital motion

The gravitational forces between two masses are equal in size but opposite in direction. However, if one of the masses is very much bigger than the other, we often assume that the force on the larger mass has negligible effect, while the same force acting on the much smaller mass produces a significant acceleration. If the smaller mass is already moving then the gravitational force can provide the centripetal force to make it **orbit** (move continuously around) the larger mass. It is then described as a **satellite** of the larger mass. The Earth and the other planets orbiting the Sun, and moons orbiting planets, are all examples of **natural satellites**. In the modern world we are becoming more and more dependent on the **artificial satellites** that orbit around the Earth.

Artificial satellites around the Earth

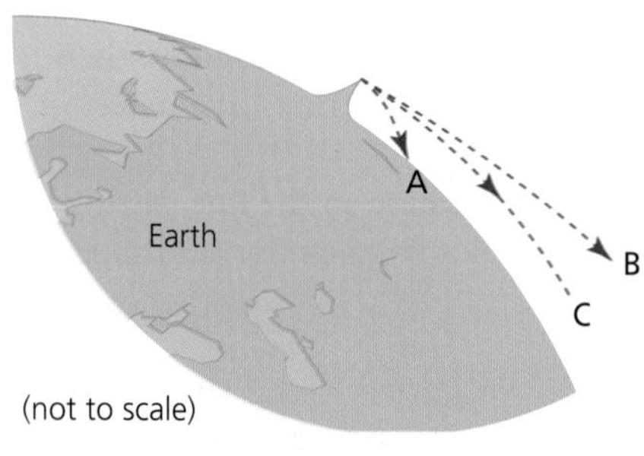

■ **Figure 6.26** The path of objects projected at different speeds from a mountain top

If we assume that there is no air resistance, a bullet that is fired 'horizontally' from the top of a mountain would move in a parabolic path and hit the ground some distance away, as shown by path A in Figure 6.26. If the bullet was travelling fast enough, it could move as shown in path B, and 'escape' from the Earth. Path C shows the path of an object moving with exactly the right speed and direction so that it remains at the same distance above the Earth's surface (remember that we are assuming that there is no air resistance); that is, it is in orbit around the Earth.

Gravity is the only force acting on the bullet and it is acting continuously and perpendicularly to its instantaneous velocity. As we have discussed, this is a necessary condition for circular motion. The force of gravity (weight) is providing the centripetal force. Remembering the equation for centripetal acceleration, we can write:

$$\frac{v^2}{r} = g \qquad \text{or} \qquad \left(\frac{mv^2}{r} = mg\right)$$

This enables us to calculate the theoretical speed necessary for an orbit very close to the Earth's surface (radius = 6.37×10^6 m).

$$v^2 = gr = 9.81 \times (6.37 \times 10^6)$$
$$v = 7910\,\text{m s}^{-1}$$

Without air resistance, an object moving horizontally with a speed of $7910\,\text{m s}^{-1}$ close to the Earth's surface would orbit the Earth.

To avoid air resistance, a satellite needs to be above the Earth's atmosphere. Most satellite orbits are at a height of at least 300 km. In order to maintain a satellite in a circular orbit around the Earth, it needs to be given the correct speed for its particular height. If there is no air resistance, a satellite in a circular orbit will continue to orbit the Earth without the need for any engine. The force of gravity acts perpendicularly to motion so that no work is done by that force.

Knowing the value of g at any particular height enables us to calculate the speed necessary for a circular orbit at that height. The speed does not depend on the mass. All satellites at the same height move with the same speed and have the same orbital period.

If the speed of a satellite is faster than the speed necessary for a circular orbit (but slower than the escape speed) it will move in an elliptical path. The orbits of planets, moons and satellites are not perfectly circular, but the difference is often insignificant.

32 a Calculate values for the gravitational field strengths at heights above the Earth of 250 km, 1000 km, 10 000 km and 30 000 km.
b Calculate the necessary speeds for circular orbits at these heights.
c Use a compass to draw a scale diagram of the Earth with these orbits around it.
d Use $v = 2\pi r/T$ to determine the times for complete orbits (periods), T, at these heights and mark them on your diagram.

33 Two satellites of equal mass orbit the same planet as shown in Figure 6.27. Satellite B is twice as far away from the centre of the planet as satellite A. Copy the table and complete it to show the properties of the orbit of satellite B.

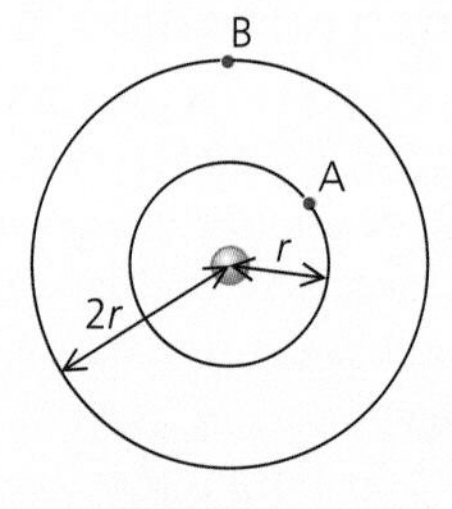

■ Figure 6.27

	Satellite A	Satellite B
Distance from planet's centre	r	$2r$
Gravitational field strength	g	
Gravitational force	F	
Circumference of orbit	c	
Speed	v	
Time period	T	

34 The Earth is an average distance of 1.50×10^{11} m from the Sun.
a Calculate the average orbital speed of the Earth around the Sun.
b Determine the centripetal acceleration of the Earth towards the Sun.
c Use your answers to calculate a value for the mass of the Sun.

Gravity controls the motions of moons, planets, stars and galaxies

Gravity is the only significant force controlling the motions of all the large objects in space. Although perfect circular motion is an idealized example, Newton's laws of motion and gravitation can be combined with the concept of centripetal force to make reliable predictions of the motions of moons, planets and stars, and the rotation of galaxies etc.

Nature of Science

Can the laws of physics predict the future?

Newton is rightly famed for formulating the 'laws' of motion and gravitation that bear his name. These laws are central to much of what is known as 'classical physics' and, by using them, future motions can be predicted from current and past observations. Newtonian physics is often described as **deterministic**, meaning that a given set of circumstances can only lead to one, certain outcome (a concept that modern physics has shown to be incorrect on the atomic scale). For example, we can use Newton's laws to predict very accurately what will happen when cars collide, or when a comet will be observable from Earth hundreds of years in the future, but not to confidently predict events within an atom. Also, although Newton's law may accurately describe gravitational events in the terms of mathematical equations, it offers us no help at all in trying to understand the true nature of gravity.

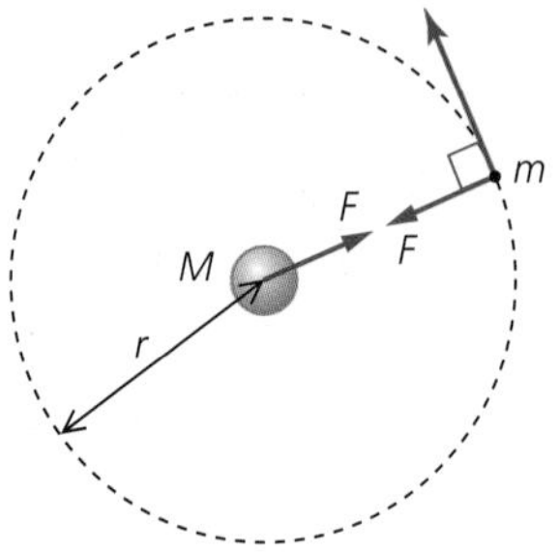

■ **Figure 6.28**

Relating the average radius of an orbit to its time period

The centripetal force (mv^2/r) required to keep any mass, m, in a circular orbit around a larger mass, M, is provided by the force of gravity (GMm/r^2), which always acts perpendicularly to the motion of the orbiting mass (see Figure 6.28).

$$\frac{mv^2}{r} = \frac{GMm}{r^2}$$

$$v^2 = \frac{GM}{r}$$

(This equation is equivalent to $v^2 = gr$ as used earlier in this chapter.)

If we replace v with $2\pi r/T$ (circumference/period), we get an important equation that shows us directly how the time period of an orbit depends on its radius. This equation can also be applied to elliptical orbits if the average radius is used.

$$\left(\frac{2\pi r}{T}\right)^2 = \frac{GM}{r}$$

Rearranging, we get a general equation linking the radius to the period for all satellites orbiting the same mass, M.

$$\frac{r^3}{T^2} = \frac{GM}{4\pi^2}$$

$GM/4\pi^2$ is a constant for all masses in orbit around the same mass, M.

This means that r^3/T^2 must also be a constant. This was first discovered by the German mathematician Johannes Kepler (1571–1630, Figure 6.29). Kepler was using observations and data on the planets of our solar system, but his law can be applied wherever different smaller masses orbit the same larger mass, M. (For example, the moons around Jupiter, or the artificial satellites around Earth). His law is empirical (based only on observation, rather than theory) and the derivation shown above was proposed by Isaac Newton many years later.

■ **Figure 6.29** Johannes Kepler

Worked example

11 Io is a moon of Jupiter. It is an average distance of 422 000 km from the centre of Jupiter and takes 1.77 days to complete an orbit. Calculate the mass of Jupiter.

$$\frac{r^3}{T^2} = \frac{GM}{4\pi^2}$$

$$\frac{(4.22 \times 10^8)^3}{(1.77 \times 24 \times 3600)^2} = (6.67 \times 10^{-11})\frac{M}{4\pi^2}$$

$$M = 1.90 \times 10^{27}\,\text{kg}$$

Artificial satellites orbiting the Earth have a wide range of uses, which is expanding every year. This requires a high level of international consultation and control; such collaboration also aims to avoid duplication of research and costs. Figure 6.31 shows the launch site for rockets of the European Space Agency, which has twenty member states.

■ **Figure 6.30** European Space Agency launch site in French Guiana

35 a Calculate the time period and orbital speed of a satellite in a low orbit 300 km above the Earth's surface.
b Suggest one advantage and one disadvantage of placing satellites at this height.

36 The mass of the Sun is 1.99×10^{30} kg. Use Kepler's third law and the time period of the Earth to determine the average distance to the Sun.

37 a Make a spreadsheet of the planets of the solar system, their average distances from the Sun, r, and their periods, T.
b Use the spreadsheet to calculate r^3, T^2 and r^3/T^2.
c Use the program to plot a graph of r^3 against T^2.
d Do your results confirm Kepler's third law? Explain your answer.
e Alternatively, a graph of log T could be plotted against log r. What could be determined from the gradient of this graph?

38 Two moons, A and B, orbit a planet at distances of 5.4×10^7 m and 8.1×10^7 m from the planet's centre. If moon A has a period of 24 days, calculate the period of moon B.

ToK Link

Are classical physics and modern physics compatible?

The laws of mechanics along with the law of gravitation create the deterministic nature of classical physics. Are classical physics and modern physics compatible? Do other areas of knowledge also have a similar division between classical and modern in their historical development?

Are the motions of all masses, ranging from stars and galaxies down to sub-atomic particles, controlled by the same 'laws' of physics, or is it possible, or acceptable, that we need totally different physics to describe motions on different scales?

Examination questions – a selection

Paper 1 IB questions and IB style questions

Q1 An object moving in a circular path of radius 40 cm completes two revolutions in 4 s. Which of the following is an accurate calculation concerning the object's motion?

A Its total angular displacement was 2π radians.
B Its average angular velocity was $\pi\,\text{rad}\,\text{s}^{-1}$.
C Its average linear velocity was $2.5\,\text{m}\,\text{s}^{-1}$.
D Its frequency was 2 Hz.

Q2 A small rubber ball on a length of thin cotton was spun faster and faster in a nearly horizontal circle until the cotton broke. Which of the following describes this situation?

A Tension in the string provides the centripetal force on the ball.
B When the cotton broke the ball initially moved radially away from the centre of the circle.
C If the experiment was repeated with a more massive ball moving in a path of the same radius, it would need to be spun faster before the cotton would break.
D If the experiment was repeated with the original ball moving in a path of larger radius, it would need to be spun faster before the cotton would break.

Q3 The Moon orbits the Earth.

Which of the following diagrams correctly represents the force(s) acting on the Moon?

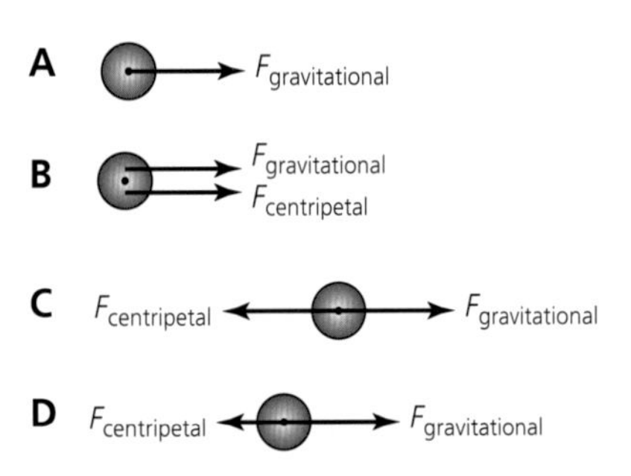

Q4 Which of the following correctly describes the constant, G?

A It is a vector quantity.
B It has a larger magnitude on the Moon than on the Earth.
C It has a smaller magnitude on the Moon than on the Earth.
D It is a fundamental constant.

Q5 There is a place between the Earth and the Moon where the gravitational field strength is zero. This is because:

A The Earth's gravitational field is stronger than the Moon's gravitational field.
B Gravitational fields only exist close to the surfaces of planets and moons.
C The fields from the Earth and the Moon act in opposite directions.
D The Moon does not have a gravitational field.

Q6 Newton's universal law of gravitation concerns the force between two masses. These masses are:
A planets
B stars
C spherical masses
D point masses.

Q7 Which of the following is a correct statement about on object moving in a circle at a constant linear speed?
A There is no force acting.
B Its angular velocity increases uniformly.
C A force is needed in the direction of instantaneous motion.
D The object is accelerating.

Paper 2 IB questions and IB style questions

Q1 This question is about gravitational fields.
a Define gravitational field strength. (2)
b The gravitational field strength at the surface of Jupiter is 25 N kg^{-1} and the radius of Jupiter is 7.1×10^7 m.
i Derive an expression for the gravitational field strength at the surface of a planet in terms of its mass *M*, its radius *R* and the gravitational constant *G*. (2)
ii Use your expression in b i above to estimate the mass of Jupiter. (2)

Q2 **a** Name the type of force that enables a car to travel around a curve on a horizontal road. (1)
b A 1200 kg car is travelling around a road that is the arc of a circle of radius 80 m. What is the maximum possible speed of the car if the available centripetal force cannot exceed 4500 N? (3)
c Explain how a banked road surface can enable a car to travel around a corner faster than on a similar horizontal surface. (3)

7 Atomic, nuclear and particle physics

ESSENTIAL IDEAS

- In the microscopic world energy is discrete.
- Energy can be released in nuclear decays and reactions as a result of the relationship between mass and energy.
- It is believed that all the matter around us is made up of fundamental particles called quarks and leptons. It is known that matter has a hierarchical structure with quarks making up nucleons, nucleons making up nuclei, nuclei and electrons making up atoms and atoms making up molecules. In this hierarchical structure, the smallest scale is seen for quarks and leptons (10^{-18} m).

7.1 Discrete energy and radioactivity – *in the microscopic world energy is discrete*

Light is electromagnetic energy that has been emitted from atoms – we can learn a lot about the energy inside atoms by examining the light (spectra) that atoms emit.

Emission and absorption spectra

When white light from the Sun passes through a prism, the light is dispersed into its component colours. The band of different colours merge into a continuous spectrum as shown in Figure 4.22.

We can make some substances *emit* light by supplying them with energy by heating them. Many elements emit different colours from which they can be identified in *flame tests*. We also commonly produce light using electric currents to supply energy in a variety of ways (see Chapter 5). If an electric current (at high voltage) is passed through an element in the form of a low-pressure gas in a *discharge tube* it will produce its own unique emission spectrum. See Figure 7.1.

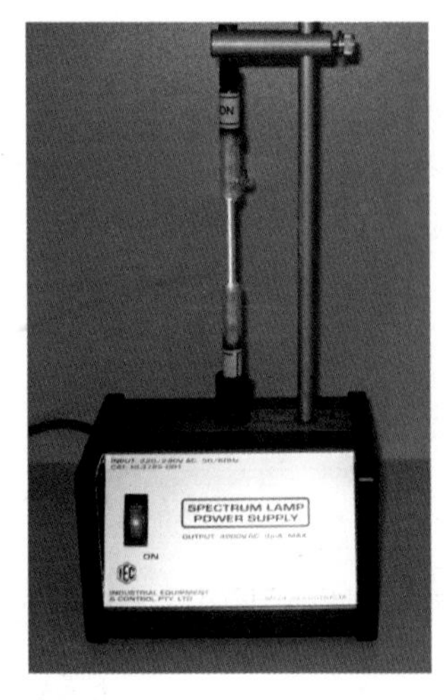

Figure 7.1 A discharge tube containing atoms of neon

When a high potential difference is applied across the two electrodes in the tube, energy is transferred to the gas atoms or molecules and light is emitted. Examination of the light with a spectroscope shows that the emitted spectrum is not continuous but consists of a number of bright lines, as shown in Figure 7.2. (A simple hand spectroscope can be made using just a black tube with a slit and a *diffraction grating* – see page 397).

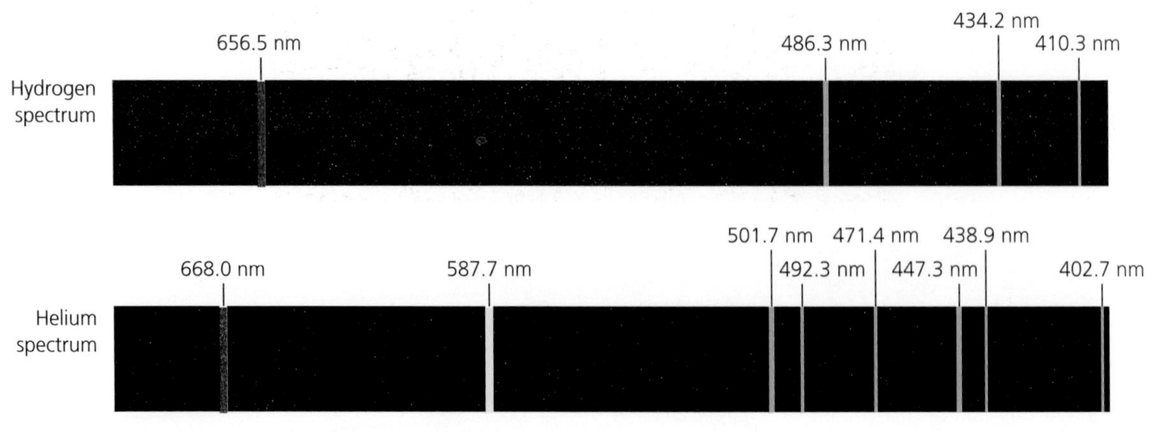

Figure 7.2 Emission spectra of hydrogen and helium

The spectrum produced by the gaseous element in the discharge tube is called a **line spectrum**. It consists of a number of *discrete* (separate) colours; each colour is the image of the slit in front of the light source. The range of wavelengths corresponding to the lines of the emission spectrum is characteristic of, and unique to, the element in the discharge tube. If broad spectrum white light is passed through a sample of gaseous atoms or molecules maintained at low pressure and the spectrum is analysed, it is found that light of certain wavelengths is *missing*. In their place are a series of sharp dark lines. This pattern of lines is identical to that seen in the emission spectrum of the same gas. It is called an **absorption spectrum**.

Figure 7.3 compares the emission and absorption spectra of the same element, and how they can be observed using a prism. (The prism could be replaced by a diffraction grating, as discussed in Chapter 9.)

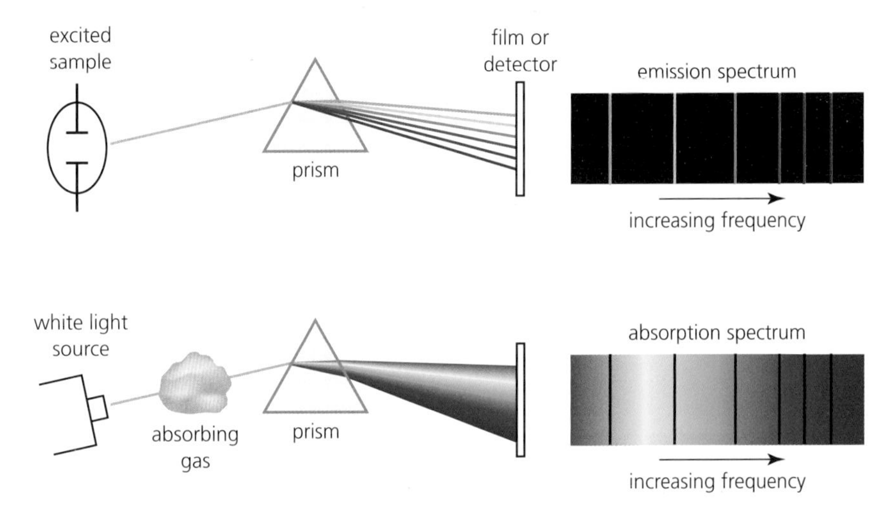

■ **Figure 7.3** Production of emission and absorption spectra of the same element

The study of spectra is called **spectroscopy**, and instruments used to measure the wavelengths of spectra are called *spectrometers* (see Figure 7.4).

■ **Figure 7.4** A spectrometer

Discrete energy and discrete energy levels

Photons

Before we can explain the origin of line spectra you need to understand a very important concept in physics: all electromagnetic radiations, including light, transfer energy in *discrete* (separate) amounts called **photons**. The transfer of energy is not continuous. More generally, we use the term **quantum** to describe the smallest possible quantity of any entity that can only have discrete values. We can say that light is **quantized**.

The fact that light can behave as if it is 'lumpy', as if it were composed of 'particles', contradicts the *wave theory of light* used in Chapter 4 to explain diffraction and interference. It is worth stressing that there is no single, simple theory of light that explains all of its properties, and this dilemma is often referred to as the *wave–particle duality* of light. Photons are discussed in much more detail in Chapter 12.

The energy, E, carried by one photon of electromagnetic radiation depends only on its frequency, f, as follows:

$$E = hf$$

This equation is given in the *Physics data booklet*.

h is a very important fundamental constant that controls the properties of electromagnetic radiations. It is called **Planck's constant**. Its value of 6.63×10^{-34} J s is given in the *Physics data booklet*.

Since we know from Chapter 4 that $c = f\lambda$, this equation can be rewritten as $E = hc/\lambda$, although it is written in the *Physics data booklet* as:

$$\lambda = \frac{hc}{E}$$

Worked example

1 Calculate the energy carried by one photon of microwaves of wavelength 10 cm (as might be used in a mobile phone):
a in J
b in eV.

a $E = \frac{hc}{\lambda}$

$E = 6.63 \times 10^{-34} \times \frac{3.00 \times 10^{8}}{0.10}$

$E = 2.0 \times 10^{-24}$ J

b $\frac{2.0 \times 10^{-24}}{1.6 \times 10^{-19}} = 1.2 \times 10^{-5}$ eV

1 a What frequency of electromagnetic radiation has photons of energy 1.0×10^{5} eV?
b What name do we give to that kind of radiation?

2 A microwave oven uses electromagnetic photons of energy 1.6×10^{-24} J. What is the wavelength of this radiation?

3 The greenhouse gas carbon dioxide absorbs radiation of frequency of 1600 nm.
a How much energy is carried by the absorbed photons?
b In what part of the electromagnetic spectrum is this radiation?

4 A particular visible line in the spectrum of oxygen has a wavelength of 5.13×10^{-7} m. What is the energy transferred by one photon of this radiation?

5 A light bulb emits light of power 4 W. Estimate the number of photons emitted every second.

6 Light has a typical wavelength of 5×10^{-7} m, and X-rays have a typical wavelength of 5×10^{-11} m.
a Draw a small square of sides 2 mm to represent the energy carried by a light photon.
b Assuming that photon energy is represented by the area, draw another square to represent the energy carried by an X-ray photon.
c Suggest why X-rays are more dangerous than light.

Transitions between energy levels

The emission of a light photon transfers energy away from an atom and this must mean that the amount of energy inside the atom has decreased. The fact that the spectra from elements is in the form of separate lines tells us that energy can only be emitted (as photons) from atoms in definite, *discrete* amounts. We may conclude that the emission of each photon occurs when an atom changes from one precise *energy level* to another precise, but lower, energy level. We refer to *transitions* between energy levels.

The emission of photons of different energies from atoms of the same element must mean that each atom has many possible energy levels. Consider Figure 7.5, which shows, with a simplified example, that an atom with four different energy levels could have six possible transitions between energy levels and, therefore, atoms with these energy levels might emit photons of six different energies (wavelengths).

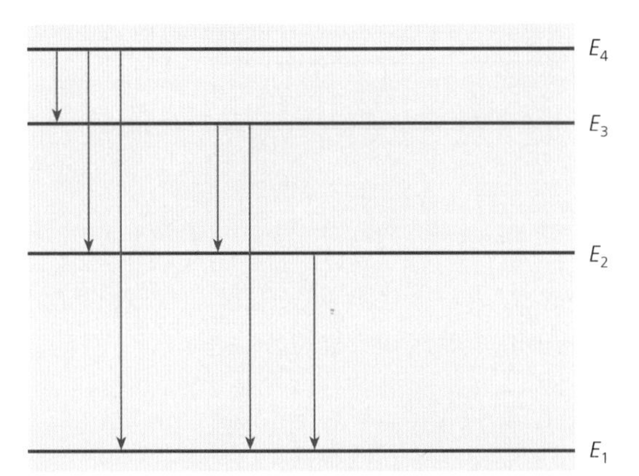

Figure 7.5 Energy transitions between four energy levels in an atom

Each of the transitions results in the emission of a photon with a particular wavelength. The transition from E_4 to E_1 involves the highest amount of energy, and so results in light with the highest frequency and shortest wavelength. The transition from E_4 to E_3, which involves the least energy, gives light with the lowest frequency and longest wavelength. In general, if there is a difference in energy, ΔE, between two levels then the frequency, f, of a photon involved with a transition between these two levels is given by:

$$\Delta E = hf$$

The lowest possible energy level is called the ground state of the atom and it is, logically, the lowest level shown in the diagram. Atoms are usually in their ground states and they need to be *excited* (given energy) by heating, or from an electric current, to raise them to a higher energy level, which is called an **excited state.** After which, an atom usually returns very quickly to a lower energy level by emitting a photon.

The different energy levels of atoms are explained by changes to the way electrons are arranged around the nucleus, so it is common to alternatively refer an atom's energy levels as *electron energy levels*.

When energy is absorbed by an atom, its electric potential energy increases. This is because of the forces between the positively charged nucleus and negatively charged electrons. When a photon is emitted, the electric potential energy decreases.

Absorption spectra are explained by atoms being raised to higher energy levels by absorbing photons with the same energy as those that can be emitted by the same atom, as they undergo transitions between identical energy levels. The atom will quickly re-emit a photon, but in a random direction, thereby decreasing the intensity of radiation travelling in the original direction.

Atoms of different elements have different energy levels. This means that the differences between energy levels are unique to each element, so that each element (in the gaseous state) produces a different and characteristic spectrum. These spectra can be used to identify the presence of a particular element in a sample that has been vaporized. Analysis of the light emitted from stars is used to determine the elements present in them (this is covered in further detail in Option D: Astrophysics).

Measurement of the numerous wavelengths of the spectral lines from an element (using a spectrometer) can be used to determine the magnitudes of the large number of energy level transitions occurring within the atoms. From this information a detailed energy level diagram can be constructed. Figure 7.6 shows energy levels within the simplest atom: hydrogen. Note that the energy levels are all shown with negative values; this represents the fact that energy must be *supplied* to the atom to raise an electron to a higher level of potential energy, and that a free electron is considered to have zero electric potential energy. The energy that needs to be supplied to an atom to take an electron from its ground state to the highest excited state is called the ionization energy of the atom because it is the energy needed to free the electron from the attraction of the nucleus, the energy needed to *ionize* the atom.

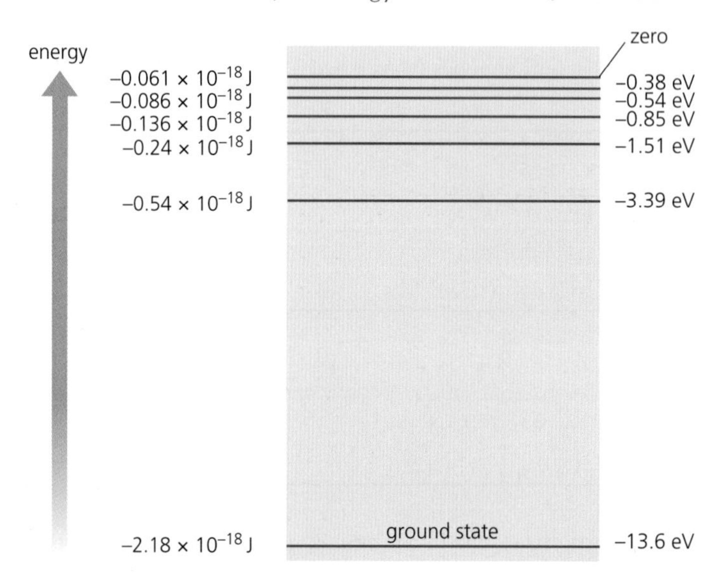

■ Figure 7.6 Some of the energy levels of the hydrogen atom

Figure 7.7 represents an electron in a hydrogen atom moving to the next higher energy level and then quickly returning to the ground state as a photon is released.

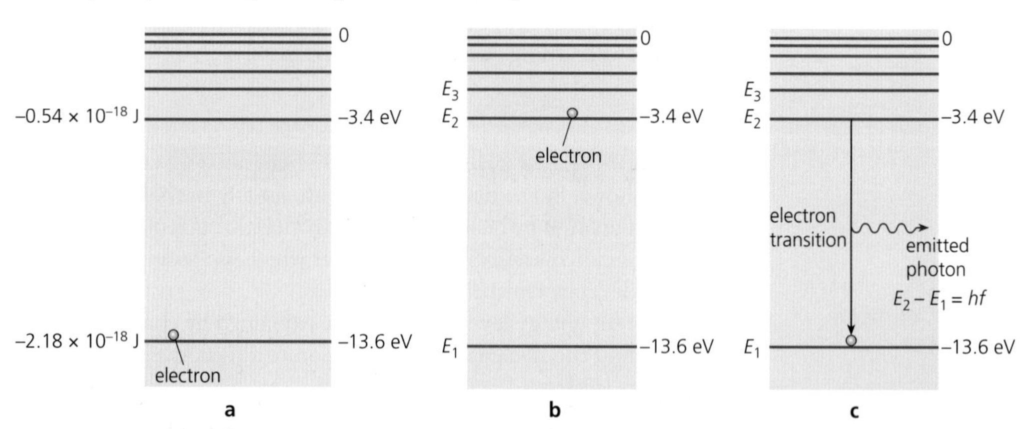

■ Figure 7.7 A hydrogen atom and its electron: **a** in its ground state, **b** in a short-lived excited state, **c** returning to the ground state

Worked example

2 Consider Figure 7.7.

a What is the ionization energy of the hydrogen atom?

b What is the frequency of a photon emitted when the atom falls from the −3.4 eV energy level down to the ground state?

c Hydrogen absorbs radiation of wavelength 6.7×10^{-7} m. Identify the transition involved.

d How many energy transitions are theoretically possible between the levels shown?

a 2.18×10^{-18} J

b $E = hf$

Difference in energy levels (J) = $(2.18 - 0.54) \times 10^{-18} = (6.63 \times 10^{-34})f$

$f = 2.5 \times 10^{15}$ Hz

c $E = \frac{hc}{\lambda}$

$E = 6.63 \times 10^{-34} \times \frac{3.0 \times 10^{8}}{6.7 \times 10^{-7}}$

$E = 3.0 \times 10^{-19}\,\text{J}$

This corresponds to the transition up from the second to the third energy level.

d $6 + 5 + 4 + 3 + 2 + 1 = 21$

7 Consider Figure 7.8.
 a What wavelength of radiation is emitted by the transition shown?
 b In what part of the electromagnetic spectrum is this radiation?
 c When radiation of frequency 1.18×10^{15} Hz passes through cool mercury vapour it is absorbed. Identify the transition involved in this process.
 d What is the longest wavelength of radiation that could be emitted by a transition between the levels shown?

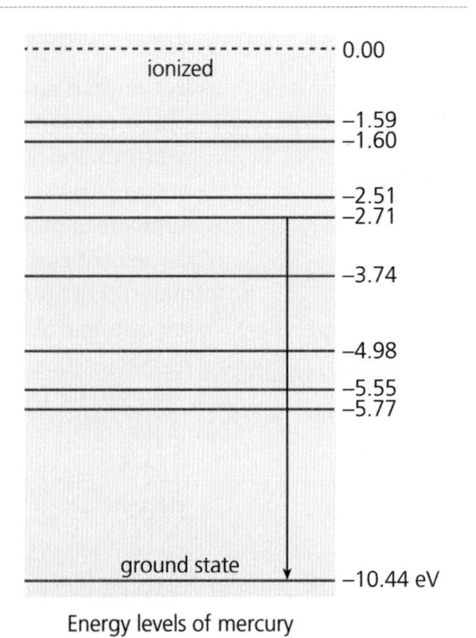

■ **Figure 7.8** Some of the energy levels of mercury

8 When the spectrum emitted by the Sun is observed closely using a spectrometer, by looking at a white surface – *not* the sun directly, it is found that light of certain frequencies is missing and, in their place, are dark lines.
 a Explain how the cool outer gaseous atmosphere of the Sun is responsible for the absence of these frequencies.
 b Suggest how an analysis of the solar absorption spectrum could be used to determine which elements are present in the Sun's atmosphere.

9 Rydberg discovered that the frequencies of all the lines in the spectrum of hydrogen could be predicted by the following expression:

$f = 3.29 \times 10^{15}\left(\frac{1}{n_1^2} - \frac{1}{n_2^2}\right)$, where $n_1 < n_2$ (both integers)

 a Use a spreadsheet to calculate the frequencies predicted by using $n_1 = 1$. Let n_2 vary from 2 to a very large number.
 b Show that the series converges to a numerical limit. What is that limit? (These frequencies are in the ultraviolet part of the spectrum. They are known as the Lyman series.)

■ Atomic structure

Before we can discuss radioactivity we need to have some understanding of the structure of atoms.

Nature of Science

What is matter?

The question 'what is matter made of?' has been one of the central issues in science and philosophy for centuries. As recently as 2012 a major discovery (the 'Higgs boson') confirmed a significant new development (see Section 7.3), but we need to go back more than 2400 years for the beginnings of the story – Democritus is often credited as being the first person to propose that matter was made of atoms (about 400 BCE). However, it is only about 100 years since the first proposal that the atom had a central nucleus.

The basic *nuclear* model of the atom was first proposed by Ernest Rutherford in 1911 following the famous Geiger and Marsden experiment (see Section 7.3). This model describes the atom as consisting of a very small and dense central **nucleus** surrounded by orbiting electrons. Two years later (1913) Niels Bohr proposed that the existence of energy levels could be explained if electrons could exist only in specific orbits (known as 'shells' in chemistry). More details can be found in Chapter 12.

The diameter of an atom is typically about 10^{-10} m and the diameter of a nucleus is typically about 10^{-15} m.

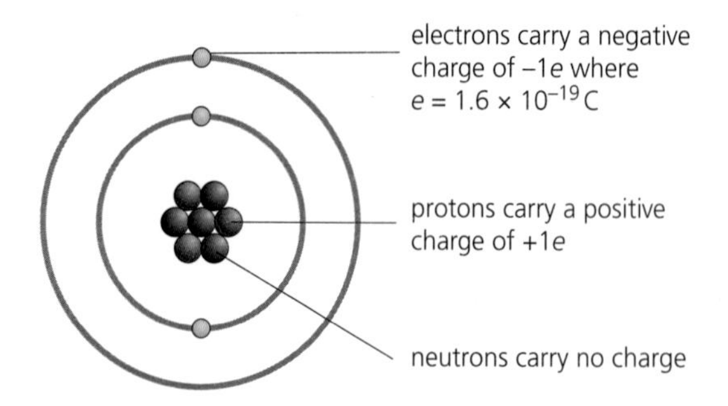

■ **Figure 7.9** Simple nuclear model of a lithium atom (not to scale)

In the years that followed it was confirmed that the nucleus consists of **protons** and **neutrons**, which contain almost all of the mass of the atom. The protons are positively charged and the neutrons are electrically neutral. The electrons are negatively charged but have very little mass in comparison to protons and neutrons. Atoms are electrically neutral because there are equal numbers of protons and electrons. In this model, the electrons orbit the nucleus because of the centripetal force provided by the electrical attraction between opposite charges. The vast majority of an atom is empty space, a vacuum. The properties of protons, neutrons and electrons are summarized in Table 7.1.

■ **Table 7.1** Properties of sub-atomic particles

Name of particle	Approximate relative mass	Relative charge
proton	1	+1
neutron	1	0
electron	$\frac{1}{1840}$	−1

This model of the atom was *never* fully satisfactory because electromagnetic theory predicts that the centripetally accelerating electrons would radiate away energy (and spiral inwards). Also, there were many properties of atoms that the theory could not explain, such as the *reason* for energy levels, and why the protons were not repelled from each other.

This electrostatic model of the atom is still a common visualization but it has been replaced by a quantum mechanical model based on the wave behaviour of electrons – see Chapter 12. There are many more sub-atomic particles than protons, electrons and neutrons, and details are given later in this chapter (in Section 7.3).

Nuclear structure

Proton number, Z

The number of protons in the nucleus of an atom determines which element it is. So, atoms of a particular element are identified by their **proton number** (sometimes called *atomic number*), which is given the symbol Z. The periodic table of the elements arranges the elements in order of increasing proton number.

The **proton number**, Z, is the number of protons in the nucleus of an atom.

Because atoms are electrically neutral, the number of protons is equal to the number of electrons orbiting the nucleus.

Nucleon number, A

Neutrons and protons are both found in the nucleus of the atom. The term **nucleon** is used for both these particles.

The **nucleon number**, A, is defined as the total number of protons and neutrons in a nucleus.

The nucleon number represents the mass of an atom, because the mass of an electron is negligible. Nucleon number is sometimes referred to as the *mass number*.

Neutron number, N

The **neutron number,** N, is defined as the number of neutrons in a nucleus.

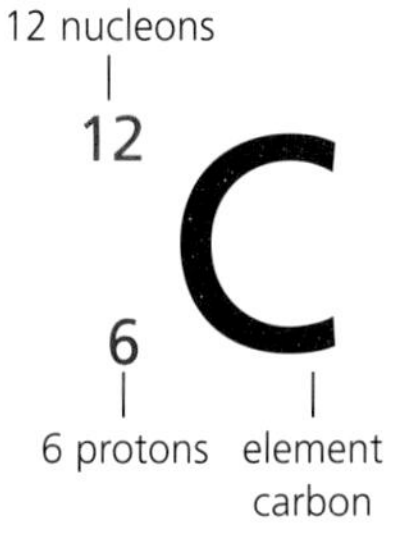

■ **Figure 7.10** Standard notation for specifying a nuclide

The difference between a nucleon number and the proton number gives the number of neutrons in the nucleus:

$$N = A - Z$$

Isotopes

Two or more atoms with the same proton number may have *different* numbers of neutrons. The atoms are the same element but have different nucleon numbers. These atoms are called **isotopes.**

Isotopes of an element have the same chemical properties. Most elements have several isotopes.

Nuclides

The term nuclide is used to specify one particular species (type) of atom, as defined by the structure of its nucleus.

All atoms with the same nucleon number *and* the same proton number are described as atoms of the same *nuclide.*

There is a standard notation (Figure 7.10) used to represent a *nuclide* by identifying its proton number and nucleon number.

It is important to make clear the difference between nuclides and isotopes. When referring to different types of atom we call them nuclides, however, when referring specifically to atoms of the same element with different nuclei, we call them isotopes.

Some elements have many isotopes, but others have very few or even one. For example, the most common isotope of hydrogen is hydrogen-1, ^1_1H. Its nucleus is a single proton. Hydrogen-2, ^2_1H, is called *deuterium*; its nucleus contains one proton and one neutron. Hydrogen-3, ^3_1H, with one proton and two neutrons, is called *tritium*. Hydrogen isotopes (Figure 7.11) are involved in fusion reactions (see Section 7.2 and Chapter 8).

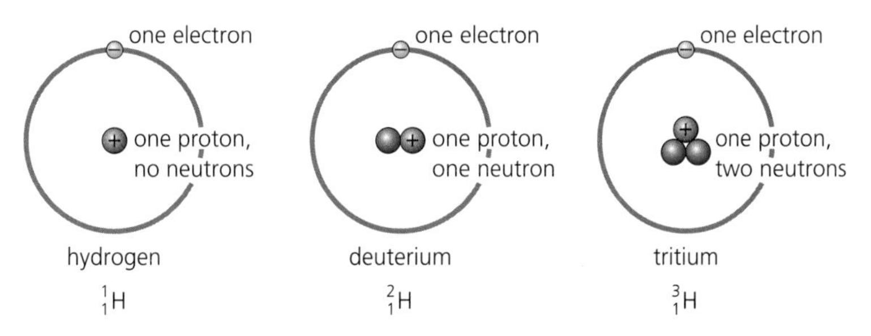

■ **Figure 7.11** The three isotopes of hydrogen

As a further example, the following nuclides are three isotopes of carbon:

$^{12}_{6}C$ (six protons, six neutrons)

$^{13}_{6}C$ (six protons, seven neutrons)

$^{14}_{6}C$ (six protons, eight neutrons)

Samples of elements are usually mixtures of isotopes. Isotopes cannot be separated by chemical means. Separation can be achieved by processes that depend on the difference in masses of the isotopes, for example the diffusion rate of gaseous compounds.

The notation for describing nuclides can also be applied to the nucleons. For example, a proton can be written as ^1_1p and a neutron as ^1_0n. The electron can also be represented using this notation: its charge is −1 compared to the +1 charge on a proton, so an electron can be represented by $^{0}_{-1}e$, remembering that the mass (number) of the electron is effectively zero compared to the proton and neutron.

Worked examples

3 Deduce the nucleon, proton and neutron numbers for these two isotopes of lithium:

$^{6}_{3}Li$ and $^{7}_{3}Li$

Draw simple diagrams showing the structure of these atoms.

Lithium-6: 6 nucleons, 3 protons, and 6 – 3 = 3 neutrons

Lithium-7: 7 nucleons, 3 protons, and 7 – 3 = 4 neutrons

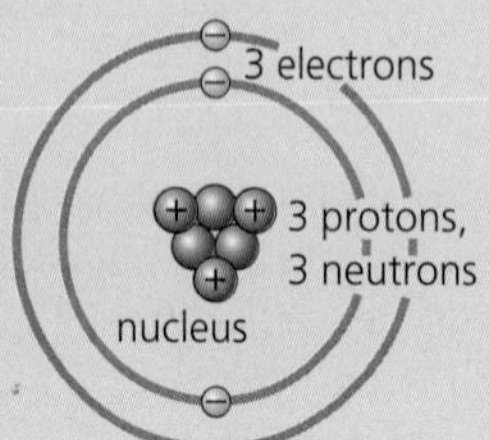

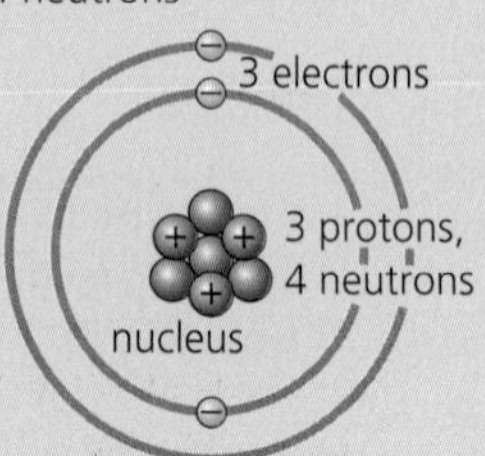

■ **Figure 7.12**

4 An oxygen atom nucleus is represented by $^{16}_{8}O$. Describe its atomic structure.

The nucleus has a proton number of 8 and a nucleon number of 16. Thus, its nucleus contains eight protons and eight neutrons (16 – 8). There are also eight electrons.

5 A potassium atom contains 19 protons, 19 electrons and 20 neutrons. Deduce its nuclide notation.

Proton number Z = 19, and nucleon number A = 19 + 20 = 39.

$^{39}_{19}K$

6 State the numbers of protons and neutrons in an atom of the isotope plutonium-239, $^{239}_{94}Pu$.

number of protons (proton number) Z = 94

nucleon number (A) = 239

neutron number $N = A - Z = 239 - 94 = 145$

10 The nuclides $^{129}_{53}I$, $^{137}_{55}Cs$ and $^{90}_{38}Sr$ were all formed during atomic weapons testing. State the number of neutrons, protons and electrons in the atoms of these nuclides.

11 What is the electric charge of the nucleus $^{4}_{2}He$?

12 The number of electrons, protons and neutrons in an ion of sulfur are equal to 18, 16 and 16, respectively. What is the correct nuclide symbol for the sulfur ion?

13 State the number of nucleons in one carbon-13 atom, $^{13}_{6}C$.

14 Chlorine is an element that has 17 protons in its nucleus. The two most common isotopes of chlorine are chlorine-35 and chlorine-37.
 a What are the nuclide symbols for these two isotopes?
 b Suggest why the periodic table shows an atomic mass for chlorine of 35.45.

■ Fundamental forces and their properties

We know from Chapter 5 that there are electrical forces between charged particles and that Coulomb's law can be used to calculate the magnitude of those forces, as shown in this worked example.

Worked example

7 a Estimate an order of magnitude for the force between a helium nucleus and one of the electrons in the atom (assume that the separation is 10^{-10} m).

b Estimate an order of magnitude for the force between the two protons in the helium nucleus.

a $F = k\dfrac{q_1q_2}{r^2}$

$$F = (8.99 \times 10^9) \times \frac{(-1.6 \times 10^{-19}) \times (+1.6 \times 2 \times 10^{-19})}{(10^{-10})^2} \approx -10^{-7}\,\text{N (attractive force)}$$

b $F = k\dfrac{q_1q_2}{r^2}$

$$F = (8.99 \times 10^9) \times \frac{(+1.6 \times 10^{-19}) \times (+1.6 \times 10^{-19})}{(10^{-15})^2} \approx +10^{2}\,\text{N (repulsive force)}$$

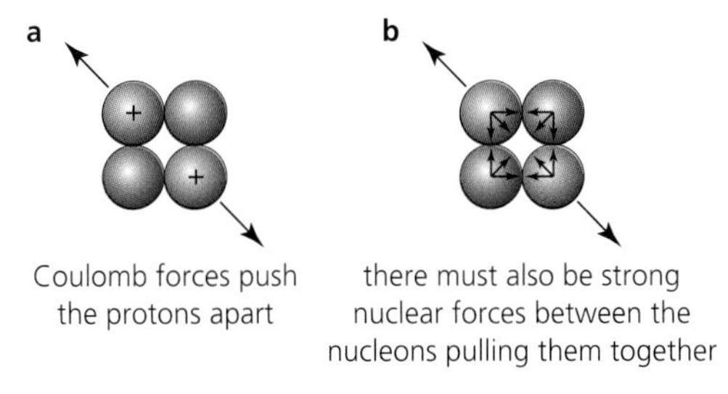

■ **Figure 7.13** Repulsive and attractive forces in a helium nucleus

In a simple model of the atom the attractive forces between the nucleus and the electrons keep the electrons in orbit moving around the nucleus, but the *much larger* forces (Worked example 7b) between the protons within the nucleus suggests that they should be repelled apart (see Figure 7.13a).

It is clear that there must be another strong force acting, which opposes the electric repulsion and attracts the protons together. This attractive, very short-range force acts between nucleons (including neutrons) and is known as the strong nuclear force.

The electrical force is proportional to 1/separation^2 (as shown in the equations above), but the strong nuclear force must decrease much more quickly with distance because it is negligible or zero except when the separation of the nucleons is very small, typically less than 10^{-15} m. In other words, the strong nuclear force extends effectively only to its immediate neighbours – it does not extend to all the nucleons in the nucleus.

The strong nuclear force is considered to be one of the *four fundamental forces (interactions)* in the universe:

- *Strong nuclear force*: only acts on sub-atomic particles known as quarks (from which protons and neutrons are constructed). 'Short-range', within the nucleus.
- *Electromagnetic force*: causes both magnetic and electrostatic effects and exists between two charges. It is this force that binds electrons to the positively charged nucleus to form an atom. It is also the force that binds atoms together to form solids, liquids and gases. It is attractive for unlike charges but repulsive for like charges. Its range is infinite but its magnitude obeys the inverse square law.
- *Gravitational force*: it exists between any two objects with mass. It is this force that binds stars together to form galaxies (see Option D: Astrophysics). It is always attractive and its range is infinite. A galaxy can experience the gravitational pull by another galaxy many billions of kilometres away. Its magnitude obeys the inverse square law and is the weakest of the four fundamental forces.
- *Weak nuclear force*: involved with radioactive decay (even shorter range than the strong nuclear force).

These fundamental forces are discussed in more detail in Section 7.3.

■ Radioactive decay

ToK Link

Fortune favours the prepared mind

The role of luck/serendipity in successful scientific discovery is almost inevitably accompanied by a scientifically curious mind that will pursue the outcome of the 'lucky' event. To what extent might scientific discoveries that have been described as being the result of luck actually be better described as being the result of reason or intuition?

Radioactivity was discovered by accident in 1896 by Henri Becquerel who was carrying out a series of experiments on fluorescence. In the course of his experiments he put some potassium uranium sulfate (which emits radioactivity, but nobody knew that then) on a sheet of photographic film that was between two pieces of black paper in a drawer. When Becquerel developed the film he found that it had the same appearance as if it had been exposed to sunlight, although that was not possible. In fact, radiation had passed from the uranium compound through the black paper and exposed the film. In this rather strange and serendipitous way Becquerel had discovered radioactivity. ('**Serendipity**' means unplanned and unexpected good luck.)

There are many similar examples of accidental discoveries in science, but these should not simply be viewed as 'luck' because a spirit of investigation is a prerequisite, as well as excellent experimental skills. And, most importantly, the scientist needs to appreciate the significance of what may appear at the time to be a minor effect (although Becquerel was uncertain of the explanation at first). The French microbiologist Louis Pasteur is widely quoted as saying 'fortune favours the prepared mind'.

An understanding of the causes of radioactivity begins with a consideration of the forces in the nucleus.

The relative sizes of the Coulomb repulsion and the attractive strong nuclear force depends on the ratio of neutrons to protons (N/Z) in a nucleus; the balance between these two forces controls *nuclear stability*. Big nuclei with large numbers of protons and neutrons (typically $Z > 82$) are unstable because the longer-range repulsive electrical forces (from a large number of protons) can be greater than the short-range strong nuclear forces. But many nuclides of lower mass are also unstable. Figure 7.14 indicates how nuclear stability depends on the N/Z ratio. For low-mass atoms, stable nuclei have approximately equal numbers of protons and neutrons ($N = Z$), but larger nuclei need more neutrons (compared to protons) to achieve stability.

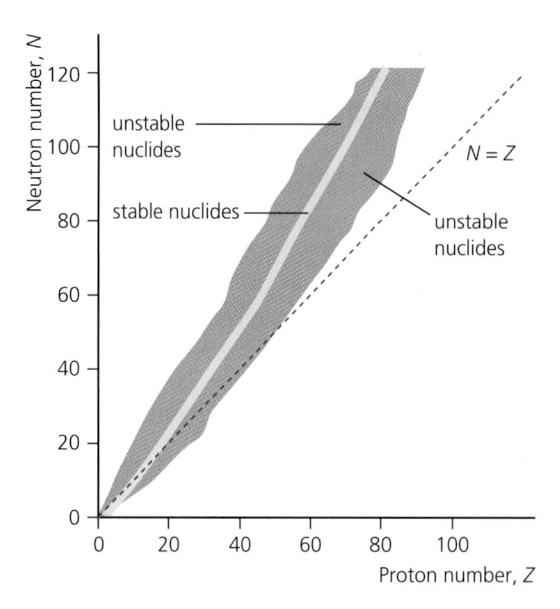

■ **Figure 7.14** Stable nuclei shown on a chart of neutron number plotted against proton number

In an attempt to become more stable, unstable nuclei may release **nuclear radiation** in the form of small particles and/or gamma rays. This is also described as **ionizing radiation** because it causes atoms in the surrounding materials to lose electrons and form ions. The release of a particle from a nucleus results in the formation of a different nuclide of a different element. This is known as **transmutation**.

Radioactivity is the emission of ionizing radiation caused by changes in the nuclei of unstable atoms. The process by which radioactive atoms change into other elements is known as **radioactive decay.**

Radioactive decay should not be confused with chemical or biological decay. The decay of a radioactive material will often not involve any obvious change in appearance.

A material that emits measurable amounts of radioactivity may be described as *radioactive*, while an unstable atom may be referred to as a **radioisotope** or **radionuclide**.

It should be stressed that the decay of an unstable nucleus is spontaneous, random, unpredictable and uncontrollable.

Radioactive decay and the emission of nuclear radiation is unaffected by factors such as chemical composition, temperature and pressure. We cannot control when radioactive nuclei will decay; however, we can control our exposure to the emitted radiation.

Additional Perspectives

The effect of ionizing radiation on humans

All radiation that can cause ionization is dangerous to humans. Ionizing radiation comes in many forms and from many sources. It includes the radiation from natural radioactive substances and radiation from artificial sources, such as X-ray machines in hospitals and nuclear reactors.

People are exposed to a variety of ionizing radiations that are health risks. The danger is due to the absorption of energy from the radiation by the tissues of the body. This results in the formation of ions, which can kill or change living cells.

Very high doses of radiation can cause cells to stop functioning and prevent them from undergoing cell division, resulting in the death of the cell. Widespread damage of cells in different tissues can result in death. There are also possible long-term delayed effects of ionizing radiation, such as sterility and cancers, especially leukaemia (cancer of the white blood cells) and inherited genetic defects (mutations) in the children of people who have been exposed to radiation.

The short-term effects of exposure to large doses of radiation include radiation burns (redness and sores on the skin) mainly caused by beta and gamma radiations. There may also be blindness and the formation of cataracts in the eyes. Radiation sickness occurs when a person has a single large exposure to radiation. Nausea and vomiting are the main symptoms, together with fever, hair loss and poor wound healing.

Medical experts are uncertain about the long-term effects of exposure to low doses of radiation. Estimates of the risk of radiation damage are based on the assumption that as the dose of radiation becomes smaller, so the risk is reduced in proportion.

The main risks from alpha and beta radiation come from sources that get inside a person. Because alpha and beta particles do not penetrate very far into the body, the risks from external solid or liquid sources are fairly small. However, care must be taken to stop radioactive materials from being eaten or inhaled (breathed in) from the air. So no eating, drinking or smoking is allowed where any radioactive materials are handled, and disposable gloves and protective clothing are worn. Masks are worn in mines where radioactive dust particles are airborne.

Gamma radiation and X-rays from external sources can be absorbed deep inside the body, and people who are exposed to sources of X-rays and gamma radiation must be protected as much as possible. The dose a person receives can be limited in a number of ways: by using lead shielding, by keeping a large distance between the person and source, and by keeping exposure times as short as possible.

People who work with ionizing radiation may wear a film badge or detector that gives a permanent record of the radiation dose received. Workers may also be checked for radiation contamination by using sensitive radiation monitors before they leave their place of work. A worker handling radioactive materials may use remote-controlled tools and sit behind a shielding wall made of thick lead and concrete.

When radioactive materials are used in medicine they are chosen carefully to have the least damaging effects on the body.

Nuclear reactors produce large amounts of high-frequency electromagnetic radiation (gamma rays) and radioactive waste materials. The safe storage of nuclear waste for thousands of years is a very important safety issue that requires international debate. Nuclear reactors also produce large numbers of neutrons, which are released from the nuclei of uranium atoms. High-speed neutrons are also a form of dangerous ionizing radiation.

1 Explain why nuclear radiation is used extensively in hospitals, even though it is considered to be a serious health risk.

Alpha particles, beta particles and gamma rays

Radioactive nuclides may emit three kinds of radiation:

- alpha particles
- beta particles (positive and negative)
- gamma rays (often associated with alpha or beta emission).

Atoms of the same radionuclide always emit the same kind of radiation.

Alpha radiation

An **alpha particle** is the same as a helium-4 nucleus: the combination of two protons and two neutrons. It has a nucleon number of 4 and a proton number of +2. Alpha particles are represented by the symbols ${}^{4}_{2}\alpha$ or ${}^{4}_{2}\text{He}$.

Clearly the emission of an alpha particle results in the loss of two protons and two neutrons from a nucleus, so that the proton number of the nuclide decreases by two and a new element is formed (transmutation). This can be represented in a **nuclear equation** as follows:

$${}^{A}_{Z}\text{X} \rightarrow {}^{A-4}_{Z-2}\text{H} + {}^{4}_{2}\text{He}$$

parent nuclide → daughter nuclide + alpha particle

Nuclear equations must balance: the sum of the nucleon numbers and proton numbers must be equal on both sides of the equation. The terms *parent nuclide* and *daughter nuclide* are commonly used to describe the nuclides before and after the decay, respectively.

The change to a more stable nucleus is equivalent to a decrease in nuclear potential energy; this energy is transferred to the kinetic energy of the alpha particle (and the daughter nucleus). Because there are only two particles after the decay, they must initially move in exactly opposite directions. This is because of the law of conservation of momentum, which also predicts that the alpha particle will have a much faster speed (if it is emitted from a massive nucleus). Emitted alpha particles from the same type of nuclide usually have the same speed and the same energy, typically about 5 MeV (reminder: energy in J = energy in eV/1.6 × 10^{-19}). (But some radionuclides can emit alpha particles of one of several different energies.)

For example, an alpha particle is emitted when a radium-226 nucleus decays, resulting in the formation of a radon-222 nucleus. This is described by the nuclear equation:

$${}^{226}_{88}\text{Ra} \rightarrow {}^{222}_{86}\text{Rn} + {}^{4}_{2}\text{He}$$

The alpha particle produced in this decay has a kinetic energy of 4.7 MeV.

Beta radiation

In an unstable nucleus it is possible for an uncharged neutron to be converted into a positive proton and a negative electron. This also involves the creation of another particle called an (electron) **antineutrino**, $\bar{\nu}_e$. Antineutrinos (and neutrinos) are *very* small particles with no charge, so they are very difficult to detect.

$${}^{1}_{0}\text{n} \rightarrow {}^{1}_{1}\text{p} + {}^{0}_{-1}\text{e} + \bar{\nu}_e$$

After this nuclear reaction happens it is not possible for the newly formed electron to remain within the nucleus and it is ejected from the atom at a very high speed (close to the speed of light). It is then called a **beta-negative particle** and it is represented by the symbol ${}^{0}_{-1}\beta$ or ${}^{0}_{-1}\text{e}$.

When beta-negative decay occurs, the number of nucleons in the nucleus remains the same, but the number of protons increases by one, so that a new element is formed (transmutation). This can be represented in a nuclear equation:

$${}^{A}_{Z}\text{X} \rightarrow {}^{A}_{Z+1}\text{X} + {}^{0}_{-1}\text{e} + \bar{\nu}_e$$

parent nuclide → daughter nuclide + beta particle

As with alpha decay, the change to a more stable nucleus is equivalent to a decrease in nuclear potential energy and this is transferred to the kinetic energy of the particles. Because there are three (rather than two) particles after the decay, which may move in a variety of different directions, the beta particles emitted from the same type of nuclide will *not* all have the same speeds and energies (consider the law of conservation of momentum). There will be a well-defined maximum energy however, typically about 1 MeV.

For example, beta-negative particles are emitted when a strontium-90 nucleus undergoes beta decay, resulting in the formation of an yttrium-90 nucleus:

$$^{90}_{38}\text{Sr} \rightarrow {}^{90}_{39}\text{Y} + {}^{0}_{-1}\text{e} + \overline{\nu}_e$$

The beta-negative particle in this decay has a kinetic energy of 0.55 MeV.

In a similar process called beta-*positive* decay, a positively charged proton in a nucleus is converted into neutron and a positively charged electron, called a positron, which is then ejected from the atom (after which it is called a beta-positive particle). An (electron) neutrino, ν_e, is created at the same time. The electron and the positron are considered to be matter-antimatter. This is discussed in more detail in Section 7.3.

$$^{1}_{1}\text{p} \rightarrow {}^{1}_{0}\text{n} + {}^{0}_{+1}\text{e} + \nu_e$$

The following equation represents a typical beta-positive decay:

$$^{23}_{12}\text{Mg} \rightarrow {}^{23}_{11}\text{Na} + {}^{0}_{+1}\text{e} + \nu_e$$

A full understanding of beta decay requires knowledge of quarks and the weak nuclear force (see Section 7.3).

Gamma radiation

Gamma rays are high-frequency, high-energy electromagnetic radiation (photons) released from unstable nuclei. A typical wavelength is about 10^{-12} m. This corresponds to an energy of about 1 MeV (use $E = hc/\lambda$). Gamma rays are often emitted after an unstable nucleus has emitted an alpha or beta particle. Gamma rays may be represented by the symbol $^{0}_{0}\gamma$.

For example, when a thorium-234 nucleus is formed from a uranium-238 nucleus by alpha decay, the thorium nucleus contains excess energy and is said to be in an *excited state*. The excited thorium nucleus (shown by the symbol * in an equation) returns to a more stable state by emitting a gamma ray:

$$^{234}_{90}\text{Th}^* \rightarrow {}^{234}_{90}\text{Th} + {}^{0}_{0}\gamma$$

Because gamma rays have no mass or charge, the composition of the emitting nucleus does not change. There is no transmutation.

15 Complete the following nuclear equations and name the decay processes involved.

a $^{131}_{53}\text{I} \rightarrow \text{Xe} + {}^{0}_{-1}\text{e}$

b $^{241}_{95}\text{Am} \rightarrow \text{Np} + {}^{4}_{2}\text{He}$

16 A gamma ray has a wavelength of 5.76×10^{-12} m. What is the energy (in MeV) of a photon of this radiation?

17 Write nuclear equations for the following nuclear reactions:

a alpha decay of $^{184}_{78}\text{Pt}$

b beta-negative decay of $^{24}_{11}\text{Na}$

c gamma emission from $^{60}_{27}\text{Co}$

Use a periodic table to identify any new elements formed.

18 Calculate the speed of an alpha particle with kinetic energy of 3.79 MeV (use the *Physics data booklet* to obtain the masses involved).

19 Use the law of conservation of momentum to explain why beta particles emitted from the same source must have a range (spectrum) of different speeds and energies.

20 When the unstable nucleus magnesium-23 decays, a positron is emitted (beta-positive decay). Write a nuclear equation for this.

Radioactivity experiments

The properties of nuclear radiations can be investigated in a school laboratory if suitable radioactive sources are available. Alternatively, it is possible to use apparatus or computer programs that simulate radioactive experiments.

If ionizing radiation from a radioactive source enters the human body it is hazardous to health (see Additional Perspectives, page 293), so it is important that any radioactive sources used have a very low power, and that they are treated with the appropriate safety precautions. These include:

- Sources should be kept in thick, lead-lined containers. When not in use they should be locked away securely.
- Sources should be used for as short a time as possible.
- Sources need to be well-labelled, with appropriate warnings.
- Sources should be handled with tongs and always directed away from people.
- Lead shielding can be used to stop radiation spreading outside of the apparatus.

When nuclear radiation enters a detector it is usually detected as single events, and these are 'counted', so that we may refer to the **count** or the **count rate** (count per second) from a detector and (Geiger) *counter*. The unpredictable nature of individual radioactive decays means that significant variation in count may be observed. This is not an error, but more consistent measurements will be obtained with higher counts.

Background radiation

Traces of radioactive isotopes occur *naturally* in almost all substances. We are therefore all exposed to very low levels of radiation all the time. This cannot be avoided and it is called **background radiation.** Of course, some locations have a higher background count than others. Figure 7.15 shows where the background radiation typically comes from in the UK.

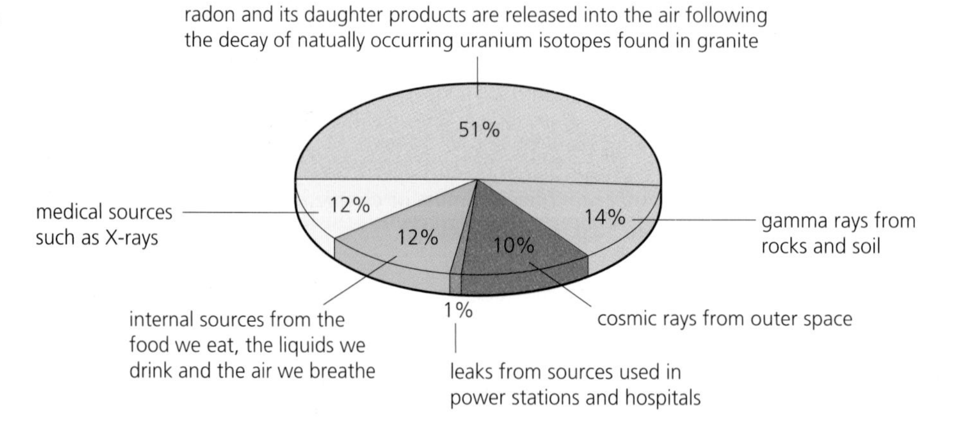

Figure 7.15 Sources of background radiation in the UK

The internal sources are radioactive nuclides, such as potassium-40, $^{40}_{19}K$, which are found in our bodies. The radon gas in the air comes from naturally occurring radioactive materials such as radium, thorium and uranium in the Earth's crust. Most of the cosmic radiation from space is absorbed by the Earth's atmosphere but some reaches the ground. However, people in high-flying aircraft and astronauts receive a much larger cosmic radiation dose. Cosmic rays include fast-moving nuclei of hydrogen and helium atoms.

Measurements of radiation counts from sources in the laboratory should normally be adjusted for the background count. For example, if the background count was $30\,\text{min}^{-1}$ at the time that a count rate of $387\,\text{s}^{-1}$ was being measured from a source, then the adjusted rate would be $386.5\,\text{s}^{-1}$. Clearly the background count will be more significant when lower readings are being taken.

Absorption characteristics of decay particles

Testing the absorption of ionizing radiation in various materials is one of the most common radioactivity investigations. However, their absorption (range) in air should be understood first.

Alpha particles are absorbed by a few centimetres of air (typically four or five). Beta particles penetrate further – up to about 30 cm. Gamma rays are very penetrating because there is very little absorption. But note that a gamma ray count rate will fall rapidly with increasing distance from the source because, even without any absorption, as the rays spread out the intensity will decrease with an inverse square relationship.

As alpha particles and beta particles travel through matter, they collide with atoms causing them to lose one or more electrons. The ionized atom (ion) and the resulting free electron are called an *ion pair* (Figure 7.16). When the kinetic energy of the particle has been reduced to a low value (comparable to the surrounding atoms), it will stop moving forward and can be considered as 'absorbed'.

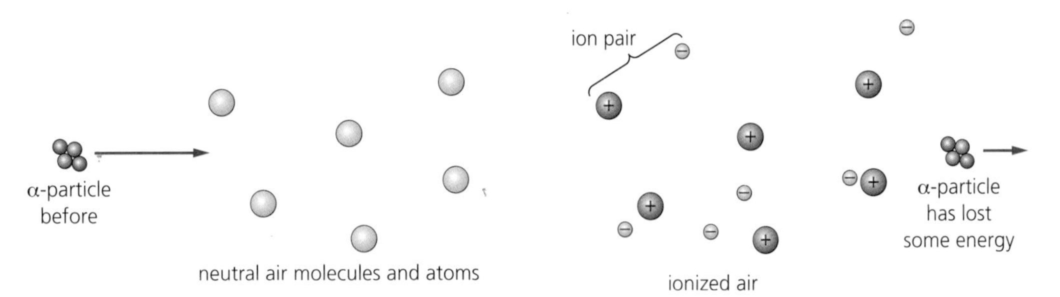

■ **Figure 7.16** Formation of ion pairs by alpha particles from molecules in the air

The production of an ion pair requires the separation of unlike charges, hence this process requires energy. Alpha particles have a relatively large mass, so that they transfer energy quite effectively in collisions with molecules typically of greater mass. This makes alpha particles efficient ionizers. They may produce as many as 10^5 ion pairs for every centimetre of air through which they pass. As a result they lose energy relatively quickly and have low *penetrating power*.

Beta particles have considerably less mass than alpha particles. Consequently they are much less efficient at transferring energy in collisions and produce much fewer ion pairs per centimetre. They are, therefore, far more penetrating than alpha particles.

Because gamma rays have no charge or mass, their ionizing power is much lower than that of alpha or beta particles. Their greater penetrating power makes gamma rays the most dangerous of the radiations entering the body from outside. Their energy is also transferred all in one interaction, rather than gradually during many collisions, and this can cause more harm to individual body cells.

Generally we would expect absorption of any radiation to be more extensive in denser materials. Figure 7.17 illustrates the absorption of nuclear radiations in paper, aluminium and lead. These materials have become the standards used to compare absorptions. Note that because alpha particles are easily absorbed in air, their absorption in other materials must be tested within a few centimetres of the source.

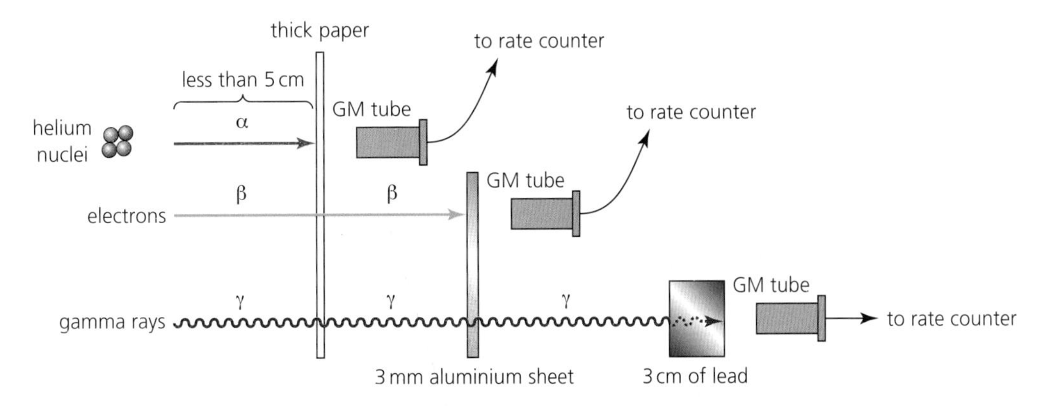

■ **Figure 7.17** Absorption of ionizing radiations

An unknown radiation can be identified from the materials that will absorb it. Alternatively, a radiation can be identified from its behaviour in electric or magnetic fields, as discussed in the next section.

Deflecting radiation in fields

Alpha and beta radiation will be emitted in random directions from their sources, but they can be formed into narrow beams (collimated) by passing the radiation through slits. Because a beam of alpha particles, or beta particles, is a flow of charge they will be deflected if they pass perpendicularly across a magnetic field (as discussed in Chapter 5). Gamma radiation is uncharged, so it cannot be deflected.

Figure 7.18 shows the passage of the three types of ionizing radiation across a strong magnetic field. Fleming's left-hand rule can be applied to confirm the deflection of the alpha and beta particles into circular paths, the magnetic force providing the centripetal force. The radius of the path of a charged particle moving perpendicularly across a magnetic field can be calculated from $r = mv/qB$ (page 252).

An alpha particle has twice the charge and about 8000 times the mass of a beta particle, although a typical beta particle may be moving ten-times faster. Taking all three factors into consideration, we can predict that the radius of an alpha particle's path may be about 400 times the radius of a beta particle in the same magnetic field.

(Note that observation of the deflection of alpha particles will usually require a vacuum.)

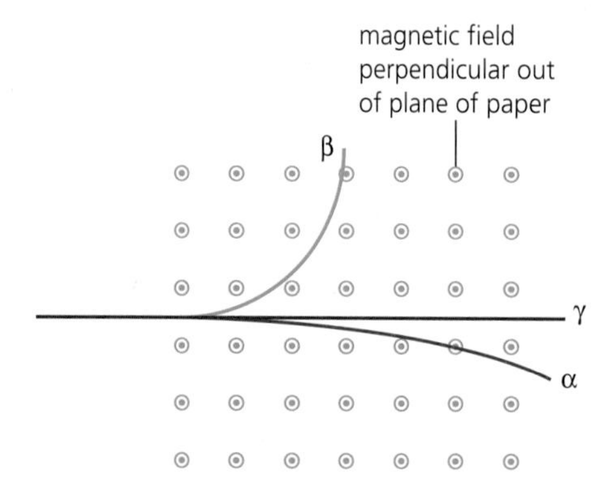

■ **Figure 7.18** Behaviour of ionizing radiations in a magnetic field

Alpha and beta radiation can also be deflected by electric fields, as shown in Figure 7.19. Alpha particles are attracted to the negative plate; beta particles are attracted to the positive plate. The combination of constant speed in one direction, with a constant perpendicular force and acceleration, produces a parabolic trajectory. This is similar to the projectile movement discussed in Chapter 2. The deflection of the alpha particles is small in comparison to beta particles, due to the same factors as discussed for magnetic deflection.

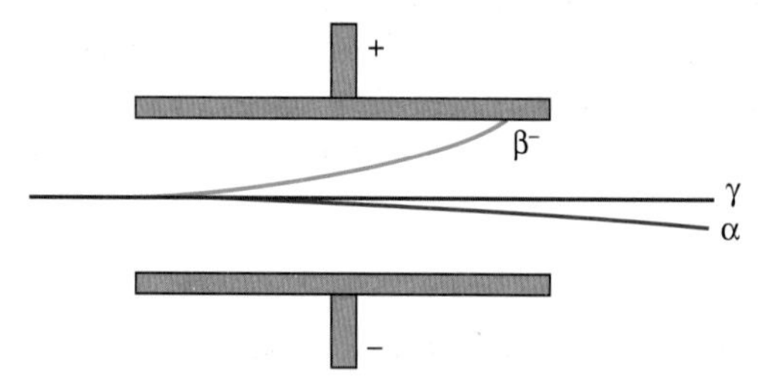

■ **Figure 7.19** Behaviour of ionizing radiations in an electric field

21 Alpha particles lose about 5×10^{-18} J of kinetic energy in each collision with an atom or molecule in the air. An alpha particle travelling through air makes 10^5 ionizing collisions with molecules or atoms in the air for each centimetre of travel. Calculate the approximate range of an alpha particle if the particle begins with an energy of 4.7×10^{-13} J.

22 Alpha particles are absorbed much more easily than beta particles with the same energy, yet alpha particles are more massive than beta particles. Explain this.

23 Why does ionizing radiation travel further in air if the pressure is reduced?

24 Explain why a source of alpha particles *outside* the human body may be considered to be very low risk, but a source *inside* the body is a high risk.

25 Radiation from a gamma ray source is detected at a distance of 12 cm from a source and a count rate of $373\,s^{-1}$ is recorded. Estimate the count rate if the distance was:
a reduced to 6 cm
b increased to 39 cm.

26 During a radioactivity investigation the background count was measured three times, giving the results $38\,min^{-1}$, $41\,min^{-1}$ and $36\,min^{-1}$.
a Explain why these measurements are all different.
b Calculate an average background count rate in s^{-1}.
c Explain why background radiation causes problems in experiments that involve measuring low count rates.

27 Explain how beta particles can be identified using experiments that observe their ability to penetrate different materials.

Summary of properties of alpha and beta particles and gamma rays

Table 7.2 Summary of the properties of alpha, beta and gamma radiations

Property	Alpha (α)	Beta (β^+ or β^-)	Gamma (γ)
Relative charge	+2	+1 or −1	0
Relative mass	4	$\frac{1}{1840}$	0
Typical penetration	5 cm air; thin paper	30 cm air; few millimetres of aluminium	highly penetrating; partially absorbed by thick, dense materials
Nature	helium nucleus	positron or electron	electromagnetic wave/ photon
Typical speed	$10^7\,m\,s^{-1}$ $\approx 0.1c$	$\approx 2.5 \times 10^8\,m\,s^{-1}$ $\approx 0.9c$	$3.00 \times 10^8\,m\,s^{-1}$ c
Notation	${}^4_2He$ or ${}^4_2\alpha$	${}^{0}_{-1}e$ or ${}^{0}_{-1}\beta$	γ or ${}^0_0\gamma$
Ionizing ability	very high	light	very low
Absorbed by	piece of paper	3 mm aluminium	intensity halved by 2 cm lead

Patterns of radioactive decay

The decay of a single unstable nucleus is a spontaneous and random event. It is unpredictable in a similar way that the result of tossing a single coin is unpredictable. However, we can be sure that if we toss a large number of coins, about 50 per cent will be heads and about 50 per cent will be tails. Similarly, we can say that when we have a large number of undecayed nuclei, about 50 per cent of them will decay in a certain time. Even the tiniest sample of a radioactive material contains a very large number of nuclei, and this means that the random process of radioactive decay can be predicted.

Half-life

The half-life of a radioactive nuclide is the time taken for half of any sample of unstable nuclei to decay.

Half-life is given the symbol $T_{\frac{1}{2}}$. The half-lives of different radioactive nuclides can vary enormously, from fractions of a second to billions of years. See Table 7.3 for some examples.

■ **Table 7.3** Examples of half-lives

Radioactive nuclide	Half-life
uranium-238	4.5×10^9 years
radium-226	1.6×10^3 years
radon-222	3.8 days
francium-221	4.8 minutes
astatine-217	0.03 seconds

After each half-life has elapsed, half of the remaining unstable nuclei will decay in the next half-life. As a much simplified example, if there were 64 (billion) nuclei at the start, after each successive half-life, the number remaining would be about 32, 16, 8, 4, 2 ... Theoretically the number will never reduce to zero.

■ **Figure 7.20** The pattern of radioactive decay can be simulated by throwing dice

The process of radioactive decay can be simulated by the throwing of dice (Figure 7.20). Consider the following experiment in which a group of students start by throwing 6000 dice. Each time a six is thrown, that dice is removed. The results for the number of dice after each throw are shown in Table 7.4.

■ **Table 7.4** Results of dice-throwing experiment

Number of throws	Number of dice remaining	Number of dice removed
0	6000	0
1	5020	980
2	4163	857
3	3485	678
4	2887	598
5	2420	467
6	2009	411
7	1674	335
8	1399	75

Figure 7.21 is a graph of the number of dice remaining plotted against the number of throws. This is an example of a *decay curve*. The rate at which dice (with a six) are removed decreases because fewer sixes occur when fewer dice are thrown. Reading values from the graph shows that approximately 3.8 throws would be required to halve the number of dice. After another 3.8 throws the number of dice would halve again. The 'half-life' of the process is 3.8 throws.

Alternatively a graph of the number of dice removed against number of throws could be drawn. This graph will have the same shape and same half-life as the graph shown in Figure 7.21.

The dice experiment is a useful model (analogy) for representing radioactive decay, with the dice representing 6000 radioactive nuclei and the rate of decay decreasing with time.

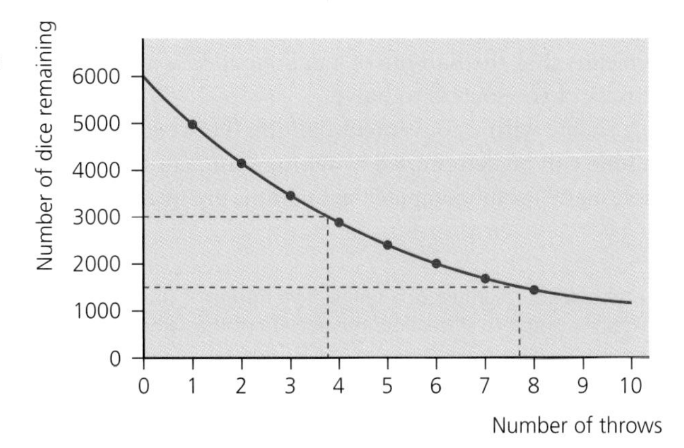

■ **Figure 7.21** Graph of the number of dice remaining against the number of throws

The graph in Figure 7.21 is an example of a quantity undergoing decay in which the rate of decay (modelled by the dice) is directly proportional to the amount remaining (number of dice). This is known as an *exponential decay*. The mathematics of exponential decay is not needed here, but it is described in Chapter 12 and in IB mathematics courses.

Figure 7.22 shows a typical graph representing a radioactive decay with a half-life of $T_{\frac{1}{2}}$. It has the same shape as Figure 7.21. A value for the half-life can be determined by choosing *any* value and finding the time taken for that value to halve. Accuracy can be improved by choosing several pairs of values and calculating an average.

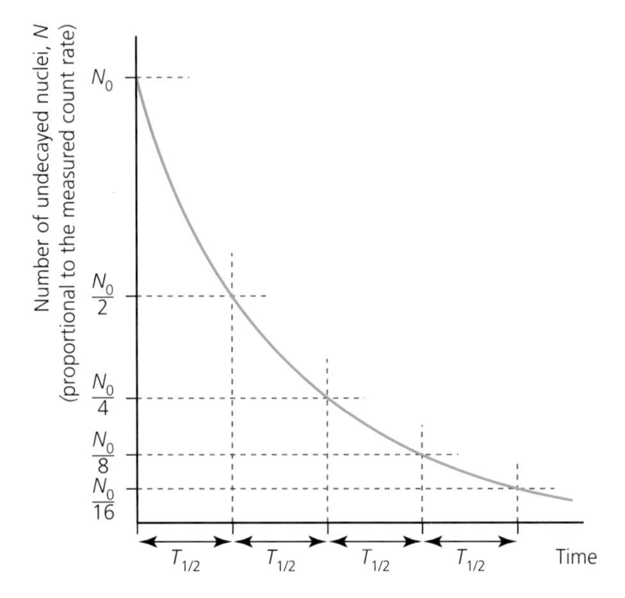

■ **Figure 7.22** Radioactive decay curve

For example, radium-226 (an alpha emitter) has a half-life of 1620 years. This means that if we start with 1 g of pure radium-226, then 0.5 g of it will have undergone radioactive decay in 1620 years. After another 1620 years (another half-life), half of the radium-226 atoms remaining will have decayed, leaving 0.25 g, and so on.

Experimental determination of half-life

Counting numbers of undecayed nuclei in a school's radioactive source is not possible, so we need to use changes in count rates to determine half-lives. The count rate from a radiation counter placed close to a radioactive source will be an indication of the rate of decay in the source. (The actual number of decays occurring every second in the source is called its *activity* and has the unit *Bequerel, Bq*. This is discussed in Chapter 12.)

As stated earlier, the rate of radioactive decay is proportional to the number of atoms still undecayed. This means that the half-life of a radionuclide is also equal to the time taken for the count rate (or activity of the source) to halve.

If a radioactive source with a convenient half-life (between a few minutes and a few hours) is available, the half-life can be determined by taking sufficient count rate measurements to plot a decay curve. If not, many useful computer simulations are available.

28 The initial count rate from a sample of a radioactive nuclide is $8000\,s^{-1}$. The half-life of the nuclide is 5 minutes. Sketch a graph to show how the activity of the sample changes over a time interval of 25 minutes.

29 The table below shows the variation with time, t, of the count rate of a sample of a radioactive nuclide X. The average background count during the experiment was $36\,min^{-1}$.

t/hour	0	1	2	3	4	5	6	7	8	9	10
Count rate/min^{-1}	854	752	688	576	544	486	448	396	362	334	284

a Plot a graph to show the variation with time of the corrected count rate.
b Use the graph to determine the half-life of nuclide X.

30 Explain why it might be difficult for a laboratory to provide a radioisotope with a half-life of, for example, 10 minutes.

Radioactive decay problems

If the half-life of a nuclide is known then the number of undecayed nuclei, or the count rate, after successive half-lives can be determined if the initial values are known.

Figure 7.23 illustrates the radioactive decay of a sample of americium-242, which has a half-life of 16 hours. Initially there are 40 million undecayed nuclei. The grey circles represent undecayed nuclei and the orange circles represent decayed nuclei. With each half-life, half of the remaining undecayed nuclei decay.

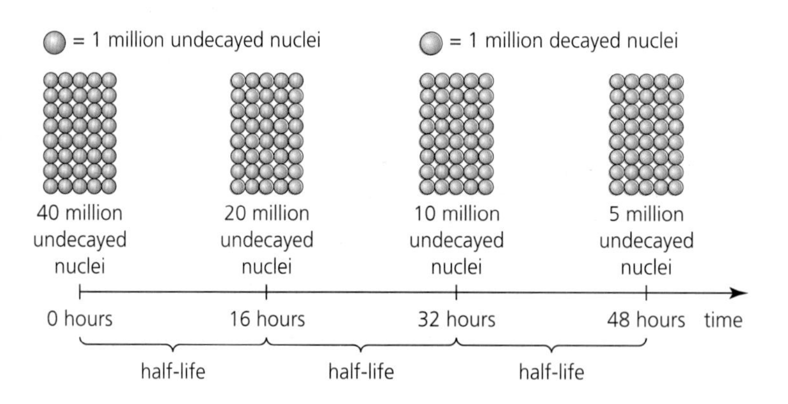

■ **Figure 7.23** The radioactive decay of a sample of americium-242

This process of radioactive decay is summarized in numerical form in Table 7.5.

■ **Table 7.5** The radioactive decay of a sample of americium-242

Number of undecayed nuclei	Fraction of original undecayed nuclei remaining	Number of decayed nuclei	Number of half-lives elapsed	Number of hours elapsed
40×10^6	1	0	0	0
20×10^6	½	20×10^6	1	16
10×10^6	¼	30×10^6	2	32
5.0×10^6	⅛	35×10^6	3	48
2.5×10^6	$\frac{1}{16}$	37.5×10^6	4	64

Worked examples

8 The half-life of francium-221 is 4.8 minutes.

a Calculate the fraction of a sample of pure francium-221 remaining undecayed after a time of 14.4 minutes.

b What fraction of the original sample will have decayed?

a The half-life of francium-221 is 4.8 minutes, so 14.4 minutes is three half-lives (14.4 ÷ 4.8). After one half-life, half of the sample will remain undecayed.

After two half-lives, 0.5 × 0.5 = 0.25 of the sample will remain undecayed.

After three half-lives, 0.5 × 0.25 = 0.125 of the sample will remain undecayed.

Hence after 14.4 minutes the fraction of francium-221 remaining undecayed is 0.125, or $\frac{1}{8}$.

b $1 - \frac{1}{8} = \frac{7}{8}$

9 Caesium-137 has a half-life of 30 years. If the initial count rate from a source was $10^4\,s^{-1}$, what is the count rate after:

a 120 years

b 135 years?

a 120 years is equivalent to four half-lives, so the count rate will have decreased by a factor of 16 (2^4) and will be $10^4/16 = 630\,s^{-1}$.

b 135 years is equivalent to 4.5 half-lives, so we know that the count will be between 625 and 313, but it is not midway between these values because decay is not linear. The mathematics needed to make this calculation is given in Section 12.2. (The correct answer is $442\,s^{-1}$.)

---

31 A radioactive substance has a half-life of five days and the initial count rate is $500\,min^{-1}$. If the background count was found to be $20\,min^{-1}$, what will be the count rate after 15 days?

32 a The half-life of francium-221 is 4.8 minutes. Calculate the fraction of a sample of francium-221 remaining undecayed after a time of 24.0 minutes.

b The half-life radon-222 is 3.8 days. Calculate the fraction of a sample of radon-222 that has decayed after 15.2 days.

c Cobalt-60 is used in many applications in which gamma radiation is required. The half-life of cobalt-60 is 5.26 years. A cobalt-60 source has an initial activity of 2.00×10^{15} decays s^{-1}. What will be its activity after 26.30 years?

d A radioactive element has a half-life of 80 minutes. How long will it take for the count rate to decrease to 250 per minute if the initial count rate is 1000 per minute?

e The half-life of radium-226 is 1601 years. For an initial sample:

i what fraction has decayed after 4803 years?

ii what fraction remains undecayed after 6404 years?

33 Technetium-99 is a radioactive waste product from nuclear power stations. It has a half-life of 212 000 years.

a Estimate the percentage of technetium that is still radioactive after one million years.

b Approximately how many years are needed for the activity from the technetium to fall to 1 per cent of its original value?

Decay series

Heavy radioactive nuclides, such as radium-226 and uranium-238, cannot become stable by emitting just one particle. They undergo a radioactive **decay series**, producing either an alpha or a beta particle and maybe gamma radiation during each step, until a stable nuclide is formed. For example, uranium-238 undergoes a decay series (see Figure 7.24) to eventually form stable lead-206. Each decay will have its own particular half-life.

If it was possible to have a source of pure uranium-238, for example, it would immediately start decaying into other nuclides and, after some time, all the nuclides in the decay series would be present in the sample. The relative proportions of different nuclides depends on their half-lives. After a *very* long time most of the source will have turned into lead. In other words, we should expect many radioactive sources (of the heavier elements particularly) to contain a range of different nuclides.

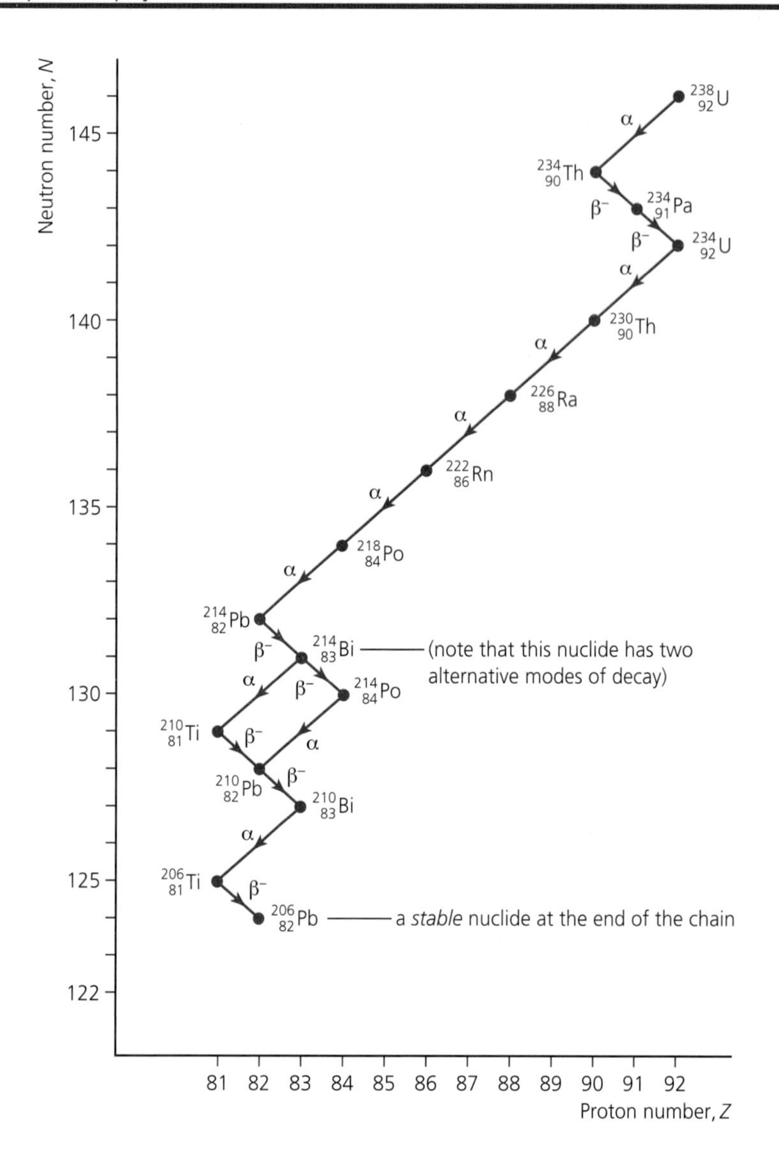

■ Figure 7.24 An example of a decay series (uranium-238) on a chart of the nuclides

Figure 7.24 shows a small part of an overall N–Z plot. Readers are advised to inspect a full neutron number–proton number plot, which is commonly known as a *chart of the nuclides* because it has a space for every known nuclide. Stable and unstable nuclides are clearly indicated and the kind of radiation emitted is often shown.

Figure 7.25 represents how the chart position of a nuclide changes during alpha decay and beta-negative decay (beta-positive decay is in the opposite direction). The movement on the chart can be seen as an attempt to move towards the line of stability. Gamma rays do not cause transmutation.

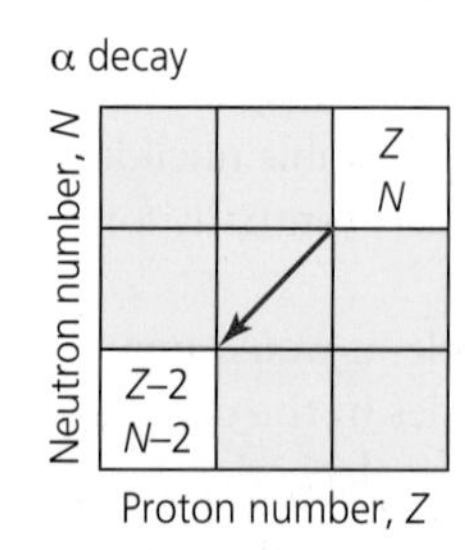

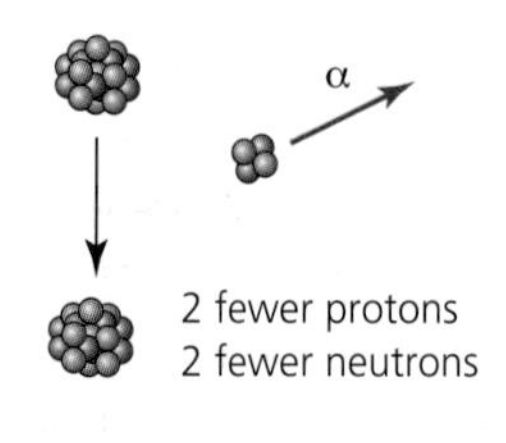

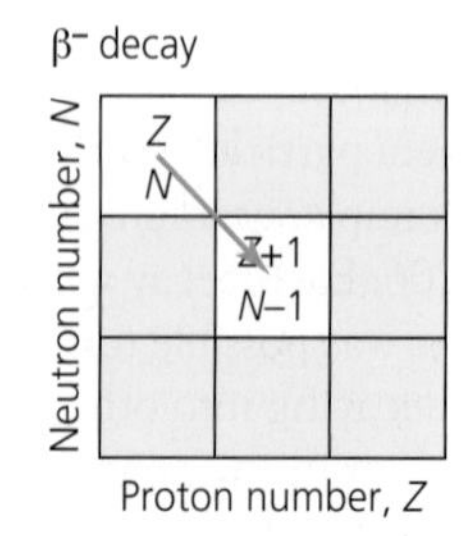

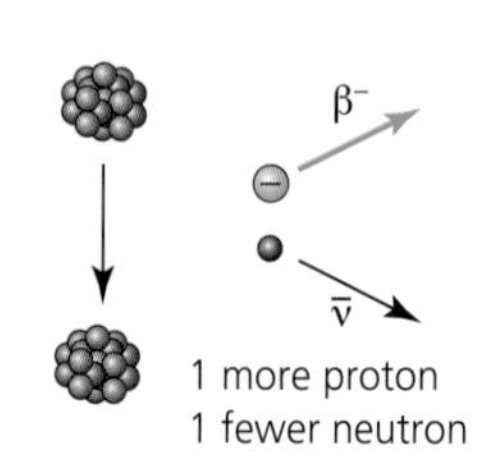

■ Figure 7.25 Transmutations on a chart of the nuclides

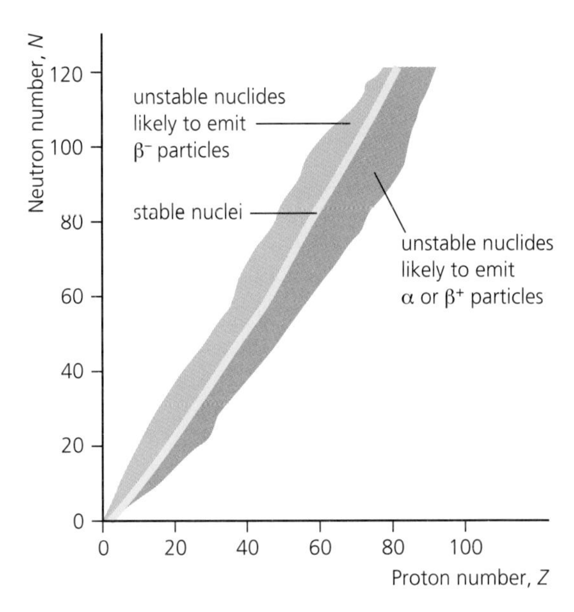

Figure 7.26 Stability and transmutations on a chart of the nuclides

Uses of radionuclides

Radioactive substances have a wide range of uses and, although these are not directly included in the IB course, students are recommended to research one or two of the following applications:

- diagnosis of illness
- treatment of disease
- 'tracers'
- food preservation
- sterilization of medical equipment
- determining the age of rocks
- locating faults in metal structures, such as pipes
- carbon dating

Utilizations

Carbon dating

Archaeological specimens (which were previously living plants or animals) with ages up to about 50 000 years can be dated using the isotope carbon-14. Carbon-14, which has a half-life of 5730 years, is constantly being formed in the upper atmosphere. When cosmic rays enter the atmosphere they can produce neutrons. The nuclear reaction shown below can then take place:

$$^{14}_{7}N + ^{1}_{0}n \rightarrow ^{14}_{6}C + ^{1}_{1}H$$

This type of reaction is an example of a natural transmutation in which a nuclide absorbs another smaller particle into its nucleus and undergoes a nuclear reaction to form a new nuclide.

The carbon-14 (present in the atmosphere as carbon dioxide, $^{14}CO_2$) is taken in by plants during photosynthesis. The plants are eaten by animals, which may then be eaten by other animals. However, once the animal or plant dies no more radioactive carbon-14 is taken in and the percentage in the remains starts to decrease due to radioactive decay (Figure 7.27).

This means that if the ratio of carbon-14 to carbon-12 is known then the age of the specimen can be determined. The activity of carbon-14 in living materials is about 19 counts per minute for each gram of specimen. This technique of carbon dating can be used to determine the ages of animal remains and wood, paper and cloth. It has also provided supporting evidence that there has been change (evolution) in groups of organisms over time.

■ **Figure 7.27** Carbon-14 is incorporated into living tissue

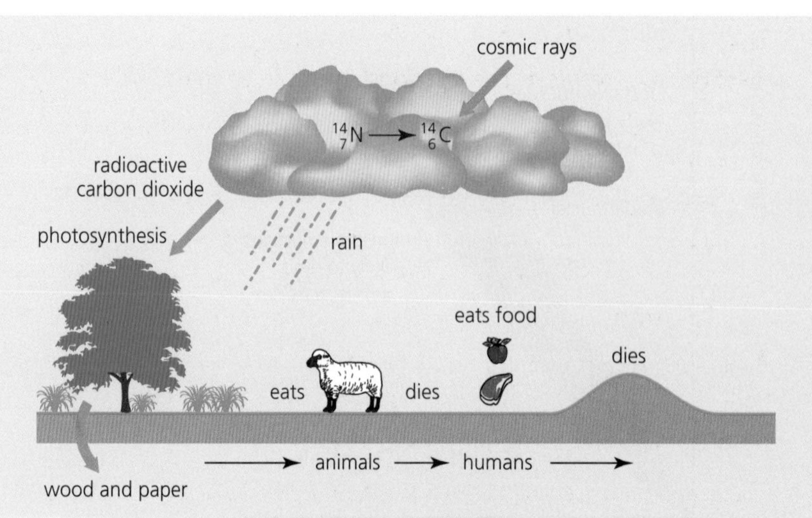

1 Find out about at least one specific example of the use of carbon dating to determine the age of an historical artefact.

2 Sketch a graph to show how the carbon-14 count rate from a sample might vary over about 20 000 years.

Utilizations

Medical tracers

Nuclear medicine involves the use of radioactive substances to detect (or treat) abnormalities in the function of particular organs in the body. Substances introduced into the body for this purpose are called *tracers*. They may be injected or ingested (see Figure 7.28). The radioactive substance most commonly used is technetium-99m. This is an excited atom produced from molybdenum-99 by beta decay. The decay process has a half-life of six hours and the gamma photons have an energy of 0.14 MeV. Both of these properties make technetium-99m a good choice for tracer studies. The amount of energy carried by the gamma photons makes them easy to detect (using a gamma camera). The half-life of six hours is a good compromise between activity lasting long enough to be useful in nuclear medicine and the wish to expose the patient to ionizing radiation for as short a time period as possible. The technetium is usually used to 'label' a molecule that is preferentially taken up by the tissue. This illustrates how knowledge of radioactivity, radioactive substances and the radioactive decay law are crucial in modern nuclear medicine.

■ **Figure 7.28** Injecting a radioactive tracer

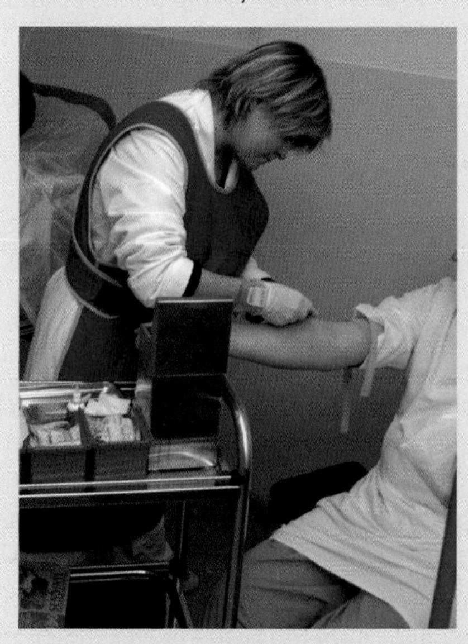

1 Write a nuclear equation showing the formation of technetium-99m.

7.2 Nuclear reactions – *energy can be released in nuclear decays and reactions as a result of the relationship between mass and energy*

In this section we will concentrate on the energy that can be released when there are changes in the structure of nuclei. These changes include:

- spontaneous natural radioactive decays, as discussed in Section 7.1
- nuclear fission (splitting a nucleus into two smaller nuclei)
- nuclear fusion (joining two small nuclei to make one larger nucleus)
- other artificially induced reactions.

As we shall see, when energy is released the masses involved must decrease slightly, so we will begin by introducing the unified atomic mass unit.

The unified atomic mass unit

The kilogram is an inconveniently large unit for the masses of atoms and sub-atomic particles. The *unified atomic mass unit*, u, is the most widely used unit for mass on the microscopic scale. The unified atomic mass unit is approximately equal to the mass of a proton or neutron, but it is very precisely defined as follows:

The unified atomic mass unit is defined as one-twelfth of the mass of an isolated carbon-12 atom.

From Chapter 3 we know that, by definition, one mole of carbon-12 has a mass of 12.000 00 g and contains $6.022\,141 \times 10^{23}$ atoms (this is the value for N_A to seven significant figures).

Therefore, the mass of one carbon-12 atom $= \dfrac{12.000\,00}{6.022\,141 \times 10^{23}} = 1.992\,647 \times 10^{-23}$ and one unified atomic mass unit, $u = \dfrac{1}{12}(1.992\,647 \times 10^{-23}\,\text{g}) = 0.166\,054 \times 10^{-23}\,\text{g}$

To four significant figures, $1\,\text{u} = 1.661 \times 10^{-27}\,\text{kg}$.

This value is given in the *Physics data booklet*.

Expressed in unified atomic mass units, the *rest masses* of the proton, neutron and electron are:

- mass of the proton, $m_p = 1.007\,276\,\text{u}$ ($= 1.673 \times 10^{-27}\,\text{kg}$)
- mass of the neutron, $m_n = 1.008\,665\,\text{u}$ ($= 1.675 \times 10^{-27}\,\text{kg}$)
- mass of the electron, $m_e = 0.000\,549\,\text{u}$ ($= 9.110 \times 10^{-31}\,\text{kg}$)

All these values are given in the *Physics data booklet*.

The **rest mass** of a particle is the mass of an isolated particle that is at rest relative to the observer.

The relationship between mass and energy

According to Einstein's special theory of relativity, *any* mass, *m*, is equivalent to an amount of energy, *E*. Einstein said 'mass and energy are different manifestations of the same thing'. If the energy of any system increases (in any form), then its mass increases too. This is explained in more detail in Option A: Relativity.

The mass and the energy of a system are proportional, E/m = constant. The constant is the speed of light, *c*, squared, so that:

$$E = mc^2$$

This famous equation shows us that a very small mass is equivalent to an *enormous* amount of energy. For example, a mass of 1 g is equivalent to about $10^{14}\,\text{J}$. If we increase or decrease the energy of an everyday object by changing its potential energy, kinetic energy or internal energy, there will always be an accompanying change in mass, but it will be much too small to measure.

But on the atomic scale, relatively large amounts of energy are transferred from relatively small masses, so that the **energy–mass equivalence** becomes much more relevant and useful.

Whenever a nuclear reaction results in the release of energy from an atom, there must be a corresponding decrease in mass of the nucleus, Δm, which is equivalent to its loss of energy, ΔE. The nucleus has less energy, so it must have less mass.

$$\Delta E = \Delta mc^2$$

This equation is given in the *Physics data booklet*.

The energy equivalent of 1 u = $(1.6605 \times 10^{-27}) \times (2.9979 \times 10^8)^2 = 1.4924 \times 10^{-10}$ J

Converting joules to electronvolts: $(1.4924 \times 10^{-10})/(1.6022 \times 10^{-19}) = 9.315 \times 10^8$ eV

The energy equivalent of a mass of 1 u is 931.5 MeV.

For a larger scale example, when one kilogram of uranium-235 undergoes fission in a nuclear reactor, the total energy released from the nuclei is approximately 8×10^{13} J. This corresponds to a small, but measurable, decrease in overall mass:

$$\Delta m = \frac{\Delta E}{c^2} = \frac{8 \times 10^{13}}{(3.00 \times 10^8)^2} \approx 9 \times 10^{-4}\,\text{kg}$$

ToK Link

Paradigm shift in physics

The acceptance that mass and energy are equivalent was a major paradigm shift in physics. How have other paradigm shifts changed the direction of science? Have there been similar paradigm shifts in other areas of knowledge?

A physics **paradigm** is a widely accepted model and a way of thinking by which an aspect of the physical world is viewed and described. A **paradigm shift** is when one paradigm is discarded in favour of another. It often involves radical change.

For many centuries scientists had considered energy and mass to be distinct and unrelated to each other. But Einstein's proposal that mass and energy are equivalent to each other completely changed that way of thinking. It was a completely new paradigm. *Mass–energy equivalence* is fundamental to most important studies and research in modern physics, from particle physics and nuclear physics to cosmology.

One of the most famous paradigm shifts in science was the change from an Earth-centred model of the universe to a Sun-centred model (and then to a model without a centre). Each model rejected the one that preceded it as fundamentally flawed, and it seems that science cannot willingly have conflicting models to describe the same thing. The same may not be true of other areas of knowledge.

An alternative unit for atomic-scale mass

Considering the equivalence of mass and energy, we can reconsider the unit we use for mass. Since $m = Ec^{-2}$, a unit for mass can be expressed in terms of energy divided by the speed of light squared.

The SI unit for mass is then seen as $\text{J}/(\text{m s}^{-1})^2$, which is known as the kilogram.

But, as we have seen in atomic and nuclear physics, energy is often measured in electron volts, eV (keV or MeV). In this measurement system the unit of mass becomes MeV c^{-2}. This unit is not given any other name.

Worked example

10 The rest mass of a proton is 1.007 276 u.

a What is this mass in units of MeV c^{-2}?

b What is the energy equivalent of this mass?

a $1.007\,276 \times 931.5 = 938.3\,\text{MeV c}^{-2}$

b 938.3 MeV

34 Use Einstein's mass–energy equivalence relationship to calculate the energy equivalent of 500 g of matter.

35 An atom loses a mass of 2.2×10^{-30} kg after a nuclear reaction. Determine the energy obtained in joules due to this loss of nuclear mass.

36 Calculate the increase in mass when 1.00 kg of water absorbs 4.20×10^{4} J of energy to produce a temperature rise of 10.0° C.

37 The mass decrease for the decay of one radium atom is 8.5×10^{-30} kg. Calculate the energy equivalent in MeV.

Mass defect and nuclear binding energy

In any nucleus, strong nuclear forces hold the nucleons together, so it can be described as a *bound system*. If we wanted to separate any or all of the nucleons (in a particle physics thought experiment) we would have to supply energy to the system. See Figure 7.29.

The **binding energy** of a nucleus is the amount of energy needed to completely separate *all* of its nucleons.

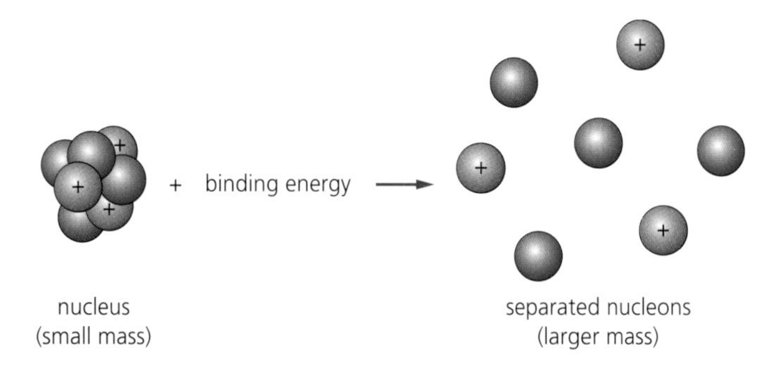

Figure 7.29 Binding energy is needed to separate nucleons; this example is lithium-7

Alternatively, binding energy can be interpreted as the energy *released* when a nucleus is formed from its nucleons. The binding energy is equivalent to the potential energy in the nucleus.

The energy needed to remove one nucleon from the nucleus is approximately equal to the binding energy divided by the number of nucleons. This is known as the **average binding energy per nucleon**. Nuclei with high values for their average binding energy per nucleon are considered to be the most stable. (Nuclear stability should not be confused with chemical stability.) Nickel-62 is the nuclide with the highest average binding energy per nucleon.

Supplying energy to a system means that the mass of the system must increase (energy–mass equivalence), so the total mass of the separated nucleons must be higher than their combined mass when they were together in the nucleus.

Consider a numerical example: the mass of a carbon-12 atom is (by definition) 12.000 000 u. The atom is composed of six protons, six neutrons and six electrons. However, the sum of the separate rest masses of the sub-atomic particles is more than 12.000 000 u as follows:

$$(6 \times 1.007\,276\,\text{u}) + (6 \times 1.008\,665\,\text{u}) + (6 \times 0.000\,549\,\text{u}) = 12.098\,94\,\text{u}$$

When these particles joined together to make a carbon-12 atom there was a reduction in mass of 0.09894 u and an equivalent increase in nuclear potential energy. This difference in mass is called the *mass defect* of the atom. More usually this term is just applied to a nucleus.

The **mass defect** of a nucleus is the difference between the mass of the nucleus and the sum of the masses of its nucleons if separated.

The binding energy and mass defect of a nucleus are equivalent to each other.

Worked examples

11 A particular nucleus has a mass defect of 0.369 u.
a What is its binding energy in MeV?
b If it contains 40 nucleons, what is the average binding energy per nucleon?

a $0.369 \times 931.5 = 344\,\text{MeV}$

b $\frac{344}{40} = 8.6\,\text{MeV}$

12 Calculate the mass defect (in electronvolts) and binding energy of a helium atom (4.00260 u). It consists of two protons (each of mass 1.007276 u), two neutrons (each of mass 1.008665 u) and two electrons (each of mass 0.000549 u). 1 u = 931.5 MeV c^{-2}.

The total mass of the individual particles = $(2 \times 1.007\,276) + (2 \times 1.008\,665) + (2 \times 0.000\,549)$
$= 4.032\,98\,\text{u}$

Mass defect = $4.032\,98 - 4.002\,60 = 0.030\,38\,\text{u}$

$\Delta E = 0.030\,38 \times 931.5 = 28.30\,\text{MeV}$

Variations in nuclear stability (average binding energy per nucleon)

A plot of binding energy per nucleon against nucleon number shows some important variations. Clearly larger nuclei have higher overall binding energies, and we might expect that the binding energy was approximately proportional to the number of nucleons, so that the average binding energy per nucleon was constant. However Figure 7.30 shows that, although a value of between 7 and 9 is typical for most nuclides, a clear pattern is seen with a peak at nickel-62.

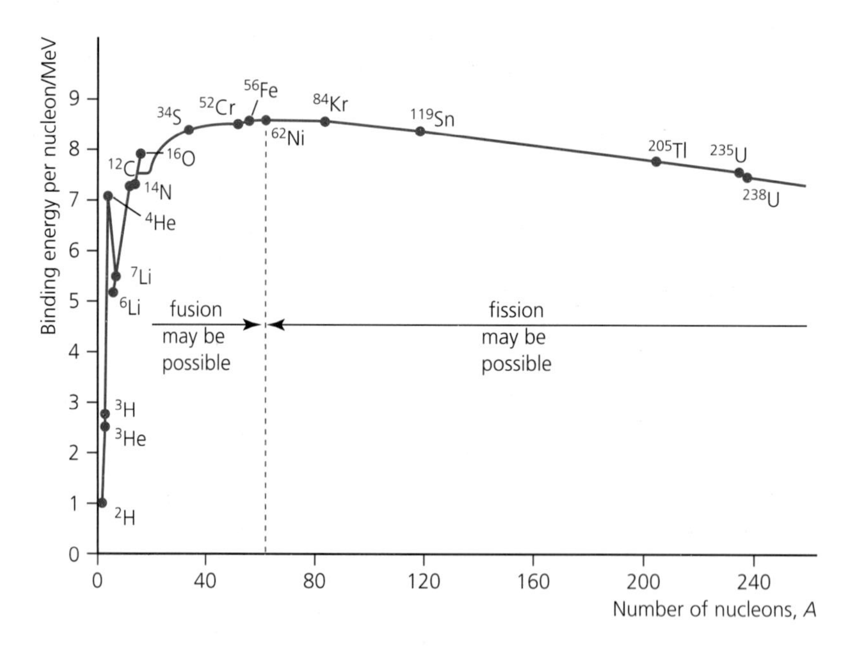

■ **Figure 7.30** A plot of binding energy per nucleon against number of nucleons

This chart is sometimes drawn the other way up, with *negative* binding energies. The use of negative binding energies is consistent with systems in which separated particles have zero energy, and bound particles have negative binding energy, representing the fact that energy must be provided to the system in order for it to reach zero energy. As other examples, remember that the electron energy levels within atoms are given negative values and that gravitational potential energies are always negative.

Worked example

13 Determine the binding energy per nucleon of the nucleus of $^{12}_{6}C$.

Total rest mass of individual protons and neutrons:

$(6 \times 1.007276\,u) + (6 \times 1.008665\,u) = 12.095646\,u$

Mass defect:

$12.095646\,u - 12.000000\,u = 0.0956465\,u$

Binding energy per nucleon:

$$\frac{0.0956465 \times 931.5}{12} = 7.425\,\text{MeV per nucleon}$$

38 The binding energy of a carbon-14 nucleus is 102 MeV. Calculate the binding energy per nucleon.

39 Calculate the binding energy, in MeV, of a nitrogen-14 nucleus with a mass defect of 0.108517 u.

40 The mass of a nickel-62 atom is 61.92835 u. Calculate its average binding energy per nucleon (Proton number of nickel is 28).

41 a Estimate the average binding energy per nucleon of uranium-235 from Figure 7.30.
b Calculate a value for the total binding energy of the nuclide.

Nature of Science

Patterns, trends and discrepancies

Searching for patterns in measurements and observations is a major preoccupation of scientists and there are two excellent examples in this chapter, both of which have been represented graphically: Figures 7.14 and 7.30. These two charts are presented as empirical results, without any theoretical background. Of course, such patterns demand explanations but, even without those, they provide the information needed to make very useful predictions about nuclide stability. As a further example of the usefulness of patterns, if we check further into the observed patterns of nuclear stability, we find that, for example:

- Nuclides containing odd numbers of both protons and neutrons are the least stable.
- Nuclides containing even numbers of both protons and neutrons are the most stable.
- Nuclides that contain odd numbers of protons and even numbers of neutrons are less stable than nuclides that contain even numbers of protons and odd numbers of neutrons.

Nuclear fission and nuclear fusion

Nuclear fission

Nuclear fission is the splitting of a heavy nucleus into two lighter nuclei. If fission can be made to happen, it is accompanied by the release of a large amount of energy (kinetic energy of the particles and the electromagnetic energy of photons). The two newly-formed nuclei are more stable than the previous single nucleus, and this is shown by the fact that they have a higher average binding energy per nucleon. See Figure 7.31a.

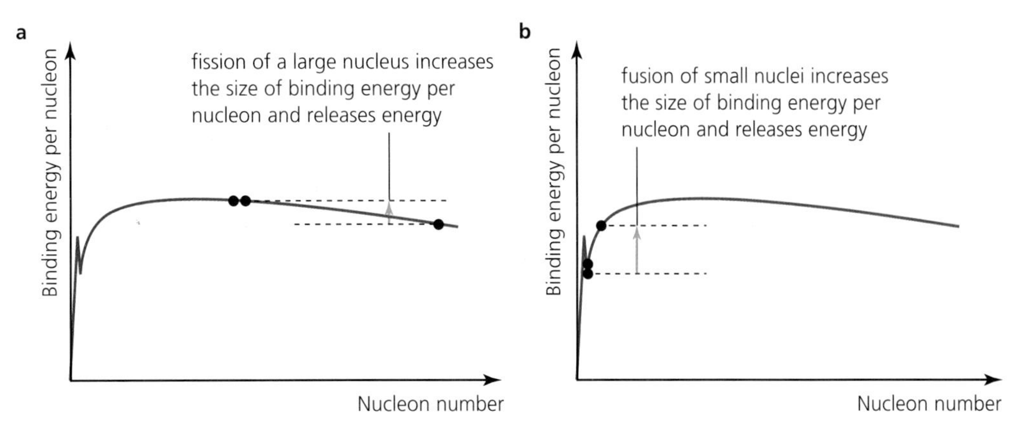

Figure 7.31 Any nuclear process that increases the size of average binding energy per nucleon releases energy

The most common fission reaction is instigated by the bombardment of uranium-235 nuclei by slow-moving neutrons. This can result in the capture of a neutron and the formation of uranium-236. This is an unstable nuclide that undergoes fission readily. A typical fission reaction is as follows:

$$^{1}_{0}n + ^{235}_{92}U \rightarrow ^{236}_{92}U \rightarrow ^{141}_{56}Ba + ^{92}_{36}Kr + 3^{1}_{0}n + \text{photons}$$

The neutrons released may be absorbed by other uranium-235 nuclei and induce additional fission reactions. Nuclear fission reactors make use of *controlled* fission (see Chapter 8). Atomic bombs make use of *uncontrolled* fission.

The risks associated with nuclear power

Because of the possible risks involved, the use of nuclear power and the possible development of nuclear weapons has been a major issue worldwide for the past 70 years. Most people have strong opinions on this subject and a few countries have nuclear policies that can cause major disagreements with their neighbours and other countries. In a world with an ever-increasing demand for energy supplies, nuclear power has significant advantages, but many people consider that the risks are too high. International discussion and collaboration are needed to avoid conflict, but who makes the final decisions? Nuclear power is discussed in much greater detail in Chapter 8.

Nuclear fusion

Nuclear fusion is the combination of two light nuclei to produce a heavier nucleus. Energy is released in the process. As with fission, the newly-formed nucleus is more stable than the previous two nuclei, as shown by the fact that it has a higher average binding energy per nucleon. See Figure 7.31b.

An example is the fusion of two hydrogen-2 nuclei (deuterium) to produce a helium-3 nucleus:

$$^{2}_{1}H + ^{2}_{1}H \rightarrow ^{3}_{2}He + ^{1}_{0}n$$

Temperatures in the region of 10^8 kelvin are required to provide the fusing nuclei with sufficient kinetic energy to overcome their mutual repulsion. Experimental nuclear fusion reactors make use of controlled nuclear fusion, but they have not been able to produce a sustainable power output. There is more about this in Chapter 8.

Calculating energy released during nuclear reactions

Radioactive decay

As an example, consider the alpha decay of radium-226. We will assume that the radium atom has no significant kinetic energy.

$$^{226}_{88}Ra \rightarrow ^{222}_{86}Rn + ^{4}_{2}He$$

Consider the masses on both sides of this equation:

- Rest mass of radium = 226.0254 u
- Rest mass of radon = 222.0176 u
- Rest mass of alpha particle = 4.0026 u

A simple calculation shows that the mass after the reaction is 0.0052 u less than before. The same result will be obtained whether we consider the masses of atoms or the masses of nuclei, because the masses of the electrons remain unchanged. We know that the total mass–energy must be the same before and after the decay (conservation of energy), so the difference in mass must have been converted to the kinetic energy of the alpha particle and the radon nucleus); 0052 u is equivalent to 4.84 MeV.

Because the alpha particle has a much smaller mass than the radium nucleus we know, from the law of conservation of momentum, that it will move much faster and carry most of the kinetic energy. The actual energy of the alpha particle is 4.7 MeV.

One mole of radium (226 g) will release a total energy of:

$$6.02 \times 10^{23} \times 4.71 = 2.84 \times 10^{24}\,\text{MeV}$$

This is a lot of energy (4.54×10^{11} J), but remember that the energy will be released over a very long time because the half-life of radium-226 is about 1600 years. Radionuclides are not generally used to transfer large amounts of energy because they are both low power and very expensive, but they can provide energy for a long period of time. Alpha sources can be used to generate small amounts of electrical energy in places that are difficult to access, such that replacing a power source would be problematic. This includes some uses on satellites and space probes.

But another kind of nuclear reaction is capable of transferring large amounts of energy quickly – nuclear fission.

Nuclear fission

As an example, consider this fission of uranium-235 into nuclides of barium and krypton, similar to that discussed before:

$$^{1}_{0}n + ^{235}_{92}U \rightarrow ^{236}_{92}U \rightarrow ^{144}_{56}Ba + ^{89}_{36}Kr + 3^{1}_{0}n + \text{photons}$$

- Rest mass of neutron = 1.0087 u
- Rest mass of uranium-235 = 235.0439 u
- Rest mass of barium-144 = 144.9229 u
- Rest mass of krypton-89 = 88.9178 u

Rest mass on left-hand side of the equation = 236.0526 u
Rest mass on right-hand side of the equation = 235.8668 u
Mass difference = 0.1858 u

This mass difference is equivalent to 173.1 MeV (0.1858×931.5). This amount of energy is released from each fission in the form of kinetic energy of the resulting nuclei and neutrons, and as photons. The fission of one mole of uranium-235 nuclei (235 g) in this way will release:

$$(173.1 \times 10^{6}) \times (6.02 \times 10^{23}) \times (1.6 \times 10^{-19}) = 1.7 \times 10^{13}\,\text{J}$$

As we will discuss in Chapter 8, the release of such a large amount of energy from a small amount of uranium can be controlled. The energy is mostly absorbed within the source and this raises its temperature such that steam can be produced and used to turn turbines that generate electricity.

Nuclear fusion

As an example, consider the example of a fusion reaction given before:

$$^{2}_{1}H + ^{2}_{1}H \rightarrow ^{3}_{2}He + ^{1}_{0}n$$

- Rest mass of hydrogen-2 = 2.0141 u
- Rest mass of helium-3 = 3.0161 u
- Rest mass of neutron = 1.0086 u

Rest mass on left-hand side of the equation = 4.0282 u
Rest mass on right-hand side of the equation = 4.0247 u
Mass difference = 0.0035 u

This mass difference is equivalent to 3.26 MeV (0.0035×931.5). This amount of energy is released from each fusion in the form of kinetic energy of the resulting nucleus and neutron. The fission of one mole of hydrogen-2 nuclei (2 g) in this way will release:

$$\frac{(3.26 \times 10^{6}) \times (6.02 \times 10^{23}) \times (1.6 \times 10^{-19})}{2} = 1.6 \times 10^{11}\,\text{J}$$

The energy that would be released from the fusion of one *kilogram* of hydrogen-2 is of the same order of magnitude as that released from the fission of one kilogram of uranium-235.

In order for two positively charged nuclei to fuse together they must have enough kinetic energy to overcome the repulsive electric forces between them. This requires extremely high temperatures. Much research continues into the technology that will allow this to happen in a controlled and sustained way. See Chapter 8.

Although nuclear fusion has only been achieved for relatively short times on Earth, it is the dominant energy transfer taking place in stars. The fusion of hydrogen into helium is the origin of the radiated energy that is released from stars, including the Sun. See Option D: Astrophysics.

42 Thorium-227 decays by alpha emission to radium-223 (proton number of thorium = 90, mass of thorium nucleus = 227.0278 u, mass of radon nucleus = 223.0186 u, mass of alpha particle = 4.0026 u).
a Write a nuclear equation for this decay.
b Calculate the energy (MeV) released in the decay.

43 Use data from the previous page to confirm that 'the energy that would be released from the fusion of one *kilogram* of hydrogen-2 is of the same order of magnitude as released from the fission of one kilogram of uranium-235'.

44 An alpha particle is emitted from radon-222 with energy of 5.49 MeV.
a Calculate the speed of the alpha particle.
b Determine the speed of the radon nucleus immediately after the decay – assume that it was at rest before. (Masses: alpha particle = 6.645×10^{-27} kg, radon nucleus = 3.687×10^{-25} kg)

45 One possible reaction taking place in the core of a nuclear reactor is:

$${}^{235}_{92}\text{U} + {}^{1}_{0}\text{n} \rightarrow {}^{95}_{42}\text{Mo} + {}^{139}_{57}\text{La} + 2{}^{1}_{0}\text{n} + 7{}^{0}_{-1}\text{e}$$

(Masses: uranium-235, 235.123 u; molybdenum-95, 94.945 u; lanthanum-139, 138.955 u; neutron, 1.009 u. 0.235 kg of uranium-235 contains 6×10^{23} atoms; 1 u = 931.5 MeV.) For this fission reaction, calculate:
a the mass (in atomic mass units) on each side of the equation (ignore the mass of electrons)
b the change in mass (in atomic mass units)
c the energy released per fission of a uranium-235 nucleus (in MeV)
d the energy available from the fission of 0.100 g of uranium-235 (in joules).

46 Two fusion reactions that take place in the Sun are described below.
- A hydrogen-2 nucleus absorbs a proton to form a helium-3 nucleus.
- Two helium-3 nuclei fuse to form a helium-4 nucleus plus two free protons.

For each nuclear reaction, write down the appropriate nuclear equation and calculate the energy released in MeV.

(Masses: hydrogen-2, 2.01410 u; helium-3, 3.01605 u; helium-4, 4.00260 u; proton, 1.00728 u and neutron, 1.00867 u; 1 u = 931.5 MeV)

Additional Perspectives

Comparing chemical and nuclear reactions

A nuclear reaction is said to occur whenever there is any change to the nucleus of an atom. Nuclear reactions will produce different elements if the number of protons changes. Chemical reactions always produce new substances, but they never involve changes in nuclei, so they do not create new elements. Chemical reactions involve the rearrangement of electrons and this typically involves much smaller amounts of energy than nuclear changes. Rates of chemical reactions are influenced by temperature and catalysts. Rates of nuclear reactions are unaffected by such factors, or the physical form of the substances involved.

Both chemical and nuclear reactions are involved in the production of nuclear power. The extraction of uranium from the ground and the processes preparing the fuel for the reactors necessitate chemical changes, whereas nuclear fission is obviously a nuclear reaction.

1 Find out why uranium hexafluoride is needed in the production of nuclear fuel.

2 Use the internet to compare (an order of magnitude) the energy that can be transferred when 1 kg of oil is burned in a conventional power station to when 1 kg of uranium-235 reacts in a nuclear power station.

7.3 The structure of matter – *it is believed that all the matter around us is made up of fundamental particles called quarks and leptons. It is known that matter has a hierarchical structure with quarks making up nucleons, nucleons making up nuclei, nuclei and electrons making up atoms and atoms making up molecules. In this hierarchical structure, the smallest scale is seen for quarks and leptons (10^{-18} m)*

The basic nature of matter and the different particles within it (and the forces between them) have always been a central theme of physics. With such knowledge, scientists are also able to develop an understanding of the beginnings of the universe and make predictions about how it will change in the future. This kind of research has been a source of fascination and motivation for many generations of scientists.

The basic proton–neutron–electron structure of the atom outlined at the beginning of this chapter was developed in the early years of the twentieth century and it still remains the most common visualization of an atom. This final section of the chapter will outline how this model has been refined in the light of more recent experimental evidence, up to and including the discovery of the Higgs boson in 2012. But we will begin with a discussion of one of the most famous of all physics experiments, the series of investigations that first led to the concept of the atom with a central nucleus – the Geiger–Marsden experiment.

■ The Geiger–Marsden experiment

The most common method of finding out about atomic structure is to direct a stream of fast-moving tiny particles at much larger atoms and see how the atoms and particles are affected by the 'collisions' (if at all). Modern nuclear research laboratories accelerate charged particles with electric fields, to give them as much energy as possible, before they are made to collide with other particles. This branch of physics is generally known as *high-energy physics*.

The first experiments investigating the structure of the atom took place just over 100 years ago but, at that time, modern particle accelerators were obviously not available and the most energetic particle beams were those produced in radioactive decay.

In 1909, Ernest Rutherford and two of his research students, Geiger and Marsden, working at the University of Manchester, UK, directed a narrow beam of *alpha particles* from a radioactive source at a *very thin* gold foil. A zinc sulfide detector was moved around the foil to determine the directions in which alpha particles travelled after striking the foil (Figure 7.32). The alpha particles have enough energy to emit a tiny flash of light when they are stopped by the zinc sulfide.

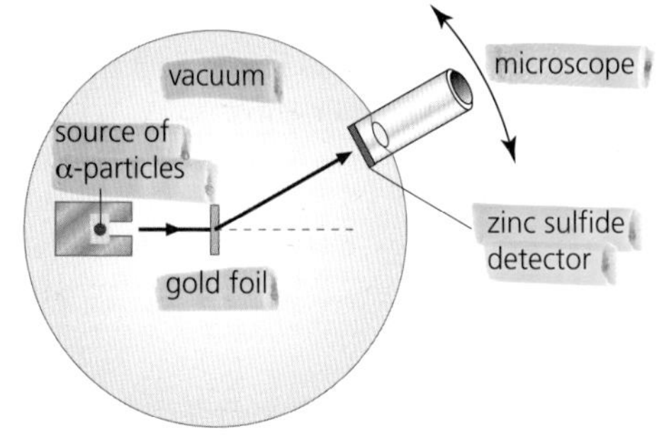

■ **Figure 7.32** The alpha particle scattering experiment

Alpha particles carry a very large amount of energy relative to their small size and, at that time, it was expected that the alpha particles would not be affected much by passing through such thin gold. Gold foil can be made very thin (less than 10^{-6} m) and may then only have about 3000 layers of atoms. Although alpha particles only travel about 4 cm in air, they would encounter many more molecules travelling that distance in air than passing through gold atoms in very thin foil. But Geiger and Marsden's results were surprising.

Rutherford published the results in 1911. He reported that:

- most of the alpha particles passed through the foil with very little or no deviation from their original path
- a small number of particles (about 1 in 1800) were deviated through an angle of more than about 10°
- an extremely small number of particles (about 1 in 10 000) were deflected through an angle larger than 90°.

At that time, physicists knew that atoms were not elementary particles because they contained electrons and emitted radiation. Many atomic structure theories were being put forward but none were really satisfactory until Rutherford had the insight to propose the nuclear model of the atom.

From alpha particle scattering experimental results Rutherford drew the following conclusions:

- Most of the mass of an atom is concentrated in a very small volume at the centre of the atom. Most alpha articles would therefore pass through the foil undeviated (continuing in a straight line) because most of the atom was empty space (a vacuum).
- The centre (he called it the *nucleus*) of an atom must be positively charged in order to repel the positively charged alpha particles. Alpha particles that pass close to a nucleus will experience a strong electrostatic repulsive force, causing them to change direction.
- Only alpha particles that pass very close to the nucleus, almost striking it head-on, will experience electrostatic repulsion large enough to cause them to deviate through angles larger than 90°. The fact that so few particles did so confirms that the nucleus is very small, and that most of the atom is empty space (a vacuum).

Figure 7.33 shows some of the possible trajectories (paths) of the alpha particles. Rutherford used his new nuclear model of the atom and Coulomb's inverse square law (covered in Chapter 5) to explain the repulsive force between the positively charged particles. He used the forces to calculate the fraction of alpha particles expected to be deviated through various angles. Rutherford's calculations agreed very closely with the results from the experiment, supporting his proposal of a nuclear model of the atom.

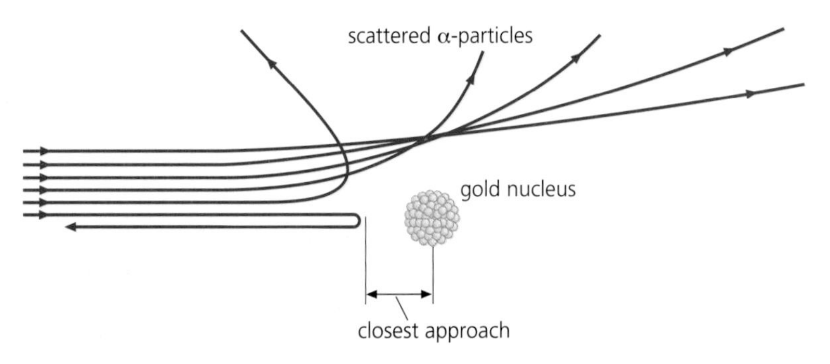

Figure 7.33 Alpha particle trajectories in the gold foil experiment

From his results, Rutherford calculated that the diameter of the nucleus is in the order of 10^{-15} m, compared to the diameter of the whole atom, which is in the order of 10^{-10} m. Figure 7.34 shows the features of the nuclear model of a nitrogen atom with approximate dimensions.

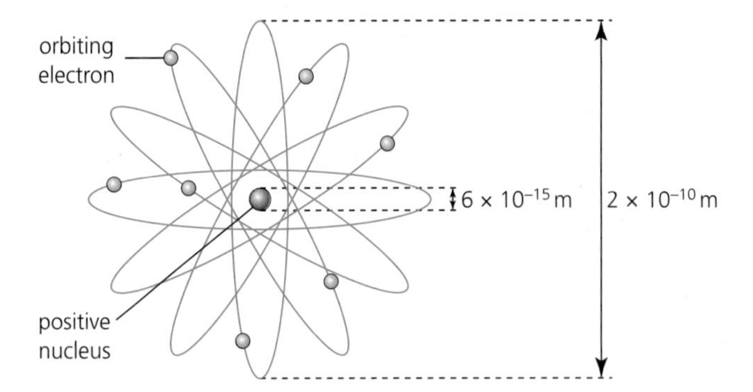

Figure 7.34 The structure and dimensions of a nitrogen atom, with seven electrons orbiting the positive nucleus. Note that the diameter of the atom is more than 30000 times larger than that of the nucleus

Alpha particle scattering can be modelled using simple apparatus such as that shown in Figure 7.35, in which a small ball rolls down a wooden ramp onto a specially shaped metal 'hill'. The shape of the hill is made so that, when viewed from above, the ball moves as if it was being repelled from the centre of the hill by an inverse square law of repulsion. In other words, gravitational forces are used to model electric forces. Using this apparatus it is possible to investigate how the direction in which a ball travels after leaving the hill (the scattering angle) depends on its initial direction ('aiming error') and/or its energy. In Geiger and Marsden's experiment it was not possible to observe the scattering paths of individual alpha particles, but the observed scattering pattern of large numbers of alpha particles in a beam is found to be in very close agreement with modelling based on individual balls rolling on hills.

■ **Figure 7.35** Alpha particle scattering analogue

47 Explain in fewer than 100 words (without using a diagram) why Rutherford concluded that atoms contain a small, positively charged central nucleus.

48 a Make a sketch of an alpha particle being deflected through 90° by a gold nucleus.
b On the same sketch draw the path of an alpha particle (approaching along the same path as before) being scattered by a copper nucleus (which has a lower proton number).
c Show how an alpha particle of higher energy would be affected if it approached a gold nucleus along the same path.

49 An alpha particle nearly collides with a gold nucleus and returns along the same path. Sketch a graph showing how the electric potential energy and kinetic energy possessed by the alpha particle vary with the distance of the alpha particle from the gold nucleus.

50 What would have happened if neutrons had been used in Rutherford's experiment instead of alpha particles? Explain your answer.

Additional Perspectives

The plum-pudding model

The first scientist to propose an internal structure for the atom was Joseph John ('J.J.') Thomson in 1904. This followed his discovery of the negatively charged electron a few years earlier from which it became clear that the atom was not an elementary, indivisible particle. Thomson proposed that an atom consists of a uniform sphere of positive charge in which a number of negatively charged electrons were randomly distributed. This was popularly known as the 'plum-pudding' model (Figure 7.36) and it assumed that the mass and positive charge of the atom was spread evenly over the entire atom.

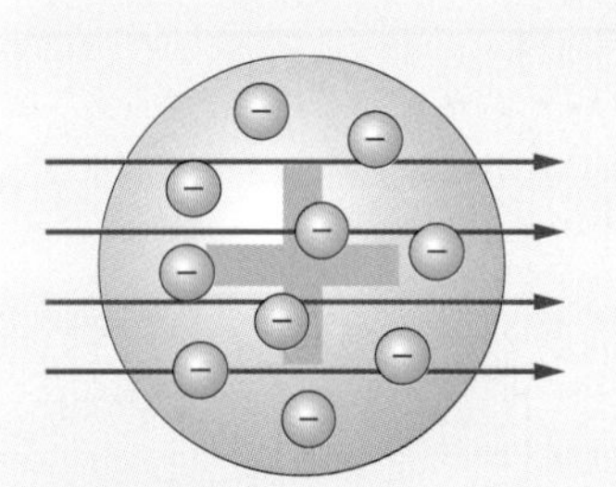

■ **Figure 7.36** Thomson's 'plum-pudding' model of atomic structure showing undeflected alpha particles

Thomson's model could explain the electrical neutrality of the atom, but could not be reconciled with Rutherford's scattering of alpha particles, discovered years later. Rutherford's nuclear model was a paradigm shift in thinking about atomic structure and Thomson's model was then abandoned.

If Thomson's model had been correct, the maximum deflecting force on the alpha particle as it passed through an atom would be small, because the positive and negative charges and mass are uniformly distributed in the sphere of the atom. The highly energized alpha particles would pass through mostly undeflected (Figure 7.36), as had been originally predicted.

1 Explain why the results of Geiger and Marsden's alpha particle scattering experiment do not support Thomson's 'plum-pudding' model of the atom.

■ Particle classifications

So far in this chapter we have referred to the following sub-atomic particles and radiation:

- protons
- neutrons
- electrons
- (electron) neutrinos
- photons.

And the following antiparticles (see comment below):

- positrons
- (electron) antineutrinos.

We will now discuss whether or not these are *elementary* particles and what other types of particles are found inside atoms. There are three types of sub-atomic particle – and this is the basis of the Standard Model of elementary particles. More details are provided later in this chapter.

- **Leptons:** electrons, neutrinos, antineutrinos and positrons are particles of low mass known as leptons. These particles have never been split apart into smaller particles and behave as point-like fundamental particles. Particles such as these, which have no detectable internal structure, are known as **elementary particles**. Two or more elementary particles of the same kind, for example two electrons, are completely identical.
- **Quarks:** the proton and the neutron are *not* elementary particles. They are made up of smaller elementary particles known as quarks and, as such, they are examples of **composite particles**.
- **Exchange particles**, also known as *gauge bosons*. As we shall see, the four fundamental forces (interactions) that were listed in Section 7.1 can be explained by the exchange of particles. For example, the photon is responsible for the electromagnetic interaction.

Within these three classifications there are a large number of composite, non-elementary particles, only some of which are mentioned in this book. Most of these can only be created in high-energy particle physics laboratories and they may be difficult to detect and/or are short lived (decay quickly into other particles). There is no requirement to remember their specific names or properties.

The naming of the different types of sub-atomic particles can be confusing. Figure 7.37 is presented for easy reference. It summarizes the names that will be used in the rest of this chapter.

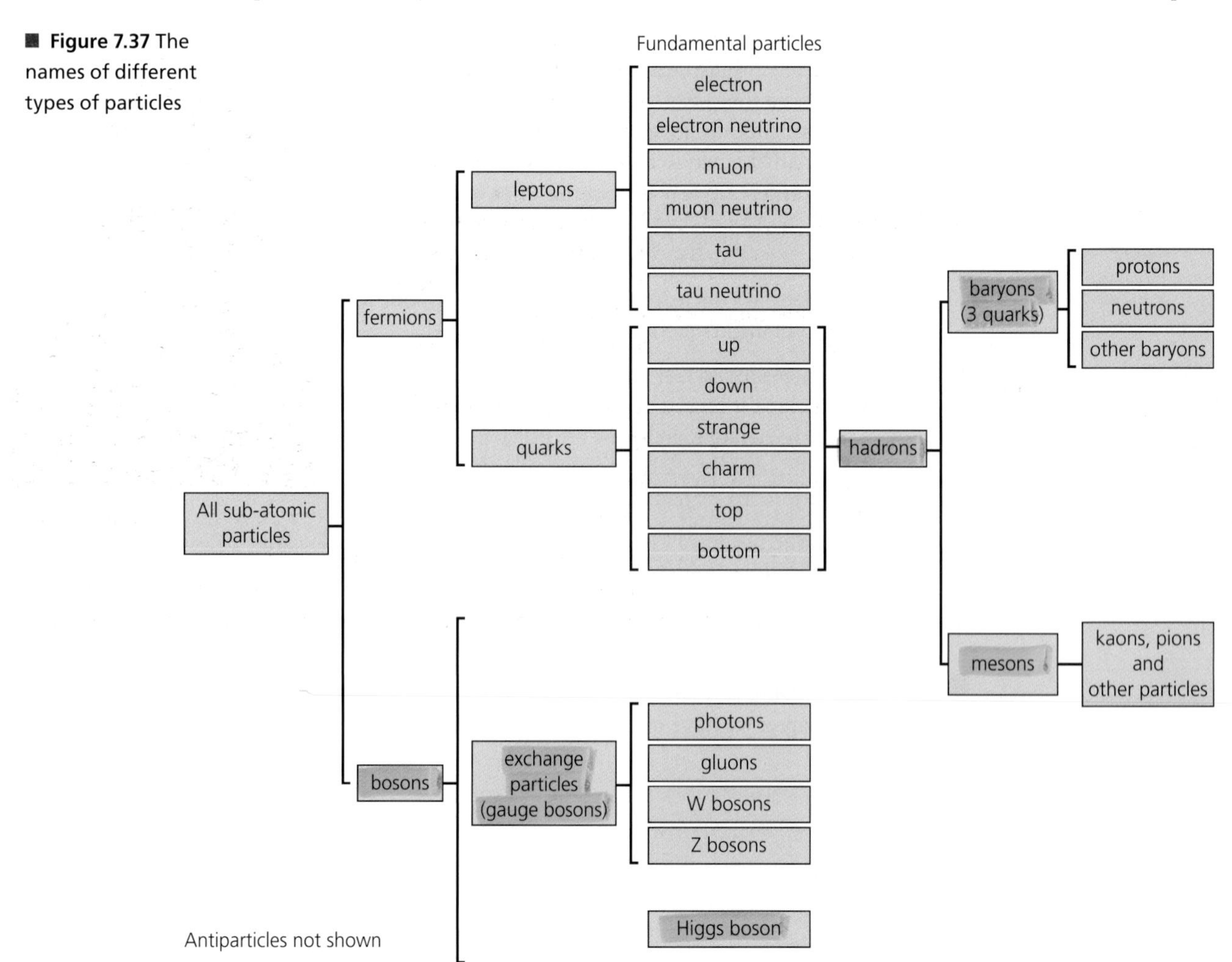

■ **Figure 7.37** The names of different types of particles

Particles and antiparticles

For every sub-atomic particle there is an **antiparticle** that has same mass as the particle, but the opposite charge (if it is charged) and opposite quantum numbers (see page 322).

For example, the electron, e^-, has a negative charge, while its antiparticle, the positron, e^+, has positive charge. Other antiparticles are typically denoted with a bar over the particle's symbol. In some cases, however, particle and antiparticle coincide; for example, the photon is its own antiparticle.

In 1932, the positron was the first antiparticle to be discovered when a particle with the mass of an electron was observed to be deflected by a magnetic field in the opposite direction (to an electron). The antiproton was discovered more than 20 years later.

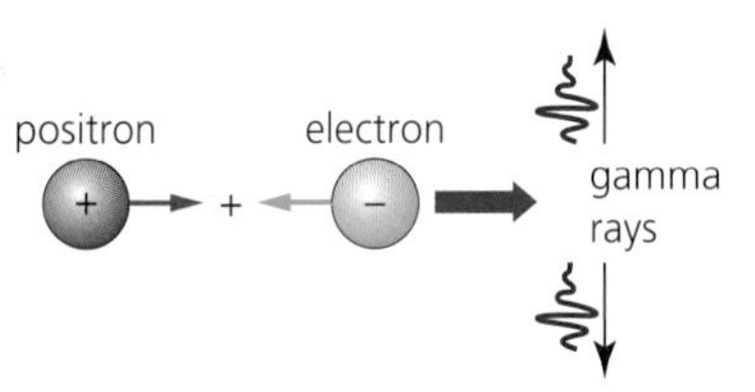

Figure 7.38 Annihilation of an electron–positron pair

Antimatter is material that is made up of antiparticles. If antimatter comes into contact with matter, they will **annihilate** (Figure 7.38), turning all the mass into energy (according to $E = mc^2$). For example, in a collision of an electron and a positron, a pair of high-energy photons (gamma rays) are created. The reverse of this process is also possible and is an example of *pair production* (see page 517).

Particle physics theories based on perfect symmetry predict that the early universe after the Big Bang would contain equal numbers of particles and antiparticles. Today, however, physicists observe a universe that consists almost entirely of matter and very little antimatter. The reasons for this are still not fully understood.

Quarks, leptons and their antiparticles

Quarks

By the late 1960s a large number of different sub-atomic particles had been identified and physicists were questioning whether they were all elementary or made of a much lower number of smaller particles combined in different ways. Murray Gell-Mann (see Figure 7.39) and George Zweig proposed the existence of quarks in 1964 but their discovery was not confirmed until about ten years later, when the scattering of high-energy electrons off nucleons suggested that protons and neutrons had some kind of inner structure and, therefore, that they were *not* elementary.

The name 'quark' was selected from *Finnegan's Wake* written by the Irish author James Joyce and published in 1939. At that time only three quarks had been proposed:

Three quarks for Muster Mark!
Sure he has not got much of a bark
And sure any he has it's all beside the mark.

Figure 7.39 Murray Gell-Mann

In order to explain the properties of the composite particles that contain quarks, six different types, or **flavours**, of quark are needed. They are given the names *up, down, strange, charm, bottom and top*. The *flavour* of quarks has no meaning other than as a way of distinguishing them (Table 7.6).

All six types of quark have an electric charge that is either ⅓ or ⅔ of the magnitude of an electron's charge, *e*.

■ Table 7.6 Properties of quarks

Flavour	Symbol	Electric charge	Baryon number	Strangeness
up	*u*	$+\frac{2}{3}e$	$\frac{1}{3}$	0
down	*d*	$-\frac{1}{3}e$	$\frac{1}{3}$	0
strange	*s*	$-\frac{1}{3}e$	$\frac{1}{3}$	−1
charm	*c*	$+\frac{2}{3}e$	$\frac{1}{3}$	0
bottom	*b*	$-\frac{1}{3}e$	$\frac{1}{3}$	0
top	*t*	$+\frac{2}{3}e$	$\frac{1}{3}$	0

Baryon number and strangeness are explained later. A simplified version of this chart is included in the *Physics data booklet*.

Quarks are *never* found as free particles. They are *always* combined together. Particles composed of combinations of quarks are called hadrons.

Antiquarks have the opposite charge to quarks. For example, the anti-down quark has the symbol $\bar{d}$ and charge $+\frac{1}{3}e$.

Up and down quarks are generally stable (although they are involved in beta decay) and they are by far the most common quarks in the universe, whereas charm, strange, top and bottom quarks can only be produced in high-energy collisions (such as those involving *cosmic rays* and in *particle accelerators*). For the rest of this chapter, for simplicity, we will only discuss up, down and strange quarks. It is, however, interesting to mention that the top quark is the heaviest sub-atomic particle ever observed, with a mass that is nearly as heavy as an entire atom of gold and, together with the recently-discovered Higgs boson, it is one the elementary particles receiving the greatest attention.

The concept of strangeness was first introduced when some mesons (see below) were found to have 'strange' properties (they took much longer to decay than was expected). This unusual behaviour was then explained by proposing that the particle contained a different kind of quark: a *strange quark*. The concept of strangeness was extremely important from an historical perspective; however, it is no more fundamental than any of the other quark flavours – and we can talk similarly of topness, bottomness etc.

Hadrons, baryons and mesons

There are two types of hadron (particles containing quarks) – baryons and mesons.

Baryons

Baryons contain three quarks. A *proton* is a baryon composed of two up quarks and one down quark (uud), such that it has total charge of +*e*. A *neutron* is a baryon composed of two down quarks and one up quark (udd), such that it has no overall charge. See Figure 7.40.

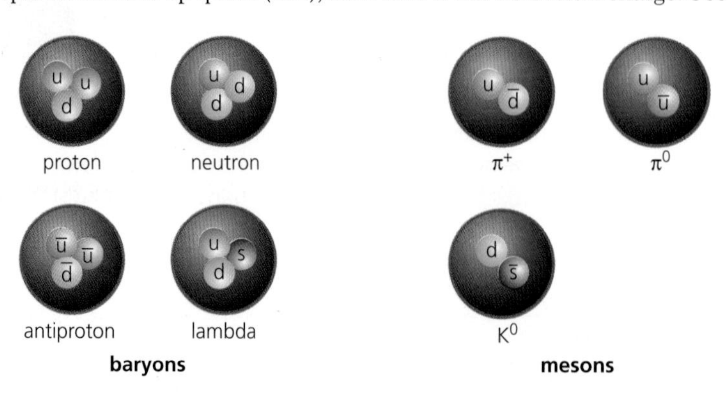

■ Figure 7.40 Combinations of quarks in some baryons and mesons

But there are other baryons apart from protons and neutrons. For example, the *lambda zero* particle contains an up quark, a down quark and a strange quark (uds). This particle is uncharged. It is described as having a strangeness of −1 (meaning that it contains a strange quark).

Mesons

The combination of a quark with an antiquark is called a **meson**. For example, the π^+ meson (*pion*) is a combination of an up quark and a down antiquark ($u\bar{d}$). See Figure 7.40. It has a charge of +1*e*. Its antiparticle, the π^- meson, contains a down quark and an up antiquark ($\bar{u}d$).

A K^0 *kaon* is a meson contains an anti-strange quark, ($d\bar{s}$). It is uncharged and has a strangeness of +1 (meaning that it contains an antistrange quark).

Mesons occur only in high-energy interactions in matter. They have short lifetimes before they decay.

Confinement

The fact that quarks are only ever found bound together as hadrons and never exist as free particles is called **quark confinement.**

To explain this we need to consider the nature of the strong force between two quarks: this force does *not* vary considerably with distance (unlike other forces, which decrease with distance; for example the electric force between two charges decreases with distance with an inverse square relationship). This means that the work (= force × distance) that would have to be done to completely separately quarks would be infinite. If a large amount of energy is supplied to two quarks there is the possibility that the energy could be converted to mass, with the formation of another pair of quarks. In other words, the energy supplied in an attempt to separate them produces a quark–antiquark pair long before they are far enough apart to observe separately.

51 Summarize the essential difference between baryons and mesons.

52 What is the quark structure of:
a an antiproton
b an antineutron?

53 Use the internet to determine the structure of eta mesons.

54 When a proton collided with a negative pion ($\bar{u}d$), a neutron was formed and one other particle.
a What was the quark structure of the other particle?
b What assumption did you make?

Leptons

There are six leptons: the **electron** and its **neutrino**, the muon and its neutrino, and the tau and its neutrino. See Table 7.7.

■ **Table 7.7** Properties of leptons

Lepton	Symbol	Electric charge/*e*	Rest mass (MeV c^{-2})	Lepton number
electron	e^-	−1	0.511	1
electron neutrino	ν_e	0	very small, not confirmed	1
muon	μ^-	−1	106	1
muon neutrino	ν_μ	0	very small, not confirmed	1
tau	τ^-	−1	1780	1
tau neutrino	ν_τ	0	very small, not confirmed	1

Lepton number is explained later (page 322). A simplified version of this chart is included in the *Physics data booklet*.

The three neutrinos of the electron, muon and tau all have very small masses. The muon is present in cosmic rays, but the tau has only been detected in high-energy collisions in a particle collider. The muon is unstable and quickly decays into an electron, an electron antineutrino and a muon neutrino. The electron is the charge carrier in metallic conductors and semi-conductors. Electron neutrinos are formed during beta-positive decay, as we saw earlier in this chapter.

The conservation laws of charge, baryon number, lepton number and strangeness

Particles can have a range of properties. Apart from mass–energy equivalence, we have already referred to charge, flavour and spin. In any particle decay or interaction, many of these properties are conserved. This is a similar principle to the conservation of proton and nucleon numbers in radioactive decay.

Quantum numbers are used to describe quantized values of conserved quantities:

- The quantum number for *charge*, $\frac{1}{3}$, $\frac{2}{3}$, 1, $1\frac{1}{3}$ ... is always conserved.
- The quantum **baryon number** is the number of baryons; it is always conserved (antibaryons have a baryon number of −1). Quarks have a baryon number of $\frac{1}{3}$.
- The quantum **lepton number** is the number of leptons; it is always conserved (antileptons have a lepton number of −1)
- Strangeness is conserved in electromagnetic and strong interactions, but not always in weak interactions. All quarks have a strangeness of 0 except the strange quark (−1).

Worked example

14 Apply the conservation laws to the following possible interactions:

a $n \rightarrow p + e^- + \bar{\nu}_e$

b $K^0 \rightarrow \pi^+ + \pi^-$

a Baryon number: 1 = 1 + 0 + 0
Lepton number: 0 = 0 + 1 + (−1)
Charge: 0 = 1 + (−1) + 0
Strangeness: 0 = 0 + 0 + 0

The conservations laws are all correct, so the interaction can occur. It is beta-negative radioactive decay.

b Baryon number: 0 = 0 + 0
Lepton number: 0 = 0 + 0
Charge: 0 = 1 + (−1)
Strangeness: 1 = 0 + 0
Strangeness is not conserved, so the interaction cannot occur.

55 A neutral kaon ($d\bar{s}$) can interact with a proton to produce a positive kaon ($u\bar{s}$) and another particle, *x*:
$K^0 + p \rightarrow K^+ + x$
Deduce the nature of *x*.

56 Use the conservation rules to determine if the following nuclear reactions are possible. Research the properties of the particles where necessary.
a $n + p \rightarrow e^+ + \bar{\nu}_e$
b $p + \bar{p} \rightarrow \pi^- + \pi^+$
c $e^+ + e^- \rightarrow \gamma + \gamma$
d $\pi^- + p \rightarrow K^0 + \Lambda^0$
e $\pi^0 + n \rightarrow K^+ + \Sigma^+$

The nature and range of the strong nuclear force, weak nuclear force and electromagnetic force

The four fundamental interactions in nature were briefly mentioned earlier in this chapter. They are summarized in Table 7.8. (The weak and the electromagnetic interactions are sometimes considered as combined in the *electro–weak interaction*.)

Table 7.8 the four fundamental interactions

Fundamental interaction	Acts on	Approximate relative size	Range/m	Exchange particle
strong nuclear	quarks and gluons	1	10^{-15}	gluon (*g*)
electromagnetic	all charged particles	10^{-2}	Infinite, but reduces with an inverse square law	photon (γ)
weak nuclear	quarks and leptons	10^{-6}	10^{-18}	W or Z bosons
gravitational	all masses	10^{-38} (not significant for individual particles)	Infinite, but reduces with an inverse square law	graviton (theoretical)

A simplified version of this is included in the *Physics data booklet*.

Exchange particles

At the fundamental level of particle physics, forces are explained in term of the transfer of **exchange particles (gauge bosons)** between the two particles experiencing the force.

For example, there is a force between two charged particles because there is an exchange of a photon between them. The photon is the exchange particle for the electromagnetic interaction.

We can begin to explain exchange particles with an everyday example, as follows. Consider Figure 7.41. The forward force needed to throw the ball from A to B is equal and opposite to the backward force on A. When B catches the ball there will be a backward force on her. In other words, the exchange of the ball results in A and B being 'repelled' from each other. If the ball is thrown back the process is repeated. If the ball was not visible we would simply observe that the boats were repelled.

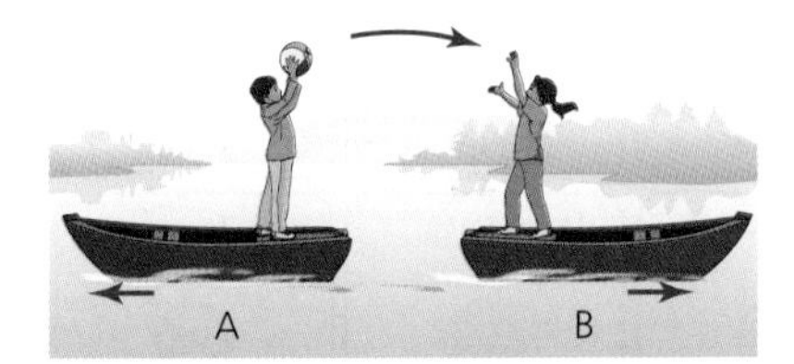

Figure 7.41 A ball being exchanged involves forces

Of course, this macroscopic example is very different from the action of exchange particles in fundamental interactions and it does not easily explain attractive forces (some books use a boomerang to explain this). Nevertheless, the visualization helps our understanding.

Physicists believe that the fundamental interactions occur because of the exchange of particles. These exchanges are very fast and are not observable. Because of this, the exchange particles are often described as **virtual particles**.

The short range of nuclear forces can be explained if the virtual particles have a limited lifetime and do not have enough time to be transferred between the other two particles unless they are very close together. (Heisenberg's uncertainty principle explains why this can occur. It is discussed in Chapter 12.)

57 Suggest how the use of a boomerang analogy could help in explaining attractive forces between particles.

58 Which fundamental forces act on:
a protons
b neutrons
c electrons
d quarks
e leptons?

59 In Table 7.8 electromagnetic forces and gravitational force are stated to be in the approximate ratio 10^{-2} to 10^{-38}. As an example, calculate the electric force between a proton and an electron separated by a distance of 5×10^{-11} m and compare it to the gravitational force between the same two particles.

60 Describe the characteristic features of the strong nuclear force that distinguish it from the electromagnetic force.

The Standard Model

The Standard Model of particle physics is a theoretical framework that reflects scientists' current understanding of the sub-atomic world. The model has been extremely successful in accounting for most observed sub-atomic phenomena. However, it is not believed to be the complete picture. For example, gravity cannot be incorporated in the model. There are other theoretical shortcomings. As a result, the focus of current high-energy experiments is on attempting to discover what physics lies beyond the Standard Model.

Figure 7.42 shows a chart of the Standard Model of elementary particles. The simplicity of the chart does not truly reflect the extent of the efforts that preceded it: considerable research, both experimental and theoretical, in many countries over the last 50 years has been necessary to reach this point. In the latest particle-colliding experiments the quantity of data produced is huge, requiring enormous computing power for its analysis. However, there are still many unanswered questions, such as where does gravity fit into the model, what is dark matter (in space) and why is the universe mostly matter and not antimatter? The chart also includes the Higgs boson, which will be discussed later.

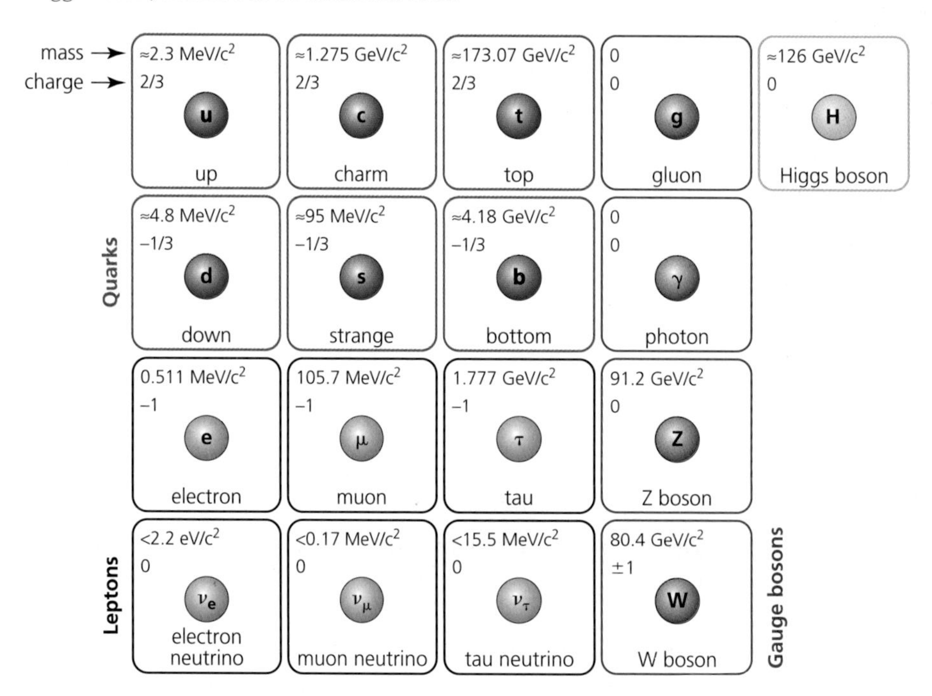

Figure 7.42 The Standard Model of atomic physics

Nature of Science

Predictions: belief in 'simplicity'

The concept of a proton–neutron–electron model for atoms had an elegant simplicity, so the discovery of a large number of other sub-atomic particles was problematic for physicists who tend to believe that it is the fundamental nature of science that the structure of the universe should be 'simple' and composed of a relatively small number of elementary particles (and forces).

Despite the hundreds of particles that have been discovered or theorized, the assumption has always been made that most of them were *not* elementary and that there had to be an underlying simplicity awaiting confirmation. That faith has been restored by the development of the Standard Model.

The Standard Model also reflects scientists' belief in the importance of *symmetry*. For example, there are positive and negative charges, particles and antiparticles, and six leptons and six baryons.

ToK Link

The contribution of physics to human knowledge

Does the belief in the existence of fundamental particles mean that it is justifiable to see physics as being more important than other areas of knowledge?

What individuals choose to believe is important is for them to decide, but knowledge about the structure of matter and the universe is undoubtedly of considerable intrinsic interest to most people and such knowledge is undeniably about as basic and fundamental as it is possible to be. However, science has progressed to the stage where such understanding is not easily attained by anyone who does not have a scientific background, and to many people, advanced science is totally disconnected from their everyday lives.

It is certainly possible to argue that the application of other areas of scientific knowledge (medicine for example) is more 'important' because they directly affect the quality of our everyday lives. It should be added that we do not know what benefits may come from high-energy physics research; for example, CERN played a key role in the development of the internet.

■ **Figure 7.43** Richard Feynman: 'I have approximate answers and possible beliefs and different degrees of certainty about different things, but I'm not absolutely sure about anything'

■ Feynman diagrams

Feynman diagrams are a way of visualizing any *elementary* particle interaction using a **vertex** (or two or more *vertices*).

Figure 7.44 is a simple example. The vertex is the point in the middle of the diagram.

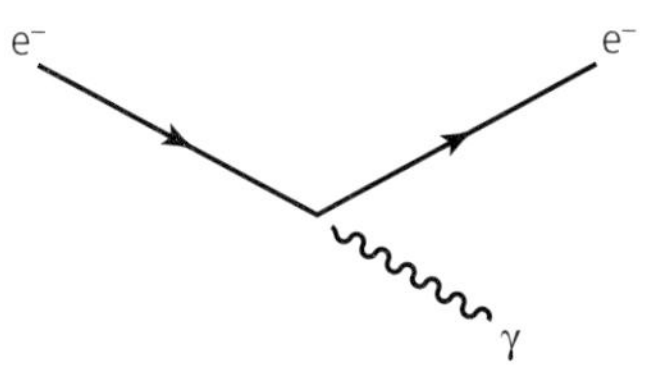

■ **Figure 7.44** Feynman diagram of a simple interaction

- The area to the left of the vertex represents *before* the interaction, the area to the right represents *after* the interaction.
- Observable particles are represented by straight lines with arrows pointing to the right. Antiparticles are represented by arrows to the left, even though *all particles are always considered to be moving from left to right.*
- There is always one arrow pointing into a vertex and one arrow pointing away.
- Unobservable exchange particles are represented by curly lines, without arrows.
- The change in orientation of a line signifies that the motion of the particle has changed.
- Each vertex joins two straight lines and typically one curly line.
- The conservation laws can be applied to the interaction represented by a single vertex.

With this information we can see that the vertex in Figure 7.44 represents an electron emitting an exchange particle (a photon) and changing direction.

It is important to understand that these diagrams are *not* representing the spatial arrangement of particles, or the vectors in a collision. Feynman diagrams are not displacement–time graphs. In fact, they are a simplification of a complex mathematical treatment of unobservable particle interactions, which is not included in the IB course.

In this book we are representing increasing time by moving from *left to right*, but other sources may use *bottom to top* to represent time. This is easily checked by seeing which way the arrows point for particles.

Feynman diagrams provide a powerful tool in representing, analysing and predicting particle interactions. The diagrams in Figure 7.45 show some variations to Figure 7.44. Despite their obvious similarities they represent different situations.

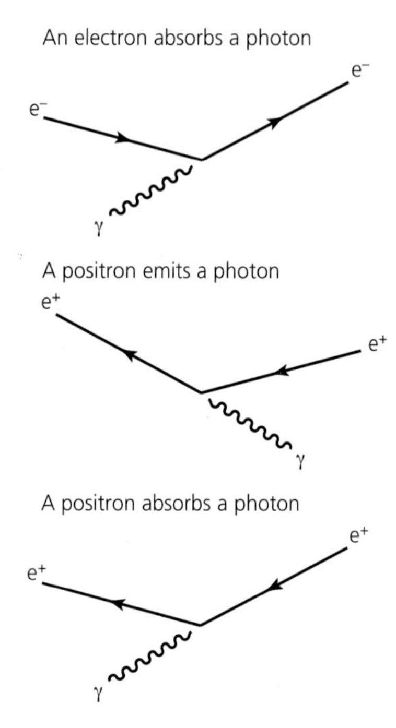

■ **Figure 7.45** Electron and positron vertices

We will now expand our discussion of Feynman diagrams to include interactions between different particles. Students will not be expected to remember individual Feynman diagrams, rather they will need to be able to interpret diagrams provided on an examination paper, or construct simple diagrams from information in the question.

Examples of Feynman diagrams

Figure 7.46a represents a possible interaction between two electrons, which we would previously have interpreted as the electric repulsion between similar charges. However, there are many other Feynman diagrams that could be drawn to achieve the same result of overall repulsion between two electrons. One example is shown in Figure 7.46b. A full mathematical analysis would be needed to consider all possibilities.

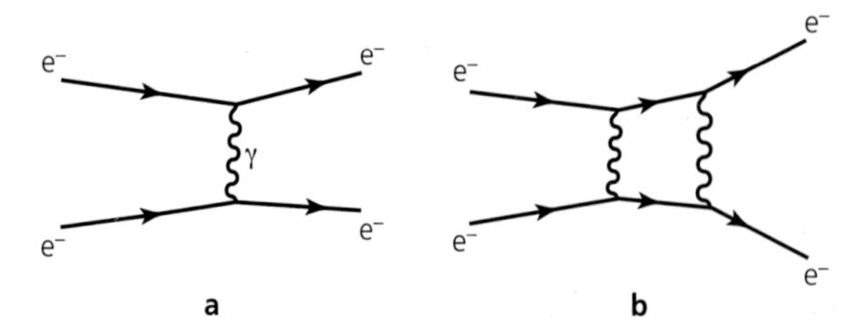

■ **Figure 7.46** Two Feynman diagrams for electron–electron scattering

Figure 7.47 represents an interaction in which two photons scatter off each other due to the exchange of electrons and positrons.

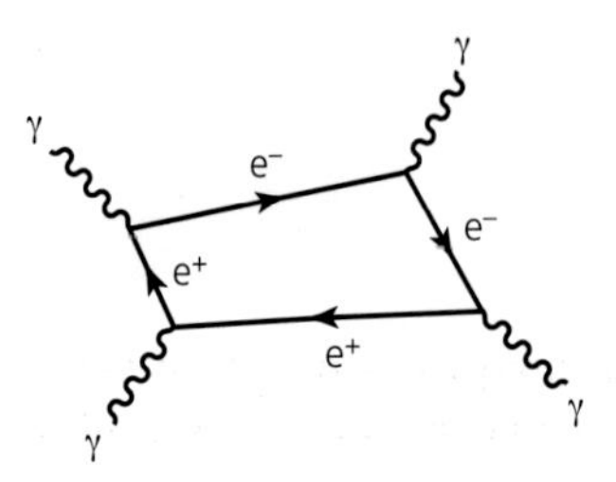

■ **Figure 7.47** A Feynman diagram for photon–photon scattering

Figure 7.48a represents the annihilation of an electron–positron pair. (This is an unusual example, normally two photons will be created). Figure 7.48b represents pair production.

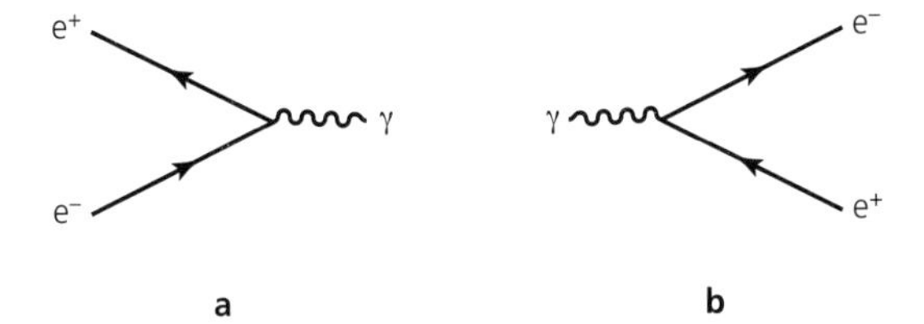

■ **Figure 7.48** Annihilation and pair production

Figure 7.49a shows a Feynman diagram representing beta-negative decay, while Figure 7.49b shows a particle representation.

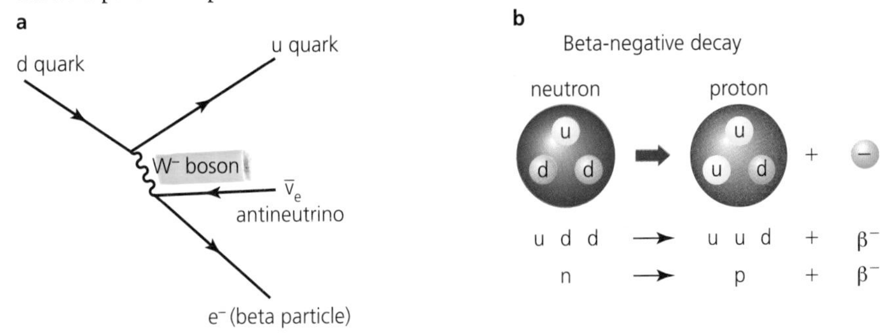

■ **Figure 7.49** Beta-negative decay

Figure 7.50 shows the formation of a quark–antiquark pair involving the strong nuclear force exchange particle, the gluon.

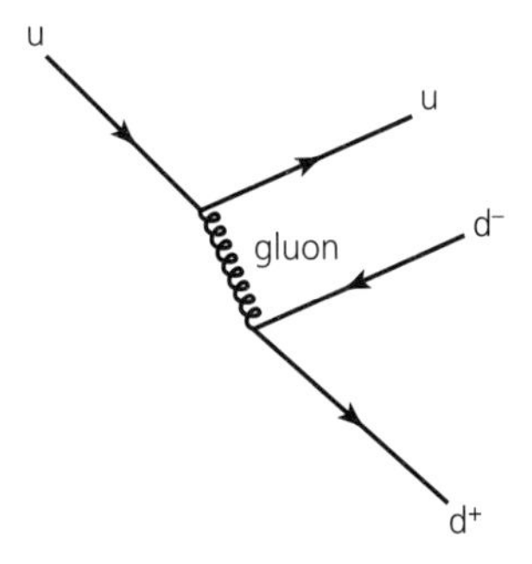

■ **Figure 7.50** Exchange of a gluon in the formation of quarks

61 Draw a Feynman diagram to represent beta-positive decay, which involves a W^+ boson.

62 Research the meaning of the 'eightfold way'.

■ Higgs boson

■ **Figure 7.51** François Englert and Peter Higgs

The existence of the Higgs boson – an elementary particle – was first proposed by a number of physicists in the early 1960s but it took about 50 years until it was finally identified on 4 July 2012.

On that date a particle with the correct properties (the only elementary particle to have zero charge *and* zero spin) and with a large energy–mass of $126\,GeVc^{-2}$ (135 u) was discovered. The search for the Higgs boson took so long because it required the extremely high-particle energies that only became achievable at CERN's Large Hadron Collider (see Figure 5.90).

Peter Higgs and François Englert were then awarded the Nobel prize for physics in Oslo in 2013 'for the theoretical discovery of a mechanism that contributes to our understanding of the origin of mass of sub-atomic particles, and which recently was confirmed through the discovery of the predicted fundamental particle, by the ATLAS and CMS experiments at CERN's Large Hadron Collider'.

The Higgs particle was the vital missing link in the Standard Model and its importance was reflected in newspaper and internet headlines.

Higgs and his colleagues proposed that particles do not have intrinsic mass but acquire the properties of mass from interacting with a field – the **Higgs field** – that occupies the entire universe. As with the other fields we have discussed, a boson – the Higgs boson – is needed for this interaction to occur. Clearly the discovery of the Higgs boson was the fundamental evidence needed to support this revolutionary theory that explains, for the first time, why different particles have different masses.

Nature of Science

International collaboration at CERN

Nuclear physics experiments that are aimed at increasing our knowledge of sub-atomic particles and the beginnings of the universe require ever increasing particle energies, and these can only be provided by extremely expensive particle accelerators, such as at CERN. It has been beneficial for countries to collaborate in sharing knowledge and costs, but the allocation of considerable funds to nuclear research has been a matter of much debate. Many people consider that countries have more important priorities (social needs) than spending money on specialized scientific knowledge that may, or may not, have any long-term benefits to society.

One reason for the establishment of CERN in 1952 was political. During the Second World War, scientists and engineers from European countries had been competing to develop the nuclear bomb and missiles. After the war it was hoped that a European centre could be constructed so that scientists could work together for non-military research. CERN staff design and build particle accelerators and help to prepare, carry out, analyse and interpret data from complex particle physics experiments. Half of the world's particle physicists visit CERN for their research. They represent almost 600 universities and 100 nationalities. Particle physics and the discovery of new particles intensified when there was large-scale national and international **collaboration**. Large particle accelerators, such as the Large Hadron Collider at CERN are so expensive that they can only be operated if many countries contribute.

Examination questions – a selection

Paper 1 IB questions and IB style questions

Q1 Why are gamma rays not deflected by a strong magnetic field?

A They have no mass.
B They are weakly ionising.
C They are strongly penetrating.
D They are electrically neutral.

Q2 A freshly prepared sample contains 40 µg of the isotope iodine-131. The half-life of iodine-131 is 8 days Which of the following is the best estimate for the mass of the iodine-131 remaining after 24 days?

A 10 µg
B 13 µg
C 5 µg
D zero

Q3 Two neutrons are captured by a nucleus. Which of the following gives the changes in the proton number and nucleon number of the nucleus?

	Proton number	Nucleon number
A	unchanged	unchanged
B	unchanged	increases by 2
C	increases by 2	unchanged
D	increases by 2	increases by 2

Q4 What is particle X in the fusion reaction shown below?

$${}^{7}_{3}\text{Li} + {}^{2}_{1}\text{H} \rightarrow 2\,{}^{4}_{2}\text{He} + \text{X}$$

A a proton
B an electron (beta particle)
C an alpha particle
D a neutron

Q5 The binding energy of a helium-3 nucleus is defined to be:

A the energy released when a helium-3 nucleus is formed from its individual constituents
B the energy released when the helium-3 nucleus is separated into its individual constituents
C the total energy of the protons inside the helium-3 nucleus
D the total energy of the helium-3 nucleus.

Q6 The diagram shows four possible electron energy levels in the hydrogen atom.

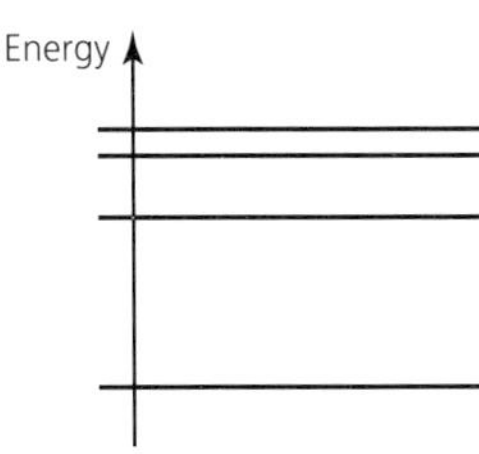

The number of different frequencies in the emission spectrum of atomic hydrogen that arise from electron transitions between these levels is:

A 0
B 2
C 4
D 6

Q7 The radiation emitted when an electron changed energy levels in an atom had a wavelength of 6×10^{-7} m. Which of the following correctly describes this energy change?

A The electron moved to a higher energy level within the atom.
B The frequency of the emitted radiation was 2×10^{-15} Hz.
C The energy of the emitted photon was approximately 3×10^{-19} J.
D 3×10^{-19} J is approximately equivalent to 0.5 eV.

Q8 Which of the following statements about sub-atomic particles is correct?

A Protons, neutrons and electrons are all elementary particles.
B Electrons belong to a class of particles called leptons.
C Electromagnetic forces are explained by the exchange of particles called gluons.
D Protons and neutrons each contain two quarks.

Q9 The Standard Model has three classes of particle. These are:

A hadrons, leptons and quarks
B hadrons, leptons and exchange bosons
C hadrons, quarks and exchange bosons
D leptons, quarks and exchange bosons.

Q10 Which of the following is not a conserved quantum number in some, or all, nuclear reactions?

A mass **B** baryon number **C** charge **D** strangeness

Q11 Which of the following is true about beta minus (β–) decay?

A An antineutrino is absorbed.
B The charge of the daughter nuclide is less than that of the parent nuclide.
C An antineutrino is emitted.
D The mass number of the daughter nuclide is less than that of the parent nuclide.

Q12 In the Geiger-Marsden experiment α particles are scattered by gold nuclei. The experimental results provide evidence that:

A α particles have discrete amounts of kinetic energy.
B Most of the mass and positive charge of an atom is concentrated in a small volume.
C The nucleus contains protons and neutrons.
D Gold atoms have a high binding energy per nucleon.

Q13 Protons and neutrons are held together in the nucleus by the:

A electrostatic force
B gravitational force
C weak nuclear force
D strong nuclear force.

Paper 2 IB questions and IB style questions

Q1 **a** **i** Describe the phenomenon of natural radioactive decay. (3)

ii Ionising radiation is emitted during radioactive decay. Explain what is meant by the term *ionising*. (2)

b The sketch graph below shows the variation with mass number (nucleon number) A of the binding energy per nucleon E of nuclei.

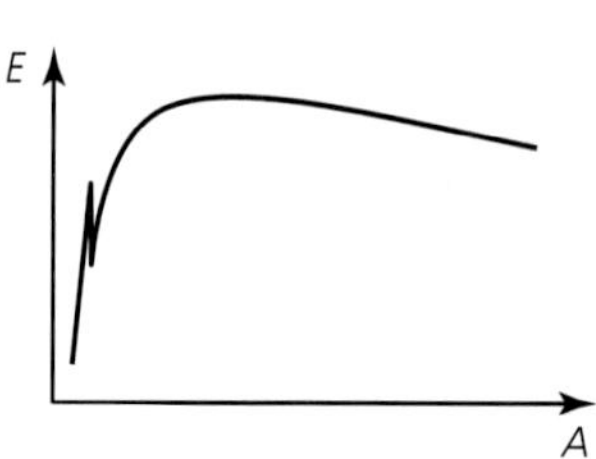

One possible nuclear reaction that occurs when uranium-235 is bombarded by a neutron to form xenon-142 and strontium-90 is represented as:

$^{235}_{92}U + ^{1}_{0}n \rightarrow ^{142}_{54}Xe + ^{90}_{38}Sr + 4^{1}_{0}n$

i Identify the type of nuclear reaction represented above. (1)

ii Copy the sketch graph above. On your copy, identify with their symbols the approximate positions of the uranium (U), the xenon (Xe) and the strontium (Sr) nuclei. (2)

iii Data for the binding energies of xenon-142 and strontium-90 are given below.

Isotope	Binding energy/MeV
xenon-142	1189
strontium-90	784.8

The total energy released during the reaction is 187.9 MeV. Determine the binding energy per nucleon of uranium-235. (2)

iv State why the binding energy of the neutrons formed in the reaction is not quoted. (1)

Q2 A stationary radon-220 ($^{220}_{86}Rn$) nucleus undergoes α-decay to form a nucleus of polonium (Po). The α-particle has kinetic energy of 6.29 MeV.

a **i** Complete the nuclear equation for this decay. (2)

$^{220}_{86}Rn \rightarrow Po +$

ii Calculate the kinetic energy, in joules, of the α-particle. (2)

iii Deduce that the speed of the α-particle is $1.74 \times 10^7\,m\,s^{-1}$. (1)

The diagram below shows the α-particle and the polonium nucleus immediately after the decay. The direction of the velocity of the α-particle is indicated.

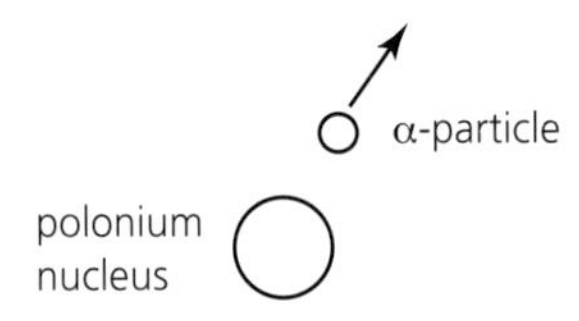

b **i** On a copy of the diagram above, draw an arrow to show the initial direction of motion of the polonium nucleus immediately after the decay. (1)

ii Determine the speed of the polonium nucleus immediately after the decay. (3)

iii In the decay of another radon nucleus, the nucleus is moving before the decay. Without any further calculation, suggest the effect, if any, of this initial speed on the paths shown in b i. (2)

The half-life of the decay of radon-220 is 55 s.

c **i** Explain why it is not possible to state a time for the life of a radon-220 nucleus. (2)

ii Define *half-life*. (2)

8 Energy production

ESSENTIAL IDEAS

- The constant need for new energy sources implies decisions that may have a serious effect on the environment. The finite quantity of fossil fuels and their implication in global warming has led to the development of alternative sources of energy. This continues to be an area of rapidly changing technological innovation.
- For simplified modelling purposes the Earth can be treated as a black-body radiator and the atmosphere treated as a grey body.

8.1 Energy sources – *the constant need for new energy sources implies decisions that may have a serious effect on the environment. The finite quantity of fossil fuels and their implication in global warming has led to the development of alternative sources of energy. This continues to be an area of rapidly changing technological innovation*

Data concerning the use of energy sources throughout the world is changeable over time and can vary in reliability. For this reason the figures quoted throughout this chapter are for guidance only and should not necessarily be assumed to be indisputably precise. It is wise to consult reputable websites for the latest information.

A **fuel** is a widely used term for any substance from which energy from changes within its atoms and molecules (that is, from chemical or nuclear energy) is used to do useful work. Examples include coal and uranium. Coal is an example of a **fossil fuel**. Oil and natural gas are also fossil fuels. Fossil fuels are formed underground by the action of high pressure and temperature (in the absence of air) over many millions of years. They are formed mostly from dead plants and marine creatures. Although the processes that form fossil fuels are still happening today, because of the very long time needed for their formation, they are known as **non-renewable energy sources**: we are using them at a very much quicker rate than they are being formed.

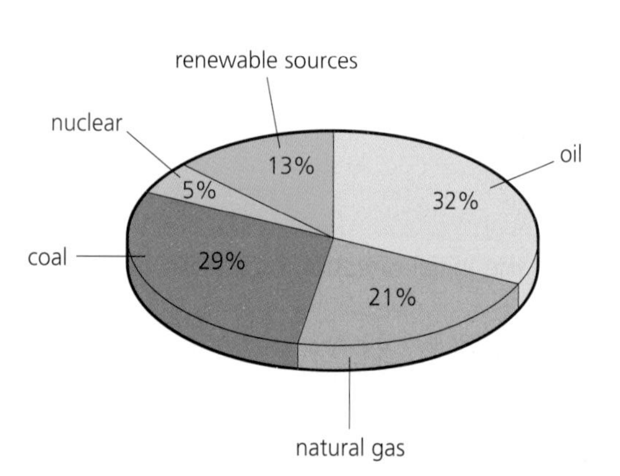

■ **Figure 8.1** The proportion of different energy sources contributing to the total energy consumption of the world

In 2012, the total annual world primary energy consumption was about 5.5×10^{20} J. Although this figure cannot be known with a high degree of certainty, it amounts to a global average power consumption of over 2000 W for every person on the planet. Of course, this is only an average and so does not take into account the wide differences in energy consumption between richer and poorer countries. This figure includes energy sources used for transport, industry and the generation of electricity, but does not include the use of small-scale individual resources such as firewood. About one-quarter of the people on Earth do not have electricity in their homes. Sometimes figures quoted on websites or in books can be misleading if they only take into account energy sources used for the generation of electricity.

The proportions of different energy sources that are used around the world are shown in Figure 8.1.

- **Fossil fuels** provide about 82 per cent of the total energy consumed in the world. The products of crude oil are the most widely used (32 per cent), followed by coal and natural gas (methane) at 29 per cent and 21 per cent, respectively. Despite the considerable advantages of using fossil fuels, they have at least two major disadvantages: they release carbon dioxide when they are burned (and almost certainly lead to climate change) and they are non-renewable (see page 333).
- **Nuclear power** provides approximately five per cent of the total energy consumed in the world. This is also a non-renewable energy source.
- **Renewable energy sources** supply about 13 per cent of the total energy consumed in the world.

Countries need fuel supplies to maintain the living standards of their citizens. Because most countries do not have enough energy resources for themselves, they need to import energy from those few countries that have more than they need. This requires international understanding and co-operation between countries if conflicts, and possibly wars, are to be avoided.

Any threat to international fuel supplies can have a rapid and dramatic effect on world economics; the price of a barrel of oil on international markets is a very important financial indicator that affects the price of many other items that we buy.

Primary energy sources

A **primary energy source** is a source that occurs naturally and that has not been processed in any way. For example, crude oil is a primary energy source.

When a primary source is converted into another, more useful, resource it is then described as a **secondary energy source**. For example, crude oil refined into other fuels, uranium processed into plutonium or the generation of electricity.

The original source of most of the energy consumed on the Earth is the Sun. Plants get their energy directly from the Sun's radiation, and plants and animals that died hundreds of millions of years ago are the source of all fossil fuels. Wind and rain are a result of temperature changes caused by the Sun's radiation and waves are formed by the wind.

The main sources from the above lists that do not get their energy from the Sun are nuclear energy (which comes from changes inside the nuclei of atoms) and geothermal energy (from radioactivity, which is also from nuclear changes). A third example is tidal energy, which gets its energy from very slight changes in the gravitational potential energy of the Earth–Moon–Sun system.

Advantages and disadvantages of different energy sources

The most important factors to consider when discussing the advantages and disadvantages of individual energy sources are:

- greenhouse gas emissions and possible effects on global warming
- risks to human health/life
- possible pollution and environmental effects, including problems caused by the transportation and storage of fuels and waste products
- whether the source is renewable or non-renewable
- cost
- how much energy can be transferred from a given mass (or volume) of the source
- whether the energy is continuously available or dependent on factors such as weather conditions and the time of day/night.

Renewable and non-renewable energy sources

A **renewable energy source** is continuously replaced by natural processes and will not be used up (become depleted/run out). It will continue to be available for our use for a very long time.

Wind and waves, for example, will always be available as long as there is energy coming from the Sun to warm the planet. The main renewable sources are:

- biomass/biofuels (energy transferred from wood and other plants, and also from animal waste)
- hydroelectricity
- solar heating
- wind power
- geothermal (energy transferred from hot rocks under the ground)
- photovoltaic cells
- wave power.

■ **Figure 8.2** Brazil uses a large amount of ethanol (produced from sugar cane) as a fuel.

This list is in the approximate (decreasing) order of total energy use. Hydroelectric power is the most widely used renewable energy source for the generation of electricity. These are fast-developing technologies and the latest information can easily be researched on the internet.

When we call an energy source non-renewable we mean that the source cannot be replaced once it has been used up, and so supplies will run out (be depleted) sometime in the not-too-distant future. The main non-renewable sources are:

- fossil fuels
- nuclear power.

Specific energy and energy density of fuel sources

It is useful to know how much energy can be transferred from a certain amount of fuel so that different fuels can be compared. This can be determined in a laboratory by burning a known mass or volume of fuel and transferring the resulting thermal energy to raise the temperature of a known mass of water. (Remember that $Q = mc\Delta T$). Figure 8.3 shows how a chemist might quickly do this for a liquid fuel, although this simple method will not be very accurate (students should be able to explain why using ideas from Chapter 3).

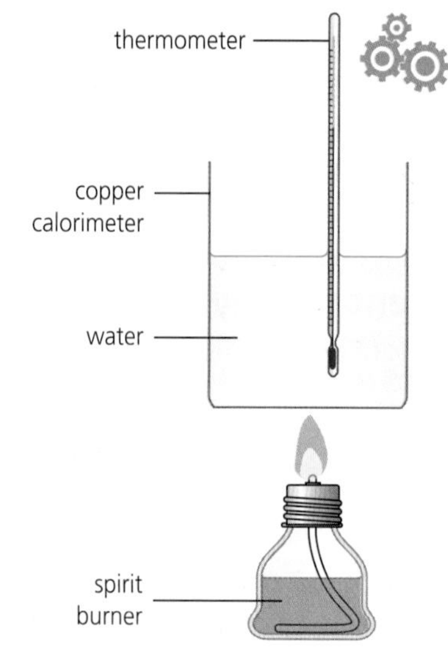

■ **Figure 8.3** Estimating the energy density or specific energy of a fuel

Specific energy = energy transferred from unit mass. Unit: J kg^{-1}

Energy density = energy transferred from unit volume. Unit: J m^{-3}

For many renewable energy sources, such as wind power, it may not be sensible to refer to a numerical value of energy density, although sometimes wind power may be described in general terms as having a low energy density because it requires a relatively large area of land to achieve a moderate power output.

A major advantage of nuclear power is its high specific energy (and energy density). Of the fossil fuels, natural gas (methane) has the highest specific energy.

When using a fuel with a high specific energy, less mass of fuel will be needed to provide a certain amount of energy compared to using a fuel with a lower specific energy. Table 8.1 gives some examples.

■ **Table 8.1** Specific energies

Material	**Specific energy/MJ kg^{-1}**
nuclear fusion (of deuterium and tritium)	340 000 000
uranium used in nuclear reactors	83 000 000
hydrogen	143
natural gas (methane)	55
gasoline/petrol	46
crude oil	46
vegetable oil	42
ethanol	30
coal	28
wood	17
typical carbohydrate (food)	17
cow dung	15
household waste	9
torch battery	0.1
water at height of 100 m	0.001

1 In an experiment similar to that shown in Figure 8.3, when $1.24\,cm^3$ of fuel was completely burned in air the temperature of 54.8 g of water rose from 18.0°C to 83.4°C. (Specific heat capacity of water is $4180\,J\,kg^{-1}K^{-1}$.)
 a Calculate a value for the energy density of the fuel.
 b Would you expect the actual value to be higher or lower than your answer? Explain.

2 A country has a population of 46 million people and a total energy consumption of $5.2 \times 10^{18}\,J$ in one year.
 a Calculate the average energy consumption per person in one year.
 b What is the average power consumption per person?
 c If an average home has 2.8 people and a total power consumption of 2.7 kW, what percentage of the country's energy consumption occurs in people's homes?
 d Suggest what the rest of the energy is used for.

3 a How much gasoline is needed to accelerate a 1500 kg car from rest to $20\,m\,s^{-1}$ if the overall efficiency of the process is 32 per cent?
 b Estimate the rate at which the same car consumes gasoline if it is travelling at a constant speed of $20\,m\,s^{-1}$ against a total resistive force of 2000 N.
 c What is the specific energy of a torch battery containing 7.9 g of chemicals that supplies enough energy to run a 0.85 W light bulb for 110 minutes?

4 A coal-burning power station has an efficiency of 36 per cent and an output power of 312 MW.
 a Calculate the mass of coal burned:
  i every second
  ii every week.
 b Approximately how much uranium would be needed each week for a nuclear power station of the same output but with an efficiency of 42 per cent?

5 a Find out which countries of the world have the highest average power consumptions (per person/capita).
 b Suggest possible reasons why the people of those countries use so much energy.

6 Suggest in what ways a list of the energy sources used in large-scale electrical power generation might be different from a list of overall energy sources used in the world.

7 Use the internet to find out the latest figures for the use of renewable energy sources:
 a in the world
 b in the country where you live.

8 Suggest circumstances under which it might be more useful to refer to the energy density of a fuel, rather than its specific energy.

Electricity as a secondary and versatile form of energy

It was not until the end of the nineteenth century that science and technology had advanced to the point where power stations could be built in order to burn fuels to generate electricity for large numbers of people (see Figure 8.11). Before that time, coal and wood were used for heating, and gas, oil or candles were used for lighting. Even today almost half of the world's population still uses local wood or other plant or animal materials for providing some, or most, of their energy.

In a period of just over 130 years most of the world has been completely transformed by the widespread use of electricity. This is because:

- large amounts of electrical energy can be generated quickly and economically in power stations
- electric currents can transfer energy efficiently to where it is needed by using metal wires and cables
- a wide variety of devices have been invented that transfer electrical energy to other very useful forms.

But electric current has to be generated when it is needed. It cannot be stored directly.

Nature of Science

All scientific advances have associated risks

Electricity has transformed the world as a result of the collaboration between generations of scientists and engineers across the world on a massive scale. (Although there is still a long way to go before everyone has access to electricity; it is estimated that about 1.2 billion people still live in homes without electricity.) This has come at a cost – the Earth's resources have been depleted;

we are threatened by climate change and pollution. But it is the very nature of science to solve problems, especially when, as in the hope of providing inexpensive and environmentally friendly energy, there may be tremendous financial advantages to be gained.

Of course, the scientists who carried out the pioneering investigations of current electricity would never have dreamed where their discoveries would lead in the following 200 years! The full consequences of a scientific or technological development are often not foreseeable but, when they are, it can be argued that it should be part of scientists' roles to undertake and present risk–benefit analyses, rather than simply pursue knowledge regardless of the implications. Where there is any doubt, it is now generally accepted that scientists must convince their colleagues, and the public, that none of their intended actions/research could be harmful in any way. (This is called the '**precautionary principle**'.) Although it must be admitted that, by the very nature of science and the unpredictability of research, this is a very difficult thing to achieve.

Using energy sources to generate electricity in power stations

Most power stations that produce electricity transfer the energy from burning fossil fuels or from fission reactions in nuclear fuels. The thermal energy is used to raise the temperature of water in a boiler and turn it into high-pressure steam. The steam causes the rotation of turbines, shown in Figure 8.4, which are connected to coils of wire. The process of *electromagnetic induction* (Chapter 11) produces electrical energy as the coils rotate in strong magnetic fields. When the steam exits the turbine it is cooled, causing it to condense, and the water is pumped back into the boiler. Figure 8.5 shows a simplified representation of a fossil-fuel power station.

The same basic ideas apply to *all* power stations that transfer energy from a fuel to electricity using the flow of thermal energy. Machines that are powered by a flow of thermal energy are called **heat engines**. Examples include power stations, cars, planes and trains.

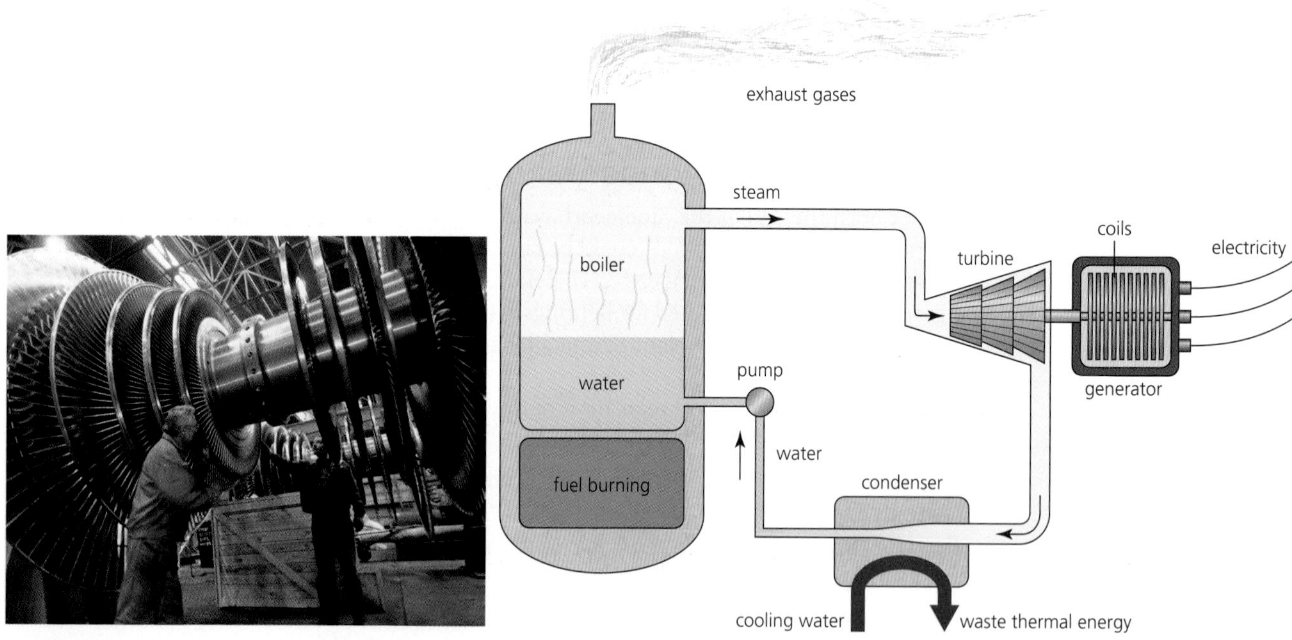

■ **Figure 8.4** A turbine used for generating electricity

■ **Figure 8.5** A schematic diagram of a fossil-fuel power station

■ Energy conversions and energy degradation

Heat engines use the principle that when there is a temperature difference between two places, thermal energy will flow between them. There is then the possibility of using some of that energy to do useful work, for example turning turbines to generate electricity. Figure 8.6 illustrates the principle by showing an example in which, for every 100 J of energy flowing between a 'hot' and a 'cold' area, 30 J of energy can be used to do useful work and 70 J is transferred to the cold area, from where it spreads into the surroundings (environment). Of course, 30 J + 70 J = 100 J because of the law of conservation of energy. The efficiency of this process is 30/100 = 0.3, or 30 per cent.

Energy that spreads into the surroundings **(dissipates)** cannot be recovered to do any useful work and it is known as **degraded energy.**

A typical power station may be only 35 per cent efficient in producing electricity (a disappointingly low figure), which means that 65 per cent of the energy obtained from the fuel is degraded and transferred to the surroundings at the power station. Of course, we want power stations to be as efficient as possible and it is therefore important to understand why energy becomes degraded. As we saw in Chapter 2, in all mechanical processes friction will result in the dissipation of energy, but this is *not* the major reason for power station inefficiency.

To understand the inefficiency of all heat engines, consider the following simplified example. In Figure 8.7, thermal energy is supplied to gas sealed in a cylinder. The increasing pressure inside the cylinder forces the movable piston (assumed to be friction-free) to the right. In this example it is possible, in theory, for all of the thermal energy supplied to the gas to be converted into the kinetic energy of the moving piston. The system can be 100 per cent efficient.

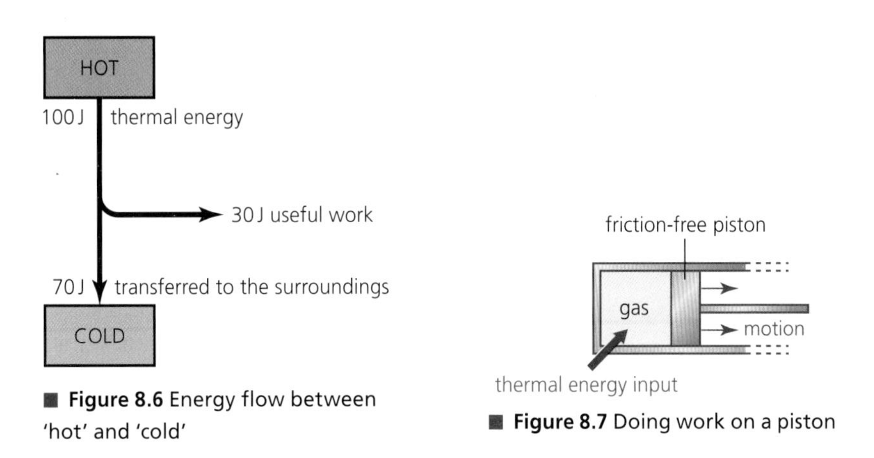

■ **Figure 8.6** Energy flow between 'hot' and 'cold'

■ **Figure 8.7** Doing work on a piston

But a heated gas in a cylinder cannot expand forever. This means that any practical heat engine must involve a process that repeats itself over and over again in order for it to work continuously. A practical heat engine must work in *cycles* in a confined space, with an expansion followed by compression, followed by expansion etc. However, it is not possible for any cyclical process to happen without a lot of the thermal energy being transferred to the surroundings (out of the system). This is the main reason why power stations are so inefficient. The efficiency of a power station (or any other heat engine) improves with a higher temperature input and a lower temperature output.

Sankey diagrams

Energy transformations can be usefully represented in flow diagrams, such as Figure 8.8. This is another way of representing the simple situation described earlier in Figure 8.6. The width of

■ **Figure 8.8** A simple Sankey diagram

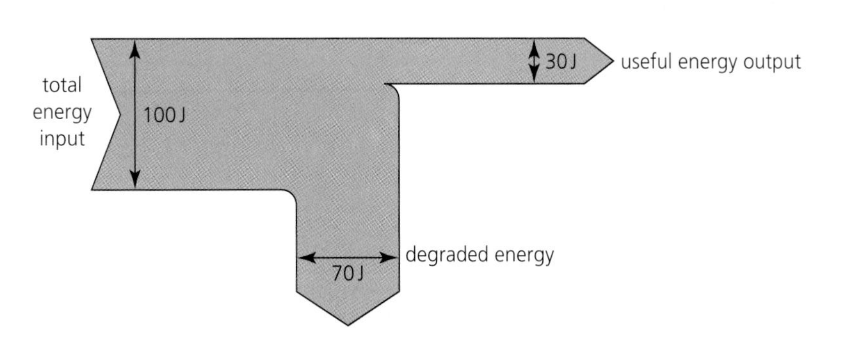

each section is proportional to the amount of energy (or power), starting with the energy input shown at the left of the diagram. Degraded energy is shown with downwards arrows and useful energy flows to the right.

Diagrams like these are known as **Sankey diagrams** and they can be used to help represent many energy transformations.

Figure 8.9 represents the useful energy transformations in a fossil-fuel power station (like that shown in Figure 8.5) and Figure 8.10 shows a Sankey diagram representing the energy flow through the same system, including the degraded energy.

Computer simulations are very helpful for observing and comparing Sankey diagrams representing various physical processes.

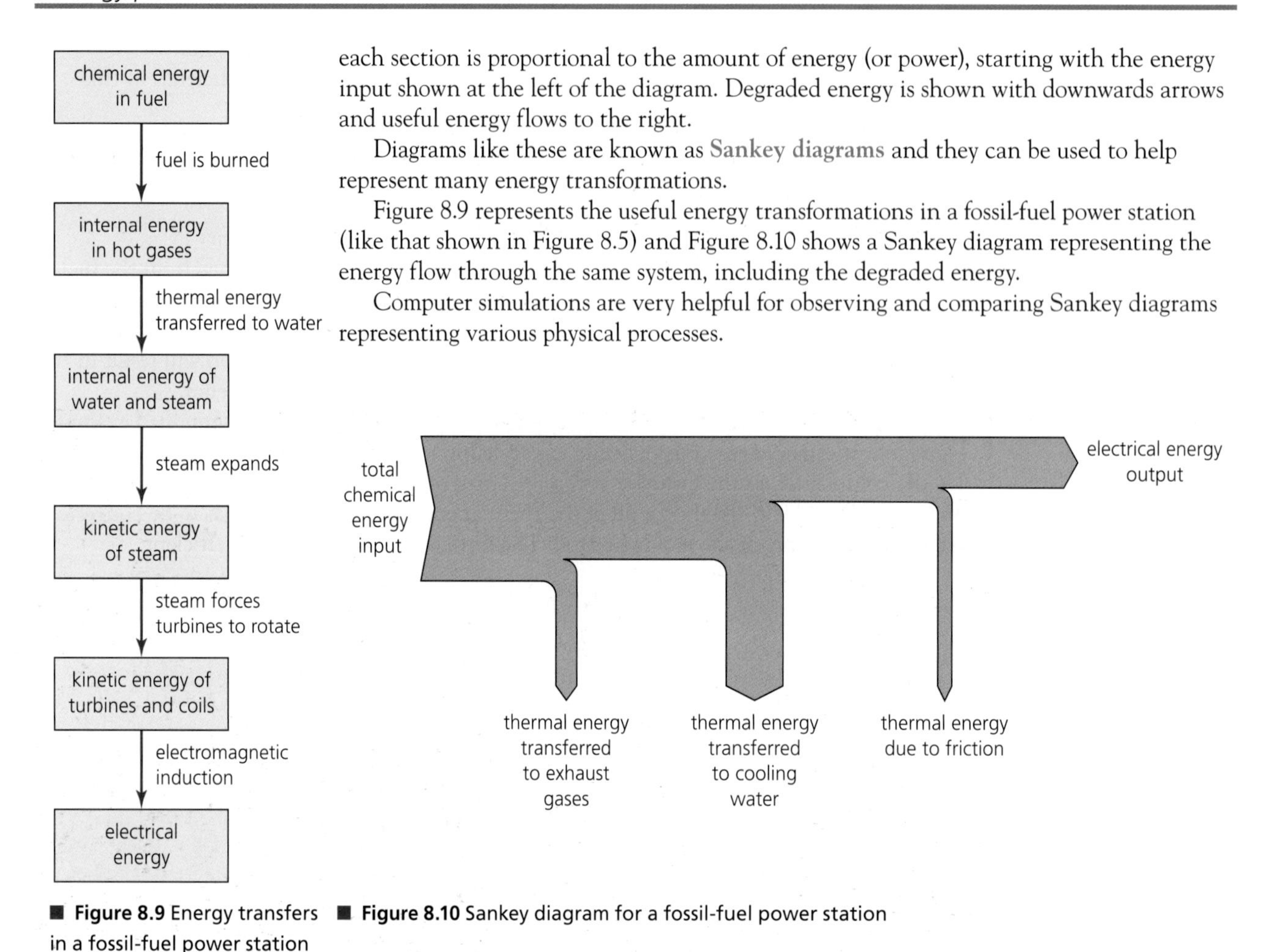

■ **Figure 8.9** Energy transfers in a fossil-fuel power station

■ **Figure 8.10** Sankey diagram for a fossil-fuel power station

9 Estimate the overall efficiency of the power station represented in Figure 8.10.

10 Draw Sankey diagrams to represent the energy flows for:
  a a car travelling at a constant speed
  b a small torch bulb powered by a battery.

11 a Which has more internal energy: 1 kg of water at 25°C or 1 kg of water at 35°C?
  b Explain why it is theoretically possible to transfer energy from water at 25°C to do useful work if it is in a room at 15°C, but useful energy cannot be transferred from the same water if it is in a room at 35°C.
  c Everything around us contains very large amounts of internal energy. Explain why we are not able to extract this energy to do useful work.

■ Energy choices

In the rest of this section we will describe the basic features, and some of the advantages and disadvantages, of using the following energy sources:

- fossil-fuel power stations
- nuclear power stations
- wind power generators
- hydroelectric systems
- solar power.

Fossil-fuel power stations

Fossil fuels currently provide between 80 per cent and 85 per cent of the world's energy needs. Oil, coal and natural gas are used worldwide for the generation of electricity and for transportation. However, the burning of fossil fuels has two very important disadvantages:

- The gases released during the combustion of fossil fuels enter the Earth's atmosphere and almost certainly lead to increased global temperature and *climate change*. This vital topic is discussed in more detail in Section 8.2.
- Fossil fuels are non-renewable and we cannot continue to use them at the current rate. Our supplies will become depleted.

Figure 8.11 show the outside of a typical fossil-fuel power station. Figure 8.5 shows an outline sketch of a typical interior. Figure 8.9 uses a Sankey diagram to represent the energy flow in a fossil-fuel power station.

Figure 8.11 A fossil-fuel power station

All fossil fuels have high specific energies compared to most other energy sources except nuclear energy (as shown in Table 8.1). This is one of the main reasons that the world has become so dependent on them. Worked example 1 shows a typical calculation involving the fuel consumption of a gas-fired power station.

Worked example

1 Estimate the useful output power of a gas-fired power station that uses natural gas at a rate of $15.0\,\text{kg}\,\text{s}^{-1}$. Assume that the specific energy of the gas is $55\,\text{MJ}\,\text{kg}^{-1}$ and the efficiency of the power station is 45 per cent.

Power output, P = efficiency × mass of fuel burned every second × specific energy

$P = 0.45 \times 15.0 \times (55 \times 10^6)$

$P = 3.7 \times 10^8\,\text{W}$ (370 MW or 0.37 GW)

Natural gas power stations are the most efficient type of power station. They are able to convert nearly half of the chemical potential energy in the gas into electrical energy (approximately 45 per cent efficient). Coal-fired power stations are typically no more than 35 per cent efficient. Oil-fired power stations have efficiencies of about 40 per cent. (These figures are for rough guidance only – they have been rounded-off to the nearest five per cent to make them easy to remember.) Option B: Engineering physics uses the laws of thermodynamics to explain why it is impossible to achieve much higher efficiencies than these values.

Because of its high specific energy, the cost of transporting coal over quite large distances by rail is not too high when compared with the other expenses of power generation. For similar reasons, oil and natural gas (being fluids) are commonly moved around the world through very long pipelines. Large quantities of oil are also moved around the world in large ships (tankers). The advantages of moving fossil fuels from where they are extracted from the ground to where they are used are considerable, but so too are the disadvantages, especially when accidents result in oil being spread widely into the environment. This kind of pollution can be particularly harmful when an oil spill occurs at sea, such as the accident off the coast of the USA in 2010 (see Figure 8.12). Because of their high energy density, it is possible to store enough fossil fuels to provide for the energy demands for many weeks ahead.

■ **Figure 8.12** The oil spill in the Gulf of Mexico, 2010

How long will fossil-fuel reserves last?

There are many reasons why it is difficult to predict how many years fossil fuel reserves will last. However, it seems probable that sometime within the next 100 years the world will have to face the major problem of very significant reductions in the availability of coal, crude oil and natural gas as the Earth comes to the end of the 'fossil-fuel age' and cheap energy. However, at the same time there will be a beneficial decrease in the emission of greenhouses gases.

At the *present rate of usage*, existing coal reserves may last about another 150 years, natural gas about 60 years and crude oil about 50 years. But these figures can be very misleading because the world continues to change and we cannot be sure about many factors, including:

- how much fossil fuel remains to be discovered
- whether existing sources that are currently considered uneconomic to extract from the ground will later become more viable as technologies improve and economic conditions change (e.g. *fracking*)
- to what extent the use of renewable sources and nuclear power will increase
- by how much the world's consumption of energy will increase.

On top of these factors, economic and political pressures will probably have unforeseen and all-important effects. Of course, some countries are 'rich' in fossil-fuel reserves (for example, the USA has the highest coal reserves, see Figure 8.13), while other countries have very little or none.

Table 8.2 summarizes the main advantages and disadvantages of using fossil fuels.

■ **Figure 8.13** The USA has the highest reserves of coal

Advantages	Disadvantages
• High energy density. • Fuel is relatively cheap (although economic and political factors may result in significant and sudden changes in price). • Power stations are relatively inexpensive to construct and maintain (when considering their high power outputs). • Power stations can be built in almost any location (that has good transport links and a plentiful water supply). • These are established technologies – power stations, transport and storage systems already exist.	• Greenhouse gas emissions and global warming. • Chemical pollution during mining and burning (including acid rain). • Non-renewable sources. • Extraction/mining can damage the environment and be hazardous to health. • Leakage from oil tankers or pipelines can cause considerable harm to the environment.

■ **Table 8.2** Advantages and disadvantages of using fossil fuels

12 An old-fashioned steam train had an output power of 2.3 MW but an efficiency of only 8.4 per cent. If the coal used had a specific energy of 29 MJ kg^{-1}, how much coal had to be burned every minute?

13 a How much fuel (kerosene) does a jet airliner consume every second when travelling at a constant speed of 240 m s^{-1} with a power output of 89 MW? Assume that the fuel has an energy density of 37 MJ m^{-3} and the engines have an efficiency of 39 per cent.
b What is the resistive force acting on the airliner?
c Estimate the amount of fuel needed to travel a distance of 5000 km at this constant height and speed.

14 A 220 MW oil-fired power station has an efficiency of 40 per cent and uses fuel at a rate of 13 kg s^{-1}. What is the specific energy of the fuel?

15 a What is the efficiency of a coal-fired power station that produces an average 560 MW of output power when burning fuel at a rate of 4.8×10^6 kg every day? (Assume that the specific energy of the coal is 30 MJ kg^{-1}.)
b If the discarded thermal energy was removed from the power station by cooling water, which should not increase in temperature by more than 4.0 °C, calculate the minimum rate of flow of cooling water that would be needed. The specific heat capacity of water is 4180 J kg^{-1} °C^{-1}.
c Estimate the mass of coal that must be burned every year to supply the needs of a large home that uses an average power of 3 kW.

16 Use the internet to find out the latest information on how long fossil-fuel reserves are expected to last.

Nuclear power stations

There are a total of approximately 500 nuclear power stations in about 30 different countries around the world. The USA has the highest number of reactors but France has the biggest percentage of the total electrical energy generated from **nuclear reactors**. China and Russia are planning to build the largest number of reactors. The use of nuclear power is a controversial issue that stirs up strong feelings in many people, but it is not sensible to form an opinion without first understanding some of the physics involved.

■ **Figure 8.14** The 3.6 GW nuclear power station in Cruas, France, supplies about five per cent of the country's electricity needs

Chain reactions and energy transfers

The energy released in a nuclear reactor comes from the *fission* of nuclei, not from radioactive decay. When a fission reaction in a nucleus of uranium-235 is initiated (started) by the *capture* of a neutron, the process will release about 200 MeV (3.2×10^{-11} J) of energy (for more details see Chapter 7). The nuclear potential energy is transferred to the kinetic energy of the resulting nuclei (the **fission fragments**) as well as to the neutrons and photons. This is an enormous amount of energy from one tiny nuclear reaction.

If the same process can be repeated with a very large number of uranium nuclei, the resulting increased kinetic energy of so many particles will effectively be a very large increase in the internal energy of the material.

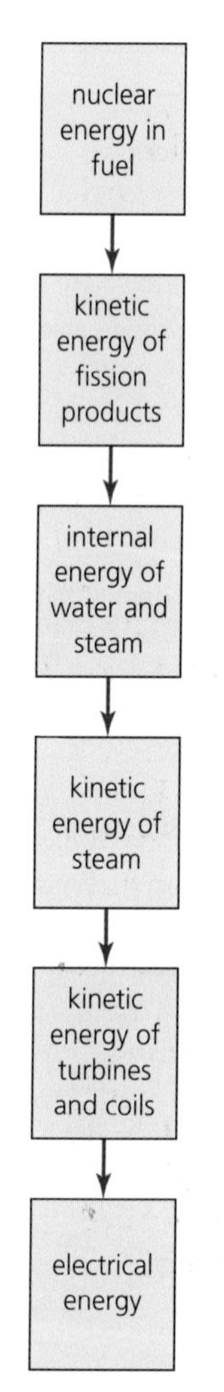

■ **Figure 8.15** Energy flow in a nuclear power station

The uranium will get much hotter with the possibility of large-scale production of electrical energy in much the same a way as in a conventional fossil-fuelled power station. The material used is called the **nuclear fuel** and it is usually in the shape of *rods*. Obviously, a nuclear fuel is different from a fossil fuel because it is not burned to transfer energy. Figure 8.15 shows the main energy transformations in a nuclear power station.

The following equation is just one of several possible fission reactions that nuclei of uranium-235 may undergo when they interact with slow-moving neutrons. It is also shown diagrammatically in Figure 8.16.

$$^{1}_{0}n + ^{235}_{92}U \rightarrow ^{144}_{56}Ba + ^{89}_{36}Kr + 3^{1}_{0}n + \text{about } 200\,\text{MeV energy}$$

When a neutron collides with a nucleus of uranium-235, fission is not certain; in fact it is unlikely and typically only a small percentage of interactions will initiate nuclear fission. In the equation above it is clear that, in this example, each nuclear fission produces three more neutrons (the average for all uranium-235 fissions is about 2.5) and these three neutrons may themselves then go on to cause further fissions. If, on average, at least one of the neutrons causes further fission, then the process will continue and it is described as being a **chain reaction**. This type of reaction is difficult to achieve.

If, on average, each fission reaction leads to one further fission reaction (and so on), then the number of fissions per second remains *constant* and the process can be described as *controlled* or **self-sustained nuclear fission**. This process provides the basis for nuclear reactors generating electricity in power stations. However, if each fission produces more than one further fission, then the reaction is *uncontrolled* and the number of fissions per second will rise rapidly. This will result in the release of an enormous amount of energy very quickly. This is what happens in nuclear weapons.

Figure 8.17 shows an idealized example of the start of a chain reaction in which all neutrons cause further fissions.

A relatively small number of fission reactions occur randomly in any sample of uranium-235 all the time. These reactions provide the neutrons needed for further fission reactions. In order to encourage and control nuclear fission, scientists need to understand the main factors that affect the probability of fission occurring. These are as follows:

1 Neutrons are penetrating particles (because they are uncharged) and many or most of them will usually pass out of the material without interacting with any nuclei.
2 Uranium-235 atoms make up only a small percentage of all the atoms in the material.
3 When a neutron does interact with a uranium-235 nucleus, it will only cause fission if it is travelling relatively slowly. Fast neutrons do not cause fission.

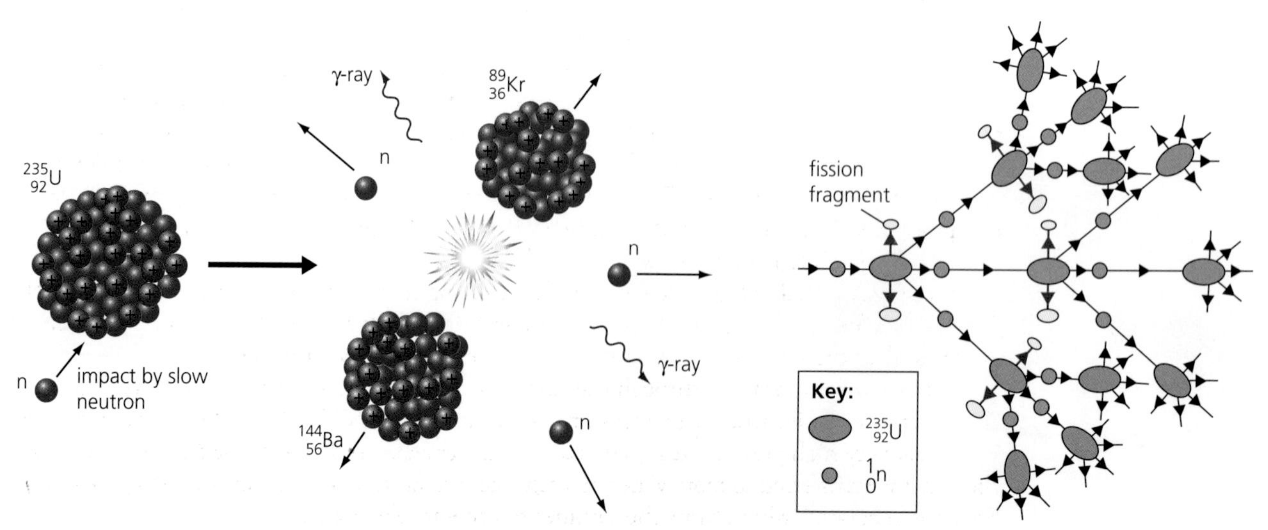

■ **Figure 8.16** Fission of uranium-235

■ **Figure 8.17** The start of a chain reaction

Critical mass

The ratio of volume to surface area of a solid increases as it gets larger (consider solid cubes of different sizes). This means that the more massive a material is, the smaller the percentage of neutrons that will reach the surface and escape. That is, a higher percentage may cause fission. The critical mass of a material is the minimum mass needed for a self-sustaining chain reaction. (Uranium that contains 20 per cent of uranium-235 has a critical mass of over 400 kg.)

Fuel enrichment – increasing the percentage of uranium-235

Uranium-235 is the only nuclide on the planet that occurs in significant quantities that can sustain a chain reaction. However, when uranium ore is extracted from the ground and refined, the uranium atoms are approximately in the ratio 99.3 per cent uranium-238 and only 0.7 per cent uranium-235 (with traces of other uranium isotopes). All the isotopes of uranium are radioactive, but the half-life of uranium-238 is very long (4.5×10^9 years), similar to the age of the Earth, whereas uranium-235 has a half-life of 7.0×10^8 years.

■ **Figure 8.18** Uranium fuel rods

For a chain reaction and power generation, the percentage of uranium-235 has to be increased to about three per cent at the very least, although higher percentages are preferable. (Nuclear weapons require a much higher percentage.) This process is called fuel enrichment. Figure 8.18 shows a photograph of enriched uranium fuel rods.

Uranium-238 nuclei can absorb neutrons without causing fission, so too much uranium-238 will discourage a chain reaction. Enrichment cannot be done chemically because isotopes of the same element have identical chemical properties, so physical processes need to be involved (for example, using the *diffusion* of gaseous uranium compounds) but these are difficult and expensive technologies. The remaining uranium is called *depleted uranium*; it has physical properties, especially its high density, that have made it useful in military engineering but this has been controversial (because it is radioactive).

Moderator, control rods and heat exchanger

The neutrons released in nuclear fission have typical energies of about 1 MeV, which means that they travel very fast. This is usually too fast to initiate another fission reaction. The slower a neutron travels, the higher the probability it has of it being captured by a uranium nucleus. Therefore, before a chain reaction can occur the neutrons need to be slowed down to energies of less than 1 eV (they are often then described as *thermal* neutrons). This process is called *moderating* the neutrons and the material used is called a moderator.

In order for the fast neutrons to lose so much of their kinetic energy, they need to collide many times with the nuclei of atoms. In general, when particles collide there is a higher rate of transfer of kinetic energy between them if they have approximately the same mass (see Chapter 2). The mass of a neutron is always lighter than the mass of a whole nucleus, but the difference is less for nuclei with low mass. This is why atoms with nuclei of small mass are preferable for this process of moderation, but it is also important that the nuclei do not absorb neutrons. Commonly, graphite (carbon) or water is used as a moderator.

In the pressurized water-cooled reactor shown in Figure 8.19, the nuclear fission occurs in the fuel rods in the reactor vessel. As the cold water flows past the hot fuel rods it removes thermal energy and, at the same time, acts as the moderator, slowing down the neutrons.

The hot water that flows through the reactor vessel is in a sealed system and it never leaves the concrete containment. Its temperature may be as high as 300°C, but it does not boil because it is under very high pressure. It is pumped to a heat exchanger in which the thermal energy in the water is transferred to more water in a *separate* system (this is an important safety measure). Steam is generated, which turns the turbines to generate electricity.

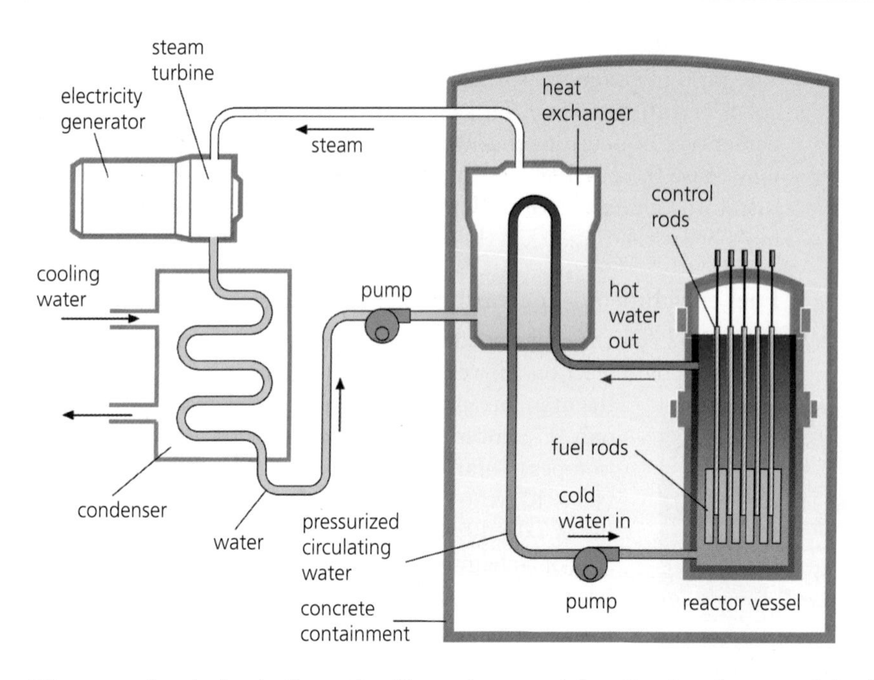

Figure 8.19 A pressurized water-cooled reactor

The control rods (typically made of boron) are used for adjusting the rate of the fission reactions by absorbing neutrons. This is done by moving the rods into or out of the system as necessary. Figure 8.20 represents the use of control rods to produce controlled fission.

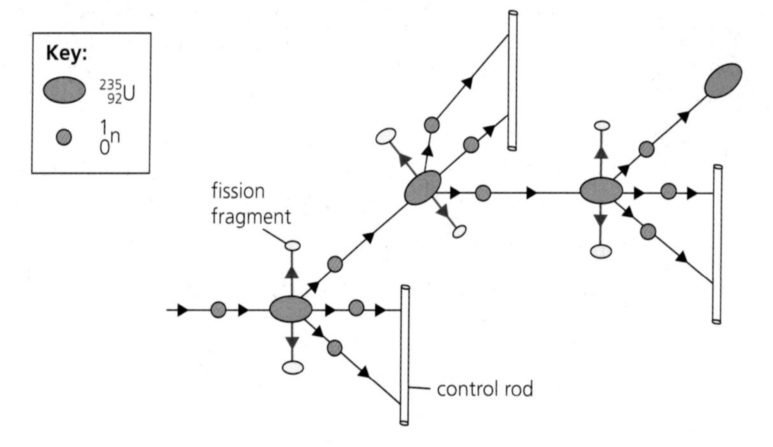

Figure 8.20 How control rods are used to control a chain reaction

Students are advised to use computer simulations to model the working of nuclear power stations and other nuclear processes.

17 Calculate the speed of: a 1 MeV neutron; b a 1 eV neutron.

18 Suggest why isotopes of uranium are so difficult to separate by physical methods.

19 A nuclear power station using uranium-235 has an efficiency of 43 per cent and a useful power output of 1 GW.
  a If the average energy released per fission is 190 MeV, calculate how many fissions occur every second.
  b What mass of uranium-235 is needed every day?

20 a Calculate the specific energy of pure uranium-235 (in $J\,kg^{-1}$).
  b If uranium-235 is only three per cent of the total mass of the fuel, calculate the energy density of the fuel.
  c Compare your answer for b with a type of coal that can transfer $29\,MJ\,kg^{-1}$.

21 Another possible fission of uranium-235 results in the formation of $^{138}_{55}Cs$ and $^{96}_{37}Rb$. Write a nuclear equation for this reaction.

22 a If an individual uses electrical energy at an average rate of 1 kW, what is their annual energy consumption?
  b What mass of uranium-235 atoms has to undergo fission to provide the energy needed for a year? (Assume 100 per cent efficiency.)

Safety issues

Nuclear accidents

It is fair to say that the risks of nuclear power are very well understood and, as a result, safety standards are very high. But, for many people, this is not reassuring enough because no matter how careful nuclear engineers are, accidents and natural disasters can happen. Safety standards do vary from country to country, but the consequences of a nuclear accident anywhere could be really disastrous.

The world's worst nuclear accidents at Chernobyl in Ukraine in 1986 and Fukushima (see Figure 8.21) in Japan in March 2011 remain as vivid warnings about the possible risks of nuclear power. The radiation leaks and explosions at Fukushima followed the damage caused by a tsunami, whereas the 'meltdown' at Chernobyl is widely recognized to be the result of poor design and management at that time. Modern designs and safety procedures have been vastly improved but may still be insufficient in the face of an extreme natural disaster. Hundreds of thousands of people had to be evacuated because of these accidents, and the number of long-term illnesses and deaths caused by them will take many years to be confirmed.

A **thermal meltdown** is probably the most serious possible consequence of a nuclear accident. If, for some reason (for example, the loss of coolant or the control rods failing to work properly), the core of the reactor gets too hot or even melts, the reactor vessel may be badly damaged. Fires and explosions may happen as extremely hot materials are suddenly exposed to the air. Highly concentrated and dangerous radioactive materials may then be released into the ground, water or air so that they are spread over large distances by geographic and weather conditions.

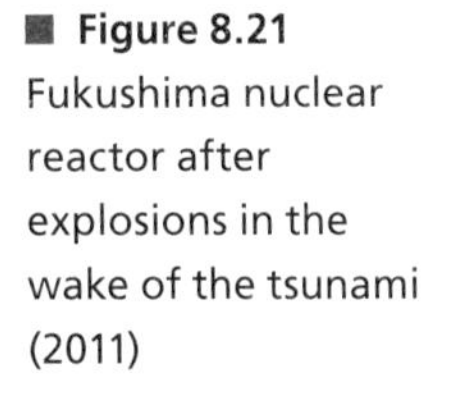

■ Figure 8.21 Fukushima nuclear reactor after explosions in the wake of the tsunami (2011)

When considering the dangers of nuclear power, it should always be remembered that the uses of *all* other kinds of energy resources have their various risks. In particular, coal mining has been responsible for an extremely high number of serious injuries, long-term health problems and deaths over the past 200 years.

Nuclear waste

The products of fission reactions are highly radioactive and can therefore be dangerous if safety standards are not high enough. Anything associated with the reactor must be considered as a potential health risk, although the biggest dangers are obviously posed by the fuel rods themselves. Nuclear waste products, such as the fuel rods after they are no longer useful, are highly radioactive and contain some isotopes with very long half-lives, which means that they will be dangerous for many thousands of years. As an example, ^{99}Tc has a half-life of 210 000 years. It should be noted, however, that if a radioisotope has a very long half-life, the activity from each gram will be very low.

■ **Figure 8.22** Storage of nuclear waste

■ **Figure 8.23** McClean Lake uranium mine in Saskatchewan, Canada

There is no way to stop a radioactive material from emitting nuclear radiation, so safety measures have to concentrate on preventing people from being exposed to the radiation. Radioactive substances need to be surrounded by materials that will absorb the radiation safely. Thick concrete, water and lead are widely used for this purpose, depending on the circumstances. Exposure to radiation is also reduced by keeping people as far away as possible or, if they need to be close, by wearing protective clothing and limiting the length of time of their exposure.

The disposal of highly dangerous nuclear waste remains a serious problem that we are leaving for generations to come. At present, some of this waste is stored deep underground in very strong and thick containers designed to prevent leakage, while some is stored above ground where it can be more easily monitored (see Figure 8.22).

Mining

Uranium is widespread as a trace element in many minerals around the world, but more than half of the world's uranium mines are located in only three countries: Canada (see Figure 8.23), Australia and Kazakhstan. A lot of rock has to be extracted from large, open-cast mines to obtain only a small amount of uranium, which is usually in the form of uranium oxide.

One of the products from the decay of uranium nuclei is the radioactive gas radon. This makes uranium mines potentially dangerous places. Because it is a gas, any radon present in the air can enter the lungs and, once it is inside the body, the radiation that it emits is far more dangerous than similar radiation sources outside the body.

Utilizations

Nuclear weapons

The first nuclear bomb was tested in 1945 in the USA. The latest country to develop nuclear weapons was North Korea in 2013. It is believed that nine different countries have nuclear weapons; they all claim that these are just for national defence and that the weapons would never be used except under the extreme circumstances of being attacked by another country.

The destructive power of just a single nuclear bomb was horrifyingly demonstrated at both Hiroshima and Nagasaki in Japan in 1945. The basic requirement is a critical mass (but not all together in one piece) of plutonium-239 or highly enriched uranium-235 (enriched to 20 per cent at least, but preferably much more). If the bomb is to be detonated, the two separate halves of the critical mass have to be brought together by a relatively small conventional explosion, as shown in Figure 8.24. This then produces an uncontrolled chain reaction.

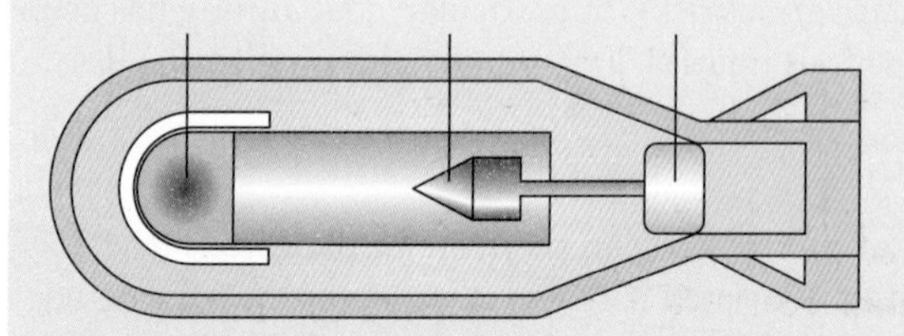

■ **Figure 8.24** Sub-critical masses in a nuclear bomb – a conventional explosion forces the two sub-critical masses together

The production of enriched 'weapons-grade' uranium is technically extremely difficult and expensive. The alternative, plutonium-239, is a product of neutron capture by uranium-238 nuclei in a nuclear power station, which can later be recovered from the fuel rods. This is why there is concern that a few countries may use their development of a nuclear power programme to (secretly) develop nuclear weapons.

Many people claim that the threat of nuclear weapons becoming available to the 'wrong' countries/organizations/people is one of the biggest threats facing the world in the twenty-first century. However, some people believe that the threat of the use of nuclear weapons has actually helped prevent major wars, although many others believe that there could never be any justification for the construction or use of nuclear weapons under *any* circumstances. Regretfully, in the modern world we need to be aware of the possibilities of terrorist attacks on nuclear power stations, or the transportation of nuclear materials, or the use of nuclear or 'dirty' bombs by terrorists.

1. It might be agreed that a world without nuclear weapons would be a better place but, once they had been invented, is it ever going to be possible to remove them all?
2. Research the *Nuclear Non-Proliferation Treaty* and which countries have not signed it.

Table 8.3 Summarizes the advantages and disadvantages of the use of nuclear capability.

Advantages	Disadvantages
• Extremely high energy density. • No greenhouse gases emitted during routine operation. (Some scientists think that nuclear power may be the only realistic solution to global warming.) • No chemical pollution during operation. • Reasonably large amount of nuclear fuels are still available. • Despite a few serious incidents, statistically over the last 50 years, nuclear power has overall proven to be a reasonably safe energy technology.	• Dangerous and very long-lasting radioactive waste products. • Expensive. • Efficiency is not high when the whole process is taken into account. • Threat of serious accidents. • Possible target for terrorists. • Linked with nuclear weapons. • Not a renewable source.

■ **Table 8.3** Advantages and disadvantages of nuclear power

■ **Figure 8.25** Protestors try to stop the shipment and storage of nuclear waste

ToK Link

To what extent should public opinion influence scientific research and development?

The use of nuclear energy inspires a range of emotional responses from scientists and society. How can accurate scientific risk assessment be undertaken in emotionally charged areas?

There can be few areas in science where knowledge and emotions are so intertwined as in any discussion about nuclear power. Very few people are undecided; they tend to be strongly for or strongly against the use of nuclear energy. The perceived worst consequences of a nuclear accident at a particular location are, no doubt, terrifying, but a realistic **risk assessment** needs to compare that carefully with the long-term and more widespread risks associated with using other energy resources, or indeed the other risks of everyday life.

When we try to form an honest opinion about anything, our own previous experiences, prejudices and emotions will influence our thoughts, no matter how hard we try to be **objective**. There is no doubt that the shared processes of science try very hard to eliminate such bias, but 'the knowledge' that we each carry around in our heads is highly personalized.

Nuclear fusion

Nuclear fusion is the source of the energy in all stars, including the Sun (see Chapter 7). A typical fusion reaction like that shown in the following equation would transfer about 18 MeV:

$$^{2}_{1}\text{H} + ^{3}_{1}\text{H} \rightarrow ^{4}_{2}\text{He} + ^{1}_{0}\text{n}$$

The two nuclei of hydrogen (known as deuterium and tritium) are both positively charged and they will only fuse together if they have enough kinetic energy to overcome the repulsive forces between them. Achieving this high kinetic energy requires temperatures of more than 10^8 K. At this high temperature electrons move around separately from hydrogen ions (protons) and the gas is called a plasma.

■ **Figure 8.26** Russian Tokamak thermonuclear reactor

Nuclear fusion at these temperatures has been achieved on Earth, but the main problem is continuing (*sustaining*) the reactions over any significant length of time. The hot plasma cannot be allowed to come in contact with the walls of a container because that would cause it to cool down and the container would also become contaminated.

To avoid this problem the plasma is confined (contained) using a very strong magnetic field, such as in the Russian designed Tokamak thermonuclear reactor, which uses a doughnut (torus) shaped field (Figure 8.26).

If energy could be released on a large scale by nuclear fusion, it would have the advantage of being a more plentiful fuel supply with fewer radioactive waste problems, compared to nuclear fission. These advantages are enormous, but despite years of considerable scientific efforts, the technology needed to sustain nuclear fusion reactions is still proving too difficult.

23 All the isotopes of uranium are radioactive and decay into other elements. Explain why there is still a significant amount of uranium left on Earth.

24 If, after a nuclear accident involving an isotope with a half-life of 12 years, the radiation level in the environment was 33 times higher than the normal background count, approximately how much time would have to pass before people could re-enter the area?

25 a Dangerous nuclear waste is to be stored underground for more than 100 000 years. Suggest ways in which this plan could go wrong.
b Is putting the waste on a rocket and firing it into the Sun a sensible alternative? Explain your answer.

26 a Use Coulomb's law to calculate the repulsive force between a tritium nucleus and a deuterium nucleus when they are separated by a distance of 5.0×10^{-15} m.
b Calculate the acceleration with which the deuterium nucleus would start to move away.

27 a Use the internet to gather data on the total number of deaths and serious illnesses that have been attributed to the Chernobyl accident since it occurred in 1986.
b You will probably find that the information from various websites is different. Suggest reasons why consistent information is difficult to obtain.
c How can you decide which website gives you the most reliable information?

28 Use the internet to find out about the latest developments in nuclear fusion: what power was achieved, and for how long? What nuclides were used? Where did it take place? In particular, what and where is ITER?

29 a What is your personal opinion about the use of nuclear power? In 100 words or fewer explain why you think that your country should, or should not, operate nuclear power stations.
b Conduct a survey of the opinions of your fellow students. Are opinions of physics students different from those of non-physics students?

30 a Under what circumstances (if any) do you think that it would be acceptable for one country to use a nuclear weapon to bomb another?
b Do you believe that every country has the right to decide for itself whether to build nuclear weapons or not? If not, who should control any development of the world's nuclear weapons?

Wind power generators

Radiation from the Sun causes differences in temperature that result in changes in air density. These changes produce convection currents and differences in air pressure. Wind is air moving between areas of high pressure and low pressure. For thousands of years the energy transferred from windmills has been used around the world for work such as grinding crops (see Figure 8.27). There is an enormous amount of kinetic energy in the winds across the planet, but it has low energy 'density' and it is only in recent years that technological advances have encouraged the widespread use of large wind-driven electrical generators. It has now become a very rapidly expanding technology.

■ **Figure 8.27** Traditional windmills in Greece

■ **Figure 8.28** A wind farm in Denmark

There are many different designs of wind generators (which generate electricity, not wind!) but we will concentrate on the type most commonly used for large-scale electrical power generation. They have a horizontal axis of rotation, as shown in Figure 8.28. However, there is an interesting variety of designs for wind generators with vertical axes, most of which function equally well whatever the direction of the wind. Figure 8.29 shows an example.

■ **Figure 8.29** Wind generators with a vertical axis

The blades of a horizontal axis generator (usually there are three) move in a vertical circle and are designed to be struck by wind moving parallel to the Earth's surface. The kinetic energy of the blades is then transferred to make coils spin in electrical generators. The number of blades, their width and angle to the wind are all carefully chosen to get the maximum amount of power transferred from the wind. For the same reason, the whole blade mechanism is often able to rotate in the horizontal plane so that the wind always strikes the blades perpendicularly. The blades are designed to rotate at a particular speed with winds within a certain range, typically about 5 to 15 $m s^{-1}$. Figure 8.30 shows the energy transformations involved in wind power.

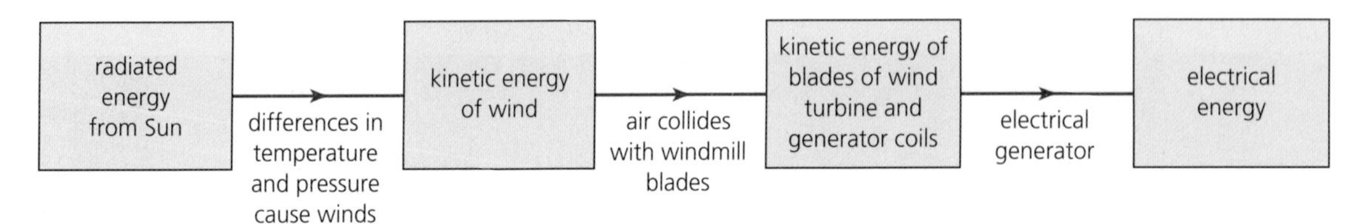

■ **Figure 8.30** Energy flow diagram for wind generators

Winds flowing close to the ground will lose energy due to friction and by striking obstacles, so generators are usually located near to coasts or on treeless hills, and the towers supporting the rotating blades are as tall as possible. Constructing wind generators in shallow water off-shore has obvious advantages, but these are expensive to construct. The largest off-shore wind farms (already operational and those in construction) are off the shores of the UK, Denmark and Germany.

Smaller wind generators may be ideal for remote locations, for example on boats or farmhouses that have no access to mains electricity. The advantages and disadvantages of using wind power are shown in Table 8.4.

Advantages	Disadvantages
• No greenhouse gas emissions. • Renewable source. • Free source of energy. • No pollution during use. • Small generators are ideal for remote locations.	• Expensive to construct. • Low energy density; large area needed (although the land around them can still be used for farming). • Wind speed (and power output) is unreliable. • Emits some noise. • Best locations are often far away from cities and towns. • Many people consider that they are ugly and spoil the environment.

■ **Table 8.4** Advantages and disadvantages of wind power

Calculations involving wind power

An equation for the power of a wind generator can be obtained by considering the loss of kinetic energy of the wind striking the blades.

The *maximum theoretical* power, P, available from a wind power generator with cross-sectional area A (see Figure 8.31) when struck by air of density ρ moving with a speed v is given by:

$$P = \frac{1}{2}A\rho v^3$$

This equation is given in the *Physics data booklet*.

It is not surprising that the power is proportional to the area and density, but the fact that it depends on the wind speed *cubed* should be noted. A doubling of the wind speed could theoretically produce eight (2^3) times the power.

The transfer of kinetic energy of the air to kinetic energy of the blades cannot be 100 per cent efficient, considering that some of the wind must pass between the blades and that the air that strikes the blades will not come to rest.

Additional Perspectives

Deriving an equation for wind power

The volume of air passing the blades every second, V, is equal to the speed of the wind, v, multiplied by the area 'swept out' by the rotating blades, A.

$$V = vA$$

Because mass = volume × density, ρ, the mass, m, passing every second is given by:

$$m = vA\rho$$

Using the equation for kinetic energy ($E_K = \frac{1}{2}mv^2$), the kinetic energy of this moving mass of air passing the blades every second is:

$$E_{K(air)} = \frac{1}{2} \times vA\rho \times v^2$$

We know that:

$$\text{power} = \frac{\text{energy transferred}}{\text{time}}$$

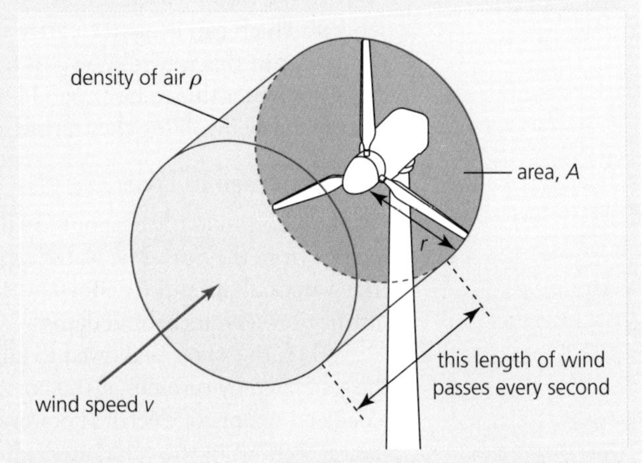

■ **Figure 8.31** Cross-section of wind striking the blades of a wind turbine

So, if all of the wind's kinetic energy is transferred to the wind generator every second (which would mean that the air behind the generator is not moving), then its maximum theoretical power, P, is given by:

$$P = \frac{1}{2}A\rho v^3$$

1 Find out why most horizontal-axis wind generators have three blades.

31 Explain what it means to say that wind power has a low 'energy density'.

32 A wind generator has blades of length 18 m and operates with an efficiency of 21 per cent. What power output would you expect with a $12\,\text{m s}^{-1}$ wind? (Density of air = $1.3\,\text{kg m}^{-3}$.)

33 A wind generator has an output of 10 kW when the wind speed is $8\,\text{m s}^{-1}$.
- a What wind speed would be needed for an output of 20 kW?
- b What power output would you predict for another generator of similar design, but with blades of twice the length, when the wind speed was $16\,\text{m s}^{-1}$?

34 A small farmhouse needs an average power of 4 kW. A wind generator is to be placed on the top of a tower where the effective average wind speed is $7.5\,\text{m s}^{-1}$. Calculate the length of blades needed to generate this power if the system is 24 per cent efficient.

35 Large wind generators can produce 5 MW or more of electrical power.
- a Make an estimate of the diameter of the circle through which the blades rotate. List the assumptions you made.
- b A total output of 20 MW is required. Discuss whether it might be better to build four similar generators or one generator of twice the size.

36
- a How many large wind generators (as described in question 35) would be needed to provide 2 GW (equivalent to a large fossil-fuelled power station)?
- b If the wind generators have to be at least 300 m apart, what is the minimum area of land needed?
- c Suggest a reason why the generators need to be so far apart.

37 Wind speeds vary considerably and quoting an 'average' value may be misleading, especially if a value for the average speed is used directly to predict an average power, as in question 34. Explain why this would almost certainly lead to an underestimate of the available power.

38 Draw a Sankey diagram to represent the energy transfers involved with a wind generator.

Hydroelectric power

When water is free to flow downwards, gravitational potential energy is transferred to kinetic energy, which can be used to drive turbines and generate electrical energy. Electric power generated in this way is known as **hydroelectric power**.

In order for this to be a useful long-term source of electricity, there must be a process (*cycle*) continuously involving the transfer of the energy needed to return the water to a higher level.

Water storage in lakes

Water is continuously evaporating from the oceans and seas due to the transfer of thermal energy from the Sun. The water vapour rises due to convection currents, forms clouds and later the water falls as rain (or snow), which can be collected and stored at high altitude in lakes, or in artificial reservoirs behind dams.

When the water is allowed to fall downwards, often nearly vertically, electrical energy can be generated by turbines at the bottom of the fall (see Figure 8.32). It is also possible to generate smaller amounts of electrical power directly from fast-flowing rivers. Some of the most powerful power stations in the world are hydroelectric, for example the Itaipu hydroelectric power station on the Brazil–Paraguay border (Figure 8.33), but there are also a very large number of small-scale schemes providing electric power for small communities in remote locations. Approximately 20 per cent of the world's electricity is generated by hydroelectric power stations and it is the world's most widely used source of renewable energy (for electricity production).

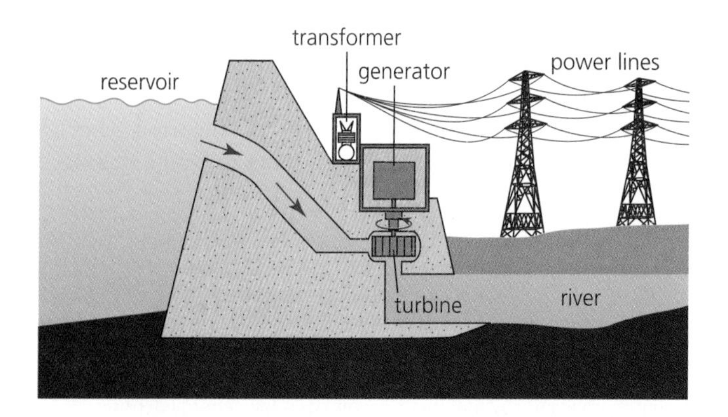

■ **Figure 8.32** Cross-section of a hydroelectric power station

■ **Figure 8.33** The Itaipu hydroelectric power station on the Brazil–Paraguay border

Pumped storage

In hilly locations, hydroelectric power offers a large-scale solution to the problem of storing excess energy from an electricity power station. Electrical power stations are designed to be operated at a particular power output at which they have their maximum efficiency but at night, when most people are asleep, there is normally less demand for electrical energy (see Figure 8.34). Some electricity companies try to encourage people to use electricity at night by offering cheaper rates. If power stations are operated at a lower power at night, their efficiency is reduced and a higher proportion of energy is wasted as thermal energy transferred into the surroundings. It is better to keep a power station running at approximately the same power 24-hours a day by storing the excess energy generated at night and transferring it back into the system during the day at time when the energy can be sold at a higher price. The most common way of doing this is to use the excess power to pump water up to a lake at a greater height during the night and then release it to generate extra power the next day.

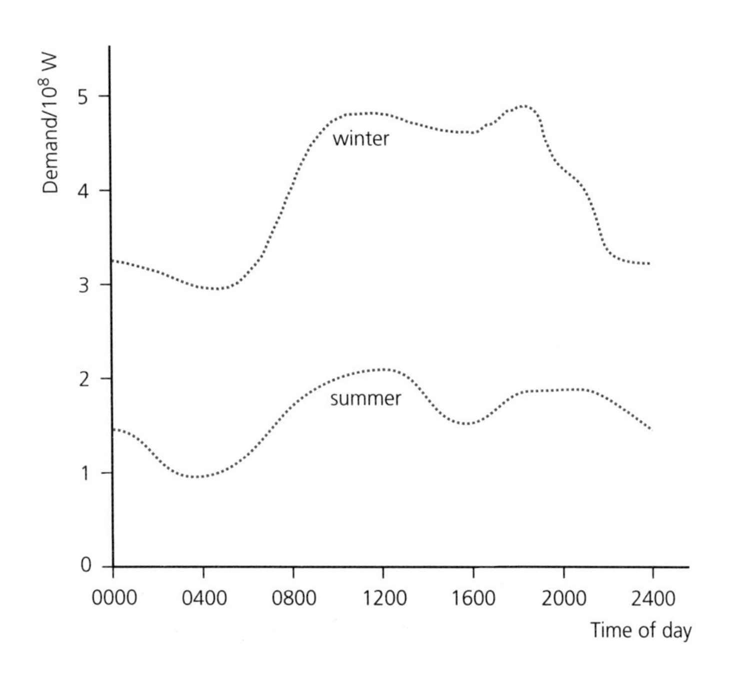

■ **Figure 8.34** Typical daily variation in energy consumption from a large power station in summer and winter

The energy transformations that take place in a hydroelectric power station are shown in Figure 8.35. *Tides*, caused by the gravitational attraction of the Moon, cause rises in sea level twice a day. The water can be trapped behind dams and released later when the sea level falls.

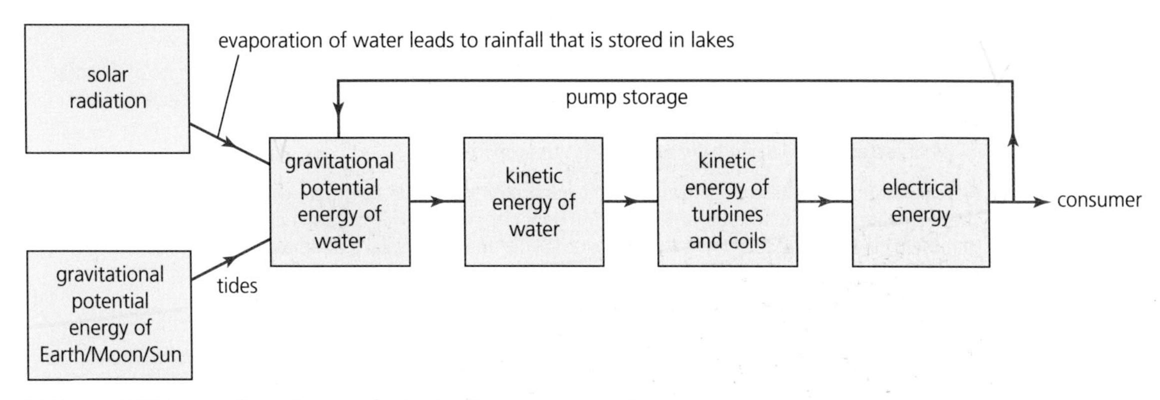

■ **Figure 8.35** Energy flow diagram for hydroelectric power stations

The advantages and disadvantages of using hydroelectric power are shown in Table 8.5.

Advantages	Disadvantages
• Renewable source of energy. • No greenhouse gas emissions (although there will probably be some increase in the release of methane from the enlarged lakes). • Relatively high efficiency. • Free source of energy. • No significant pollution during operation. • Dams may also be used to control river flow, improve irrigation and prevent flooding. • Newly created lakes can be a recreational resource and provide a new habitat for some plants and animals. • Can be the ideal energy resource for remote, hilly locations.	• The environment will be affected and the natural habitat of many plants and animals may be destroyed. • Newly built dams will form lakes that may cover land that was previously villages, towns, farmland etc. • Can only be used in certain locations (mountainous/hilly places). • Large-scale projects can be very expensive to construct, although maintenance is more reasonable. • The natural flow of rivers may be interrupted, which can have many undesirable consequences. • Hydroelectric power stations may be a long way from centres of population, so the power may need to be transmitted over long distances. • If a dam bursts it can cause considerable damage and loss of life.

Table 8.5 Advantages and disadvantages of hydroelectric power

Calculations involving hydroelectric schemes

The loss of gravitational energy when a mass, m, of water falls a distance, Δh, can be determined from:

$$\Delta E_P = mg\Delta h$$

The average theoretical power output during time Δt will be $\frac{\Delta E_P}{\Delta t}$.

Because they do not rely on the transfer of thermal energy, hydroelectric power stations will be much more efficient than fuel-burning power stations. Typically, the larger the power station, the more efficient it will be.

Worked example

2 What is the average output power from a small hydroelectric power station operating at 92.0 per cent efficiency if 242 000 kg of water passes through its turbines every hour, having fallen a vertical distance of 82.4 m?

Energy transferred from falling water in one hour:

$$\Delta E_p = mg\Delta h$$

$$\Delta E_p = 242\,000 \times 9.81 \times 82.4 = 1.96 \times 10^8\,\text{J}$$

$$\text{Maximum theoretical power} = \frac{\Delta E_p}{\Delta t} = \frac{1.96 \times 10^8}{3600} = 5.43 \times 10^4\,\text{W}$$

$$\text{Output power} = \text{input power} \times \text{efficiency} = 5.43 \times 10^4 \times 0.920 = 5.00 \times 10^4\,\text{W}$$

39 Calculate the maximum theoretical power available from a hydroelectric power station in which 2.0 m^3 of water falls a vertical distance of 112 m every second. Assume that the density of water is 1000 kg m^{-3}.

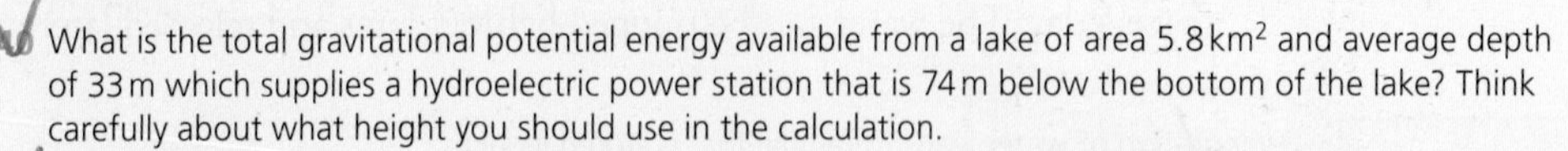

40 What is the total gravitational potential energy available from a lake of area 5.8 km^2 and average depth of 33 m which supplies a hydroelectric power station that is 74 m below the bottom of the lake? Think carefully about what height you should use in the calculation.

41 Water flows from a lake through a certain hydroelectric power station at an average annual rate of 4.3 $m^3 s^{-1}$.
  a If the water in the lake comes from the rainfall over an area of 138 km^2, what is the average rainfall (in cm y^{-1}) over that area?
  b What assumption did you have to make? Is it reasonable?

42 A family wants to install a small hydroelectric generator to provide their own electrical needs, which they estimate to be an average of 3.0 kW. The generator's efficiency is 85 per cent. If their home is at an altitude of 1420 m and they want to use water falling from a small lake at an altitude of 1479 m, what average mass of water must pass through the generator every second?

43 Make a list of the desirable features of a possible location for a new hydroelectric power station.

44 Suggest some 'undesirable consequences' that may occur if the natural flow of a large river is interrupted by one or more dams.

45 In a pumped storage system, water is pumped up to a lake/reservoir using electrical energy and then later allowed to fall back to generate electricity again. Of course some energy is wasted in this process. Explain why this process can still be a good idea.

Solar power

The intensity of solar radiation falling on a surface can be calculated using:

$$\text{intensity} = \frac{\text{power}}{\text{area}}$$

$$I = \frac{P}{A}$$

This equation is included in the *Physics data booklet*.

The intensity of the Sun's radiation arriving at the Earth is approximately 1400 W m^{-2}. Although this figure can vary by approximately seven per cent, it is known as the **solar constant**. It is the power passing through each square metre perpendicular to the Sun's 'rays' as they reach the outer limits of the atmosphere, as shown in Figure 8.36. (This power is across all wavelengths of the Sun's emission spectrum, although about half is in the visible part of the spectrum.)

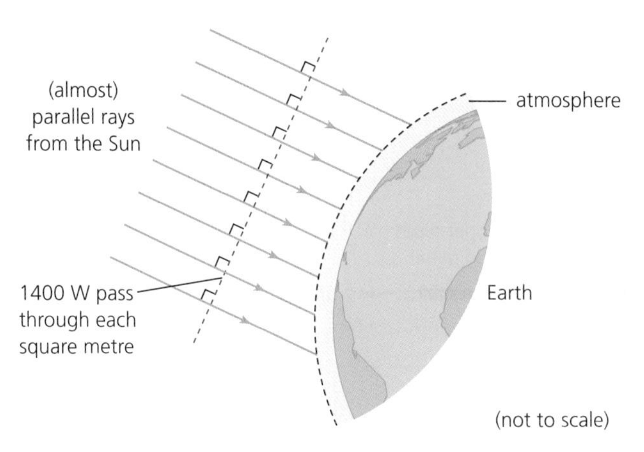

■ **Figure 8.36** Solar radiation arriving at the Earth

This is an enormous amount of energy incident on the Earth, although some of the energy from the Sun is absorbed, reflected and scattered (reflected randomly) in the atmosphere. For example, on a clear day with the Sun directly overhead, a figure for the intensity on the Earth's surface is sometimes quoted to be about 1 kW m^{-2}.

A more useful value, averaged over 24 hours and over the whole planet, is about 230 W m^{-2}. Of course, the energy transferred by radiation from the Sun is essential for all life on Earth and it is important to remember that it is, or has been, the indirect origin of the majority of the energy resources discussed in this topic. However, surprisingly little use has been made of the *direct* transfer of solar energy to other energy forms.

Solar radiation can be used directly in two different ways: in **solar heating panels** the radiant (solar) energy is transferred to raise the internal energy of water; in **photovoltaic cells** (also called photo cells or solar cells) the solar energy is transferred directly to electrical energy. Groups of photovoltaic cells are commonly called solar panels, but this can cause confusion with solar *heating* panels.

■ **Figure 8.37** Photovoltaic cells

Photovoltaic cells

When radiation falls on the semi-conducting material of a photovoltaic cell, free electrons are released within the material and a potential difference is produced across it.

When used in a circuit, both the current and the potential difference from a single cell tends to be low and there is also significant internal resistance. So, individually, they are only sources of small amounts of electrical power (see Figure 8.37). A large number can be connected together to provide much greater power. Figure 8.38 shows the energy transformation that occurs in a photovoltaic cell.

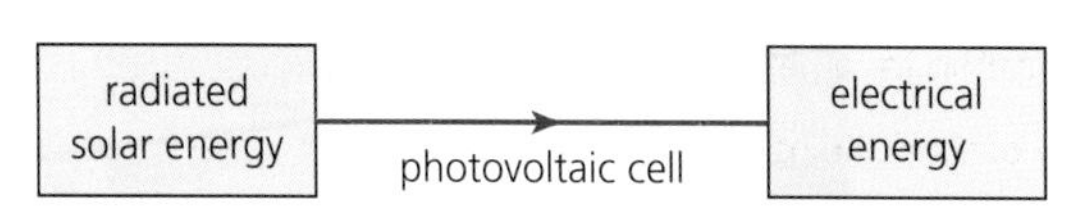

■ **Figure 8.38** Energy flow for a photovoltaic cell

Photovoltaic cells are probably the world's fastest growing renewable energy technology. Costs are falling quickly and efficiencies are rising. Most solar cells operate at less than 20 per cent efficiency, although the best are about 25 per cent, with future hopes aiming at over 40 per cent. Because of their potential convenience and low cost, a lot of research is going into the design of thin-film photovoltaic cells.

■ **Figure 8.39** Germany has made significant investments in photovoltaic power

Photovoltaic cells are used (together with rechargeable batteries) for low-power devices such as calculators and for isolated single devices (such as emergency phones), but improving technology and lower costs now mean that they are also being used widely to provide supplementary electric power to homes and offices. Figure 8.39 shows a typical example. Solar power stations are also being built to provide 100 MW (or more) of power to electric grid systems around the world.

Solar heating panels

In a solar heating panel the radiation from the Sun is absorbed by black-painted copper pipes through which water flows (Figure 8.40). The hot water is then pumped through a heat exchanger in which thermal energy is transferred from the hot water to a larger amount of water stored in an insulated tank. When the water in the tank is hot enough it can be used in the home for washing etc. Homes with solar heating panels will usually also have some other means of heating water for the times when the water is not hot enough. Figure 8.41 shows the energy transfer in a solar heating panel.

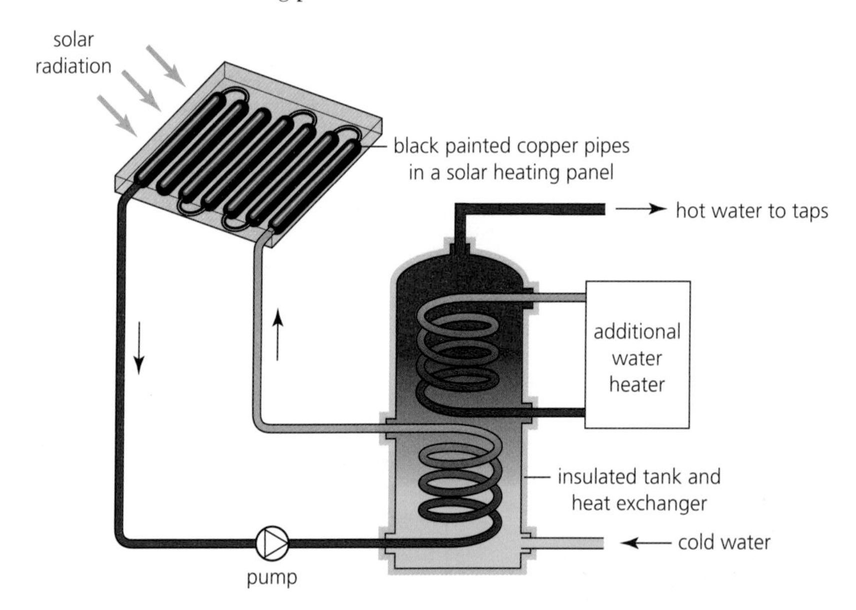

■ **Figure 8.40** A solar heating system

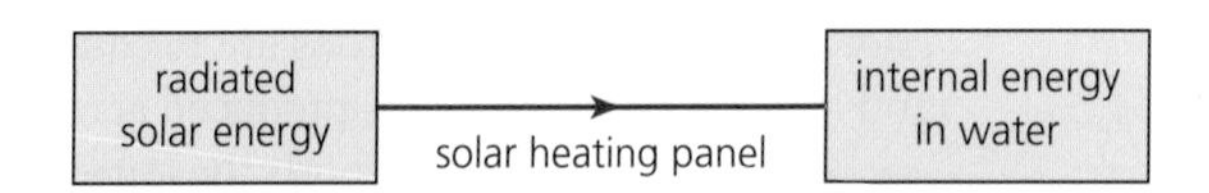

■ **Figure 8.41** Energy flow for a solar heating panel

Solar heating panels are mostly used for individual homes and they are typically placed on a roof. They are used widely in many countries, most usefully where there is a hot and/or a sunny climate (Figure 8.42). Many governments encourage their use, or even require that all new homes have solar heating panels.

■ **Figure 8.42** Solar heating panels on homes in Turkey

Variations in radiation intensity reaching the Earth's surface

The intensity of the radiation reaching a particular place on the Earth's surface depends on:

- the weather/climate
- the geographical latitude of the location
- time of day/night
- the angle of the surface to the horizontal at that location.

Figure 8.43 illustrates that if the same radiated solar power falls on areas A and B, the intensity (power/area) will be less at A because the area is larger. This diagram could be representing places at different latitudes on the Earth or, for a given location, it could be representing variations in intensity at different times of the day or at different times of the year.

To reduce the effect of these variations, photovoltaic cells or solar heating panels are not usually placed horizontally on the ground but are positioned so that they receive radiation perpendicularly at a carefully chosen (average) time of the day and year. A more expensive option is to provide the machinery necessary for the devices to move and keep aligned.

It is important to realize that if a panel were positioned so that it received the maximum radiation at midday in winter, the received power would still be less than that received at midday in summer. This is because the radiation has to pass through a greater length of the Earth's atmosphere. This can be seen by comparing the lengths P and Q in Figure 8.44.

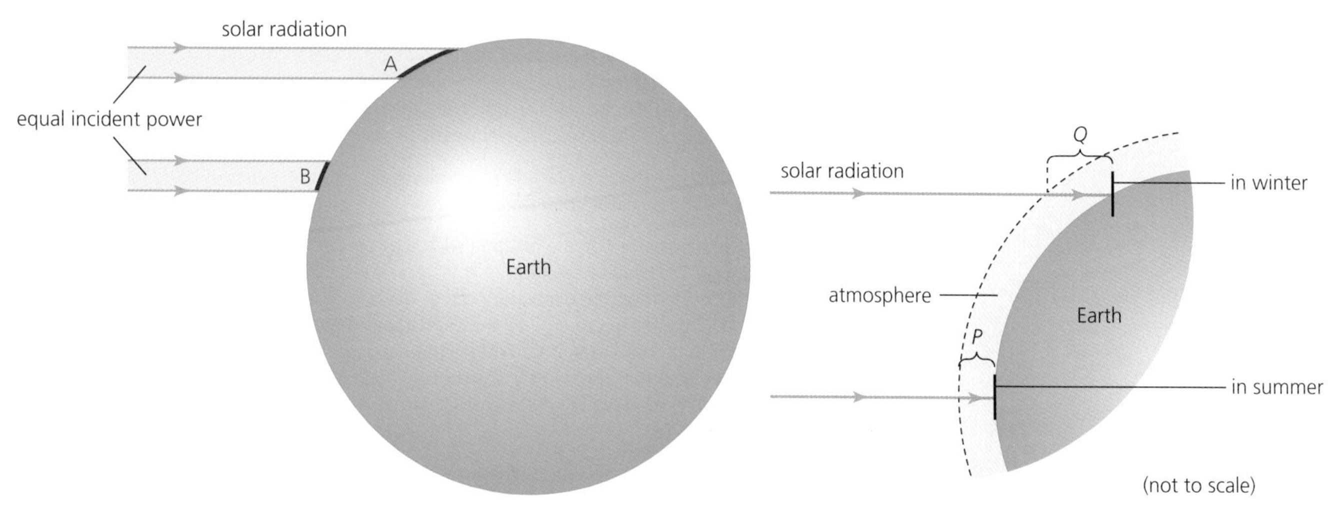

Figure 8.43 The received intensity of solar radiation varies with the angle of incidence

Figure 8.44 Effect of the atmosphere on received intensity of radiation at different times of year

The advantages and disadvantages of using solar radiation are summarized in Table 8.6.

Advantages	Disadvantages
• Renewable source of energy. • Free source of energy. • Pollution-free during use. • Low maintenance costs. • Typical 20+ years' lifetime.	• Variation of output with time of day/night, weather, time of year etc. (meaning that individual systems need supplementary power/batteries). • Photovoltaic cells and solar heating panels are expensive to construct and install (but they are becoming significantly cheaper). • Photovoltaic cells have some pollution issues during construction and end-of-use recycling. • Low energy density – large areas needed for photovoltaic power stations.

Table 8.6 Advantages and disadvantages of solar power

46 a Calculate the total amount of energy radiated from the Sun that falls on the Earth's upper atmosphere in one year. (Assume that the distance from the centre of the Earth to the limit of the atmosphere is 6.5×10^6 m.)
b Compare your answer with the world's total annual energy consumption, estimated to be approximately 5×10^{20} J.

47 Figures 8.36 and 8.44 are not drawn to scale. The radius of the Earth is 6.4×10^6 m and the height of the atmosphere is often assumed to be about 100 km. Use a compass to draw a scale diagram of the Earth surrounded by its atmosphere.

48 Make two copies of Figure 8.43 and label them (a) variation of intensity with time of the day, and (b) variation of intensity with time of the year. On each diagram indicate the position of the Earth's axis and the north pole. On (b) also label the direction of rotation.

49 Suggest two reasons why the value of the solar constant varies (by up to seven per cent).

50 Explain the ways in which the design of a solar heating system (with black-painted copper pipes) as shown in Figure 8.40 tries to maximize the amount of thermal energy transferred to the water in the tank.

51 a Draw an electric circuit that would enable you to investigate how light intensity affects the potential difference across a photovoltaic cell and the current through it.
b How could you use the data to determine the internal resistance of the cell?

52 A photovoltaic cell of area 1.8 cm^2 is placed where it receives radiation at a rate of 700 W m^{-2}.
a What electrical power is produced if its efficiency is 18 per cent?
b If the voltage across it is 0.74 V, calculate the current through the cell.
c How many cells would be needed for a power output of 50 W?
d What would be the total area of the cells?

53 a Use the internet to get an up-to-date estimate of the world's total electrical power consumption.
b Use estimates of the average radiated power received on the Earth's surface, and typical areas and efficiencies of photovoltaic cells, to calculate an approximate value for the area of cells that would be needed to supply all of the world's electrical power needs.

54 Imagine that a solar heating panel is being placed in a fixed position on the roof of a home in your town/city. Discuss which would be the best direction for it to face.

55 A solar heating panel of area 3.4 m^2 is put in a position such that on one day it received an average intensity of 640 W m^{-2} during 12 hours of daylight.
a What is the total energy incident on the panel during that day?
b The system is designed to transfer this energy to a 0.73 m^3 tank of water, and at the start of the day the temperature of the water in the tank was 2°C. What is the maximum possible temperature of the water at the end of the day? (Assume that the density of water is 1000 kg m^{-3} and that the specific heat capacity of water is 4200 J kg^{-1} °C^{-1}.)
c What assumption did you make?

56 a Make a copy of Figure 8.45, which shows how the total energy received per day by a fixed solar panel (or solar cell) is predicted to vary during the year at a location with a latitude of 30° in the northern hemisphere.
b Add lines to your graph to show the energy/day that would be predicted for the same panel if it was:
i at a latitude of 60° N
ii at a latitude of zero, on the equator
iii at a latitude of 30° S.
c Give two reasons why the total energy received per day decreases with increasing latitude.

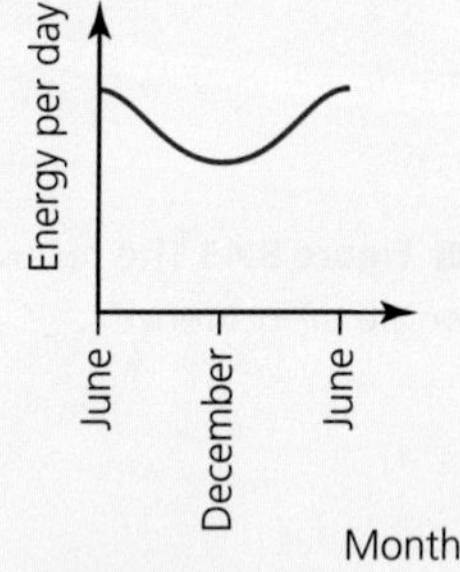

■ **Figure 8.45**

Utilizations

Geothermal energy

At a number of locations around the world, naturally occurring hot water has been used for thousands of years for bathing and for heating rooms. In recent years, attention has turned to extracting more of this *geothermal energy*, with the aim of reducing the adverse effects of using fossil fuels.

Since its formation, the core of the Earth has been extremely hot. It has remained hot partly due to energy released in the decay of radioactive materials. Some of this internal energy is continuously transferred as thermal energy to rocks close to the surface of the Earth.

■ **Figure 8.46** Iceland gets about 30 per cent of its energy needs from geothermal power

A volcano is a dramatic example of this, and the use of geothermal energy is mostly (but not exclusively) to be found in countries known for their volcanic activity.

Currently, geothermal power provides much less than one per cent of the total world's needs, but three countries each generate more than 25 per cent of their electrical power using geothermal energy: Iceland (Figure 8.46), the Philippines and El Salvador.

In some locations the energy is directly and conveniently available in the form of hot water, but in other places water has to be pumped down to hot rocks a long way beneath the surface. The most common and efficient use of geothermal energy is achieved with the direct use of hot water for heating, but where temperatures are high enough electricity can be generated.

1 Where are the nearest 'hot springs'/spas to where you live? How hot is the water and what is it used for?

2 Use the internet to find out as much information as you can about any one particular geothermal power station from around the world. Prepare a presentation for the rest of your group. (Answer questions such as: Why is it at that location? Did engineers have to dig into the Earth's crust or was the energy accessible close to the surface? Is the internal energy mostly in rocks or water? How is the energy transferred for use? Is there any pollution? What is the useful power output? What is the efficiency?). Make sure that you are using the terms 'thermal energy' and 'internal energy' correctly. Try not to use the word 'heat'.

Energy research

The world's current dependency on non-renewable fossil fuels cannot continue indefinitely. It has become a matter of great urgency that new sources of energy are developed, especially because the world's population is increasing and many people in developing countries are hoping for improved standards of living. At the same time, climate change and other environmental issues are of very great concern, and it is vital that energy sources do not make those problems worse.

These problems are now generally well known and accepted around the world, but they should never be underestimated or forgotten because of overfamiliarization. A great deal of scientific and technological research is being carried out in many different countries by scientists and engineers collaborating from almost all branches of science. But obviously this is not just a matter of science and technology – individuals and societies must accept responsibility and make changes, led by politicians and governments.

The introduction of new technologies is also driven by economics. Companies, large and small, are usually happy to invest in research if they anticipate making profits later. This is a rapidly changing business situation; a new idea may seem uneconomical when first conceived but become profitable when introduced on a large scale.

These are some of the continuing areas of development:

- discovery of new reserves of fossil fuels
- development of new technologies (e.g. fracking) to extract fossil fuels that were previously uneconomical
- imaginative new ideas and technologies for renewable energy sources
- providing education and information to people around the world (including the need for population control)

- more efficient use of energy by individuals, homes and organizations
- continuing nuclear fusion research
- increasing use of nuclear (fission) power, if this can be acceptable to the public.

The following list contains some of the many questions that might be asked about different types of power station.

- Is the energy source renewable?
- Can it provide large amounts of power to large numbers of people?
- Are there any greenhouse gases emitted?
- Is the source of energy free or expensive?
- Is it expensive or difficult to transfer the energy source to the power station?
- Does the source have a high energy density?
- Is the supply of energy reasonably constant and predictable?
- Is the process energy efficient?
- Is there any pollution during operation?
- Is there any significant pollution during construction or disposal?
- Are there any other adverse effects on the environment or people?
- Is it difficult or expensive to construct and install?
- Is the equipment expensive to maintain?
- How long before the equipment needs replacing?
- Are there plenty of suitable locations for power stations of this type?
- Are the probable locations of power stations close to towns and cities?

8.2 Thermal energy transfer – *for simplified modelling purposes the Earth can be treated as a black-body radiator and the atmosphere treated as a grey body*

Everything emits electromagnetic waves, most commonly in the infrared part of the spectrum. The hotter and bigger an object is, the more radiation is emits. The Sun, of course, is both very big and very hot, so it emits a very large amount of energy (3.84×10^{26} W). Most of the energy that we use on Earth came originally from the Sun in the form of electromagnetic waves. In this section we will consider the energy arriving at the Earth from the Sun, the energy absorbed, the energy reflected and the effect all of this has on the average temperature of the planet.

But first we will summarize the three principal means of thermal energy (heat) transfer and then describe how thermal radiation can be quantified.

Conduction, convection and thermal radiation

Thermal conduction

If a substance contains parts at different temperatures, the hottest part contains particles with the highest random kinetic energy. **Thermal conduction** is the process by which this energy spreads out and is transferred from particle to particle, by 'collisions', to a cooler part of the substance (or another substance). See Figure 8.47.

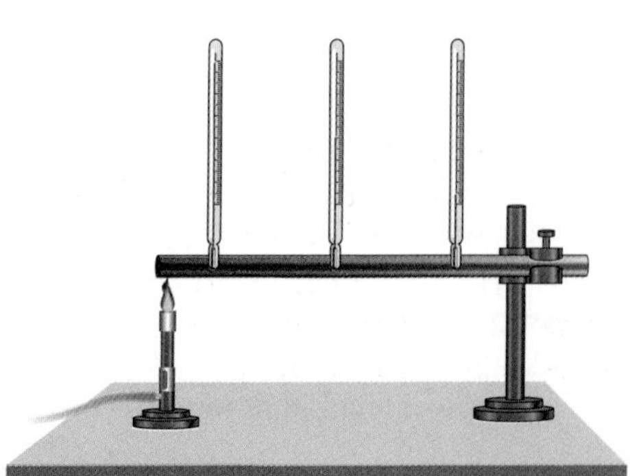

Figure 8.47 Thermal energy being conducted along a metal rod

Conduction can occur in solids, liquids and gases, but the process is quicker when particles have strong forces between them and are close together. Therefore conduction in solids is usually better than in liquids, and conduction in liquids is better than conduction in gases. The water trapped in the wetsuit of the windsurfer shown in Figure 8.48 is a poor thermal conductor and so keeps him warm.

Metals are good *conductors* of heat because the energy is also transferred by their free electrons. Most non-metals are poor conductors of heat (they are good *insulators*).

■ **Figure 8.48** Wetsuits provide thermal insulation because trapped water is a poor conductor

Nature of Science

The usefulness of analogies

Electrical conduction and thermal conduction have many similarities. In Chapter 5 we saw how a potential difference is needed to conduct a flow of charge through an electrical resistance. Simply by **analogy**, we may suggest (correctly) that a temperature difference is needed to conduct a flow of thermal energy through a thermal resistance. Thermal resistivity can be considered analogous to electrical resistivity. Similar equations can be used to represent both kinds of conduction, and even the flow of water along pipes.

In Chapters 5 and 6 we saw another powerful analogy – how gravitational and electrical forces between points were represented by similar equations.

Thermal convection

When a liquid is heated the molecules have greater random kinetic energy and they will move further apart from each other. We say that the liquid expands. Similarly, a gas may expand when it is heated, but *only if* it is not in a container that prevents expansion.

When *part* of a fluid (gas or liquid) is heated, the molecules will move further apart and this will result in a localized decrease in density. The warmer part of the fluid will then rise and flow above the cooler part of the fluid, which has a higher density. This movement of thermal energy and fluid is called thermal convection. It is common for convection to produce a *circulation* of a gas or liquid. Convection currents in air and water are very important in the study of weather patterns and climate. Figure 8.49 shows a simple example: during the day the land heats up quicker than the sea because it has a lower specific heat capacity; thermal energy is transferred to the air, which then rises.

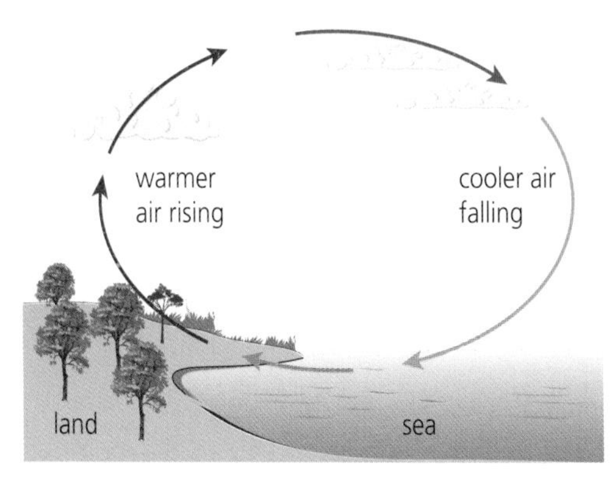

■ **Figure 8.49** Convection currents near the beach during the day

There are many important examples of convection currents. These include ocean currents and the movement of molten rocks under the Earth's surface.

Thermal radiation

The various interactions of the particles contained within all matter results in the emission of electromagnetic radiation. This occurs for *all* matter at *all* temperatures above absolute zero (0 K). The radiation (most commonly infrared) is called thermal radiation. Figure 8.50 shows a simple laboratory experiment designed to compare how the surfaces of identical containers affect cooling. Figure 8.51 show another experiment to compare how the surfaces of identical containers affect their rises in temperature.

Experiments like these should show that dark surfaces are good at *emitting* thermal radiation and good at *absorbing* thermal radiation. Similarly, white surfaces are poor emitters and absorbers: they are good *reflectors* of radiation. Although simple in principle, in practice both of these basic experiments may produce unconvincing results.

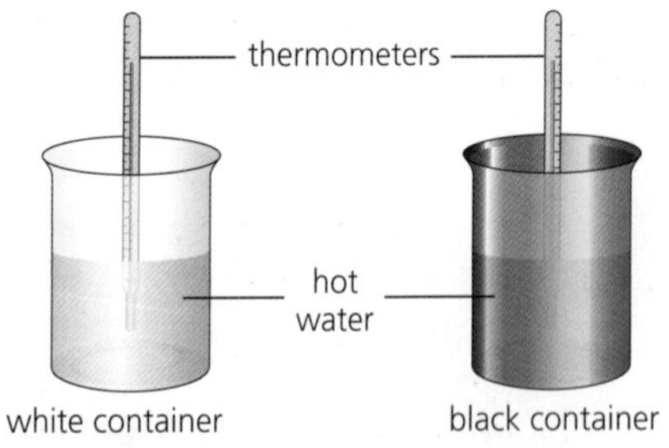

■ **Figure 8.50** Comparing thermal radiation losses

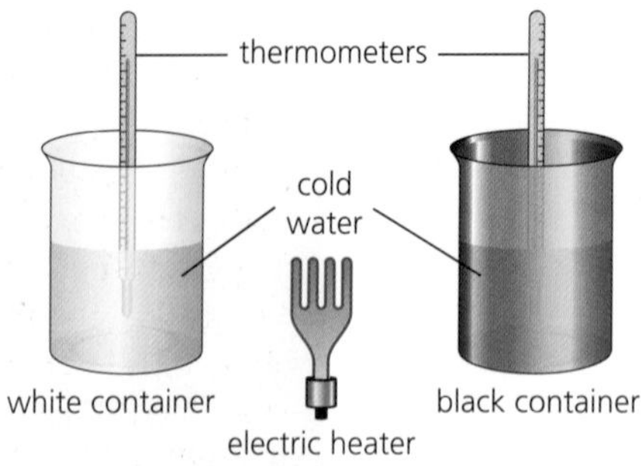

■ **Figure 8.51** Comparing thermal radiation gains

Thermal energy transfers between everyday objects often involve conduction, convection *and* radiation at the same time. Conduction is best through solids, convection only occurs in fluids and radiation can occur across a vacuum. Thermal radiation becomes *much* more significant at higher temperatures.

Black-body radiation

The concept of a *black body* is central to an understanding of thermal radiation.

> A perfect **black body** is an idealized object that absorbs *all* the electromagnetic radiation that falls on it.

Because it does not reflect light, it will appear black (unless it is so hot, that it emits visible light).

Any good absorber of radiation is also a good emitter of radiation. The radiation emitted from a 'perfect' emitter is called black-body radiation. Remember that, if the object is hot enough to emit visible light, it will not appear black.

An object cannot emit all of its energy over all frequencies in an instant, so we can define (perfect) black-body radiation by using graphs that show the intensity distribution for the different frequencies emitted. Figure 8.52 compares the radiation emitted from black bodies at different temperatures.

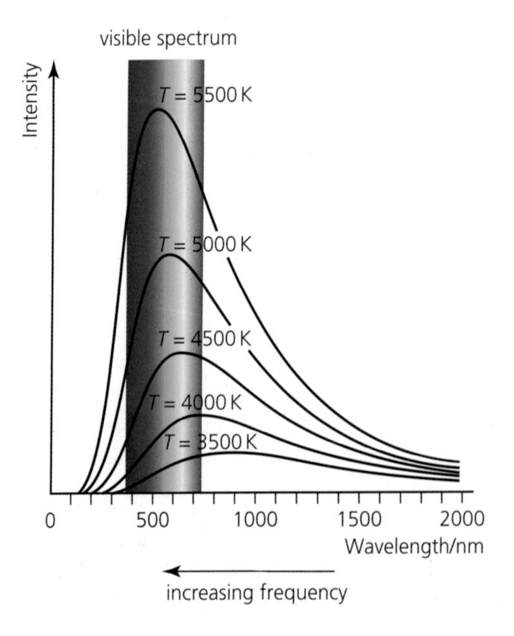

■ **Figure 8.52** Spectra of black-body radiation at different temperatures

As any object gets hotter it emits more radiation at all frequencies, so that the graph is shifted upwards for higher temperatures. However, it is very important to realize that the distribution of frequencies/wavelengths also changes, with bigger increases occurring at lower wavelengths. The peak wavelength moves to a lower value (higher frequency).

The vertical axis on the graph represents the intensity (power emitted per unit area) from a surface over a small range of wavelengths. The area under the graph is an indication of the total power emitted per unit area.

The surface temperature of the Sun is about 5800 K and the Earth's average surface temperature is approximately 288 K (too low to be shown clearly on the scale used in Figure 8.52). If we assume that they both behave approximately like perfect black bodies, we can compare the radiation emitted from the Sun and from the Earth. The solar radiation arriving at the Earth from the Sun is approximately 50 per cent visible light and 50 per cent infrared, with small amounts of ultraviolet and other radiations. The radiation going away from the Earth is much lower in intensity and it is all in the infrared part of the electromagnetic spectrum. As we will see later in this chapter, it is because of this difference in the spectra of the incoming and outgoing radiations that the greenhouse gases in the Earth's atmosphere will absorb a much bigger proportion of the radiation emitted from the Earth than of the radiation arriving from the Sun.

Calculating radiated power emitted

The total power radiated from *any* surface depends on only three variables:

- surface area, A
- surface temperature, T
- nature of the surface.

Note that the power emitted depends on the nature of the surface, but *not* the substance itself.

Emitted power varies considerably with changes in temperature and this is shown by the Stefan–Boltzmann law. This states that for a perfect black body, the total power emitted per unit area is proportional to the fourth power of the (absolute) temperature:

$$\frac{\text{power}}{\text{area}} \propto T^4$$

The power, P, emitted from a perfect black body is given by the equation:

$$P = \sigma A T^4$$

Here σ is a constant. It is called the Stefan–Boltzmann constant and has the value $5.67 \times 10^{-8}\,\text{W m}^{-2}\,\text{K}^{-4}$. (This value is given in the *Physics data booklet*.)

The term *grey body* is sometimes used to describe objects that are not perfect black bodies; this equation can be easily adapted for use with any such bodies by using the concept of the emissivity, e, of a surface.

Emissivity is a number between zero and one: a body with an emissivity of 1 represents a perfect black body emitting the maximum power; an emissivity of 0.5 represents a surface that emits 50 per cent of the maximum power.

Emissivity is defined as the power radiated by a surface divided by the power radiated from a black body of the same surface area and temperature. Emissivity is a ratio and therefore does not have a unit.

$$\text{emissivity} = \frac{\text{power radiated by a surface}}{\text{power radiated from a black body of the same temperature and area}}$$

Dark surfaces have high emissivity, which means that they radiate heat well, but they also absorb heat well. Light surfaces have a low emissivity, which means that they radiate heat poorly, but they also absorb heat poorly. The average emissivity of the Earth (and its atmosphere) is 0.61.

The Stefan–Boltzmann law for the power emitted from *any* surface then becomes:

$$P = e\sigma A T^4$$

This equation is given in the *Physics data booklet*.

Worked example

3 The power of our Sun is 3.8×10^{26} W and its surface temperature is 5780 K. Assuming that it behaves like a perfect black body, calculate the radius of the Sun.

$P = e\sigma AT^4$

$3.8 \times 10^{26} = 1 \times (5.67 \times 10^{-8}) \times (4\pi r^2) \times 5780^4$

$r^2 = \frac{3.8 \times 10^{26}}{7.95 \times 10^{8}}$

$r = 6.9 \times 10^8$ m

Wien's displacement law

The variation in the wavelength at which the maximum power is emitted by a black body (see Figure 8.52) can be represented mathematically by the following equation (known as Wien's displacement law):

$$\lambda_{max} \text{ (metres)} = \frac{2.90 \times 10^{-3}}{T \text{ (kelvin)}}$$

This equation is given in the *Physics data booklet*. (Remember that the temperature must be in kelvins.)

This is an *empirical* law, determined by investigation.

Worked example

4 The surface temperature of the Sun is 5780 K.
a At what wavelength is the emitted radiation maximized?
b What assumption did you make?
c In what part of the visible spectrum is this wavelength?

a $\lambda_{max} \text{ (metres)} = \frac{2.90 \times 10^{-3}}{T \text{ (kelvin)}}$

$\lambda_{max} = \frac{2.90 \times 10^{-3}}{5780}$

$\lambda_{max} = 5.02 \times 10^{-7}$ m

b That the Sun is a perfect black body (this is almost true).
c Green

57 Describe simple laboratory experiments that demonstrate the flow of thermal energy by **a** conduction, and **b** convection.

58 Suggest reasons why the experiments shown in Figure 8.50 and Figure 8.51 'may produce unconvincing results'.

59 Give an everyday example of:
a a dark surface being good at absorbing thermal energy
b a dark surface being good at emitting thermal energy
c a white or shiny surface being good at reflecting thermal energy
d a white or shiny surface being poor at emitting thermal energy.

60 A star has a radius of 8.3×10^7 m and a surface temperature of 7500°C.
a Calculate the power it emits.
b What assumption did you make?
c What is the intensity at a distance of 2.4×10^{12} m from the star?

61 a Estimate an average surface area for a naked adult human body.
b If skin has an emissivity of 0.95, calculate the power radiated away from a naked body with a skin temperature of 33°C.
c In what other ways will the body lose heat?
d Your answers may suggest that a naked body would lose heat very quickly. Explain why this is probably not true.

62 a If the temperature of an object increases from 285 K to 312 K, what is the percentage increase in its emitted radiation?
b An object is at 20°C. To what temperature would it have to be heated to double the power of the radiation emitted?

63 When the current through an incandescent light bulb is turned on, the temperature of the filament will rise until the thermal energy is radiated away at a rate equal to the electrical power provided. Calculate the operating temperature of a 40 W light bulb if the filament has a surface area of $1.2 \times 10^{-5}\,m^2$. (Assume it behaves as a perfect black body.)

64 A tungsten-halogen bulb operates at a temperature of 3500 K.
a At what wavelength is the emitted radiation maximized?
b Suggest how the light from a tungsten-halogen bulb will appear different from the light from an incandescent light bulb operating at 2800 K.

Utilizations

Using emitted thermal radiation to determine temperatures

1 If an object is hot enough to emit visible light its temperature can be estimated from its colour. For example, when steel is heated to the temperature at which it just gives out visible (red) light, it is about 600°C. When its temperature is raised further it will emit orange light as well, and then yellow. An overall effect of yellow light is produced at a temperature of about 1000°C. The emission of radiation across all of the visible spectrum produces white light at about 1300°C.

2 The surface temperature of stars, including our Sun, can be determined from detailed analysis of the radiation they emit. If the wavelength of maximum intensity is determined, then Wein's law can be used to determine the temperature. This is explained in more detail in Option D: Astrophysics.

3 Infrared cameras can be used to produce images in some similar ways to visible light cameras. Because all objects emit infrared, they have the advantage of being able to produce images from objects even if there is no visible light. (Any colour(s) in the images are added artificially.) The quality of the images is not as good as with light because longer wavelengths are used. If the power radiated by any point on an object can be determined then, in principle, its temperature can be *estimated* by using $P = e\sigma AT^4$, although an assumption would need to be made about emissivity. In this way it is possible to produce *thermographs*, which have a wide variety of uses such as scanning at airports (Figure 8.53).

■ **Figure 8.53** Thermal scanner at Incheon airport, South Korea

1 Use the internet to make a list of at least five different uses for infrared thermography.

2 Research how infrared cameras detect different intensities of radiation.

3 Suggest the advantages and disadvantages of using infrared thermometers to measure human body temperatures.

Radiation from the Sun

The solar constant

If we assume that thermal radiation from the Sun spreads out equally in all directions without absorption, then at a distance r from the Sun, the same power is passing through an area $4\pi r^2$ (the surface area of an imaginary sphere), as shown in Figure 8.54.

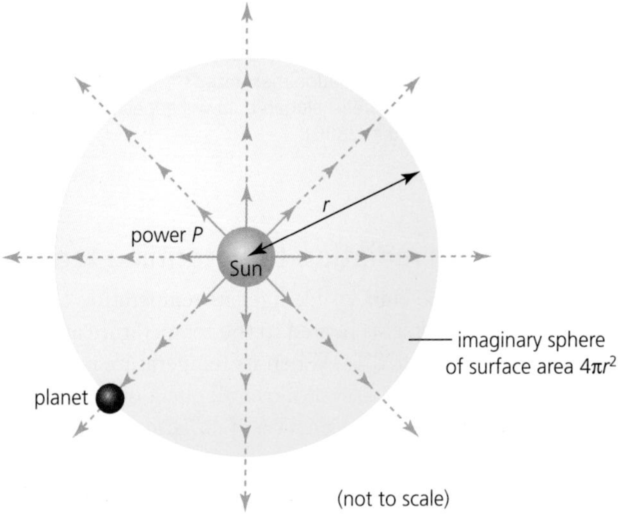

Figure 8.54 The Sun's radiation spreading out

Because:

$$\text{intensity, } I = \frac{\text{power, } P}{\text{area, } A}$$

the intensity at different distances, r, from the Sun can be calculated from:

$$I = \frac{P}{4\pi r^2}$$

Intensity follows an inverse square law with distance from the source (as discussed in Chapter 4). This equation is *not* given in the *Physics data booklet*.

Worked example

5 Calculate the intensity of the Sun's radiation arriving at:
a the Earth
b Mercury.

a Earth is an average distance of 150×10^6 km from the Sun, so:

$$I = \frac{P}{4\pi r^2} = \frac{3.84 \times 10^{26}}{4 \times \pi \times (150 \times 10^9)^2}$$

$I = 1.36 \times 10^3\,\text{W m}^{-2}$

b Mercury is an average distance of 58×10^6 km from the Sun, so:

$$I = \frac{P}{4\pi r^2} = \frac{3.84 \times 10^{26}}{4 \times \pi \times (58 \times 10^9)^2}$$

$I = 9.1 \times 10^3\,\text{W m}^{-2}$

The result for Earth calculated in Worked example 5 was used earlier in this chapter when discussing solar power. It is called the *solar constant*, *S*, and it represents the average intensity falling on an area above the Earth's atmosphere that is perpendicular to the direction in which the radiation is travelling. It has been measured on orbiting satellites to have an average value of 1361 W m^{-2}. The value given in the *Physics data booklet* is:

solar constant, $S = 1.36 \times 10^3\ \text{W m}^{-2}$

There are *slight* variations in the solar constant due to periodic change in the power output from the Sun, and differences in the distance from the Earth to the Sun at different times of the year.

Albedo and emissivity

On average about 30 per cent of the radiation from the Sun that is incident on the Earth and its atmosphere is directly reflected back, or scattered, into space, but there are considerable variations from place to place and at different times of the day and year.

The ratio of the total scattered or reflected power to the total incident power is known as the albedo.

$$\text{albedo} = \frac{\text{total scattered power}}{\text{total incident power}}$$

This equation is given in the *Physics data booklet*.

Albedo is sometimes given the symbol α. An albedo of 1 would mean that *all* radiation was scattered or reflected. An albedo of zero would mean that all the radiation was absorbed and *none* scattered or reflected. White and bright surfaces (such as snow, ice and clouds) have a high albedo because they reflect a lot of the incident radiation. They will also have a low emissivity. Darker surfaces (such as water), which have higher emissivity, will have a lower albedo because they absorb more of the incident radiation (Figure 8.55).

Good emitters of radiation (high emissivity) are also good absorbers of radiation. A surface that absorbs radiation well, and so reflects poorly (low albedo), will also have a high emissivity.

Albedo also varies with the angle of incidence (which changes with inclination of the surface, season, latitude and time of the day). The Earth's average albedo is approximately 0.30 (30 per cent), as mentioned above.

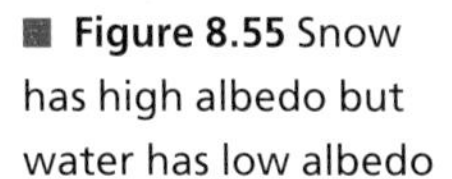

Figure 8.55 Snow has high albedo but water has low albedo

65 Use the solar constant and the radius of the Earth (6.37×10^6 m) to show that the total radiated energy arriving at the Earth from the Sun is 1.74×10^{17} J every second.

66 The total radiated power arriving from a sun at a planet is 2.82×10^{17} W. The orbit of the planet around the sun has a radius of 1.23×10^7 m.
a Calculate the intensity of the radiation arriving at the planet (perpendicular to the direction of the radiation).
b If the average albedo of the planet is 0.22, calculate the average intensity of radiation radiated away from the planet in all directions.

67 One of the planets in our solar system receives solar radiation with an intensity of about $15\,\text{W m}^{-2}$. Calculate its distance from the Sun and hence find out which planet it is.

68 A small light bulb is the only lamp in an otherwise dark room. It emits visible radiation with a power of 5 W. What is the intensity received at a book 90 cm from the bulb?

69 Suggest possible reasons why the average albedo at a particular place is:
a different in winter than in summer
b different at different times of the same day.

70 Two planets are orbiting a distant star. Planet A is 100 million kilometres from the star and receives an intensity of $860\,\text{W m}^{-2}$. What is the intensity received at planet B if it is 120 million kilometres from the star?

The greenhouse effect

The Earth receives an average of about 1.74×10^{17} J of radiated energy from the Sun every second (Figure 8.56). If the Earth reflects or radiates energy back into space at the same rate, then it will be in *thermal equilibrium* and its average temperature will remain constant. If the Earth reflects or radiates less energy than it receives it will get hotter. If it reflects or radiates more than it receives, it will become cooler.

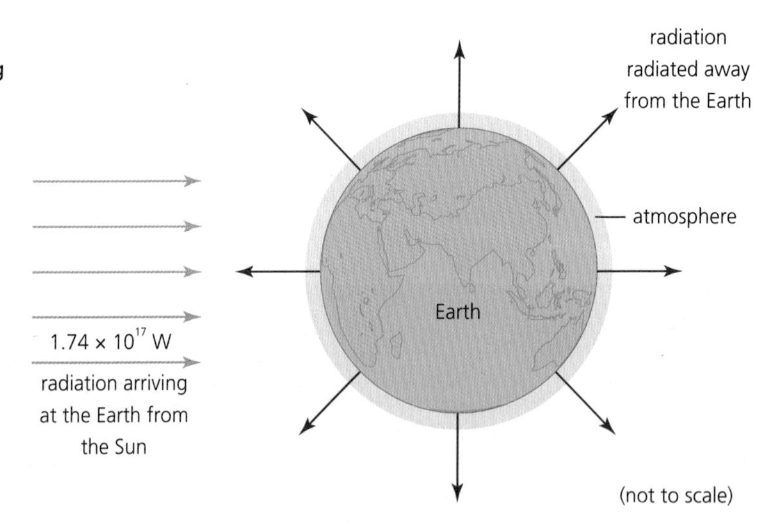

Figure 8.56 Earth receiving and emitting radiation

The **greenhouse effect** is the name that has been given to the *natural* effect a planet's atmosphere has in increasing the temperature of the planet to a value higher than it would be without an atmosphere.

The Earth has had a *beneficial* greenhouse effect for as long as it has had an atmosphere; if the Earth did not have an atmosphere it would be about 30°C cooler and without life. The Moon and the Earth are approximately the same distance from the Sun, but the surface of the Earth is hotter because the Earth has an atmosphere.

However, most scientists now believe that the greenhouse effect has become *enhanced* (increased). It is believed that this is mainly because human activity has changed the composition of the atmosphere. The enhanced greenhouse effect is discussed later in this chapter.

If the Earth *did not* have an atmosphere, the average intensity of radiation received from the Sun over the whole of the Earth's surface (over an extended period of time) could be calculated as follows:

$$I = \frac{\text{total power received by Earth from the Sun}}{\text{total surface area of the Earth}} = \frac{1361 \times \pi r_E^2}{4\pi r_E^2}$$

where r_E is the radius of the Earth, 6.37×10^6 m.

$$I = 340\,\mathrm{W\,m^{-2}}$$

But the Earth *does* have an atmosphere. Of the $340\,\mathrm{W\,m^{-2}}$ incident on the Earth and its atmosphere, $107\,\mathrm{W\,m^{-2}}$ is immediately reflected back into space (about $77\,\mathrm{W\,m^{-2}}$ from the atmosphere and $30\,\mathrm{W\,m^{-2}}$ from the Earth's surface). These values give the albedo of the Earth as $107/340 \approx 30$ per cent, which agrees with the figure quoted earlier for the average albedo of the Earth.

This leaves a global average of $233\,\mathrm{W\,m^{-2}}$ absorbed by the Earth and its atmosphere. Figure 8.57 shows what happens to this energy: $66\,\mathrm{W\,m^{-2}}$ is absorbed by the atmosphere and $167\,\mathrm{W\,m^{-2}}$ is absorbed by the Earth's surface. (Although actual values are quoted here, figures may vary slightly depending on the sources of information used; these do not need to be learnt.)

For the Earth to remain in thermal equilibrium it is necessary for the incident $233\,\mathrm{W\,m^{-2}}$ to be matched by an equal intensity radiated back into space.

Radiation is the only way that thermal energy can travel away from the Earth through space. (But thermal energy is transferred from the Earth's surface to the atmosphere by a number of other processes as well, involving conduction, convection and evaporation.)

The atmosphere absorbs a much higher percentage of this outgoing radiant energy than it does incoming radiant energy from the Sun. This is because the Earth is much cooler than the Sun and the radiation it emits is mostly at lower wavelengths (all infrared, rather than visible). See Figure 8.52. The (greenhouse) gases in the atmosphere absorb some of these infrared wavelengths, but do not absorb so much of the incoming radiation.

The energy absorbed by the atmosphere is also transferred away as radiated energy, which is emitted in all directions, some of which goes directly into space. However, most of the energy is transferred back down to the Earth's surface (by radiation and other processes) and so the process continues, and thermal equilibrium is established at a certain temperature. (On average this is about 15°C at present on the Earth's surface.) Figure 8.57 attempts to simplify and summarize these complex processes.

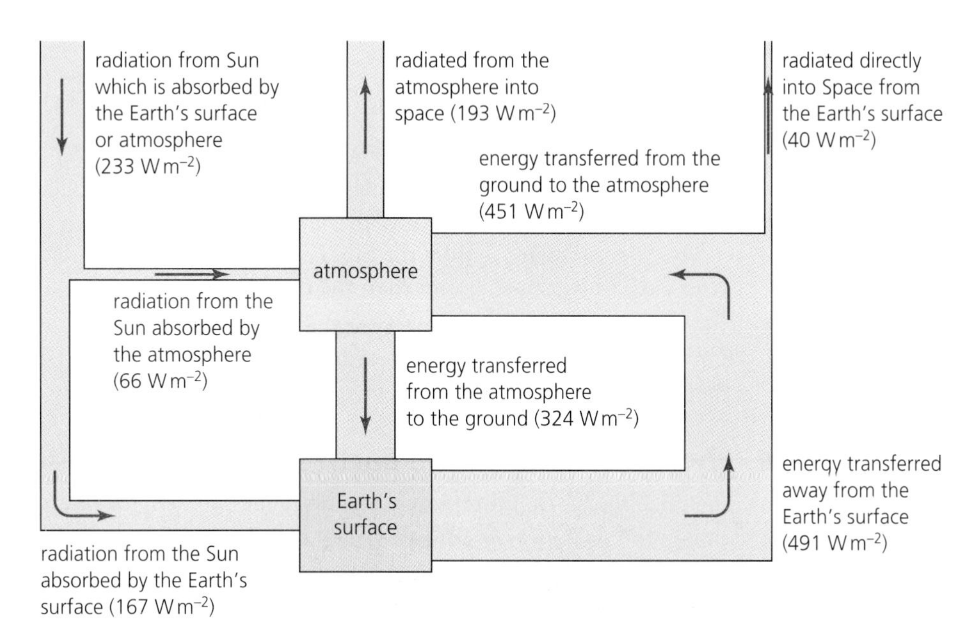

■ **Figure 8.57** Energy flow through the Earth's atmosphere

Greenhouse gases

The Earth's atmosphere has been formed over millions of years by naturally occurring volcanic and biological processes and from collisions with comets and asteroids. The air in the atmosphere contains approximately (by volume) 78 per cent nitrogen, 21 per cent oxygen and 0.9 per cent argon. There are also naturally occurring traces of many other gases, including carbon dioxide and water vapour. Some of these trace gases are called **greenhouse gases** because they play a very important part in controlling the temperature of the Earth and the greenhouse effect. Greenhouse gases absorb (and then re-emit) infrared radiation. Nitrogen, oxygen and argon have no greenhouse effect (because they have non-polar molecules).

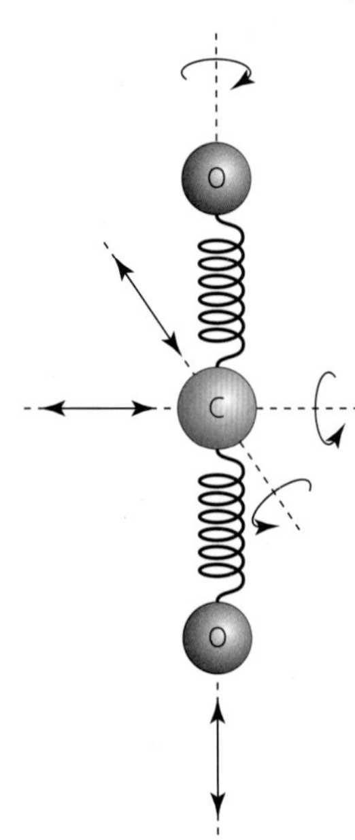

■ **Figure 8.58** A few possible molecular vibrations in carbon dioxide

There are many greenhouse gases but the four most important, in decreasing order of their contribution to the greenhouse effect, are:

- water vapour
- carbon dioxide
- methane
- nitrous oxide (dinitrogen monoxide).

The relative importance of these gases in causing the greenhouse effect depends on their relative abundance in the atmosphere as well as their ability to absorb infrared radiation. Each of the gases has natural as well as man-made origins.

Water vapour is by far the most abundant of the greenhouse gases. Of the other gases, carbon dioxide contributes between 15 and 20 per cent of the overall greenhouse effect.

Methane and nitrous oxide absorb infrared radiation more strongly than carbon dioxide but their concentrations in the atmosphere are very much lower.

Molecules of greenhouse gases absorb infrared radiation because the atoms in their molecule are not at rest – they oscillate with simple harmonic motion, like masses connected by springs. Figure 8.58 shows a much simplified example of possible modes of vibration of a carbon dioxide molecule.

If the atoms in a molecule oscillate at the same frequency as the radiation passing through the greenhouse gas, then energy can be absorbed (an example of an effect known as *resonance*), raising the molecule to a higher vibrational energy level. The energy is quickly released again but the released energy is emitted in random directions.

The behaviour of the greenhouse gases is also an important topic in chemistry.

Summary of the greenhouse effect

- Some of the radiant infrared energy emitted by the Earth's surface (and atmosphere) is absorbed by molecules of 'greenhouse' gases in the atmosphere.
- All gas molecules oscillate. Molecules of greenhouse gases oscillate at the same frequencies as infrared radiation. This is why they absorb energy.
- Greenhouse gas molecules also absorb some of the energy arriving from the Sun, but they absorb a lower percentage from the Sun because the distribution of wavelengths is different – because the Sun is much hotter than the Earth.
- The molecules re-radiate the energy at the same wavelengths, but in random directions, so that some of the energy that was previously moving away from the Earth is redirected back towards the planet's surface.

■ Energy balance in the Earth surface–atmosphere system

A planet, like the Earth, is in thermal equilibrium *if* the total incoming power equals the total outgoing power. A planet can only receive or emit energy by radiation, so:

incoming radiated power (from the Sun) = outgoing radiated power (from the Earth)

$$\text{solar constant} \times \text{cross-sectional area of Earth} \times (1 - \text{albedo}) = e\sigma A T^4$$

For the Earth and its atmosphere, as we have seen, the solar constant is 1361 W m^{-2}, the average albedo is 0.30 and the average emissivity is 0.61, so that:

$$1361 \times \pi r_E^2 \times (1 - 0.30) = 0.61 \times (5.67 \times 10^{-8}) \times 4\pi r_E^2 \times T^4$$

This calculation confirms an average surface temperature of about 288 K (290 K).

Although this calculation uses a much simplified model of the Earth's energy balance, it can be used to predict the effect of changing emissivity and/or albedo.

71 Recalculate a temperature for the Earth (for the calculation above) assuming that the albedo rose by ten per cent and the emissivity decreased by ten per cent.

72 The planet Uranus is an average distance of 2.9×10^{12} m from the Sun.
  a Calculate the solar constant for this planet.
  b Estimate its surface temperature assuming it has values for albedo and emissivity that are the same as for the Earth.
  c Research the true value of the surface temperature of Uranus.

73 Suppose that a new planet is discovered orbiting a distant star at an average radius of 2.7×10^{12} m.
  a If the power emitted by the star is 1.3×10^{28} W, estimate the surface temperature of the star. State the assumptions you made.
  b Do you think there could be life on this planet?

74 Explain why a planèt of radius $2r$ would have approximately the same temperature as a planet of radius r at the same distance from a star (assume equal albedos and emissivities).

Global warming and climate change

The Earth has changed temperature in the past, but increases in the past 100 years appear to be relatively sudden. Figure 8.59 shows data representing global average temperatures over the last 130 years. Although there has been some doubt over the accuracy of some of this information (especially the older data), scientists generally agree that the planet is getting warmer, that the average temperature rose by about 0.7°C (± 0.2°C) in the hundred years between 1905 and 2005, and that the rise in temperature in the twenty-first century will be greater. From 2004 to 2013 there was no significant change in the (already high) global average temperature, but the time scale is too short to draw any firm conclusions from these data.

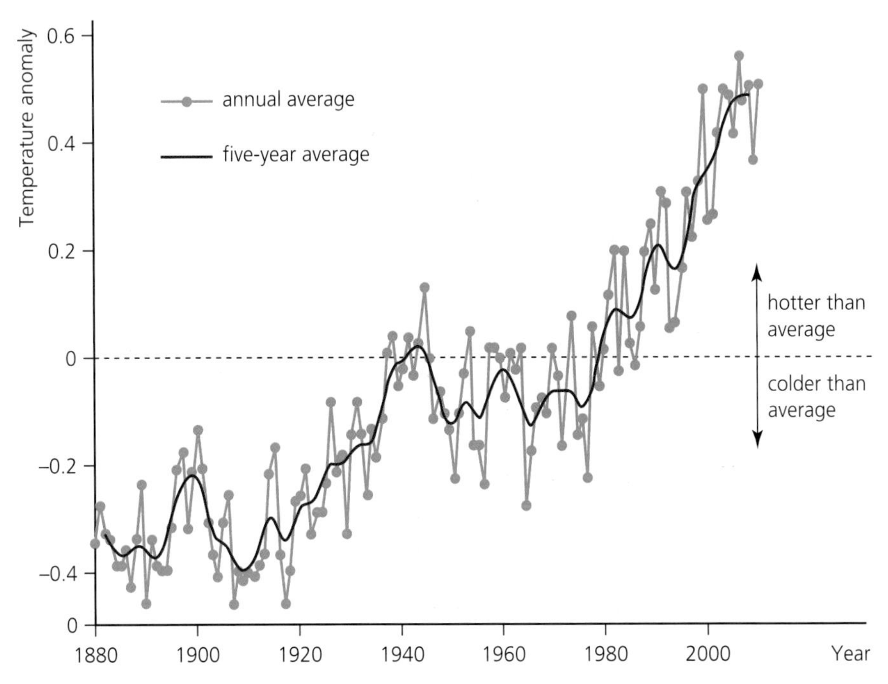

■ **Figure 8.59** Changes in global temperatures. The *temperature anomaly* is the difference between the measured temperatures and the long-term average

■ **Figure 8.60** The Maldives are a low-lying country threatened by rising sea-levels

Almost all scientists now agree that the cause of the rises in global temperatures is an **enhanced greenhouse effect** due to the increasing concentration of greenhouse gases (especially carbon dioxide) caused by human activities. This reduces the intensity of the radiation emitted from the Earth. The Intergovernmental Panel on Climate Change (IPCC) concluded that rises in global temperature were 'very likely due to the observed increase in anthropogenic greenhouse gas concentrations'. ('Anthropogenic' means caused by humans.) This means that the increased combustion of fossil fuels has released extra carbon dioxide into the atmosphere and is probably the most important cause of the current global warming.

The two main consequences of global warming are *climate change* and *rising sea levels*, but there are many other interconnected factors as well, including:

- Snow and ice have a high albedo; global warming will result in less snow and ice covering the Earth's surface, therefore lowering the Earth's average albedo and increasing the absorption of solar radiation in those places.
- If the temperatures of the oceans increase, less carbon dioxide will be dissolved in the water (the solubility of carbon dioxide reduces as the temperature rises) meaning that more gas is released into the atmosphere to further increase the greenhouse effect.
- Changing land-use has an effect on the concentration of greenhouse gases, especially carbon dioxide. Trees and other plants remove carbon dioxide from the air in the process of photosynthesis. This is an example of *carbon fixation* – the removal of carbon dioxide from the air and the formation of solid carbon compounds. Any widespread changes of land-use (particularly deforestation) may lead to an increase in the atmospheric concentration of carbon dioxide. Changes of land use may also lead to changes in albedo and, if the land is cleared by burning, more carbon dioxide will be released in the process.

Evidence linking global warming to the levels of greenhouse gases

Ice extracted from deep below the Antarctic surface, at the Russian base at Vostok, has been studied in great detail (isotopic analysis) for evidence of atmospheric carbon dioxide concentrations. These concentrations have been linked to global temperatures going back hundreds of thousands of years.

■ **Figure 8.61** Are hurricanes (typhoons) getting worse because of global warming?

The results are presented in Figure 8.62 and it seems clear that there is a definite *correlation* between average global temperatures and the amount of carbon dioxide in the atmosphere. This, in itself, does not prove that changing levels of carbon dioxide *caused* the changes in temperature.

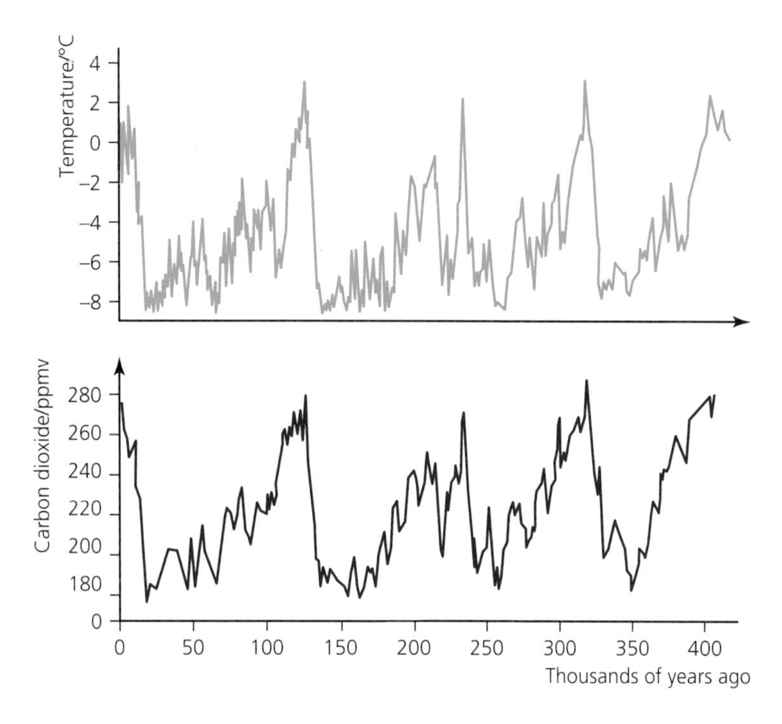

Figure 8.62 There is a correlation between average global temperatures and the amount of carbon dioxide in the atmosphere (ppmv = parts per million by volume)

More recently, and over a much shorter time scale, significant increases in the concentrations of greenhouse gases in the atmosphere have occurred at a time of global warming, as indicated approximately in Figure 8.63.

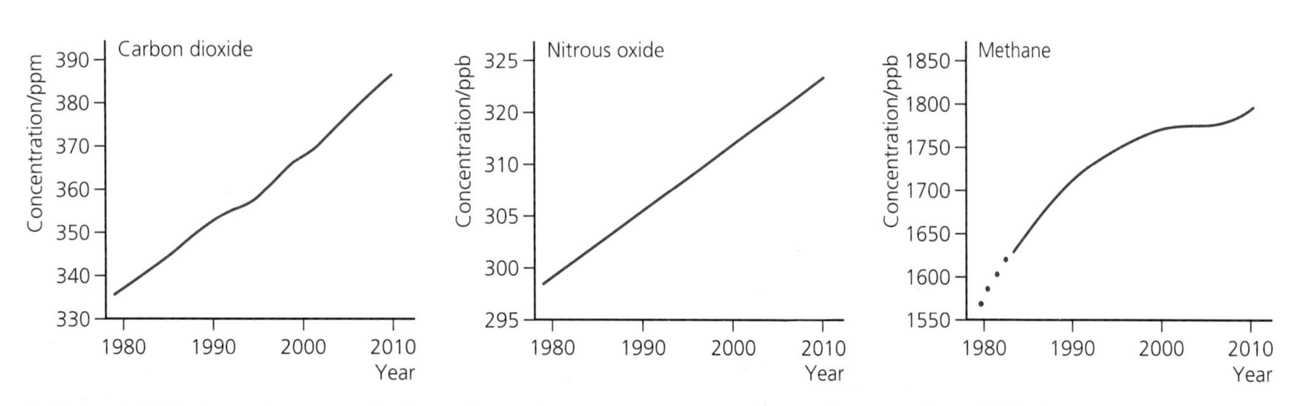

Figure 8.63 The increasing concentrations of greenhouse gases over recent years (ppb = parts per billion)

ToK Link

Correlation and cause

The **correlation** (link) between increasing concentrations of greenhouse gases and rising global temperatures is well established and accepted by (almost) everybody. However, that does not mean that we can be sure that global warming is *caused* by the release of more greenhouse gases. Obviously, controlled experiments to test such a theory cannot be carried out and we must rely on limited statistical evidence, computer modelling and scientific reasoning. In such cases, 100 per cent certainty is never possible and individuals and societies must make informed judgements based on the best possible scientific evidence. Of course, some people will always choose to disagree with, or ignore, the opinions of the majority.

Nature of Science

Simple and complex modelling of the behaviour of gaseous systems

The kinetic theory of gases is a relatively simple mathematical model that accurately predicts the behaviour of most gases that are confined in limited spaces. The Earth's atmosphere is also a gaseous system, but its size and complexity defy detailed mathematical modelling.

Weather forecasting has always been a difficult science and accurately predicting next week's weather is often impossible, although modern advances in the collection of vast quantities of accurate data and the use of powerful computers are making this more reliable. But reliably predicting the long-term climate changes that will probably be produced by global warming (of an uncertain amount) is a task that is a much bigger challenge for the very best scientists and computer programmers.

It should be clear that the world's climate is controlled by a large number of interconnected variables, any or all of which are subject to considerable uncertainty. Complex computer modelling of the world's future climate has been based on a vast amount of current and past data. Such research continues to be carried out around the world, but the predictions are all subject to significant uncertainty, and there is no worldwide agreement on the results: the situation is simply too complex and not fully understood.

Computer simulations of energy exchanges and climate models are available and can be useful educational resources although, inevitably, they greatly simplify the situation.

ToK Link

The future can never be predicted with certainty

The debate about global warming illustrates the difficulties that arise when scientists cannot always agree on the interpretation of the data, especially because the solution would involve large-scale action through international government co-operation. When scientists disagree, how do we decide between competing theories?

There can never be certainty, or agreement, in predicting events as complex as the future climate, yet governments and organizations need to make vital long-term decisions that affect us all. But should they plan for the worst that might happen, because to do otherwise would be irresponsible, or should they look for some level of consensus among most scientists?

Nature of Science

Using computers to expand human knowledge

Trying to predict the future, or to answer the question 'what would happen if …' has always been a common and enjoyable human activity, but it seems that our predictions are usually much more likely to be wrong than right. This is partly because, in all but the simplest of examples, there are just too many variables and unknown factors. Of course, the delightful inconsistencies of human nature play an important part when we are dealing with people's behaviour, but accurately predicting events governed mostly by the laws of physics, such as next week's weather, can also be extremely difficult.

When a situation can be represented by equations and numbers, mathematical modelling is a powerful tool to help understanding. But even in the simplest situations, there are nearly always simplifications and assumptions that result in uncertainty in predictions. When dealing with complex situations – such as predicting next month's weather, the value of a financial stock next year or the climate in 50 years' time – even the most able people in the world will struggle with the complexity and amount of data. The rapid increase in computing power in recent years has changed this.

Modern computers have computing power and memory far in advance of human beings. They are able to handle masses of data and make enormous numbers of calculations that would never be possible without them. They are ideal for making predictions about the future climate, but that does not necessarily mean that the predictions will turn out to be correct. Computer predictions are limited by the input data provided to them and, more particularly, by the specific tasks that human beings have asked them to perform. To check the accuracy of predictions, computer models can be used to model known complex situations from the past to see if they are able to predict what actually happened next. But predicting the past is always much easier than predicting the future.

75 Use data from Figure 8.63 to estimate by what percentage the concentrations in the atmosphere of the three greenhouse gases has increased between the years 1980 and 2010.

76 a Estimate the amount of energy needed to raise the average temperature of the world's oceans by 1°C.
b How long would it take the Sun to radiate that amount of energy to Earth?

77 a The specific latent heat of vaporization of water is $2.26 \times 10^6\,\mathrm{J\,kg^{-1}}$. Explain what this means.
b Use this figure to calculate the thermal energy that must have been supplied to a 500 000 kg cloud to evaporate it from water on the surface of an ocean.
c If the cloud is 5 km above the ocean, compare your answer in b to the gravitational potential energy that has also been gained by the water in the cloud.
d Assuming that the thermal energy for evaporation was supplied by radiation from the Sun at an average rate of $25\,\mathrm{W\,m^{-2}}$ over an area of $10\,\mathrm{km^2}$, how long would it have taken for the water in the cloud to evaporate from the ocean (give your answer to the nearest hour)?

78 Use the internet to find out about the West Antarctic Ice Sheet and why some scientists are worried about it.

79 Research into the origins and uses of the gas SF_6, which is possibly the most potent of all greenhouse gases.

80 Research into:
a the precise reasons why increasing global temperatures result in rising sea levels
b how much average sea levels have changed in the last:
i 100 years
ii 10 years.

A great deal has been done in recent years to combat the effects of global warming, but is it enough? Because climate change has been a major issue around the world for many years, it may have become too familiar and accepted, with the result that many people are perhaps no longer taking the consequences seriously enough.

However, many scientists are convinced that we need to take greater action now. In April 2014 a UN report from the Intergovernmental Panel on Climate Change (IPCC) was blunt, describing the effects of climate change as 'severe, pervasive and irreversible'. The amount of carbon dioxide in the atmosphere was reconfirmed as the dominant factor driving climate change, and perhaps the single most important recommendation (in detailed reports) was to switch from carbon based fuels to renewable sources of energy (and maybe nuclear power) 'sooner rather than later'. The overall economic costs of delaying and having to cope with the consequences of climate change in the future were predicted to be greater than the economic costs of widespread action in the near future. Such dramatic changes will not occur without the intervention and leadership of governments, and meaningful international agreements, to a much greater extent that has been evident in recent years.

Despite the apparent inevitability of increased global warming and climate change, the 2014 report offered some encouragement, saying that solving the problems offered challenges that could be met and opportunities for a different kind of future.

Additional Perspectives

Reducing global warming

Reducing global warming has become a major worldwide issue and many people claim that this is the biggest problem facing the modern world. It receives an enormous amount of media attention and considerable efforts are being made at all levels of society to try to solve the problems.

However, the increasing demand for energy in the developing nations and a world population that is still rising mean that the overall demand for electrical energy (and the need to burn fossil fuels) is still rising. Many people would also claim that there is a fundamental conflict between the need to reduce energy demands and modern lifestyles in consumer-driven developed countries.

While it is true that most informed scientists believe firmly in the enhanced greenhouse effect, 100 per cent proof is impossible, so there is still an element of doubt and there are a significant number of people who prefer to believe, for a variety of reasons, that the problem has been exaggerated (or even that the scientists are wrong). From the vast amount of information and disinformation that is available on this subject, individuals and organizations are able to select whatever data they want in order to promote their own views or interests.

Possible solutions

- Encouraging organizations and individuals to reduce their energy needs. It is never easy to make people change their habits and any government that tried to introduce large price rises or new laws in order to reduce energy consumption would be very unpopular. Some people claim that there should be trends in society (especially in the developed countries) towards the use of smaller and fewer cars, and better, more energy-efficient public transport, travelling less (to work and for leisure), reduced use of heating or air-conditioning and generally a life of sufficiency rather than consumerism and excess. Reducing a person's or an organization's need for the energy transferred from burning fossil fuels has become known as reducing their *carbon footprint*.
- Developing and greatly increasing the use of renewable energy resources and nuclear power.
- Setting and enforcing strict limits and controls on greenhouse gas emissions, such as those imposed by international agreements (such as the Kyoto Protocol). Countries (or large organizations) would then have no choice but to stick to the agreed total limits, although it may not be fair or sensible to set the same limits on different countries. To encourage enterprise and flexibility within these limits, *carbon trading* (also known as 'cap and trade') can be introduced. This is when one country or organization can exceed its limit by buying the right to extra emissions from another country or organization. People and organizations can be encouraged to become *carbon neutral*, so that any of their actions that result in the emission of greenhouse gases are balanced by other actions that are beneficial to the environment, for example planting trees.
- Using the latest technology to ensure that power stations are as efficient as possible. This may mean closing older power stations and using natural gas as the fuel wherever possible. Using natural gas is more efficient and creates lower emissions than coal or fuel oil. As discussed, a large and unavoidable percentage of the input energy to a power station is transferred to thermal energy, which is often just spread into the surroundings. In *combined heat* and *power systems* (CHP) some of that thermal energy is transferred using hot water to keep local homes and offices warm. Helsinki, in Finland, is a very good example of the widespread use of this idea (Figure 8.64).
- Encouraging the use of electric and *hybrid vehicles*. Vehicles that are powered by petrol (gasoline) are convenient and powerful, but they are very inefficient and emit greenhouse gases into the atmosphere. Electric vehicles have considerable advantages: they are quiet, efficient and emit no pollution or greenhouse gases while they are being driven. However, there are also disadvantages:
  - The energy that they use is stored in a battery, and even the very best batteries have a much lower energy density than petrol/gasoline. This means that, even with a large battery, it is not possible to store enough energy for long journeys, and a large battery also adds considerable mass to the car.
  - Electric vehicles are usually less powerful, and have less acceleration, than gasoline-driven vehicles.
  - It takes hours to recharge the batteries.
  - The electrical energy used to charge the battery had to be generated in a power station (probably burning a fossil fuel), and then transferred to the place where the vehicle was charged.

■ **Figure 8.64** Homes in the local community can be heated by thermal energy from this power station in Helsinki

■ **Figure 8.65** A hybrid vehicle

The overall energy efficiency and greenhouse emissions of electric vehicles may be only slightly better than using gasoline engines, especially when the manufacture of the vehicles and batteries are also considered.

Hybrid vehicles (Figure 8.65) have been developed to overcome the disadvantages of both electric and gasoline-powered vehicles, at the same time as combining their advantages. In a hybrid vehicle, an electric motor provides the power efficiently for driving at slower speeds and an internal combustion engine is used, at its greatest efficiency, when a higher power is needed for faster speeds or accelerations. The use of a regenerative braking systems means that when brakes are used to reduce the speed of a vehicle, kinetic energy is transferred to chemical energy in the battery using a generator (rather than thermal energy in the brakes, tyres and road). The whole system is computer-controlled and the vehicle has a smaller mass than a gasoline-powered vehicle.

- *Carbon capture* and *storage*. 'Carbon capture' is the term used for the removal of carbon dioxide from the gases released when fossil fuels are burned at power stations. Various methods are available and effective, but they considerably reduce the efficiency of the power station, so that more fuel has to be burned to produce the same output power. This would mean a sharp rise in electricity prices. Once the carbon dioxide has been removed it has to be permanently stored (also known as *sequestering*). There are several possibilities, including storage as a gas and storage in the form of carbonates. Increasing the area of the Earth covered by trees and other plants (and stopping deforestation) captures carbon in the plant material and is an appealing way of increasing the amount of carbon dioxide removed from the atmosphere by photosynthesis.

1 a Do you think that most of the people you know are happy to change their habits in an effort to reduce global warming?

b If yes, explain how. If not, explain why not.

2 Hybrid cars are significantly more expensive to buy than conventional gasoline-powered cars, so not so many are sold. Explain why some people are still happy to pay the extra money for hybrid vehicles.

3 Use the internet to research the latest information on carbon sequestering.

4 Find a suitable site on the internet that will enable you to calculate the carbon footprint of your own lifestyle.

Examination questions – a selection

Paper 1 IB questions and IB style questions

Q1 Which of the following does not affect the rate at which electromagnetic radiation is transferred away from a hot object kept at a constant temperature?

A the material from which it is made
B surface temperature
C surface area
D emissivity of the surface

Q2 Which parts of a nuclear reactor are responsible for ensuring that the rate of the reaction is kept constant?

A heat exchanger and control rods
B fuel rods and heat exchanger
C fuel rods and moderator
D moderator and control rods

Q3 Two black bodies X and Y are at different temperatures. The temperature of body Y is higher than that of body X. Which of the following shows the black-body spectra for the two bodies?

A

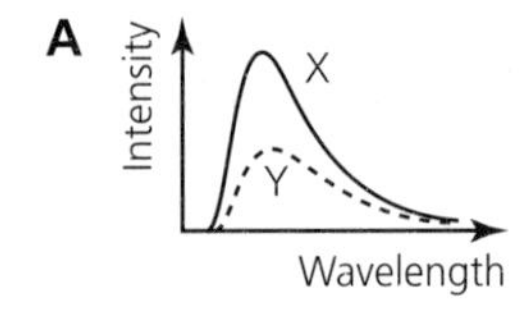

B

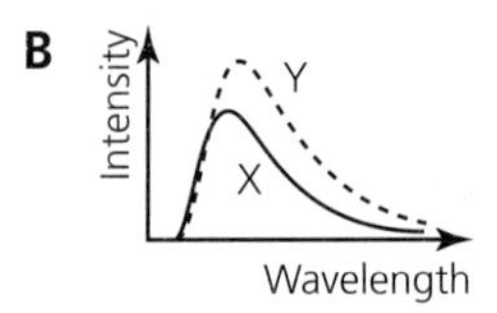

C

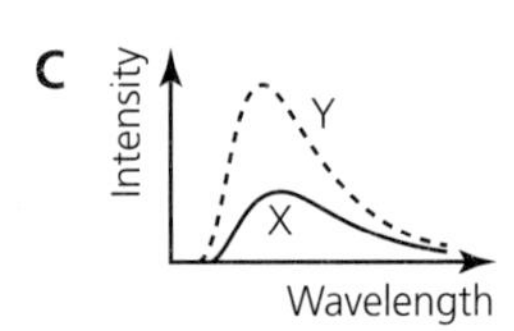

D

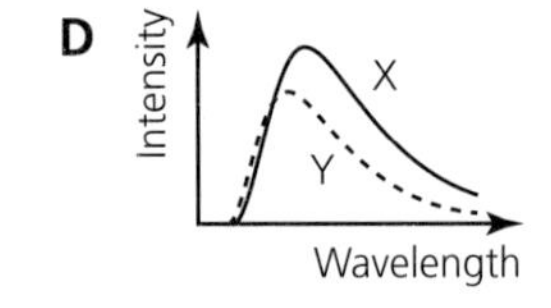

© IB Organization

Q4 The output power from a wind generator is 64 kW when the wind speed is 8 m s^{-1}. If the wind speed falls to 4 ms^{-1}, which of the following is the best estimate for the reduced power output?

A 32 kW
B 16 kW
C 8 kW
D 4 kW

Q5 Plutonium-239 is produced in nuclear power stations that use uranium as their fuel. Which of the following statements about this process is correct?

A The plutonium is fissile and can be used as a nuclear fuel.
B The plutonium is produced by nuclear fission.
C The plutonium is created in the moderator.
D Plutonium does not undergo radioactive decay.

Q6 A planet with a surface temperature of 30°C radiates power at a rate P per square metre. Another planet has a similar surface area and its temperature is 90°C. Which of the following is the best estimate of the power radiated per square metre from the second planet?

A $1.2P$
B $2P$
C $3P$
D $9P$

Q7 Which of the following is the correct meaning of the enrichment of uranium?

A increasing the percentage of uranium-237 in the fuel
B increasing the percentage of uranium-235 in the fuel
C increasing the density of the fuel
D increasing the percentage of plutonium in the fuel

Q8 Which of the following is *not* considered to be an advantage of using hydroelectric power generated from water stored in lakes?

A The source of energy is renewable.
B It does not emit significant amounts of greenhouse gases.
C The power generation is more efficient than in fossil-fuelled power stations.
D There is a negligible effect on the environment.

Q9 If global warming results in less snow and ice on the Earth's surface, how will this affect the Earth's albedo and energy absorption?

A The albedo and the rate of energy absorption will both increase.
B The albedo and the rate of energy absorption will both decrease.
C The albedo will decrease and the rate of energy absorption will increase.
D The albedo will increase and the rate of energy absorption will decrease.

Q10 Which of the following is *not* a possible way to reduce the rate of global warming?

A replacing coal-fired power stations with the use of natural gas
B increasing the use of nuclear power
C using hybrid vehicles
D operating power stations at lower temperatures

Paper 2 IB questions and IB style questions

Q1 **a** State two examples of fossil fuels. (2)
b Explain why fossil fuels are said to be non-renewable. (2)
c A Sankey diagram for the generation of electrical energy using fossil fuel as the primary energy source is shown.

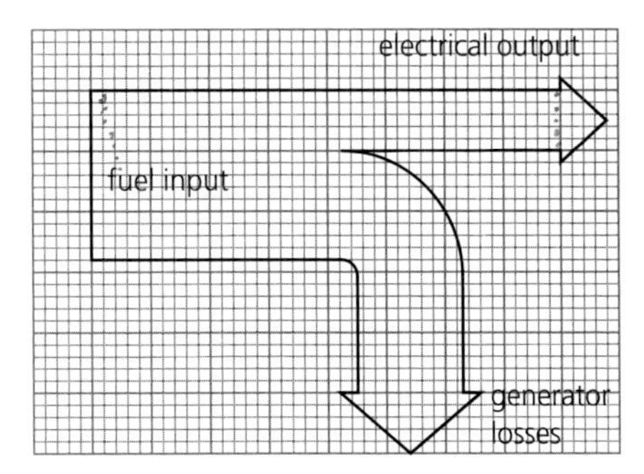

Use the Sankey diagram to estimate the efficiency of production of electrical energy and explain your answer. (2)

d Despite the fact that fossil fuels are non-renewable and contribute to atmospheric pollution there is widespread use of such fuels. Suggest three reasons for this widespread use. (3)

Q2 **a** The graph shows part of the absorption spectrum of nitrogen oxide (N_2O) in which the intensity of absorbed radiation A is plotted against frequency f.

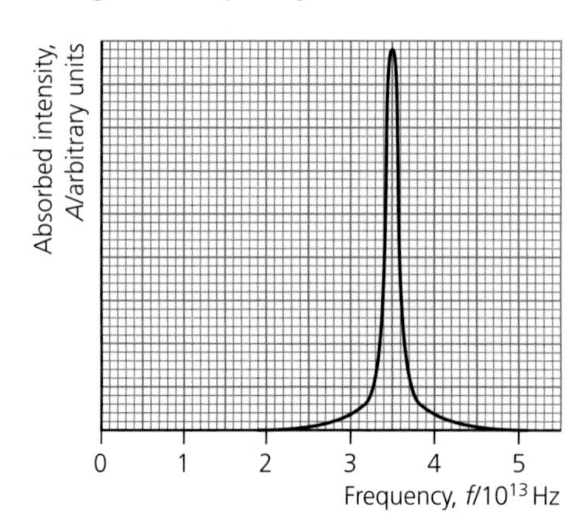

i State the region of the electromagnetic spectrum to which the resonant (peak) frequency of nitrogen oxide belongs. (1)

ii Using your answer to a i, explain why nitrogen oxide is classified as a greenhouse gas. (2)

b Define *emissivity* and *albedo*. (3)

c The diagram shows a simple energy balance climate model in which the atmosphere and the surface of the Earth are two bodies each at a constant temperature. The surface of the Earth receives both solar radiation and radiation emitted from the atmosphere. Assume that the Earth's surface behaves as a black body.

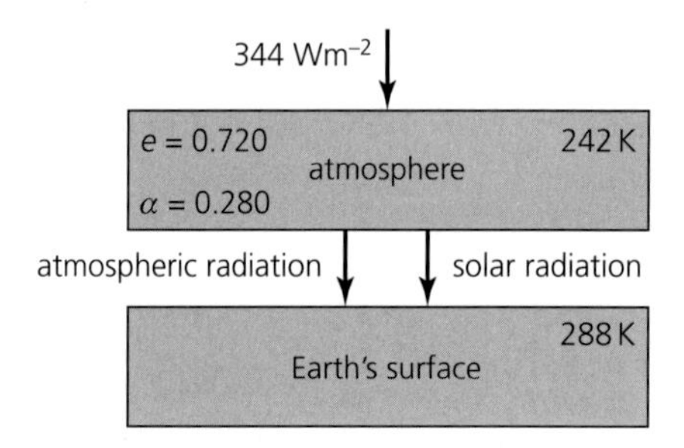

The following data are available for this model:
average temperature of the atmosphere of the Earth = 242 K
emissivity, e of the atmosphere of Earth = 0.720
average albedo, α of the atmosphere of the Earth = 0.280
solar intensity at top of atmosphere = $344\,\text{W m}^{-2}$
average temperature of the surface of the Earth = 288 K

Use the data to show that the

i power radiated per unit area of the atmosphere is $140\,\text{W m}^{-2}$ (2)

ii solar power absorbed per unit area at the surface of the Earth is $248\,\text{W m}^{-2}$. (1)

d It is hypothesised that, if the production of greenhouse gases were to stay at its present level, then the temperature of the Earth's atmosphere would eventually rise by 6.0 K. Calculate the power per unit area that would then be:

i radiated by the atmosphere (1)

ii absorbed by the Earth's surface. (1)

e Estimate, using your answer to d ii, the increase in temperature of the Earth's surface. (3)

© IB Organization

9 Wave phenomena

ESSENTIAL IDEAS

- The solution of the harmonic oscillator can be framed around the variation of kinetic and potential energy in the system.
- Single-slit diffraction occurs when a wave is incident on a slit of approximately the same size as the wavelength.
- Interference patterns from multiple slits and thin films produce accurately repeatable patterns.
- Resolution places an absolute limit on the extent to which an optical or other system can separate images of objects.
- The Doppler effect describes the phenomenon of wavelength/frequency shift when relative motion occurs.

9.1 Simple harmonic motion – *the solution of the harmonic oscillator can be framed around the variation of kinetic and potential energy in the system*

Simple harmonic motion (SHM) was introduced in Chapter 4. We will now look at this very important concept in more mathematical detail, starting by re-examining two basic oscillators, and then explaining that SHM and circular motion have many mathematical similarities.

Equations for the periods of two common simple harmonic oscillators

The period, T, of a *simple* pendulum depends only its length, l (and the value of the gravitational field strength, g):

$$T = 2\pi\sqrt{\frac{l}{g}}$$

This equation is given in the *Physics data booklet*.

The period, T, of a mass oscillating on the end of a spring depends on the magnitude of the mass, m, and the force constant, k, of the spring:

$$T = 2\pi\sqrt{\frac{m}{k}}$$

This equation is given in the *Physics data booklet*.

Both equations are idealized and make simplifying assumptions about the oscillating system involved. However, careful laboratory investigations, and the drawing of appropriate graphs to represent the results, confirm that these equations can make accurate predictions of time periods.

It is possible to investigate a wide variety of simple oscillating systems in a laboratory. Many of them have some similarities with one of the two systems referred to above. For example, there are many different kinds of pendulum, including those that twist (*torsion pendulums*). The results can be represented graphically and assessed to see if the period of the system and the variables that control it are connected by any simple mathematical relationship. A *tuning fork* is a simple mechanical oscillator, which has one dominant frequency, and the waveform produced can be inspected using a microphone connected to an oscilloscope.

There are also many computer simulations of various kinds of SHM that enable quick manipulation of the variables in order to observe their effect on the period.

Worked example

1 **a** Using the results from an investigation of a simple pendulum, what graph would you draw to get a straight line through the origin?
b How could a value for the acceleration due to gravity (gravitational field strength) be determined from the graph?

a $T = 2\pi\sqrt{\frac{l}{g}}$

$T = \frac{2\pi}{\sqrt{g}}\sqrt{l}$

Comparing this to the equation for a straight line, $y = mx + c$, a graph of T against $\sqrt{l}$ (or T^2 against l) should result in a straight line that extends to the origin.

b g can be found from the gradient of the graph ($= 2\pi/\sqrt{g}$).

1 a A simple pendulum completed 20 oscillations in 30.9 s. What was its length?
b Sketch the pendulum and clearly label the length calculated in a.

2 a A 200 g mass was fixed on the end of a long spring and it increased its length by 5.4 cm. What is the force constant of the spring?
b If the mass is displaced a small distance from its equilibrium position and undergoes SHM, calculate the frequency of oscillations.
c Suggest a possible reason why oscillations with a larger amplitude may not be simple harmonic.

3 a Using the results from an investigation of the oscillations of a mass–spring system (varying the mass), what graph would you draw to get a straight line through the origin?
b How could a value for the force constant of the spring be determined from the graph?

Oscillations and circular motion

There is a close connection between oscillations and circular motion. Indeed, viewed from the side, motion in a circle has the same pattern of movement as a simple oscillation.

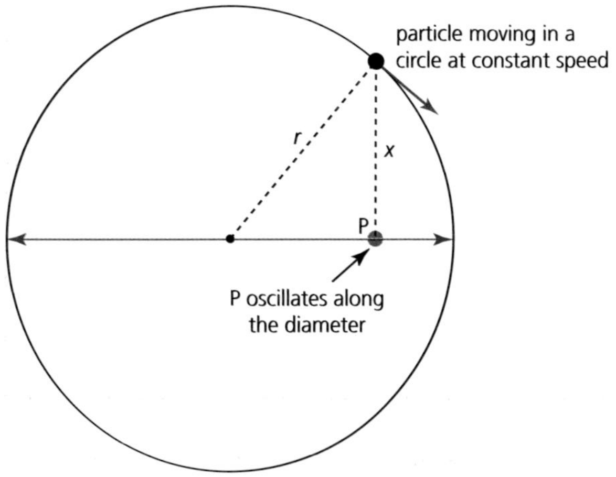

Figure 9.1 shows a particle moving in a circle at constant speed. Point P is the projection of the particle's position onto the diameter of the circle.

As the particle moves in a circle, point P oscillates backwards and forwards along the diameter with the same frequency as the particle's circular motion, and with an amplitude equal to the radius, r, of the circle.

One complete oscillation of point P can be considered as equivalent to the particle moving through an angle of 2π radians (or 360°).

If a graph of the perpendicular displacement from the diameter, x, against time is plotted for point P, it will look exactly like an SHM graph (Figure 9.2) with the amplitude, x_0, being equal to r.

The concept of *angular velocity*, ω was introduced in Chapter 6 as a key quantity in the mathematical description of motion in a circle. Angular velocity is also used widely in the description of SHM. Remember that, although the terms period, frequency and angular velocity are all used to describe oscillations, they are just different ways of representing the same information:

$$\omega = \frac{2\pi}{T} = 2\pi f$$

■ **Figure 9.1** Comparing motion in a circle to an oscillation

4 Using a slow-motion video replay of the movement of a bird's wings, it was calculated that they were waving up and down at a frequency of 22 Hz. What was the time period of the oscillation and its angular velocity?

5 A car engine was measured at 4300 rpm (revolutions per minute). What was its angular velocity?

6 The Earth spins once on its axis in 23 hours and 56 minutes.
a What is its angular velocity?
b Through what total angle will it rotate in 1 day (24 hours) 3 hours and 24 minutes?

7 An object is oscillating with an angular velocity of 48.1 rad s^{-1}. What is its period?

The defining equation of SHM

We saw in Chapter 4 that simple harmonic motion (SHM) is defined as an oscillation in which the acceleration, a, of a body is proportional to its displacement, x, from its equilibrium position and in the opposite direction:

$$a \propto -x$$

We can rewrite the equation as:

$$a = -\text{constant} \times x$$

The constant must involve frequency because the magnitude of the acceleration is bigger if the frequency is higher. The constant can be shown to be ω^2. So, the defining equation of SHM can be written as:

$$a = -\omega^2 x$$

This equation is given in the *Physics data booklet*. It allows us to calculate the instantaneous acceleration of an oscillator of known frequency at any given displacement, as shown in Worked example 2.

Worked example

2 A mass oscillates between two springs with a frequency of 1.4 Hz.
a What is its angular velocity?
b What is its acceleration when:
i its displacement is 1.0 cm
ii its displacement is 4.0 cm
iii it passes through its equilibrium position?

a $\omega = 2\pi f = 2\pi(1.4) = 8.8\,\text{rad}\,\text{s}^{-1}$

b i $a = -\omega^2 x = -(8.8)^2 \times 0.010 = -0.77\,\text{m}\,\text{s}^{-2}$
ii $a = -\omega^2 x = -(8.8)^2 \times 0.040 = -3.1\,\text{m}\,\text{s}^{-2}$
iii $a = -\omega^2 x = -(8.8)^2 \times 0.0 = 0.0\,\text{m}\,\text{s}^{-2}$

8 A pendulum has a period of 2.34 s. How far must it be displaced to make its acceleration 1.00 m s^{-2}?

9 During SHM a mass moves with an acceleration of 3.4 m s^{-2} when its displacement is 4.0 cm. Calculate:
a its angular velocity
b its period.

10 A mass oscillating on a spring performs exactly 20 oscillations in 15.8 s.
a What is its acceleration when it is displaced 62.3 mm from its equilibrium position?
b What assumption did you make?

Solutions to the SHM equation ($a = -\omega^2 x$)

There are still many things about an oscillator that we cannot determine directly from the equation defining SHM, $a = -\omega^2 x$. For example, what are the values of displacement and velocity at any given time? To answer such questions we need either accurate graphs or other mathematical 'solutions' to the SHM equation. (Here, 'solutions' means either graphs or equations that will present the same information in more useful forms.)

Graphical solutions

Figure 9.2 shows the waveform of an SHM starting with zero displacement at time $t = 0$. A detailed graph like this could be used to find the displacement at any time.

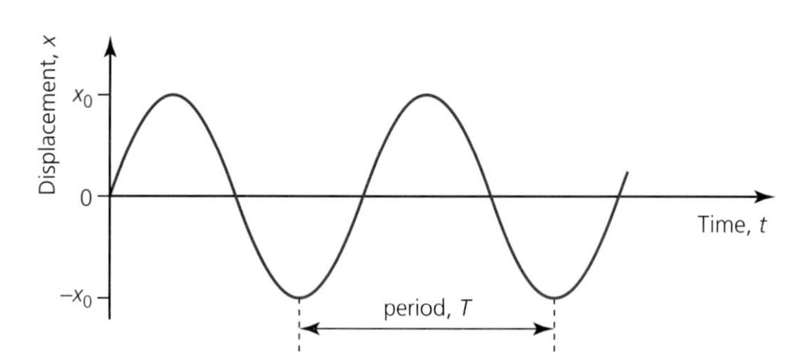

■ **Figure 9.2** Displacement–time graph for SHM, represented by a sine wave; timing was started when the particle had zero displacement

The velocity at a particular time, or displacement, can be found from the gradient of the displacement–time graph at that point (velocity, v = change in displacement/change in time).

Similarly, the acceleration at a given time, or displacement, is determined from the gradient of the velocity graph (acceleration, $a = \Delta v/\Delta t$).

So, in theory, the data from the displacement–time graph can be processed to calculate values for velocity and acceleration, and the corresponding graphs can be drawn, as shown in Figure 9.3. Algebraic solutions using trigonometric equations for the lines are usually easier. The equations shown in Figure 9.3 are explained in the next section.

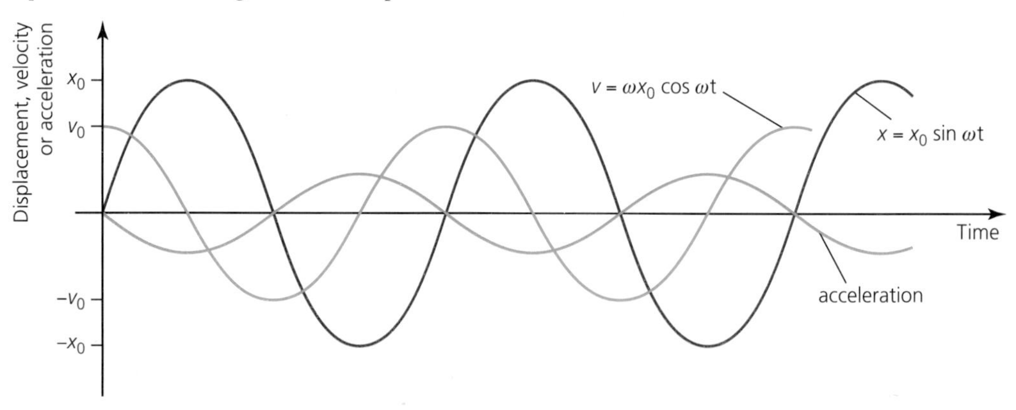

■ **Figure 9.3** Comparing displacement, velocity and acceleration for SHM, with timing starting at displacement $x = 0$

The displacement and acceleration graphs are π out of phase with each other and $\pi/2$ out of phase with the velocity graph.

Iteration

Graphs like those shown in Figure 9.3 can be produced using the laws of basic mechanics (Chapter 2) from first principles, but it can be a lengthy process. Given the mass, its position and the force involved at any moment it is possible to calculate what will happen in the next interval of time, assuming that the force does *not* change. But, of course, as the mass moves during the oscillation, the force on it does change. However, if the time interval is chosen to be small enough, the calculation can still be accurate. The results from each calculation are then used as the input data for the next calculation, and so on. This is called an **iterative process**. It can produce very accurate results if the time intervals are very small and a very large number of repeated calculations are performed. Ideal for a computer!

Algebraic solutions

The displacement–time graph shown in Figure 9.2 has a simple sinusoidal shape and it can be described mathematically by an equation using a sine:

$$x = x_0 \sin\theta$$

where θ is the angular displacement at time t.

Because, by definition, $\theta = \omega t$, this equation for displacement can be expressed much more usefully in terms of t and ω (or T, or f):

$$x = x_0 \sin \omega t$$

The velocity graph, which starts with maximum velocity, v_0 at $t = 0$, can be represented by the following *cosine* equation:

$$v = v_0 \cos \omega t$$

But we know from Chapter 6 that, in general for circular motion/oscillations, speed $v = \omega r$, so that:

$$v_0 = \omega x_0$$

Then the equation for v can more usefully (not needing knowledge of v_0) be rewritten as:

$$v = \omega x_0 \cos \omega t$$

The two equations for x and v highlighted above are given in the *Physics data booklet*. They show how displacement and velocity vary with time for SHM with zero displacement at the start of the timing ($x = 0$ when $t = 0$). We can say that they are *solutions to* the SHM equation $a = -\omega^2 x$.

We could choose to start timing the SHM when the particle is at its maximum displacement from the equilibrium position during the oscillation, that is, $x = x_0$ when $t = 0$. The graphs shown in Figure 9.3 would then appear as shown in Figure 9.4.

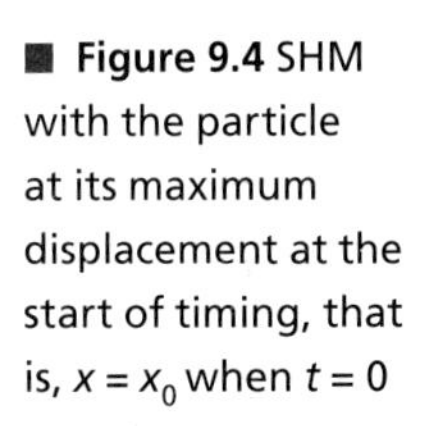

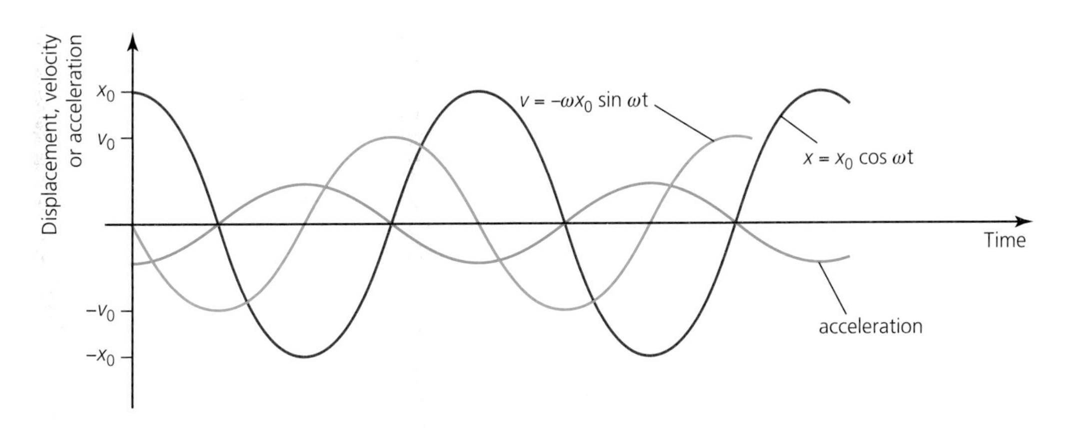

■ **Figure 9.4** SHM with the particle at its maximum displacement at the start of timing, that is, $x = x_0$ when $t = 0$

In this case, the solutions to the SHM equation are as follows:

$$x = x_0 \cos \omega t$$

$$v = -\omega x_0 \sin \omega t$$

These equations are given in the *Physics data booklet*.

It can also be shown that the velocity at any known displacement can be calculated from the following equation:

$$v = \pm\omega\sqrt{x_0^2 - x^2}$$

This equation is also listed in the *Physics data booklet*.

This equation confirms that the maximum velocity, v_0, will occur when the particle is moving through its equilibrium position (at zero displacement, $x = 0$), so that:

$$v_0 = \pm\omega x_0$$

This equation is *not* given in the *Physics data booklet*.

Worked example

3 An oscillating mass is set in motion with SHM. It is at its maximum displacement of 12 cm when a timer is started, and its period of oscillation is 2.4 s. Calculate:

a the displacement after 3.3 s
b its maximum speed
c its speed after 5.6 s
d its speed when its displacement is 8.8 cm.

a $x = x_0 \cos \omega t$ and $\omega = \frac{2\pi}{T}$, with $x_0 = 0.12$ m, $T = 2.4$ s and $t = 3.3$ s

$$x = 0.12 \times \cos\left(\frac{2\pi}{2.4} \times 3.3\right) = 0.12 \times -0.704 = -0.084\,\text{m}$$

b $v_0 = \omega x_0 = \left(\frac{2\pi}{2.4}\right) \times 0.12 = 0.31\,\text{m s}^{-1}$

c $v = -v_0 \sin \omega t$ with $v_0 = 0.31\,\text{m s}^{-1}$ and $t = 5.6$ s

$$v = 0.31 \times \sin\left(\frac{2\pi}{2.4} \times 5.6\right) = 0.27\,\text{m s}^{-1}$$

d $v = \pm\omega\sqrt{x_0^2 - x^2}$ with $x = 0.088$ m

$$v = \frac{2\pi}{2.4} \times \sqrt{0.12^2 - 0.088^2} = 0.21\,\text{m s}^{-1}$$

11 A mass is oscillating between two springs with a frequency of 1.5 Hz and amplitude of 3.7 cm. It has a speed of 34 cm s^{-1} as it passes through its equilibrium position and a timer is started. What are its displacement and velocity 1.8 s later?

12 An object of mass 45 g undergoes SHM with a frequency of 12 Hz and an amplitude of 3.1 mm.
a What is its maximum speed and kinetic energy?
b What is the object's displacement 120 ms after it is released from its maximum displacement?

13 A mass is oscillating with SHM and has an amplitude of 3.84 cm. Its displacement is 2.76 cm at 0.0217 s after it is released from its maximum displacement. Calculate a possible value for its frequency.

14 A simple harmonic oscillator has a time period of 0.84 s and its speed is 0.53 m s^{-1} as it passes through its mean (equilibrium) position.
a What is its speed exactly 2.0 s later?
b If the amplitude of the oscillation is 8.9 cm, what was the displacement after 2.0 s?

15 The water level in a harbour rises and falls with the tides, with 12 h 32 min for a complete cycle. The high tide level is 8.20 m above the low tide level, which occurred at 4.10 am. If the tides rise and fall with SHM, what will be the level of the water at 6.00 am?

16 What does the area under a velocity–time graph of an oscillation represent?

Energy changes during SHM

All mechanical oscillations involve a continual interchange between kinetic energy and some form of potential energy. For example, the potential energy involved in a pendulum swinging is gravitational, whereas for a mass oscillating horizontally between springs the potential energy is stored in the form of elastic strain energy in the spring:

total energy, E_T = kinetic energy, E_K + potential energy, E_P

We know that kinetic energy, $E_K = \frac{1}{2}mv^2$. Combining this with the equation for v given earlier leads to the following equation, which shows how KE varies with displacement during SHM:

$$E_K = \tfrac{1}{2}m\omega^2(x_0^2 - x^2)$$

This equation is given in the *Physics data booklet.*

This relationship is represented graphically in Figure 9.5. Also included on the same graph are the lines for total energy and for potential energy (= total energy − kinetic energy).

The total energy is equal to the *maximum* value of KE, which is the value of KE with zero displacement and zero potential energy. That is:

$$E_T = \tfrac{1}{2}m\omega^2 x_0^2$$

This equation is given in the *Physics data booklet.*

The potential energy can be determined from total energy less the kinetic energy:

$$E_P = \tfrac{1}{2}m\omega^2 x^2$$

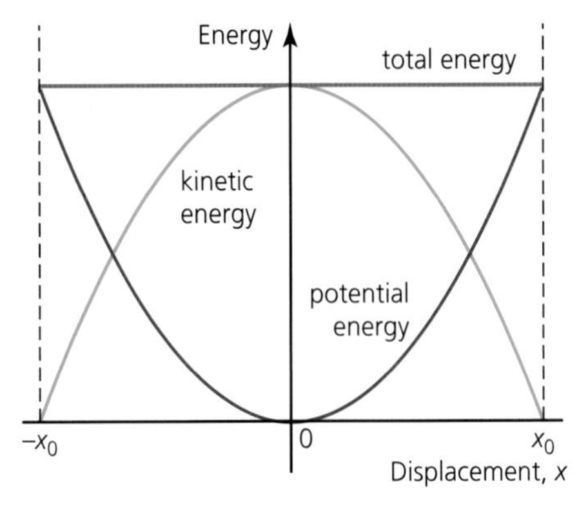

■ **Figure 9.5** Variation of energies of a simple harmonic oscillator with displacement

This equation (which is not in the *Physics data booklet*) confirms what we saw in Chapter 4 (page 161): the energy of an oscillation is proportional to its amplitude *squared*. For example, if pendulum A swings with three times the amplitude of an identical pendulum B, then pendulum A has nine (3^2) times the potential energy of pendulum B.

In this chapter we have been concerned with the interchange between kinetic energy and some form of potential energy for *mechanical* oscillators, but the usefulness of SHM theory reaches further. For example, an understanding of oscillations in *electric* circuits is an essential aspect of learning about radio and phone communication.

Worked example

4 A mass of 1250 g oscillates with a period of 0.56 s and an amplitude of 32 cm.

a What is its total energy?

b What is its potential energy when its displacement is 12 cm?

c What is its kinetic energy when displaced 12 cm?

d What is its kinetic energy when it has a displacement of 4.3 cm?

a $E_T = \frac{1}{2}m\omega^2 x_0^2$

$= 0.5 \times 1.25 \times \left(\frac{2\pi}{0.56}\right)^2 \times 0.32^2$

$= 8.1\,\text{J}$

b $E_P = \frac{1}{2}m\omega^2 x^2$

$= 0.5 \times 1.25 \times \left(\frac{2\pi}{0.56}\right)^2 \times 0.12^2$

$= 1.13\,\text{J}$

c $E_T = E_P + E_K$

$8.1 = 1.13 + E_K$

$E_K = 7.0\,\text{J}$

d $E_K = \frac{1}{2}m\omega^2(x_0^2 - x^2)$

$= 0.5 \times 1.25 \times \left(\frac{2\pi}{0.56}\right)^2 \times (0.32^2 - 0.043^2)$

$= 7.9\,\text{J}$

17 A large pendulum of mass 2.6 kg has a length, l, of 4.63 m.
 a Determine its period, T.
 b Explain why a pendulum of twice the mass has the same period (if it has the same length).
 c The motion of a pendulum is a close approximation to SHM. Explain why the equation for the time period of an oscillator (like the pendulum) does not include the amplitude.
 d What is the angular velocity of this pendulum?
 e What is the oscillator's total energy if it has an amplitude of 1.2 m?
 f What is its speed as it passes through the equilibrium position?

18 A 435 g mass is oscillating horizontally between two springs with a frequency of 0.849 Hz. If its total energy is 4.28 J, what are its amplitude and its maximum speed?

19 An oscillator of mass 786 g has a total energy of 2.4 J.
 a Calculate its period if it is moving with an amplitude of 23 cm.
 b What is its speed when it has a displacement of 10 cm?

20 Real oscillators spread energy into the surroundings. Sketch a graph to show how the potential and kinetic energies might change with time over several oscillations of a real oscillator.

Nature of Science

Analysis of more complicated oscillations

The analysis of oscillations, of many and various kinds, plays a very important part in physics and technology. This chapter has been mainly concerned with an understanding of the basic SHM of oscillators moving with constant frequency and amplitude: oscillations that can be represented by *single* sine and cosine functions. SHM is the starting point for the study of all oscillations, but most real-life oscillators are not so simple and their waveforms are much more complex.

However, mathematics has a solution: *any* complicated, but periodic, waveform can be reproduced by the addition (superposition) of an appropriate number of simple trigonometric wave functions (sine and cosine waves). This is known as **Fourier analysis**. We have already met this concept briefly in Chapter 4: sounds played on musical instruments are the addition of a series of mathematically related *harmonics*. Computer programs can be used to add trigonometric wave functions together and show the results graphically or, conversely, a complicated waveform can be analysed into component mathematical functions. Combining simple trigonometric functions to make a more complicated waveform is called **Fourier synthesis**.

9.2 Single-slit diffraction – *single-slit diffraction occurs when a wave is incident on a slit of approximately the same size as the wavelength*

Single-slit diffraction was discussed briefly in Chapter 4. Figure 9.6 shows how a laser can be used to produce a single-slit diffraction pattern on a screen.

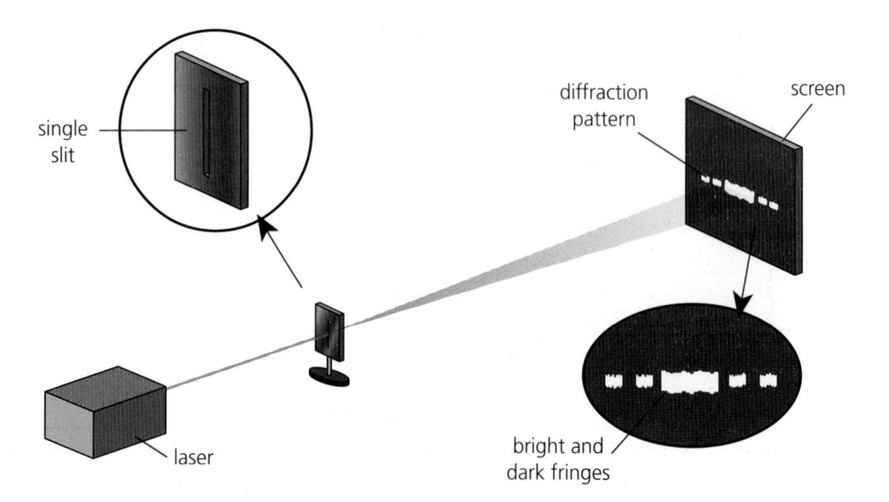

■ **Figure 9.6** Diffraction of light through a single slit

Figure 9.7a shows a diffraction pattern for a beam of monochromatic light passing through a narrow vertical slit. This is the type of aperture (opening) usually used to produce simple diffraction patterns. The diffraction pattern consists of a series of bands (fringes) of light and dark. *The central band is brighter and about twice the width of the others.* If an adjustable slit is used, the pattern can be observed getting broader as the slit width is reduced.

Figure 9.8 shows the diffraction pattern produced when monochromatic light passes through a small *circular* aperture.

By comparing Figures 9.7a and 9.8, we can see that the shape of the pattern depends on the shape of the aperture. The light used to produce these patterns was *monochromatic*, which means that it has only one wavelength (and frequency), or a very narrow range of wavelengths. Light from most sources is not monochromatic and will consist of different wavelengths. These will not produce such a simple pattern because different wavelengths are sent to different places on the screen. The diffraction pattern produced by white light passing through a vertical narrow slit will have a mostly white central band, but the other bands will show various coloured effects. The lower image in Figure 9.7b shows the pattern produced by white light passing through a slit of the same width as used to produce Figure 9.7a.

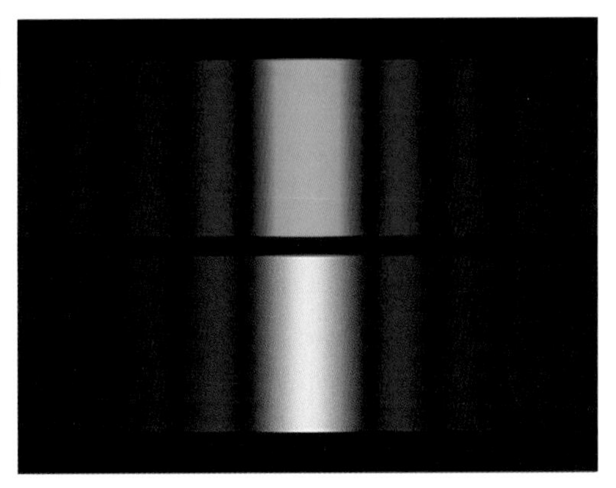

■ **Figure 9.7** Diffraction of **a** monochromatic light and **b** white light by a narrow vertical slit

■ **Figure 9.8** Monochromatic light diffracted by a very small *circular* aperture

The nature of single-slit diffraction

In order to produce these diffraction patterns for light, the waves must travel away from the aperture in some directions, but not in others. The simple diffraction theory covered in Chapter 4 does not explain this. Figure 9.9a shows the diffracted waves coming away from a gap of about the same width as the wavelength (as discussed in Chapter 4). The gap acts effectively like a point source, spreading waves forward equally in *all* directions. This *cannot* be used to help to explain the diffraction of light because the wavelengths of light are so much smaller than the width of even a very small hole. For example, a hole of diameter 0.5 mm is 1000 times wider than the wavelength of green light (approximately 5×10^{-7} m).

To produce a diffraction pattern for light, such as that shown in Figure 9.7a, we must develop a theory that explains why waves spread away from a gap, as shown in Figure 9.9b when the gap is much bigger than the wavelength.

Diffraction patterns can be explained by assuming that each point on a wavefront passing through the slit acts as a point source of **secondary waves**. (This idea was first suggested in 1678 by the Dutch physicist Christiaan Huygens, when he proposed that light waves propagate forwards because *all* the points on any wavefront act as sources of

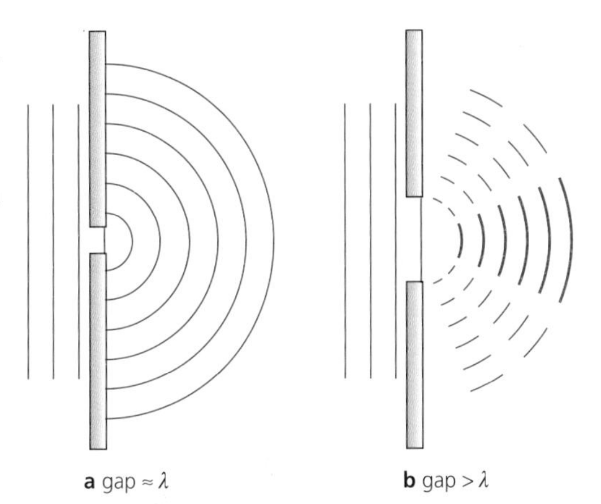

■ **Figure 9.9** Diffraction of waves by different sized apertures

secondary waves.) What happens after the wave has passed through the slit depends on how these secondary waves *interfere* with each other.

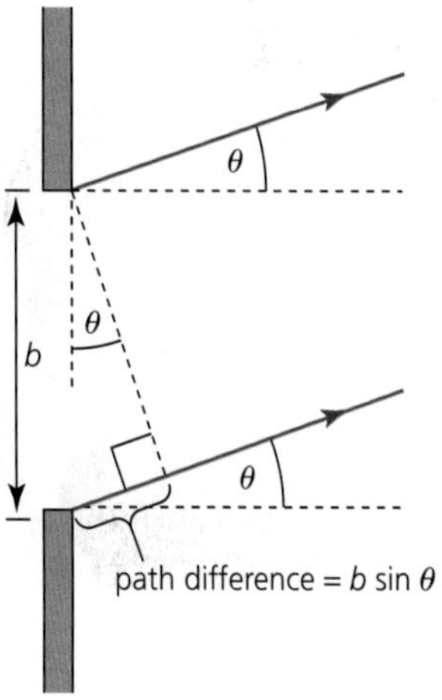

Figure 9.10 shows a direction, θ, in which secondary waves travel away from a slit of width, b. If θ is zero, all the secondary waves will interfere constructively in this direction (straight through the slit) because there is no *path difference* between them. (Of course, in theory, waves travelling parallel to each other in the same direction cannot meet and interfere, so we will assume that the waves' directions are *very nearly* parallel.)

Consider what happens for angles increasingly larger than zero. The path difference, as shown in Figure 9.10, equals $b \sin \theta$ and this increases as the angle θ increases. There will be an angle at which the waves from the two edges of the slit interfere constructively because the path difference has increased to become one wavelength, λ.

■ **Figure 9.10** Path differences and interference

But if secondary waves from the edges of the slit interfere constructively, what about interference between all the other secondary waves? Consider Figure 9.11 in which the slit has been divided into a number of point sources of secondary waves. (Ten points have been chosen, but it could be many more.)

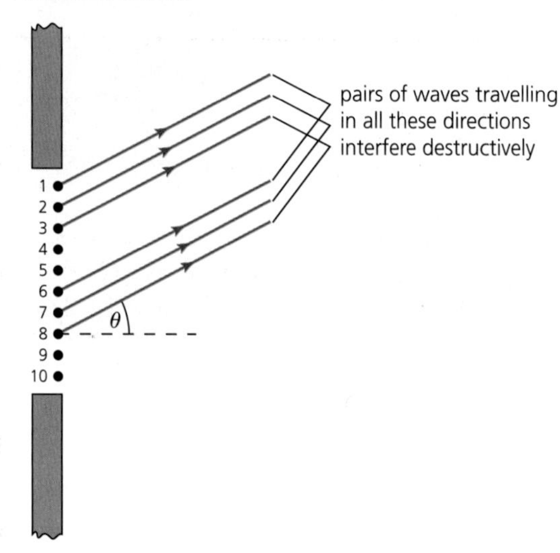

■ **Figure 9.11** Secondary waves that will interfere destructively can be 'paired off'

If the angle, θ, is such that secondary waves from points 1 and 10 would interfere constructively because the path difference is one wavelength, then secondary waves from 1 and 6 must have a path difference of half a wavelength and interfere destructively. Similarly, waves from points 2 and 7, points 3 and 8, points 4 and 9 and points 5 and 10 must all interfere destructively. In this way waves from *all* points can be 'paired off' with others, so that the first *minimum* of the diffraction pattern occurs at such an angle that waves from the edges of the slit would otherwise interfere constructively.

The first minimum of the diffraction pattern occurs when the path difference between secondary waves from the edge of the slit is equal to one wavelength. That is, if $b \sin \theta = \lambda$.

For the diffraction of light through a narrow slit, the angle θ is usually small and approximately equal to sin θ if the angle is expressed in radians (valid for angles less than about 10°).

The angle for the first minimum of a single slit diffraction pattern is $\theta = \dfrac{\lambda}{b}$.

This equation is given in the *Physics data booklet*.

ToK Link

Comparing science with other areas of knowledge

Are explanations in science different from explanations in other areas of knowledge such as history?

For example, would it be possible in principle to develop a mathematical model to describe the political and/or economic situation in Europe before the start of World War 1? And could such a model be used to predict what happened? Is there something fundamentally different between knowledge in physics and history, or is any historical situation just too complicated, or dependent on human personalities, for mathematical analysis?

Relative intensity–angle graph for single-slit diffraction

Similar reasoning to that discussed above can be used to show that further diffraction minima occur at angles $2\lambda/b$, $3\lambda/b$, $4\lambda/b$ etc. Figure 9.12 shows a graphical interpretation of these figures and approximately how it corresponds to a drawing of a single-slit diffraction pattern.

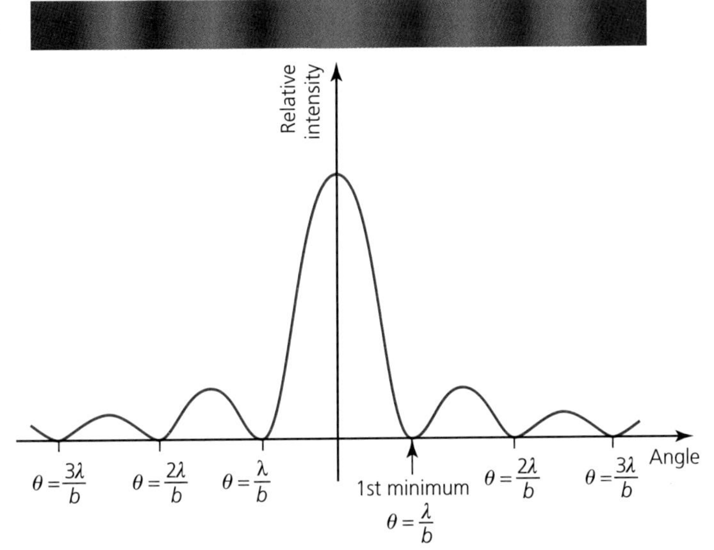

■ **Figure 9.12** Variation of intensity with angle for single-slit diffraction

Worked example

5 Monochromatic light of wavelength 663 nm is shone through a gap of width 0.0730 mm.

a At what angle is the first minimum of the diffraction pattern formed?

b If the pattern is observed on a screen that is 2.83 m from the slit, what is the width of the central maximum?

a $\theta = \frac{\lambda}{b}$

$\theta = \frac{663 \times 10^{-9}}{7.30 \times 10^{-5}}$

$= 9.08 \times 10^{-3}$ radians

b See Figure 9.13.

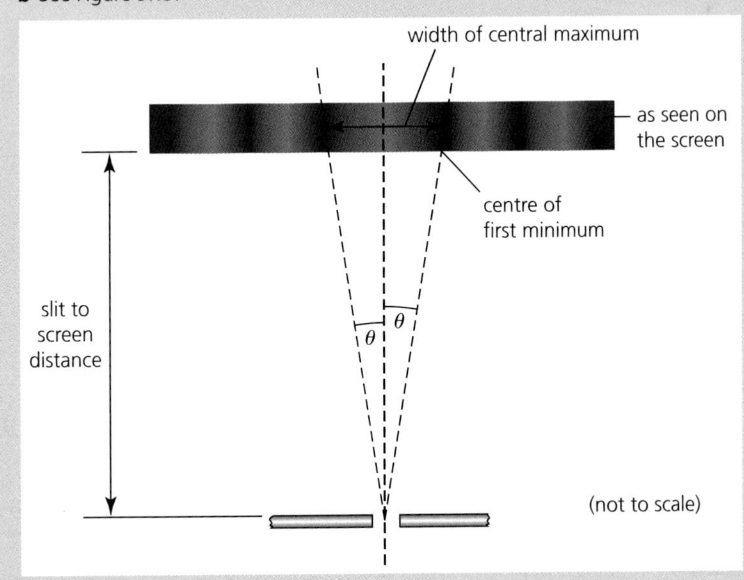

■ **Figure 9.13**

$\theta \approx \sin \theta$

$= \frac{\text{half width of central maximum}}{\text{slit to screen distance}}$

half width of central maximum = $(9.08 \times 10^{-3}) \times 2.83 = 0.0257$ m

width of central maximum = $0.0257 \times 2 = 0.0514$ m

21 Electromagnetic radiation of wavelength 2.37×10^{-7} m passes through a narrow slit of width 4.70×10^{-5} m.
a In what part of the electromagnetic spectrum is this radiation?
b Suggest how it could be detected.
c Calculate the angle of the first minimum of the diffraction pattern.

22 What is the wavelength of light that has a first diffraction minimum at an angle of 0.0038 radians when it passes through a slit of width 0.15 mm?

23 When light of wavelength 6.2×10^{-7} m was diffracted through a narrow slit, the central maximum had a width of 2.8 cm on a screen that was 1.92 m from the slit. What was the width of the slit?

24 a Sketch and label a relative intensity against angle graph for the diffraction of red light of wavelength 6.4×10^{-7} m through a slit of width 0.082 mm. Include at least five peaks of intensity.
b Add to your graph a sketch to show how monochromatic blue light would be affected by the same slit.

The importance of diffraction

The diffraction of light is evidence confirming the wave nature of light, and when it was discovered that electrons could be diffracted, their wave nature was also understood for the first time (Chapter 12). Other electromagnetic radiations will be diffracted significantly by objects and gaps that have a size comparable to their wavelengths. For example, the diffraction of X-rays by atoms, ions and molecules can be used to provide detailed information about solids with regular structure. The diffraction of radio waves used in communication is also of considerable importance when choosing which wavelength to use in order to get information efficiently from the transmitter to the receiver.

Waves travelling from place to place can have their direction changed by diffraction, but diffraction can also be important when waves are emitted or received. They can then be considered to be passing through an aperture, and the ratio λ/b will be important when considering how the waves spread away from a source, or what direction they appear to come from when they are received.

The fundamental difference between diffraction and interference can be confusing. Diffraction is the spreading of waves around obstacles and corners, and though gaps, but when we refer to a diffraction *pattern*, it is the result of interference between wavefronts caused by diffraction.

Nature of Science

Changing theories of diffraction

The first detailed observations of the diffraction patterns produced in the shadows by light passing through apertures were made more than 350 years ago. At that time the phenomenon defied any simple explanation in terms of light 'particles', and the wave nature of light was not understood. After the wave theory of light became established, a theory of diffraction could be developed that involved the adding together of waveforms arriving at the same point from different places within the aperture. (Depending on the context of the discussion, the addition of waveforms can be variously described as *superposition*, *interference* or *Fourier synthesis*.) The more recent photon theory of light (Chapter 7) returns to a 'particle' explanation, but particles involving 'probability waves' (Chapter 12).

9.3 Interference – *interference patterns from multiple slits and thin films produce accurately repeatable patterns*

The interference of waves was introduced in Chapter 4, where it was explained that the necessary condition for interference is that the sources are **coherent**. The important principles will be repeated here quickly. Students are recommended to make use of some of the many computer simulations available on interference.

Consider Figure 9.14, which shows two coherent sources, S_1 and S_2, and the waves they have emitted.

The solid blue lines represent the crests of waves and the troughs of the waves will be midway between them. At a point like P, two crests are coming together, but at the same point a short time later, two troughs will come together. The waves arriving at points like P will always arrive in phase and, using the principle of superposition, we can determine that constructive interference occurs.

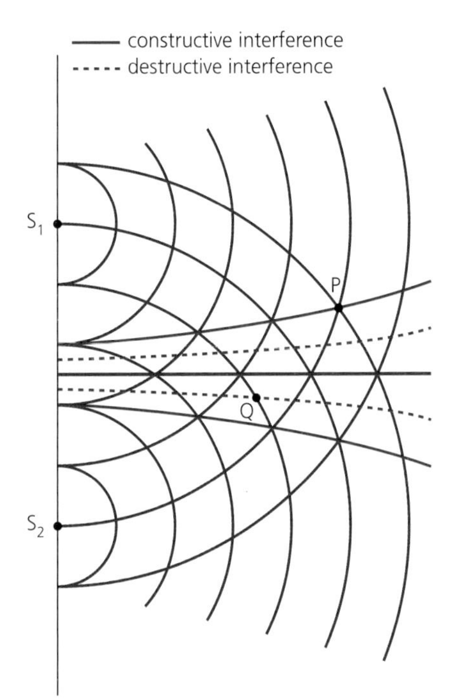

■ **Figure 9.14** Two sets of coherent waves crossing each other to produce an interference pattern

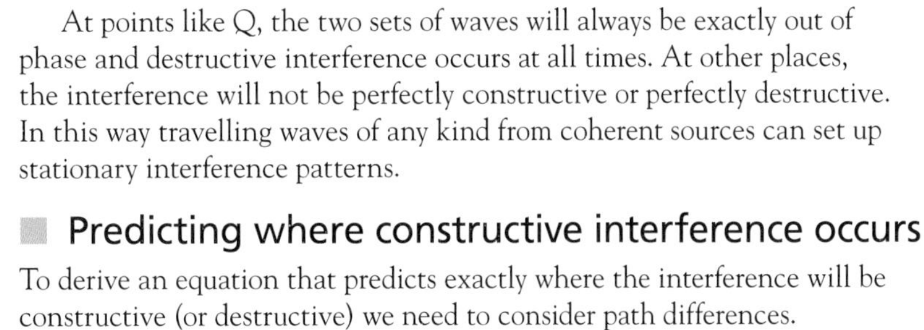

At points like Q, the two sets of waves will always be exactly out of phase and destructive interference occurs at all times. At other places, the interference will not be perfectly constructive or perfectly destructive. In this way travelling waves of any kind from coherent sources can set up stationary interference patterns.

Predicting where constructive interference occurs

To derive an equation that predicts exactly where the interference will be constructive (or destructive) we need to consider path differences.

In Figure 9.15, rays are used to represent the directions of wave travel and it is clear that the waves arriving at a point P from S_2 have travelled further than the waves arriving from S_1. We say that there is a *path difference* between the waves ($S_2P - S_1P$).

Constructive interference occurs if the path difference between the waves is zero, or one wavelength, or two wavelengths, or three wavelengths etc.

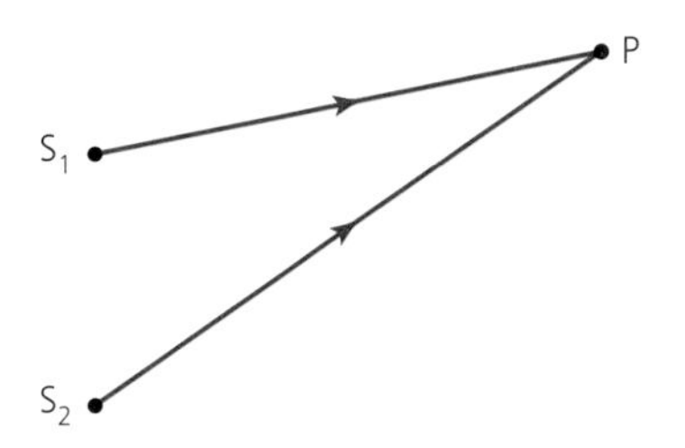

■ **Figure 9.15** Path difference

The condition for *constructive* interference is that the path difference = $n\lambda$ (where n is an integer: 1, 2, 3, 4, 5 etc.).

The condition for *destructive* interference is that the path difference equals an odd number of half wavelengths. Path difference = $(n + \frac{1}{2})\lambda$.

These equations for path difference are given in the *Physics data booklet* (for Chapter 4).

Interference of light waves

Light waves are not usually coherent, so the interference of light is not a phenomenon of which we are usually aware. Light from a laser (which is both monochromatic and coherent) is the ideal source for producing interference patterns, but interference can also be produced without lasers.

Rather than use two separate sources, light from a single source is split into two parts using two narrow slits, which must be close together. The waves passing through each slit then act like separate coherent sources as the waves spread away from the slits because of diffraction (Figure 9.16). In general, we know that for diffraction to be significant, the size of the gap needs to be comparable to the size of the wavelength, so the gaps must be very narrow (because the wavelength of light is so small).

■ **Figure 9.16** Using two slits to produce an interference pattern

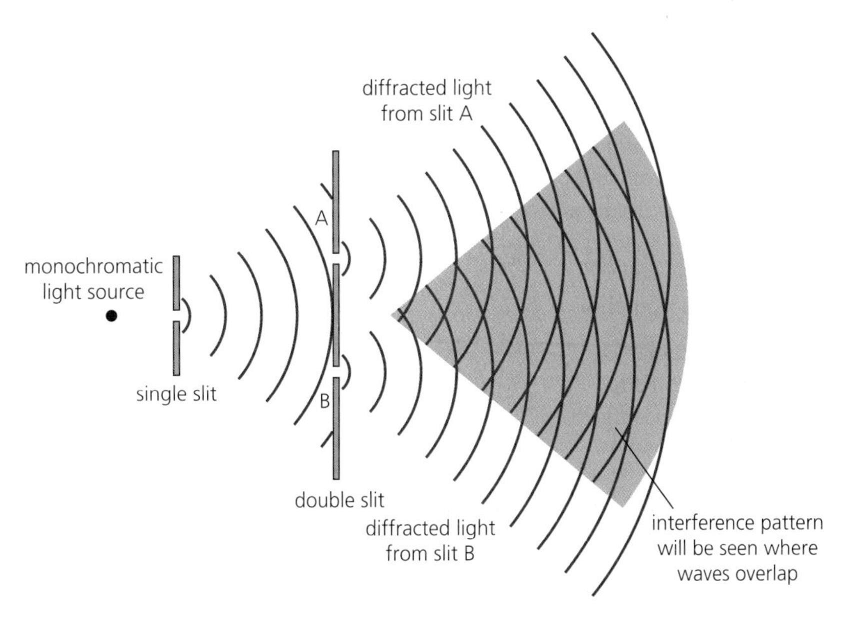

The original source of light should be as bright as possible and as small as possible (that is why a slit is also placed in front of the source). Ideally the light should be monochromatic, but a white light source can also be used, perhaps with a filter across the slit to reduce the range of wavelengths.

Young's double-slit experiment

With an arrangement such as that shown in Figure 4.86 (page 187) interference patterns like those shown in Figure 9.17a can be seen. This experiment is similar in principle to that first performed by Thomas Young in the early nineteenth century. Measurements taken from the experiment can be used to determine a value for the wavelength of light used.

If we want to predict the exact locations of constructive and destructive interference we must link path differences to the dimensions of the apparatus. Consider Figure 9.18, which shows two identical, very narrow slits separated by a distance d (between the centres of the slits).

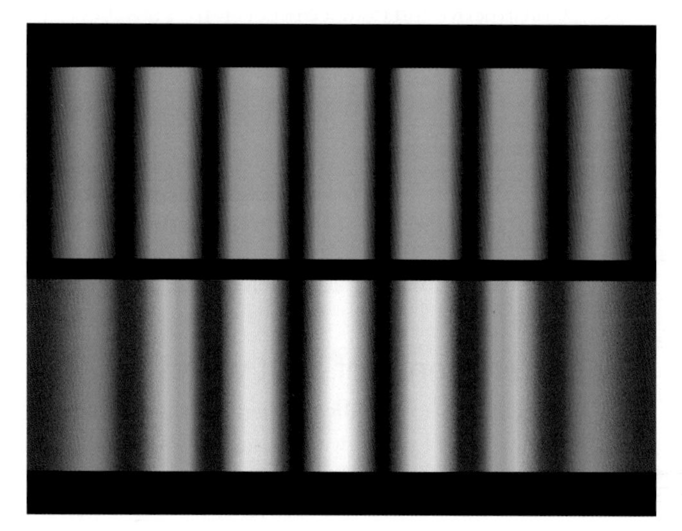

■ **Figure 9.17** Double-slit interference patterns of **a** monochromatic light and **b** white light

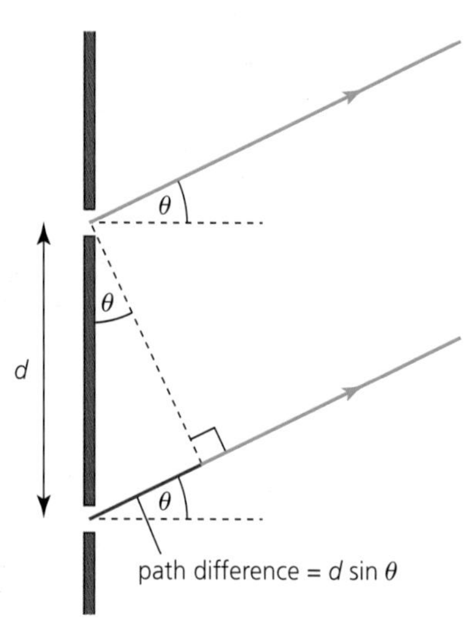

■ **Figure 9.18** Explaining path difference = $d\sin\theta$

Note that d is significantly greater than the individual widths of the two slits. The arrowed lines represent rays showing the direction of waves leaving both slits at an angle θ to the normal. Suppose that these waves meet and interfere constructively at a distant point. Of course, if the rays are perfectly parallel they cannot meet, so we must assume that because the point is a long way away (compared with the slit separation) the rays are very nearly parallel.

From the triangle in Figure 9.18 we can see that the path difference between the rays is $d\sin\theta$ and we know that this must equal $n\lambda$ because the interference is constructive:

$$n\lambda = d\sin\theta$$

This equation is given in the *Physics data booklet*. Note that the angle θ is usually small in double-slit experiments, so that $\sin\theta \approx \theta$ in radians.

Destructive interference will occur at angles such that $(n + \frac{1}{2})\lambda = d\sin\theta$.

Figure 9.19 links these equations to a sketch of how the intensity varies across the centre of a fringe pattern. For small angles, the horizontal axis could also represent the distance along the screen (as an alternative to $\sin\theta$ or θ in radians). (This equation and graph can be easily confused with similar work on single-slit diffraction, covered earlier in this chapter.)

In an actual experiment we cannot usually measure angles directly, so measurements are taken from the screen on which the interference pattern is observed. In Figure 9.20 s represents the distance between the centres of adjacent interference maxima (or minima). s can be

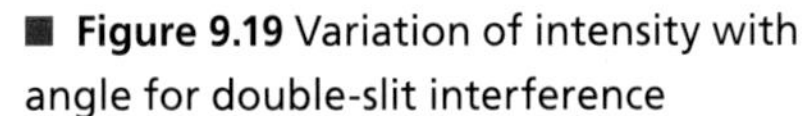

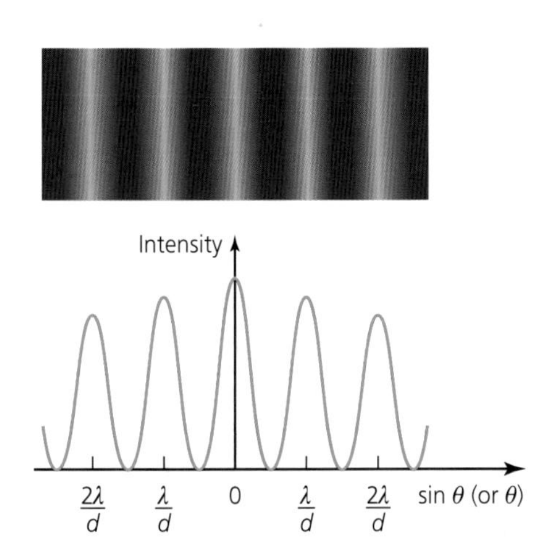

■ **Figure 9.19** Variation of intensity with angle for double-slit interference

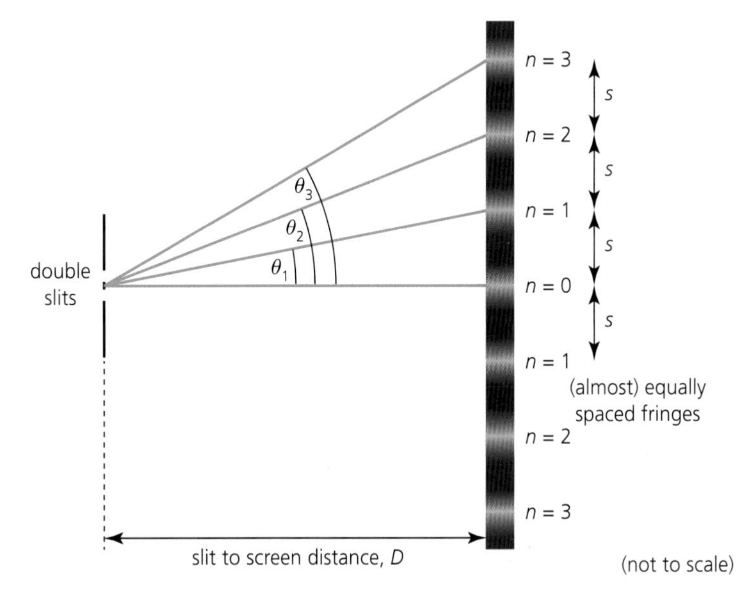

■ **Figure 9.20** Separation and numbering of fringes seen on a screen

considered to have a constant value along the screen, assuming that the angle θ is small enough for $\sin\theta$ to be approximately equal to θ and $\tan\theta$.

From $n\lambda = d\sin\theta$, we see that:

$$1\lambda = d\sin\theta_1$$

$$2\lambda = d\sin\theta_2$$

$$3\lambda = d\sin\theta_3$$

etc.

For small angles:

$$\sin\theta_1 = \tan\theta_1 = \theta_1 = \frac{s}{D}$$

So that $1\lambda = d\left(\frac{s}{D}\right)$ can be rewritten in the form used in Chapter 4:

$$s = \frac{\lambda D}{d}$$

Worked example

6 When monochromatic light of wavelength 5.9×10^{-7} m was directed through two narrow slits at a screen 2.4 m away, a series of light and dark fringes was seen. If the distance between the centres of adjacent bright fringes (and adjacent dark fringes) was 1.1 cm, what was separation of the slits?

$$s = \frac{\lambda D}{d}$$

$$s = 1.1 \times 10^{-2} = \frac{(5.9 \times 10^{-7}) \times 2.4}{d}$$

$$d = 1.3 \times 10^{-4}\,\text{m}$$

Modulation of two-slit interference pattern by one-slit diffraction effect

The analysis of a two-slit interference pattern presented above has, for the purpose of simplification, ignored one important factor – the light emerging from each slit is itself diffracted into a pattern, as described earlier in this chapter. If the diffraction of light emerging from each of two identical slits has a minimum intensity at a certain angle, then there will be a minimum at that angle in the interference pattern, even if the two-slit equation predicts a maximum.

The overall effect can be summarized by saying that the single-slit diffraction pattern *modulates* the two-slit interference pattern. This is shown in Figure 9.21, in which the double-slit intensity graph is drawn underneath an *envelope* shape of the single-slit pattern. Remember that the angles at which *minima* occur for a single slit are predicted by $n\lambda/b$, whereas the *maxima* of a two-slit pattern are predicted by $n\lambda/d$. Note also that the separation of the centres of slits, d, must *always* be less than the width of each slit, b.

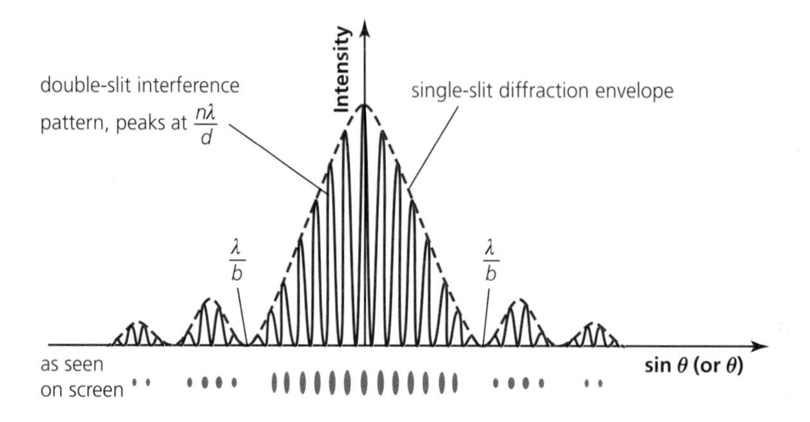

Figure 9.21 Modulated double-slit interference pattern

ToK Link

Justifying simplifications

Most two-slit interference descriptions can be made without reference to the one-slit modulation effect. To what level can scientists ignore parts of a model for simplicity and clarity?

When introducing the theory for two-slit interference in a book like this, it is much easier to discuss interference between light from the separate slits and to ignore the effects of the diffraction pattern from each slit. After the basic theory is explained and understood, any further complications can be discussed. This is typical of the educational process in science, in which it is common to be told that 'knowledge' gained from earlier lessons has been simplified and is not the whole truth.

25 a Light of wavelength 450 nm passes through two slits that are 0.10 mm apart. Calculate the angle to the first maximum.
b Sketch a graph of the intensity variation across the interference pattern.
c How far apart will the fringes be on a screen placed 3.0 m from the slits?

26 Give reasons why it is difficult to observe the interference of light from a household light bulb using double slits.

27 The angular separation between the centres of the first and the fourth bright fringes in an interference pattern is 1.23×10^{-3} rad. If the wavelength used was 633 nm, calculate the separation of the two slits that produced the pattern.

28 A Young's double-slit experiment was set up to determine the wavelength of the light used. The slits were 0.14 mm apart and the bright fringes on a screen 1.8 m away were measured to be 6.15 mm apart.
a What was the result of the experiment?
b Explain what would happen to the separation of the fringes if the experiment was repeated in water (refractive index = 1.3).

29 In a school experiment to demonstrate interference, two slits each of width 3 cm are placed in front of a source of microwaves. The distance between the slits is 12 cm.
a Suggest why the slit widths were chosen to be 3 cm.
b Some distance away, a microwave detector is moved along a line that is parallel to a line joining the slits. The detector moves 24 cm between two adjacent maxima. Estimate the approximate distance between the detector and the slits.

30 The interference pattern produced by double slits with monochromatic light is observed on a screen. What will happen to the pattern if each of the following changes is made (separately)?
a Light of a longer wavelength is used.
b The slit to screen distance is increased.
c The slits are made closer together.
d White light is used.

31 A plane is flying at altitude directly between two radio towers emitting coherent electromagnetic waves of frequency 6.0 MHz. If the strength of the signal detected on the plane rises to a maximum every 0.096 s, what is the speed of the plane?

32 Monochromatic light of wavelength 5.8×10^{-7} m passes through two parallel slits of width 0.032 mm, the centres of which are 0.27 mm apart. Show, with appropriate calculations, why the eighth and ninth fringes either side of the centre of the pattern will not be clearly visible.

Multiple-slit and diffraction grating interference patterns

2 slits

5 slits

10 slits

■ **Figure 9.22** How an interference pattern changes as more slits are involved

Multiple slits

As we have seen, a pattern of interference fringes can be formed using a beam of light incident on just two slits. If the number of slits (of the same width and separation) is increased, the observed diffraction/interference pattern will have an increasing intensity, but the maxima will still occur at the same angles because the wavelength and slit separation have not changed. However, the most important feature to note is that, if more slits are involved, the maxima become much sharper and more accurately located at those particular angles. This is shown in Figure 9.22, which compares the patterns seen using monochromatic light for two, five and ten slits.

Diffraction gratings

If the light incident normally on slits is not monochromatic but contains a range of different wavelengths, the maxima of the different wavelengths will occur at slightly different angles, so that the light will be *dispersed*. The different maxima will overlap if a small number of slits is used, but the maxima can be separated (resolved) by using a large number of slits. A very large number of slits close together, which is used for dispersing light into different wavelengths, is called a **diffraction grating** (see Figure 9.23). The resolution provided by diffraction gratings is called its **resolvance**, and this is discussed in Section 9.4.

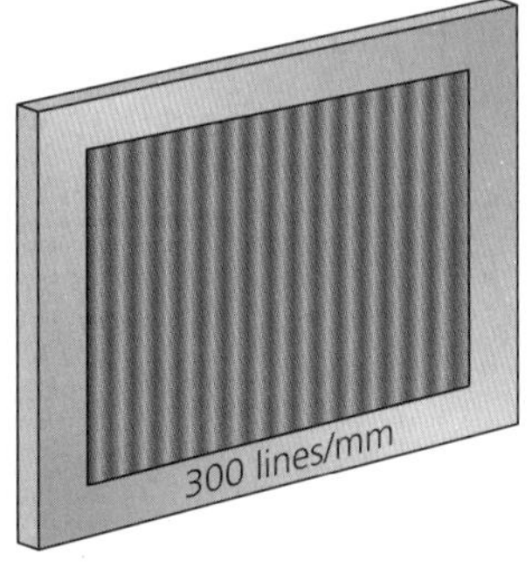

■ **Figure 9.23** Diffraction grating

The slits on a diffraction grating are usually called 'lines'. A typical grating may have 600 lines in every millimetre, which gives a spacing, d, of about 1.7×10^{-6} m. This is approximately three average wavelengths of visible light and it is therefore much smaller than the double-slit separations that were discussed earlier. This means that diffraction gratings typically deviate light through much larger angles than a pair of slits.

Diffraction gratings are widely used for analysing light and offer an alternative to glass prisms, which use the *refraction* (rather than diffraction) of light to produce dispersion. A typical application would be the investigation of emission and absorption spectra, as discussed in Chapter 7.

A simple **spectroscope** for observing spectra can be made from a lightproof tube with a diffraction grating at one end and a narrow slit at the other (Figure 9.24).

■ **Figure 9.24** Observing spectra with a spectroscope

Using a diffraction grating to measure wavelength

The equation used for predicting the angles of constructive interference with two slits of separation d can also be applied to multiple slits and gratings with separation of lines equal to d:

$$n\lambda = d\sin\theta$$

Figure 9.25 shows how a diffraction grating is used to produce an interference pattern (of monochromatic light). In many practical arrangements, a converging lens is used to focus the light on the screen, but that is not shown here. The first maximum out from the middle of the pattern is called the first diffraction order, the second is called the second diffraction order and so on.

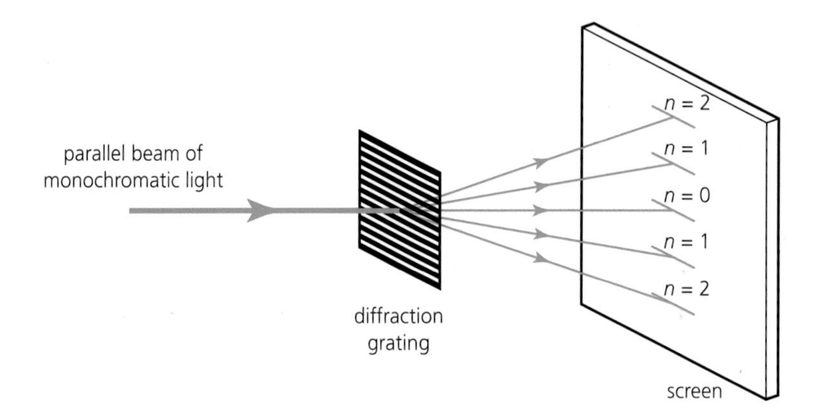

■ **Figure 9.25** Action of a diffraction grating to produce different orders

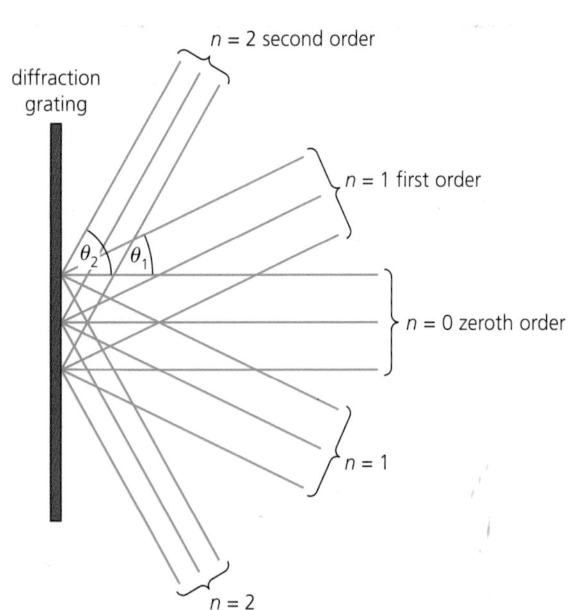

■ **Figure 9.26** Different orders in different directions

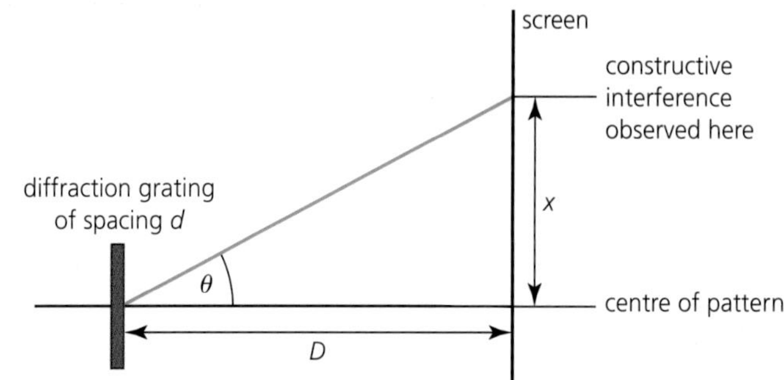

■ **Figure 9.27** Measurements needed to determine wavelength

Figure 9.26 shows sets of adjacent rays emerging in different directions. All waves transmitted along the normal will interfere constructively because there is no path difference between them. This is called the zeroth order, $n = 0$.

At angle θ_1, the rays again interfere constructively because there is a path difference of one wavelength between adjacent rays ($n = 1$). This is called the first order.

The second order occurs at an angle θ_2 when there is a path difference of two wavelengths between rays from adjacent rays ($n = 2$). Gratings are usually designed using small slit separations to spread the light out as much as possible, and our attention is normally concentrated on the lower orders. Higher orders are not usually needed.

To determine a wavelength using a diffraction grating, measurements are taken from a screen placed a known distance away so that $\sin\theta$ can be calculated (Figure 9.27). Because diffraction gratings are usually used to produce large angular separations we cannot make the approximation $\sin\theta \approx \theta$ without introducing a significant error.

Worked example

7 Monochromatic light passes through a diffraction grating with 400 lines mm^{-1} and forms an interference pattern on a screen 2.25 m away.

a If the distance between the central maximum and the first order is 57.2 cm, what is the wavelength of the light?

b If a second beam of monochromatic light of wavelength 6.87×10^{-7} m also passes through the grating, how far apart are the first orders on the screen?

a Considering Figure 9.27, $\tan\theta_1 = 0.572/2.25 = 0.254$, so that $\theta_1 = 14.3$ and $\sin\theta_1 = 0.246$.

$n\lambda = d\sin\theta$

$1 \times \lambda = d\sin\theta_1$

$\lambda = \frac{1.0 \times 10^{-3}}{400} \times 0.246$

$\lambda = 6.16 \times 10^{-7}$ m

(Using the $\sin\theta \approx \tan\theta$ approximation produces an answer of 6.36×10^{-7} m.)

b $1 \times \lambda = d\sin\theta_1$

$6.87 \times 10^{-7} = \frac{(1.0 \times 10^{-3})}{400} \times \sin\theta_1$

$\sin\theta_1 = 0.275$

So $\theta = 16.0°$ and $\tan\theta = 0.286$.

$\tan\theta$ = distance between the central maximum and the first order divided by the slit to screen distance.

Distance between the central maximum and the first order = $0.286 \times 2.25 = 0.643$ m. So the separation of the orders on the screen = (64.3 − 57.2) = 7.1 cm.

Using a diffraction grating to analyse spectra containing different wavelengths

The main reason why diffraction gratings are so useful is because they produce very sharp and intense maxima. Figure 9.28 compares the relative intensities produced by double slits (using monochromatic light) with a diffraction grating of the same spacing. Compare this with Figure 9.22, which showed the visual effect of increasing the number of slits.

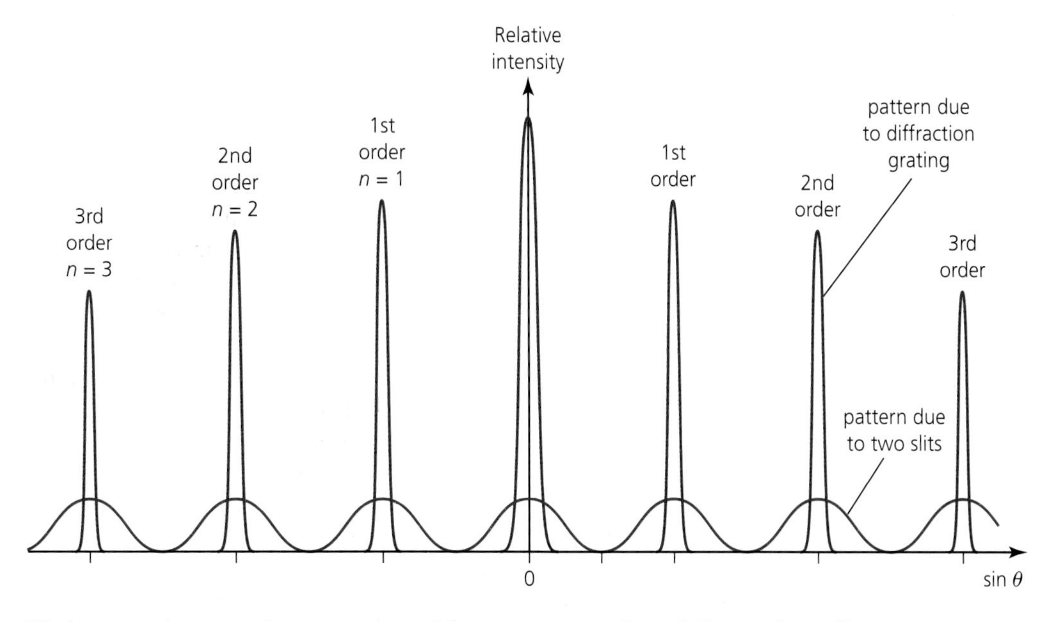

■ **Figure 9.28** Comparing the maxima produced by double slits and a diffraction grating using monochromatic light

If light is incident on a larger number of slits, more waves from different slits will arrive at any particular point on the screen. But, unless the angle is perfect for constructive interference (as represented by the equation $n\lambda = d\sin\theta$), there will be enough waves arriving at that point out of phase to result in overall destructive interference. In practice, light striking a grating will usually be incident on well over 1000 slits, so that the peaks become very much sharper and more intense than those shown in Figure 9.28.

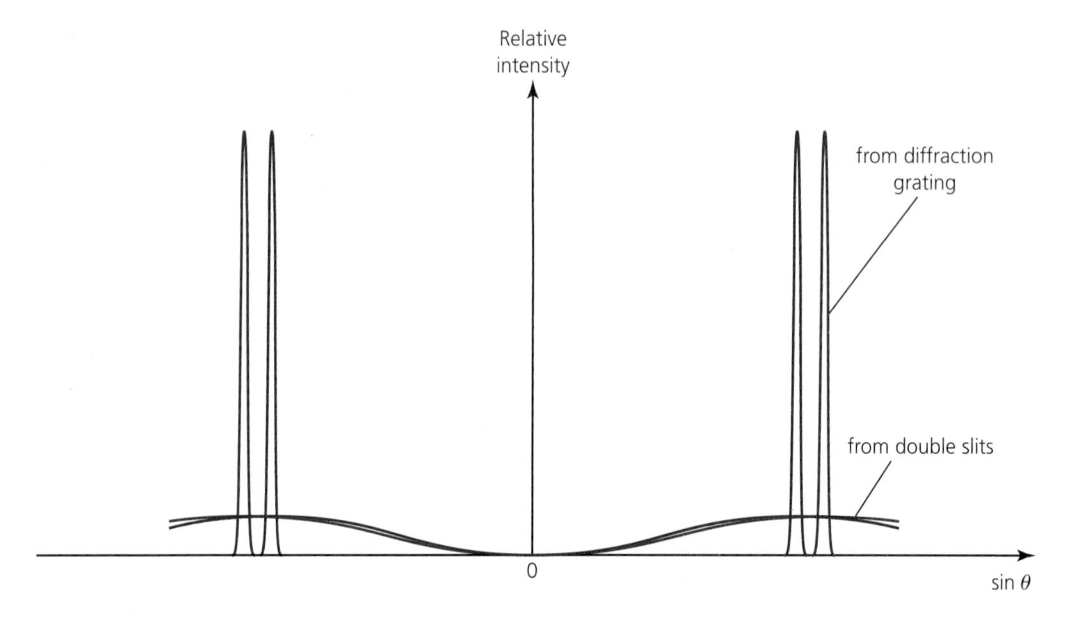

■ **Figure 9.29** High resolution produced by diffraction gratings

Diffraction gratings analyse spectra by separating (resolving) the maxima of different wavelengths. This is shown in Figure 9.29, which compares the intensities of the first-order spectra from double slits and a diffraction grating (for a spectrum containing only red and blue light).

If white light is incident on a grating, a series of continuous spectra is produced, as shown in Figure 9.30. Because the wavelength of red light is less than twice the wavelength of violet light, the first-order and the second-order visible spectra cannot overlap, although higher orders do.

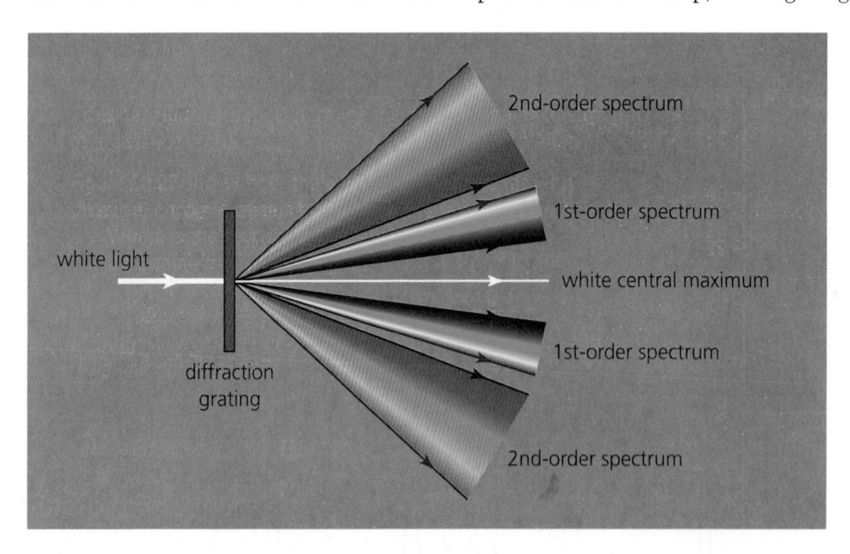

■ **Figure 9.30** White light passing through a diffraction grating

If the emission or absorption spectrum of a substance is being observed, a good-quality grating will be able to resolve the wavelengths into the different 'lines' of a line spectrum, like those shown in Figure 7.2 (page 283).

The diffraction gratings discussed so far have all been *transmission* gratings, but similar principles can be applied to light *reflected* off a series of very close lines. Reflection gratings can be useful when examining radiation that would otherwise be absorbed in the material of a transmission grating, for example ultraviolet. The manufacture of optical data storage discs (for example, DVDs) produces a regularity of structure that results in them acting as reflection gratings. The colours seen in reflections from their surfaces are a result of this 'diffraction grating effect'.

33 Light of wavelength 460 nm is incident normally on a diffraction grating with 200 lines per millimetre. Calculate the angle to the normal of the third-order maximum.

34 A diffraction grating is used with monochromatic light of wavelength 6.3×10^{-7} m and a screen a perpendicular distance of 3.75 m away. How many lines per millimetre are there on the diffraction grating if it produces a second-order maximum 38 cm from the centre of the pattern?

35 Monochromatic light of wavelength 530 nm is incident normally on a grating with 750 lines mm^{-1}. The interference pattern is seen on a screen that is 1.8 m from the grating. What is the distance between the first and second orders seen on the screen?

36 a A teacher is using a diffraction grating of 300 lines mm^{-1} to show a class a white light spectrum. She wants the first-order spectrum to be 10 cm wide. Estimate the distance that she will need between the grating and the screen.
b A prism can also be used to produce a spectrum. Explain why red light is refracted less than blue light in a prism, but diffracted more by a diffraction grating.

37 Light of wavelength 5.9×10^{-7} m is incident on a grating with 6.0×10^{5} lines per metre. How many orders will be produced?

38 When using white light, explain why red light in the second-order spectrum overlaps with blue light in the third-order spectrum.

39 A second-order maximum of blue light of wavelength 460 nm is sent to a certain point on a screen using a diffraction grating. At the same point the third-order maximum of a different wavelength is detected.
a What is the wavelength of the second radiation?
b In what section of the electromagnetic spectrum is this radiation?
c Suggest how it could be detected.

40 The theory developed in this section has all been for light incident normally on a diffraction grating. Suggest what the effect would be on an observed pattern if the grating was twisted so that the light was incident obliquely.

41 Suggest two reasons why an optical diffraction grating would not be of much use with X-rays.

42 Sketch a relative intensity against $\sin\theta$ graph for monochromatic light of wavelength 680 nm incident normally on a diffraction grating with 400 lines per mm.

43 A diffraction grating produced two first-order maxima for different wavelengths at angles of 7.46° and 7.59° to the normal through the grating. This angular separation was not enough to see the two lines separately. What is the angular separation of the same lines in the second order?

44 a Make a sketch of the variation of relative intensity with $\sin\theta$ for red light as it passes through a diffraction grating that produces two orders.
b Add to your sketch the variation in intensity of blue light passing through the same grating.
c Make a separate sketch showing the relative intensities of the same red and blue light after it passes through ten slits (with the same spacing as the grating).
d Use your sketches to help explain why diffraction gratings are so useful for analysing light.

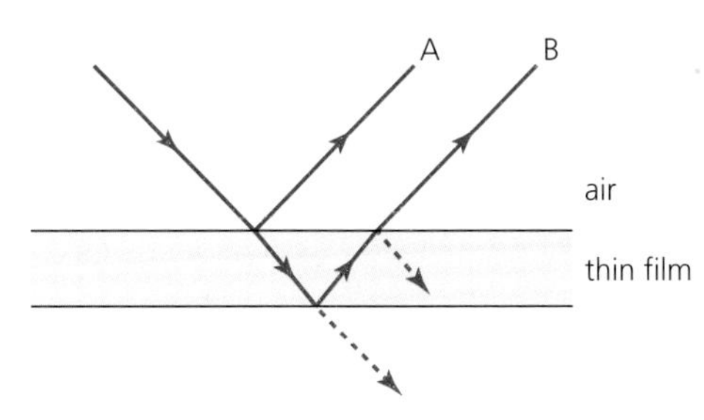

Figure 9.31 Waves reflected off two surfaces of a thin film

Thin-film interference

We know that *coherent* light waves are needed to produce any interference pattern, and another way of providing these is by the use of thin films. (The word 'film' is used here to mean a thin layer.) Figure 9.31 shows waves represented by the simplification of light rays being reflected off the top and bottom surfaces of a thin film of a transparent material, which, for example, might be water or oil.

If the film is thin enough, the rays A and B will be coherent (because they came from the same wavefront) and they will produce interference effects, depending on their path difference. If the film is not very thin the waves will lose their coherence and no interference pattern will be seen.

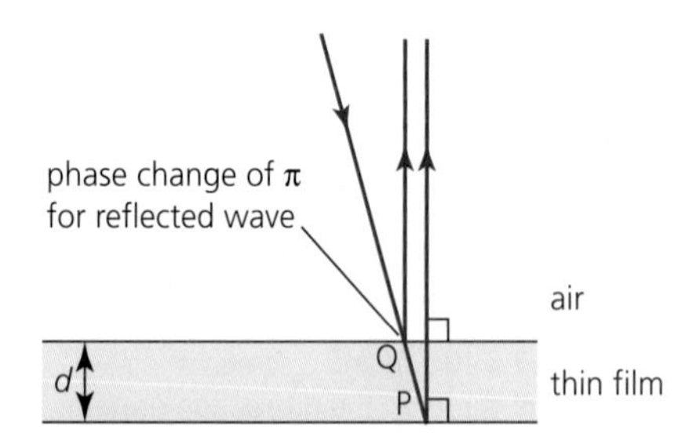

■ **Figure 9.32** There is a phase change when passing from air into the film

Equations for interference for normal incidence on a parallel film

Figure 9.32 shows the reflection of waves (represented by rays), initially in air, which are *reflected normally* from both surfaces of a parallel-sided film of thickness d. The rays are drawn at a slight angle to the perpendicular only to show them clearly. We know that the usual condition for constructive interference is that the path difference is a whole number of wavelengths, but whenever reflections are involved, we also need to consider if there are any *phase changes*.

There will be a phase change of $\lambda/2$ (π radians) at Q, when the light travelling in air reflects off a medium with a higher refractive index, n. This was discussed briefly in Chapter 4. We will assume that the medium on the other side of the thin film has a lower refractive index than the film, so that there will *not* be a phase change at P, at the second boundary. This assumption is valid for the two most common examples of thin film interference – air/water/air and air/oil/water.

Because of the introduction of a path difference of $\lambda/2$ (because of the phase change), the condition for constructive interference changes from path difference = $m\lambda$ to path difference = $(m + ½)\lambda$. (The symbol m is used here to represent an integer so as to avoid confusion with n, which is used for refractive index.)

Light will have a shorter wavelength in the film. For example, water has a refractive index of 1.33, which means that the speed of light in water is $c/1.33$ and the wavelength is reduced to $\lambda/1.33$. Therefore, the wavelength used in the equation must be divided by the refractive index and the conditions for interference become as follows:

Constructive interference:

$$2dn = (m + ½)\lambda$$

Destructive interference:

$$2dn = m\lambda$$

These equations, which apply only to normal incidence on certain parallel films, are given in the *Physics data booklet*.

Examples of thin-film interference

Determining the thickness of a film

Oil films and soap bubbles are typically very thin, comparable to the wavelength of light. Their thickness may be determined from observation of their interference effects. Because they are so thin, we may assume that m in the equations above has a value of one, and hence the thickness, d, of the film can be estimated from a knowledge of the wavelength and refractive index.

Worked example

8 An oil film appeared blue when observed normally. Estimate a possible value for its thickness if the refractive index of the oil was 1.57.

For constructive interference:

$$2dn = (m + ½)\lambda$$

with $m = 0$ and $\lambda = 4.7 \times 10^{-7}$ m.

$$2 \times d \times 1.57 = 0.5 \times (4.7 \times 10^{-7})$$

$$d = 7.5 \times 10^{-8}\,\text{m}$$

Utilizations

Quantifying oil spills

The energy-intense lifestyles of modern life demand a continuous supply of crude oil (see Chapter 8). The extraction of crude oil from the under the ocean floor, its transfer to land and its movement around the planet in large oil tankers and through pipes has caused a number of serious accidents. These accidents cause considerable pollution to the seas, and the plants and animals that live in and around them. Figure 9.33 shows a penguin covered with oil from a spill off the coast of Tauranga in New Zealand in 2011.

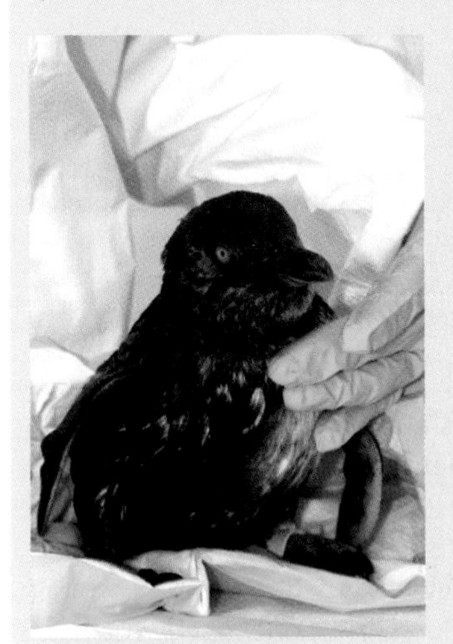

■ **Figure 9.33** Effects of an oil spill on marine life

Oil is less dense than water and therefore it floats on the surface of the sea. An oil spill can be difficult to contain because the oil tends to spread out very thinly on the surface of water (to an extent that varies with the type of oil). The phrase *oil slick* is sometimes used for a smooth layer of oil that has spread over a large area and has not been broken up by the action of the waves.

An oil spill of about 1000 litres that has spread to cover a square kilometre would be about 10^{-3} mm thick and be noticeable by the coloured interference patterns it produced. An oil film about one tenth of this thickness would have a shiny appearance. A thicker spill would appear much darker, without colours. It is often possible to estimate the volume of an oil spill from its area and appearance.

Obviously the severity of the possible pollution following an oil spill depends on the quantity of oil released into the environment and its location, but there are many other factors involved, including the type of oil and the prevailing weather, currents and waves. The temperature of the water may also be an important factor.

1 Research the various methods that can be used to contain and clean up a large oil spill on the surface of the sea.

Anti-reflection coatings

Consider a glass surface with a thin film of transparent material ('coated') on top of it, as shown in Figure 9.34. The material of the film has a refractive index lower than glass. The incident light is monochromatic, with a wavelength λ, and it is incident normally, although in the diagram it has been shown at a slight angle in order for the separate rays to be distinguished.

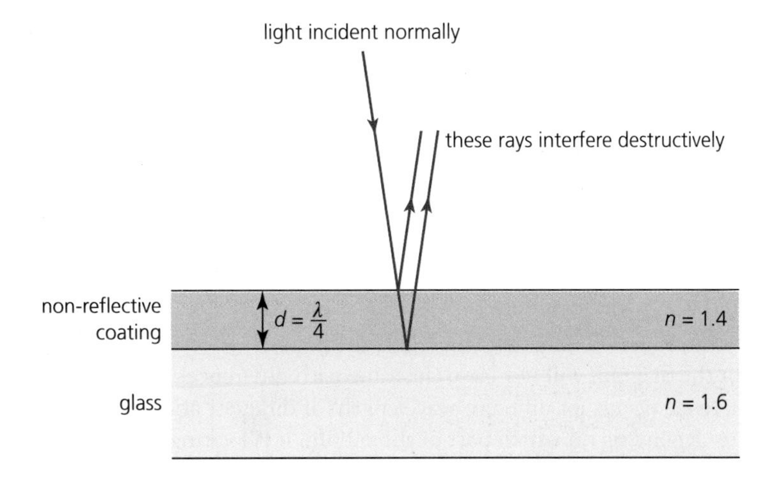

■ **Figure 9.34** Reflections from a non-reflective coating

Because both reflections occur when the light is entering a medium with a higher refractive index, we do not need to consider the effect of any phase changes involved. The minimum condition for *destructive* interference is:

$$\text{path difference} = 2d = \tfrac{1}{2}\lambda_c$$

where λ_c is the wavelength in the coating (not air), or:

$$d = \frac{\lambda_c}{4}$$

An anti-reflection coating needs to have a thickness of one quarter of a wavelength (measured in the coating, not air).

Alternatively, because $\lambda_c = \lambda/n$, where λ is the wavelength in air:

$$d = \frac{\lambda}{4n}$$

If light is not reflected, then more useful light energy will be transmitted through the glass. Coating lenses is a way of ensuring that the maximum possible light intensity is transmitted through all kinds of optical instruments (including eye glasses), as well as solar cells and solar panels. The process of putting a non-reflective coating on lenses is called **blooming**.

■ **Figure 9.35** Stealth aircraft

The thickness of a single layer will result in perfectly destructive interference for only one particular wavelength, but other wavelengths will also undergo destructive interference to some extent. By using multiple coatings of different thickness or refractive index, it is possible to effectively reduce reflections over all visible wavelengths. These effects also depend on the angle of incidence, but most of the time lenses are used with light that is incident approximately normally.

Non-reflecting coatings are also used on military aircraft to help them avoid detection by radar. Radar uses a range of possible wavelengths, typically a few centimetres. Figure 9.35 shows a Stealth aircraft, which uses a wide variety of techniques to avoid detection.

Interference at oblique angles

We will now broaden our discussion to include the possibility of observing thin-films from different angles. Figure 9.36 could represent, for example, reflections off an oil film floating on water.

■ **Figure 9.36** Observing a thin film at different angles

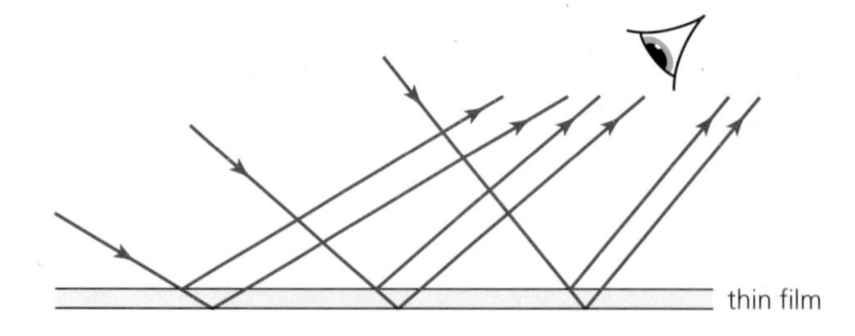

If the thin film is of constant thickness, the pairs of reflections from each of the three incident rays in the diagram will not have the same path differences and this means that constructive interference occurs for different wavelengths at different angles. The eye will see different colours depending on which part of the oil film it is looking at.

Figure 9.37 shows the typical appearance of an oil film on water. There is usually no regularity in the patterns observed because the thickness of the oil film is not constant. Similar effects are seen in soap bubbles (Figure 9.38), but their appearances change rapidly as the water moves towards the bottom of the bubble because of gravity and because the water evaporates quickly.

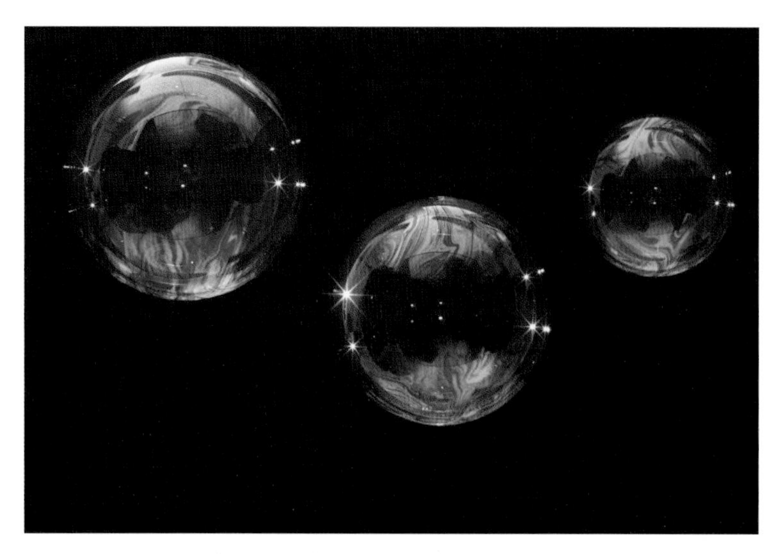

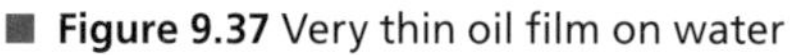

■ **Figure 9.37** Very thin oil film on water

■ **Figure 9.38** Interference effects in soap bubbles

The coloured effects seen in some bird feathers and insect wings, called **iridescence** (Figure 9.39), puzzled early scientists. We now know that this is an interference effect caused by the structures of the surfaces.

■ **Figure 9.39** Iridescence in peacock feathers

Nature of Science

Curiosity and serendipity

The beauty of iridescence is easily appreciated by everyone, and it is therefore perhaps not surprising that it stimulated the search for an explanation and the subsequent development of thin-film theory. The natural curiosity of human beings is a significant driving force in science, and many scientists see beauty and inspiration in unexplained phenomena of all kinds. But there are many examples of scientific investigations that led to surprising results.

Serendipity is the term used to describe something fortunate happening when you are not expecting it. The first laboratory production of thin films was serendipitous when Joseph Fraunhofer observed the effect of treating a glass surface with nitric acid.

45 An oil film is observed using normal incidence of light of wavelength 624 nm. If the oil has a refractive index of 1.42, what is the minimum thickness that will produce constructive interference?

46 A transparent plastic film of refractive index 1.51 is observed using monochromatic visible light at normal incidence. If destructive interference occurs when the film is 4.58×10^{-7} m thick, calculate possible values for the wavelength.

47 The blooming on a lens is 8.3×10^{-5} mm thick. What is its approximate refractive index if it is designed to be non-reflective for blue light?

48 Find out the various means by which a military aircraft might avoid detection.

49 The equation $2dn = m\lambda$ can be used to determine thicknesses of a thin film that will result in destructive interference. Give two separate examples of thin-films observation for which it would not be appropriate to use this equation.

9.4 Resolution – *resolution places an absolute limit on the extent to which an optical or other system can separate images of objects*

The importance of the size of a diffracting aperture

If you stand on a beach and look down at the sand, you probably will not be able to see the separate grains of sand. Similarly, if you look at a tree that is a long way away, you will not be able to see the separate leaves. In scientific terms, we say that you cannot **resolve** the detail.

If you assume that a person has good eyesight, the ability of their eyes to resolve detail depends largely on the amount of diffraction of light as it enters the eye through the *pupil* (aperture) and the separation of the light receptors on the *retina* of the eye. The larger the aperture, the less the diffraction and, therefore, the better the resolution. You can check this by looking at the world through a *very* small hole made in a piece of paper held in front of your eye (increasing the diffraction of light as it enters your eye).

When discussing resolution, in order to improve understanding, we usually simplify the situation by only considering waves of a single wavelength coming from two identical point sources of light.

The resolution of simple monochromatic two-source systems

Figure 9.40 shows an eye looking at two distant identical point objects, O_1 and O_2; θ is called the **angular separation** of the objects. (This is sometimes called the *angle subtended* at the eye.)

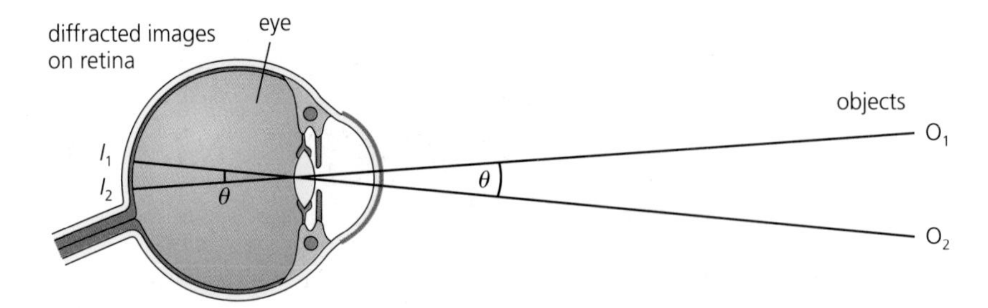

Figure 9.40 An eye receiving light from two separate objects

When the light from each of the objects enters the eye it will be diffracted and two single-slit diffraction patterns (similar to that shown in Figure 9.12) will be received by the light receptors on the retina at the back of the eye. So, point objects do not form point images. (Note that the light rays are not shown being refracted by the eye in this example because they are passing through the optical centre of the lens.)

The ability of this eye to resolve two separate images depends on how much the two diffraction patterns overlap (assuming that the receptors in the retina are close enough together). Similar comments apply to any equipment designed to form images, such as telescopes, microscopes and cameras.

Consider Figure 9.41, which shows intensities from two identical sources that might be received by the eye or other image-forming system. In Figure 9.41a the two sources can easily be resolved because their diffraction patterns do not overlap. In Figure 9.41c the sources are so close together that their diffraction patterns merge together and the eye cannot detect any fall in the resultant intensity between them; the images are not resolved. The resultant waveform is not shown, but can be determined using the principle of superposition.

Figure 9.41b represents the situation in which the sources can *just* be resolved because, although the images are close, there is a detectable fall in resultant intensity between them.

■ **Figure 9.41** Intensities received by the eye as a result of two diffraction patterns

a Two sources easily resolved

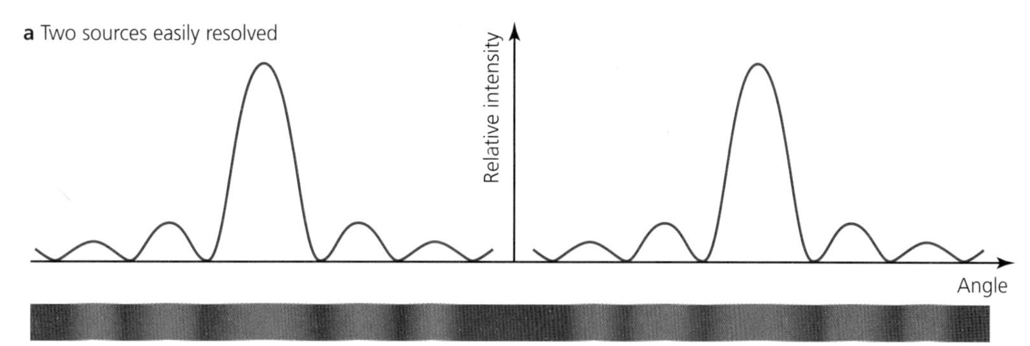

b Two sources *just* resolved

drop in intensity

$\frac{\lambda}{b}$

c Two sources not resolved

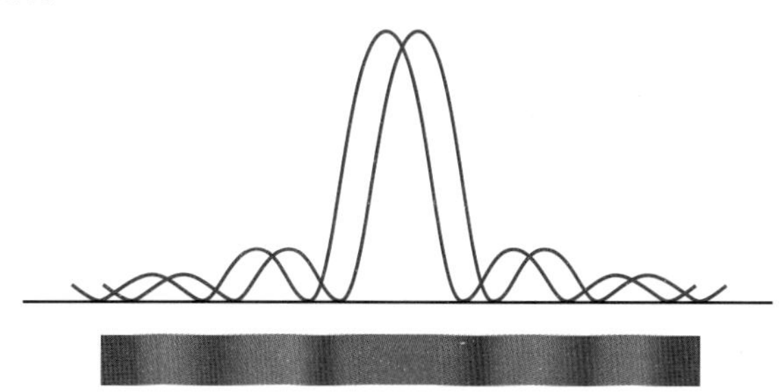

The Rayleigh criterion

Figure 9.42a shows two point sources viewed through circular apertures.

Rayleigh's criterion (similar to the *Dawes limit*) states that two point sources can *just* be resolved if the first minimum of the diffraction pattern of one occurs at the same angle as the central maximum of the other (Figures 9.41b and 9.42b).

a

b

■ **Figure 9.42** Images of two point sources observed through circular apertures that are **a** easily resolvable and **b** just resolvable

Rayleigh's criterion is a useful guide, but not a *law* of physics, and there may be factors other than diffraction that can affect resolution, so it can be regarded as a limit to resolution. The criterion can be expressed mathematically as follows.

The images of two sources may *just* be resolved through a *narrow slit*, of width b, if they have an angular separation of $\theta = \lambda/b$.

If they subtend a larger angle they will be resolvable, but if they subtend a smaller angle then they cannot be resolved.

When we consider eyes, telescopes and microscopes it is clear that light waves are usually received and detected through circular apertures, rather than slits. The resolution of a circular aperture of diameter *b* will not be as good as with a slit of constant width *b*, so the criterion is adjusted, as follows:

> The images of two sources can *just* be resolved through a *circular aperture* if they have an angular separation of $\theta = 1.22\lambda/b$.

This equation is given in the *Physics data booklet*. Although this equation relates to two identical sources and a single wavelength, λ_{av}/b can be taken as a good guide to the resolution of all images through any aperture of width *b*. The *lower* the value of λ_{av}/b, the better the resolution.

This equation shows clearly that the resolution of images provided by a certain kind of equipment – a telescope for example – can be improved by using a wider aperture or shorter wavelengths (if that is possible).

Nature of Science

Improved technology

Redesigning equipment and/or using higher-quality components can improve resolution (if it is not already as good as suggested by the criterion) and it may also improve other aspects of images. Technological advances such as these are characteristic of scientific development and they provide opportunities for further discoveries. Astronomy and microscopy are obvious examples of areas of study that are greatly influenced by the current state of technology.

Tok Link

The limits of technology are constantly advancing

The resolution limits set by Dawes and Rayleigh are capable of being surpassed by the construction of high-quality telescopes. Are we capable of breaking other limits of scientific knowledge with our advancing technology?

These limits to resolution are based on certain simplifying assumptions about the sources, the radiation and the receiver. They are very useful guidelines, but not perfect.

Worked example

9 Two small point sources of light separated by 1.1 cm are placed 3.6 m away from an observer who has a pupil diameter of 1.9 mm. Can they be seen as separate when the average wavelength of the light used was 5.0×10^{-7} m?

They will be resolvable if their angular separation is greater than or equal to $1.22\lambda/b$.

$$1.22\frac{\lambda}{b} = \frac{1.22 \times 5.0 \times 10^{-7}}{1.9 \times 10^{-3}} = 3.2 \times 10^{-4} \text{ radians}$$

$$\text{angular separation of sources} = \frac{1.1 \times 10^{-2}}{3.6} = 3.1 \times 10^{-3} \text{ radians}$$

The angular separation is much larger than $1.22\lambda/b$, so they are easily resolvable.

Other examples of resolution

Radio telescopes

Stars emit many other kinds of electromagnetic radiation apart from light and infrared.
Radio telescopes, like the one shown in Figure 9.43, detect wavelengths much longer than visible light and, therefore, would have very poor resolution if they were not made with wide diameters.

■ **Figure 9.43** The Jodrell Bank radio telescope in England, UK

Worked example

10 a The Jodrell Bank radio telescope in the UK has a diameter of 76 m. When used with radio waves of wavelength 21 cm, is it capable of resolving two stars that are 2.7×10^{11} m apart if they are both 1.23×10^{16} m from Earth?

b Compare the ability of this radio telescope to resolve with that of a typical human eye.

a $1.22\frac{\lambda}{b} = \frac{1.22 \times 21 \times 10^{-2}}{76} = 3.4 \times 10^{-3}$ radians

angular separation of sources $= \frac{2.7 \times 10^{11}}{1.23 \times 10^{16}} = 2.2 \times 10^{-5}$ radians

The angular separation is much smaller than $1.22\lambda/b$, so they cannot be resolved – they will appear like a single star.

b Using the data from Worked example 9 as an example, a human eye can resolve two objects if their angular separation is about 3×10^{-4} radians or more, but the radio telescope in this question requires the objects to be at least 3.4×10^{-3} radians apart. This angle is about ten times bigger. The eye has a much better resolution than a radio telescope mainly because it uses waves of a much smaller wavelength. The resolution of radio telescopes can be improved by making them even larger, but there are constructional limits to how big they can be made.

Utilizations

More about radio telescopes

Despite their poor resolution (compared with the human eye), radio telescopes collect a lot of information about the universe that is not available from visible light radiation. Light waves from distant stars are affected when they pass through the Earth's atmosphere and this will decrease the possible resolution. This is why optical telescopes are often constructed on mountain tops or put on orbiting satellites. Radio telescopes do not have this problem.

Hydrogen is the most common element in the universe and its atoms emit electromagnetic waves with wavelength 21 cm, in the part of the electromagnetic spectrum known as radio waves. Many other similar wavelengths (from a few centimetres to many metres) are received on Earth from space, and radio telescopes are designed to detect these waves.

In a dish telescope, such as that shown in Figure 9.43, the radio waves reflect off a parabolic reflector and are focused on a central receiver (like a reflecting optical telescope). As we have seen, the resolution of a radio telescope will be disappointing unless its dish has a wide aperture. Of course, having a wide aperture also means that the telescope is able to detect fainter and more distant objects because more energy is received from them when using a larger dish.

There are many different designs of radio telescope, but the largest with a single dish is that built at Arecibo in Puerto Rico (diameter 305 m), which uses a natural hollow in the ground to help provide support. However, a radio telescope is being built in China with a diameter of 500 m, which is due to be operational in 2016.

In radio interferometry, signals received from individual telescopes grouped together in a regular pattern (an *array*, as in Figure 9.44) are combined to produce a superposition (interference) pattern that has a much narrower spacing than the diffraction pattern of each individual telescope. This greatly improves the resolution of the system.

1 Find out what you can about China's *FAST* telescope.

■ **Figure 9.44** An array of linked radio telescopes

Electronic displays

Digital pictures are composed of a very large number of tiny picture elements (*pixels*) and information about the display is usually given in a form such as 1366 × 768, meaning that there are 1366 pixels in each horizontal row and 768 in each vertical row, making a total of 1 049 088 pixels, or approximately one megapixel (1 MPx).

If the screen size is 41 cm by 23 cm, then the centres of pixels are an average distance of 0.03 cm apart in their rows and columns. The spaces between the pixels are much smallest. (Confusingly, the number of pixels is often also called the 'resolution' of the display.)

When we look at a screen, we do not want to see individual pixels, so the angular separation as seen by a viewer must be smaller than $1.22\lambda/b$. Using 4.0×10^{-7} m as a value for the shortest wavelength produced by the screen, we can calculate a rough guide to the minimum distance at which a viewer (with a pupil of size of 2 mm) would need to be positioned so that they could not resolve individual pixels:

$$\frac{1.22 \times 4.0 \times 10^{-7}}{2 \times 10^{-3}} = \frac{\text{pixel separation}}{\text{minimum distance between viewer and screen}}$$

Using a pixel separation of 0.03 mm gives a minimum viewing distance of about 10 cm from the screen. Television screens designed to be viewed from longer distance can have their picture elements further apart.

Digital cameras

The quality of the lens and the sizes of the lens and aperture are the most important factors in determining the quality of the images produced by a camera. But for a digital camera the distance between the receptors (CCDs) on the image sensor must be small enough to resolve the detail provided by the lens. The 'resolutions' of digital cameras are also described in terms of the numbers of (light-receiving) pixels. The settings used on the camera, the way in which the data are processed and the 'resolution' of the display used will all affect the amount of detail resolvable in the final image.

Optical storage of data

Information is stored using tiny bumps (called 'lands') on a plastic disc (CD or DVD). The information is 'read' by light from a laser that reflects off the reflective coating on the lands and/or the 'pits' between them.

The closer the lands (and pits) are located to each other, the greater the amount of information that can be stored on the disc. But, if the lands are *too* close together, the diffracting laser beam will not be able to resolve the difference between them. A laser with a shorter wavelength will diffract less and enable more data to be stored on the same sized disc.

On a typical CD (which stores 700 MB of information), the pits are about 8×10^{-6} m in length and they are read by light from a laser of wavelength 7.8×10^{-7} m. DVDs store information in the same way, but they have an improved data capacity: the pits and lands can be closer together because a laser light of shorter wavelength (6.5×10^{-7} m) is used. A single-layer DVD can store about seven times as much information as a CD.

The development of blue lasers, with their smaller wavelengths (4.05×10^{-7} m), has led to Blu-ray technology and greatly increased data storage capacity on discs of the same size. This enables the storage and playback of high-definition (HD) video.

Satellites

Satellites have an ever-increasing range of uses including communications, television, mapping, navigation, spying, observing the climate, weather and land use, and astronomical research. In earlier years the launch and use of satellites was controlled by the military and governments, but the use of satellites for commercial interests is becoming much more important and widespread. This requires international consultation and agreements about satellite launches, locations of satellite orbits and the selective use of microwave frequencies.

Communication between satellites and Earth stations requires waves that are unaffected by passing through the atmosphere. Typically, microwaves are used with a wavelength of a few centimetres. Parabolic dish reflectors behind the *transmitting* aerials can produce highly directional beams if diffraction can be limited by having a large enough dish aperture.

Receiving aerials are similarly designed with reflectors that are used to focus the incoming microwaves onto an aerial. They need to be large enough to collect a signal of sufficient power.

Currently there are more than 1000 operational satellites in orbit around the Earth and it is estimated that in the coming years over 100 new satellites will be launched every year. With so many transmissions taking place, it is very important that receiving aerials are able to resolve the signal that they wish to detect from others nearby.

Worked example

11 A transmitting aerial has a diameter of 5.0 m and transmits a microwave beam of frequency 4.8 GHz. What is the approximate cross-sectional area of the central diffraction beam at a distance of 300 km from the aerial (typical distance between Earth and a low orbit satellite)?

$$\text{wavelength} = \frac{v}{f} = \frac{3.0 \times 10^8}{4.8 \times 10^9} = 6.3 \times 10^{-2}\,\text{m}$$

Angle between first diffraction minimum and centre of the diffraction pattern:

$$\theta = \frac{\lambda}{b} = \frac{6.3 \times 10^{-2}}{5.0} = 0.0126$$

Radius of beam at a distance of 300 km:

$$r = 0.0125 \times 3 \times 10^5 = 3780\,\text{m}$$

$$\text{area of beam} = \pi r^2 = 4.5 \times 10^7\,\text{m}^2\ (44\,\text{km}^2)$$

Resolution of diffraction gratings

The use of diffraction gratings was discussed earlier in this chapter as a method of analysing light. This can be considered as another form of resolution – separating wavelengths, rather than separating sources.

The ability of a grating to separate wavelengths is called its **resolvance** (or *resolving power*), *R*. It is defined as:

$$R = \frac{\lambda}{\Delta\lambda}$$

Where $\Delta\lambda$ is the smallest difference between two resolvable wavelengths, λ (either value can be taken). This equation is given in the *Physics data booklet*. Because it is a ratio, resolvance has no unit.

Worked example

12 What is the resolvance of a diffraction grating that can *just* separate wavelengths of 589.00 nm and 589.59 nm?

$$R = \frac{\lambda}{\Delta\lambda} = \frac{589.59}{0.59} = 1.0 \times 10^3$$

However, we are more likely to determine the resolvance of a diffraction grating from the number of lines mm^{-1} and the width of the light beam that is incident on it, as follows:

$$R = mN$$

Where *N* is the total number of slits illuminated and *m* is the diffraction order (the resolution is better for higher orders). This equation is also given in the *Physics data booklet*. Note that, for a given grating, increasing the width of the incident beam increases the resolvance (improves the resolution).

Worked example

13 A parallel light beam of width 5 mm falls on a diffraction grating with 80 lines mm^{-1}. What is the resolvance in the second-order spectrum?

$$R = mN$$

$$R = 2 \times (5 \times 80) = 800$$

50 a Why might you expect a camera that has a lens with a wider diameter to produce better pictures?
b Suggest a reason why bigger lenses might produce poorer images.

51 Why do astronomers sometimes prefer to take photographs with blue filters?

52 The pupils in our eyes dilate (get bigger) when the light intensity decreases. Discuss if this means that people can see better at night.

53 A car is driving towards an observer from a long way away at night. If the headlights are 1.8 m apart, estimate the maximum distance away at which the observer will see two distinct headlights. (Assume that the average wavelength of light is 5.2×10^{-7} m and the pupil diameter is 4.2 mm.)

54 A camera on a satellite orbiting at a height of 230 km above the Earth is required to take photographs that resolve objects that are 1.0 m apart. Assuming a wavelength of 5.5×10^{-7} m, what minimum diameter lens would be needed?

55 a A radio telescope with a dish of diameter 64 m is used to detect radiation of wavelength 1.4 m. How far away from the Earth are two stars separated by a distance of 3.8×10^{12} m that can just be resolved?
b What assumption(s) did you make?

56 Could an optical telescope with a (objective) lens of diameter 12 cm be used to read the writing on an advertising sign that is 5.4 km away if the letters are an average of 8.5 cm apart? Explain your answer.

57 The Moon is 3.8×10^{8} m from the Earth. Estimate the smallest distance between two features on the Moon's surface that can just be observed from the Earth by the human eye.

58 Use Figure 9.45 to measure the resolution of your own eyes.

59 a What is the second-order resolvance of a diffraction grating that has lines separated by 3.3×10^{-6} m for an incident beam of width 1 mm?
b Would this grating be able to resolve wavelengths of 6.22×10^{-7} m and 6.23×10^{-7} m?
c How could the resolution be improved using the same grating?

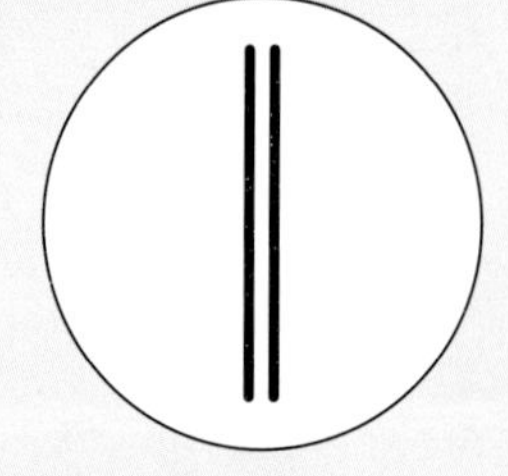

Figure 9.45

9.5 Doppler effect – *the Doppler effect describes the phenomenon of wavelength/frequency shift when relative motion occurs*

The Doppler effect for sound waves

When we hear a sound we can usually assume that the frequency (*pitch*) that is heard by our ears is the same as the frequency that was emitted by the source. But if the source of the sound is moving towards us (or away from us) we will hear a sound with a different frequency. This is usually only noticeable if the movement is fast; the most common example is the sound heard from a car or train that moves quickly past us.

This change of frequency that is detected when there is relative motion between a source and a receiver of waves is called the **Doppler effect**, named after the Austrian physicist Christian Doppler, who first proposed it in 1842. The Doppler effect can occur with any kind of wave.

Figure 9.46 shows a way in which the Doppler effect can be demonstrated with sound. A small source of sound (of a single frequency) is spun around in a circle. When the source is moving towards the listener a higher frequency is heard; when it is moving away, a lower frequency is heard.

■ **Figure 9.46** Demonstrating the Doppler effect with sound waves

ToK Link

Applying our own direct experience is important in understanding phenomena we cannot observe directly

How important is ***sense perception*** *in explaining scientific ideas such as the Doppler effect?*

The Doppler effect can occur with any kind of wave, but it seems so much easier to understand after we have heard the effect with sound waves. In a similar way, the interference and diffraction of, for example, radio waves, is much easier to understand if we have observed similar effects with light and sound.

The easiest way to explain the Doppler effect is by considering wavefronts (Figure 9.47). Figure 9.47a shows the common situation in which a stationary source, S, emits waves that travel towards a stationary detector, D. Figure 9.47b shows a detector moving towards a stationary source and Figure 9.47c shows a source moving directly towards a stationary detector. Similar diagrams can be drawn to represent the situations where the source and detector are moving apart. The Doppler effect can be better understood by observing computer simulations of moving sources and observers.

The detector in Figure 9.47b will meet more wavefronts in a given time than if it remained in the same place, so that the received frequency, f', is higher than the emitted frequency, f. In Figure 9.47c the distance between the wavefronts between the source and the observer (the received wavelength) is reduced, which again means that the received frequency will be higher than the emitted frequency. (Frequency = v/λ and the wave speed, v, is constant. The speed of sound through air does not vary with the motion of the source or detector.)

■ **Figure 9.47** Wavefront diagrams to demonstrate the Doppler effect

a Source and detector both stationary

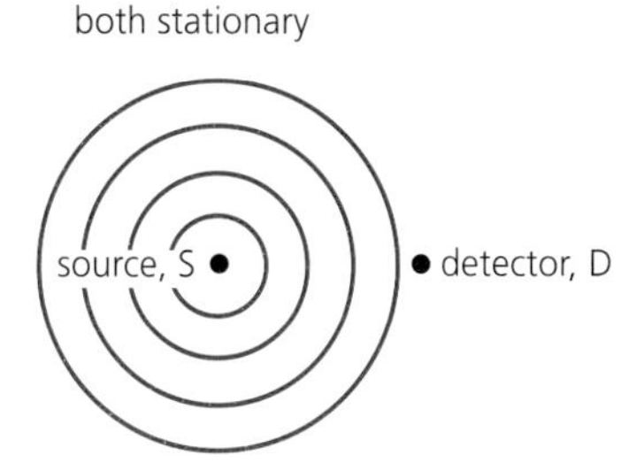

b Detector moving towards stationary source

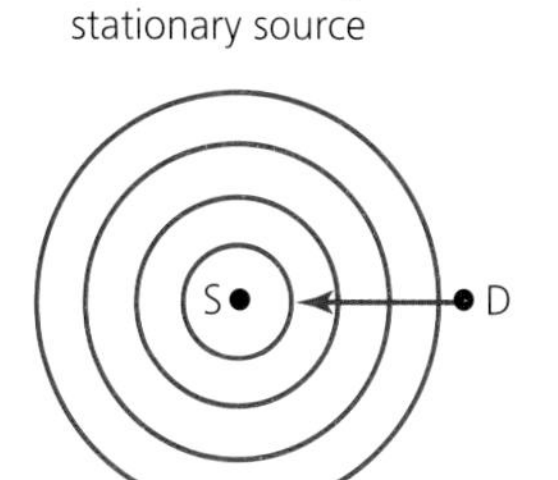

c Source moving towards stationary detector

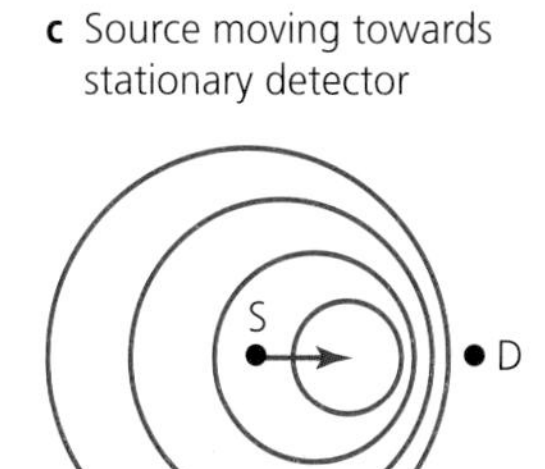

Utilizations

Shock waves: breaking the sound barrier

As an object, like a plane, flies faster and faster, the sound waves that it makes get closer and closer together in front of it. When a plane reaches the speed of sound, at about 1200 km h^{-1}, the waves superpose to create a 'shock wave'. This is shown in Figure 9.48.

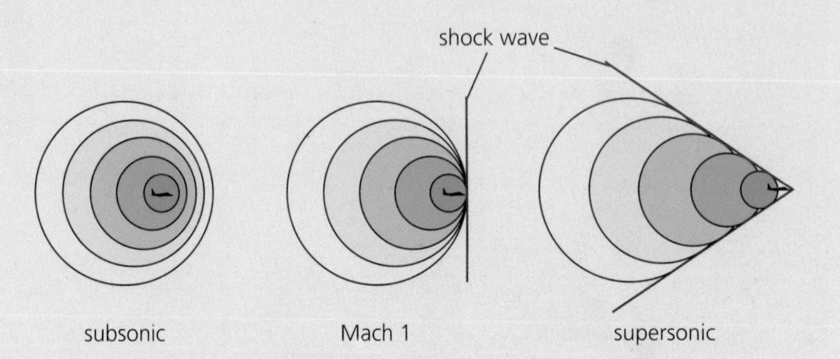

■ **Figure 9.48** Creating a shock wave in air

When a plane reaches the speed of sound it is said to be travelling at 'Mach 1' (named after the Austrian physicist, Ernst Mach). Faster speeds are described as 'supersonic' and twice the speed of sound is called Mach 2 etc. As Figure 9.49 shows, the shock wave travels away from the side of the plane and can be heard on the ground as a 'sonic boom'. Similarly, 'bow waves' can often be seen spreading from the front (bow) of a boat because boats usually travel faster than the water waves they create. The energy transferred by such waves can cause a lot of damage to the land at the edges of rivers (river banks).

■ **Figure 9.49** Plane breaking the sound barrier

For many years some engineers doubted if the sound barrier could ever be broken. The first confirmed supersonic flight (with a pilot) was in 1947. Now it is common for military aircraft to travel faster than Mach 1, but Concorde and Tupolev 144 were the only supersonic passenger aircraft in regular service.

It is possible to use a whip to break the sound barrier. If the whip gets thinner towards its end then the speed of a wave along it can increase until the tip is travelling faster than sound (in air). The sound it produces is often described as a whip 'cracking'.

1 In a world in which time is so important for so many people, suggest reasons why there are no longer supersonic aircraft in commercial operation.

Equations for the Doppler effect with sound

Figure 9.50a shows waves of frequency f and wavelength λ travelling at a speed v between a stationary source S and a stationary observer O. (The term *observer* can be used with any kind of waves, not just light.) In the time, t, that it takes the first wavefront emitted from the source to reach the observer, the wave has travelled a distance vt. The number of waves between the source and observer is ft. The wavelength, λ, equals the total distance divided by the number of waves $= vt/ft = v/f$, as we would expect.

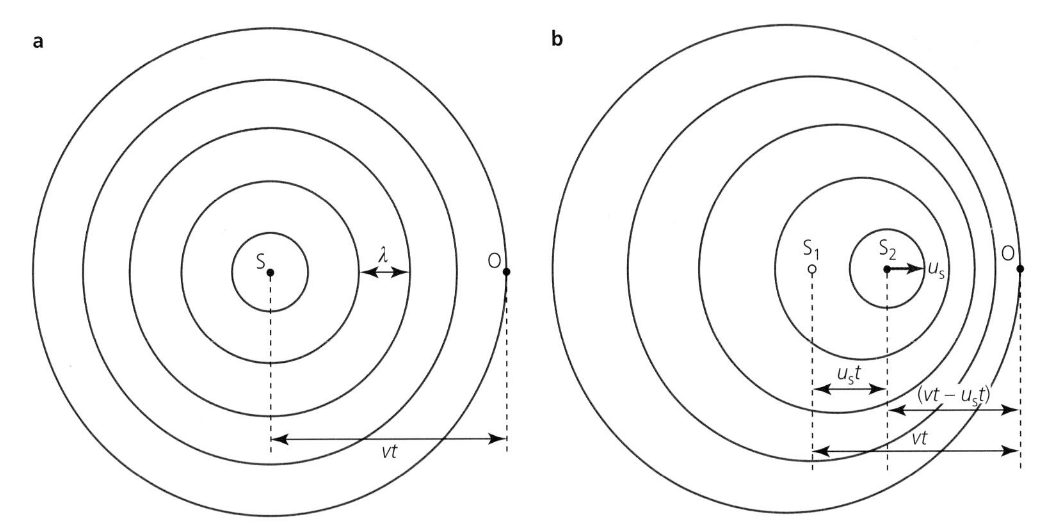

Figure 9.50
a Waves between a stationary source and a stationary observer
b Waves between a moving source and a stationary observer

Figure 9.50b represents exactly the same waves emitted in the same time from a source moving towards a stationary observer with a speed u_s. In time, t, the source has moved from S_1 to S_2. The number of waves is the same as in Figure 9.50a, but because the source has moved forwards a distance, $u_s t$, the waves between the source and the observer are now compressed within the length $vt - u_s t$.

This means that the observed (received) wavelength, λ', equals the total distance divided by the number of waves:

$$\lambda' = \frac{vt - u_s t}{ft} = \frac{v - u_s}{f}$$

The observed (received) frequency, f', is given by:

$$f' = \frac{v}{\lambda'} = \frac{vf}{v - u_s}$$

If the source is moving away from the observer, the equation becomes:

$$f' = \frac{vf}{v + u_s}$$

In general, we can write:

$$f' = f\left(\frac{v}{v \pm u_s}\right)$$

This is the equation for the Doppler effect from a *moving source* detected by a stationary observer and it is given in the *Physics data booklet*.

In a similar situation, the equation for the frequency detected by a *moving observer* from a stationary source is:

$$f' = f\left(\frac{v \pm u_O}{v}\right)$$

This equation is also given in the *Physics data booklet*.

If the source of the sound and the observer are getting closer together, along a line directly between them, a higher-frequency sound (than that emitted) will be detected by the observer and that frequency will be constant, although it will increase in intensity. Similarly, if the source and the observer are moving apart, the observed sound will have a lower, constant frequency (than that emitted) and it will decrease in intensity. The frequency must change quickly at the moment the source and detector move past each other. (If the motion is not *directly* between the source and the observer, or both the source and the observer are moving, the principles are the same, but the mathematics is more complicated and it is not included in this course.)

Worked examples

14 a A source of sound emitting a frequency of 480 Hz is moving directly towards a stationary observer at 50 m s^{-1}. If it is a hot day and the speed of sound is 350 m s^{-1}, what frequency is received?

b What frequency would be heard on a cold day when the speed of sound was 330 m s^{-1}?

c Explain why the speed of sound is less on a colder day.

a $f' = \frac{vf}{(v - u_s)}$

$f' = \frac{(350 \times 480)}{(350 - 50)} = 560\text{Hz}$

b $f' = \frac{(330 \times 480)}{(330 - 50)} = 566\text{Hz}$

c Sound is transferred though air by moving air molecules. On a colder day the molecules will have a lower average speed.

15 What frequency will be received by an observer moving at 24 m s^{-1} directly away from a stationary source of sound waves of frequency 980 Hz? (Take the speed of sound to be 342 m s^{-1}.)

$f' = f\left(\frac{v - u_o}{v}\right)$

$f' = 980 \times \left(\frac{(342 - 24)}{342}\right) = 910\,\text{Hz}$

60 An observer receives sound of frequency 436 Hz from a train moving at 18 m s^{-1} directly towards him. What was the emitted frequency? (Take the speed of sound to be 342 m s^{-1}.)

61 An observer is moving directly towards a source of sound emitted at 256.0 Hz, with a speed of 26.0 m s^{-1}. If the received sound has a frequency of 275.7 Hz, what was the speed of sound?

62 A car emitting a sound of 190 Hz is moving directly away from an observer who detects a sound of frequency 174 Hz. What was the speed of the car? (Take the speed of sound to be 342 m s^{-1}.)

63 A train is moving at constant speed along a track, as shown in Figure 9.51, and is emitting a sound of constant frequency.

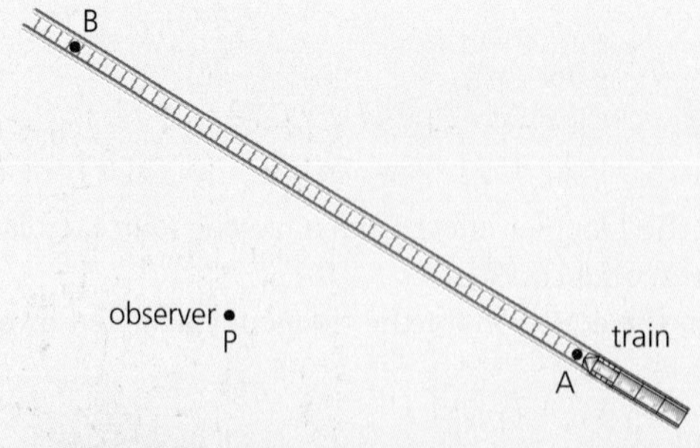

Figure 9.51

a Suggest how the sound heard by an observer at point P will change as the train moves from A to B.

b Describe the sound heard by someone sitting on the train during the same time.

64 The source of sound shown in Figure 9.46 is rotating at 4.2 revolutions per second in a circle of radius 1.3 m. If the emitted sound has a frequency of 287 Hz, what is the difference between the highest and lowest frequencies that are heard? (Take the speed of sound to be 340 m s^{-1}.)

The Doppler effect with electromagnetic waves

The Doppler effect also occurs with electromagnetic waves, but the situation is made more complicated because the received speed of electromagnetic waves is unaffected by the speed of the observer. The equations given earlier cannot be used with electromagnetic waves.

However, the following equation for the change (shift) in frequency, Δf, or wavelength, $\Delta\lambda$, can be used if the relative speed between source and observer, v, is very much slower than the speed of the electromagnetic waves, c ($v << c$).

$$\frac{\Delta f}{f} = \frac{\Delta\lambda}{\lambda} \approx \frac{v}{c}$$

This equation is given in the *Physics data booklet*.

Because the speed of electromagnetic waves is so high ($c = 3.00 \times 10^8\,\mathrm{m\,s^{-1}}$ in vacuum or air) this equation can nearly always be used with accuracy.

Worked example

16 A plane travelling at a speed of $250\,\mathrm{m\,s^{-1}}$ transmits a radio signal at a frequency of 130 MHz. What change of frequency will be detected by the airport it is travelling directly towards?

$$\Delta f = \frac{v}{c}f = \frac{250}{3.00 \times 10^8} \times 1.3 \times 10^8 = 108\,\mathrm{Hz}$$

The airport will receive a frequency 108 Hz higher than 1.3×10^8 Hz. This is a very small increase, requiring good-quality electronic circuits to be detected.

Clearly the shift in frequencies of light is not something we will observe in everyday life. It becomes more significant for very fast-moving objects like stars.

Using the Doppler effect to determine speed

Nature of Science

Using the Doppler effect requires a high level of technology

The Doppler effect may not seem to be so important when first studied, but it has a surprising number of interesting and useful applications, mostly in the determination of speeds, as outlined below. The mathematical link between speed and change of frequency is straightforward enough, but its use to determine speeds accurately depends on a high level of technology and sophisticated measuring instruments.

In order to measure the (average) velocity of a moving object we can observe the position of the object on two separate occasions over a known time interval. One way of doing this, especially if the object is inaccessible, like a plane for example, is to send waves towards it and then detect the waves when they are reflected back to the transmitter. From these data, the direction from the transmitter to the plane can be determined and the distance between the transmitter and the plane can be calculated from the time delay between the sent and the received signals. If this process is repeated, the velocity of the plane can be determined. Microwaves are used in this example of a simple primary radar system.

Figure 9.52 Air traffic control uses the Doppler effect

The Doppler effect, however, provides a better way of determining the speed of a moving object, like a plane, using the same kind of waves. This is used in air traffic control at airports around the world (see Figure 9.52). Some animals, like bats (see Figure 9.53) and dolphins, use the same principles but with sound or ultrasonic waves.

If waves of a known speed, v, and frequency, f, are directed towards a moving object and reflected back, the object will effectively be acting like a moving source of waves. The Doppler equations can be used to determine the speed of the object if the received frequency, f', can be measured.

The measurement of the speeds of cars, as well as planes and other vehicles, is a common example of the use of the Doppler effect. In many countries the police reflect electromagnetic radiation off cars in order to determine their speeds. Microwaves and infrared radiation are commonly used for this purpose. The speed of athletes or balls in sports can also be determined using the Doppler effect. The measurement of the rate of blood flow in an artery is another interesting example, which is shown in Figure 9.54.

■ **Figure 9.53** These bats in Malaysia use the Doppler effect to navigate

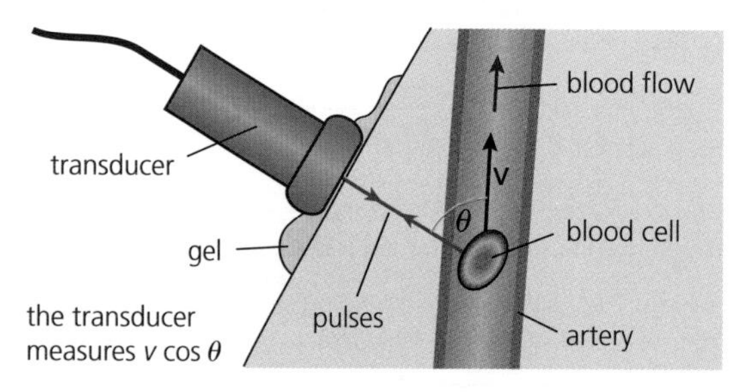

■ **Figure 9.54** Measuring blood flow rate using the Doppler effect

Pulses of ultrasonic waves are sent into the body from the **transducer** and are reflected back from blood cells flowing in an artery. The received waves have a different frequency because of the Doppler effect and the change of frequency can be used to calculate the speed of the blood. This information can be used by doctors to diagnose many medical problems. Because the waves usually cannot be directed along the line of blood flow, the calculated speed will be the component ($v \cos \theta$).

There are also some very important applications of the Doppler effect used in astronomy. The decrease in frequency of radiation received from distant galaxies is known as the 'red shift', and it provides very important information about the nature and age of the universe. This is discussed in more detail in the astronomy option. See also question 68.

Weather forecasting is another interesting application (weather radar). The location and velocity of, for example, rain can be determined using the Doppler effect with waves of wavelength in the order of 10 cm.

65 a What change of frequency will be received back from a car moving directly away at 135 km h^{-1} if the radiation used in the 'speed gun' has a frequency of 24 GHz?
b Suggest reasons why ultrasonic waves are not usually used in speed guns.

66 An airport radar system using microwaves of frequency 98.2 MHz sends out a pulse of waves that is reflected off a plane flying directly away. If the reflected signal is received back at the airport 8.43×10^{-5} s later, at a frequency 85.2 Hz lower than was emitted, what was the speed of the plane and how far away was it?

67 In the television broadcast of some sports, for example, baseball, tennis and cricket, viewers can see a replay of the trajectory (path) of a moving ball. Use the internet to find out how this is done. (Is the Doppler effect used?)

68 A star emits radiation of frequency 1.42×10^9 Hz. When received on Earth the frequency is 1.38×10^9 Hz. What is the speed of the star? Is it moving towards the Earth, or away from the Earth?

Examination questions – a selection

Paper 1 IB questions and IB style questions

Q1 A moving source of sound passes a stationary observer with a speed v. If the speed of sound in air is c, what will the speed of the waves be as they pass the observer?

A c **B** $c - v$ **C** $c + v$ **D** $v - c$

Q2 The equation $\theta = \lambda/b$ is used to describe diffraction at a single slit. In this equation:

A b is the width of the fringes seen on the screen
B b is the path difference between waves arriving at adjacent fringes
C θ is the angular width of the central maximum
D θ is half the angular width of the central maximum.

Q3 A source S, moving at constant speed, emits a sound of constant frequency. The source passes by a stationary observer O, as shown below.

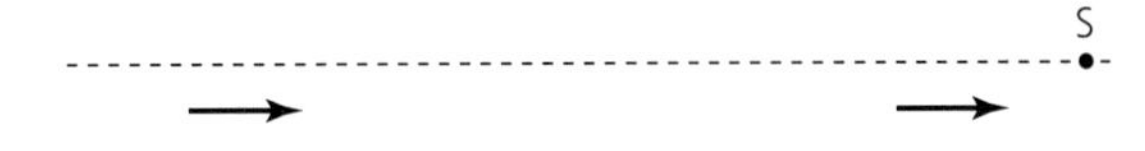

• O

Which of the following shows the variation with time t of the frequency f observed at O as the source S approaches and passes by the observer?

A

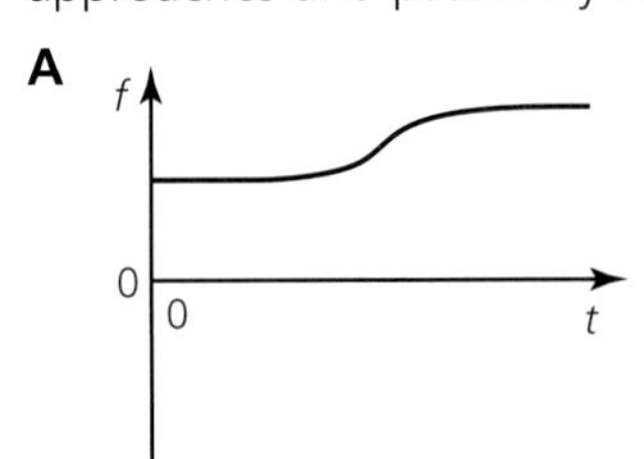

B

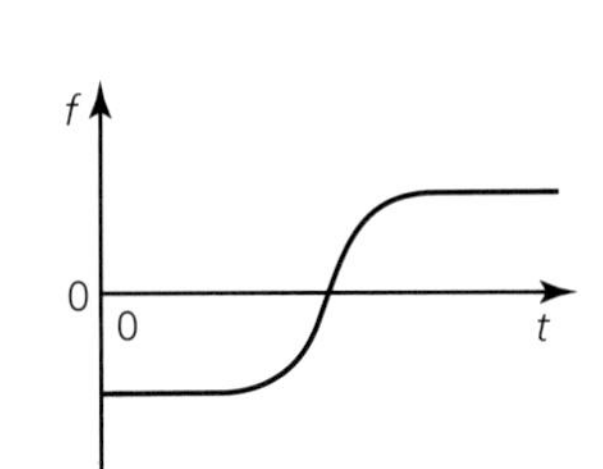

C

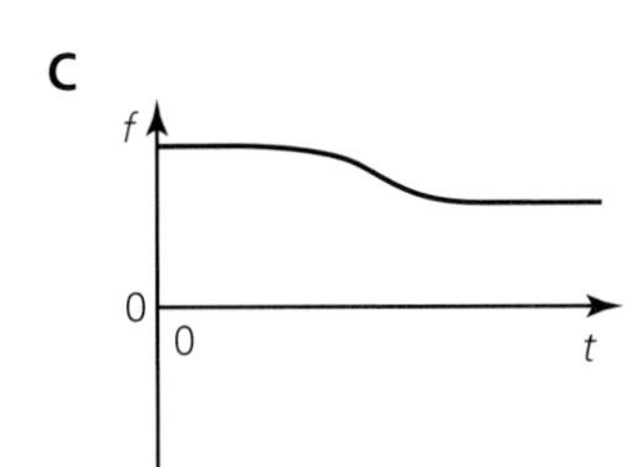

D

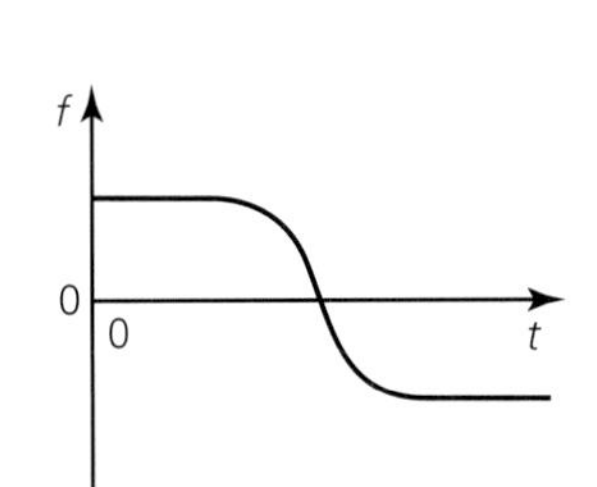

Q4 Which of the following is not (on its own) a possible way to increase the resolution of an optical telescope?

A Increase the diameter of the lenses.
B Use a blue filter.
C Use more powerful lenses to produce a bigger magnification.
D Use better quality lenses.

Q5 A particle performs simple harmonic oscillations. Which of the following quantities will be unaffected by a reduction in the amplitude of oscillations?

A the total energy
B the maximum speed
C the maximum acceleration
D the period.

Q6 A simple pendulum is preforming SHM. Which of the following statements about its period is correct?

A If it was taken to the moon it would have the same period.
B If its length was doubled, its period would increase by a factor of $\sqrt{2}$.
C If its mass was doubled, its period would be halved.
D Its period will gradually increase as energy is dissipated.

Q7 The spectrum of light received from distant galaxies is observed to have a red-shift. This is explained by the fact that:

A galaxies are moving away from each other
B galaxies are rotating
C the Earth is spinning
D the galaxies have very high temperatures.

Q8 Sound waves from two separate loudspeakers meet at a point. Which of the following is necessary for any kind of interference effect to be heard?

A The waves must arrive at the point in phase.
B The waves must have equal amplitude.
C The waves must have travelled the same distance from their sources.
D The waves must have a constant phase difference.

Q9 A 0.5 kg mass is oscillating with a frequency of 2 Hz and amplitude of 5 cm.

a Assuming the motion is simple harmonic, its total energy is approximately:

A 10^{-2} J **B** 10^{-1} J **C** 10^{1} J **D** 10^{3} J

b The maximum velocity of the oscillator is approximately:

A 10^{-1} m s^{-1} **B** 10^{0} m s^{-1} **C** 10^{1} m s^{-1} **D** 10^{2} m s^{-1}

Q10 Monochromatic, coherent light is incident normally on a double slit. The width of each slit is small compared with their separation. After passing through the slits the light is brought to a focus on a screen. Which of the following diagrams best shows the variation with distance *x* along the screen of the intensity *I* of the light?

A 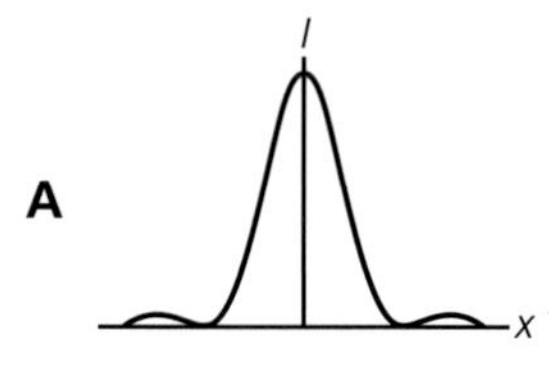

B 

C 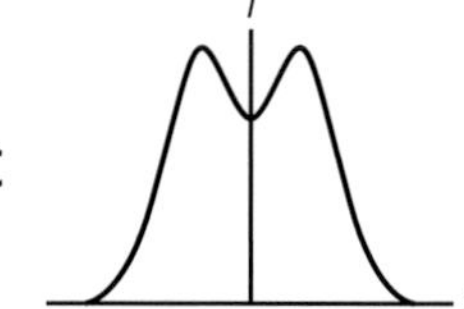

D 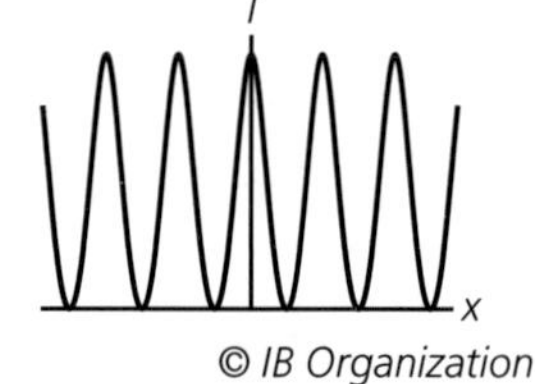

Q11 The diagram below shows an arrangement for a two-slit interference experiment. Coherent light of frequency *f* is incident on two narrow parallel slits S_1 and S_2 and an interference pattern is observed on a screen a large distance away. The speed of light is *c*. The centre bright fringe is at M and the next bright fringe is at N.

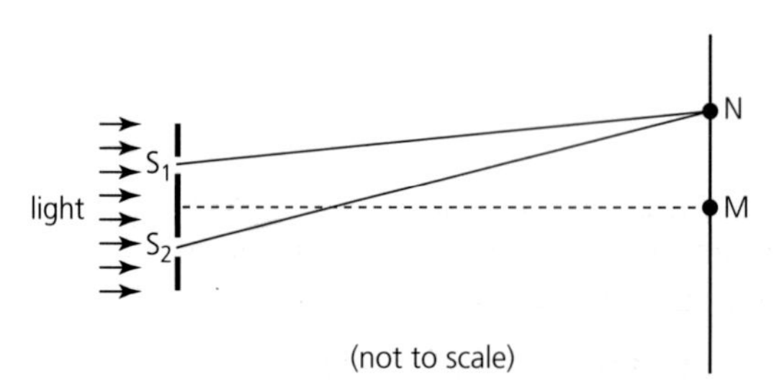

The distance S_2N-S_1N is equal to:

A $\frac{c}{2f}$ **B** $\frac{c}{f}$ **C** $\frac{f}{2c}$ **D** $\frac{f}{c}$

Q12 Which one of the following changes would result in an increase of the fringe separation in the previous question?

A Decrease the frequency of the incident light.
B Decrease the distance between the slits and the screen.
C Increase the separation of the slits.
D Increase the intensity of the incident light on the slits.

© IB Organization

Paper 2 IB questions and IB style questions

Q1 The diagram shows a source of sound, S, which is moving in the direction shown with a constant velocity. The sound emitted by S is received at stationary points P and Q.

Q S P

a Make a copy of the diagram and show a series of wavefronts that have been emitted by S. (2)
b Use your drawing to explain the differences in the sounds that are heard at points P and Q. (3)
c **i** What is the name of this effect?
ii Give one example of this effect in everyday life. (2)

© IB Organization

Q2 Radio telescopes can be used to locate distant galaxies. The ability of such telescopes to resolve the images of galaxies is increased by using two telescopes separated by a large distance *D*. The telescopes behave as a single radio telescope with a dish diameter equal to *D*.

The images of two distant galaxies G_1 and G_2 are just resolved by the two telescopes.

a State the phenomenon that limits the ability of radio telescopes to resolve images. (1)
b State the Rayleigh criterion for the images of G_1 and G_2 to be just resolved. (1)
c Determine, using the following data, the separation *d* of G_1 and G_2. (1)
effective distance of G_1 and G_2 from Earth = 2.2×10^{25} m
separation $D = 4.0 \times 10^3$ m
wavelength of radio waves received from G_1 and G_2 = 0.14 m

© IB Organization

Q3 **a** Plane wavefronts of monochromatic light of wavelength λ are incident on a narrow slit. After passing through the slit they are incident on a screen placed a large distance from the slit.

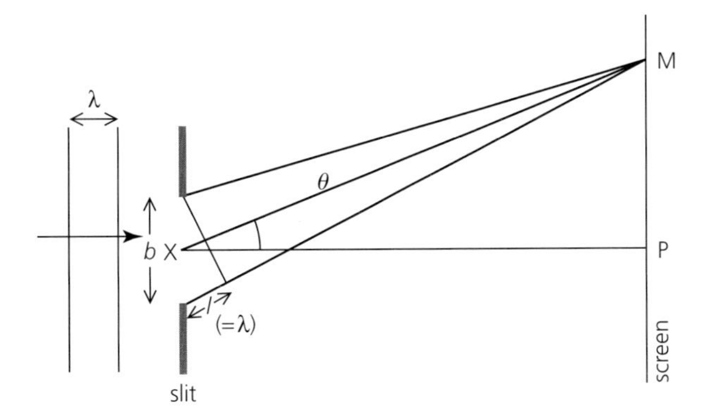

The width of the slit is *b* and the point X is at the centre of the slit. The point M on the screen is the position of the first minimum of the diffraction pattern formed on the screen. The path difference between light from the top edge of the slit and light from the bottom edge of the slit is *l*.

Use the diagram to explain why the distance *l* is equal to λ. (3)

The wavefronts in a are from monochromatic point source S_1. Diagram 1 is a sketch of how the intensity of the diffraction pattern formed by the single slit varies with the angle θ. The units on the vertical axis are arbitrary.

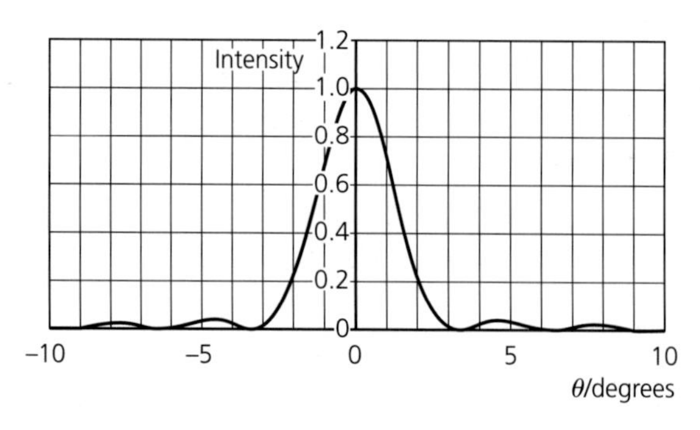

Diagram 1

Another identical point source S_2 is placed close to S_1, as shown in diagram 2.

S_1 •

S_2 •

Diagram 2

b The diffraction patterns formed by each source are just resolved. Make a copy of diagram 1 and sketch the intensity distribution of the light from source S_2. (2)

c Outline how the Rayleigh criterion affects the design of radio telescopes. (2)

d The dish of the Arecibo radio telescope has a diameter of 300 m. Two distant radio sources are 2.0×10^{12} m apart. The sources are 3.0×10^{16} m from Earth and they emit radio waves of wavelength 21 cm. Determine whether the radio telescope can resolve these sources. (3)

© IB Organization

Q4 a A particle of mass m that is attached to a light spring is executing simple harmonic motion in a *horizontal direction*. State the condition relating to the net force acting on the particle that is necessary for it to execute simple harmonic motion. (2)

b The graph shows how the kinetic energy E_K of the particle in a varies with the displacement x of the particle from equilibrium.

i Make a copy of the diagram and add a sketch graph to show how the potential energy of the particle varies with the displacement x. (2)

ii The mass of the particle is 0.30 kg. Use data from the graph to show that the frequency f of oscillation of the particle is 2.0 Hz. (2)

© IB Organization

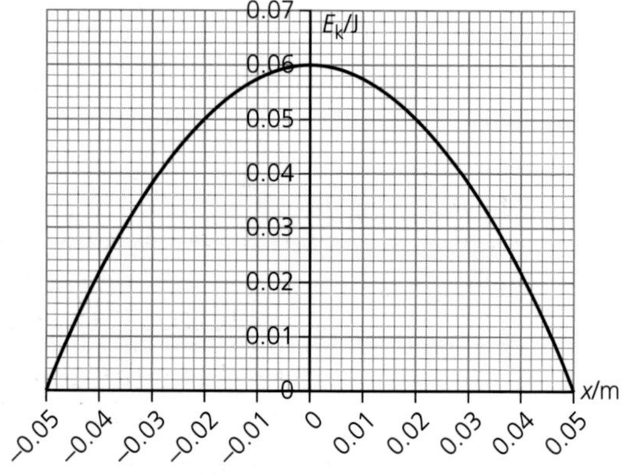

10 Fields

ESSENTIAL IDEAS

- Electric charges and masses each influence the space around them and that influence can be represented through the concept of fields.
- Similar approaches can be taken in analysing electrical and gravitational potential problems.

10.1 Describing fields *– electric charges and masses each influence the space around them and that influence can be represented through the concept of fields*

We have met the concept of fields before in this course, particularly in Chapters 5 and 6. In this chapter we will quickly review ideas about forces in gravitational and electric fields, as well as the concept of field strength. Then we will develop those ideas to include more about energy in fields, and introduce the important idea of *potential*. Section 10.2 contains more details about the mathematical modelling of fields, especially radial fields.

Physicists use field theory to help explain forces that act 'at a distance' across a space. In this chapter we will deal only with gravitational and electric fields. As explained in Chapters 5 and 6, the forces between point masses and the forces between point charges have many mathematical similarities, so that a good understanding of gravitational fields should be of benefit when studying electric fields.

Nature of Science

A paradigm shift in the understanding of forces

Although we are all used to the force of gravity, the fact that some kinds of forces (like gravity) can act across empty space is conceptually very difficult once we start to think in depth about it. That difficulty is made worse when it seems that the forces act instantaneously, regardless of the distance. Classical physics could not offer an explanation and a completely new model (a **paradigm shift**) was needed to explain 'at a distance' interactions between charges or between masses.

The concept of exchange particles was introduced in Chapter 7, but here we are more concerned with describing and using fields, rather than explaining the true nature of the interactions. We will begin with a revision of ideas covered in earlier chapters.

Gravitational fields

- A gravitational field exists where a mass experiences a gravitational force.
- Gravitational forces between two masses, M and m, separated by a distance r are described by Newton's law of gravitation: $F = GMm/r^2$ (an inverse square law). G is the universal gravitation constant.
- Gravitational forces can only be attractive.
- Gravitational field strength = gravitational force/mass: $g = F_g/m$ (unit N kg^{-1}).
- Gravitational field is a vector quantity, pointing in the direction of the force.
- Gravitational fields can be represented on paper by the use of field lines.
- Gravitational fields tend to dominate in the macroscopic world.

Electrostatic fields

- An electric field exists where a charge experiences an electric force. If the force is constant, then we can describe the field as **electrostatic**.
- Electric forces between two charges, q_1 and q_2, separated by a distance r are described by Coulomb's law: $F = kq_1q_2/r^2$ (an inverse square law). k is Coulomb's constant.
- Electric forces can be attractive (between opposite charges), or repulsive (between like charges).
- An electric field is a vector quantity, pointing in the direction of the force on a positive charge.

- Magnitude of electric field strength = electric force/charge. $E = F_q/q$. (Unit N C^{-1})
- Electric fields can be represented on paper by the use of field lines.
- Electric fields tend to dominate in the microscopic world.

Additional Perspectives

Gravitational fields and the lunar cycle

We can see the Moon because it reflects light from the Sun. As the Moon orbits the Earth its appearance (*phase*) changes because, depending on the relative positions of the Sun, Earth and Moon, the light from the Sun falls on a changing proportion of the side of the Moon that is facing the Earth. When we can see a complete circular moon, or 'full moon', the whole of the side that is facing the Earth is illuminated; this occurs every 29.53 days. In fact, the Moon orbits the Earth every 27.32 days, but the repeating lunar cycle is longer because it depends on the relative positions of the Sun, Earth and Moon, and not just the rotation of the Moon.

■ **Figure 10.1** Boats stranded at low tide in Vietnam

The regular phases of the Moon have long been the basis for calendars and the timing of annual events, festivals and holidays.

Gravitational attraction keeps the Moon in orbit around the Earth and this same force (of the order of 10^{20} N) also affects the Earth. The most noticeable effect is the tides in the Earth's oceans. As the Earth spins on its axis, water in the oceans is pulled closer to the Moon causing a high tide. At the same time, a high tide also occurs on the opposite side of the Earth, resulting in high tides (and low tides) about twice every day in any particular location. Boats that are in the water at high tide can become stranded when the water level goes down at low tide (Figure 10.1).

The Sun also affects the tides. When the Earth, Sun and Moon are in an approximate straight line (at a new moon or full moon), the gravitational effects are greatest and the tides highest.

1 Draw a diagram to illustrate the different phases of the Moon.

2 Name three different events in three different cultures or religions that occur at a particular time depending on the phase of the Moon. Explain how the exact date of one of these events is decided each year.

3 Do you believe that a full moon can have an effect on human or animal behaviour? Give a scientific reason for your answer.

4 Calculate the gravitational attraction between the Moon and the Earth.

■ Potential energy in fields

When masses (or charges) move with, or against, forces in fields there must be *changes* in potential energy because work has to be done. The work done (change in potential energy) can be calculated using $W = Fs\cos\theta$ if F is constant, or from the area under an F–s graph, for a variable force, if the way in which the force varies with distance is known (Chapter 2).

Alternatively, we can determine a *change* in potential energy by subtracting the value of potential energy at one point from the potential energy at another point. Obviously, this requires that we must know how to calculate the potential energy of a mass (or charge) *at a point*. And that requires that we must first choose a reference place where we consider that potential energy is zero.

In elementary physics courses the gravitational potential energy of a mass at a point close to the Earth's surface is often calculated from $E_P = mgh$. However, such calculations assume that the mass has zero potential energy on the Earth's surface, which may be convenient but is not the best choice if we want our calculations to be valid for everywhere in the universe (which, should be an important aim of any physics theory). In Chapter 2 we stressed the use of $\Delta E_P = mg\Delta h$ to calculate a *difference* in gravitational potential energy between two places (rather than potential energy at a point).

Logically, physicists consider potential energy to be zero when masses (or charges) are so far apart that there are no forces between them. The variation of forces with distance in both gravitational and electric fields is described by inverse square laws and so, in theory, the forces will never reduce to zero. However, in practice forces become very, very small (negligible) when the distances are large.

Infinity is chosen to be the place where gravitational and electric potential energies are zero.

Nature of Science

Infinity

Infinity is not an actual place, but an abstract concept that appears regularly in physics and mathematics.

The idea of an *infinite* quantity (field, distance, time, number etc.) is used in physics to suggest a quantity that is limitless (without end) and therefore greater than any real, measurable quantity. The opposite of infinite is **finite**, which means within limits.

The idea of an infinite series of numbers, an infinite time, or even a field that extends for ever (but becomes vanishingly small) may all seem somehow acceptable to the human mind. However, the concept of an infinite universe gives most of us problems!

We say that there is gravitational (or electric) potential energy stored in any system of masses (or charges) because, at some time in the past, work was done when the masses (or charges) moved to their present positions.

The total gravitational (or electric) potential energy of a system, E_P, is defined as the work done when bringing all the masses (or charges) of the system to their present positions, assuming that they were originally at infinity (an infinite distance apart).

The work that would be done in moving a mass (or charge) to or from infinity (Figure 10.2) would *not* be infinite, because the forces involved become very, very small (negligible) when the distances become large.

The work done in moving between two points in a field does *not* depend on the path taken.

It should be clear that potential energy is stored in a *system* of masses (or charges) rather than in a single mass (or charge). However, it is common to refer to the potential energy of a single smaller mass (or charge) if the other mass (or charge) is very much greater, so that it is almost unaffected by any changes to the smaller mass (or charge). For example, we can refer to the gravitational potential energy of a book on the Earth's surface, without considering the Earth, or we can refer to the electric potential energy of an electron in an atom without considering the nucleus.

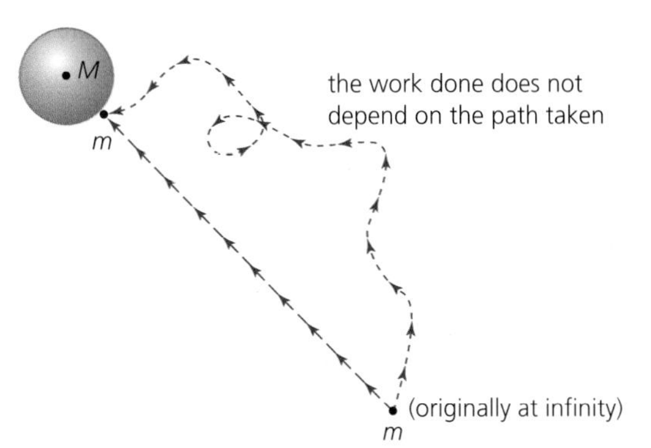

■ **Figure 10.2** Moving a mass from infinity

Examples

Gravitational potential energy

The gravitational potential energy of, for example, a 5 kg mass on the Earth's surface is approximately -3.13×10^8 J (this is explained on page 436). This statement means that a minimum energy of 3.13×10^8 J would have to be *supplied* to move the mass to where it would have zero gravitational potential energy (at infinity). Gravitational potential energy of a mass on Earth + energy transferred to move the mass to infinity = gravitational potential energy of mass at infinity (zero):

$$(-3.13 \times 10^8) + (3.13 \times 10^8) = 0$$

The fact that gravitational potential energies are *always* negative is because gravitational forces are always attractive, so that energy must always be supplied to move masses further apart.

The fact that gravitational potential energies are large is because they usually involve one or more large masses (like the Earth) and they relate to movement to or from infinity. *Changes* in gravitational potential energy are, of course, typically much smaller and can be positive or negative: moving a 5 kg mass up or down 2 m close to the Earth's surface involves about ±100 J ($\Delta E_p = mg\Delta h$).

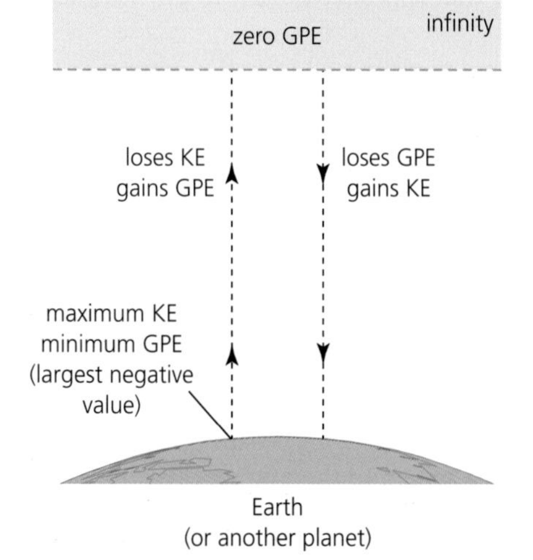

■ **Figure 10.3** Changes of energy when moving between a planet and infinity

If any mass – a ball, for example – is projected upwards from the Earth's surface, it loses kinetic energy (KE) and gains gravitational potential energy (GPE). This is also true for an unpowered spacecraft and is even true for the theoretical extreme in which a mass travels all the way 'up' to infinity. For the same reasons, a theoretical mass 'falling' from infinity would gain kinetic energy and lose gravitational potential energy (as shown in Figure 10.3), so that its minimum gravitational energy (on the Earth's surface) corresponds to its largest negative value.

Electric potential energy

The electric potential energy of an electron in a hydrogen atom at a distance of 5.3×10^{-11} m from a proton is -4.3×10^{-18} J (this is explained on page 436). This means that a moving electron would have to have a minimum kinetic energy of 4.3×10^{-18} J in order to move an infinite distance away from the proton, where it would then have zero electric potential energy. (The electron within the atom already has some kinetic energy.)

electric potential energy of hydrogen atom + kinetic energy of electron to move it to infinity = electric potential energy of electron and proton an infinite distance apart

$$(-4.3 \times 10^{-18}) + (4.3 \times 10^{-18}) = 0$$

Electric potential energy can be negative or positive, depending on the signs of the charges. If they are two similar charges, the potential energy is positive, which shows that there is a repulsive force between them, so that if they are moved apart they lose potential energy.

■ Electric potential and gravitational potential

When calculating gravitational potential energies we refer to a particular mass (or charge) in a particular place, but if we want to answer questions like 'how much energy would be needed to put a mass in a certain location?' it is better to use the important concept of **potential**. (Potential energy and potential are not the same thing, although they are closely connected.)

The gravitational potential at a point is defined as the work done *per unit mass* in bringing a small test mass from infinity to that position.

The concept of a *test mass*, or *test charge*, was introduced in Chapters 5 and 6, when defining field strengths. It means a mass so small that it will not affect the field or potential that it is measuring.

The definition of electric potential is similar, except that it needs to include reference to the sign of the charge. As before, in the definition of electric field strength, we chose to refer to *positive* charge, so that:

The electric potential at a point is defined as the work done *per unit charge* in bringing a small *positive* test charge from infinity to that position.

In practice, the Earth is considered to be capable of absorbing any amount of charge without affecting its potential, so that *connecting to earth* (grounding) is considered to be effectively connecting to zero potential. The negative terminal of a battery is often considered to be at zero volts, but it needs to be connected to earth (grounded) to be sure.

Potential is potential energy per unit mass (or charge).

Potential is given the symbol V_g or V_e and has the units of $J\,kg^{-1}$ or $J\,C^{-1}$ (commonly known as volts, V). Potential is a scalar quantity.

$$V_g = \frac{E_P}{m}$$

and

$$V_e = \frac{E_P}{q}$$

These two equations are given in the *Physics data booklet* in the form:

$$E_P = mV_g$$

$$E_P = qV_e$$

Returning to the previous examples:

- If the gravitational *potential energy* of a 5 kg mass on the Earth's surface is -3.13×10^8 J, then the gravitational *potential* on the Earth's surface is $(-3.13 \times 10^8)/5 = -6.26 \times 10^7\,J\,kg^{-1}$. Using this value for gravitational potential we can easily calculate the gravitational potential energy of any other mass on the Earth's surface. For example, the gravitational potential energy of a 52.0 kg girl is:

  $$52.0 \times (-6.26 \times 10^7) = -3.26 \times 10^9\,J$$

- If the electric potential energy of an electron in a hydrogen atom -4.3×10^{-18} J, then the electric potential at that point is:

  $$\frac{-4.3 \times 10^{-18}}{1.6 \times 10^{-19}} = -27\,J\,C^{-1} = -27 \text{ volts}$$

To understand the difference between potential energy and potential, it can be helpful to compare them to the total cost of a commodity and its price per kilogram. It might be true to say that the price of some meat is \$6, but it is generally more informative to quote a price per kg: \$14 kg^{-1}, for example. Similarly, it might be true that the potential energy of a 5 kg mass in a particular location is -3.13×10^8 J, but saying that the potential at that point is $-6.26 \times 10^7\,J\,kg^{-1}$ is generally much more informative.

Potential difference, p.d.

Although it is important to understand the concept of potential, we are usually much more concerned with the *difference* in potential (**potential difference**) between two places, than the actual potential at any location.

The potential difference, ΔV, between two points is the work done, W, when moving a unit mass (or unit positive charge) between those two points.

$$\Delta V_g = \frac{W}{m}$$

or

$$\Delta V_e = \frac{W}{q}$$

We have met the concept of electrical potential difference before, when dealing with electric circuits (Chapter 5). Then it was expressed simply as $V = W/q$, but in this chapter we are using the Δ sign to stress that we are referring to potential difference and not potential.

These equations are given in the *Physics data booklet* in the form:

$$W = m\Delta V_g$$

$$W = q\Delta V_e$$

Worked example

1 Two points in an electric field have potentials of 12.7 V and 15.3 V. How much energy will be transferred when an electron moves between these points:

a in J

b in eV?

a $W = q\Delta V_e = (1.6 \times 10^{-19}) \times (15.3 - 12.7) = 4.2 \times 10^{-19}$ J

b $15.3 - 12.7 = 2.6$ eV

1 If 4.9×10^{-5} J of energy were transferred when a charge of 5.1 μC was moved from a certain point to earth (ground), what was the magnitude of the potential of the point?

2 The gravitational potential on the surface of a planet is $-4.8\,\text{MJ kg}^{-1}$. Calculate the gravitational potential energy of an 86 kg mass on the planet.

3 The gravitational potential at a height of 500 km above the Earth's surface is $-58.0\,\text{MJ kg}^{-1}$. How much potential energy must be transferred to a mass of 5000 kg to raise it to an orbit at this height from the Earth's surface?

4 A satellite of mass 3760 kg is in a circular orbit where the potential is $-5.74 \times 10^{7}\,\text{J kg}^{-1}$.
  a What would be the change in gravitational potential energy if it were to move to an orbit where the potential is $-6.00 \times 10^{7}\,\text{J kg}^{-1}$?
  b Would the satellite have to move towards or away from the Earth?

5 How much energy is gained by an overall charge of 2 C when it passes through a battery that has a terminal p.d. of 12 V?

6 What is the gravitational potential difference between the top and bottom of a 24.5 m tall building?

7 Explain why 'infinity seems a more sensible choice for a zero of gravitational potential than for a zero of electric potential'.

8 A charge of +4.5 C is moved from a place where the potential is +0.23 V to a place where the potential is −1.44 V.
  a How much energy is transferred?
  b Is work done *on* the charge or *by* the charge?

9 Explain why there is no change in gravitational potential energy for a mass that is moved horizontally on the Earth's surface, or for a satellite in a circular orbit around the Earth.

Field lines and equipotential surfaces

A field line represents the direction of force on a mass (or positive charge). If the field lines get closer together it means that the field strength is increasing.

When drawing field lines remember that they cannot cross each other because it would mean that a force could act in two directions at the same place.

An **equipotential surface** (or line) connects places with the same potential.

Gravitational equipotentials

We would not normally expect the gravitational potential (or field) around a large mass, like a planet or star, to change and that is another reason why the concept of potential is useful: we can draw 'energy maps' to represent how the gravitational potential varies around planets and similar bodies. Figure 10.4 shows *equipotential lines* around a spherical planet, which acts like a point mass.

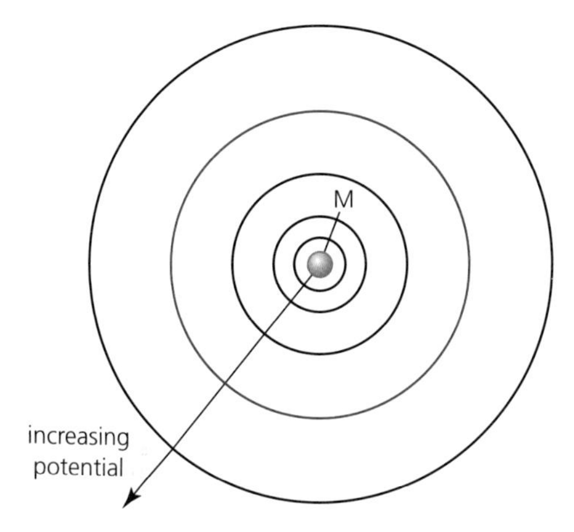

■ **Figure 10.4** Equipotential lines around a spherical mass

For example, the same amount of energy would be needed to take a certain mass located at *any* point on the red equipotential line to infinity, and no *net* energy would be needed to move the mass from any place on the red line to any other place on the same equipotential line. Because potential is a scalar quantity, there are no arrows on the lines.

The numerical difference in gravitational potential between successive equipotential lines is always kept the same, which means that the lines get further apart as the distance from the mass increases (because the gravitational field strength is getting weaker). Figure 10.4 represents a three-dimensional situation on a flat piece of paper; in three dimensions the equipotential surfaces in this example would be spherical.

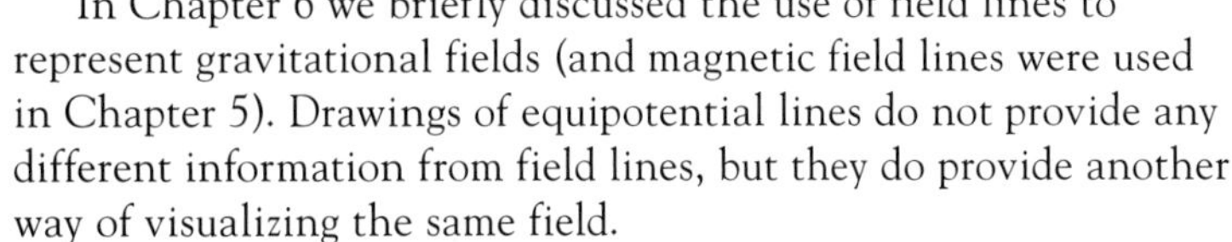

In Chapter 6 we briefly discussed the use of field lines to represent gravitational fields (and magnetic field lines were used in Chapter 5). Drawings of equipotential lines do not provide any different information from field lines, but they do provide another way of visualizing the same field.

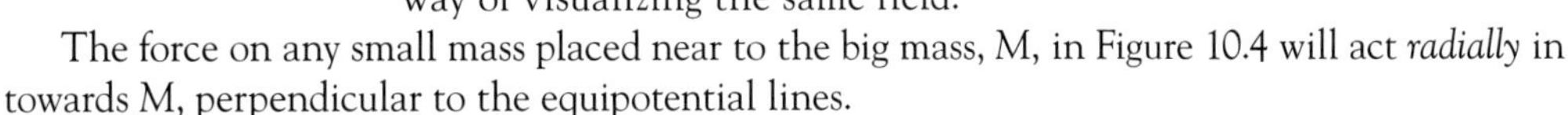

The force on any small mass placed near to the big mass, M, in Figure 10.4 will act *radially* in towards M, perpendicular to the equipotential lines.

Gravitational field lines are always perpendicular to equipotential surfaces, pointing from higher potential to lower potential.

Figure 10.5a shows a radial gravitational field and Figure 10.5b shows an (almost) uniform gravitational field, such as in a room. The field is always strongest where the field lines and the equipotential lines are closest together.

a

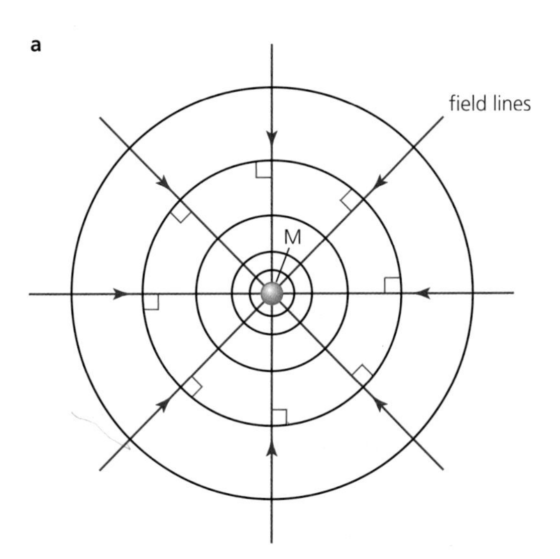

b

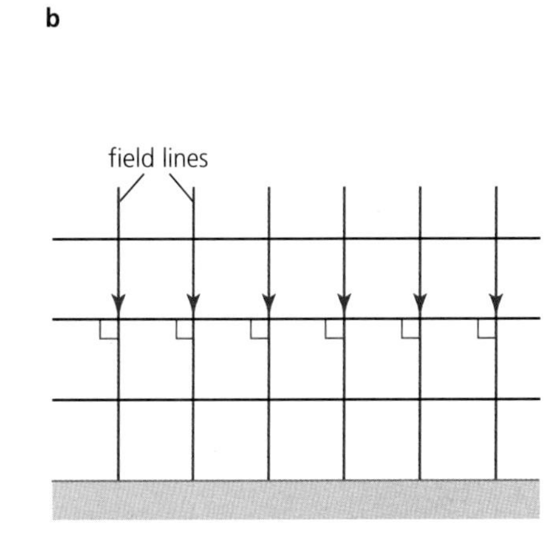

■ **Figure 10.5** Equipotential lines and field lines are perpendicular to each other

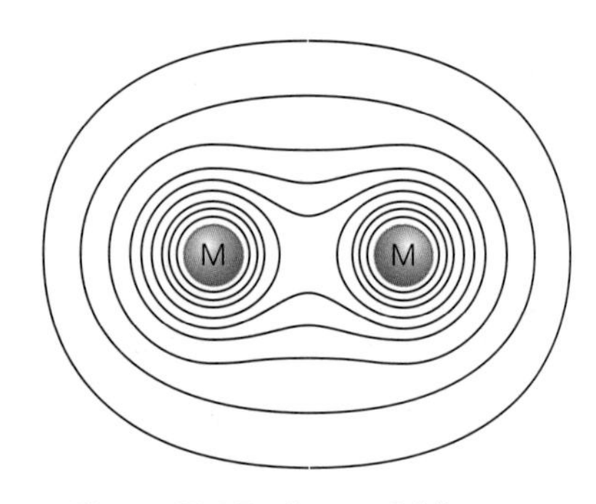

■ **Figure 10.6** Equipotential lines around two equal masses

The shape of the equipotential lines around two equal masses is shown in Figure 10.6.

Contour lines drawn on a geographical map, such as those shown in Figure 10.7, are useful in the same way that equipotential lines are useful. In fact contour lines *are* a type of equipotential line. Contour lines join up places at the same height, so that, by looking at the map, we know in which direction to move to go up (to gain gravitational energy), to go down or to stay at the same level. We also know that the ground is steepest where the lines are closest together and that anything able to move freely, like a river, will go downwards perpendicular to the contour (equipotential) lines.

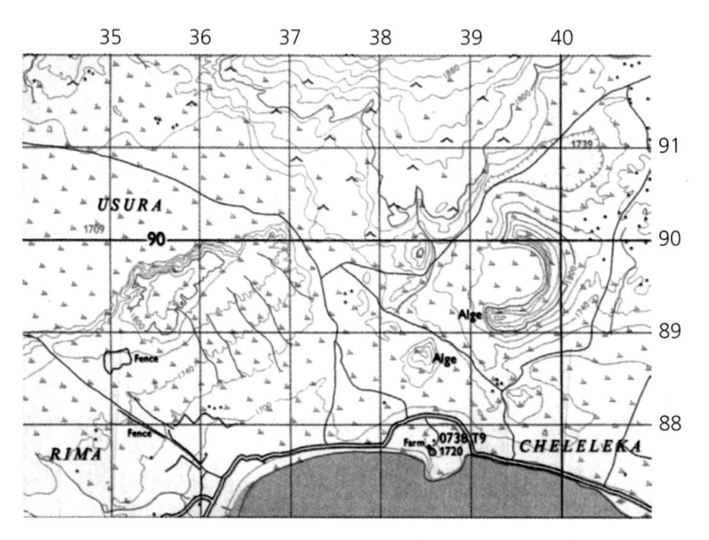

■ **Figure 10.7** Contour lines on a map

10 a Make a copy of Figure 10.6 and add to it gravitational field lines.
b Mark where the gravitational field strength is zero.

11 Draw a sketch of the equipotential lines around two masses that are not equal – for example, the Earth and the Moon.

12 Find out what is meant by the *geoid*.

Electric equipotentials

Variations of electric potential around any particular arrangement of charges are also best shown on drawings of equipotential lines or surfaces. The principles are the same as for gravitational equipotentials, but they are repeated below for clarity.

Electric field lines are always perpendicular to equipotential lines and point from higher to lower potential (the direction of force on a positive charge).

Figure 10.8 shows the equipotential lines and field lines around individual point charges and Figure 10.9 represents the uniform field between charged parallel plates. The numerical difference between adjacent equipotential lines is always kept the same. Where the lines are getting further apart the *potential gradient* is decreasing, which shows that the field is getting weaker. Field lines and equipotential lines drawn on a flat surface are a simplified way of representing a three-dimensional situation and more generally we should be referring to equipotential surfaces.

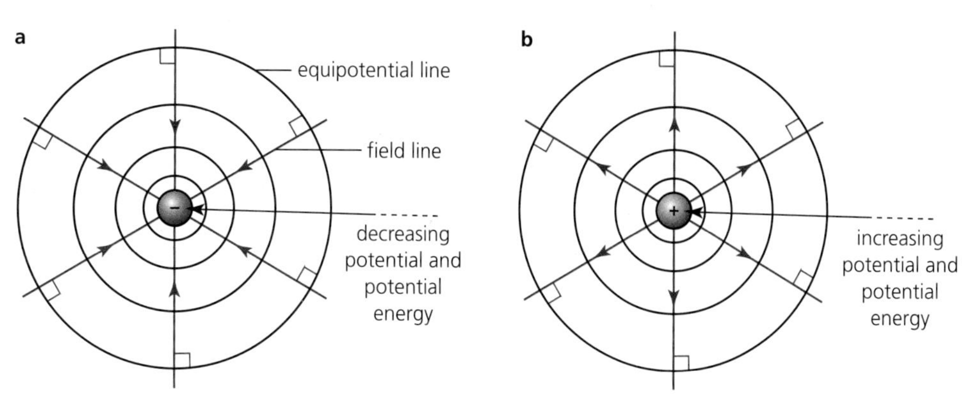

■ **Figure 10.8** Electric field and potential around point charges

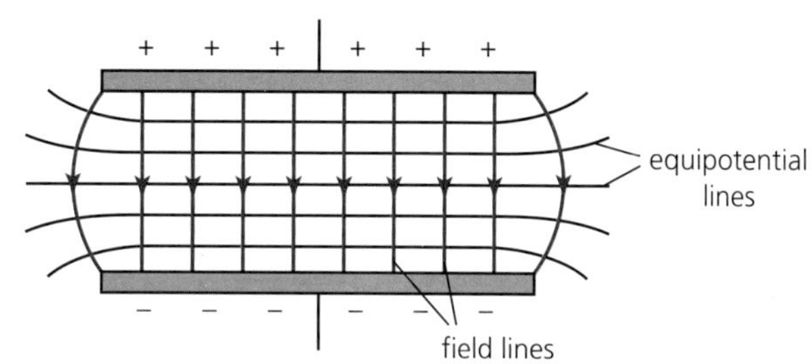

■ **Figure 10.9** Field and potential in a uniform field

By definition, no net energy is transferred when a charge is moved between different places on the same equipotential line.

Figure 10.10 shows the equipotential lines around two equally sized positive charges. Some field lines have also been included.

Figure 10.11 shows the equipotential lines around two equally sized charges of opposite sign. (A pair of equal and opposite charges separated by a small distance is called a **dipole**.)

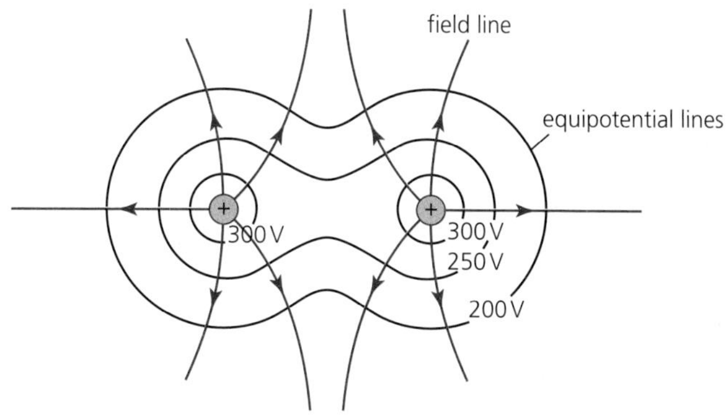

■ **Figure 10.10** Equipotentials around equal point charges

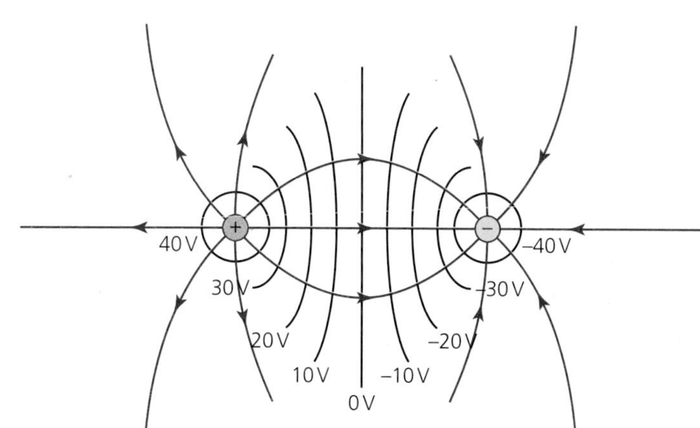

■ **Figure 10.11** Equipotentials around a dipole

Electric equipotentials can be investigated in two dimensions easily if *resistance paper* is available. See Figure 10.12. The two copper electrodes must be in good contact with the paper. The probe is moved so that the reading on the voltmeter remains constant. This means that the potential difference between one electrode and the probe is constant, so that the movement of the probe is along an equipotential line. A range of different electric fields can be investigated by using electrodes of different shape, size and separation.

Field lines representing a *static* electric field must meet any conducting surface perpendicularly, otherwise there would be a component of force acting on free electrons making them move through the conductor (it would then not be in a static field in equilibrium).

On a flat charged conductor electrons will distribute themselves evenly, but on a curved surface they will tend to be more concentrated where the surface is most curved. This means that electric field strength is highest near curved and pointed objects. See Figure 10.13.

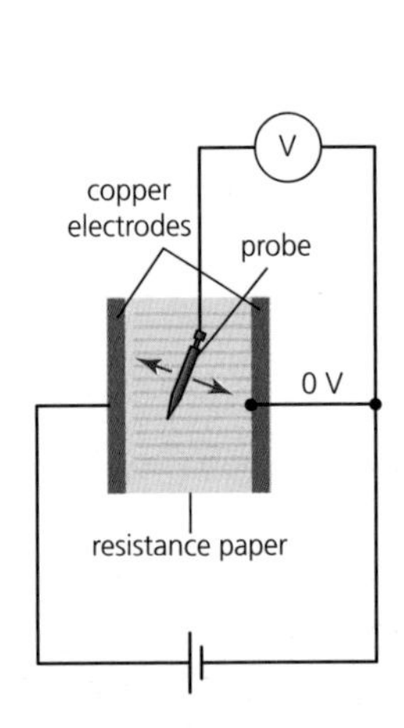

■ **Figure 10.12** Investigating equipotentials

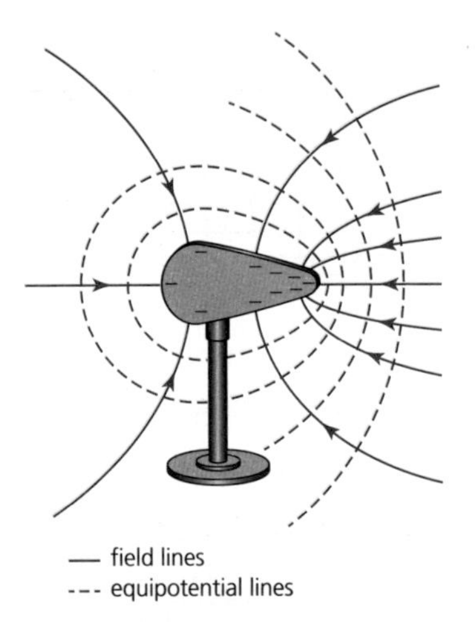

■ **Figure 10.13 Electric field lines and equipotentials near a negatively charged curved conductor**

13 Sketch the equipotential and field lines around two point charges of different magnitude if
 a they have similar signs
 b they have opposite signs.

14 Sketch the shapes of the equipotential lines that you would expect to obtain with the arrangement as shown in Figure 10.12.

15 Consider Figure 10.9. Make a sketch of the field lines and equipotential lines that would represent the situation if the p.d. across the plates were changed from $+V$ to $-V/2$.

16 Figure 10.14 shows a kind of *coaxial cable* that is widely used for transferring data, such as signals to televisions. The outer copper mesh is maintained at 0 V and the signal is sent as an electromagnetic wave in the insulator between the central copper wire and the surrounding mesh. This design, with its earthed outer mesh, helps reduce interference to and from other electromagnetic waves. Make a sketch of a cross-section of the cable and add electric field lines and equipotential lines.

■ **Figure 10.14**

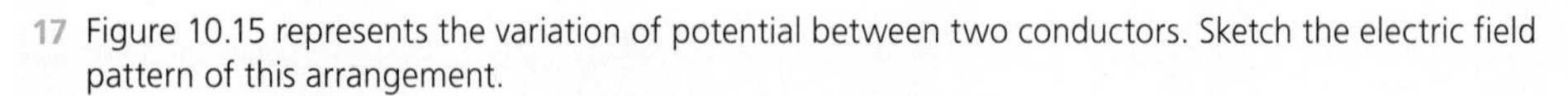

17 Figure 10.15 represents the variation of potential between two conductors. Sketch the electric field pattern of this arrangement.

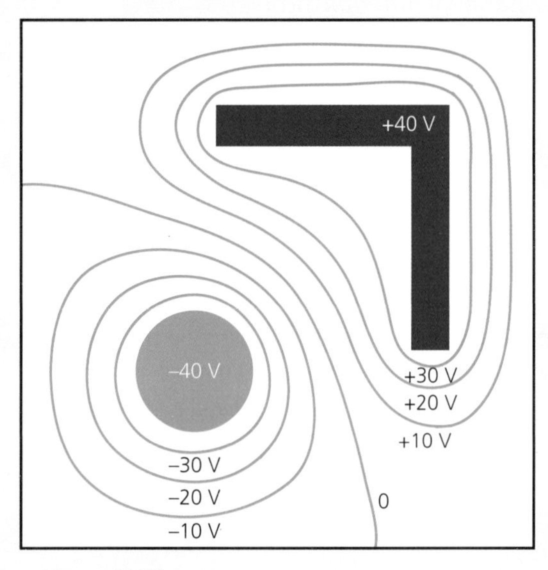

■ **Figure 10.15**

18 Draw a sketch to represent the electric field lines between a positively charged cloud and some flat ground with a single tall tree beneath it.

Utilizations

Electric fields and lightning conductors

A lightning conductor is a long, metallic object fixed pointing upwards on the top of a tall or exposed structure, such as a building, bridge or statue (Figure 10.16). A thick copper cable connects the lightning conductor down the side of the structure and into the ground. The simplest lightning conductors are just single metal rods, but many other patterns are in use. Sometimes they can be designed to be decorative or just inconspicuous.

■ **Figure 10.16** The Eiffel tower must be protected by a lightning conductor

Lightning will strike the conductor rather than the structure and the very high electric currents flow harmlessly to the ground rather than damaging the structure, or people inside or near it.

The pointed nature of the lightning conductor increases the electric field around it and helps to ionize the air to provide a conducting path for the lightning, but it is still not fully understood how important this process is.

1 Research the risks to aircraft from lightning strikes, and how these effects are minimized.

10.2 Fields at work – *similar approaches can be taken in analysing electrical and gravitational potential problems*

In this section we will use our understanding of field concepts to consider radial fields and uniform fields in more mathematical detail, and then apply that knowledge to specific examples. To begin with it may be helpful to summarize the four principal concepts and their interconnections:

- All masses exert *gravitational forces* on other masses. Similarly, all charges exert *electric forces* on other charges.
- This means that there must be *potential energy* stored in all such arrangements.
- Any calculations of the forces and energy will be specific to the particular masses (or charges) involved.
- The concepts of *field strength* and *potential* describe the properties of points in space, rather than the properties of particular masses (or charges).
- The interconnections between these four related concepts are shown in Figure 10.17.

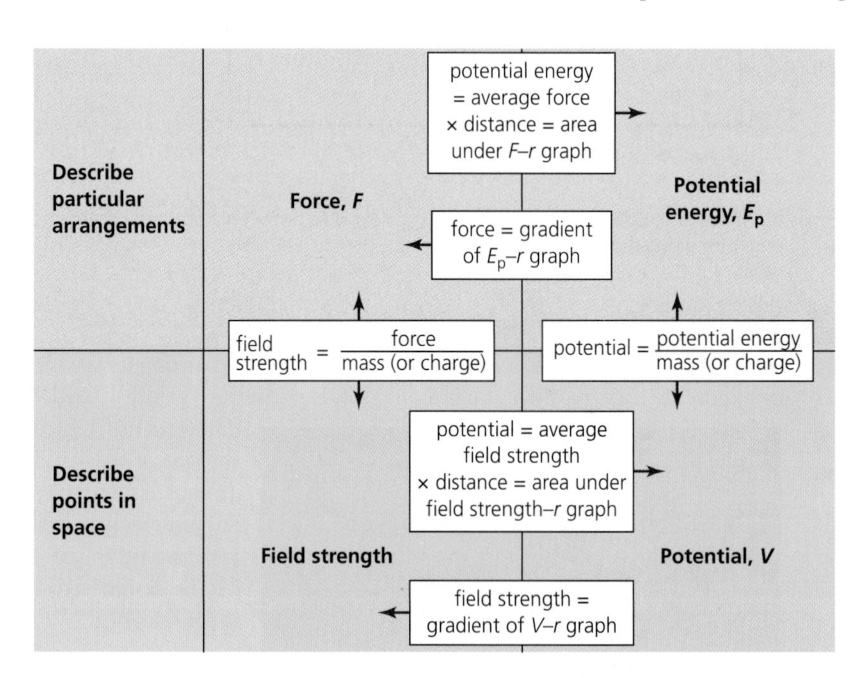

Figure 10.17 Equations relating the magnitudes of field quantities

Nature of Science

Communication with non-scientists

Gravitational fields extend throughout the universe to unimaginably large distances. Conversely, applications of electric fields include sub-atomic distances much smaller than can be seen with any microscope (see Table 1.5 on page 5). Of course, the scientists involved in these areas have specialist knowledge and are very familiar with the scale of their subject, but most of the general public will have very little understanding of these topics, which are difficult to visualize. Added to which, the concepts used in this chapter have a reputation for being difficult!

Scientists and scientific organizations cannot work in isolation; they have a responsibility to communicate the essence of their work, not only to other scientists but also to the general public. This requires communication skills that are very different from those skills used in the normal work of a scientist.

Forces and inverse-square law behaviour: radial fields

The *radial fields* around point masses (or charges) are the most basic of types field and should be well understood. All other fields can be understood as the combination of many radial fields. Remember also that spherical masses and spherical charged conductors behave as if all their mass (or charge) is concentrated at the centres of the spheres (as observed from outside the spheres).

Forces

We already know that the gravitational and electrical forces between points are described by the following equations, which are given in the *Physics data booklet*:

$F_G = G\frac{Mm}{r^2}$ (Newton's law of gravitation, sometimes written as $F_G = G\frac{m_1m_2}{r^2}$)

$F_E = k\frac{q_1q_2}{r^2}$ (Coulomb's law)

Multiplying together electric charges of the same sign will result in a positive force, which means a repulsion. The force acts in the same direction as increasing separation. Multiplying a positive charge by a negative charge will result in a negative force, which means an attraction.

For consistency, it might be considered that the equation for gravitational forces should also have a negative sign. However, this is usually considered unnecessary and omitted because gravitational forces are always attractive and we are only concerned with the magnitude of the force.

Figure 10.19 shows a graphical representation of an inverse square law of (gravitational) attraction.

ToK Link

Is there a difference between negligible and zero?

Although gravitational and electrostatic forces decrease with the square of the distance and will only become zero at infinite separation, from a practical stand point they become negligible at much smaller distances. How do scientists decide when an effect is so small that it can be ignored?

Negligible means too small to be of significance, although in this context *negligible* also implies too small to be measured using current technology. For any particular *application* of a theory it may be reasonable to assume that an effect is negligible, but the basic theory will not change.

Field strength

We know that in general $g = F_G/m$ and $E = F_E/q$, so the equations for field strengths in the radial fields around points are simply:

$$g = \frac{F_G}{m} = G\frac{m_1m_2}{r^2}$$

$$E = \frac{F_E}{q} = k\frac{q_1q_2/r^2}{q}$$

q_1 and q_2 are interchangeable here, and when discussing the field around a single point charge we can just use q.

Cancelling, we get:

$$g = G\frac{M}{r^2}$$

$$E = k\frac{q}{r^2}$$

The equation for gravitational field strength is given in the *Physics data booklet* (for Chapter 6). The equation for electric field strength is *not* given in the *Physics data booklet*.

Figure 10.18 shows the variation of gravitational field strength around the Earth. The field *inside* the Earth gets *less* as the distance below the surface increases.

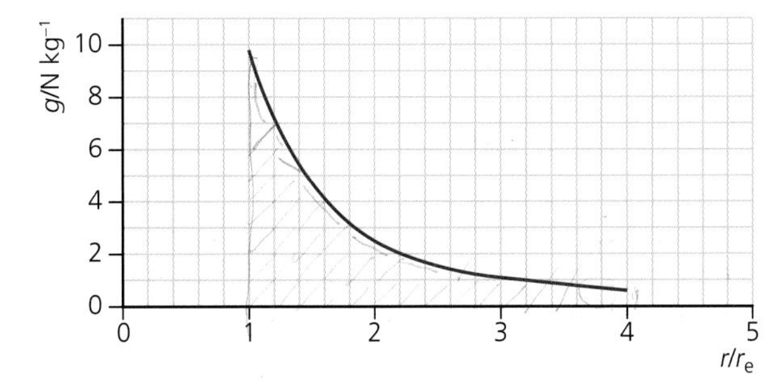

■ **Figure 10.18** Gravitational field strength above the Earth's surface (radius r_e)

28

19 If the electric field strength 34 cm from the centre of a charged sphere of radius 10 cm is $-6.0\,NC^{-1}$, what is the charge on the sphere?

20 Draw a graph of the electric field strength around an isolated metallic sphere of radius 10 cm if there is a charge of +2.6 nC on its surface.

21 a The gravitational field strength on the surface of the planet Venus is $8.9\,N\,kg^{-1}$. Calculate the mass of Venus if it has a diameter of 12 100 km.
b What would be the weight of a 5000 kg spacecraft 6050 km above the surface of Venus?

22 The area under a field strength–distance graph between infinity and a particular distance is equal to the potential at that point. Use Figure 10.18 to estimate the gravitational potential on the Earth's surface ($r_e = 6.4 \times 10^6$ m).

23 Suggest reasons why:
a the gravitational field strength inside the Earth is less than $9.8\,N\,kg^{-1}$
b the field strength at the centre of the Earth is zero.

Potential and potential energy

Potential energy, E_p

Figure 10.19 represents a variation of an attractive force between two points that is proportional to $1/r^2$. The shaded area, between $r = r_0$ and r = infinity, is equal to the work done in bringing a mass (or charge) from infinity to a distance r_0 from the point mass, or point charge (or from the centre of a sphere). (The work is done by the field because the force is attractive.)

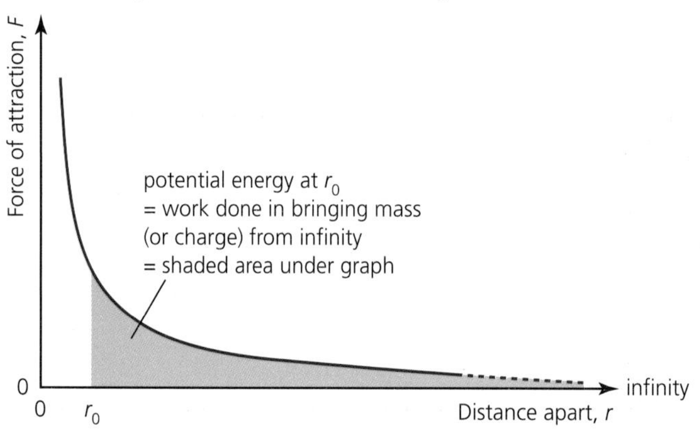

■ **Figure 10.19** Force–distance graph for an inverse square relationship

24 The gravitational force (weight) acting on a 20 kg mass on the surface of the planet Mars is 74 N. The radius of Mars is 3400 km.
a Calculate the mass of Mars.
b Draw a force–distance graph for this mass up to a distance of at least 2×10^4 km from the centre of Mars.
c Use your *graph* to estimate the gravitational potential energy of the mass when it is on the surface of Mars.

25 Figure 10.20 shows how the force acting on a −73 pC point charge varies as it moved away from a charged sphere.
a Determine the charge on the sphere.
b Use the *graph* to estimate the potential energy of the point charge when it is 5 cm from the centre of the sphere.

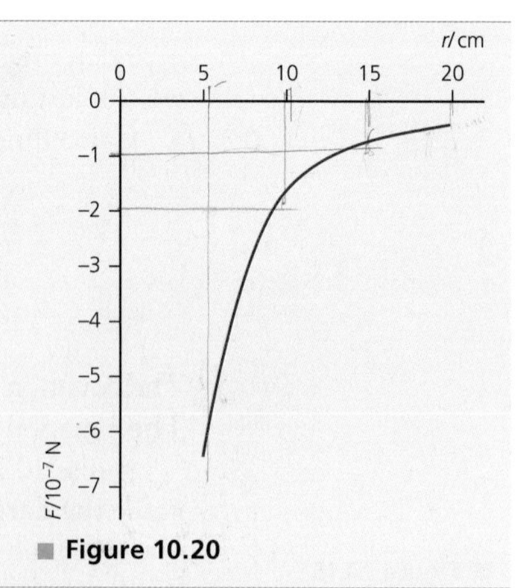

■ **Figure 10.20**

Of course, using equations is much easier than measuring areas under graphs. It can be shown that gravitational potential energy and electric potential energy can be calculated from the following equations:

$$E_P = -G\frac{Mm}{r}$$

$$E_P = k\frac{q_1q_2}{r}$$

Both equations are given in the *Physics data booklet*.

Reminder: the negative sign in the gravitational equation represents the facts that the zero of potential energy is at infinity and that gravitational forces are always attractive. The sign of the electric potential energy depends on the signs of the two charges.

Figure 10.21 shows the variation in gravitational potential energy (proportional to $1/r$) of a (test) mass, around a point mass. The gradient at any point equals the magnitude of the force at that distance. The potential energy graph for two point charges of opposite sign will be similar. For two similar charges, the potential energy will be positive because there will be a repulsive force between them.

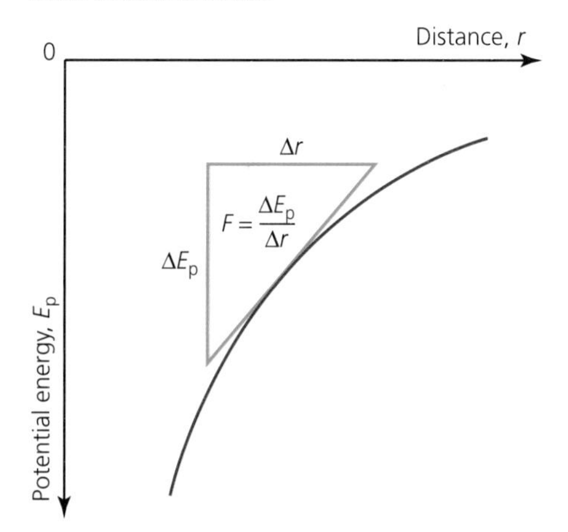

Figure 10.21 Graph of potential energy against distance in an inverse-square force field

Worked example

2 **a** Calculate a value for the gravitational potential energy of a 2000 kg spacecraft 100 km above the surface of the Moon. (mass of Moon = 7.35×10^{22} kg; radius of Moon = 1740 km)

b Explain why its actual potential energy will be bigger.

a $E_p = \frac{-GMm}{r} = \frac{(-6.67 \times 10^{-11}) \times (7.35 \times 10^{22}) \times 2000}{1.84 \times 10^6} = 5.30 \times 10^9\,J$

b The effect of the Earth's gravitational field has been ignored.

26 Figure 10.22 shows the variation of gravitational potential energy of a spacecraft close to a planet.

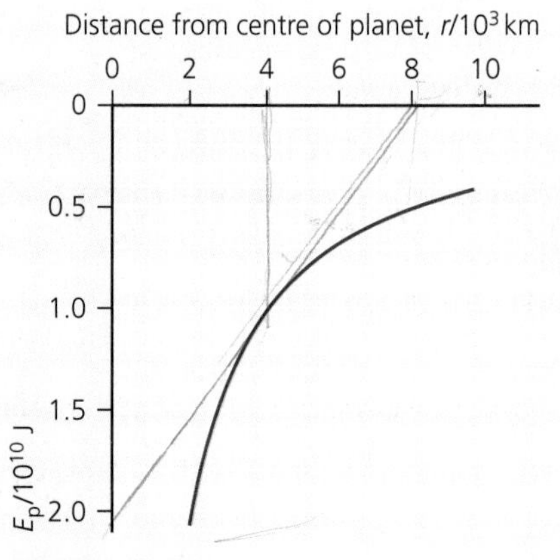

Figure 10.22

a Use the gradient of the graph to determine the magnitude of the force on the spacecraft when it is 4000 km from the centre of the planet.

b Estimate the potential energy of the spacecraft if it is 1.0×10^5 km from the centre of the planet.

27 The surfaces of two identically charged spheres are separated by 12 cm. If the radius of each sphere is 2.8 cm and the electric potential energy in the arrangement is $+3.6 \times 10^{-8}$ J, what is the charge on each sphere?

28 Estimate the gravitational potential energy stored in the Sun–Earth system (research any data needed).

Potential, V

In general, we have defined potentials as $V_g = E_p/m$ and $V_e = E_p/q$. In particular, for radial fields:

$$V_g = \frac{-(GMm/r)}{m}$$

and

$$V_e = \frac{(kq_1q_2/r)}{q_1}$$

q_1 and q_2 are interchangeable here, and when discussing the potential around a single point charge, we can just use q:

$$V_g = -\frac{GM}{r}$$

$$V_e = \frac{kq}{r}$$

Both equations are in the *Physics data booklet*.

A potential–distance graph will have the same inverse proportion shape as a potential energy–distance graph. Figure 10.23 shows the variations in potential around both positive and negative charges. (Gravitational potentials will always be negative).

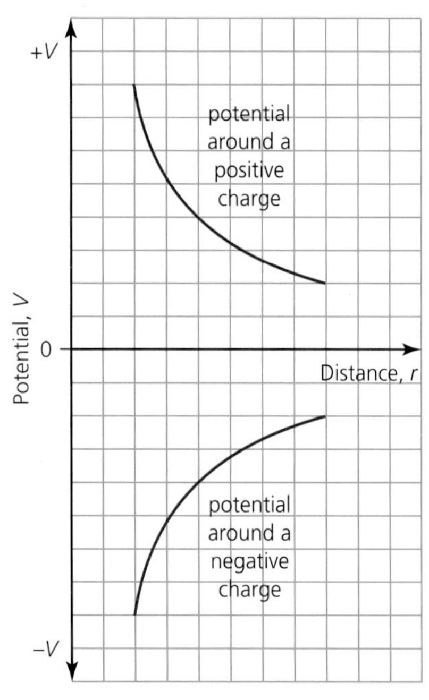

■ **Figure 10.23** Variation of potential around charges

29 a Explain exactly what we mean when we say that the gravitational potential on the Earth's surface is $-6.25 \times 10^7\,\text{J kg}^{-1}$ (at a distance r from its centre).

b Calculate the gravitational potential at the points P, Q and R in Figure 10.24.

c How much gravitational potential energy would have to be given to a 2420 kg spacecraft to move it from Q to R?

30 Use a spreadsheet to calculate the gravitational potential at different distances from the centre of the Earth (every 10 000 km from zero to 100 000 km). Use the computer program to generate a graph of the results.

31 Io is one of the moons of the planet Jupiter. It has a radius of 1820 km and a mass of 8.9×10^{22} kg.

a What is the gravitational potential on the surface of Io? (Ignore the effects of Jupiter.)

b What would be the potential on another moon if it had the same density, but twice the radius?

c Use these data to determine how much energy would be needed to transfer a 8.5 kg mass to infinity from the surface of Io.

d Why is your answer to **c** likely to be an underestimate.

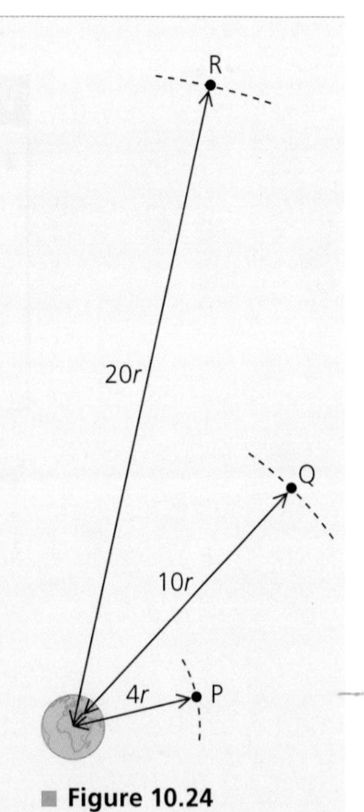

■ **Figure 10.24**

32 What is the electric potential at a distance of 12.6 cm from a point charge of 60 μC?

33 What is the electric potential 10^{-10} m away from a positive charge of 1.6×10^{-19} C? (This is the approximate distance of the nucleus from the electron in a hydrogen atom.)

34 A conducting sphere of radius 2.5 cm has a charge of −76 nC. How far from its surface will the potential be −1000 V?

Summary of the equations for radial fields

Figure 10.25 shows the equations for the key concepts in radial fields. Refer to Figure 10.17 for a reminder of how the concepts are interconnected. Seven of these eight equations can be found in the *Physics data booklet*, but the equation for electric field strength is *not* included.

■ **Figure 10.25** Equations for radial gravitational fields (in blue) and radial electric fields (in red)

Force	**Potential energy**
$F_G = G\frac{Mm}{r^2}$ $\left(\text{or } F_G = G\frac{m_1 m_2}{r_2}\right)$ $F_E = k\frac{q_1 q_2}{r^2}$	$E_p = -\frac{GMm}{r}$ $E_p = \frac{kq_1 q_2}{r}$
Field	**Potential**
$g = \frac{GM}{r^2}$ $E = \frac{kq}{r^2}$	$V_g = -\frac{GM}{r}$ $V_e = \frac{kq}{r}$

■ Potential gradient and field strength

The magnitude of the gradient of any potential–distance graph for a radial field, ($\Delta(GM/r)/\Delta r$ or $\Delta(kq/r)/\Delta r$), equals the magnitude of the field strength at that point, but the following equations apply equally to any field:

$$g = -\frac{\Delta V_g}{\Delta r}$$

$$E = -\frac{\Delta V_e}{\Delta r}$$

These equations are given in the *Physics data booklet*.

Expressed in this way we can see that an alternative unit for electric field strength is V m^{-1} (equivalent to N C^{-1}).

$\frac{\Delta V}{\Delta r}$ is known as the **potential gradient**.

Field strength is equal in magnitude to the potential gradient.

As mentioned in Chapter 5, determining an electric field strength from measurements of force and charge ($E = F/q$) is experimentally difficult, but determining electric field strength from a potential gradient ($E = -\Delta V_e/\Delta r$) is much easier, especially if the field is uniform (see below), such that only a voltmeter and a ruler are needed.

Figure 10.26 shows an example of how the gravitational potential varies around a planet. The field can be found at any distance (for example at R) by determining the gradient of the graph at that distance.

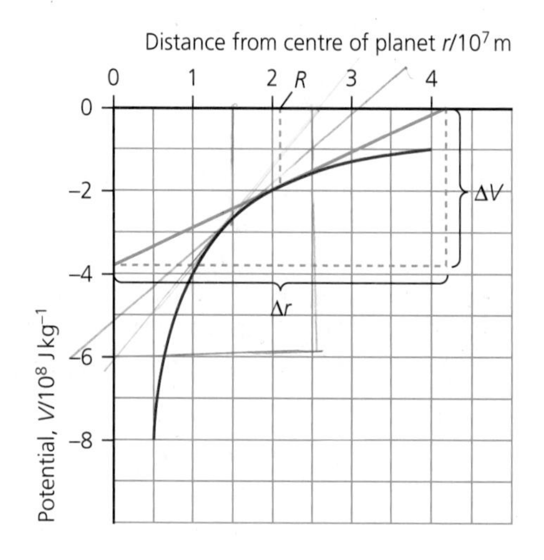

■ **Figure 10.26** Graph showing variation of gravitational potential around a planet

35 Consider Figure 10.26. Determine the gravitational field strength at a distance of 1.5×10^7 m from the centre of the planet.

36 Figure 10.27 shows how the electric potential varies with distance, r, from a certain point, P.

a Where is the electric field strength a minimum?

b Determine the field strength:

i 12 cm from P

ii 32 cm from P.

c Suggest what arrangement of charges might produce this variation in potential.

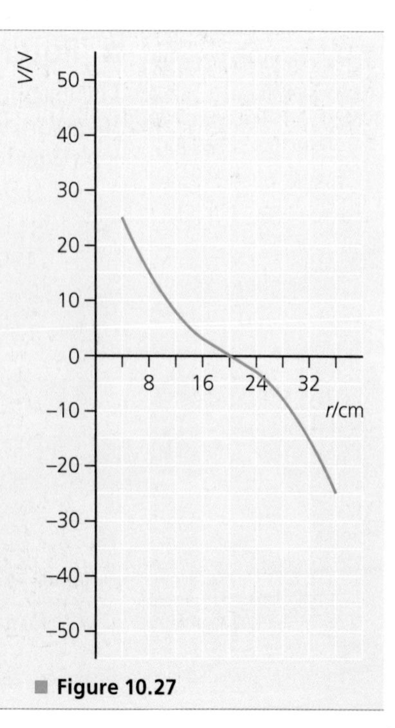

■ **Figure 10.27**

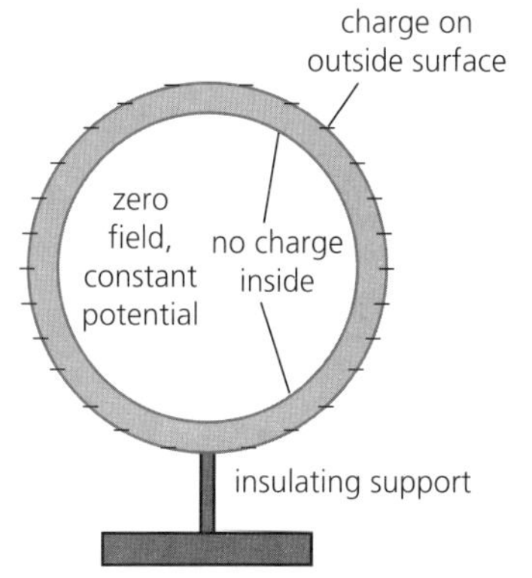

■ **Figure 10.28** Negatively charged hollow metallic sphere

Field and potential inside a hollow charged sphere

If a metallic hollow sphere is charged, there will be forces between the charges that make them separate as much as possible. This means that the charge will accumulate on the *outside* surface of the sphere (or any other shaped hollow conductor). Because the inside is uncharged, there will be zero electric field *inside* the sphere (Figure 10.28).

A zero electric field inside a conducting sphere means that the potential gradient is also zero, so that the potential is constant inside the sphere and throughout the conductor.

If there were a potential difference anywhere, charges would experience forces and move, so that any p.d. would then disappear.

Utilizations

Hollow conductors: Faraday's cage

■ **Figure 10.29** A Faraday cage, showing sparks on the outside, but with someone safe inside

If a *constant* electric field is applied to a metal conductor surrounding a space, free electrons will very quickly redistribute themselves on the outer surface depending on the strength and direction of the field, and the shape of the conductor. For a spherical conductor the charge distribution would be the same everywhere on the surface. The conductor does not need to be solid, so a mesh or 'cage' will work just as well. The inner surface will be unchanged and there will be no electric field or potential *difference* inside the cage. Figure 10.29 shows a dramatic example of a 'Faraday' cage.

The safest place to be in a lightning storm is inside a conductor, like a car or a building (that has a lightning conductor if it is in an exposed location). Electronic equipment can be protected by putting it inside a Faraday cage.

Similar ideas can be applied to the *oscillating* electric and magnetic fields of electromagnetic waves, but the situation is more complicated and the extent to which the space inside the cage remains free of electromagnetic radiation depends on a number of factors including the wavelength of the radiation, the size of any gaps in the cage and the material, thickness, size and shape of the cage. *Electromagnetic shielding* is used widely to protect important equipment from external electromagnetic waves, or to stop the emission of electromagnetic signals.

1 Investigate the extent to which a signal to, or from, a mobile phone, or the effectiveness of a remote control, are reduced by placing them inside metallic conductors, with or without holes (e.g. a *disconnected* microwave oven, mesh, aluminium foil…)

Uniform fields

Gravitational

Gravitational fields around planets, like the Earth, are radial in nature so a perfectly uniform gravitational field is not a realistic situation. However, for most purposes we can assume that in our everyday lives on the surface of Earth we move around in a constant gravitational field (Figure 10.5b). So we can use $g \approx 9.8\,\mathrm{N\,kg^{-1}}$ anywhere on Earth. Very accurate measurements of gravitational field strength will reveal slight variations in g because of a variety of factors, including the density of the nearby rocks.

Electric

Uniform electric fields are important and are easily created by connecting a potential difference across parallel conducting plates (Figure 10.9). The p.d. forces electrons off one plate and onto the other. This is discussed in much greater detail in Chapter 11. The charges spread evenly on the plates and this produces a uniform electric field between the plates. Beyond the limits of the plates the field strength reduces. This is sometimes described as the *edge effect*.

Figure 10.30 shows a charge q between parallel plates separated by a distance d. When a potential difference, V, is applied across the plates, the charge experiences a force F. The force will be constant anywhere between the plates (because the field is uniform) and the charge will accelerate along a field line.

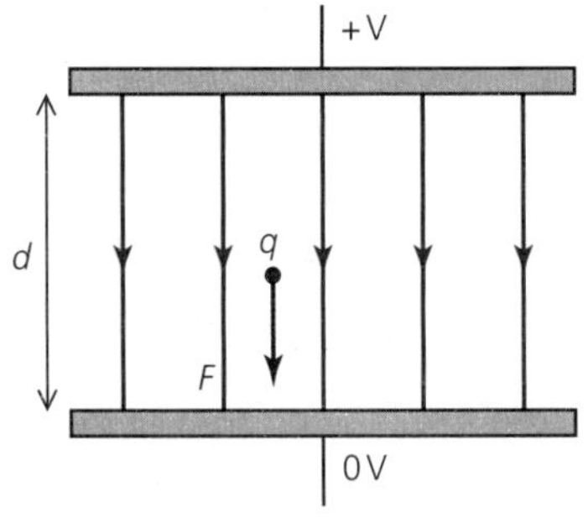

■ Figure 10.30 The force on a charge in a uniform electric field

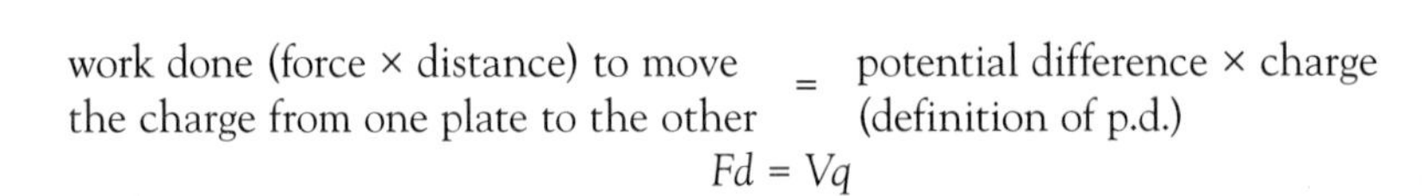

work done (force × distance) to move the charge from one plate to the other = potential difference × charge (definition of p.d.)

$$Fd = Vq$$

Or, $F/q = V/d$, and because electric field strength is defined as F/q:

strength of uniform electric field between parallel plates, $E = \dfrac{V}{d}$

This equation is *not* given in the *Physics data booklet* because it is a particular example of the more general statement field strength = potential gradient ($E = -\Delta V_e/\Delta r$).

Worked example

3 A beta-negative particle is moving perpendicular to the uniform electric field created by a p.d. of 5000 V across parallel metal plates separated by a vertical distance of 5 cm.

a What is the magnitude of the electric field between the plates?

b What is the magnitude of the force on the beta particle?

c Describe the motion of the particle between the plates.

d Calculate the acceleration of the beta particle between the plates (mass of electron = 9.1×10^{-31} kg).

a $E = \dfrac{V}{d} = \dfrac{5000}{0.050} = 1.0 \times 10^{5}\,\mathrm{V\,m^{-1}}$

b $F = Eq = (1.0 \times 10^{5}) \times (1.6 \times 10^{-19}) = 1.6 \times 10^{-14}\,\mathrm{N}$

c The horizontal speed of the electron will be unaffected, but at the same time it will be accelerated vertically by the force. These will combine to produce a parabolic path.

d $a = \dfrac{F}{m} = \dfrac{1.6 \times 10^{-14}\,\mathrm{N}}{9.1 \times 10^{-31}} = 1.8 \times 10^{16}\,\mathrm{m\,s^{-2}}$

37 We live in a *radial* gravitational field, so explain why we can assume that the gravitational field strength on Earth is approximately *constant*.

38 Suggest three reasons why the gravitational field strength varies slightly between different locations on the Earth's surface.

39 Explain why it is reasonable to assume that the electric field at the edges of a uniform field between parallel plates has about half the strength of the field at the centre.

40 a What voltage is needed across parallel metal plates, separated by 1.5 cm, to create an electric field strength between them of $3.8 \times 10^4 \, \text{V m}^{-1}$?
b What charge placed in this field will experience a force of $1.0 \times 10^{-3} \, \text{N}$?

41 If a uniform electric field of at least $2.9 \times 10^6 \, \text{V m}^{-1}$ is needed to cause electrical breakdown of dry air, estimate the voltage needed to cause a spark between parallel plates separated by 2 mm.

42 Consider the equipotential lines in the uniform electric field between the two parallel plates shown in Figure 10.31.
a Determine the electric field strength.
b If a charge of $+7.8 \times 10^{-7} \, \text{C}$ was placed at point A, what size force would it experience?
c What is the direction of that force?
d What is the potential difference between points A and B?
e How much work is done if the charge moves from A to B?
f What happens to the charge as it moves from A to B?

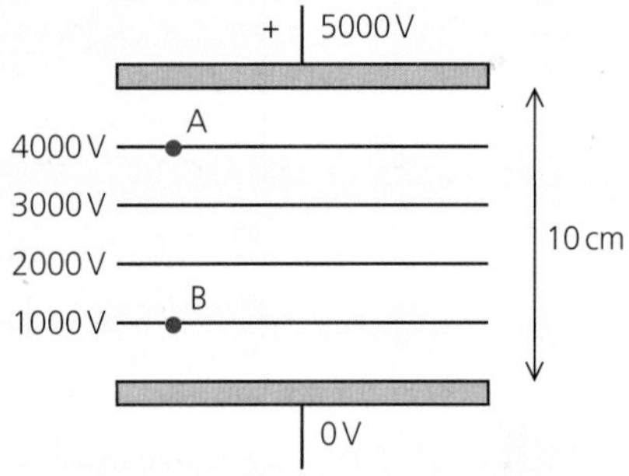

■ **Figure 10.31**

■ Using electric fields to produce beams of charged particles

Electric fields are widely used to accelerate charged particles such as electrons, protons and ions. Examples, many of which are included in this course, include:

- electron beams used in diffraction and scattering experiments
- oscilloscopes that use 'cathode rays'
- mass spectrometers
- high-energy beams used in nuclear physics experiments
- production of X-rays (see Option C).

As an example of controlling particle beams with electric fields, consider Figure 10.32, which shows a type of electron beam deflection tube that is commonly used to demonstrate the production and properties of electron beams. Electrons are emitted from the cathode when it is heated by the hot wire next to it (a process known as *thermionic emission*). The electrons are accelerated in the electric field created by a high p.d. connected between the cathode and the positive anode, and they form a beam as they travel at high speed across the *evacuated* tube. When the beam of electrons strikes the *fluorescent* screen some of their kinetic energy is transferred to a spot of visible light.

■ **Figure 10.32** Electron deflection tube

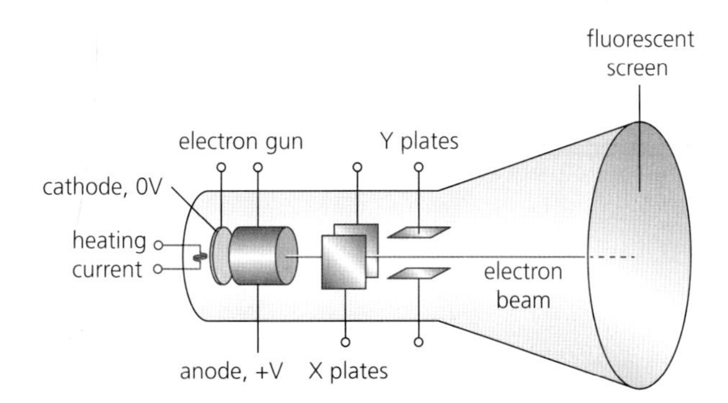

If p.d.s are applied to the X-plates and/or the Y-plates the electrons will experiences forces as they pass through the two uniform electric fields. By varying the p.d.s the spot can be moved anywhere on the screen.

Worked example

4 **a** 2500 V is used across an 'electron gun' to produce a beam of electrons. What is the kinetic energy of each electron emerging from the gun?

b What is the maximum speed of the electrons?

a 2500 eV, assuming that it has zero kinetic energy to begin with, or:

energy = p.d. × charge = $2500 \times (1.6 \times 10^{-19}) = 4.0 \times 10^{-16}$ J

b kinetic energy = $\frac{1}{2}mv^2$

$4.0 \times 10^{-16} = 0.5 \times (9.1 \times 10^{-31}) \times v^2$

$v = 3.0 \times 10^7\,\text{m s}^{-1}$

Combination of fields and potentials

First consider a gravitational example: Figure 10.33 shows two planets (or a planet and a moon), A and B, and the potentials and fields that they individually produce at a point P.

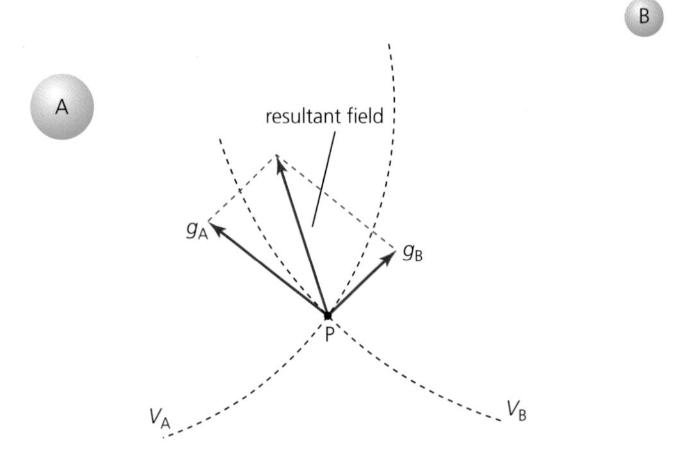

Figure 10.33 Combining gravitational fields

Field strength is a vector quantity, so vector addition is needed to determine the resultant of the two fields.

In Figure 10.33 this has been done by drawing a parallelogram, with the diagonal representing the resultant gravitational field strength in magnitude and direction.

Potential is a scalar quantity so the resultant potential at a point is just the sum of the individual potentials.

$V_{combined} = V_A + V_B$. For example, if the potential at P due to A was $-8\,\text{MJ kg}^{-1}$ and the potential at P due to B was $-3\,\text{MJ kg}^{-1}$, then the overall potential would be $-11\,\text{MJ kg}^{-1}$. To take 1 kg from P to infinity would need 8 MJ if only A was present, 3 MJ if only B was present and 11 MJ if they were both present.

Exactly the same principles can be applied to electric fields, but remember that electric potential can be either positive or negative. Consider Figure 10.34.

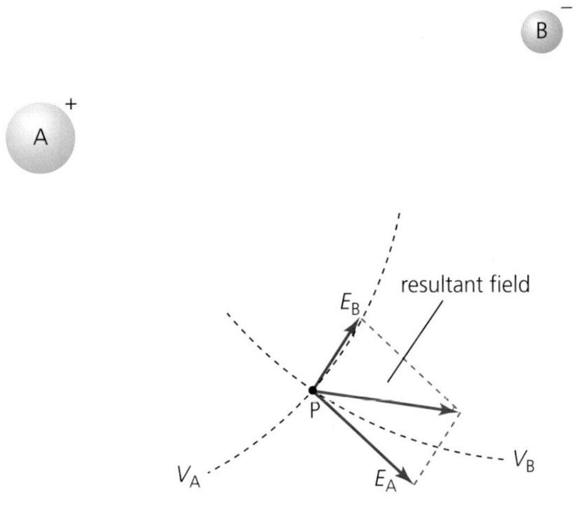

Figure 10.34 Combining electric fields

$V_{combined} = V_A + V_B$. For example, if the potential at P due to A was +8 V and the potential at P due to B was −3 V, then the overall potential would be +5 V. A charge of +1 C moving from P to infinity would *gain* 8 J if only A was present, but 3 J would be *needed* to move 1 C to infinity if only B was present. If both A and B were present, +1 C would *gain* 5 J moving from P to infinity.

43 A singly ionized atom of mass 23.0 u is accelerated by 4700 V in a mass spectrometer.
  a What is the charge on the ion?
  b What is its maximum kinetic energy (J)?
  c What is the maximum speed of the ion?

44 What are the potentials at points R, S and T as shown in Figure 10.35? (Assume that the drawing is full size.)

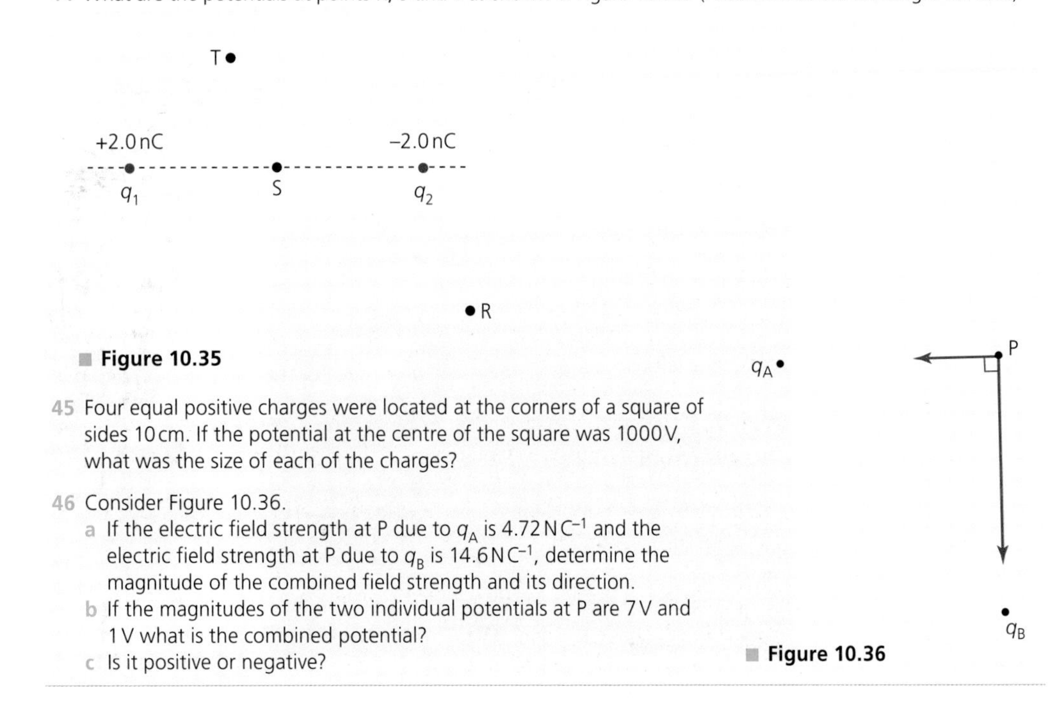

Figure 10.35

45 Four equal positive charges were located at the corners of a square of sides 10 cm. If the potential at the centre of the square was 1000 V, what was the size of each of the charges?

46 Consider Figure 10.36.
  a If the electric field strength at P due to q_A is 4.72 N C^{-1} and the electric field strength at P due to q_B is 14.6 N C^{-1}, determine the magnitude of the combined field strength and its direction.
  b If the magnitudes of the two individual potentials at P are 7 V and 1 V what is the combined potential?
  c Is it positive or negative?

Figure 10.36

47 Figure 10.37 shows how the gravitational potential, V, between a planet and its moon varies with the separation of their centres, r.

a Estimate the radius of the planet.

b Estimate the mass of the planet.

c At what distance from the centre of the planet is the gravitational field zero?

d Determine the magnitude of the gravitational field strength 1.0×10^8 m from the centre of the moon.

48 a Use a spreadsheet to calculate the combined potential of the Earth and Moon at points along a line joining their centres (research the necessary data).

b Plot a graph of your results. It should be similar in shape to Figure 10.37.

c How much energy would be needed to move a spacecraft of mass 3400 kg from the surface of the Earth to the surface of the Moon?

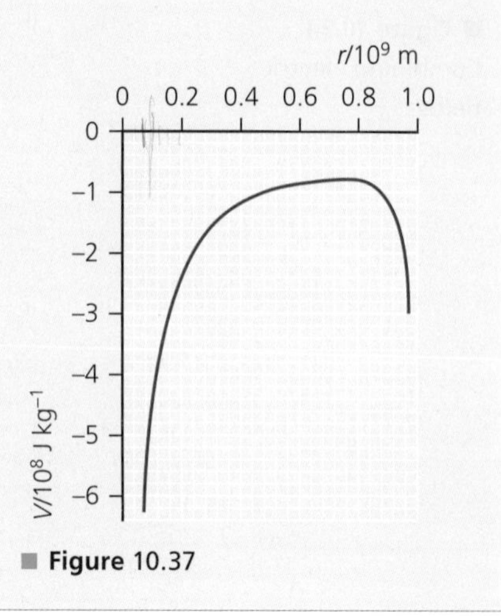

■ **Figure 10.37**

■ Spacecraft in gravitational fields

In this section we will first consider how much energy needs to be transferred to a spacecraft in order for it to travel a very great distance away from the Earth (or the Moon or another planet), and then how much energy must be transferred to a satellite to put it into orbit.

Utilizations

Gravitational fields: launching satellites

To send something up into space so that it stays up and does not come down again was a dream of scientists for centuries. The first artificial satellite was sent into orbit by Russia on 4 October 1957 and this truly historic event has rightly been called the start of the 'space age'. The satellite had a mass of 84 kg and it orbited the Earth at an approximate average height of 570 km and speed of 29 000 km h^{-1}. It was called 'Sputnik 1' (Figure 10.38). Three months after launching, Sputnik 1 burned up as it re-entered the Earth's atmosphere. Soon afterwards, the launch of Sputnik 2 carried the first animal into space, a dog called Laika (see Figure 10.39). She survived the launch but died in orbit a few hours later.

■ **Figure 10.38** Sputnik 1, the first satellite in space

■ **Figure 10.39** Laika: the first animal in space

In simple terms, the major challenge was to transfer sufficient energy to the Sputnik satellite to get it up above the Earth's atmosphere and get it moving in the correct direction at the right speed.

1 Calculate the gravitational potential energy of the Sputnik satellite on the Earth's surface and in orbit.

2 a Calculate the gravitational potential energy that had to be gained by the satellite for it to reach a height of 570 km.

  b Calculate the satellite's kinetic energy in orbit, and therefore its total increase in energy compared with being stationary on the Earth's surface.

3 Find out what fuel is used for rocket launches and the reasons why it is chosen.

4 Find out which countries currently have working satellites in orbit.

5 Find out if there is any country or organization that controls the use of the space around the Earth.

Your answer to question **2b** should be about 4×10^9 J. Much more energy than this, however, needs to be transferred because a large amount of energy will be dissipated due to air resistance and because the engines are inefficient. It is not possible to simply project a satellite into space, so the launch has to be powered by rocket engines. One of the requirements of rocket fuel is that a large amount of energy can be transferred from a relatively small mass of fuel. For example, if 35 MJ is the theoretical maximum amount of energy that can be released in the combustion of 1 kg of fuel and oxidant, then more than 100 kg of fuel and oxidant would be needed to supply 4×10^9 J. Of course, more energy would then be needed to raise the fuel itself. These calculations totally ignore the mass of the engine and the tanks needed for the fuel and oxidant. The total mass of the rocket reduces considerably as the fuel and oxidant are burned and, when they are empty, the tanks can be discarded.

A typical useful 'lifetime' of a satellite is about 10 years. Currently there are about 3000 satellites in use around the Earth. Russia and the USA have the most, but many other countries, including China (Figure 10.40), have launched their own satellites. Since 1957 about 25 000 objects have been recorded in orbit. Most have burned up after a few years as their orbits re-entered the Earth's upper atmosphere, but many are still in orbit and commonly described as 'space junk'.

■ **Figure 10.40** The launch of a satellite in China

Escape speed

In theory an object can be *projected* (not powered) upwards in such a way that it could continue to move away from the Earth forever. For this to be possible the object would need to be given a *very* high speed. To calculate that speed we need to consider energies. In general, the initial kinetic energy transferred to an object will be transferred to gravitational energy, and also dissipated due to air resistance in the Earth's atmosphere.

But if we assume that the effects of air resistance are negligible, we can calculate the *minimum* theoretical speed that a projectile of mass m needs in order to 'escape' from a planet of mass M. This is called its **escape speed,** v_{esc}. It can be calculated as follows:

initial kinetic energy = change in gravitational potential energy between location and infinity

$$\frac{1}{2}mv_{esc}^2 = 0 - \left(-\frac{GMm}{r}\right)$$

$$V_{esc} = \sqrt{\frac{2GM}{r}}$$

This equation is given in the *Physics data booklet*. Note that the theoretical value for the escape speed does not depend on the mass or the direction of motion.

Worked example

5 Calculate the escape speed of a projectile launched from the Earth's surface. (Ignore any effects due to the Moon.)

$$v_{esc} = \sqrt{\frac{2GM}{r}}$$

$$v_{esc} = \sqrt{\frac{2 \times 6.67 \times 10^{-11} \times 5.97 \times 10^{24}}{6.37 \times 10^6}}$$

$$v_{esc} = 1.12 \times 10^4\,\text{m s}^{-1}$$

This means that, *ignoring the effects of air resistance*, any mass projected in any direction from the Earth's surface with a speed of 11.2 km s^{-1} will never come back down to Earth. This speed is theoretically enough to take it to infinity (also assuming it does not pass near any other masses), where its speed would reduce to zero. If it was launched with a faster speed, the mass would still be moving when it reached infinity.

Of course, it is not sensible to assume that there is no air resistance for an object moving at high speed through the Earth's atmosphere, so this calculated value is just a theoretical minimum. However, on a planet or a moon that has no atmosphere, or if the starting point was already above the Earth's atmosphere, the calculations are valid. Such calculations also ignore any possible effects of other gravitational fields from other planets, moons and stars.

It is possible to give an object an initial speed of 11.2 km s^{-1}, but it is not possible to project any object with the much higher speed and kinetic energy needed to overcome all the resistive forces of the atmosphere. Spacecraft put into orbit around the Earth are *not* projectiles; they are transported away from the Earth's surface by continuous forces from rocket engines that act over extended periods of time.

49 a Calculate the escape speed from the Moon's surface (mass = 7.4×10^{22} kg and radius = 1.7×10^6 m).
b Explain why the actual escape speed is a little higher than your answer.
c Explain why a space vehicle launched from the Moon could reach the Earth if it was given a smaller initial speed.

50 Calculate the escape speed from an orbit 350 km above the Earth's surface.

51 How far away from Earth does the escape speed reduce to 1% of its value on the Earth's surface?

52 a The mass of the Sun is 2.0×10^{30} kg. Calculate how small its radius would have to be if the escape speed from its 'surface' was to be equal to the speed of light (3×10^8 m s^{-1}), so that it might become a *black hole*.
b Compare your answer to the actual radius of the Sun.

53 Consider the combined effect of the Earth's and the Moon's gravitational fields on the Earth's surface.
  a How much higher than $1.12 \times 10^4\,m\,s^{-1}$ is the escape speed from the Earth?
  b Can the effect of the Moon be ignored?

54 Calculate the escape speed from a planet of radius $2.2 \times 10^3\,km$ if the gravitational field strength on its surface is $3.9\,N\,kg^{-1}$.

Additional Perspectives

Atmospheres

In a gas in a container on the Earth's surface we would normally assume that the molecules are distributed randomly and move with an average speed represented by the absolute temperature (Chapter 3). For example, at 300 K nitrogen molecules have an average speed of about $500\,m\,s^{-1}$. However, if the container was really large we would also need to consider the effect of gravity, which acts on gas molecules in the same way as it acts on everything else.

As the molecules move upwards they slow down and, because of the range of molecular speeds in all gases, there will be fewer molecules per unit volume at greater heights. This results in a decrease in the macroscopic properties of the gas: pressure, density and temperature. Remember that molecules continually collide with each other, resulting in random changes of individual kinetic energies and speeds.

The Earth's atmosphere is not an isolated system and, although the principles described above are valid, the situation is actually much more complicated because energy is continually being transferred to the atmosphere from the Sun and the Earth. Energy is mainly transferred to the air from the Earth's surface, so this is where we would expect the air to be warmest and is why it gets colder when we go up a mountain. Such temperature changes cause changes in air pressure and density, setting up local and worldwide convection currents.

■ **Figure 10.41** The Moon has no significant atmosphere, so the sky is black

A very small percentage of molecules that reach the upper atmosphere can gain a speed higher than the escape speed, so they leave the atmosphere permanently. However, the geological and biological processes that created the Earth's atmosphere are still occurring and the gases are replenished.

Because of its lighter mass, the escape speed from the Moon is much slower than from the Earth. This means that, over a long period of time, almost all the gas molecules that were present in the past have escaped and there are no processes to replace them. As a result, the Moon has no significant atmosphere and that is why the sky seen from the Moon is black – there are no gas molecules to scatter the light (Figure 10.41).

1 a Show that molecules moving vertically upwards at $500\,m\,s^{-1}$ would rise to a height of about 12 km if they did not collide with other molecules.
  b Give reasons why many air molecules are found at much greater heights.

2 Find out where the Earth's atmosphere came from, and whether there could be an atmosphere on the Moon if it was larger.

3 Do Mercury, Venus and Mars have atmospheres? Are there any other planets or moons in the solar system that have atmospheres?

Orbital motion, orbital speed and orbital energy

Speed of an orbiting satellite

We have seen in Chapter 6 that a satellite in a circular orbit around a planet must have the right orbital speed, v_{orbit}, for a given height. This speed can be calculated by equating centripetal acceleration to gravitational acceleration:

$$\frac{v_{orbit}^2}{r} = g = \frac{GM}{r^2}$$

so that

$$v_{orbit} = \sqrt{gr}$$

and

$$v_{orbit} = \sqrt{\frac{GM}{r}}$$

This equation is given in the *Physics data booklet.*

Because:

$$v_{orbit} = \frac{2\pi r}{T}$$

$$\left(\frac{2\pi r}{T}\right)^2 = \frac{GM}{r}$$

Rearranging this gives an equation connecting radius directly to time period:

$$\frac{r^3}{T^2} = \frac{GM}{4\pi^2}$$

(This useful equation is a form of Kepler's third law, but it is *not* required knowledge in the IB physics course.)

Assume that all orbits are circular.

55 The radius of the Moon is 1740 km and the gravitational field strength on its surface is $1.61\,N\,kg^{-1}$.
  a What would be the necessary orbital speed for a satellite around the Moon at a height of 1740 km above the surface?
  b How many hours would such a satellite take to complete one orbit?

56 The mass of Jupiter is 1.90×10^{27} kg. Ganymede is one of its moons (satellites) at an average distance away (centre to centre) of 1.07×10^{6} km. What is the time period of its orbit in days?

57 At what height above the Earth's surface would a satellite orbiting the Earth need to have a speed of exactly $5\,km\,s^{-1}$? (mass of Earth = 5.97×10^{24} kg; radius of Earth = 6.37×10^{6} m)

58 a The average distance between the centres of the Earth and the Sun is 1.50×10^{11} m. Treating the Earth as a satellite of the Sun, estimate its time period in seconds. (mass of Sun = 1.99×10^{30} kg)
  b Suggest a reason why the answer is not exactly 1 year.

59 Use a spreadsheet to calculate orbital speeds for satellites orbiting the Earth at different heights (up to 40 000 km). Draw a graph of your results.

Energy of an orbiting satellite

We know that when a ball is thrown upwards through the air, it will gain gravitational energy and lose kinetic energy as it moves away from the Earth. The process is reversed when it moves closer to the Earth. If there were no air resistance, the *total* energy of the ball would remain constant. The same principle applies to satellites, moons, planets and comets in their orbits.

To determine the total energy, E_T, of a satellite, we add its kinetic energy, E_K, to its gravitational potential energy, E_P:

$$E_T = E_K + E_P$$

Remember that the total energy and the gravitational potential energy must both be negative because an orbiting satellite does not have enough energy to escape from the Earth's gravitational field. For a satellite of mass m, orbiting a planet of mass M with a speed v at a distance r from its centre, we know that:

$$E_P = -\frac{GMm}{r}$$

So that:

$$E_T = \frac{1}{2}mv^2 + \left(-\frac{GMm}{r}\right)$$

But we also know that:

$$v^2 = \frac{GM}{r}$$

So:

$$E_K = \frac{1}{2}mv^2 = \frac{1}{2}m\frac{GM}{r}$$

Which gives:

$$E_K = \frac{1}{2}\frac{GMm}{r}$$

This leads to:

$$E_T = \frac{1}{2}\frac{GMm}{r} + \left(-\frac{GMm}{r}\right)$$

$$E_T = -\frac{1}{2}\frac{GMm}{r}$$

These equations are *not* given in the *Physics data booklet*.

They show us that the total energy of a satellite in a stable orbit is always equal in magnitude to its kinetic energy, but the kinetic energy is positive, while the total energy is negative. The total energy is equal to half the gravitational potential energy. This is shown graphically in Figure 10.42.

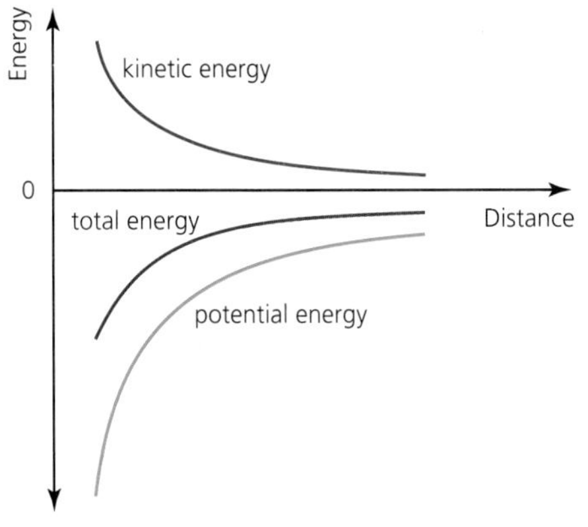

■ **Figure 10.42** Energies of a satellite

In order to 'escape' from its orbit (not from the ground) a satellite needs to have kinetic energy equal to its gravitational potential energy at that height (or more). It already has the necessary kinetic energy for its orbital motion, so it would need to gain a further equal amount of kinetic energy; its kinetic energy needs to be doubled in order to escape from the planet.

If a satellite moving in a circular orbit (at a relatively low height) experiences any air resistance, its speed at that height will decrease and the gravitational force on it will then be bigger than the centripetal force needed for that height and speed. If the satellite has no way of controlling its motion, this will cause a change of height, moving it closer to Earth, and therefore it will gain speed and kinetic energy. The total energy and the gravitational potential energy will decrease as energy is dissipated as thermal energy. As the satellite moves closer to Earth, the air resistance will increase further and its downwards trajectory will become steeper, resulting in higher speed and greater energy dissipation. Because the air resistance will be very small at great heights this process may take a long time, but low-height satellites have occasionally 'crashed' to Earth, although most of them 'burn up' in the atmosphere.

60 a What happens to the total energy of a satellite if it encounters some air resistance?
b Use the equation for total energy of a satellite to explain what must happen to a satellite that is losing energy.
c Explain why the speed of the satellite will increase and why these effects increase as the satellite gets lower.

61 a Explain what you think 'burn up' means in the paragraph above.
b Research an occasion when a satellite actually crashed on the Earth's surface and find out what happened.

62 A satellite of mass 820 kg is orbiting at a height of 320 km above the Earth's surface. Calculate:
a its gravitational potential energy
b its kinetic energy
c its total energy (Earth's radius = 6.4×10^6 m, Earth's mass = 6.0×10^{24} kg)
d the speed it would need to have in order to escape from the Earth.

Two different types of orbit

Satellites can be put into orbit at any desired height above the Earth's surface, assuming that they are high enough to avoid air resistance (above approximately 200 km). Orbits that are closer to the Earth have obvious advantages, especially if the Earth's surface is being monitored for some reason. But, apart from height, the other important (and related) factor is the period. The orientation of the orbit compared with the Earth's axis and whether it is circular or elliptical are also critical.

Polar orbits

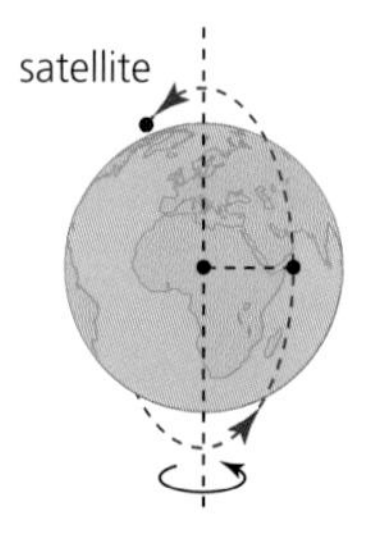

Figure 10.43 Polar-orbiting satellite

Many satellites have orbits that pass approximately over both poles of the Earth at heights of up to about 2000 km (Figure 10.43). While the orbit remains in the same plane, the Earth rotates so that the satellite passes over different parts of the planet on each orbit. These satellites make many orbits every day.

Worked example

6 **a** How many orbits of the Earth does a polar-orbiting satellite at a height of 400 km make every day?
b Suggest a use for such a satellite.

a radius of orbit = radius of Earth (6.37×10^3 km) + 400 = 6.77×10^3 km

$$v_{orbit} = \sqrt{\frac{GM}{r}} = \sqrt{\frac{(6.67 \times 10^{-11}) \times (5.97 \times 10^{24})}{6.77 \times 10^6}}$$

$v_{orbit} = 7670\,\text{m s}^{-1}$

$$\text{period, } T = \frac{2\pi r}{v_{orbit}} = \frac{2 \times \pi \times 6.77 \times 10^6}{7670} = 5546\,\text{s, or } 1.54\,\text{h}$$

$$\text{number of orbits in one day} = \frac{24}{1.54} = 15.6$$

b There are many: mapping, monitoring the weather and spying for example.

Geostationary orbits

A geostationary satellite is one that remains above the same point on the Earth's surface. This is only possible if the period of the satellite is the same as the period of rotation of the Earth (23 h 56 min). As we have seen, this requires that a geostationary satellite is at exactly the right height. Using:

$$v_{orbit} = \sqrt{\frac{GM}{r}}$$

and

$$T = \frac{2\pi r}{v_{orbit}} \text{ (or Kepler's third law)}$$

shows that the required orbital radius is 4.2×10^7 m.

Any satellite at this height will have the same period as the rotation of the Earth and they are described as *geosynchronous*. However, to be *geostationary* satellites must be made to orbit in the same plane as the equator. Both orbits shown in Figure 10.44 are geosynchronous, but only the orbit in **a** is geostationary.

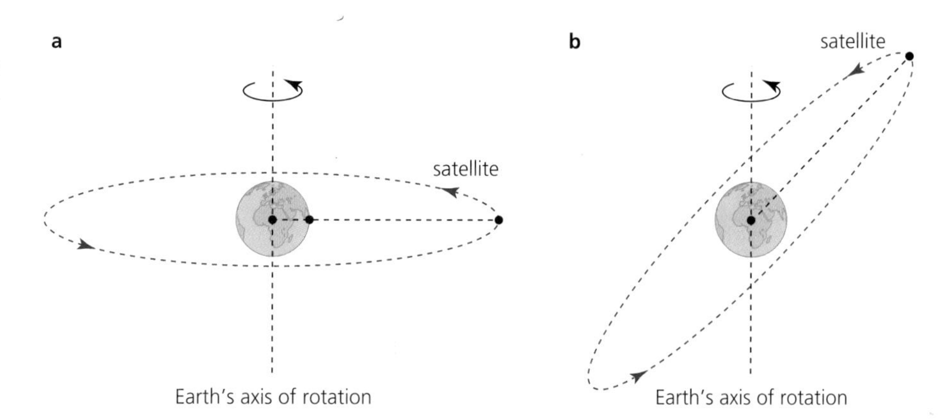

■ **Figure 10.44** Geostationary orbits must be in the plane of the equator

Geostationary satellites are especially useful because transmitting and receiving aerials on Earth can remain with the same orientations, pointing at the same place in the sky. There is no need to 'track' the satellite. This makes them especially useful for communications, TV broadcasts etc.

63 Determine the necessary height above the Earth's surface for a geostationary orbit, in kilometres to three significant figures.

64 The Moon rotates once on its axis every 27.3 days.
  a Explain why, from Earth, we always see the same side of the Moon.
  b What would be the radius of a lunar-stationary orbit? (mass of Moon = 7.35×10^{22} kg)

65 Explain why the period of the Earth's rotation is not exactly 24 h.

66 Geostationary satellites are far above the Earth's surface.
  a Does this mean that there is a significant delay when communicating through them?
  b Discuss whether the aerials need to be significantly bigger than when using closer satellites.

67 Make a list of the different uses of satellites.

Utilizations

Orbits: global positioning systems (GPS)

About 30 satellites orbiting the Earth at an approximate height of 20 000 km are needed for a GPS system. The most used system is maintained by the US government, but it is freely available to anyone. Russia has its own system and China is also developing its own. Each American GPS satellite continually transmits microwave messages of frequency 1.6 GHz (wavelength 19 cm), with the exact time at which each is transmitted. Each satellite orbits the Earth about twice a day.

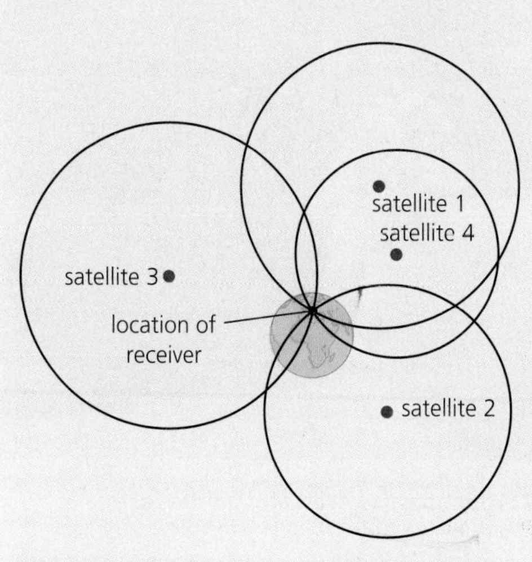

■ **Figure 10.45** Intersecting spheres around GPS satellites

When a GPS receiver detects a signal it compares the time it was received to the time it was transmitted, and then calculates the time delay (transit time) and the distance to the satellite (using the speed of electromagnetic waves). Obviously, the receiver must contain a very accurate clock. If similar calculations are made using the signals from another two, or more, satellites, the position of the receiver can be determined. Figure 10.45 shows a two-dimensional representation of the three-dimensional intersection of four spheres. All the satellites are at similar heights above the Earth's surface but their distances from any given receiver are constantly changing.

With the very best (military) equipment it is possible to locate a receiver in terms of latitude, longitude and altitude to within a few centimetres. However, international regulations limit the accuracy of civilian receivers to several metres.

If a moving receiver re-calculates its position after a known short interval of time, it can calculate its speed and direction of motion. All this information is commonly displayed on a two-dimensional electronic map of the area, but GPS can also be used to determine altitudes (in planes, for example).

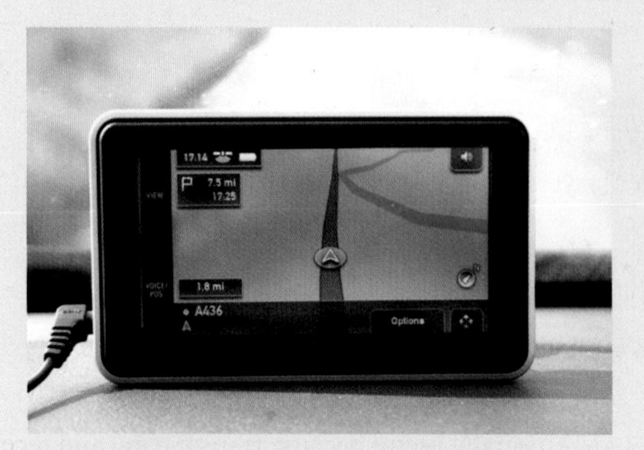

■ **Figure 10.46** A GPS system in a car

The European Union is in the process of developing its own GPS system. It is called Galileo and should be fully operational in the next few years. The intention is that the Galileo system will be for civilian use only and that Europe will have its own, independent system.

1 Suggest why using the signals from four or more satellites is better than just using three.

2 If a system is to be accurate to within 5 cm, approximately what level of accuracy is needed for the clock in the receiver?

3 Why do you think that governments may want to limit the accuracy of the GPS systems that are available to the general public?

4 Why is it important that the signals travel from the satellite to the receiver in a straight line?

Weightlessness

Because weight is the force of gravity on a mass, to be truly weightless a mass would have to be in a place where there was no significant gravitational field, such as in deep space – far, far away from the gravitational effects of any stars or planets.

Alternatively, a mass could appear weightless because it was at a point where the gravitational forces from two or more fields cancelled each other out, such as a point somewhere between the Earth and the Moon.

We are all familiar with videos of astronauts moving around freely inside, or outside, satellites orbiting the Earth (Figure 10.47). They are often described as being 'weightless', but this cannot be true because the gravitational field strength at the height of a satellite's orbit is not much lower than it is on the Earth's surface. We need to understand how and why an object can seem to be weightless when it is in a gravitational field.

■ **Figure 10.47** 'Weightless' in orbit around the Earth

We cannot feel or sense our own weight directly, but we are aware of our weight because of the contact force from the surface underneath us pushing upwards. If that contact is removed we feel weightless, although we may be more concerned with the fact that we are free falling! In an elevator, on a fairground ride or in a vehicle on a bumpy ride, the contact force between us and the supporting surface changes during any vertical change of speed. As a consequence we may have a sense of increased or decreased weight; we may even feel brief 'weightlessness' if, for a short time, there is little or no contact at all between us and the surface.

A satellite is effectively accelerating towards the centre of the Earth (while moving sideways at the same time). Astronauts in the satellite will be accelerating in exactly the same way as the satellite and there will be no need for contact between them and the satellite, so they will feel 'weightless' (Figure 10.48).

■ **Figure 10.48** Losing contact with the floor

a

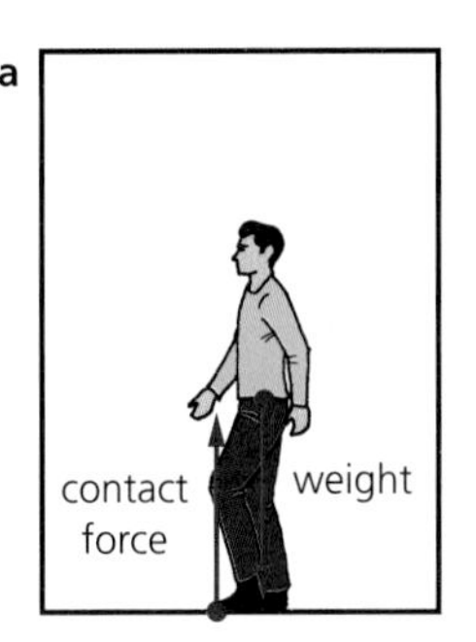

constant velocity, zero acceleration (including stationary)

No resultant force on person

b

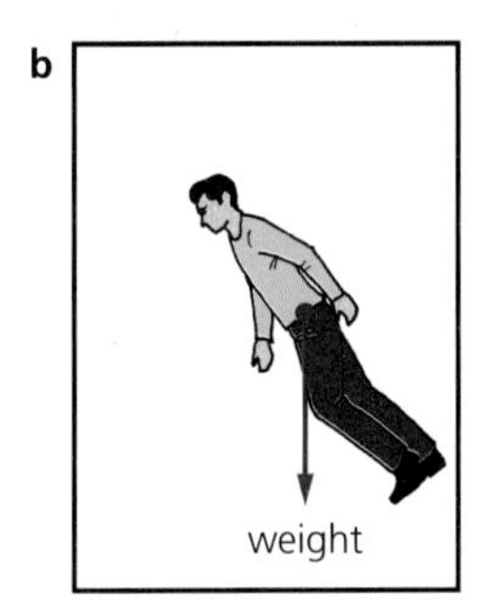

acceleration, person and container in free-fall

Resultant force downwards

Training for astronauts includes simulating weightlessness aboard flights in aircraft that are following the parabolic trajectory that would be followed by a freely moving projectile (without air resistance).

Examination questions – a selection

Paper 1 IB questions and IB style questions

Q1 A satellite is in a circular orbit around the Earth. When it is moved to an orbit that is closer to the Earth's surface, which of the following describes the energy changes that take place?

A gravitational potential energy increases, kinetic energy increases
B gravitational potential energy increases, kinetic energy decreases
C gravitational potential energy decreases, kinetic energy increases
D gravitational potential energy decreases, kinetic energy decreases

Q2 The diagram represents equipotential lines of a gravitational field.

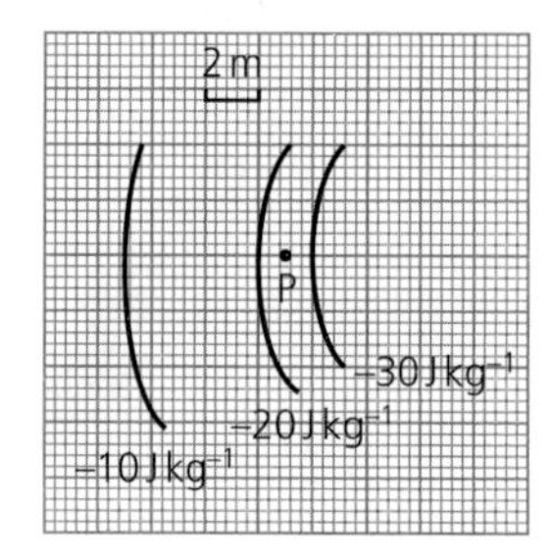

Which of the following is the direction and strength of the field at point P?

	Direction	Strength
A	←	$5.0\,N\,kg^{-1}$
B	→	$5.0\,N\,kg^{-1}$
C	←	$13\,N\,kg^{-1}$
D	→	$13\,N\,kg^{-1}$

Q3 Which of the following is a correct description of the electric field strength and electric potential around a single point charge?

A Field strength is a vector quantity; potential is a vector quantity.
B Field strength is a vector quantity; potential is a scalar quantity.
C Field strength is a scalar quantity; potential is a vector quantity.
D Field strength is a scalar quantity; potential is a scalar quantity.

Q4 Which of the following is the correct expression for the minimum kinetic energy needed by a mass m in order to escape from the gravitational attraction of a planet of mass M and radius R?

A $\frac{GM}{r}$

B $\frac{GM}{2R}$

C $\frac{GMm}{R}$

D $\frac{GM}{R^2}$

Q5 Four charges are arranged at the four corners of a square, as shown in the diagram.

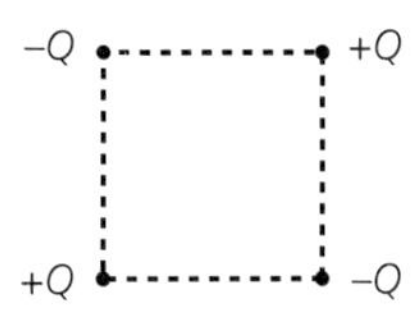

At which position is the total electric potential zero?

A nowhere
B at the centre of the square
C midway between the two negative charges
D midway between the two positive charges

Q6 The gravitational potential on the surface of a planet is V. What is the potential on the surface of another planet of the same average density, but with half the radius?

A $\frac{V}{4}$
B $\frac{V}{2}$
C $2V$
D $4V$

Q7 The two graphs below represent the variation with distance d, for $d = r$ to $d = 2r$, of the electric field and the electric potential around an isolated point charge.

Graph 1:

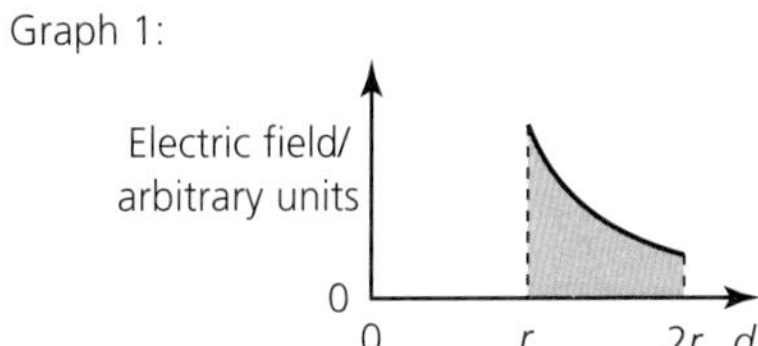

Graph 2:

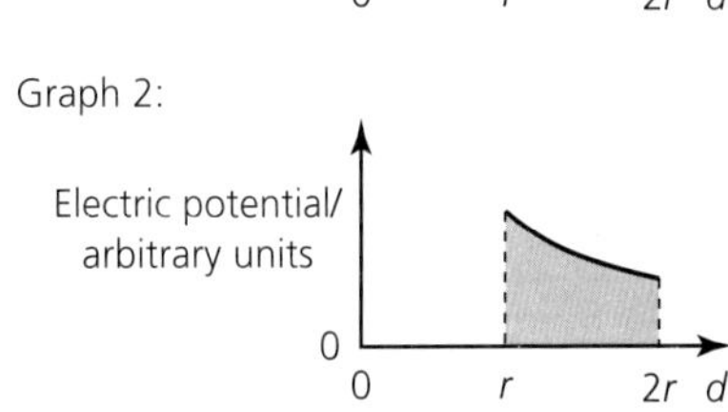

The work done by an external force in moving a test charge $+q$ from $d = 2r$ to $d = r$ is equal to q multiplied by:

A the shaded area under graph 1
B the shaded area under graph 2
C the average value of the electric field
D the average value of the electric potential.

Q8 Two satellites, P and Q, have been placed in circular orbits around the Earth. The time period of satellite P is 2 hours and the time period of satellite Q is 16 hours. Which of the following is the correct ratio of radius of orbit of satellite Q to radius of orbit of satellite P?

A 64
B 16
C 4
D 2

Paper 2 IB questions and IB style questions

Q1 **a** Define *gravitational* potential at a point in a gravitational field. (3)

b The graph below shows the variation with distance R from the centre of a planet of gravitational potential V. The radius R_0 of the planet = 5.0×10^6 m. Values of V are not shown for $R < R_0$.

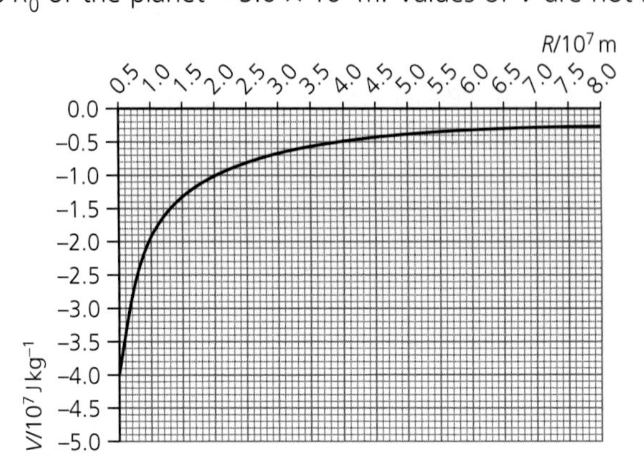

Use the graph to determine the magnitude of the gravitational field strength at the surface of the planet. (3)

c A satellite of mass 3.2×10^3 kg is launched from the surface of the planet. Use the graph to determine the minimum launch speed that the satellite must have in order to reach a height of 2.0×10^7 m above the surface of the planet. (You may assume that it reaches its maximum speed immediately after launch.) (4)

Q2 **a** Electric fields and magnetic fields may be represented by lines of force. The diagram below shows some lines of force.

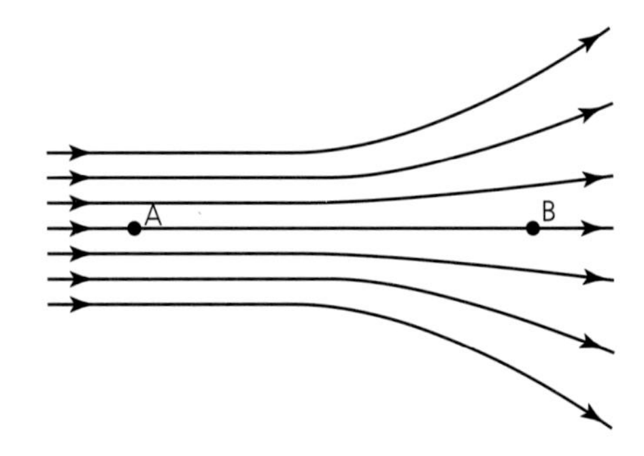

i State whether the field strength at A and B is constant, increasing or decreasing when measured in the direction from A towards B. (2)

ii Explain why field lines can never touch or cross. (2)

b The diagram below shows two insulated metal spheres. Each sphere has the same magnitude of positive charge.

Copy the diagram and in the shaded area between the spheres draw the electric field pattern due to the two spheres. (3)

Q3 a Explain why gravitational potentials (and potential energies) are always given negative values. (2)

b The electric potential 1.0 m at a point away from the surface of a hollow charged sphere of radius 10 cm is +250 V. Determine the charge on the sphere. (2)

c What is the potential on the surface of the sphere? (1)

d What is the potential at the centre of the sphere? (1)

11 Electromagnetic induction

ESSENTIAL IDEAS

- The majority of electricity generated throughout the world is generated by machines that were designed to operate using the principles of electromagnetic induction.
- Generation and transmission of alternating current (ac) electricity has transformed the world.
- Capacitors can be used to store electrical energy for later use.

11.1 Electromagnetic induction – *the majority of electricity generated throughout the world is generated by machines that were designed to operate using the principles of electromagnetic induction*

Electromotive force (emf)

The concept of electromotive force (emf) was introduced in Chapter 5 in connection with cells and batteries. In this chapter we will explain the origin of the emfs supplied to circuits from electric generators.

Inducing an emf by movement

Whenever a conductor moves across a magnetic field, or a magnetic field moves across a conductor, an emf may be *induced*. This effect is called **electromagnetic induction**. (Induction in this respect means 'the act of making something happen'.) Although these two effects are essentially the same, we will begin by describing them separately.

Electromagnetic induction by moving a conductor across a magnetic field

Figure 11.1 shows a simple experiment in which an emf is induced when a conductor (a metal wire) is moved *across* a magnetic field. The induced emf is detected because it makes a small current flow through a circuit containing a sensitive ammeter, called a **galvanometer**.

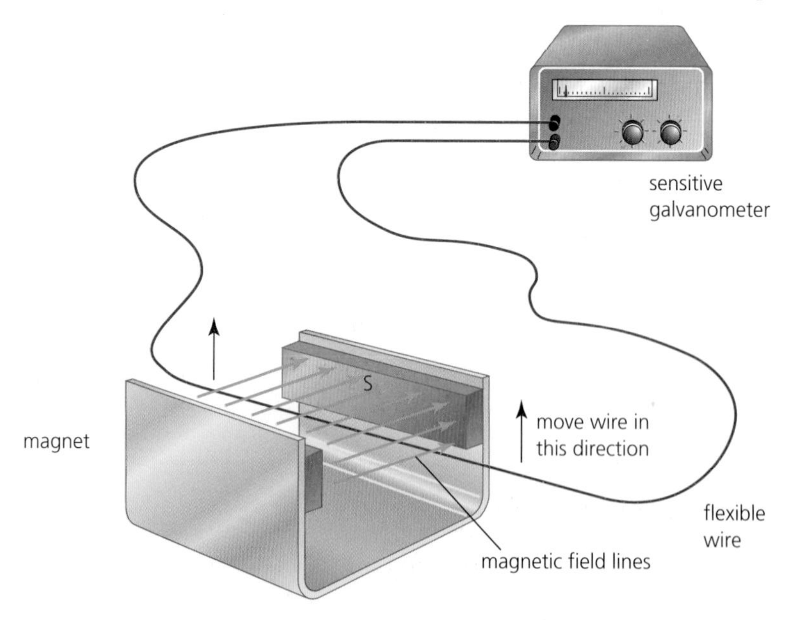

Figure 11.1 Inducing an emf by moving a wire across a magnetic field

The charged particles in the conductor experience forces (as discussed in Chapter 5) because they are moving with the wire as it crosses the magnetic field. Because it is a conductor, it contains *free electrons* that can move along the wire under the action of these forces. Other charges (protons and most of the electrons) also experience forces but are not able to move along the conductor. Moving the wire, containing free electrons, is equivalent to a conventional current of *positive* charge in the opposite direction. We can use the left-hand rule to predict the forces on the electrons, as shown in Figure 11.2. In this case the magnetic force pushes the electrons to the left, so the left-hand end of the conductor becomes negatively charged, while the other end becomes positively charged (because some electrons have flowed away). This charge separation produces a potential difference (emf) across the ends of the conductor.

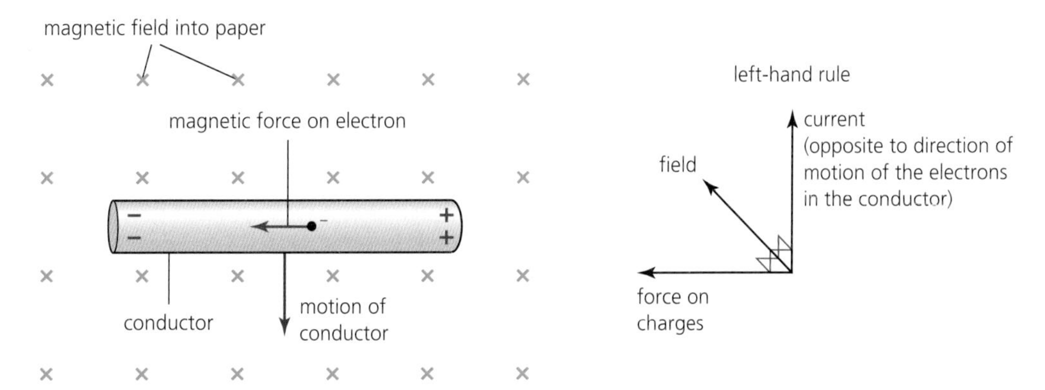

■ **Figure 11.2** Magnetic force on electrons produces charge separation

If the motion *or* the magnetic field is reversed in direction, then the emf is also reversed. (If both the motion *and* the magnetic field are reversed, then the emf is unchanged.) If the conductor and the magnetic field are both moving with the same velocity, no emf is induced. For electromagnetic induction to occur there must be *relative motion* between the conductor and the magnetic field. This effect is important and it needs to be well understood because it has many applications. Most importantly, it is the basic principle behind the generation of the world's electrical energy.

In order to induce an emf, a conductor needs to move *across* a magnetic field. Magnetic fields are represented by *field lines* and the conductor needs to be moving so that it 'cuts' across (through) the field lines. There will be no induced emf if the conductor is moving in a direction that is parallel to the magnetic field lines. Consider the examples shown in Figure 11.3. To induce an emf in **a**, the conductor has to be moved so that its motion has a component into, or out of, the plane of the paper. In **b** the magnetic field is out of the plane of the paper and the conductor would have to be moved so that its motion has a component parallel to the paper.

For similar conductors moving at the same speed, the induced emf is highest if the motion is perpendicular to the magnetic field, as shown in Figure 11.4.

■ **Figure 11.3** Direction of motion that causes electromagnetic induction

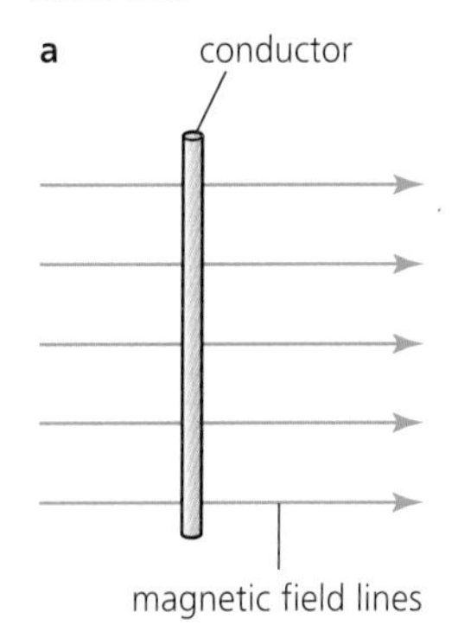

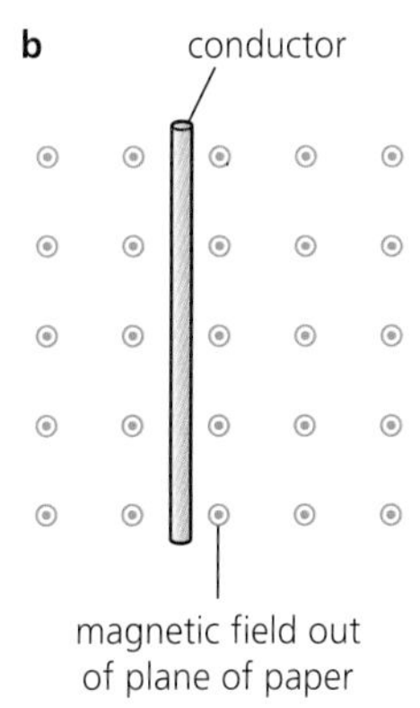

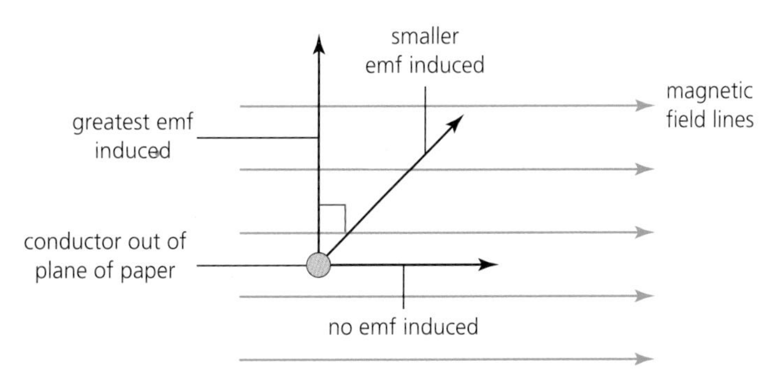

■ **Figure 11.4** The size of an induced emf depends on the direction of motion

The induced emf in Figure 11.1 is very small, but there are simple ways by which it can be increased greatly:

- faster motion
- using a magnet with a higher magnetic field strength

- wrapping a long wire into a coil with many turns (and moving *one* side of the coil through the field)
- using magnetic fields and coils with larger areas.

1 Explain why no emf is induced across a string made of plastic when it is moved through a magnetic field.

2 Figure 11.5 shows a copper wire between the poles of a permanent magnet. In which direction(s) should the wire be moved to induce:
   a the highest emf
   b zero emf?
   c Explain why no *current* can be induced in this wire as shown.

3 Draw a diagram to show how to demonstrate the induction of a current using a stationary magnet and a moving coil of wire.

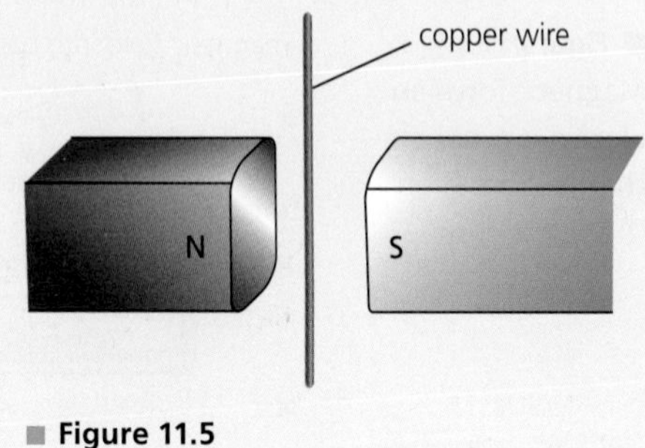

■ **Figure 11.5**

Electromagnetic induction by moving a magnetic field across a conductor

Figure 11.6 shows electromagnetic induction by moving a magnetic field (around a permanent magnet) through a conductor. Again, the induced emf and current will be very small in this basic example, but Figure 11.7 shows how the effects can be increased greatly by winding the conductor into a coil or *solenoid* with many turns. The direction of the induced current around the coil will be reversed if the magnet is reversed or the motion of the magnet is reversed.

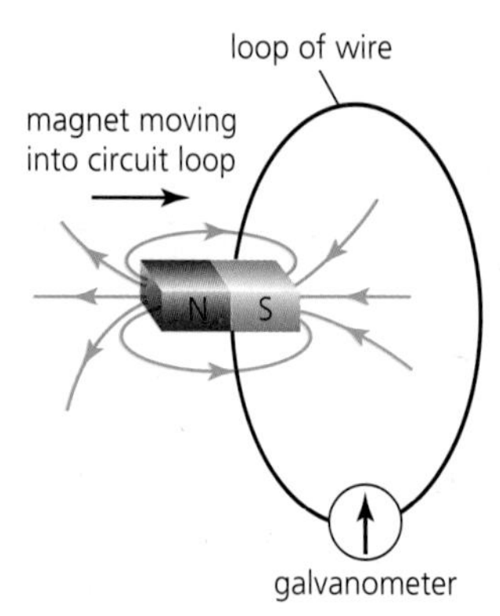

■ **Figure 11.6** Moving a magnet to induce an emf and a current

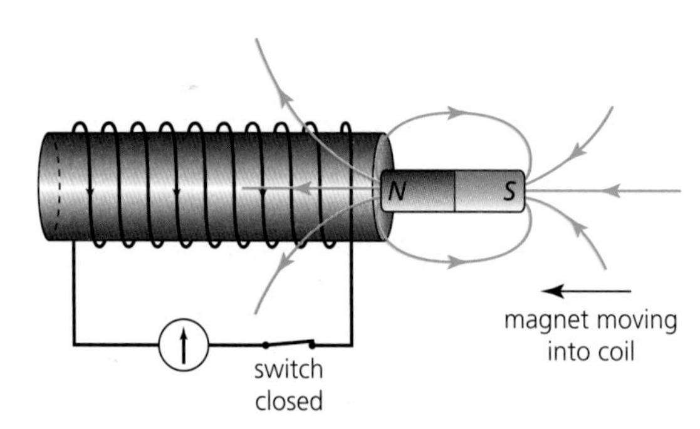

■ **Figure 11.7** Inducing an emf and a current in a coil of wire

Figure 11.8 shows an electromagnetic induction experiment recorded on a data logger and computer. The data logger records the emf being induced at regular time intervals when a magnet is dropped through a coil, and then the data are used to draw a graph.

■ **Figure 11.8** Inducing a current by dropping a magnet through a coil

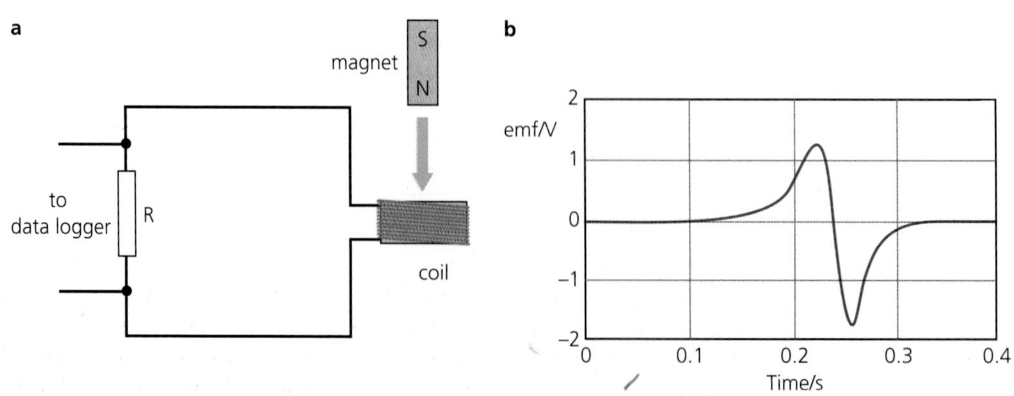

4 In a demonstration of electromagnetic induction similar to that shown in Figure 11.6, the induced current was very small.
 a Suggest two ways of increasing the induced current while still using the same single loop of wire.
 b Give two ways in which the current can be made to flow in the opposite direction around the circuit.

5 Draw a sketch similar to Figure 11.7 to show the current direction when the bar magnet comes out of the coil at the other end.

6 Suggest explanations for the shape of the graph shown in Figure 11.8.

An equation for the induced emf when a straight conductor moves across a uniform magnetic field

Figure 11.9 represents a conductor of length l moving perpendicularly across a uniform magnetic field of strength B, with speed v.

Figure 11.9 Deriving $\varepsilon = Bvl$

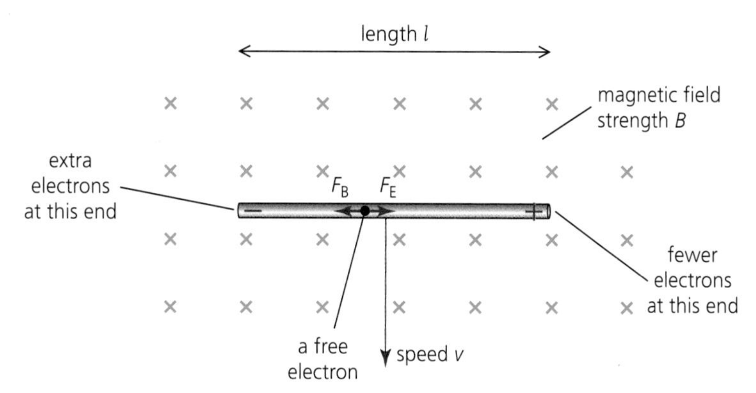

Free electrons in the conductor will each experience a magnetic force, F_B, given by the expression, $F = qvB\sin\theta$ (page 250). In this perpendicular arrangement $\sin\theta = 1$. These forces tend to move free electrons towards the left of the conductor. As more electrons move along the conductor, the increasing amount of negative charge repels the motion of further electrons to that end. The right-hand end of the conductor, which has lost electrons, will become positively charged and act as an attractive force on the electrons.

The charge separation produces an electric field along the conductor, $E = \frac{\varepsilon}{l}$, where ε is the induced emf across the length of the conductor (Chapter 10). The maximum induced potential difference will occur when the force on each free electron due to the magnetic field, F_B, is equal and opposite to the force on the electron, F_E, due to the electric field (see Figure 11.9).

$$\text{electric field} = \frac{\varepsilon}{l}$$

$$\text{electric force, } F_E = \text{electric field} \times \text{charge} = \frac{\varepsilon q}{l}$$

$$F_E = F_B$$

$$\frac{\varepsilon q}{l} = qvB$$

So that,

$$\varepsilon = Bvl$$

This equation is given in the *Physics data booklet*.

This equation shows us that the induced emf is proportional to the speed, the magnetic field strength and the length of conductor in the field. This is a useful equation, but it can only be used for a straight wire moving perpendicularly across a uniform magnetic field. (If the field was not perpendicular to the wire, the component of the field in that direction would have to be used in the calculation.)

If a conductor is wound into a rectangular coil with N turns, the induced emf when one side of the coil moves perpendicularly across a magnetic field is given by:

$$\varepsilon = BvlN$$

This equation is given in the *Physics data booklet.*

Worked example

1 What emf is produced across a 23.0 cm long conductor moving at $98.0\,\text{cm}\,\text{s}^{-1}$ perpendicularly across a magnetic field of strength 120 μT?

$\varepsilon = Bvl$

$\varepsilon = (120 \times 10^{-6}) \times 0.98 \times 0.23 = 2.7 \times 10^{-5}\,\text{V}$

7 When a conductor of length 90 cm moved perpendicularly across a uniform magnetic field of strength $4.5 \times 10^{-4}\,\text{T}$, an emf of 0.14 mV was induced. What was the speed of the conductor?

8 What strength of magnetic field is needed for a voltage of 0.12 V to be induced when a conductor of length 1.6 m moves perpendicularly across it at a speed of $2.7\,\text{m}\,\text{s}^{-1}$?

9 A plane is flying horizontally at a speed of $280\,\text{m}\,\text{s}^{-1}$ at a place where the vertical component of the Earth's magnetic field is 12 μT, as shown in Figure 11.10.
   a Calculate the emf induced across its wing tips if its wingspan is 58 m.
   b Suggest why this voltage would be larger if the plane was flying close to the North, or South, Pole.
   c Could the induced emf be used to do anything useful on the plane? Explain your answer.

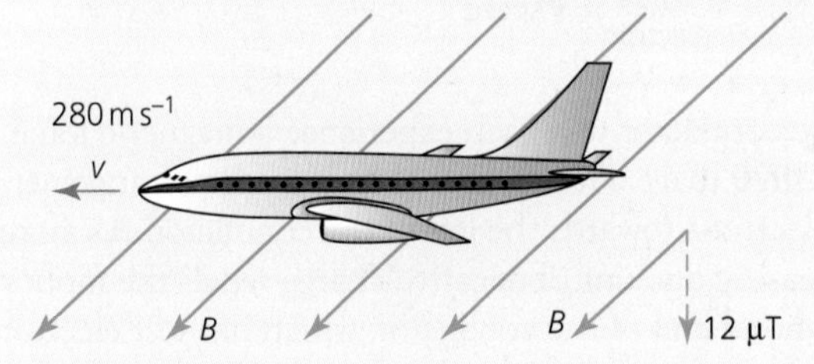

■ **Figure 11.10**

10 A rectangular coil of copper wire has 500 turns and sides of length 5.2 cm and 8.7 cm. The coil rotates at a frequency of 24 Hz about an axis that passes centrally through the shorter sides. A uniform magnetic field of 0.58 T acts on the coil in a direction perpendicular to the axis.
   a Draw a labelled sketch of this arrangement.
   b Calculate the speed of the sides of the coil.
   c What emf is induced across one of the longer sides at the instant that it is moving perpendicularly across the field?
   d What are the maximum and minimum induced emfs as the coil rotates?

Magnetic flux and magnetic flux linkage

Magnetic flux

The size of an induced emf depends not only on the strength of the magnetic field, B, but also on the area, A, of the circuit over which it is acting, and the angle at which it is passing through the circuit. For these reasons it is useful to introduce and define the important concept of **magnetic flux**.

Magnetic flux, Φ, (for a uniform magnetic field) is defined as the product of the area, A, and the component of the magnetic field strength perpendicular to that area, $B\cos\theta$. Figure 11.11 illustrates

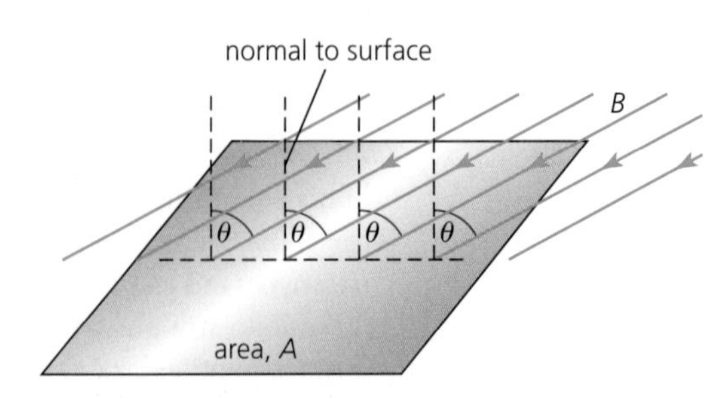

■ **Figure 11.11** Magnetic flux depends on field strength, area and angle.

an example that shows that the angle θ is measured between the normal to the area and the direction of the field.

$$\Phi = BA\cos\theta$$

This equation is given in the *Physics data booklet.*

If the magnetic field is perpendicular to the area, then $\theta = 0$ and $\cos\theta = 1$. The flux is maximized and the equation simplifies to $\Phi = BA$. If the field is parallel to the area, the flux is zero because $\cos\theta = 0$.

The unit of magnetic flux is the **weber**, Wb. One weber is equal to one tesla multiplied by one metre squared ($1\,\text{Wb} = 1\,\text{T}\,\text{m}^2$).

We can rearrange the equation to give $B\cos\theta = \Phi/A$, and this shows why magnetic field strength is widely known as **magnetic flux density** (flux/area).

Magnetic flux can be a difficult concept to understand and Figure 11.12 may help. It shows a non-mathematical interpretation of magnetic flux as the number of magnetic field lines that pass through the area. Flux also has mathematical similarities with ideas in Chapter 8, where the total power flowing through an area depends on the intensity of the electromagnetic radiation, the area and the angle of incidence.

■ Figure 11.12 Magnetic flux explained in terms of field lines

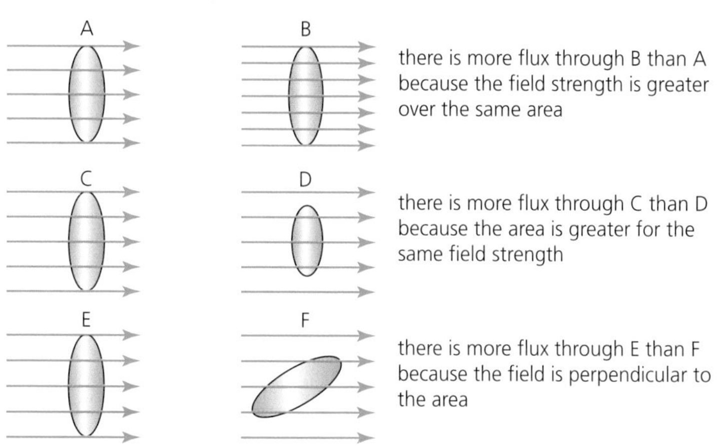

Worked examples

2 **a** Calculate the magnetic flux in a square coil of wire of sides 6.2 cm when it is placed at 45° to a magnetic field of strength 4.3×10^{-4} T.

b What would the magnetic flux become if only half of the coil was in the magnetic field?

a $\Phi = BA\cos\theta$

$\Phi = (4.3 \times 10^{-4}) \times (6.2 \times 10^{-2})^2 \times \cos 45°$

$\Phi = 1.2 \times 10^{-6}\,\text{Wb}$

b The magnetic flux would be reduced to half because the area used in the calculation is the area of the coil in the magnetic field, not the total area of the coil.

3 This example provides a solar radiation analogy to help understanding of the concept of flux. The 'solar flux density' arriving perpendicularly at the Earth's upper atmosphere is $1360\,\text{W}\,\text{m}^{-2}$. (This was called the *solar constant* in Chapter 8.) Suppose that near the Earth's surface this value has reduced to $800\,\text{W}\,\text{m}^{-2}$. Calculate the power arriving at a horizontal solar panel of area $4.0\,\text{m}^2$ if the radiation arrives at an angle of 40° to the surface (see Figure 11.13).

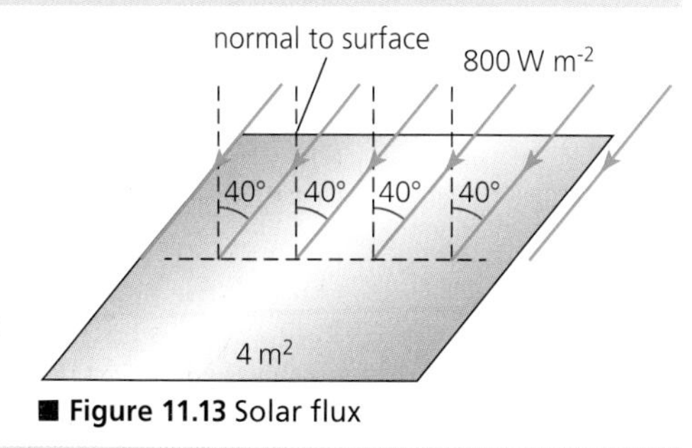

■ Figure 11.13 Solar flux

By analogy with $\Phi = BA\cos\theta$, we can write:

received power = $800 \times 4.0 \times \cos 40° = 2.4 \times 10^3\,\text{W}$

Magnetic flux linkage

Increasing the number of turns, N, in a coil increases the size of an induced emf in proportion and, for this reason, the concept of **magnetic flux linkage** becomes useful. Magnetic flux linkage is defined as the product of magnetic flux and the number of turns in the circuit. It does not have a widely used standard symbol. The units of flux linkage are the same as flux (Wb), although sometimes *Wb turns* is used.

magnetic flux linkage = $N\Phi$

This equation is *not* in the *Physics data booklet*.

In Worked example 2a, the *flux* through the coil was calculated to be 1.2×10^{-6} Wb. If the coil had 200 turns, the flux *linkage* would be $200 \times (1.2 \times 10^{-6}) = 2.4 \times 10^{-4}$ Wb.

11 Calculate the magnetic flux in a flat coil of area 48 cm^2 placed in a field of magnetic flux density 5.3×10^{-3} T if the field is at an angle of 30° to the plane of the coil.

12 A magnetic field of strength 3.4×10^{-2} T passes perpendicularly through a flat coil of 480 turns and area 4.4×10^{-5} m^2. What is the flux linkage?

13 A flat coil of 600 turns and area 8.7 cm^2 is placed in a magnetic field of strength 9.1×10^{-3} T. The axis of the coil was originally parallel to the magnetic field, but it was then rotated by 25°. What was the *change* of flux linkage through the coil?

All electromagnetic induction is caused by changing magnetic flux

Emfs can also be induced, not by movement, but by changes in the current in one circuit affecting another, separate, circuit. Figure 11.14 represents the simplest example, which should be understood in detail. First consider circuit B at a time when the switch in circuit A is open – there is no power source and no magnetic field near it, so there is no current shown by the galvanometer.

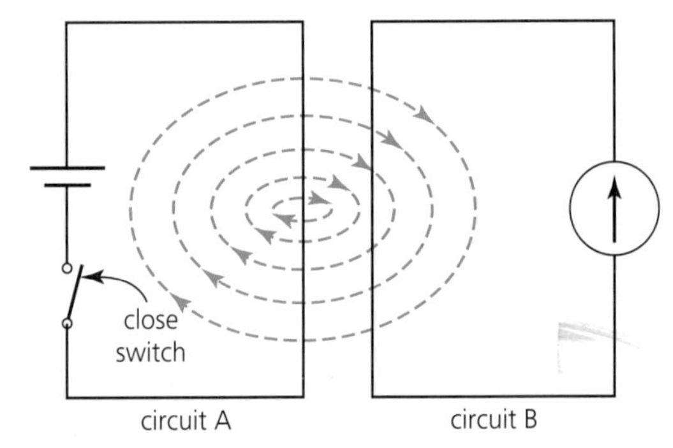

Figure 11.14 When the switch is closed, a magnetic field (flux) passes from circuit A to circuit B

When the switch in circuit A is closed, however, a current starts to flow around circuit A and this sets up a magnetic field around it. This field spreads out and passes through circuit B. The sudden *change* of magnetic flux induces an emf and a current that is detected by the galvanometer in circuit B. The *changing* current produces a *changing* magnetic flux in the same way as moving a magnet does.

This induced emf/current only lasts for a moment, while the switch in A is being turned on, because when the current in A is constant there is no longer a *changing* magnetic flux. When the switch is turned off, there is an induced emf/current for a moment in the opposite direction.

As described so far, this is a very small (but important) effect. However, the induced emf can be increased greatly by winding the conducting wires in *both* circuits into coils of many turns (to increase the flux and flux linkage) and placing them on top of each other with an iron core through the middle. This is shown in Figure 11.15. Remember from Chapter 5, that iron has high magnetic permeability and greatly increases the strength of the magnetic field.

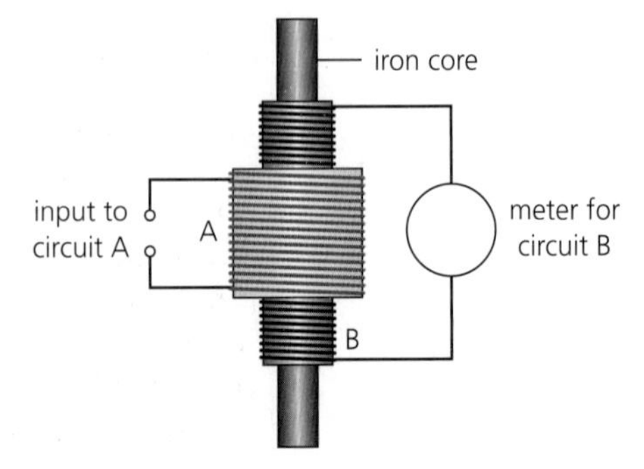

Figure 11.15 Making the induced emf larger by using an iron core and coils of many turns (increasing the flux linkage)

If the voltage source in circuit A is changed from one that gives a direct current (dc) of constant value to a source of alternating current (ac) then the magnetic flux in both circuits will

change continuously and an alternating emf will be induced continuously. This effect has a large number of important applications which are discussed later in this chapter.

14 a Figure 11.16 shows four loops of wire wrapped around a solenoid and connected to a centre-reading galvanometer (as shown on the right). Describe and explain what will be seen on the meter at the moment when:
 i the switch is closed
 ii the switch is opened.
b How would your answer to a i change if:
 i there were eight turns on the wire wrapped around the solenoid
 ii an iron core was placed inside the solenoid?
c What is seen on the meter when a constant current is flowing in the solenoid?

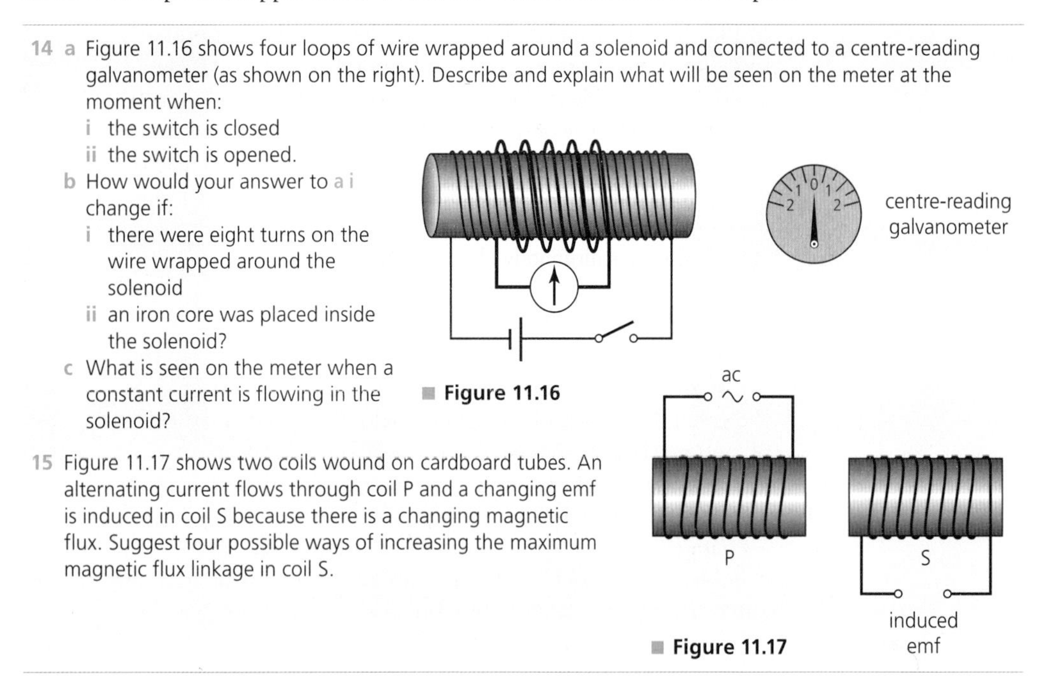

■ Figure 11.16

■ Figure 11.17

15 Figure 11.17 shows two coils wound on cardboard tubes. An alternating current flows through coil P and a changing emf is induced in coil S because there is a changing magnetic flux. Suggest four possible ways of increasing the maximum magnetic flux linkage in coil S.

Summary: three kinds of electromagnetic induction

In order to induce an emf across a circuit, a *changing* magnetic field must pass through the circuit. There are three different ways in which this can be done:

1 Part of a circuit can be moved through a stationary magnetic field (as shown in Figure 11.1).
2 A magnetic field, for example around a bar magnet, can be moved through a circuit (as shown in Figures 11.6, 11.7 and 11.8).
3 A changing current in one circuit can produce a changing magnetic field which spreads out and passes through another circuit (as shown in Figures 11.14 and 11.15).

Electromagnetically induced currents have perhaps their most important application in the generation and transmission of electrical power (transformers), but there are many other uses including radio communication, regenerative braking, metal detectors and geophones (devices that induce emfs from movement of the Earth's surface).

Utilizations

Metal detectors

'Treasure hunting' using simple metal detectors is a popular hobby. In Figure 11.18 a metal detector is being used to hunt for metal objects, like coins and rings, among the stones on the seashore. Other applications include airport security and the search for buried land-mines in war zones.

■ **Figure 11.18** Treasure hunting

An oscillating electric current is passed through a coil in the metal detector and the changing magnetic field passes through another coil in the instrument and induces a current. If there is any metal close to the coils it will affect the magnetic field passing between the coils and change the induced current. Any such change is detected by the circuit and usually an audible signal is emitted. When the original oscillating magnetic field passes through the metal it will induce a current within the solid metal called an *eddy current* and this will set up its own magnetic field.

1 Most airport metal detectors are based on a pulse induction (PI) system. Find out the principles on which this metal detection system is based.

Nature of Science

Apparently insignificant experimental observations may have enormous consequences

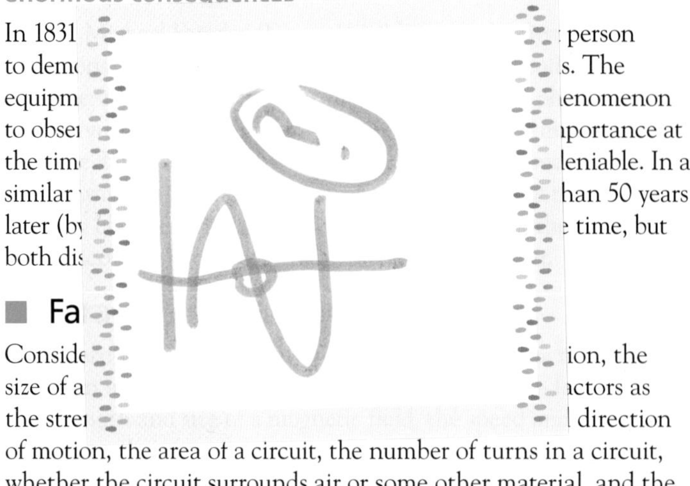

In 1831 … person
to dem… s. The
equipm… enomenon
to obse… portance at
the tim… leniable. In a
similar … han 50 years
later (b… time, but
both di…

■ **Figure 11.19** Michael Faraday (1791–1867) is considered to be one of the greatest scientists ever

■ Fa…

Consid… ion, the size of a … ctors as the stre… direction of motion, the area of a circuit, the number of turns in a circuit, whether the circuit surrounds air or some other material, and the rate at which an electric current is changing.

Faraday was the first scientist to realise that *all* these examples of electromagnetic induction could be summarized by a single law expressing the fact that *any* induced emf depends on the rate of change of magnetic flux linkage.

Faraday's law of electromagnetic induction states that the magnitude of an induced emf is equal to the rate of change of magnetic flux linkage.

$$\varepsilon = -N\frac{\Delta\Phi}{\Delta t}$$

This equation is given in the *Physics data booklet*. The negative sign is because of the conservation of energy as represented by *Lenz's law*. This will be explained later.

Remembering that $\Phi = BA\cos\theta$:

- $\varepsilon = -NB\cos\theta \times (\Delta A/\Delta t)$ can be applied to a situation in which the area of a circuit is changing in a constant field.
- $\varepsilon = -NA\cos\theta \times (\Delta B/\Delta t)$ can be applied to a situation in which the magnetic field passing through a fixed circuit is changing.

Worked example

4 The magnetic flux through a coil of 1200 turns increases from zero to 4.8×10^{-5} Wb in 2.7 ms. What is the magnitude of the induced emf during this time?

$$\varepsilon = -\frac{1200 \times (4.8 \times 10^{-5})}{2.7 \times 10^{-3}} = -21\,\text{V}$$ (The negative sign is often omitted.)

16 A coil of area 4.7 cm^2 and 480 turns is in a magnetic field of strength 3.9×10^{-2} T.
  a Calculate the maximum possible magnetic flux linkage through the coil.
  b What is the average induced emf when the coil is moved to a place where the perpendicular magnetic field strength is 9.3×10^{-3} T in a time of 0.22 s?

17 Consider the circuit shown in Figure 11.15. When the input to circuit A is an alternating current of frequency 50 Hz, the induced emf in circuit B has a maximum value of 4 V. Suggest what would happen to the induced emf in circuit B if the frequency in circuit A was doubled (with the same maximum voltage). Explain your answer.

18 Figure 11.20 shows a coil of 250 turns moving from position A, outside a strong uniform magnetic field of strength 0.12 T, to position B at the centre of the magnetic field in a time of 1.4 s.
  a Calculate the change of magnetic flux in the coil when it is moved.
  b What assumption did you make?
  c What is the change of magnetic flux linkage?
  d Calculate the average induced emf.
  e Sketch a graph to show how the induced emf changes as the coil is moved at constant speed from A to C (no values needed).

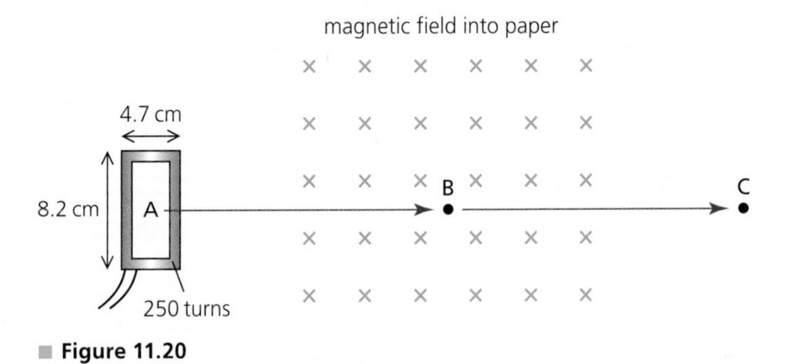

■ **Figure 11.20**

19 Imagine you are holding a flat coil of wire. You are in the Earth's magnetic field.
  a Draw a sketch to show how you would hold the coil so that there is no magnetic flux through it.
  b At a place where the magnitude of the Earth's magnetic field strength is 48 μT, what emf would be induced by moving a coil of 550 turns and area 17 cm^2 from being parallel to being perpendicular to the magnetic field in 0.50 s?

20 A small coil of area 1.2 cm^2 is placed in the centre of a long solenoid with a large cross-sectional area. A steady current of 0.5 A in the solenoid produces a magnetic field of strength 8.8×10^{-4} T.
  a How many turns would be needed on the coil if an induced emf of 2.4 mV was required when the current in the solenoid was increased to 2.0 A in a time of 0.10 s?
  b In what direction would the coil need to be positioned?

21 Refer back to the graph shown in Figure 11.8.
  a What quantity is represented by the area under the graph?
  b Compare the areas above and below the time axis.

22 Carry out some research into the work of Michael Faraday and Nikola Tesla. In your opinion, which of them contributed more to the development of the theories of electromagnetism? Explain your answer.

23 Refer back to question 10.
  a What is the maximum magnetic flux through the coil?
  b Determine the average emf induced.
  c Compare your answer to the answer for question 10 and explain the difference.

■ Energy transfers during electromagnetic induction

If a current is generated from motion by electromagnetic induction, then energy must have been transferred from outside the circuit. (We know this from an understanding of the law of conservation of energy.) The origin of this energy is the kinetic energy of the moving conductor or magnet. The moving object must therefore slow down as it loses some of its kinetic energy (unless there is a force keeping it moving).

For example, the magnets shown in Figures 11.7 and 11.8 must experience forces trying to slow them down (*opposing* their motion) as they move into the coils. This means that work has to be done to move the magnet into the coil and this energy is transferred to the electromagnetically induced current. To understand why there is a force opposing motion, we need to remember the ideas about the magnetic fields produced by currents from Chapter 5. If an (induced) current flows around the coil shown in Figure 11.7, then the coil will behave as a magnet. The direction of the current makes the end near the permanent magnet a north pole (this can be predicted using the right-hand grip rule). The two north poles repel each other and cause a repulsive force opposing motion. If the induced current flowed the other way, then the force would be attractive and the magnet would gain kinetic energy, which would conflict with the law of conservation of energy.

If the polarity of the magnet moving into the coil was reversed, the induced current would flow the other way around the coil and then two south poles would repel each other. If the magnet is taken away from the coil, the current flows in the opposite direction and a magnetic field is set up that tries to stop the magnet moving away.

If the switch in Figure 11.7 is opened, there will still be an induced emf, but no current can flow. This means that there will be no magnetic field and no force from the coil.

Lenz's law

Because of the law of conservation of energy, we know that in *all* examples of electromagnetic induction, any induced emf and current *must* set up a magnetic field that tries to stop whatever change produced the current in the first place. This is known as **Lenz's law**.

> Lenz's law states that the direction of an induced emf is such that it will *oppose* the change that produced it. This is represented mathematically by the negative sign in the Faraday's law equation.

24 Figure 11.21 shows a bar magnet on the end of a spring placed close to a coil of wire.
- a The magnet is pulled down so that its end is inside the coil and then released. Sketch a graph to show how the position of the magnet will change with time for a few seconds.
- b On the same axes indicate how the emf induced across the coil changes.
- c Suggest what differences will be observed if the switch is closed while the magnet is oscillating. Explain your answer.

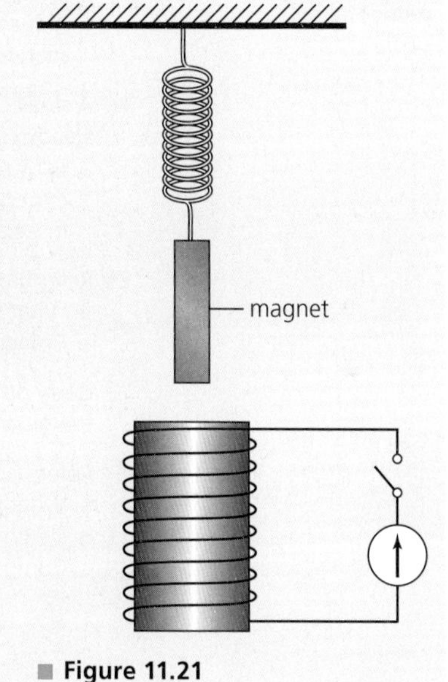

■ **Figure 11.21**

25 When a strong magnet is dropped vertically down inside a copper tube it takes significantly longer to reach the ground than when it is dropped outside the tube. Explain this observation.

26 The loop of wire shown in Figure 11.22 is travelling at constant speed when it enters the magnetic field.
- a Explain why the loop decelerates.
- b Does the top of the loop act like a north pole or a south pole?
- c In which direction does the current flow in the side of the loop first entering the magnetic field?
- d Why is there no emf or current induced while the loop is moving through the centre of the magnetic field?
- e Discuss what happens as the loop leaves the other side of the field.
- f What has happened to the kinetic energy transferred from the loop because it reduced in speed?

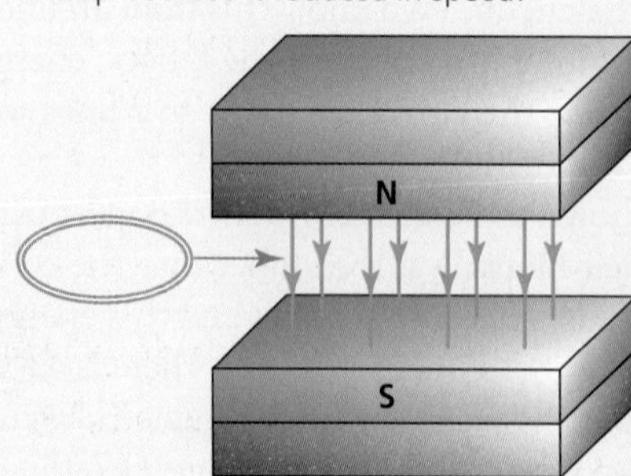

■ **Figure 11.22**

Utilizations

Eddy currents and induction cooking

In an induction cooker, like that shown in Figure 11.23a, there are coils of wire below the flat-top surface. When a high-frequency current is passed through a coil in the cooker, a strong, rapidly oscillating magnetic field is created that will pass through anything placed on or near the cooker's surface, like a cooking pot. If the material of the pot is a conductor, emfs and currents will be induced in it. Currents circulating within solid conductors, rather than around wire circuits, are known as *eddy currents*.

a

b

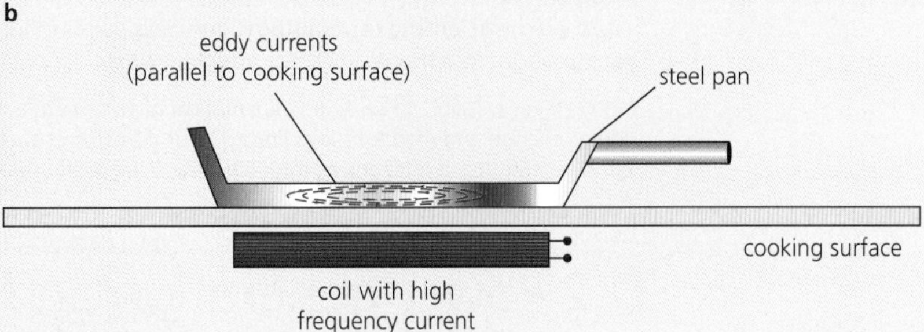

■ **Figure 11.23 a** A steel pan on an induction cooker **b** How an induction cooker works

The electrical energy in the eddy currents will be transferred to internal energy and the pan will get hotter. As in any circuit, the power generated can be calculated using $P = I^2R$. Thermal energy is then transferred by conduction to any food that is in the pan.

The choice of metal for the pan is important – for example, copper has a low resistivity and it also conducts thermal energy well. (This is not a coincidence; the ability to conduct thermal energy and the ability to conduct electricity are related. Both depend on the number of free electrons per unit volume in the material.) However, copper would *not* be a good choice for pans to be used with an induction cooker because it does not have good magnetic properties. The pans need to be made of material with high magnetic permeability, in other words a ferromagnetic material like iron or steel, so that the strength of the magnetic field in the pan is greatly increased. (Some pans made especially for use with induction cookers use a combination of different metals.)

Induction cookers have at least two significant advantages:

- The cooker itself does not get noticeably hot (except for any thermal energy transferred from the pan). This means that induction heaters are safe, quick and reasonably efficient ways of cooking.
- The heating process can be completely controlled, stopped or started by the turn of a switch.

Induction cookers are more efficient at cooking food than conventional electric hot plates and both are a lot more efficient than natural gas cookers (although this does not allow for the inefficiencies of electrical power production). The reason why they are not used more widely is that induction cookers are more expensive than gas or electric cookers (largely because of the extra electronics needed) and they also need special cooking vessels. In locations where space is limited, safety is an issue and an oven is not needed. Then small induction cookers may be the ideal choice.

1 Suggest why it is not possible to design and make an efficient induction oven.

2 Why do the heating coils in an induction cooker use high-frequency currents?

3 Explain why, although induction cookers are much more efficient than gas cookers, they both create roughly the same 'carbon footprint'.

ToK Link

The use of scientific terminology can make communication with non-scientists difficult

The terminology used in electromagnetic theory is extensive and can confuse people who are not directly involved. What effect can lack of clarity in terminology have on communicating scientific concepts to the public?

Faraday's law states that an induced emf is equal to the rate of change of magnetic flux linkage. A good student of physics will appreciate that this is an elegant and precise way of expressing a very important concept. However, for non-scientists it may seem like a foreign language. Scientists aim to express ideas as briefly and as succinctly as possible, especially when communicating with other scientists. This involves the use of precise **scientific terminology**, including the introduction of new words for new ideas, or perhaps the use of common words in precise scientific ways.

It is certainly possible to write an explanation of Faraday's law without using the phrases *induced emf, rate of change* and *magnetic flux linkage*, but it would probably require several pages instead of one line (consider the first few pages of this chapter).

11.2 Power generation and transmission – *generation and transmission of alternating current (ac) electricity has transformed the world*

Alternating current (ac) generators

Consider Figure 11.24, which shows a coil of wire between the poles of a magnet. (For simplicity, only one loop is shown, but in practice there will be a large number of turns on the coil.) If the coil is rotating, there will be a changing magnetic flux passing through it and a changing emf will be induced. As side WX moves upwards, the induced emf will make a current flow into the page, if it is connected in a circuit. At the same time any induced current in YZ will flow out of the page, because it is moving in the opposite direction. In this way, the current flows continuously around the coil.

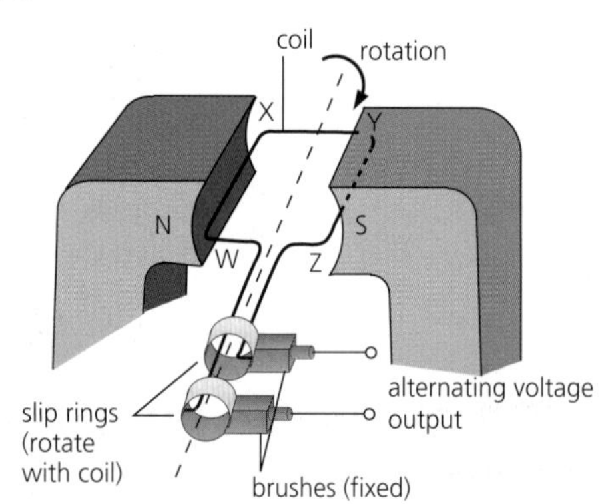

■ **Figure 11.24** A simple ac generator

The connection between the coil and the external circuit cannot be fixed and permanent because the wires would become twisted as the coil rotated. Therefore **carbon 'brushes'** are used to make the electrical contact with **slip rings** which rotate with the coil, so that the induced current can flow into an external circuit.

Figure 11.25 shows three views of the rotating coil from the side. In Figure 11.25a the plane of the coil is parallel to the magnetic field and, at that moment, the sides WX and YZ are cutting across the magnetic field at the fastest rate, so this is when the maximum emf is induced. In Figure 11.25b the sides WX and YZ are moving parallel to the magnetic field, so no emf is induced at that moment. In Figure 11.25c the induced emf is a maximum again, but the direction is reversed because the sides are moving in the opposite direction to Figure 11.25a.

a

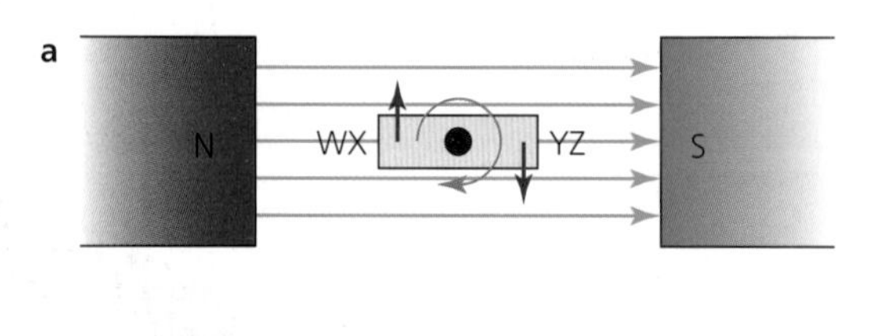

b

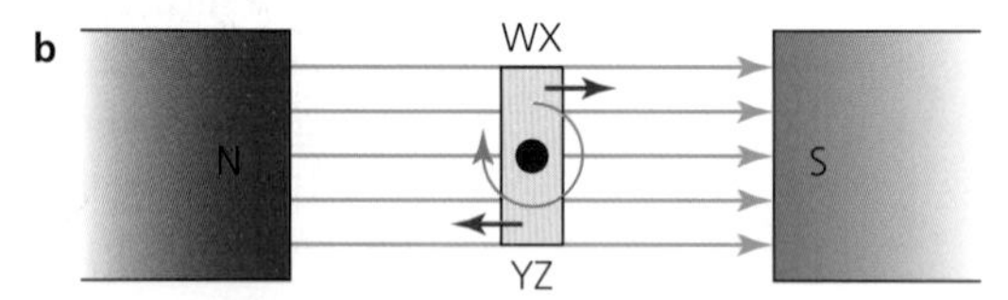

c

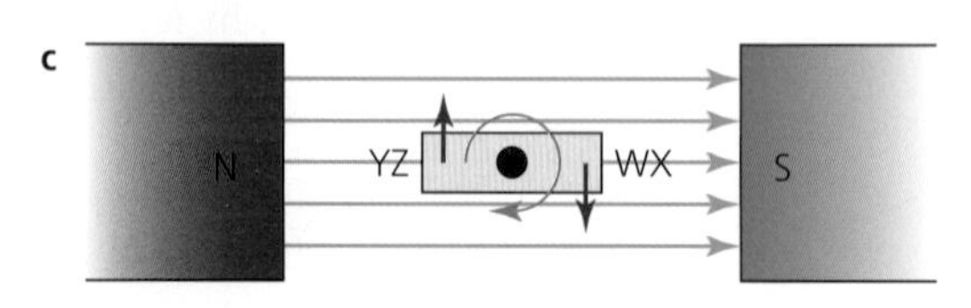

■ **Figure 11.25** The sides of the coil cut the magnetic field at different angles as they rotate, alternating the emf produced

The overall result, if the coil rotates at a constant speed in a uniform magnetic field, is to induce an emf that varies **sinusoidally**. This is shown by the green line in Figure 11.26. In positions B and D the plane of the coil is parallel to the magnetic field. At A, C and E the plane of the coil is perpendicular to the field. One complete revolution occurs in time T. Frequency, f, equals $1/T$.

If the coil rotates at a slower frequency (fewer rotations every second), then there will be a smaller rate of change of magnetic flux through it and a smaller emf will be induced. For example, halving the frequency will halve the rate of change of magnetic flux linkage and therefore halve the induced emf. The time period is doubled. This is represented by the blue line in Figure 11.26. Students are advised to watch a computer simulation of an ac generator as the coil(s) rotates slowly to aid understanding.

Throughout the world electrical energy is generated in this way using ac generators. Turbine blades can be made to rotate by the forces provided by, for example, high-pressure steam produced from burning fossil fuels, or nuclear reactions, or from falling water (as discussed in Chapter 8). Turbine blades are attached to large coils inside an ac generator. The coils – with many turns and cores with high magnetic permeability – are rotated in strong magnetic fields by the action of the turbine blades.

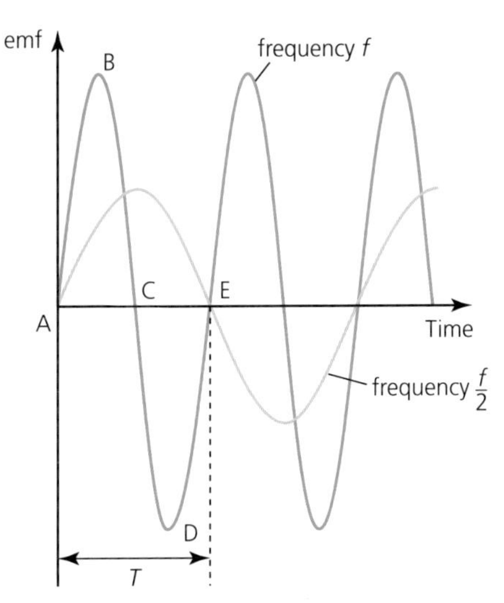

■ **Figure 11.26** Comparing induced emfs at different frequencies

Electricity can also be generated using the same principle, but with the magnetic field rotating inside the circuit, rather than the other way around. Such devices are commonly used in cars and they are often called *alternators*.

(Note that dc generators are similar to ac generators in basic design, but the connections to the external circuit need to be modified.)

27 Consider Figure 11.24.
a Explain why no emf is induced across the rotating end of the coil, XY.
b List four ways in which the induced emf could be increased.

28 An ac generator produces a sinusoidal emf of frequency 30 Hz and maximum value 9.0 V.
a Sketch an emf–time graph to show how the output of the generator changes during a time of 0.1 s.
b Add a second line to your graph to show the output from the same generator if the frequency is reduced to 10 Hz.

29 Figure 11.27 shows the basic details of the construction of a bicycle dynamo. Explain how it works.

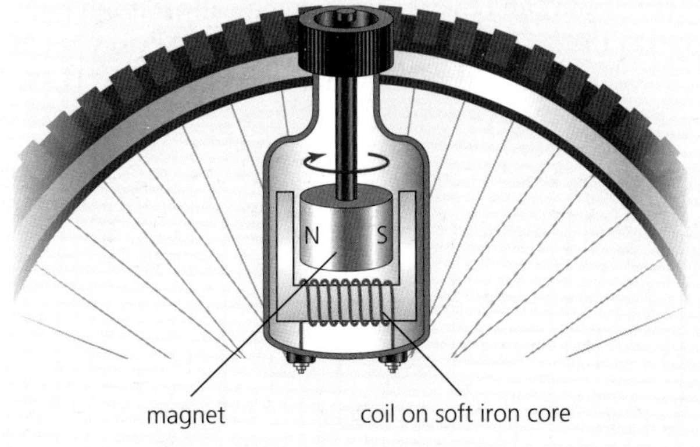

■ **Figure 11.27** A simple bicycle dynamo

Utilizations

Induced emfs: motors

Simple electric motors are similar in their basic design to generators – both consist of coils rotating in magnetic fields. We know from Chapter 5 that when a current is made to flow across a magnetic field it experiences a force. In electric motors, like the simplified version shown in Figure 11.28, forces in opposite directions on opposite sides of a current-carrying coil result in continuous rotation. Compare the simple ac generator in Figure 11.24 with the simple dc motor in Figure 11.28.

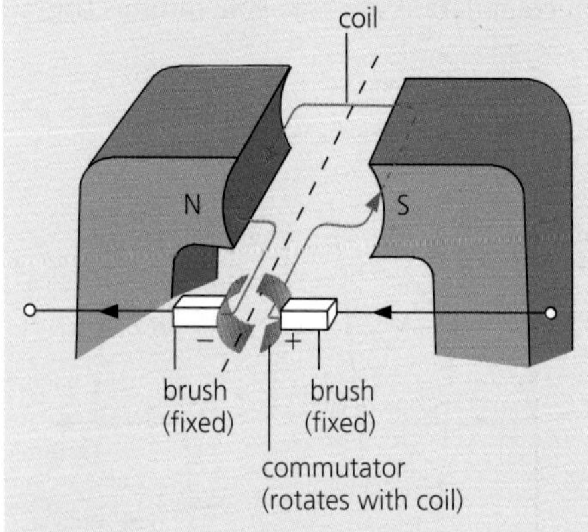

■ **Figure 11.28** A simple dc motor

When the coil in a motor is rotating, electromagnetic induction also takes place and an emf is induced across the coil which will tend to oppose its motion (Lenz's law). In other words, when a potential difference is applied to a coil in a motor to make it rotate, the motion then induces another potential difference in the opposite direction, trying to stop the rotation. This is often called a 'back emf'.

In a well-designed, efficient dc motor that is spinning freely without doing any useful work, the back emf can be almost as high as the applied p.d., so that the resultant p.d. is just enough to drive a small current through the coil and provide the necessary low power. But when the motor is doing useful mechanical work, for example lifting a load, the rate of rotation decreases and the back emf is reduced (because of the decreased rate of changing magnetic flux linkage). The resultant p.d. and current increase to provide the extra power needed.

An advantage of a dc motor is that the speed of rotation can be controlled easily, but most motors operate with alternating currents. There are two main types of motor that use alternating currents:

- A synchronous ac motor uses slip rings (as in the ac generator) and usually rotates with the same frequency as the ac supply.
- An ac induction motor uses electromagnetic induction to transfer the energy to the coil.

1 Explain why a split ring commutator and brushes, as shown in Figure 11.28, are needed for a motor to rotate when supplied with a direct current.

2 Find out some basic information on how induction motors work.

Alternating currents and direct currents

When a current is described as **alternating**, it means that the *direction* of the current changes (usually in a regular way) and this is shown graphically by both positive and negative values. We have seen that the output from ac generators is usually sinusoidal, but other waveforms are possible, like the square and triangular waves shown in Figure 11.29.

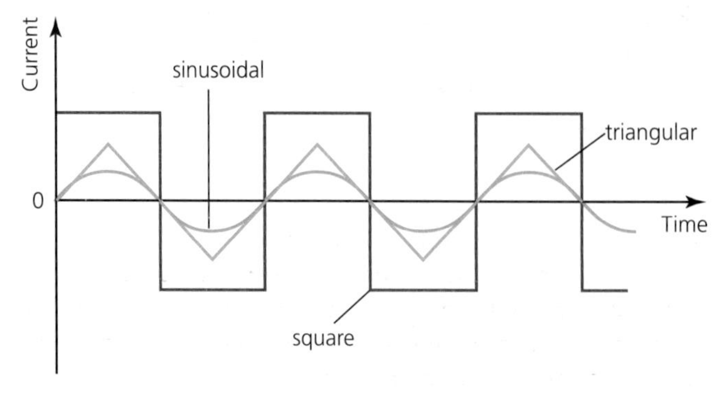

■ **Figure 11.29** Alternating currents shown graphically with different waveforms and the same frequency, but different amplitudes

A direct current always flows in the same direction around a circuit (shown graphically by always being positive or always being negative), although its magnitude may vary.

Electrical energy is mostly generated and transmitted using alternating currents because of the need to reduce energy losses. As we shall see later in this section, high voltages are needed for transmission and this requires transformers which can change (transform) voltages to higher or lower values. Simple transformers can only work using alternating current, although it is also possible to transform direct currents.

Many electrical devices that we use in our homes can be designed to use ac or dc directly, but some devices can only operate with a dc supply. For example, electronic devices need dc, so their designs need to include ac to dc convertors (rectifiers).

Average power and root mean square (rms) values of current and voltage

Figure 11.31 shows the varying voltage, current and power in a circuit with an ohmic resistor. For easy comparison all three are shown on the same axes. The peak (maximum) values of voltage and current are labelled V_0 and I_0. The voltage waveform is easily observed using an oscilloscope connected across the resistor.

The current at any moment is determined from $I = V/R$, and the instantaneous power can be determined from $P = VI$.

The maximum power is given by:

$$P_{max} = I_0V_0$$

This equation is given in the *Physics data booklet*.

The power varies but always remains positive because the voltage and current are either both positive or both negative at all times. This means that any power always flows *from* the supply to the circuit. (This is only true for circuits that can be considered to be purely resistive.) The power transferred in the resistor does not depend on the direction of current flow.

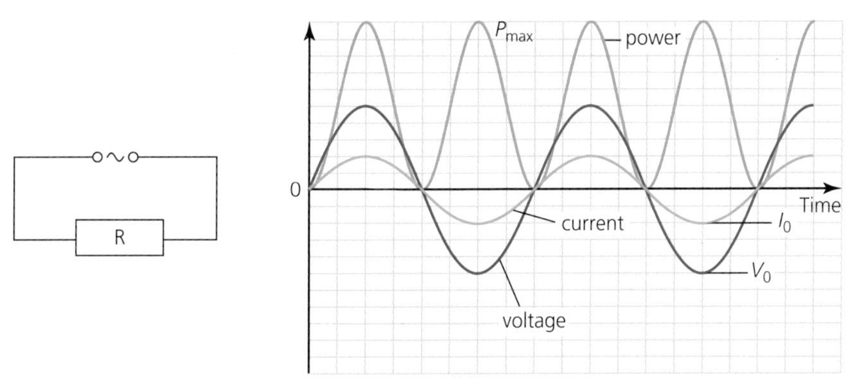

Figure 11.31 Oscillating voltage, current and power for a circuit with a resistor and an ac power supply

Both the voltage and current variations are sinusoidal, which means that when they are multiplied together, the power variation ($P = VI$) follows a *sine squared relationship*. For this kind of relationship, the average value of power, $\overline{P}$, is exactly half the peak value:

$$\overline{P} = \frac{1}{2}I_0V_0$$

This equation is given in the *Physics data booklet*.

The *effective* values of the current and voltage are the values that give the average *power*; these are not the average values of the voltage and current, which are both zero, but they are less than the peak values. The equation below can be used to determine the relationship between peak and effective values for current and voltage:

$$P_{average} = I_{effective} \times V_{effective} = \frac{1}{2}I_0V_0 = \frac{I_0}{\sqrt{2}} \times \frac{V_0}{\sqrt{2}}$$

The effective value of an alternating current (or voltage) is therefore its maximum value divided by $\sqrt{2}$. The effective value is called its **root mean square (rms)** value, or its *rating.*

The rms value (rating) given to an alternating current (or voltage) is the same as the value of a direct current (or voltage) that would dissipate power in an ohmic resistor at the same rate.

This means, for example, that 230 V ac and 230 V dc transfer energy in a resistor at the same rate, even though the alternating voltage rises to a maximum of $230 \times \sqrt{2} = 325$ V, while the dc voltage remains constant at 230 V.

$$I_{rms} = \frac{I_0}{\sqrt{2}}$$

$$V_{rms} = \frac{V_0}{\sqrt{2}}$$

These equations are given in the *Physics data booklet.*

The equations $R = V/I$ and $P = VI = I^2R = V^2/R$ can all be used with ac circuits using maximum values or rms values (but *not* a mixture), so that:

$$R = \frac{V_0}{I_0} = \frac{V_{rms}}{I_{rms}}$$

This equation is given in the *Physics data booklet.*

Additional Perspectives

Oscilloscopes

Oscilloscopes are one of the most useful pieces of equipment found in a physics or electronics laboratory. They are effectively voltmeters for observing quickly changing and repeating potential differences and they are usually used to plot voltage–time graphs over short intervals of time. But it is not just voltages that can be displayed; any physical quantity that can be *represented* by a voltage can be shown in the same way. A common example would be using the output from a microphone as the input to an oscilloscope to show the pattern of the sound waves. A much simplified representation of the front of an oscilloscope is shown in Figure 11.30.

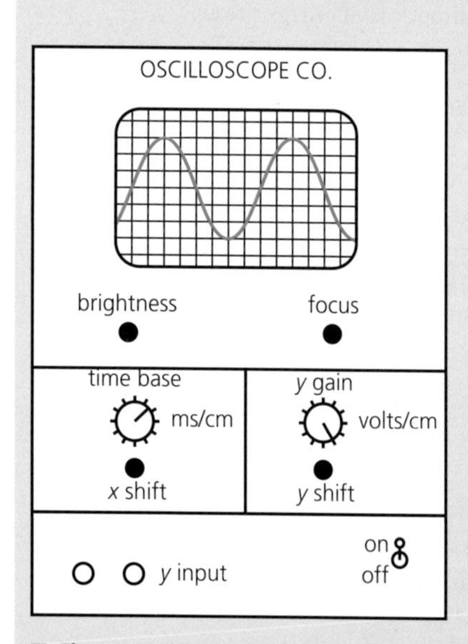

■ **Figure 11.30** A simplified oscilloscope

To begin with, a bright spot is seen on the screen and it can be moved around by voltages that make it move up and down, or left and right (y shift and x shift respectively). Circuits within the oscilloscope are usually used to make the dot move automatically from left to right across the screen at any of a very wide variety of speeds. The dot then moves back to the left of the screen and the process is repeated very quickly. This is known as using the *time base.* Unless the speed is quite slow, the moving dot will appear as a horizontal line.

Like any voltmeter, an oscilloscope is connected across the component whose p.d. is to be investigated, using the y input. If, for example, a sinusoidal voltage was applied to the y input at the same time as a suitable time base was moving the spot in the x direction, then a *trace* such as shown in Figure 11.30 would be seen on the screen, provided the y gain (amplification) was adjusted to a suitable value.

1 If the squares shown in Figure 11.30 are each 1 cm × 1 cm and the settings on the oscilloscope are $50\,\text{V}\,\text{cm}^{-1}$ and $4\,\text{ms}\,\text{cm}^{-1}$, calculate:

a the peak voltage

b the rms voltage

c the time period

d the frequency.

Worked examples

5 If the resistor in the circuit shown in Figure 11.31 is 47 Ω and the supply voltage has a peak value of 12 V, what is the rms current in the circuit?

$$I_0 = \frac{V_0}{R} = \frac{12}{47} = 0.255\text{ A}$$

$$I_{rms} = \frac{0.255}{\sqrt{2}} = 0.18\text{ A}$$

6 Consider Figure 11.32, which shows the variation with time of a supply of alternating p.d.
a What is the peak value of the voltage?
b What is the rms voltage of this supply?
c What are the time period and frequency of the supply?
d If this voltage were connected to a 64 Ω resistor, what would be the peak and rms currents?

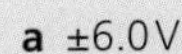

a ±6.0 V

b $\frac{6.0}{\sqrt{2}} = 4.2\text{ V}$

c $T = 8.0 \times 10^{-2}\text{ s}$

$f = \frac{1}{T} = 12.5\text{ Hz}$

d $I_0 = \frac{V_0}{R} = \frac{6.0}{64} = 0.094\text{ A}$

$I_{rms} = \frac{I_0}{\sqrt{2}} = \frac{0.094}{1.414} = 0.066\text{ A}$

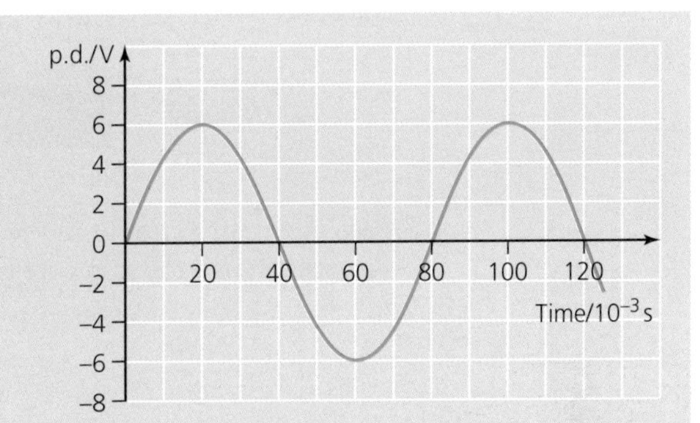

■ **Figure 11.32**

30 a If the peak value of the power in a circuit like that shown in Figure 11.31 is 14 W, what is the average power?
b If the rms value of voltage is 25 V, what is the value of the resistance?

31 a In a country where the mains electricity has a rating of 230 V, what is its peak value?
b What is the peak value of the current through an electric heater rated 2.15 kW in that country?
c What is the rms current through the heater?
d If the same heater was taken to a country where the voltage was rated at 110 V, at what average rate would it transfer energy?

32 What is the rms voltage across a light bulb that has an average power of 60 W when an alternating current of peak value 0.36 A flows through it?

33 a What is the average power dissipated in a 24.0 Ω resistor when an alternating voltage of peak value 150 V is applied across it?
b If the rms value of the voltage was doubled what would be the new power?
c What assumptions did you make?

34 A water heater rated at 2.0 kW is designed to work with a voltage rated at 250 V and frequency 60 Hz.
a Calculate the peak values of power, voltage and current.
b Sketch three graphs on the same axes to show how voltage, current and power vary over 0.03 s. Include numerical values.

■ Transformers

The transfer of electrical energy from power stations to communities requires the use of much higher voltages than is safe for use in homes (or than is produced by power stations). This means that there is the need to change (transform) voltages between different values, and the devices that do this are called **transformers**. Transformers are also needed in our homes whenever we need a voltage that is different from the mains supply. For example, we may wish to charge a mobile phone at 4 V in a country where electricity is supplied at 110 V.

Transformers use the principle of electromagnetic induction and because this requires time-changing magnetic fluxes, simple transformers can only operate with alternating current, not direct current. Figure 11.33 shows the structure of a simple transformer.

The input voltage that we want to transform is connected to a coil of wire known as the primary coil. The output voltage is taken from a separate coil known as the secondary coil. The two coils are wound on the same ferromagnetic (soft iron) core which usually forms a complete loop, even when (as is usual) the coils are wound on top of each other. The ferromagnetic core has a high magnetic permeability and this increases the magnitude of the magnetic flux passing between the coils.

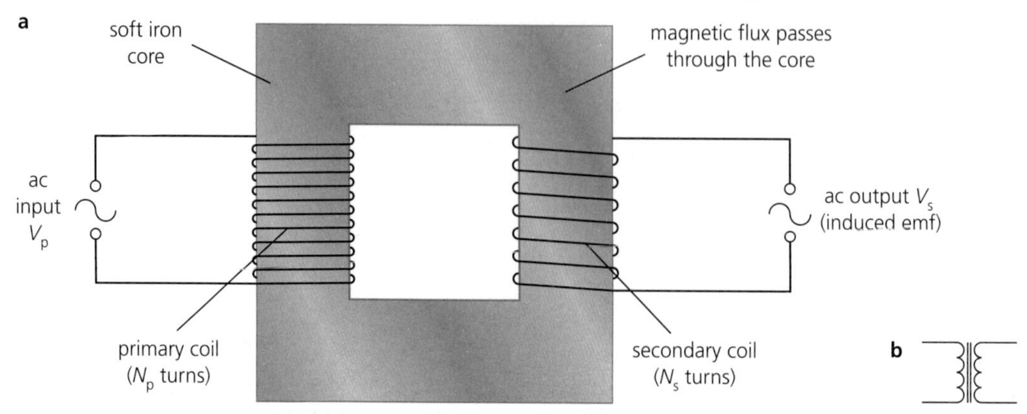

■ **Figure 11.33 a** The structure of a simple transformer and **b** the circuit symbol for a transformer

An alternating current in the primary coil creates an alternating magnetic flux, which then passes through the iron core to the secondary coil and induces an emf.

$$\varepsilon = -\frac{N\Delta\Phi}{\Delta t}$$

There are many factors that may affect the induced emf in arrangements such as that shown in Figure 11.33a and many interesting investigations are possible.

The higher the number of turns on the secondary coil, the bigger the output voltage (for a given arrangement).

For an ideal transformer, the input voltage and output voltage, and the numbers of turns are linked by the following equation:

$$\frac{\varepsilon_p}{\varepsilon_s} = \frac{N_p}{N_s}$$

This equation is given in the *Physics data booklet*.

This equation suggests that by choosing a suitable turns ratio (N_p/N_s), an ideal transformer could in theory transform any alternating voltage into any other voltage. A transformer used to increase voltages is called a step-up transformer. In contrast, step-down transformers reduce voltages.

An ideal transformer is 100% efficient, so that power into primary coil equals power out of secondary coil.

$$\varepsilon_p I_p = \varepsilon_s I_s$$

This can also be expressed as:

$$\frac{\varepsilon_p}{\varepsilon_s} = \frac{I_s}{I_p}$$

This equation is given in the *Physics data booklet*.

A transformer that steps up the voltage must step down the current. A step-down transformer must have an increased current in its secondary coil. In real transformers, such as that in Figure 11.34, there will be some energy dissipated into the environment.

■ **Figure 11.34** A transformer on a road-side pole

Worked example

7 An ideal transformer has 800 turns on its primary coil and 60 turns on its secondary coil.
 a Is this a step-up or a step-down transformer?
 b If a voltage of 110 V is applied to the primary coil, what is the output voltage?
 c Calculate the current through a 12 Ω resistor connected to the secondary coil.
 d Calculate the power in the secondary circuit.
 e What is the current in the primary coil?

a A step-down transformer, because there are fewer turns on the secondary coil.

b $\frac{\varepsilon_p}{\varepsilon_s} = \frac{N_p}{N_s}$, so $\frac{110}{\varepsilon_s} \times \frac{800}{60}$ and $\varepsilon_s = 8.3\,V$

c $I_s = \frac{\varepsilon_s}{R} = \frac{8.3}{12} = 0.69\,A$

d $P_s = \varepsilon_s \times I_s = 8.3 \times 0.69 = 5.7\,W$

e $I_p = \frac{P_p}{\varepsilon_p} = \frac{5.7}{110} = 0.052\,A$

f Because it is an ideal transformer we know that the input and output powers are equal.

Note that the input power and current can only be calculated after the power taken from the secondary circuit is known. If no current is taken out from an ideal transformer, then the input current and input power will also be zero.

Power losses in transformers

Up to this point, we have assumed that transformers are 'ideal', in other words that they are 100% efficient with no power losses. In practice, of course, this is not possible. The main reasons why some of the input power is dissipated to thermal energy in a transformer are as follows:

- The wires (*windings*) of the coils have resistance, so power ($P = I^2R$) is generated when currents flow through them. This is sometimes known as joule heating.
- Currents are induced in the *core* of the transformer as well as in the coils. These induced currents (*eddy currents*), swirling around in conductors, will result in joule heating.
- Energy is also transferred in the core as it is repeatedly magnetized and demagnetized. This is an example of an effect known as hysteresis, in which the properties of the system lag behind the effect producing the changes.
- *All* of the magnetic flux created in the primary coil will not pass through the secondary coil; there will be some 'leakage' of magnetic flux.

To reduce energy losses, thick wires of a metal with low resistivity (copper) should be used for the windings. The cores are also usually made with layers (*laminations*) of a ferromagnetic material separated by thin layers of insulation to greatly reduce eddy currents.

Larger transformers are usually the most efficient – up to 98% efficient. Even at this efficiency, large transformers may need to be cooled to prevent overheating. For example, a transformer working at a power of 100 kW and efficiency 95% must dissipate 5000 J of thermal energy safely every second. Smaller transformers are less efficient.

Most of the world uses electrical power with an alternating current frequency of 50 or 60 Hz. There are a number of conflicting factors affecting this choice, including the efficiency of transformers.

35 Make a list of the factors that could affect the induced emf in Figure 11.33a.

36 An ideal transformer has 600 turns on its primary coil.
 a If it transforms 240 V to 12 V, how many turns are on the secondary coil?
 b When the secondary coil was connected to a resistance, the current was 480 mA. What was the value of the resistance?
 c Calculate the power developed in the secondary circuit.
 d What was the current in the primary coil?

37 a Draw a fully labelled diagram of an ideal transformer capable of operating a 10 V, 16 W lamp from a 220 V ac supply. Include the output circuit.
 b What current flows in the primary circuit?

38 An ideal transformer with 36 turns on its primary coil is needed to step up a voltage from 110 V to 5.0 kV.
a How many turns should be wound on the secondary coil?
b If the input current was measured to be 0.55 A, what was the output current and power?

39 A turns ratio of 10 : 1 for a step-down transformer could be chosen from 10 : 1 or 100 : 10 or 1000 : 100 or 10 000 : 1000 etc. Suggest factors that might affect the choice of the actual numbers of turns on the transformer coils.

40 A mobile phone charger transforms 110 V to 3.8 V.
a If there are 2000 turns on the primary coil, how many turns should there be on the secondary coil?
b If the output power of the charger is 0.58 W, what is the current in the secondary coil?
c If the primary coil current is 5.9×10^{-3} A, what is the efficiency of the charger?
d What power would be dissipated by the transformer?
e How would you know if a mobile phone charger was inefficient?

41 Suggest possible reasons why larger transformers (and other electromagnetic devices) are usually more efficient than smaller ones.

Electrical power distribution

Larger power stations tend to be more efficient than smaller ones, and it is also less expensive to build fewer larger power stations than many smaller ones. This means that most people do not live or work near the power station that supplies them with electrical energy. Consequently, electrical power often has to be *transmitted* significant distances between the place where it is generated and the place where it is used. The cables that transfer electricity around the country are called **transmission lines** (or sometimes **power lines**).

Figure 11.35 Transmission lines transfer electrical power around countries

Power losses in transmission lines

We know from Chapter 5 that when a current, I, flows through a conductor of resistance R, some of the electrical energy is transformed to thermal energy. The power loss, P, can be calculated from $P = I^2R$.

Obviously, every effort must be made to keep the power loss in transmission lines as low as possible. This can be done by keeping the resistance of the cables and the current through them as low as possible.

The metal chosen for the conductor needs to have a low resistivity. However, resistivity needs to be balanced with economy, and using thicker cables of a cheaper metal rather than using thinner cables of a more expensive metal can cost less overall, assuming that they conduct equally well. (Remember $R = \rho L/A$ from Chapter 5.)

Other properties of the metal will also be important, such as its density, tensile strength and chemical reactivity. Overall, aluminium is usually considered the best choice for transmission lines, often with a steel core to increase strength (see Figure 11.36).

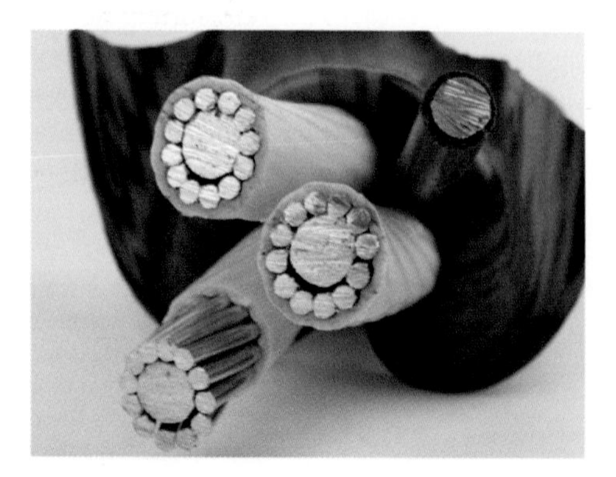

Figure 11.36 Part of a power line with separate inner cables

Why transformers are needed for the transmission of electrical power

In order to deliver a certain electrical power, P, to a community, in theory it is possible to use any current, I, and any voltage, V, for the transmission lines, as long as $P = VI$. In other words, if the current is to be reduced (to reduce energy dissipation), then the voltage must be increased by the same factor in order to deliver the power required.

Transformers are used to step up the voltages across transmission lines to as high a value as possible, so that the current can be decreased and the thermal energy wasted (I^2R) greatly reduced.

■ **Figure 11.37** Step-up transformers at a power station

Step-up transformers are sited as close as possible to power stations (see Figure 11.37). Decisions on the maximum value of voltages used with transmission lines are based largely on a combination of economic and risk factors, and there is no ideal value. The cost of the extra precautions needed to make extremely high voltages safe has to be balanced against the financial and environmental costs of the extra power losses that occur at lower voltages. In theory, stepping up the voltage by a factor of, for example, 1000 can reduce the current by the same factor; at the same time the power loss (I^2R) is reduced by a factor of $1000^2 = 10^6$.

Following transmission, the voltage has to be stepped down again to a value that is reasonably safe for use, for example, in homes and schools. In most countries this is in the range 200–250 V, although industries usually use a higher voltage. Step-down transformers are situated as close as possible to consumers.

(Note that the equation $P = V^2/R$ can be used with transmission lines to calculate the voltage drop *along* a cable. For example, if the power dissipated in a cable of resistance 2 Ω is 10^4 W, then the voltage drop along the cable is 140 V, which is not really significant if very high voltages are being used.)

The need to transform voltages explains why alternating currents are used – only ac currents induce the changing magnetic flux needed for transformers to work.

Figure 11.38 shows a block diagram of a simplified transmission system with some typical voltages. In reality, the voltages are stepped up and down by a series of transformers. The maximum voltage may rise to 750 kV or even higher.

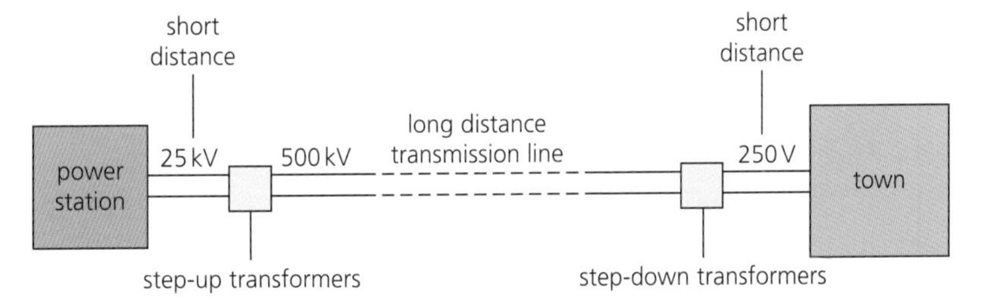

■ **Figure 11.38** A simplified power transmission system

Transmission lines in different parts of a country are usually all connected together in what is known as a **transmission grid**. The computer-controlled grid (Figure 11.39) helps balance supply and demand for electrical energy in different parts of a country, and sometimes also across international borders. Most people have become dependent on a continuous and reliable supply of electrical power and it is a high priority for electricity companies and governments to maintain that supply.

■ Figure 11.39 The Transmission Control Center of transmission network operator 50Hertz, Neuenhagen, Germany

Nature of Science

Computer modelling

Computer modelling (of power transmission, for example) must make simplifying assumptions, and it will not be able to take every inefficiency and variation into consideration. If the modelling of power transmission is based on an idealized system, then reality will involve lower powers because of inefficiencies that will never be 100% predictable. It is a major function of the system operators to appreciate the limitations of any computer model used and recognize any anomalies.

Worked example

8 **a** What is the total current supplied to 2000 homes each using an average of 2.4 kW of electrical energy, if it is supplied using a voltage of 230 V?

b Calculate the resistance of a transmission cable made of pure aluminium ($\rho = 2.8 \times 10^{-8}\,\Omega\,m$) that has a radius of 3.0 cm and a length of 15 km.

c At what rate would electrical energy be transferred to thermal energy if the current in **a** flowed through the resistance in **b**?

d Comment on your answer to **c**.

a $P = VI$

$2000 \times 2400 = 230 \times I$

$I = 21\,000\,A$

b $R = \rho L/A$

$$R = \frac{(2.8 \times 10^{-8}) \times (1.5 \times 10^4)}{(\pi \times 0.030^2)}$$

$R = 0.15\,\Omega$

(This value certainly does not seem like much resistance for a cable 15 km long!)

c $P = I^2R$

$P = 21\,000^2 \times 0.15 = 6.6 \times 10^7\,W$

d This power loss is far too high. It would be much more than the useful power delivered to the homes, making the process very inefficient. The cable would get very hot and melt.

42 Suggest reasons why copper is used for the wiring in houses but not for transmission lines.

43 A small town requires an average of 635 kW of electrical power and is supplied by a transmission line of total resistance 0.76 Ω. Calculate the power loss in the cables if the voltage used is:
a 1000 V
b 250 000 V.

44 Suggest why industries may prefer to be supplied with electrical energy at a higher voltage than homes.

45 Explain why transformers should be situated as close as possible to power stations and to homes.

46 a Suggest why it is desirable for the metal used in transmission lines to have a low density.
b What chemical properties may be needed of the metal?

47 a If the primary windings (of a coil) in a step-up transformer have 120 turns, how many turns are needed on the secondary coil if it is to transform 20.0 kV into 380 kV?
b What voltage is the output from a step-down transformer at the other end of the transmission line if it has 6800 turns on its primary coil and 80 turns on its secondary coil?
c What assumptions did you make?

48 a Suggest a reason why a transformer might work less efficiently at high frequencies.
b Suggest a reason why a transformer might work less efficiently at low frequencies.

49 A step-up transformer is used as part of a transmission system. It has 160 turns in its primary coil and 1920 turns in its secondary coil.
a If the input voltage is 1200 V, what is the maximum theoretical output voltage?
b If a current of 160 A flows in the primary coil, what is the input power?
c The step-up transformer is 96% efficient. Use your answer to a to calculate the current that flows through the secondary coil.
d How much power is dissipated to thermal energy in the transformer?
e If the output transmission line has a resistance of 0.18 Ω, what is the power loss in the cables?

Nature of Science

Competition and bias

We have explained the advantages of using alternating currents for the transmission of electrical power, which is now by far the most common system used in the world, but in the early years of electrical power production (the 1880s), direct current systems were also considered.

George Westinghouse and Thomas Edison were two famous American businessmen/inventors/engineers and their companies took opposing views on which was the best system for power transmission. This has since become known as the **'battle of the currents'**. Edison favoured a dc system in which many (smaller) power stations were to be much closer to homes, so that there would be no need for high voltages and transformers. (At that time there was no easy way to transform dc.) The whole system was intended to operate at approximately the same voltage, although a small drop would be expected between supplier and user. Because there was no need for high voltages, it was claimed that a dc system would be safer.

For the reasons already stated in this chapter, the ac systems proposed by Westinghouse have become normal practice around the world. But although such systems have predominated since those times, there have been some dc systems in operation in various countries around the world, especially in small areas of high population density.

There can be problems with the use of ac power transmissions when different systems are connected together (for example between countries operating at different frequencies). There are also significant power losses when ac is used to transfer energy under water. For these reasons, *high voltage direct current (HVDC) transmissions* are becoming more popular, although the conversion of voltages for dc is a more complicated and expensive process than for ac.

Additional Perspectives

Are the electromagnetic fields around power lines and in our homes a health risk?

Absorption of electromagnetic energy by the human body

Any oscillating current in any circuit will create changing electric and magnetic fields, which will spread away from the conductor. Mobile phones, wi-fi, TVs and radios, for example, all use this effect to great advantage with circuits and aerials designed to transmit and receive high-frequency oscillating electromagnetic fields efficiently (a typical mobile phone frequency is 900 MHz).

Human (and animal) bodies are conductors of electricity. So, if any changing electromagnetic fields produced by circuits that use alternating currents pass into a body, it is scientifically reasonable to expect that very small currents may be electromagnetically induced, much like eddy currents in a piece of metal, like a transformer core.

Some electronic equipment used in our homes (for example, computers) involves the use of high frequencies, but most of the electrical appliances in our homes use only the mains frequency of

50 or 60 Hz; the electromagnetic radiation from these can be described as *extra-low-frequency (ELF)*. Because the oscillating electromagnetic fields from mains electricity are ELF, the rate of change of magnetic flux is relatively small so that any induced currents will also tend to be very small. The corresponding wavelength is very long (confirmed by using $c = f\lambda$) and this also means that relatively little electromagnetic energy actually spreads away from the low-frequency circuits.

Possible health risks

Much research has gone on around the world into the effects of ELF radiation on the human body. Most scientists believe that the effects are negligible and that there is no evidence of any significant health risk, such as damage to genetic material. Typically, the strength of the magnetic fields a person might experience from electrical devices in their home is much smaller than the Earth's magnetic field strength.

Understandably, people tend to be more worried about the health risks of being close to transmission lines. Concerns about living near high-voltage power lines include leukaemia (especially in children), other cancers, depression and heart disease. However, there is currently no conclusive evidence making a link between these illnesses and living close to power lines.

Of course, it is certainly possible that, despite the current lack of definitive evidence, there *is* a health risk associated with living close to a power line (which could be confirmed by scientific research in the future). However, the fact that extensive research over many years has failed to make any significant connection suggests that any link, as yet unknown, may not be a great cause for concern.

Despite the lack of convincing evidence, a significant number of people still believe that there is a real risk to health, mostly because of reports they have read on the internet or in newspapers, or seen on the television. Such reports often tend to exaggerate any possibility of risks to make an interesting story, or fail to discuss any alternative scientific opinions. Of course, once the possibility of a risk is placed in their minds, many people believe that it is sensible to take precautions, even if that risk is very small or negligible. There are risks associated with everything we do as individuals or as a society, and before we worry too much about any particular activity it would be sensible to compare any possible risk with the risks of our other activities like crossing the road, taking a plane trip, smoking or eating a lot of fast-food.

Research is continuing into the possible effects of tiny induced currents in the human body. If there are any such effects, it would be reasonable to assume that they would be more pronounced with higher currents, higher frequencies or a longer time in the field. However, at present, there are no known biophysical processes in the human body that would make scientists expect that exposure to such ELF electromagnetic fields could affect genetic material (DNA) or otherwise be a danger to health.

If we consider that the electromagnetic energy is transferred by photons, the energy carried by the photons is believed to be too small to cause harmful reactions. Similar comments may be applied to the electromagnetic radiation from mobile phones.

1 a Calculate the wavelength of electromagnetic radiation of frequency 50 Hz.

  b Suggest a reason why electromagnetic wavelengths of this size do not spread away efficiently from household appliances (see Chapter 4).

2 a What amount of energy is carried by a single photon of frequency 50 Hz?

  b Suggest a reason why electromagnetic radiation of this frequency could be considered to be unlikely to affect the human body, regardless of the intensity or duration of the radiation.

  c The previous two questions use two different and conflicting theories about electromagnetic radiation which are linked by the concept of frequency. What name is given to this dilemma?

3 Use the website of the World Health Organization (WHO) to find out what their opinion is on the effects of:

a living close to sources of ELF radiation

b using mobile phones for long periods of time.

4 Explain why it would be reasonable to assume that if there were any health risks associated with exposure to ELF radiation then they would be worse with higher currents, higher frequencies or a longer length of time in the electromagnetic radiation.

ToK Link

The nature of statistical evidence

There is continued debate of the effect of electromagnetic waves on the health of humans, especially children. Is it justifiable to make use of scientific advances even if we do not know what their long-term consequences may be?

'Long-term consequences' may be very difficult to assess. Suppose, for example, after extensive research, it was shown in a report on transmission lines that for children between the ages of 5 and 15 who lived within 200 m of a high-voltage overhead line for longer than five years, there was, on average, a 10%-greater-than-average chance of developing a certain kind of cancer. Such information would be understandably very alarming for other people in similar situations and it would certainly receive a lot of publicity, with headlines such as 'Cancer risk confirmed'. The scientists and statisticians who produced the report may have stated clearly that people should not reach the conclusion that the transmission lines were definitely a health risk – but that comment would have undoubtedly received less attention.

Collecting statistical data like this is, by its nature, selective. Researchers have to choose what information they are looking for and it is not easy to be completely unbiased, especially if they start with the definite intention of looking for a certain connection. Other, relevant, information may have been overlooked, or perhaps the same data could be interpreted in a different way. For example, maybe over a period of 10 years the increased risk was much less than 10%, or maybe there was a decreased risk for people over the age of 15 – but such information will receive much less attention. Most scientists try to act with integrity and are well aware of the sometimes misleading nature of statistical evidence; there is rarely any intention to mislead.

Even if a connection between two sets of data is firmly established (that is, if there is a definite *correlation*), scientists may still be a long way from proving that one thing (for example, living near to a high-voltage power line) *causes* another (for example, an increased cancer risk). There is always uncertainty in the accuracy of any data and statistical data like this is subject to unpredictable variations and limitations. So, any correlation may be purely coincidental, although good research and analysis should reduce this possibility.

Even if statistical evidence is convincing, it may still be misleading. For example, it is a confirmed fact that, on average, taller children tend to be better at learning physics than shorter children. (Explanation at the end of this section.*)

There could be other explanations of a possible correlation. It might arise because the people living near the power lines are not typical of the general population in some way, or perhaps there is something else about the environment of the power lines that causes an increased cancer risk, or maybe there is a completely unknown factor having an influence. Such uncertainty means that confirming a definite cause and effect can be very difficult, and society and individuals have to make judgements based on the best evidence available. In Chapter 8, the increasing levels of carbon dioxide were correlated against increasing average global temperatures, leading most scientists to believe that global warming is mainly caused by gases released when fossil fuels are burned, but there are still some people who have doubts about the link.

(*Explanation: in this case, being taller is not the cause of this correlation. Growing taller and achieving a better understanding of physics are both consequences of getting older (in children). It is common for a correlation to exist between two facts because they are both caused by something else, which may not be known.)

Utilizations

Overhead or underground cables?

Almost all long-distance, high-voltage transmission lines are overhead – held well above the ground on pylons for safety reasons (see Figure 11.35). The cables themselves may be bare wires, without any insulation between them and the air, but very good insulators, like glass or ceramics, are needed where the cables are supported by the pylons (see Figure 11.40).

■ **Figure 11.40** Ceramic insulation on a pylon

The breakdown electric field strength for dry air is usually quoted to be about 3000 V mm^{-1}. This suggests that a potential difference of 3×10^6 V would be needed for a dangerous spark to jump across from a high-voltage transmission cable to a person or object 1 m away, and 6×10^6 V at 2 m apart, and so on. However, you are misled if you think that it might be safe to be within a few metres of an overhead cable. This is because the spark risk is unpredictable, depending very much on the shape and nature of the object close to the cable and the humidity of the air.

Many people think that pylons and transmission lines are ugly and spoil the beauty of the countryside, but the cost of burying cables underground is many times more expensive, partly because of the considerable extra insulation needed. When faults occur, the repair of underground cables is also slower and more expensive.

Similar considerations apply to the power cables used in towns and cities but because the voltages are lower, and because people are more concerned about safety and the appearance of their local surroundings, many communities prefer to pay the extra cost of burying cables underground. Alternatively, cables can be supported on poles at the side of the road. Although they may not be pretty to look at, they are quickly and easily repaired and extra cables can be added swiftly.

1 Look at Figure 11.41. Make a list of the advantages and disadvantages of delivering electrical power like this, compared with putting the cables underground.

■ **Figure 11.41** Delivering electrical power to homes

■ Rectifying alternating currents

Alternating currents are not suitable for use in electronic devices, like computers and televisions. This means that the ac power delivered to a home must be converted into dc for such uses. Converting ac to dc is called **rectification** and many electronic devices contain **rectifiers** inside them. Typically, electronic devices use voltages much lower than the mains supply to the home, so a transformer is also needed to step-down the voltage. The simplest kind of rectification is achieved with a single *diode*.

A **diode** is a semi-conducting component that allows current to pass through it in only one direction.

The circuit symbol for a diode is used in Figure 11.42, which shows a diode connected in series with the output of a transformer and a load resistor. Note that the direction of flow of conventional current through the diode is represented by the arrow in the symbol. We will assume that the resistance of a diode is very small for currents flowing forward (the p.d. at that moment is providing a **forward bias**) and very large for currents attempting to flow in the opposite direction (**reverse bias**).

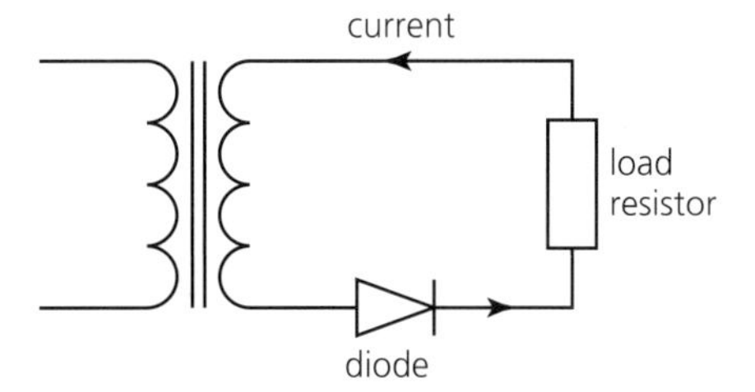

■ **Figure 11.42** A single diode used for rectification

Figure 11.43 shows a typical diode, but diode designs and sizes vary considerably, depending on their intended use.

■ **Figure 11.43** A typical semi-conductor diode (conventional current flows out of the marked end)

An oscilloscope connected across the load resistor will display voltage waveforms such as those shown in Figure 11.44. (The current waveforms will be proportional to the voltage for ohmic resistances.)

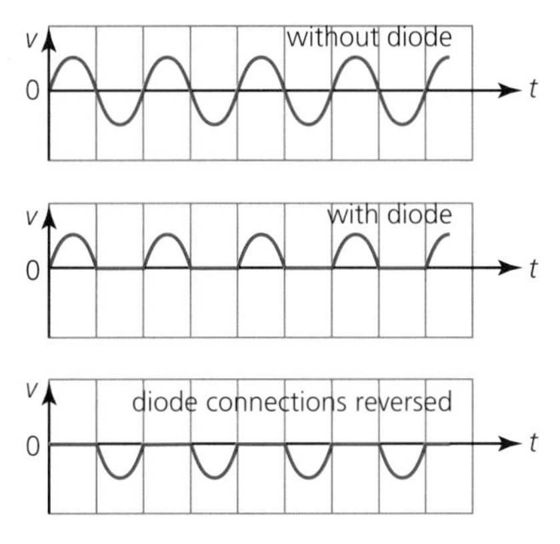

■ **Figure 11.44** Voltage waveforms for simple rectification

■ Half-wave and full-wave rectification

The lower waveforms in Figure 11.44 represent rectification, but the voltage is continuously changing and it is zero for half of the time! This is known as **half-wave rectification**. An effective numerical value for the voltage could be used to describe this output and it might be satisfactory for the operation of devices like simple lights and heaters, but for many other purposes this limited rectification is unacceptable. **Full-wave rectification** can be achieved easily with *four* diodes in a **bridge circuit**.

Diode bridges

Figure 11.45 shows a circuit for *full-wave rectification*. If the output from the transformer at any one moment is such that point A is positive (compared to F), the current will follow the route ABCDEF. A short time later the output of the transformer will make point F become positive, but the current will still flow through the load resistor in the same direction. Figure 11.46 shows the variation in voltage (proportional to the current) that would be observed by connecting an oscilloscope across the ohmic load resistor.

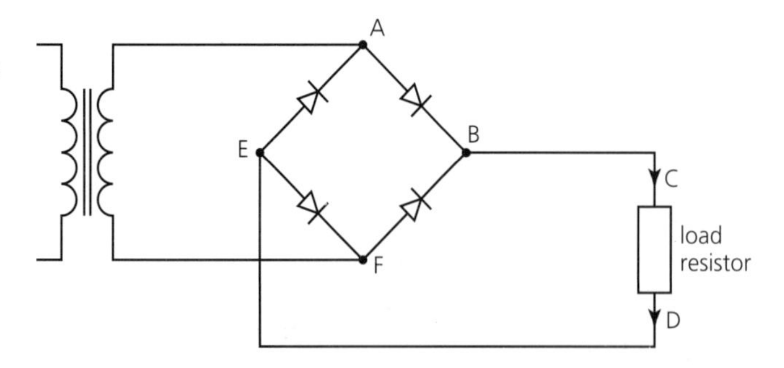

■ **Figure 11.45** Four diodes connected in a bridge circuit

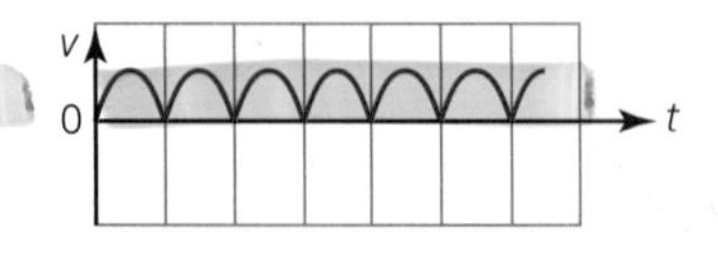

■ **Figure 11.46** Full-wave rectification (alternatively, the waveform may be inverted)

Smoothing the output

Full-wave rectification is an improvement on half-wave rectification, but the output voltage is still variable and will be unsatisfactory for many uses. The output can be smoothed by connecting a suitable *capacitor* in parallel across the resistor as shown in Figure 11.47.

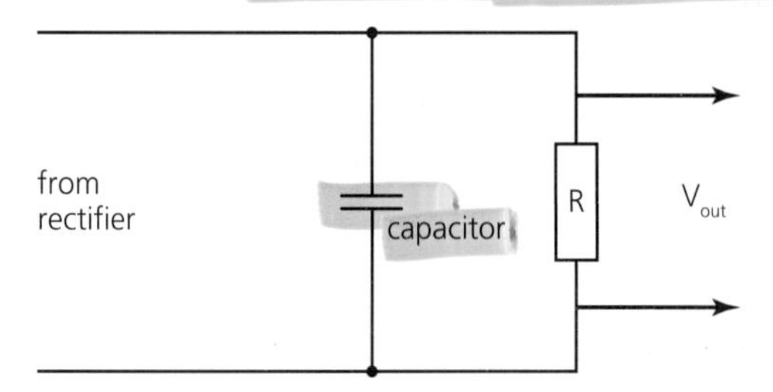

■ **Figure 11.47** Smoothing a voltage with a capacitor

Capacitors are explained in detail in Section 11.3 and a better understanding of capacitor smoothing may have to wait until then. In essence: when the output voltage from the rectifier circuit is rising, the capacitor becomes *charged* and the voltage across it rises. When the output from the rectifier is falling, the voltage across the charged capacitor falls much more slowly and this helps to maintain the voltage across *R* (see Figure 11.48). The values of C and *R* will affect the rate at which the voltage across the capacitor rises and falls.

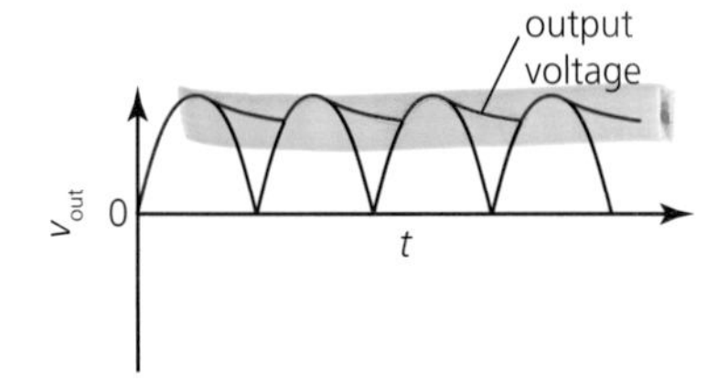

■ **Figure 11.48** Capacitor smoothing of full-wave rectification

50 Sketch a full V–I characteristic graph to show the behaviour of an ideal diode, including forward and reverse biasing.

51 Use the lettering on Figure 11.45 to indicate the flow of current when there is a positive voltage on point F (compared with A).

52 a Draw a circuit that will produce a smoothed low-voltage half-wave rectified output from a mains transformer.
b Make a sketch of how the output voltage varies with time.

53 Figure 11.49 shows a typical multi-purpose, low-voltage power supply used in school science experiments. The output is 'stepped' rather than continuously variable and it has full-wave rectification, but the output is not smoothed.
a What is the principle electrical component in this power supply?
b Suggest how different output voltages are made available.
c Should an electric heater be connected to the dc or ac output?

d Suggest other experiments for which this power supply may be:
  i suitable
  ii unsuitable.

e The power supply has a digital display for the output voltage. Why is it unlikely that the display will indicate 12 V when the switch is set to 12 V and an electric heater is connected to the output?

f The power supply will turn off automatically if the current becomes too large. Suggest how this might be achieved (the connection can be 'reset').

■ **Figure 11.49**

Additional Perspectives

Other bridge circuits

Other *bridge circuits* can make interesting investigations, but they are *not* part of the IB syllabus. The full-wave rectification circuit that we have discussed is a typical bridge circuit. Bridge circuits consists of two pairs of components (for example diodes for the rectification bridge) which are connected in parallel with each other and a voltage supply. They are called 'bridge' circuits because another component is connected between the four components, as in the further example shown in Figure 11.50. This is an historically important circuit used for measuring an unknown resistance, R_x. Its importance lies in the fact that no current is flowing through the galvanometer when the final measurement is made. This circuit is known as a *Wheatstone bridge*. A similar circuit involving capacitors is called a *Wien bridge*.

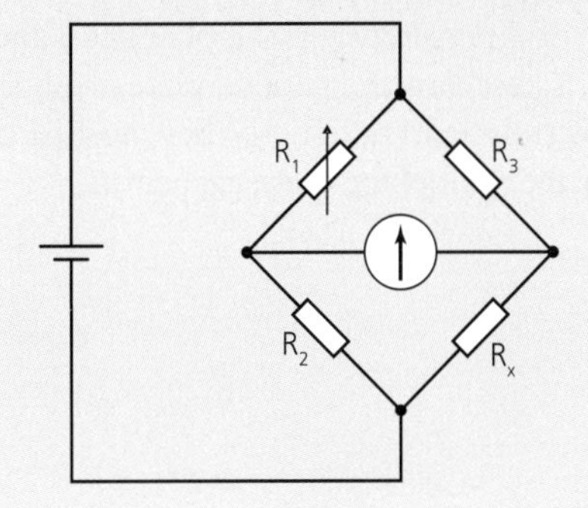

■ **Figure 11.50** A Wheatstone bridge circuit

11.3 Capacitance – *capacitors can be used to store electrical energy for later use*

The ability of any object to 'store' electric charge is described as its capacitance. An electrical component that is designed to store electric potential energy using charge separation is called a capacitor.

The simplest kind of capacitor comprises two parallel metal plates with an insulator between them.

■ Parallel plate capacitor

Consider Figure 11.51. When the switch is connected to A, electrons will be attracted to the positive terminal of the battery and repelled away from the negative terminal, so that one of the two parallel plates becomes positively charged while the other becomes equally negatively charged (as shown). This process is called **charging** the capacitor. As the charge on each plate increases, the addition or removal of more electrons becomes increasingly difficult because the charged plates will oppose their movement. A steady charge is reached when the potential difference across the plates due to the charges becomes equal to the terminal potential difference across the battery. The capacitor can be **discharged** by connecting the switch to B, so that the electrons reverse their movements. In this simple arrangement, charging and discharging are *very* quick processes because there is no resistance in the circuit. Note that charge never flows through the insulator between the plates (unless there is a fault, called a '*breakdown*').

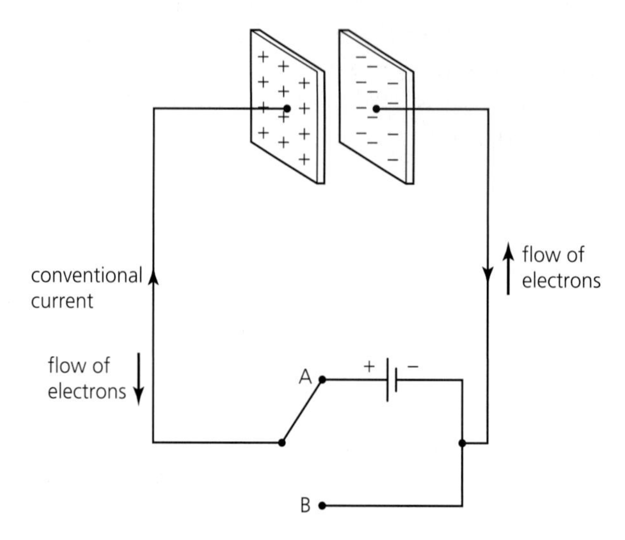

■ **Figure 11.51** Parallel metal plates connected to a battery

The amount of charge, q, on plates like these will be very small (typically picocoulombs), but how that charge depends on the size of the potential difference across the plates, V, or their area, A, or their separation, d, can be investigated with a high-voltage supply and a sensitive coulomb meter. The effect of putting different insulators between the plates can also be tested.

The circuit symbol for a capacitor is based on the parallel-plate arrangement (see Figure 11.52). (Some capacitors are electrolytic in design, so they must be connected correctly. For this reason a + sign is included in the symbol for these capacitors.)

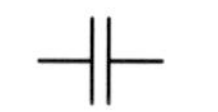

■ **Figure 11.52** The circuit symbol for a capacitor

Experimental results show that:

$$q \propto V$$

$$q \propto A$$

$$q \propto \frac{1}{d}$$

It can also be shown that $q \propto$ *electric permittivity*, ε, of the insulator between the plates.

Electric permittivity was introduced in Chapter 5 as a constant representing the electrical properties of an insulator. Relative permittivity, $\varepsilon_r = \varepsilon/\varepsilon_0$, where ε_0 is the permittivity of free space (a vacuum). In summary:

$$q = \varepsilon \frac{VA}{d}$$

This equation is *not* given in the *Physics data booklet.*

If the insulator between the plates is air or a vacuum, we can write:

$$q = \varepsilon_0 \frac{VA}{d}$$

This equation is *not* given in the *Physics data booklet.*

Capacitance

The capacitance, C, of a parallel-plate arrangement is defined as follows:

$$C = \frac{\text{the charge on the plates}}{\text{the potential difference between the plates}}$$

$$= \frac{q}{V}$$

This equation is given in the *Physics data booklet.*

The unit of capacitance is the **farad**, F (1 F = 1 C/1 V). Because this is a large unit, the smaller units millifarad (mF), microfarad (μF), nanofarad (nF) and picofarad (pF) are in common use. Note that the magnitude of the charge, q, is the same on *both* plates, because effectively the charge has been moved from one plate to the other.

(The symbol C, in italics, for capacitance should not be confused with C for coulomb.)

Comparing $C = \frac{q}{V}$ with $q = \frac{\varepsilon VA}{d}$ gives:

$$C = \varepsilon \frac{A}{d}$$

This equation is given in the *Physics data booklet.*

If a vacuum (or air) is between the plates we can use $C = \varepsilon_0 \frac{A}{d}$.

Dielectric materials

Capacitance can be increased by putting an insulator with the right electrical properties between the plates. Such materials contain polar molecules and are called **dielectrics** (or **dielectric materials**). Dielectrics are insulators with molecules that can be polarized in electric fields (see Figure 11.53).

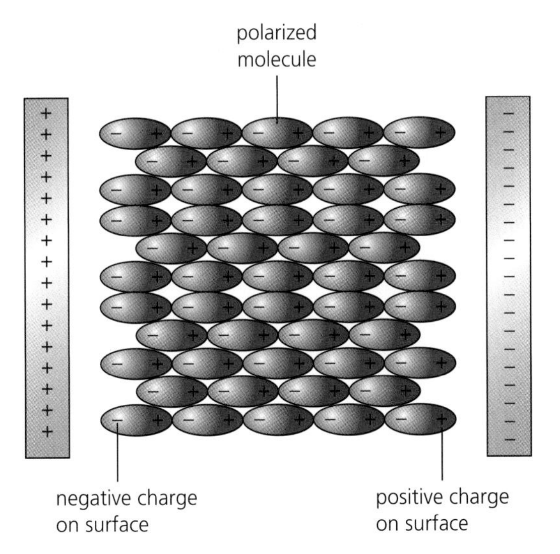

Figure 11.53 Polarized molecules in a dielectric between two charged plates

The overall effect of a dielectric is to increase the amount of charge on the plates of a capacitor (for a given potential difference). The polarized molecules create their own electric field which opposes the electric field between the plates, so that the net field between the plates tends to be reduced because of the dielectric. However, to maintain the capacitor at the potential difference provided by the source, V, more charge, q, must flow onto the plates. Because $C = q/V$, the capacitance, C, increases.

Relative permittivity is also sometimes called the **dielectric constant**.

Worked example

9 **a** Calculate the capacitance of two parallel metal plates, each of area 15 cm^2, when they are separated by 2.0 mm of air.

b What charge will flow onto each of the plates when a potential difference of 12 V is connected across them?

a $C = \varepsilon_0 A/d = (8.85 \times 10^{-12}) \times (1.5 \times 10^{-3})/(2.0 \times 10^{-3}) = 6.6 \times 10^{-12}$ F (or 6.6 pF)

b $C = q/V$

$6.6 \times 10^{-12} = q/12$

$q = 8.0 \times 10^{-11}$ C (or 80 pC)

54 A standard value for a capacitor used in electronic circuits is rated at 470 μF, 15 V. What will be the charge on this capacitor if it is connected to a 12 V supply?

55 Calculate the p.d. needed between parallel plates of area 47.6 cm^2 in order that a charge of 1.00 nC will accumulate on each plate when the separation of the plates is 2.46 mm in air.

56 Determine the capacitance of parallel plates of area 4.9 cm^2 when polythene of relative permittivity 2.25 and thickness 0.18 mm is sandwiched between them.

57 When a pair of parallel plates with an insulator sandwiched between them had a p.d. of 1550 V applied across them, a charge of 3.7 μC accumulated on each of the plates. Determine the permittivity of the insulator if it had a thickness of 0.074 mm and the area of the plates was 15.8 cm^2.

58 Find out which dielectric materials are the best choices for increasing the capacitance of capacitors.

59 A layer of aluminium oxide used to make a capacitor is 2.3×10^{-7} m thick. If the plates of a 1.0 μF capacitor have an area of 24 cm^2, what is the dielectric constant of the aluminium oxide?

60 A flow of charge around a circuit is sometimes compared to the flow of water around pipes produced by the action of a pump. Suggest how a piece of *flexible* rubber used to block the flow of water through a pipe can be considered analogous to the action of a capacitor.

61 Figure 11.54 shows a variable capacitor. Suggest what causes the capacitance to change when the knob is rotated.

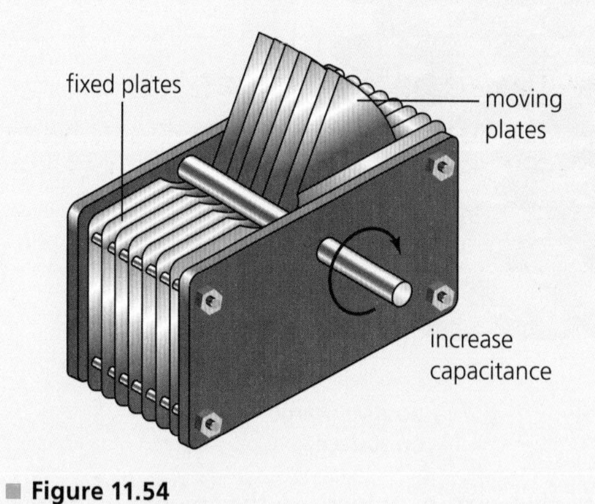

Figure 11.54

Practical capacitors and their uses

Capacitors have a very wide range of uses, including

- smoothing voltages (as discussed in Section 11.2)
- storing relatively small amounts of energy, often followed by rapid release
- tuning circuits (for example, in radios)
- timing circuits

- filtering out direct currents in an ac circuit
- maintaining voltages in the event of power failures (such as to laptop computers).

Capacitors are available in many different designs (see Figure 11.55), but the ability to store more charge for a given voltage can only be achieved by a larger physical size (consider $C = \varepsilon_0 A/d$). Designing capacitors with significant capacitance requires that the dielectric layers are very thin (which can sometimes result in breakdown faults occurring), and imaginative ways of increasing the areas of the plates are needed. This is most commonly achieved by rolling very thin plates and dielectric into the shape of a cylinder.

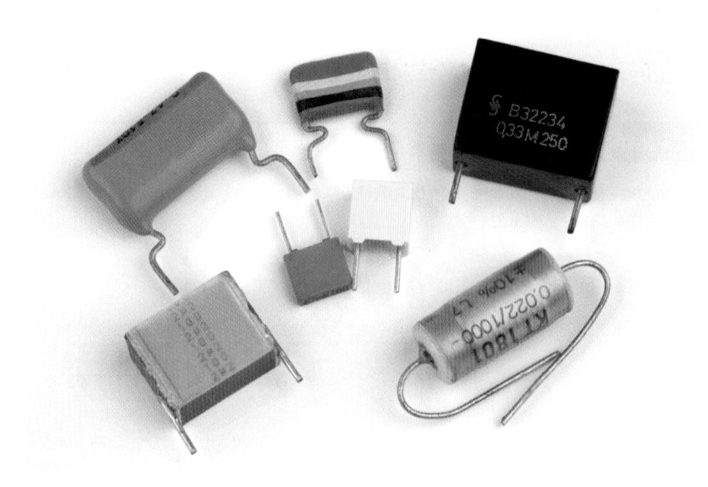

Figure 11.55 A range of different capacitors

62 Explain why large-value resistances can be very small in size, but large-value capacitors must be large in size.

63 Suggest why a capacitor is used as the energy source for the flash in a camera.

64 What problems may arise in a capacitor that uses a very thin layer of dielectric?

Energy stored in a charged capacitor

As we have seen, the charge on a capacitor is proportional to the p.d. across it. This relationship is represented in the graph shown in Figure 11.56. The gradient of this graph is equal to the capacitance, $C = \Delta q/\Delta V$.

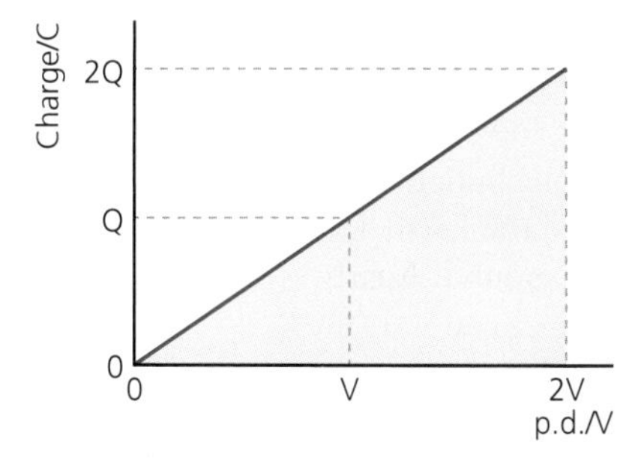

Figure 11.56 The charge–voltage relationship for a capacitor

Work has to be done to move electrons to or from the plates because the forces from the charges already on the plates have to be overcome. We know that, in general, work done = p.d. × charge (from the definition of p.d.), but neither the p.d. nor the charge are constant. Because the p.d. and charge are proportional to each other we can use the following:

$$\text{energy transferred to a capacitor} = \text{average p.d. (between 0 and } V) \times \text{charge on capacitor when p.d. is } V$$

$$E = \frac{1}{2}V \times q$$

but note also that:

Energy stored in a capacitor equals the area under a q–V graph.

Note that if V is doubled, q is also doubled, so that the energy stored is $\times 2^2$. This is shown more clearly in the following equation, which is obtained using $C = q/V$.

$$E = \frac{1}{2}CV^2$$

This equation is given in the *Physics data booklet*.

Worked example

10 Calculate the energy stored in a 6800 μF, 1.5 V capacitor and compare it to the energy stored in an AA battery of the same voltage.

Capacitor: $E = \frac{1}{2}CV^2 = \frac{1}{2} \times (6800 \times 10^{-6}) \times 1.5^2 = 7.7 \times 10^{-3}$ J

Battery: $E = VIt = 1.5 \times 1.0 \times 7200$ (an estimate made assuming that the battery can provide 1.0 A for 2 hours)

$E \approx 10^4$ J

Assuming that the volume of the battery is about ten times greater than the capacitor, the energy density in the battery is about 10^5 times greater than the energy density in the capacitor. Clearly, capacitors are not suitable for storing large amounts of energy. However, capacitors can release their stored energy very quickly (unlike batteries which have significant internal resistance and involve more time-consuming chemical processes).

The 'flash' used in a camera, or mobile phone, requires a p.d. of a few hundred volts (so the camera must include a small 'transformer') and the light emitted needs to have a high intensity. This means that the electrical power and current need to be relatively large. A capacitor is capable of providing this power because the duration of the flash is small, so the overall energy is not large. The charging of capacitors in cameras takes a few seconds because the current flows through a resistance in the circuit. Capacitors that are charged by large voltages can be dangerous because of the large initial currents they can deliver.

The energy stored in a large capacitor can be used to light a small bulb or turn a motor for a short time. If a joulemeter is available, it can be used to measure directly the energy that flows off a capacitor.

Additional Perspectives

Thunder and lightning

There are estimated to be more than one billion lightning strikes every year on or above the Earth. Of these about 25% involve strikes to the ground. Apart from being a risk to human life, lightning can start fires and cause considerable damage to buildings and electrical systems of any kind. Tall buildings and important electrical installations are designed to be protected against lightning.

Figure 11.57 A lightning flash

Lightning, like that shown in Figure 11.57, will strike from a cloud to the ground (or another cloud) when the strength of the electric field between them becomes large enough to make the air conduct. The charged cloud and the Earth below it may be considered as a very large capacitor. The air between acts as the dielectric material, but when the p.d. gets too large it breaks down and conducts in dramatic style.

Clouds contain droplets of water and ice particles, all of various sizes. These are continuously moving around and colliding with each other. During this process, charge (electrons) can be transferred between particles. This can result in a large-scale charge separation, with the top of a cloud becoming positively charged and the bottom of the same cloud becoming negatively charged as shown in Figure 11.58. This is just a simplification of what actually happens because the exact details are still not fully understood.

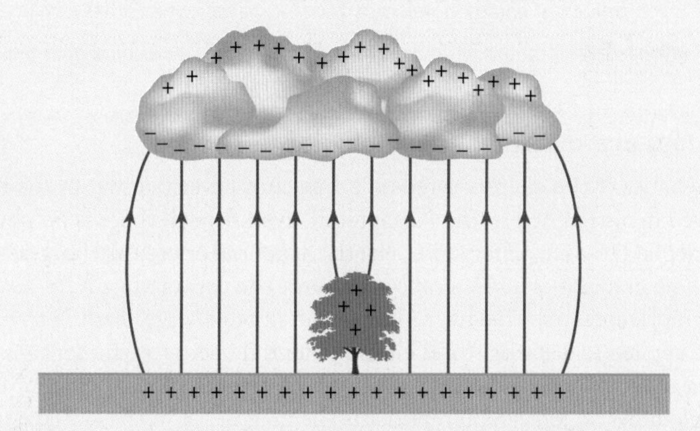

■ **Figure 11.58** Electric field under a cloud

The large negative charge at the bottom of the cloud will repel electrons on the Earth's surface and objects on it, leaving them positively charged. A very large electric field is created between the ground and the bottom of the cloud. When the charge separation and the resulting field become large enough, a conducting path is created and lightning will strike along the path of least resistance. This is a complicated process, with charge flowing from both the Earth and the cloud. Because of the high current involved in lightning, this is a very powerful process and a lot of energy is transferred to radiation (including light) and internal energy. The extremely rapid heating of the air by thousands of degrees causes rapid expansion and big pressure changes, resulting in the emission of sound (thunder).

Electric fields are stronger near pointed objects, so anything that 'sticks out' from its surroundings generally has a greater risk of being struck by lightning. But, the electric field inside a conducting surface is always zero, so people inside cars and houses, for example, should be safe from direct lightning strikes.

Apart from the large electric fields created by charged clouds, there is a permanent, but weak, electric field of about $150\,\mathrm{N\,C^{-1}}$ above the surface of the Earth, directed downwards. This is because of a layer of positive particles surrounding the Earth (the ionosphere), which is caused by radiation from the Sun.

1 What safety advice would you give to someone caught outside in a thunder storm?

2 Research the damage that can be done to buildings that are struck by lightning. How can such damage be prevented?

3 Research stories of people who have survived lightning strikes.

4 Find out which parts of the world have the most lightning strikes every year.

65 How much energy can be stored on two parallel plates of area 130 cm^2 and separated by 1.5 mm of air when a p.d. of 24 V is connected across them?

66 a What voltage would be required across the terminals of a 0.15 F capacitor in order to store 100 J of energy?
b What problems may occur if a voltage of 40 V is applied across the capacitor when it is recommended for use only for voltages up to 10 V?

67 The capacitor used in a camera has a value of 100 µF. It delivers energy to the flash at a p.d. of 280 V for a time of 0.0017 s. Determine:
a the total energy stored on the capacitor
b the power of the flash (assuming it is constant during the flash and all the energy in the capacitor is used).

68 Sketch a q–V graph for a 470 µF, 20 V capacitor.

69 Determine the amount of energy stored on a 2.7 mF capacitor when it has a charge of 5.9×10^{-3} C on it.

70 Use the internet to learn about ultra-capacitors. Write a brief summary of your findings.

Nature of Science

Relationships and mathematical analogies

The mathematics of the energy stored on a capacitor is very similar to the mathematics of the energy stored in a stretched spring (Chapter 2), even though there is no physical similarity.

- A p.d. applied to a capacitor stores electric potential energy as charges are moved onto the plates. The gradient of a q–V graph represents the capacitance ($C = \Delta q/\Delta V$). The area under a q–V graph represents the magnitude of the stored energy, leading to $E = \frac{1}{2}CV^2$.
- A force applied to a spring stores elastic potential energy as the spring is stretched. The gradient of an extension–force (x–F) graph represents 1/force constant ($k = \Delta F/\Delta x$). The area under an x–F graph represents the magnitude of the stored energy, leading to $E = \frac{1}{2}kx^2$.

There are many mathematical analogies in physics and we will soon meet another in this chapter – the mathematics of exponential changes has many applications, including the theories of capacitors and radioactivity.

Capacitors in series and parallel

There are times when we may need to work out the total capacitance of two or more capacitors connected together in a circuit. For example, we may want to know how to connect capacitors of standard values to achieve a required overall capacitance.

Capacitors in parallel

Consider three (or more) capacitors connected in parallel, as shown in Figure 11.59. The single capacitor, $C_{parallel}$, which has the same capacitance as the combined capacitance of the others, will have a total charge of $q_1 + q_2 + q_3$ on its plates. That is:

$q_{parallel} = q_1 + q_2 + q_3$, but $q = VC$, so that:

$VC_{parallel} = VC_1 + VC_2 + VC_3$

Dividing by V, which is the same for all:

$$C_{parallel} = C_1 + C_2 + C_3 \ldots$$

This equation is given in the *Physics data booklet*.

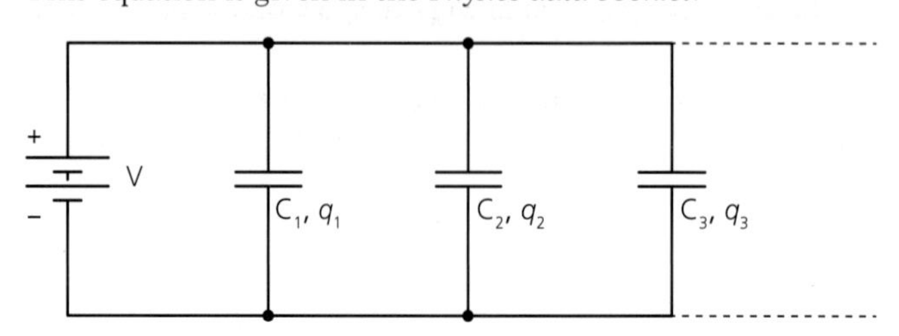

Figure 11.59 Capacitors connected in parallel

Capacitors in series

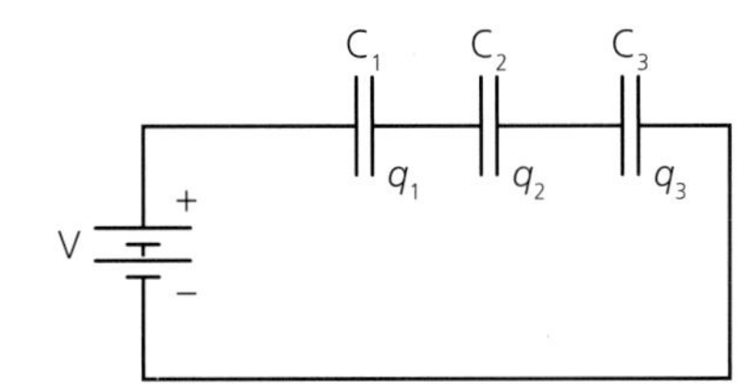

■ **Figure 11.60** Capacitors connected in series

Figure 11.60 shows three capacitors connected in series. We know that the same amount of charge, q, flows onto each capacitor and $V = V_1 + V_2 + V_3$, but $V = q/C$, so that the single capacitor, C_{series}, which has the same capacitance as the combined capacitance of the others, can be found from:

$$\frac{q}{C_{series}} = \frac{q}{C_1} + \frac{q}{C_2} + \frac{q}{C_3}$$

Cancelling q gives:

$$\frac{1}{C_{series}} = \frac{1}{C_1} + \frac{1}{C_2} + \frac{1}{C_3} \ldots$$

This equation is given in the *Physics data booklet.*

The value of an unknown, uncharged capacitor (or combination of capacitors) can be determined by connecting it in parallel with a charged, known capacitance, as shown in Worked example 11.

Alternatively, an unknown capacitance may be determined from the rate at which the discharge current (or p.d. across the capacitor) decreases with time. This will be covered in greater depth next.

Worked example

11 A capacitor of 0.10 F was charged by connecting it to a 5.9 V supply. When it was disconnected from the supply and then connected in parallel with another uncharged capacitor of unknown capacitance, the voltage across them fell to 0.71 V.

a Calculate the charge on the first capacitor before the second capacitor was connected.
b Determine the combined capacitance of the two capacitors in parallel.
c What was the unknown capacitance?

a $Q = CV = 0.10 \times 5.9 = 0.59\,\text{C}$

b $C = Q/V = 0.59/0.71 = 0.83\,\text{F}$

c $C_{parallel} = C_1 + C_2$

$0.83 = 0.1 + C_2$

$C_2 = 0.73\,\text{F}$

71 Calculate the combined capacitance of three capacitors of values 1000 μF, 2000 μF and 3000 μF connected in:
a parallel
b series.

72 If only 470 μF capacitors are available, suggest how you could combine four of them to make an overall capacitance with a value within 1% of 285 μF.

73 Two 680 μF capacitors were connected in parallel with each other and this combination was connected to a third capacitor of the same value in series. What was the combined capacitance?

74 A 470 μF capacitor was fully charged by connecting it to a 15.1 V supply. When it was removed from the voltage supply and connected in parallel across an uncharged capacitor, the voltage fell to 11.7 V.

a What was the capacitance of the second capacitor?

b Calculate the energy stored on the capacitors before and after they were connected.

c Account for the difference.

Discharging a capacitor through a resistance

The rate at which a capacitor discharges (or charges) depends on the size of the resistance in the circuit. (Up to now we have assumed that there is no resistance, so that charging and discharging occur very quickly.) As we will explain, the size of the discharge current and the rate at which the capacitor discharges can be controlled easily, and this is one reason why capacitors are such important circuit components. The currents and voltages associated with capacitors change with time, so that capacitors are essential components for circuits that have time-varying outputs, like timers and signal generators.

Resistor–capacitor (RC) series circuits

Consider the circuit shown in Figure 11.61. The capacitor can be very quickly *charged* by connecting the switch to A because there is no resistor involved. When the switch is then connected to B, the capacitor will *discharge* at a much slower rate, depending on the values of C and *R*. If the discharge rate is not too fast, measurements of current can be taken at regular intervals to enable a graph to be drawn (see Figure 11.62 for an example). Alternatively, an oscilloscope connected across the resistance can display how V_R (proportional to current) varies with time.

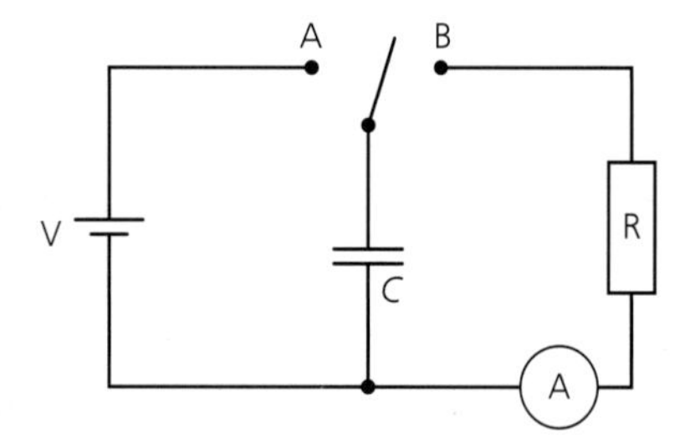

Figure 11.61 Circuit to investigate capacitor discharge current

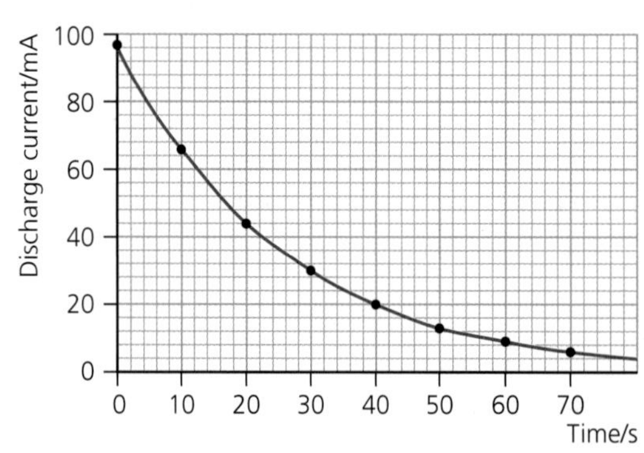

Figure 11.62 Typical variation of capacitor discharge current with time

The current drops most quickly to begin with, but then the rate of change of current (which could be determined from the instantaneous gradient of the line) becomes less and less. This is because, as the amount of charge remaining on the capacitor decreases, the forces on the electrons decrease in proportion.

This kind of change, in which a rate of decrease in a quantity is proportional to the amount remaining, is called an **exponential change** (in this case, an exponential decrease or 'decay').

One consequence of an exponential change is that the ratio of the quantities at the beginning and end of any specified time interval is always the same. This can be used to check data quickly to see if they represent an exponential relationship.

From Figure 11.62, the successive currents (mA) every ten seconds are 97, 66, 44, 30, 20, … If this is an exponential relationship 97/66 = 66/44 = 44/30 = 30/20… These are all approximately the same (1.5), so an exponential decrease is confirmed.

Graphs representing how the charge on the capacitor, and the p.d. across the capacitor ($V = q/C$), vary with time will both have the same exponential shape as that shown in Figure 11.62.

All exponential decreases can be represented by equations of the form:

$$x = x_0 e^{-kt}$$

where:

x_0 is the value of a quantity at the start of time t

x is the value at the end of time t

e is a number which occurs naturally when dealing with exponential changes; it has the value 2.718

k is the constant that characterizes the particular exponential decrease being considered. Larger values of k mean quicker changes.

Time constant

The value of a quantity that is decreasing exponentially with time will (in theory) never become zero. This means that it is impossible to quote a time for the capacitor to discharge completely. Instead, we quote the time interval that is needed for the quantity to decrease to a certain fraction (or percentage) of its value at the start of that time interval. For example, we might choose to characterize a particular exponential decrease by saying that it always takes 17 s for the quantity to fall to half of its value. The concept of *half-life* was introduced in Chapter 7 in order to characterize the rate of a radioactive decay, which is also exponential in nature. However, the concept of half-life is *not* usually used to describe capacitor circuits.

For mathematical convenience, the time chosen to characterize an exponential decrease is the time taken to fall to 1/e (= 1/2.718) or approximately 37% of its earlier value. (There is no need to be able to explain this.) This is called the **time constant**, τ, of an exponential decrease.

The time constant of the decrease shown in Figure 11.62 is approximately 26 s.

As we have said, the rate at which a capacitor discharges depends on the capacitance, C, and resistance, R, in the circuit, and it can be shown that:

time constant, $\tau = RC$

This equation is given in the *Physics data booklet.*

The unit of a time constant is s.

The meaning of time constant is shown on the charge–time graphs given in Figures 11.63 and 11.64. Graphs of I–t or V–t would show the same characteristics.

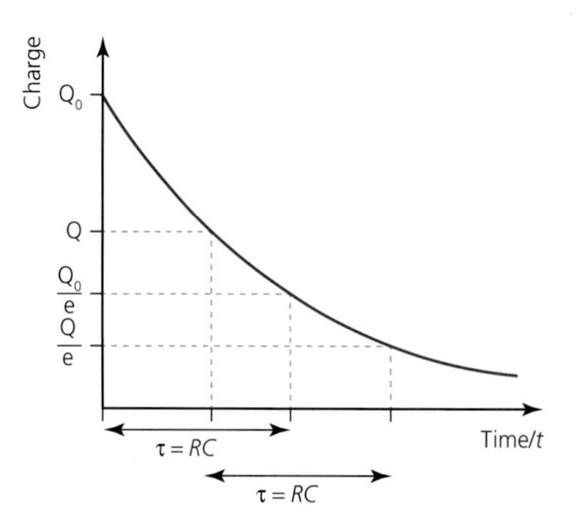

■ **Figure 11.63** Capacitor discharge graph showing the meaning of time constant

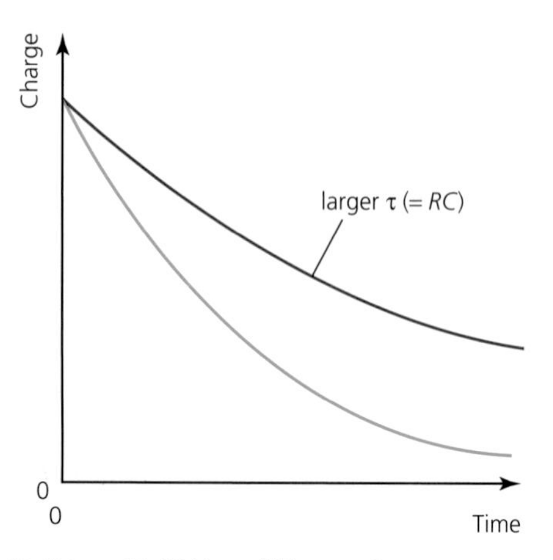

■ **Figure 11.64** How different time constants affect the discharge

If the time constant can be determined from a graph of a capacitor discharge through a known resistor, then a value for an unknown capacitance can be calculated.

Worked example

12 The current at the beginning of a capacitor discharge (as Figure 11.63) is 45.6 mA. If the capacitance is 470 μF and it is discharging through a 2.0 kΩ resistor.

a What is the time constant?

b What will the current be after:

i one time constant

ii two time constants?

c What was the initial voltage across the resistor (and capacitor)?

a $\tau = RC = (470 \times 10^{-6}) \times (2.0 \times 10^{3}) = 0.94\,\text{s}$

b Because the current falls to 37% for every time constant period:

i $0.37 \times 45.6 = 17\,\text{mA}$

ii $0.37^2 \times 45.6 = 6.2\,\text{mA}$

c $V = IR = (45.6 \times 10^{-3}) \times (2.0 \times 10^{3}) = 91\,\text{V}$

75 The voltage across a discharging capacitor falls from 10 V to 3.7 V in a time of 76 s. If the resistance of the circuit was $2.8 \times 10^3\,\Omega$, what was the value of the capacitor?

76 The charge on a capacitor falls from 4.2×10^{-5} C to 2.8×10^{-5} C in 10 s as it discharges through a 68 kΩ resistor.

a Sketch a charge–time graph for the discharge over the course of 40 s.

b Use the graph to estimate a value for the capacitor.

77 The discharge current of a 470 μF capacitor through an unknown resistance falls from 87 mA to 12 mA in exactly two minutes. Estimate the value of the unknown resistance.

78 Show that the SI unit for the time constant is s.

79 a Use Figure 11.62 to determine the rate of change of current with time at 10 s, 30 s, 50 s and 70 s.

b Check whether the ratios of successive rates are constant, which would be a further confirmation of the exponential nature of the graph.

80 A teacher wants to use a 0.05 F capacitor and a 0–10 mA ammeter to demonstrate capacitor discharge in a time of no more than 5 minutes. Suggest suitable values for the charging voltage and the resistance.

81 Suggest a reason why half-life is considered suitable for radioactive changes but not for capacitors.

Equations for capacitor discharge

The equations describing the exponential decrease of charge on a capacitor and the discharge current and the p.d. across the capacitor are as follows. All three are listed in the *Physics data booklet*.

$$q = q_0\, e^{-t/\tau}$$
$$I = I_0\, e^{-t/\tau}$$
$$V = V_0\, e^{-t/\tau}$$

These equations enable us to calculate a value for a quantity at *any time* if we know a previous value and the time constant (RC).

Taking natural logarithms of the current equation, we get the following (which is *not* in the *Physics data booklet*).

$$\ln I = \ln I_0 - \frac{t}{\tau}$$

(Or $\ln I_0/I = t/\tau$. Similarly, $\ln Q_0/Q = t/\tau$ and $\ln V_0/V = t/\tau$.)

If a graph of $\ln I$ (or $\ln Q$ or $\ln V$) is drawn against t it will appear as shown in Figure 11.65.

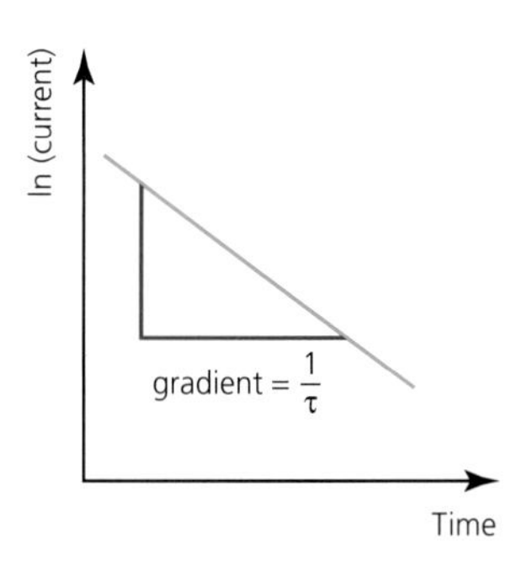

■ **Figure 11.65** Graph of natural logarithm of current against time

Comparing the equation of the line, $\ln I = -t/\tau + \ln I_0$, to the general equation for a straight line ($y = mx + c$), we see that the magnitude of the gradient of the line is equal to $1/\tau$. This is the most accurate way of determining the time constant when using the results from an experiment investigating a discharge current or voltage (see question 8).

Worked examples

13 The voltage across a 2000 μF capacitor is 12.4 V when it starts discharging through a 3250 Ω resistance. How long will it be before the voltage has fallen to exactly 1 V?

$\tau = RC = (2000 \times 10^{-6}) \times 3250 = 6.500$

$V = V_0 e^{-t/\tau}$

$1.00 = 12.4 e^{-t/6.500}$

$\ln\left(\frac{12.4}{1.00}\right) = \frac{t}{6.500}$

$t = 16.4\,\text{s}$

14 In an experiment to determine the value of an unknown capacitor, a discharge current through it was found to fall to 10% of a previous value in a time of 39 s. If the discharge was through a resistor of value 57.6 kΩ, what was the unknown capacitance?

$I = I_0 e^{-t/\tau}$

$0.1 I_0 = I_0 e^{-39/(57.6 \times 10^3)C}$

$\ln 10 = \frac{39}{57.6 \times 10^3}C$

$C = 2.9 \times 10^{-4}\,\text{F}$

82 The voltage across a discharging capacitor was measured every 10 seconds for 1 minute. The recorded values (V) were: 9.7, 8.5, 7.5, 6.6, 5.8, 5.1 and 4.5.
a Plot a graph of the natural logarithm of the voltage against time.
b Use your graph to find a value for the time constant.
c Confirm your result by calculation from the original data.
d Why is the graphical method usually considered better?
e If the circuit resistance was 1.0 kΩ, what was the capacitance in the circuit?

83 When a capacitor of 0.01 F discharges through a 10 kΩ resistor, how much of the original charge of 5.7×10^{-3} C remains on the plates after 54 s?

84 The p.d. across a 1000 μF capacitor was observed and after 80 s it had fallen to 2.3 V. If the resistance of the circuit was 0.1 MΩ, what was the initial value of the p.d.?

85 When a 680 μF capacitor discharged through an unknown resistance, the current fell from 3.3 mA to 1.9 mA in 30 s. Determine the value of the resistance.

Nature of Science

Exponential changes are common

The mathematics of exponential changes is used for two major topics in the Higher level core of this course (capacitor discharge and radioactive decay), but there are many other places where it could be applied. For example, the rate of cooling of a hot object (like a cup of coffee) may be considered to be proportional to the temperature difference from its surroundings. This could be investigated by checking if the temperature decreases (approximately) exponentially with time. It is unlikely that the relationship would be perfectly exponential because the mathematical model is a simplified representation of reality.

Exponential relationships (both growths and decreases) are found in most other branches of science, and they are also useful for modelling processes in many numerate areas of knowledge like business, geography and economics.

Charging a capacitor through a resistance

When an uncharged capacitor is connected in series with a resistance and p.d., the current is high to begin with, but then *falls* exponentially, for the same reasons as with discharging (although the current will flow in the opposite direction). The charge on the capacitor and the voltage across it will increase to limits at decreasing rates, as shown in Figure 11.66, which compares charging and discharging graphs.

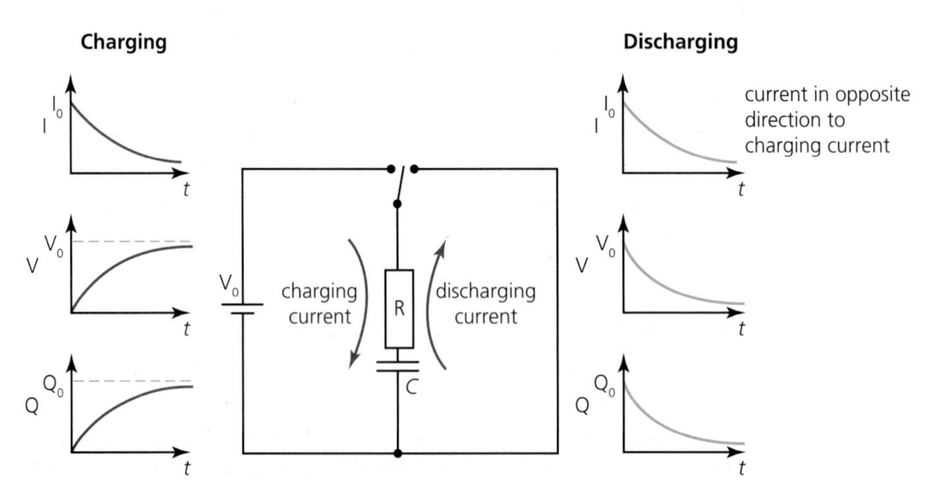

Figure 11.66 Charging and discharging circuit and graphs

The rate of the charging process is represented by the same time constant ($\tau = RC$) as for discharging:

- After one time constant, V and Q rise to $(1 - 0.37) = 0.63$ (or 63%) of their maximum values.
- After two time constants, V and Q rise to $(1 - 0.37^2) = 0.86$ (or 86%) of their maximum values.
- After three time constants, V and Q rise to $(1 - 0.37^3) = 0.95$ (or 95%) of their maximum values...

86 A 100 μF capacitor is charged through a 100 kΩ resistor using a battery with a terminal p.d. of 1.5 V. Determine:
a the p.d. and
b the charge on the capacitor after 20 s.
c What is the current through the resistor at this time?

87 Sketch a charge–time graph (for 60 s) for a 50 mF capacitor being charged through a 1200 Ω resistor by a 12 V battery.

88 Sketch a current–time graph (for 10 s) for a 360 μF capacitor being charged through a 8.2 kΩ resistor by a 6 V battery.

Additional Perspectives

Measuring capacitance with a reed switch

In this experiment an electromagnetic *reed switch* controlled by a high-frequency alternating current is used to repeatedly charge and discharge a capacitor (see Figure 11.67). Because the frequency is high, the microammeter will show a steady current and the total charge flowing through it every second is equal to the current. Dividing the current by the frequency will give the average charge flowing off the capacitor in each discharge. The capacitance can then be determined from $C = q/V$. The capacitor is assumed to discharge (almost) completely every cycle.

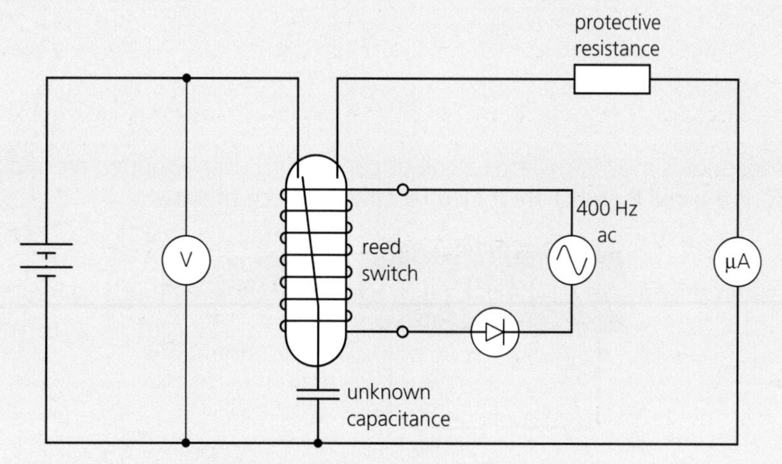

■ **Figure 11.67** Using a reed switch to measure capacitance

These questions all refer to Figure 11.67.

1 Suggest why a diode is necessary in the circuit.
2 With a 3.1 V battery a current of 480 μA was measured when using 400 Hz to control the reed switch. Determine the value of the capacitance.
3 Explain why the protective resistance is necessary and discuss whether its value should be large or small.

■ Smoothing the output of a rectifier

With a greater knowledge about the behaviour of capacitors, we can now review the action of a capacitor in smoothing the output of a rectifier. Consider again Figures 11.47 and 11.48. When the voltage from the rectifier is higher than the voltage across the capacitor, the capacitor will charge. When the voltage from the rectifier quickly falls below the voltage across the capacitor, the capacitor will maintain a voltage across the load as it discharges.

The time constant (RC) of the circuit needs to be compared to the frequency of the supply. For full-wave rectification of a 50 Hz mains supply voltage, successive peaks will occur every 0.01 s. A typical smoothing capacitor may have a value of about 10 000 μF. For a load of 100 Ω, the time constant would be 1 s, so that the fall in voltage in 0.01 s would be minimal.

Examination questions – a selection

Paper 1 IB questions and IB style questions

Q1 Power losses in transmission lines can be reduced by the use of step-up and/or step-down transformers. Which of the following describes the correct positions for the transformers?

A step-up transformer near the power station; step-down transformer near the consumers

B step-up transformer near the power station; step-up transformer near the consumers

C step-down transformer near the power station; step-down transformer near the consumers

D step-down transformer near the power station; step-up transformer near the consumers

Q2 Which of the following is *not* a possible reason for power losses in a transformer?

A eddy currents in the core

B resistive heating of the windings of the coils

C friction between the coil and the core

D 'leakage' of magnetic flux from the core

Q3 An increasing magnetic flux passes through a metal ring of cross-sectional area A. If the magnetic flux increases in value by F in a time t, the emf induced in the ring is:

A F

B $\frac{F}{t}$

C Ft

D FAt

Q4 A permanent bar magnet is moved towards a coil of conducting wire wrapped around a non-conducting cylinder. The ends of the coil P and Q are joined by straight piece of wire.

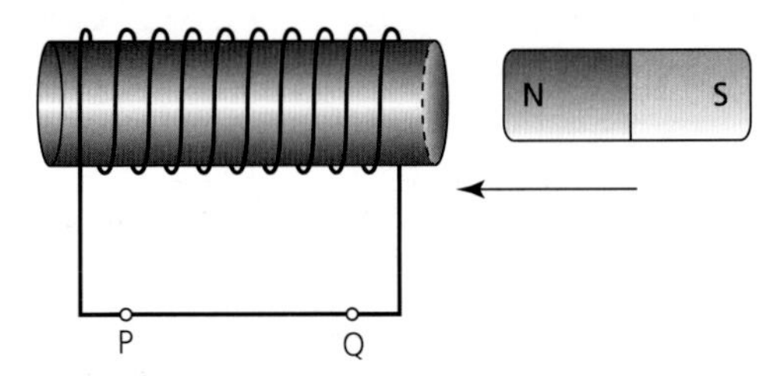

The induced current in the straight piece of wire is

A alternating

B zero

C from P to Q

D from Q to P.

Q5 The equation $\varepsilon = Bvl$ describes electromagnetic induction. Which of the following is *not* true?

A v is the speed of a conductor moving perpendicularly across a magnetic field.

B B is the magnetic flux.

C l is the length of the conductor.

D ε is the magnitude of the induced emf.

Q6 If a current is described as alternating (ac), which of the following *must* be changing?

A frequency of the current

B magnitude of the maximum current

C effective value of the current

D direction of the current

Q7 The frequency of rotation of the coil in an ac generator is doubled. The rms voltage output:

A remains the same
B is multiplied by 2
C is multiplied by $\sqrt{2}$
D is divided by $\sqrt{2}$.

Q8 The amount of charge that can be stored on a capacitor can be increased by:

A increasing the separation of the parallel plates
B increasing the electric permittivity of the dielectric medium between the plates
C decreasing the area of the plates
D decreasing the potential difference across the plates.

Q9 In order to achieve half-wave rectification of an alternating current it is necessary to use:

A a diode
B a capacitor
C a transformer
D all of the above.

Q10 Which of the following is the best description of magnetic flux?

A magnetic field strength per unit area
B magnetic field strength per unit length of conductor
C magnetic field strength normal to an area multiplied by the area
D magnetic field strength multiplied by the number of turns

Q11 High voltages are used for the transmission of electric power over long distances because:

A high voltages can be stepped down to any required value
B larger currents can be used
C power losses during transmission are minimised
D transformers have a high efficiency.

Q12 A loop of metal wire encloses an area *A*. If there are *N* turns of wire on the loop and the magnetic flux in the loop rises from a value of Φ to 3Φ in 4s, the induced emf across the loop is:

A $N\Phi$
B $2NA\Phi$
C $\frac{1}{2}NA\Phi$
D $\frac{1}{2}N\Phi$

Q13 Three similar capacitors, each with a capacitance *C* are available. Which of the following values of overall capacitance cannot be achieved by combining two or three of the capacitors?

A 0.5*C*
B *C*
C 1.5*C*
D 2*C*

Paper 2 IB questions and IB style questions

Q1 **a** In order to measure the rms value of an alternating current in a cable, a small coil of wire is placed close to the cable.

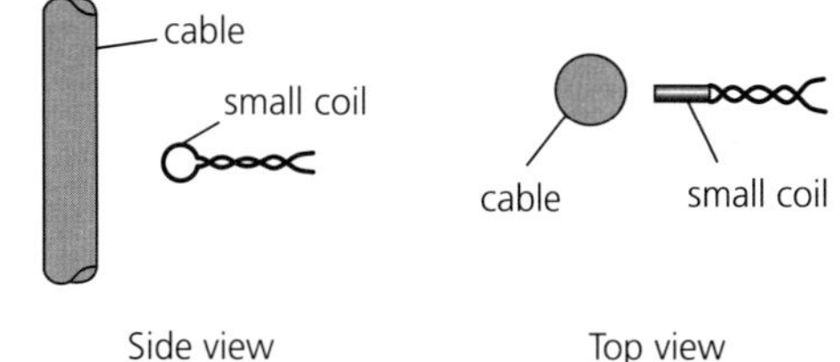

The plane of the small coil is parallel to the direction of the cable. The ends of the small coil are connected to a high-resistance ac voltmeter.

Use Faraday's law to explain why an emf is induced in the small coil. (3)

b The graph below shows the variation with time t of the current in the cable.

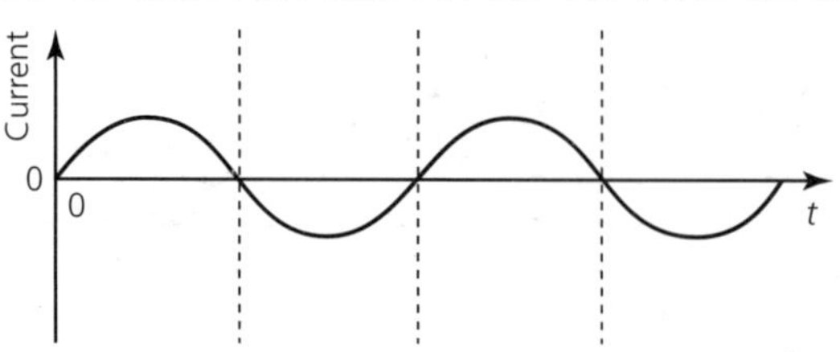

Copy the axes below. On your copy, draw the variation with time of the emf induced in the small coil. (2)

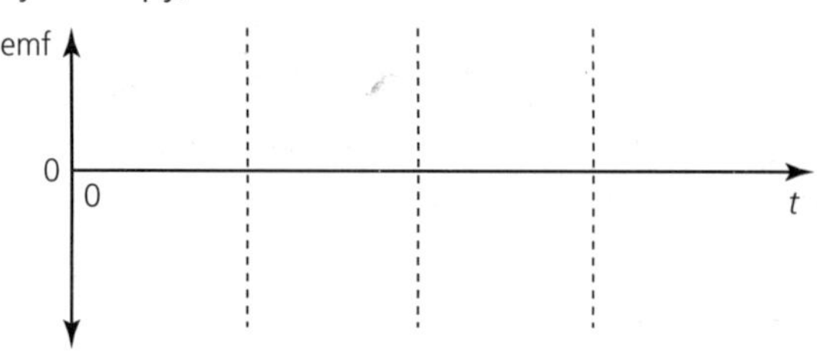

c Explain how readings on the high-resistance ac voltmeter can be used to compare the rms values of alternating current in different cables. (3)

Q2 This question is about induced electromotive force (emf).

a A rod of conducting material is in a region of uniform magnetic field. It is moved horizontally along two parallel conducting rails X and Y. The other ends of the rails are connected by a thin conducting wire.

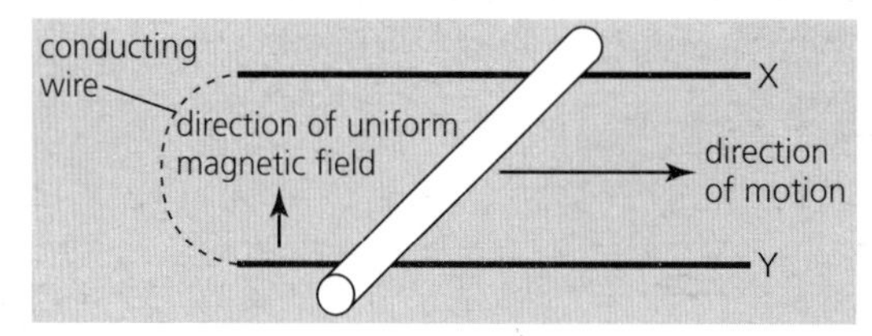

The speed of the rod is constant and is also at right angles to the direction of the uniform magnetic field.

i Describe, with reference to the forces acting on the conduction electrons in the rod, how an emf is induced in the rod. (3)

ii An induced emf is produced by a rate of change of flux. State what is meant by a rate of change of flux in this situation. (1)

b The length of the rod in a is 1.2 m and its speed is 6.2 m s^{-1}. The induced emf is 15 mV.

i Determine the magnitude of the magnetic field strength through which the rod is moving. (2)

ii Explain how Lenz's law relates to the situation described in a. (2)

Q3 A 200 μF capacitor was fully charged by connecting it to a potential difference of 12 V. A student then placed the capacitor in a circuit with a 500 kΩ resistor, a voltmeter and ammeter in order to observe the behaviour of the capacitor over a period of time.

a Draw a diagram to represent a suitable circuit. (2)

b Calculate the time constant of this circuit. (2)

c Sketch a voltage–time graph for the discharge for a total time of 4 minutes. (3)

12 Quantum and nuclear physics

ESSENTIAL IDEAS

- The microscopic quantum world offers a range of phenomena, the interpretation and explanation of which require new ideas and concepts not found in the classical world.
- The idea of discreteness that we met in the atomic world continues to exist in the nuclear world as well.

12.1 The interaction of matter with radiation – *the microscopic quantum world offers a range of phenomena, the interpretation and explanation of which require new ideas and concepts not found in the classical world*

In 1900 it was believed that physics was almost fully understood, although there were a few knowledge 'gaps', such as the nature of atoms and molecules, and the ways in which radiation interacted with matter. Many physicists believed that filling in these gaps was unlikely to involve any new theories. However, within a few years the 'small gaps' were seen to be fundamental and radically new theories were required. One important discovery was that energy only comes in certain *discrete* (separate) amounts known as **quanta** (singular: **quantum**). The implications of this discovery were enormous and collectively they are known as *quantum physics*.

As quantum physics developed, some classical concepts had to be abandoned. There was no longer a clear distinction between particles and waves. The most fundamental change was the discovery that systems change in ways that cannot be predicted precisely; only the probability of events can be predicted.

Photons

The German physicist Max Planck was the first to propose (in 1900) that the energy transferred by electromagnetic radiation was in the form of a very large number of separate, individual amounts of energy (rather than continuous waves). These discrete 'packets' of energy are called *quanta*. Quanta are also commonly called **photons**. The concept of photons was introduced in Chapter 7 as a way of explaining why atoms emit or absorb radiation in quantized amounts, and why this produces characteristic line spectra.

This very important theory, developed further by Albert Einstein in the following years, essentially describes the nature of electromagnetic radiation as being *particles*, rather than *waves*. ('Wave–particle duality' is discussed on page 516.)

The energy, E, carried by each photon (quantum) depends on its frequency, f, and is given by the relationship:

$$E = hf$$

h is Planck's constant. Its value (6.63×10^{-34} J s) and the equation are given in the *Physics data booklet*.

The following questions revise these ideas. We will then consider the experimental evidence that supports the photon model of electromagnetic radiation (the photoelectric effect).

1. Calculate the number of photons emitted every second from a mobile phone operating at a frequency of 850 MHz and at a radiated power of 780 mW.
2. What is the approximate ratio of the energy of a photon of blue light to the energy of a photon of red light?
3. A detector of very low intensity light receives a total of 3.32×10^{-17} J from light of wavelength 600 nm. Calculate the number of photons received by the detector.
4. a Calculate an approximate value for the energy of an X-ray photon in joules and in electronvolts.
   b Suggest a reason why exposure to X-rays of low intensity for a short time is dangerous, but relatively high continuous intensities of visible light causes us no harm.

The photoelectric effect

When electromagnetic radiation is directed on to a clean surface of some metals, electrons may be ejected. This is called the **photoelectric effect** (Figure 12.1) and the ejected electrons are known as **photoelectrons**. Under suitable circumstances the photoelectric effect can occur with visible light, X-rays and gamma rays, but it is most often demonstrated with ultraviolet radiation and zinc. Figure 12.2 shows a typical arrangement.

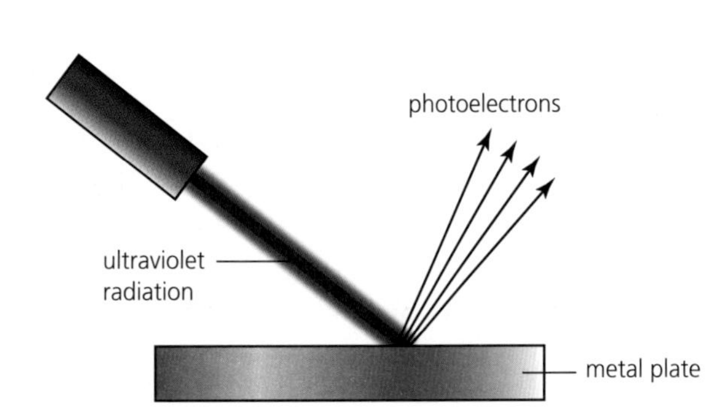

Figure 12.1 The photoelectric effect – a stream of photoelectrons is emitted from a metal surface illuminated with ultraviolet radiation

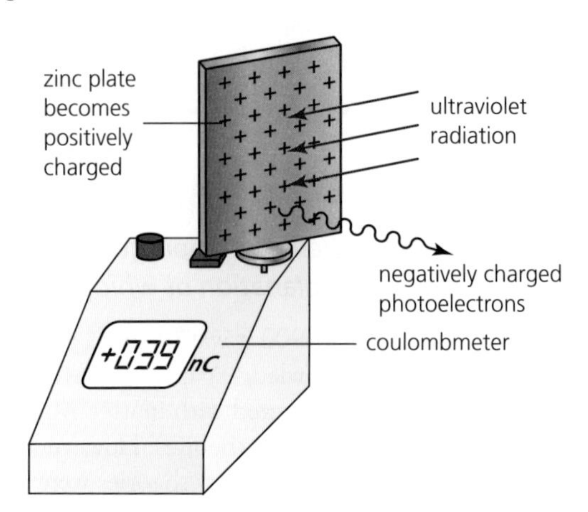

Figure 12.2 Demonstration of the photoelectric effect

Ultraviolet radiation is shone on to a zinc plate attached to a coulombmeter (which measures very small quantities of charge). The ultraviolet radiation causes the zinc plate to become positively charged because some negatively charged electrons on the (previously neutral) zinc plate have gained enough kinetic energy to escape from the surface.

Simple investigations of the photoelectric effect show a number of key features.

- If the intensity of the radiation is increased, the charge on the plate increases more quickly (because more photoelectrons are being released every second).
- There is no time delay between the radiation reaching the metal surface and the emission of photoelectrons. The release of photoelectrons from the surface is *instantaneous*.
- The photoelectric effect can only occur if the frequency of the radiation is above a certain *minimum* value. The lowest frequency for emission is called the **threshold frequency**, f_0. (Alternatively, we could say that there is *maximum* wavelength above which the effect will not occur.) If the frequency used is lower than the threshold frequency, the effect will not occur *even if the intensity of the radiation is greatly increased*. The threshold frequency of zinc, for example, is 1.04×10^{15} Hz, which is in the ultraviolet part of the spectrum. Visible light will not release photoelectrons from zinc (or most other common metals).
- For a given incident frequency the photoelectric effect occurs with some metals but not with others. This is because different metals have different threshold frequencies.

Explaining the photoelectric effect: the Einstein model

If we tried to use the wave theory of radiation to make predictions about the photoelectric effect, we would expect the following. (1) Radiation of *any* frequency will cause the photoelectric effect if the intensity is made high enough. (2) There may be a delay before the effect begins because it needs time for enough energy to be provided (like heating water until it boils).

These predictions are *wrong*, so an alternative theory is needed. Einstein realized that we cannot explain the photoelectric effect without first understanding the quantum nature of radiation.

The Einstein model explains the photoelectric effect using the concept of photons. When a photon in the incident radiation interacts with an electron in the metal surface, it transfers *all* of its energy to that electron. It should be stressed that a *single* photon can only interact with a *single* electron and that this transfer of energy is *instantaneous*; there is no need to wait for a build-up of energy. If a photoelectric effect is occurring, increasing the intensity of the radiation only increases the number of photoelectrons, not their energies.

Einstein realized that some of the energy carried by the photon (and then given to the electron) was used to overcome the attractive forces that normally keep an electron within the metal surface. The remaining energy is transferred to the kinetic energy of the newly released (photo) electron. Using the principle of conservation of energy, we can write:

$$\text{energy carried by photon} = \text{work done in removing the electron from the surface} + \text{kinetic energy of photoelectron}$$

But the energy required to remove different electrons is *not* always the same. It will vary with the position of the electron with respect to the surface. Electrons closer to the surface will require less energy to remove them. However, there is a well-defined *minimum* amount of energy needed to remove an electron, and this is called the **work function**, Φ, of the metal surface. Different metals have different values for their work functions. For example, the work function of a clean zinc surface is 4.3 eV. This means that *at least* 4.3 eV (= 6.9×10^{-19} J) of work has to be done to remove an electron from zinc.

To understand the photoelectric effect we need to compare the photon's energy, hf, to the work function, Φ, of the metal:

- $hf < \Phi$

  If an incident photon has less energy than the work function of the metal, the photoelectric effect cannot occur. Radiation that may cause the photoelectric effect with one metal may not have the same effect with another (which has a different work function).

- $hf\ (= hf_0) = \Phi$

  At the *threshold frequency*, f_0, the incident photon has exactly the same energy as the work function of the metal. We may assume that the photoelectric effect occurs but the released photoelectron will have zero kinetic energy.

- $hf > \Phi$

  If an incident photon has more energy than the work function of the metal, the photoelectric effect occurs and a photoelectron will be released. Photoelectrons produced by different photons (of the same frequency) will have a range of different kinetic energies because different amounts of work will have been done to release them.

It is important to consider the situation in which the *minimum* amount of work is done to remove an electron (equal to the work function):

energy carried by photon = work function + *maximum* kinetic energy of photoelectron

Or in symbols:

$$hf = \Phi + E_{max}$$

Or:

$$E_{max} = hf - \Phi$$

This equation is often called Einstein's photoelectric equation and it is given in the *Physics data booklet*. Because $hf_0 = \Phi$, we can also write this as:

$$hf = hf_0 + E_{max}$$

Figure 12.3 shows a graphical representation of how the maximum kinetic energy of the emitted photons varies with the frequency of the incident photons. The equation of the line is $E_{max} = hf - \Phi$, as above.

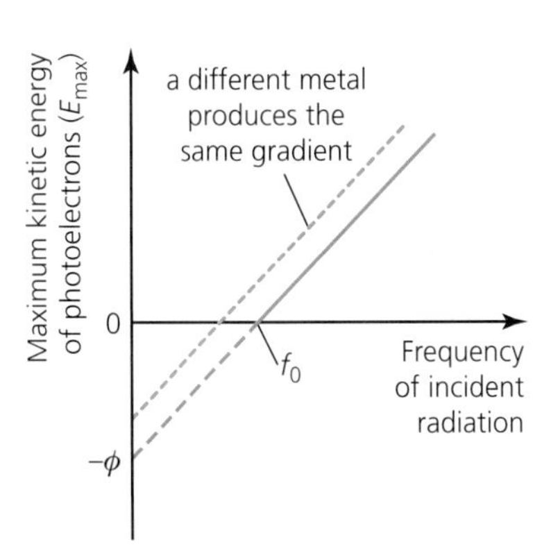

■ **Figure 12.3** Theoretical variation of maximum kinetic energy of photoelectrons with incident frequency (for two different metals)

We can take the following measurements from this graph:

- The gradient of the line is equal to Planck's constant, h. (Compare the equation of the line to $y = mx + c$.) Clearly the gradient is the same for all circumstances because it does not depend on photon frequencies or the metal used.
- The intercept on the frequency axis gives us the value of the threshold frequency, f_0.
- A value for the work function can be determined from:
  i when $E_{max} = 0$, $\Phi = hf_0$; or
  ii when $f = 0$, $\Phi = -E_{max}$

Worked example

1 Radiation of wavelength 5.59×10^{-8} m was incident on a metal surface that had a work function of 2.71 eV.

a What was the frequency of the radiation?
b How much energy is carried by one photon of the radiation?
c What is the value of the work function expressed in joules?
d Did the photoelectric effect occur under these circumstances?
e What was the maximum kinetic energy of the photoelectrons?
f What is the threshold frequency for this metal?
g Sketch a fully labelled graph to show how the maximum kinetic energy of the photoelectrons would change if the frequency of the incident radiation was varied.

a $f = \frac{c}{\lambda} = \frac{3.00 \times 10^8}{5.59 \times 10^{-8}} = 5.37 \times 10^{15}$ Hz

b $E = hf$

$E = (6.63 \times 10^{-34}) \times (5.37 \times 10^{15})$

$E = 3.56 \times 10^{-18}$ J

c $2.71 \times (1.60 \times 10^{-19}) = 4.34 \times 10^{-19}$ J

d Yes, because the energy of each photon is greater than the work function.

e $E_{max} = hf - \Phi$

$E_{max} = (3.56 \times 10^{-18}) - (4.34 \times 10^{-19}) = 3.13 \times 10^{-18}$ J

f $\Phi = hf_0$ so $f_0 = \frac{\Phi}{h} = \frac{4.34 \times 10^{-19}}{6.63 \times 10^{-34}} = 6.54 \times 10^{-14}$ Hz

g The graph should be similar to Figure 12.3, with numerical values provided for the intercepts.

5 Repeat the worked example above but for radiation of wavelength 6.11×10^{-7} m incident on a metal with a work function of 2.21 eV. Omit e.

6 a Explain how Einstein used the concept of photons to explain the photoelectric effect.
b Explain why a wave model of electromagnetic radiation is unable to explain the photoelectric effect.

7 The threshold frequency of a metal is 7.0×10^{14} Hz. Calculate the maximum kinetic energy of the electrons emitted when the frequency of the radiation incident on the metal is 1.0×10^{15} Hz.

8 a The longest wavelength that emits photoelectrons from potassium is 550 nm. Calculate the work function (in joules).
b What is the threshold wavelength for potassium? What is the name for this kind of radiation?
c Name one colour of visible light that will *not* produce the photoelectric effect with potassium.

9 When electromagnetic radiation of frequency 2.90×10^{15} Hz is incident on a metal surface, the emitted photoelectrons have a maximum kinetic energy of 9.70×10^{-19} J. Calculate the threshold frequency of the metal.

Experiments to test the Einstein model

Investigating stopping potentials

To test Einstein's equation (model) for the photoelectric effect, it is necessary to measure the maximum kinetic energy of the photoelectrons emitted under a variety of different circumstances. In order to do this the kinetic energy must be transferred to another (measurable) form of energy.

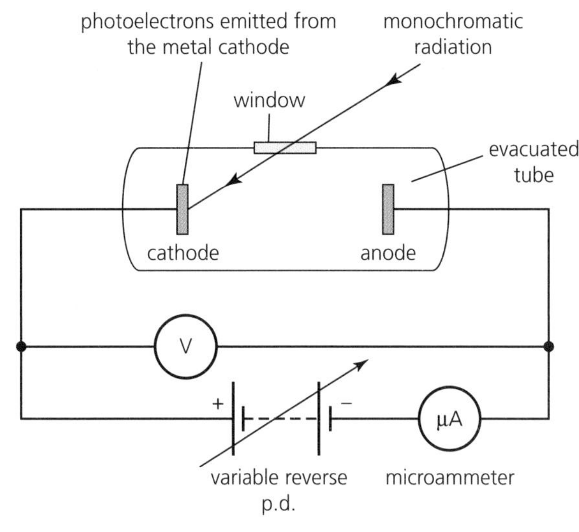

■ **Figure 12.4** Experiment to test Einstein's model of photoelectricity

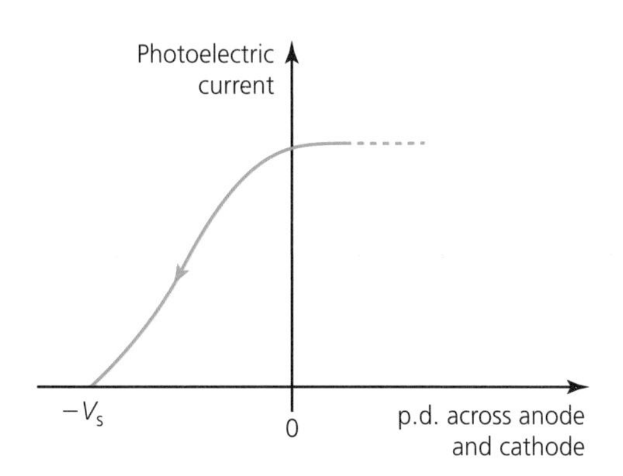

■ **Figure 12.5** Increasing the reverse p.d. decreases the photoelectric current

The kinetic energy of the photoelectrons can be transferred to electric potential energy if they are repelled by a negative voltage (potential). This experiment was first performed by the American physicist Robert Millikan and a simplified version is shown in Figure 12.4.

Ideally *monochromatic* radiation should used, but it is also possible to use a narrow range of frequencies such as those obtained by using coloured filters with white light.

When radiation is incident on a suitable emitting surface, photoelectrons will be released with a range of different energies, as explained previously. Because it is emitting negative charge, this surface can be described as a cathode (the direction of conventional current flow will be out of the cathode and around the circuit). Any photoelectrons that have enough kinetic energy will be able to move across the tube and reach the other electrode, the anode. The tube is *evacuated* (the air is removed to create a vacuum) so that the electrons do not collide with air molecules during their movement across the tube.

The most important thing to note about this circuit is that the (variable) source of p.d. is connected the 'wrong way around'. We say that it is supplying a reverse potential difference across the tube. This means that there is a negative voltage (potential) on the anode that will *repel* the photoelectrons. Photoelectrons moving towards the anode will have their kinetic energy reduced as it is transferred to electrical potential energy. (Measurements for positive potential differences can be made by reconnecting the battery the 'correct' way.)

Any flow of charge across the tube and around the circuit can be measured by a sensitive ammeter (microammeter or picoammeter). When the reverse potential on the anode is increased from zero, more and more photoelectrons will be prevented from reaching the anode and this will decrease the current. (Remember that the photoelectrons have a range of different energies.) Eventually the potential will be big enough to stop even the most energetic of photoelectrons, and the current will fall to zero (Figure 12.5).

The potential on the anode needed to just stop all photoelectrons reaching it is called the stopping potential, V_s.

Because, by definition, potential difference = energy transferred/charge, after measuring V_s we can use the following equation to calculate values for the maximum kinetic energy of photoelectrons under a range of different circumstances:

$$E_{max} = eV_s$$

For convenience, it is common to quote all energies associated with the photoelectric effect in electronvolts (eV). In which case, the maximum kinetic energy of the photoelectrons is

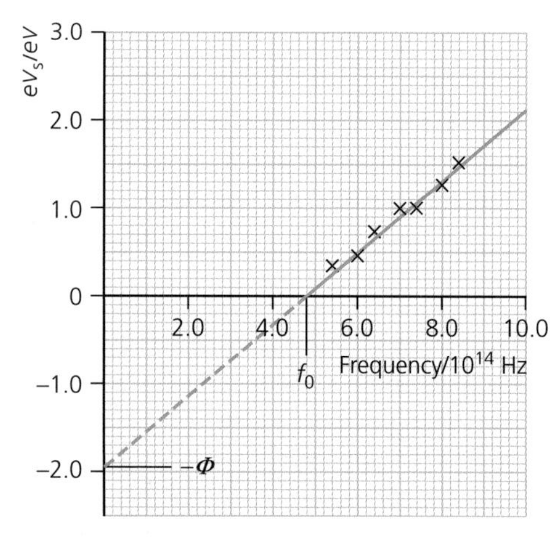

■ **Figure 12.6** Experimental results showing variation of maximum potential energy (eV_s) of photoelectrons with incident frequency

numerically equal to the stopping potential. That is, if the stopping potential is, say, 3 V, then E_{max} = 3 eV.

Einstein's equation ($E_{max} = hf - \Phi$) can be rewritten as:

$$eV_s = hf - \Phi$$

By *experimentally* determining the stopping potential for a range of different frequencies, the *theoretical* graph shown previously in Figure 12.3 can now be confirmed by plotting a graph from actual data, as shown in Figure 12.6.

- The threshold frequency, f_0, can be determined from the intercept on the frequency axis (4.8×10^{14} Hz).
- The work function, Φ, can be calculated from $\Phi = hf_0$, or from the intercept on the eV_s axis (1.9 eV).
- A value for Planck's constant, h, can be determined from the gradient (6.5×10^{-34} J s).

10 Use Figure 12.6 to confirm the values given above for f_0, Φ and h.

11 Calculate the maximum kinetic energy of photoelectrons emitted from a metal if the stopping potential was 2.4 eV. Give your answer in joules and in electronvolts.

12 Make a copy of Figure 12.5 and add lines to show the results that would be obtained with:
a the same radiation, but with a metal of higher work function
b the original metal and the same frequency of radiation, but using radiation with a greater intensity.

13 In an experiment using monochromatic radiation of frequency 7.93×10^{14} Hz with a metal that had a threshold frequency of 6.11×10^{14} Hz, it was found that the stopping potential was 0.775 V. Calculate a value for Planck's constant from these results.

Investigating photoelectric currents

Using apparatus similar to that shown in Figure 12.4, it is also possible to investigate quantitatively the effects on the photoelectric current of changing the intensity, the frequency, and the metal used in the cathode.

- **Intensity** – Figure 12.7 shows the photoelectric currents produced by monochromatic radiation of the same frequency at three different intensities.

  For positive potentials, each of the photoelectric currents remain constant because the photoelectrons are reaching the anode at the same rate as they are being produced at the cathode, and this does not depend on the size of the positive potential on the anode.

  Greater intensities (of the same frequency) produce higher photoelectric currents because there are more photons releasing more photoelectrons (of the same range of energies).

  Because the maximum kinetic energy of photoelectrons depends only on frequency and not intensity, all these graphs have the same value for stopping potential, V_s.

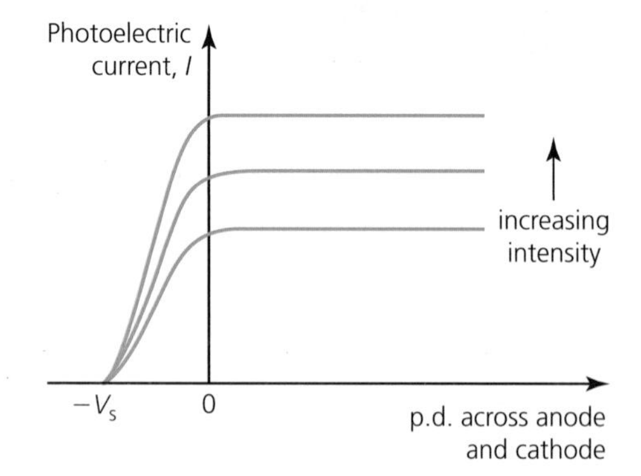

■ **Figure 12.7** Variation of photoelectric current with p.d. for radiation of three different intensities (same frequency)

- **Frequency** – Figure 12.8 shows the photoelectric currents produced by radiation from two monochromatic sources of different frequencies, A and B, incident on the same metal.

The individual photons in radiation A must have more energy (than B) and produce photoelectrons with a higher maximum kinetic energy. We know this because a bigger reverse potential is needed to stop the more energetic photoelectrons produced by A.

No conclusion can be drawn from the fact that the current for A has been drawn higher than for B, because the intensities of the two radiations are not known. In the unlikely circumstances that the two intensities were equal, the maximum current for B would have to be higher than for A because the radiation from B must have more photons, because each photon has less energy than in A.

- **Metal used in the cathode** – Experiments confirm that when different metals are tested using the same frequency, the photoelectric effect is observed with some metals but not with others (for which the metal's work function is higher than the energy of the photons).

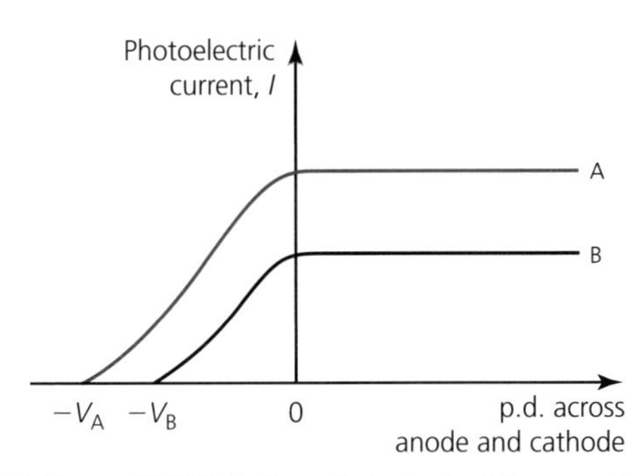

■ **Figure 12.8** Variation of photoelectric current with p.d. for radiation of two different frequencies

Light-emitting diodes (LEDs) can also be used to investigate the photoelectric effect (in 'reverse'). Electrons in an electric current passing through an LED undergo precise and identical energy transitions resulting in the emission of photons of the same single wavelength. If the potential difference across the LED is measured when it just begins to emit light, it can be used to determine a value for Planck's constant.

14 Make a copy of Figure 12.5 and add the results that would be obtained using radiation of a higher intensity (of the same frequency) incident on a metal that has a smaller work function.

15 Make a copy of Figure 12.8, line A only. Add to it a line showing the results that would be obtained with radiation of a higher frequency but with same number of photons every second incident on the metal.

16 a Select five different metallic elements and then use the internet to research their work functions.
b Calculate the threshold frequencies of the five metals.

17 The voltage across an LED was increased until it just began to emit green light of wavelength 5.6×10^{-7} m. Determine a value for Planck's constant if the voltage was 2.2 V.

Additional Perspectives

The photon nature of electromagnetic radiation: the Compton effect

This effect, discovered in the USA in 1923 by Arthur Compton, provided further evidence for the photon nature of electromagnetic radiation.

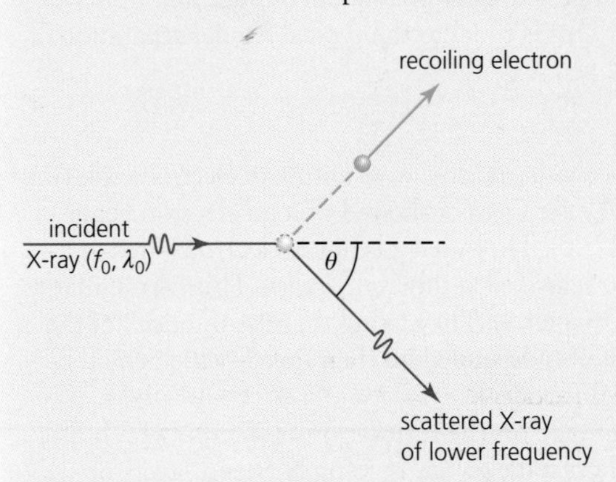

■ **Figure 12.9** Compton scattering of X-rays

It was shown that X-rays could be scattered by electrons resulting in a reduced frequency because some of the photons' energy was passed to the scattered electron. See Figure 12.9.

Momentum (as always) is conserved in this interaction. This experiment shows that an X-ray behaves as a particle carrying momentum, $p = h/\lambda$, although it has no mass. This is explained by more advanced theory and it is only possible because photons travel at the speed of light. Similar comments apply to all other photons (light for example).

1 Use the Compton equation $\Delta\lambda = (h/m_e c)(1 - \cos\theta)$ to predict the change in wavelength associated with a scattering angle of 30°.

2 Use the equation to explain why Compton scattering of light is not significant.

Matter waves

The fact that light and other electromagnetic radiations could behave as particles (photons) raises an obvious question: can particles behave like waves?

In 1924 the French physicist Louis de Broglie proposed that electrons, which were thought of as particles, might also have a wave-like character. He later generalized his hypothesis to suggest that *all* moving particles have a wave-like nature.

According to de Broglie, the wavelength, λ, of a moving particle is related to its momentum, p, by the equation:

$$\lambda = \frac{h}{p}$$

This important equation, called the **de Broglie hypothesis**, is *not* in the *Physics data booklet*.

h is Planck's constant. Once again, we can see the importance of this constant in predicting the size of quantum phenomena. The very small value of Planck's constant shows us that wave properties of particles are only significant for those with tiny momenta, as the following examples illustrate.

Worked example

2 Calculate the momentum of a moving particle that has a de Broglie wavelength of 200 pm (1 pm = 1×10^{-12} m).

$$p = \frac{h}{\lambda}$$

$$p = \frac{6.63 \times 10^{-34}}{2.00 \times 10^{-10}} = 3.32 \times 10^{-24}\,\text{kg m s}^{-1}$$

This shows us that, if there are de Broglie wavelengths for everyday objects (which are clearly not single particles), they would be immeasurably small. However, some sub-atomic particles, such as electrons, have such small masses and momenta that their wave-like properties can be detected experimentally.

In order to demonstrate that particles have wave properties it is necessary to show that they exhibit typical wave behaviour and, in particular, that they can diffract. In the 1920s electron beams were easily produced, so they were the obvious choice for an investigation into possible particle diffraction.

De Broglie's hypothesis predicts that an electron travelling in a beam with a typical speed of $5 \times 10^6\,\text{m s}^{-1}$ (and therefore momentum of about $9 \times 10^{-25}\,\text{kg m s}^{-1}$) will have a wavelength of about 1×10^{-10} m. We know from Chapter 4 that diffraction effects are greatest if the diffracting apertures are about the same size as the wavelength, so electrons should diffract well from structures with a spacing of about 1×10^{-10} m. This is equal to the typical regular separation of ions in a crystalline material.

The Davisson–Germer experiment

Experimental confirmation of de Broglie's hypothesis and the wave nature of electrons was first obtained in 1927 when Clinton Davisson and Lester Germer showed that an electron beam could be diffracted by a metal crystal (Figure 12.10). They used a beam of electrons fired at a nickel crystal target and recorded the electrons scattered at different angles. This was similar to the action of a reflection diffraction grating (Chapter 9). They found that the intensity of the scattered electrons varied with the angle and it also depended on their speed (which could be altered by changing the accelerating potential difference).

Many investigations had previously been made on the diffraction of X-rays (which also have wavelengths of about 10^{-10} m) by metal crystals, so the spacing between layers of atoms in the nickel crystal was already known. The angle at which the maximum scattered electron intensity was recorded agreed with the angle predicted by constructive interference of waves by layers of atoms in the surface of the metal crystal and was used to confirm the de Broglie hypothesis.

The diffraction of electrons can be demonstrated in school laboratories using a high-voltage evacuated tube. A beam of electrons is accelerated in an *electron gun* by a high potential difference, and then directed on to a very thin sheet of graphite (carbon atoms) (Figure 12.11). The electrons are diffracted by the regular structure of carbon atoms and form a pattern on the screen of a series of concentric rings.

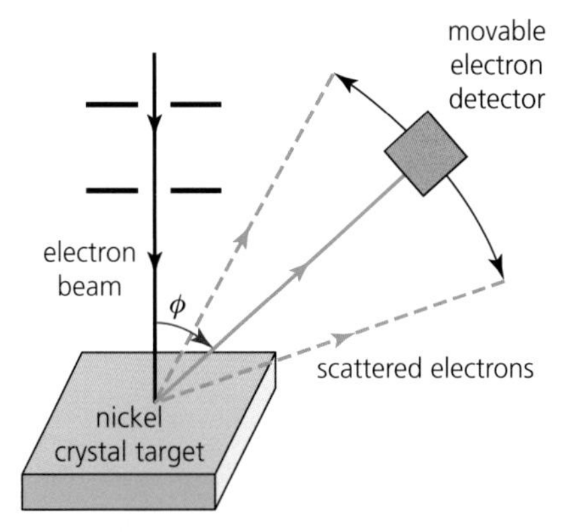

■ **Figure 12.10** The principle behind the Davisson and Germer experiment. The results of this experiment can only be explained by the fact that the electrons have wave properties and, therefore, diffract and interfere

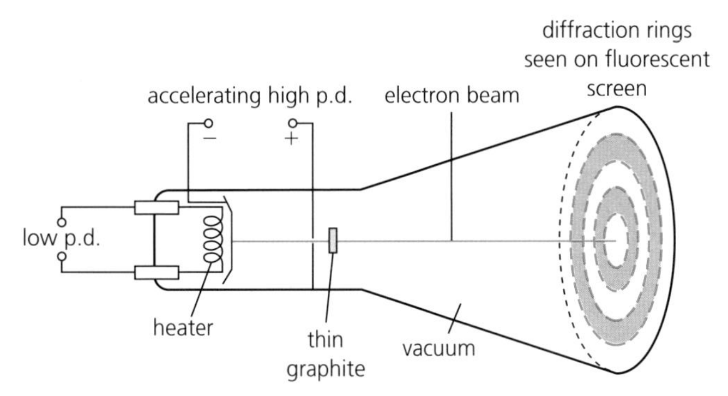

■ **Figure 12.11** An electron diffraction apparatus

If the accelerating p.d. used in Figure 12.11 is increased, the electrons gain higher kinetic energy and momentum, and de Broglie's hypothesis correctly predicts that the electrons will then have a shorter wavelength and will therefore be diffracted through smaller angles. The rings seen on the screen in Figure 12.11 would get closer together (and brighter).

Apart from electrons, the diffraction of protons, neutrons and ions of small mass has been observed.

Worked examples

3 Calculate the de Broglie wavelength of an electron travelling with a speed of $1.0 \times 10^7\,\text{m s}^{-1}$.

$$\lambda = \frac{h}{p} = \frac{6.63 \times 10^{-34}}{(9.110 \times 10^{-31}) \times (1.0 \times 10^7)} = 7.3 \times 10^{-11}\,\text{m}$$

4 Estimate a possible de Broglie wavelength of a ball of mass 0.058 kg moving with a velocity of $10^2\,\text{m s}^{-1}$.

$p = mv$; $p = 0.058 \times 10^2 = 5.8\,\text{kg m s}^{-1}$

$$\lambda = \frac{h}{p} = \frac{6.63 \times 10^{-34}}{5.8} \approx 10^{-34}\,\text{m}$$

To exhibit wave-like properties the ball would need to interact that has an object that has dimensions of the order of 10^{-34} m (over one million, million, million times smaller than the nucleus of an atom). Hence, the ball does not exhibit any measurable or detectable wave-like properties.

5 Calculate the de Broglie wavelength of an electron which has been accelerated through a potential difference of:

a 500 V

b 1000 V.

a The speed of the electrons can be calculated using the law of conservation of energy: loss of electrical potential energy = gain in kinetic energy.

$$Ve = \frac{1}{2}mv^2$$

$$v = \sqrt{\frac{2Ve}{m}} = \sqrt{\frac{2 \times 500 \times 1.60 \times 10^{-19}}{9.110 \times 10^{-31}}} = 1.32 \times 10^7\,\text{m s}^{-1}$$

$$p = mv = 1.2 \times 10^{-23}\,\text{kg m s}^{-1}$$

$$\lambda = \frac{h}{p} = 5.5 \times 10^{-11}\,\text{m}$$

b Doubling the p.d. will result in speed × √2, momentum × √2 and wavelength × 1/√2, that is, 3.9×10^{-11} m.

6 Calculate the de Broglie wavelength of a lithium nucleus ($^7_3$Li) that has been accelerated through a potential difference of 5.00 MV. (The mass of a lithium-7 nucleus is 1.165×10^{-26} kg.)

$$\frac{1}{2}mv^2 = 3e \times V$$

$$v^2 = \frac{6 \times (1.60 \times 10^{-19}) \times (5.00 \times 10^6)}{1.165 \times 10^{-26}} = 4.12 \times 10^{14}$$

$$v = 2.03 \times 10^7\,\text{m s}^{-1}$$

$$\lambda = \frac{6.63 \times 10^{-34}}{(1.165 \times 10^{-26}) \times (2.03 \times 10^7)} = 2.80 \times 10^{-15}\,\text{m}$$

18 Calculate the wavelength of an electron moving with a velocity of $2.05 \times 10^7\,\text{m s}^{-1}$.

19 A neutron has a de Broglie wavelength of 800 pm. Calculate the velocity of the neutron. The mass of the neutron is 1.675×10^{-27} kg.

20 The mass of an electron is 9.1×10^{-31} kg. If its kinetic energy is 3.0×10^{-25} J, calculate its de Broglie wavelength in metres.

21 Over what potential difference do you have to accelerate electrons from rest for them to have a wavelength of 1.2×10^{-10} m?

22 a Which is associated with a de Broglie wavelength of longer wavelength – a proton or an electron travelling at the same velocity? Explain your answer.
b The following (non-relativistic particles) all have the same kinetic energy. Rank them in order of their de Broglie wavelengths, longest first: electron, alpha particle, neutron and gold nucleus.

23 Explain why a moving aeroplane has no detectable wave properties.

24 Construct a spreadsheet for electrons, protons and alpha particles that converts values of mass and velocity into momentum and wavelength (using the de Broglie equation).

25 Calculate the wavelength of a proton accelerated (from rest) by a potential difference of 1850 V (mass of proton = 1.673×10^{-27} kg).

Nature of Science

Wave–particle duality

The discovery that radiation can behave like particles, and that particles can behave like waves, inevitably caused physicists to completely rethink their understanding of many basic ideas. This was a dramatic paradigm shift in scientists' perception of the basic ingredients of the universe. It became clear that simply labelling something as a wave or a particle, in accordance with our perception of the large-scale world around us, was not an adequate description. There is no confusion or ambiguity here; simple models and visualizations are not able to fully describe the nature of the microscopic (quantum-scale) properties of matter and radiation. Why should they?

At this level in physics sometimes we need a wave explanation and sometimes we need a particle explanation. Sometimes we need both: as when using the concept of frequency to determine the energy carried by a particle (photon). With increasing mathematical knowledge (beyond IB level), it becomes possible to develop more sophisticated models, but these do not offer more help in providing mental images to aid understanding.

Antimatter

The concept of matter and antimatter was briefly introduced in Chapter 7. We will now take a closer look at how antiparticles can be created, and what happens when a particle meets its antiparticle.

Pair production and pair annihilation

For every elementary particle there is an **antiparticle** that has same mass as the particle, but with the opposite charge (if it is charged), and opposite quantum numbers.

(A few uncharged particles (e.g. photons) are considered to be their own antiparticles.)
Some examples:

- An *antielectron*, e^+ (known as a positron), has the same mass as an *electron*, e^- and a charge of equal magnitude, but of opposite sign (positive instead of negative). The positron has a lepton number of −1 and an electron has a lepton number of +1.
- An *antineutron*, $\bar{n}$, has the same mass as a *neutron*, n, and is also not charged. The antineutron has a baryon number of −1 and the neutron has a baryon number of +1.
- A *negative pi meson* (pion), π^-, contains an anti-up quark and a down quark, and has a charge of −1. It is the antiparticle of the *positive pion*, π^+, which has a charge of +1 because it contains an anti-down quark and an up quark. The *neutral pion*, π^0, contains a down quark and an anti-down quark, or an up quark and an anti-up quark, and it is considered to be its own antiparticle.
- An *antineutrino* is uncharged and has a very small mass. A *neutrino* is also uncharged and has the same mass, but has the opposite lepton number (+1).

Pair production

Antiparticles can be produced in a process called **pair production**, in which a gamma ray photon is converted into a particle and its antiparticle. Figure 12.12 represents the pair production of an electron and a positron close to a nucleus. A Feynman diagram of this process was also shown in Chapter 7, Figure 7.48.

In order for this process to be possible, the energy carried by the photon must be greater than the sum of the rest mass-energies of the electron and positron:

$$2m_e = 2 \times 0.511\,\text{MeV}c^{-2} = 1.022\,\text{MeV}c^{-2}$$

Using energy of photon, $E = hf$:

$$(1.022 \times 10^6) \times (1.60 \times 10^{-19}) = (6.63 \times 10^{-34})f$$

$$f = 2.47 \times 10^{20}\,\text{Hz}$$

This is the minimum frequency required for a photon to be able to instigate pair production of a positron and an electron. It corresponds to a gamma ray of wavelength of 1.21×10^{-12} m. If a photon carries greater energy, it can be transferred to the kinetic energy of the particles created.

Gamma ray photons of considerably greater energy may create pairs of more massive particles, for example muons and antimuons.

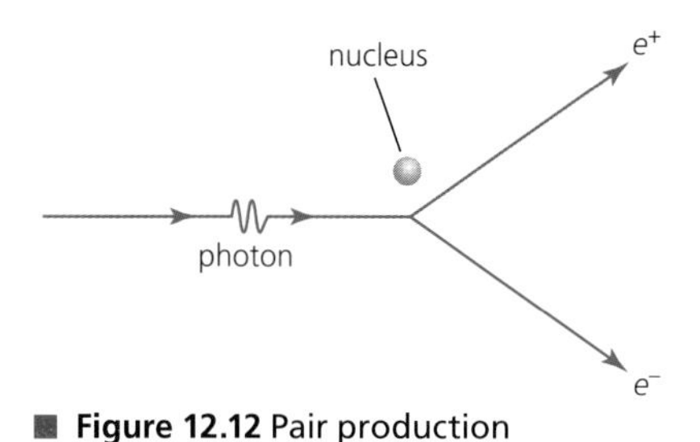

Figure 12.12 Pair production

The calculation above used the principle of conservation of energy-mass, but momentum must also be conserved in the interaction. It is for this reason that there is a need for some other mass to be involved (in Figure 12.12 the pair production occurs in the electromagnetic field near a nucleus): without another mass gaining momentum, momentum could not be conserved. Suppose that the created particles had zero velocity, how could we explain the loss of the momentum carried by the photon ($p = h/\lambda$) if there was no other particle involved? The probability of pair production increases with higher photon energy and larger nuclei.

Pair annihilation

If a particle and its antiparticle combine, the sum of their charge and quantum numbers must be zero (by definition). So, if the laws of conservation of mass–energy and momentum are obeyed, there is no physics principle that states that the reaction between two such particles cannot result in the transfer of all their masses to energy. This may be considered as the opposite of pair production. The **pair annihilation** of an electron and a positron is the most common example.

$$e^- + e^+ \rightarrow \gamma + \gamma$$

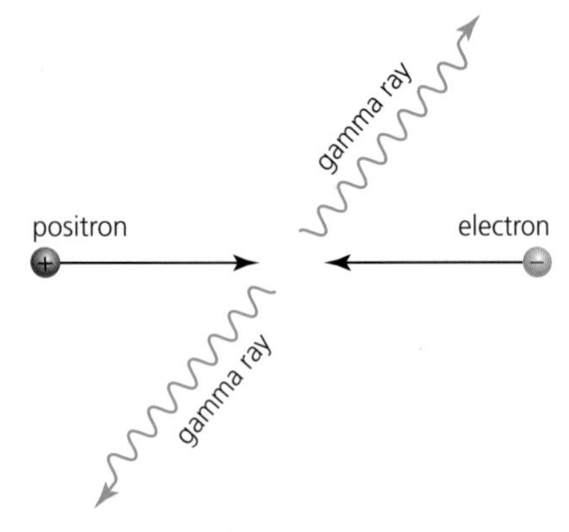

■ **Figure 12.13** Electron–positron annihilation. Each gamma ray photon will carry a minimum energy equal to the rest mass of one of the particles (0.511 MeV)

If we assume that the particles have no significant kinetic energy, or momentum, then the simplest result is a pair of gamma ray photons moving in exactly opposite directions, which is necessary if momentum is to be conserved (Figure 12.13). The situation is more complicated if the particles have higher kinetic energy.

If more massive particles (than electrons and positrons) annihilate, a variety of outcomes are possible.

We live in a universe that is predominantly made of matter, rather than antimatter, although some antimatter is created by natural processes (for example by cosmic rays interacting with the upper atmosphere, or beta-positive decay). Antimatter is also created in high-energy physics experiments. Although matter and antimatter cannot co-exist, the reasons why the universe is made of matter, rather than antimatter (or vice versa), is not fully understood.

26 If the frequency of the photon in Figure 12.12 was 3.47×10^{20} Hz, calculate the kinetic energy (MeV) of each of the particles produced (assume they are equal).

27 If a proton and an antiproton annihilate, what is the maximum possible wavelength of the photons created?

28 Consider Figure 12.13. Explain why two photons each of energy 0.511 MeV are created, rather than one of energy 1.022 MeV.

29 Explain in what way an antiproton differs from a proton.

30 Use the internet to research the latest theories concerning the balance of antimatter/matter in the early universe.

31 Draw a Feynman diagram of an electron–positron annihilation.

■ Quantization of angular momentum in the Bohr model for hydrogen

The discovery of the wave properties of particles, and in particular electrons, had enormous consequences for physicists' models of the atom. Before we look at this in more detail, we will review the prevailing model of the atom at that time (early 1920s).

In Chapter 7 the concept of quantized, discrete atomic energy levels was introduced as a means of explaining the spectra emitted or absorbed by atoms, but that discussion did not include much reference to the internal structure of atoms. Figure 12.14 is similar to Figure 7.6 but shows some transitions between electron energy levels in hydrogen atoms. The transitions have been arranged into three series. These three series of transitions produce sets of spectral lines that are distinct and do not overlap because smallest transition in one series is bigger than the biggest transition in the next series.

The ground state is labelled $n = 1$, and then successive higher energy levels $n = 2$, $n = 3$ and so on.

It was easily demonstrated that the electron energy levels themselves formed a converging series described by the following equation, although at the time the reason for this was not known:

$$E = \frac{-13.6}{n^2} \text{ eV}$$

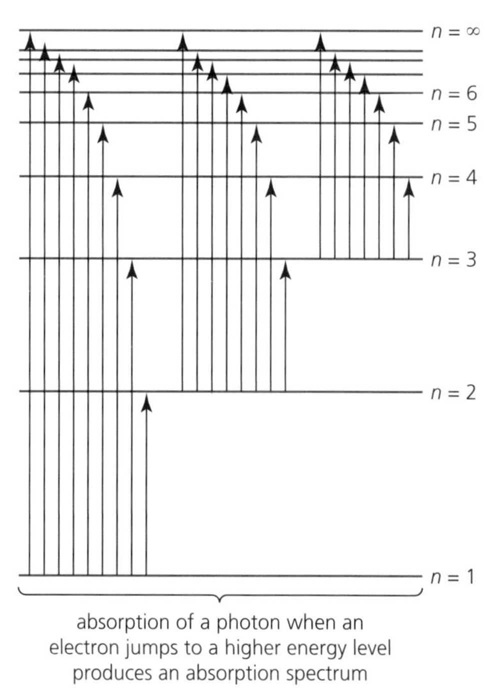

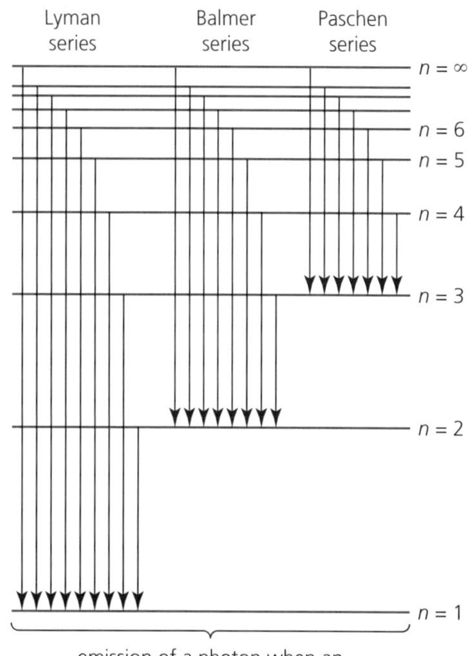

■ **Figure 12.14** The Lyman, Balmer and Paschen series of the hydrogen atom

This equation is given in the *Physics data booklet*.

- for $n = 1$, $E = -13.6/1^2 = -13.60\,\text{eV}$
- for $n = 2$, $E = -13.6/2^2 = -3.40\,\text{eV}$
- for $n = 3$, $E = -13.6/3^2 = -1.51\,\text{eV}$
- for $n = 4$, $E = -13.6/4^2 = -0.85\,\text{eV}$

and so on – an infinite series of energy levels.

After Rutherford's new model of the atom with a central nucleus in 1911, a visualization of the atom was proposed by the Danish physicist Niels Bohr that attempted to explain the quantization of energy levels and the series seen on the hydrogen spectrum.

Nature of Science

Observations: the importance of spectra

The patterns observed in the spectral lines of hydrogen may seem of no great importance, but they provided a highly significant clue in the quest to learn more about the structure of atoms (starting with the simplest, hydrogen). This was a major preoccupation of physicists in the first decades of the twentieth century, a time of great advances in atomic science.

■ **Figure 12.15** Niels Bohr

The **Bohr model** of the atom, first proposed in 1913, has electrons orbiting around the nucleus due to the centripetal force provided by electric attraction between opposite charges. But the Bohr model restricted the orbits to only certain distances from the nucleus and, most importantly, because they remained in that orbit, they did not emit electromagnetic radiation and lose energy. Each orbit had a definite and precise energy, and intermediate energies were not allowed. Photons were emitted or absorbed when electrons moved between these energy levels. See Figure 12.16, in which the distances are indicative of the radii of the orbits and *not* the energy levels (the distances are not drawn to scale).

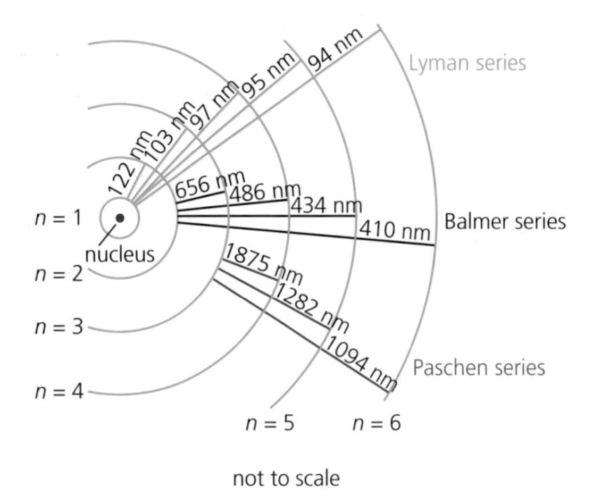

Figure 12.16 The Bohr model explains the spectrum of hydrogen using possible orbits of different radius for the electron

Using this model Bohr was able to predict that the radii, r, of possible orbits of an electron of mass, m, moving with speed, v, would be given by the equation:

$$mvr = \frac{nh}{2\pi}$$

This equation is given in the *Physics data booklet*.

The product of linear momentum and radius, mvr, is known as the **angular momentum**, of the electron (explained fully in Option B: Engineering physics). In this quantum model n is an integer known as the **principal quantum number**.

As the following mathematics shows, the Bohr model combined the classical physics of circular motion and force of electric attraction with quantum concepts in order to predict radii of orbits and energy levels in hydrogen atoms.

Equating the centripetal force on the electron to the electric attraction between it and the proton in a hydrogen atom, we get:

$$\frac{mv^2}{r} = \frac{kee}{r^2}$$

(k is Coulomb's constant), which leads to:

$$v = \sqrt{\frac{ke^2}{mr}}$$

Putting this expression for v in the equation for angular momentum (highlighted above), we get:

$$\sqrt{ke^2mr} = \frac{nh}{2\pi}$$

and rearranging enables us to obtain an expression for r:

$$r = \frac{n^2h^2}{4\pi^2ke^2m}$$

Putting $n = 1$ enabled Bohr to predict the radius of the electron's orbit when the hydrogen atom is in its ground state:

$$r = 5.3 \times 10^{-11}\,\text{m}$$

Then, using equations for electric potential energy and kinetic energy, the total energy associated with the ground state could be calculated, and this confirmed the value of −13.6 eV already known. The other energy levels of the hydrogen atom were also predicted by Bohr's model.

Although Bohr's quantized model of the atom was a big step forward in understanding the structure of the atom, and it was very accurate in predicting the energy levels of one-electron atoms such as a hydrogen atom or a helium ion, it was less successful with atoms containing more electrons. Furthermore, the reasons for the existence of energy levels were still not understood. The Bohr model still remains an important initial step for students learning about quantization in atoms, but the discovery of the wave properties of electrons quickly led to dramatic changes in physicists' understanding of the atom.

Quantum mechanics is the name given to the important branch of physics that deals with events on the atomic and sub-atomic scales that involve quantities that can only have discrete (quantized) values. In the quantum world the laws of classical physics are often of little use; trying to apply knowledge and intuition gained from observing the macroscopic world around us often only leads to confusion.

32 Determine the angular momentum of an electron in the fifth orbit of the hydrogen atom a in terms of h/π, and b in SI units.

33 a Calculate the radius of the third orbit above the ground state in the hydrogen atom.
   b What radius is predicted for the largest orbit?

34 a Calculate the kinetic energy (J) of an electron in the ground state of the hydrogen atom.
   b Calculate the electric potential (J) of the atom in its ground state.
   c Determine the overall energy of the atom (J).
   d Convert your answer to electronvolts.
   e Explain why the answer is negative.

35 a Use data from Figure 12.14 to determine the lowest frequency of the Balmer series.
   b In what part of the electromagnetic spectrum is this frequency?
   c Calculate the smallest energy (eV) carried by any of the photons involved with the five transitions shown in the Paschen series.

Fitting electron waves into atoms

After the discovery of the wave properties of electrons, the model of the structure of the atom needed to be changed significantly. However, we have already met similar ideas: in Chapter 4 we discussed waves that are 'trapped' in confined spaces and the formation of mechanical *standing waves*, such as those seen on stretched strings.

If we consider that an electron exists as a whole number of standing waves confined to the circumference of one of the Bohr orbits (radius r), then possible standing wavelengths are described by the equation:

$$n\lambda = 2\pi r$$

Comparing this to $\lambda = h/p$ (with $p = mv$) it is easy to show that:

$$mvr = \frac{nh}{2\pi}$$

as before.

In other words, the concept of an electron existing as a standing wave in a hydrogen atom successfully confirms Bohr's energy levels.

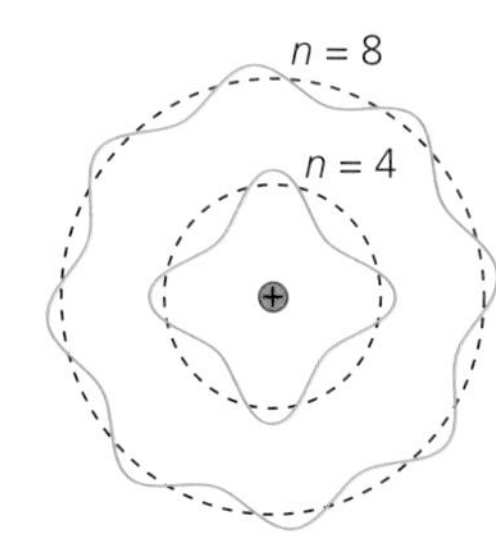

Figure 12.17 Electron standing waves in the fourth and eighth energy levels of the hydrogen atom

Figure 12.17 represents two of the possible standing wave patterns for an electron in the hydrogen atom.

However, this is a much simplified model, but before it can be taken any further we need to develop a deeper understanding of matter waves.

36 Show how the equation $mvr = \frac{nh}{2\pi}$ arises as a consequence of fitting electron standing waves into the circumferences of their orbits around the nucleus.

37 Calculate the angular momentum of an electron in its ground state in the hydrogen atom.

38 Make a sketch to represent an electron in the $n = 3$ energy level of the hydrogen atom.

■ **Figure 12.18** Erwin Schrödinger

■ The wave function

The idea that light and sound travel as waves required that we explain the nature of those waves (Chapter 4). Similarly we now need to explain the wave nature of electrons and other particle waves. However, this is not something for which we can draw a simple picture or describe in a few words. A full explanation requires a high level of mathematics, *which is not required in this course.*

In 1926 the Austrian physicist Erwin Schrödinger proposed a quantum mechanical model describing the behaviour of the electron within a hydrogen atom. In Schrödinger's theory, the electron in the atom is described by a mathematical **wave function**, $\Psi(x, t)$ as a function of position, x, and time, t. The wave function is a description of the (quantum) state of an electron; it represents everything there is to know about the particle. The wave function of a particle cannot be observed directly. It is important to understand that when the condition of the electron changes, so too does its wave function. So that, for example, the wave function of a free electron is different from the wave function of an electron bound in a hydrogen atom.

In classical physics the position of a particle at any moment can be known precisely, but on the microscopic (quantum physics) scale a particle's wave properties become important and we can then only refer to the *probability* of a particle being in a given position at any given time.

According to quantum mechanics, a particle cannot have a precise path; instead, there is only a probability that it may be found at a particular place at any particular time. The wave function of an electron may be viewed as a blurred version of its path (Figure 12.19). The darker the area of shading, the higher the probability of finding an electron there.

■ **Figure 12.19** Classical and quantum mechanical descriptions of a moving electron

The wave function of the electron provides a means of representing this probability, but not directly:

> $|\psi(x, t)|^2$ is the square of the absolute value of the amplitude of the wave function. It is a measure of the probability that an electron will be found near a point x at time t.

The probability of finding a particle at a distance r from a reference point, in unit volume, V, is known as its **probability density**, $P(r)$. It is related to $|\psi|^2$ as follows (this equation is given in the Physics data booklet); ΔV is the small volume being considered:

$$P(r) = |\psi|^2 \Delta V$$

As an example, consider Figure 12.20, which shows how the probability density for an electron varies with distance r from the centre of a hydrogen atom in its ground state ($n = 1$). The peak indicates the most probable distance of the electron from the nucleus, which is about 0.50×10^{-10} m. This is where the probability density is highest. But the essential point to understand is that the electron is *not* located in a precise orbit and could be found at any other distances (except in the nucleus).

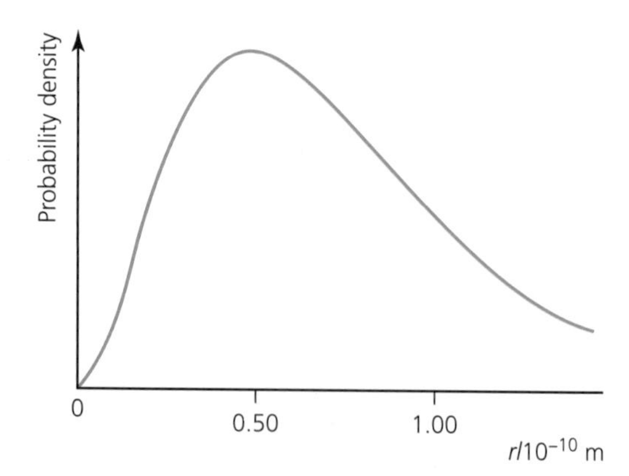

■ **Figure 12.20** Probability density for an electron in a hydrogen atom

In more advanced work, the wave function is applied to three dimensions and often the electron is pictured as an 'electron cloud' (Figure 12.21) of varying probability density. The electron density distribution may be considered to be like a photograph of the atom with a long exposure time. The electron is more likely to be in the positions where the probability density (represented by dots) is highest. For the first main energy level ($n = 1$) the orbital takes the form of a fuzzy sphere.

Interpreting diffraction and interference patterns

Earlier in this book we have described light and electron interference patterns and explained them in terms of wave superposition. Now we should consider how such patterns can be interpreted in terms of matter waves.

When a single electron wave (or photon) passes through slits its exact path is unknown (we know only its probability distribution), and it is only when significant numbers are *detected* on a screen that a pattern emerges. Figure 12.22 uses individual dots to represent the arrival of individual electrons that have been detected on a screen.

The positions of an individual electron (or photon) between its source, through the slits, to where it is detected may be described as 'everywhere' – although it is certainly more likely to be found in some places than others (as described by its wave function). When it arrives on the screen, its wave function changes ('collapses') and it can then be detected. Quantum behaviour is sometimes summarized in the quote 'nothing is real unless it is observed'.

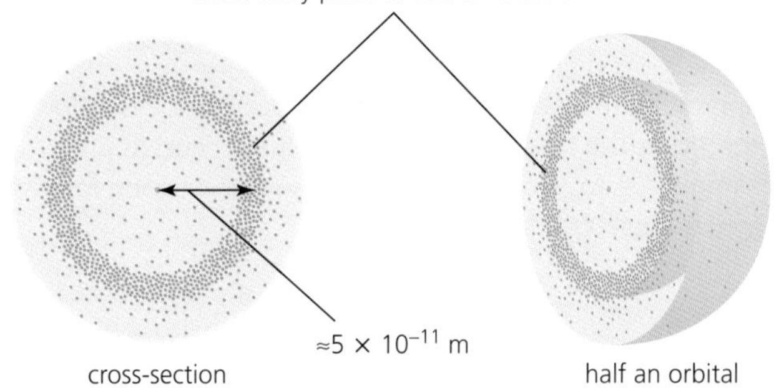

■ **Figure 12.21** Electron cloud for the first orbital in the hydrogen atom

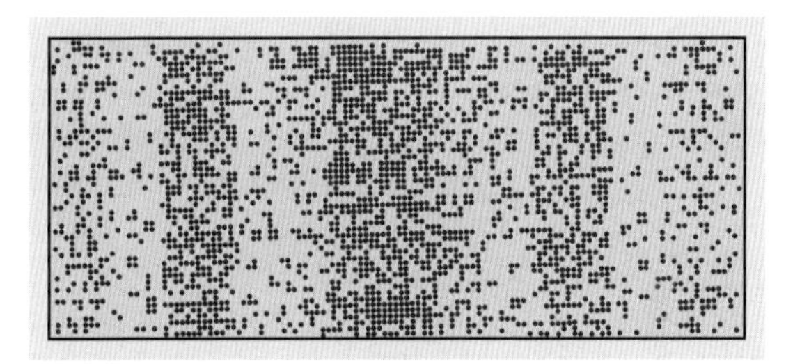

■ **Figure 12.22** Distribution of electrons arriving on a screen in an interference pattern (this would have to be a *very* low-intensity beam!)

Nature of Science

The nature of probability

Einstein is famously quoted as saying: 'God does not play dice', but the nature of probability involved in quantum theory is much more profound than tossing a die or a coin. The way a die lands could, in principle, be predicted if there was sufficient information about the die and the mechanics of how it was tossed. It is not the same in quantum theory. In quantum mechanics exactly the same initial conditions can lead to many different outcomes, so the future is not determined by the past.

■ The Heisenberg uncertainty principle

Because, on the quantum scale, particles have wave properties such that their positions are described in terms of probabilities, it is impossible to fully observe all the properties of a particle at one instant. This concept is clarified by the **Heisenberg uncertainty principle**, which links pairs of variables (such as position and momentum): the more precisely one is known, the less precisely the other is known. We cannot simultaneously measure both properties to unlimited precision.

For example, in everyday life we may commonly observe that a car is moving with a certain momentum when it is located at a certain position, but no such certainty is possible in the quantum world.

It should be stressed that *uncertainty* in atomic physics is not caused by poor experimentation skills or apparatus. It is a fundamental feature of quantum physics.

The uncertainty principle for energy and time, and position and momentum

Uncertainty in position, Δx, and uncertainty in momentum, Δp, of a particle are linked by the equation:

$$\Delta x \Delta p \geq \frac{h}{4\pi}$$

This equation is given in the *Physics data booklet*.

This shows us that the more accurately we know the momentum of a particle (smaller Δp), the less we know about where it is (larger Δx). The more accurately we know the position of a particle, the less we know about its momentum. Because of this, momentum and position are described as *linked variables* and they are called **conjugate quantities**. If one is made zero, the other must be infinite, which means that if we wanted to specify an exact location for a particle, then its momentum would have to be infinite.

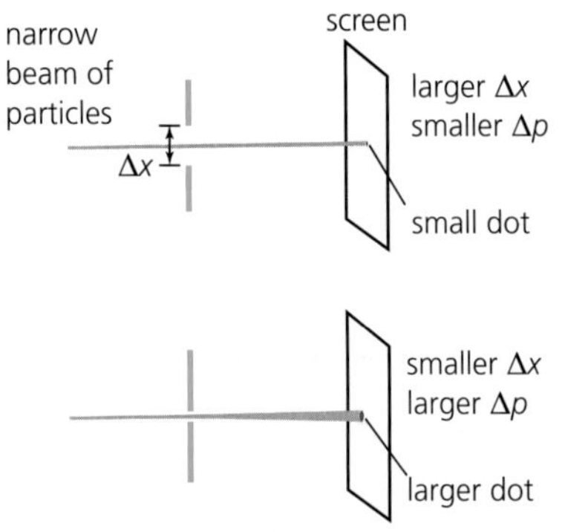

Returning to the diffraction of a narrow beam of photons, or electrons, through small gaps may help to explain the principle: we can interpret diffraction in terms of the uncertainty of position of particles as they pass through the gap. Observation shows us that if a gap is much larger than the wavelength, then diffraction is insignificant (Chapters 4 and 9). Under these circumstances the uncertainty in the position is large (because the gap is relatively large), so that the uncertainty in momentum (speed and direction) is small and that the particles' paths produce a small dot on the screen. If the gap is much smaller, the position of the particles is more precisely known and so their momentum becomes less well defined and the diffraction pattern detected on a screen spreads out (Figure 12.23).

■ **Figure 12.23** Diffraction and uncertainty

We can make an approximate mathematical link to uncertainty as follows. We know from Chapter 9 that the first diffraction minimum occurs at an angle θ such that $\theta = \lambda/b$, and in this example $b = \Delta x$, so that $\theta = \lambda/\Delta x$. A vector diagram for momentum would show that $\theta = \Delta p/p$, which suggests that:

$$\frac{\lambda}{\Delta x} \approx \frac{\Delta p}{p}$$

or

$$\Delta x \Delta p \approx p\lambda = h$$

This is in broad agreement with the more precise uncertainty principle given in the equation above.

The uncertainty principle can be linked to Schrödinger's *wave function*: the more accurately the wavelength of an electron is known, the more precisely its momentum ($p = h/\lambda$) is determined and the greater the uncertainty in its position. For example, if the wave function of a free electron was a perfect sine wave, its wavelength and momentum would be known with certainty, so there would be infinite uncertainty in its position.

Worked examples

7 An electron moves in a straight line with a constant speed ($1.30 \times 10^6\,\text{m s}^{-1}$). The speed can be measured to a precision of 0.10%. What is the maximum possible precision (minimum uncertainty) in its position if it is measured simultaneously?

$p = mv$

$p = (9.110 \times 10^{-31}) \times (1.30 \times 10^6) = 1.18 \times 10^{-24}\,\text{kg m s}^{-1}$

$\Delta p = 0.0010p = 1.18 \times 10^{-27}\,\text{kg m s}^{-1}$

$$\Delta x \geq \frac{h}{4\pi \times \Delta p} = \frac{6.63 \times 10^{-34}}{4 \times \pi \times 1.18 \times 10^{-27}}$$

$\Delta x_{min} = 4.45 \times 10^{-8}\,\text{m}$

Maximum possible precision of position (minimum uncertainty), $\Delta x = 44.5\,\text{nm}$.

8 The position of an electron is measured to the nearest $0.50 \times 10^{-10}\,\text{m}$. Calculate its minimum uncertainty in momentum.

$$\Delta p \geq \frac{h}{4\pi \times \Delta x} = \frac{6.63 \times 10^{-34}}{4 \times \pi \times 0.50 \times 10^{-10}}$$

$\Delta p_{min} = 1.1 \times 10^{-24}\,\text{kg m s}^{-1}$

9 Assume that an electron was confined within a nucleus of diameter $1 \times 10^{-15}\,\text{m}$. Use the uncertainty principle to estimate the minimum uncertainties **a** its momentum and **b** its kinetic energy. **c** Suppose the electron was in a hydrogen nucleus. Compare your answer in **b** to the electric potential energy of a proton–electron system separated by $1 \times 10^{-15}\,\text{m}$.

a

$$\Delta p = \frac{h}{4\pi \times \Delta x} = \frac{6.63 \times 10^{-34}}{4 \times \pi \times 1 \times 10^{-15}}$$

$\Delta p \approx 5.3 \times 10^{-20}\,\mathrm{kg\,m\,s^{-1}}$

b $\Delta E_k = \frac{\Delta p^2}{2m} = \frac{(5.3 \times 10^{-20})^2}{2 \times 9.1 \times 10^{-31}} = 1.5 \times 10^{-9}\,\mathrm{J}$

c $E_p = \frac{ke^2}{r} = \frac{(8.99 \times 10^9)(1.6 \times 10^{-19})^2}{1 \times 10^{-15}} = 2 \times 10^{-13}\,\mathrm{J}$

This calculation shows us that *if* an electron was confined within a nucleus it would have to have a kinetic energy many times higher than the minimum energy needed to overcome the electrical attraction and escape (= potential energy of the system). In other words, the uncertainty principle confirms that an electron *cannot* exist within the nucleus. In beta particle radioactive decay, when electrons are created within the nucleus, they are immediately ejected.

Measurements of time and energy are also linked variables and are described by the energy–time uncertainty principle:

$$\Delta E \Delta t \geq \frac{h}{4\pi}$$

This equation is in the *Physics data booklet*.

ΔE represents the uncertainty in the measurement of energy, while Δt represents the uncertainty in the measurement of time when the measurement was made.

This equation tells us that the shorter the time taken for a measurement, the greater the uncertainty in energy. This has some very important consequences – see Worked example 12.

Worked examples

10 An electron rises from the ground state of an atom to a higher level, but then returns after $5 \times 10^{-8}\,\mathrm{s}$. What is the minimum uncertainty in the energy of the excited electron?

$\Delta E \Delta t \geq \frac{h}{4\pi}$

$\Delta E \times (5 \times 10^{-8}) \geq \frac{h}{4\pi} = 5.28 \times 10^{-35}$

$\Delta E \geq 1 \times 10^{-27}\,\mathrm{J}\ (= 7 \times 10^{-6}\,\mathrm{eV})$

which is a tiny percentage of the energy levels of atoms.

11 An electron has a total energy of 2 eV but would require 5 eV to 'escape' from the nucleus that attracts it. Classical physics tells us that the electron cannot escape because it does not have enough energy, but the uncertainty principle tells us that it could acquire the extra energy (3 eV) *if* the time interval was short enough. The uncertainty in energy can be large if the uncertainty in time is small. Calculate the maximum time for the electron to escape.

$\Delta E \Delta t = \frac{h}{4\pi}$

$(3 \times 1.6 \times 10^{-19}) \times \Delta t = \frac{h}{4\pi} = 5.28 \times 10^{-35}$

$\Delta t = 1 \times 10^{-16}\,\mathrm{s}$

It is possible for an electron to escape from the atom if the time interval is $1 \times 10^{-16}\,\mathrm{s}$ or shorter. This means that, because of uncertainty, the classical law of conservation of energy can be broken if the time is short enough. There are other important examples of this, including some aspects of radioactive decay, nuclear fusion and the existence of exchange particles (see the next section).

39 Calculate the minimum uncertainty in velocity for an electron in a $1.0 \times 10^{-10}\,\mathrm{m}$ radius orbit in which the positional minimum uncertainty is 1% of the radius.

40 A mass of 40.00 g is moving at a velocity of $45.00\,\mathrm{m\,s^{-1}}$. If the velocity can be calculated with an accuracy of 2%, calculate the minimum uncertainty in the position.

41 The lifetime of a free neutron is 15 minutes. How uncertain is its energy (in eV)?

42 Use the principle of uncertainty to estimate values for the momentum and kinetic energy (eV) of an electron confined somewhere within an atom of radius 1×10^{-10} m.

43 Figure 12.24 shows the exchange of a W boson (mass $80.4\,\text{GeV}c^{-2}$) between two quarks. R represents the distance between the quarks at the time of the exchange.
  a What is the energy equivalent (J) of the mass of the boson?
  b The time for the interaction may be assumed to be R/c (distance/speed), assuming that the particle moves at a speed close to the speed of light, c. Use the uncertainty equation to estimate a maximum value for R.
  c Explain why the weak nuclear force has such a short range.

■ **Figure 12.24**

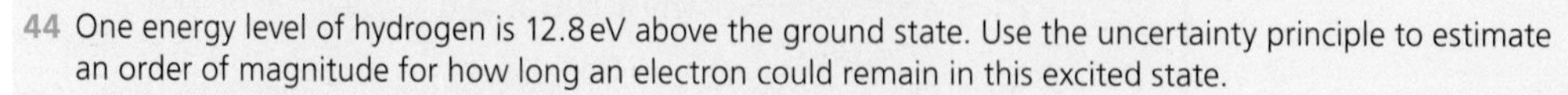

44 One energy level of hydrogen is 12.8 eV above the ground state. Use the uncertainty principle to estimate an order of magnitude for how long an electron could remain in this excited state.

45 Find out about Schrödinger's hypothetical 'cat in the box' experiment and explain its connection with quantum theory.

■ Tunnelling, potential barrier and factors affecting tunnelling probability

One of the most significant and interesting implications of quantum mechanics is that particles can do things that are considered to be impossible using the principles of classical physics. Consider Figure 12.25, which shows a particle of energy, E, trapped between two rectangular (for simplicity) *barriers* of energy, E_B. The particle seems to be trapped because $E < E_B$. A gravitational analogy could be a ball trapped with insufficient kinetic energy to convert to potential energy to get over the top of the hills. The simplest example from atomic physics would be an electron in a hydrogen atom with insufficient energy to 'escape' the electrical attraction of the proton.

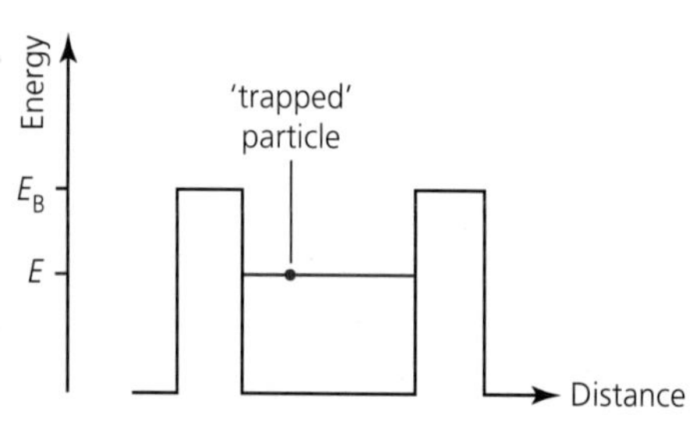

■ **Figure 12.25** A particle trapped between barriers

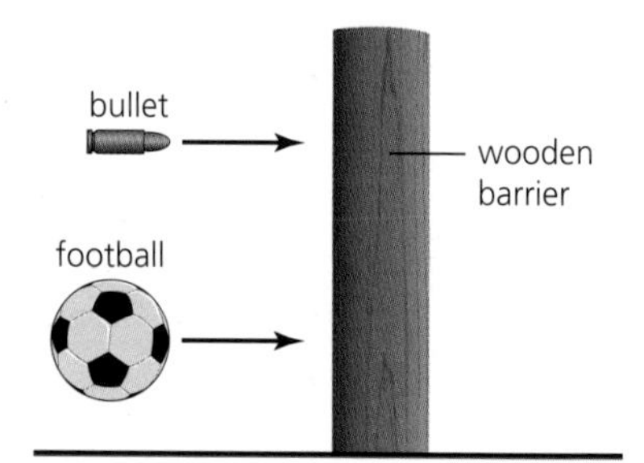

■ **Figure 12.26** A bullet and a football striking wooden barrier horizontally

Let us consider what might happen when an object meets a barrier. Figure 12.26 shows the classical physics situations of a bullet and a football striking a wooden barrier horizontally. In classical physics, the bullet and the football will not be able to pass through the thick barrier, although if the bullet is travelling fast enough, there is a possibility it may penetrate through the barrier, but only by making a physical and permanent hole in it.

Potential barriers

Now consider a particle constrained by one of the fundamental forces. As mentioned before, the simplest example from atomic physics would be an electron in an atom with insufficient kinetic energy to overcome the attraction of the nucleus in order to 'escape'. The 'barrier' in this case is one of electric potential energy because the electron needs to gain enough potential energy to escape. In order to generalize the discussion, it is usual to refer to *potential barriers* (in this example, electric potential = energy/charge) rather than barriers of potential energy. Gravitational analogies are useful in this kind of discussion and it is common to also refer to potential 'hills' and potential 'wells' in descriptions of potential barriers.

Quantum tunnelling

Figure 12.27 (top) shows an electron approaching a potential barrier. Once again the situation has been simplified by the use of a rectangular representation for the barrier (which offers no indication of how the potential varies with distance). The lower part of the figure represents the electron's probability density, which appears on *both* sides of the barrier. Because of the wave nature of the electron there is a clear probability that an electron can be detected on the other side of the barrier. In other words, some electrons will tunnel through the barrier.

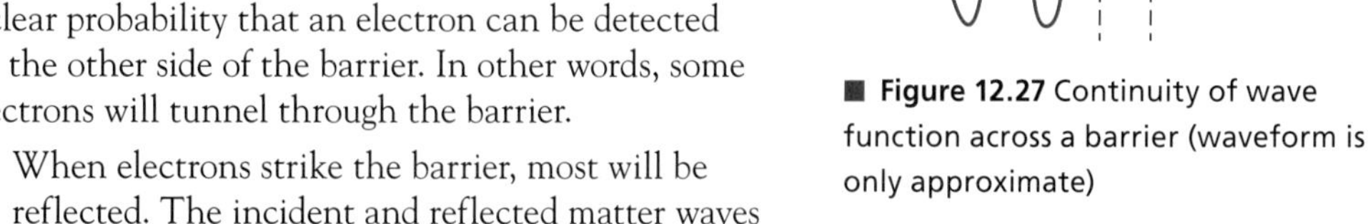

■ **Figure 12.27** Continuity of wave function across a barrier (waveform is only approximate)

- When electrons strike the barrier, most will be reflected. The incident and reflected matter waves interfere and a stationary wave pattern is obtained (giving rise to specific energy levels).
- The wave function and probability density are continuous at the interfaces of the potential barrier.
- The probability density decreases exponentially in the rectangular barrier.
- There is small probability of detecting an electron outside the rectangular barrier.

Tunnelling probability

The number of particles that will tunnel through any given potential barrier depends on the probability with which an approaching particle will penetrate to the other side of a barrier. It depends on the following factors:

- the mass of the particle
- the thickness of the barrier
- the 'height' of the potential barrier
- the energy carried by the particle.

Atomic and nuclear physics deals with enormously large numbers of particles and interactions. This means that, even when the probability of a certain interaction (such as quantum tunnelling) happening is very, very low, it may still occur in significant numbers. For comparison, is said that the worldwide chances of being killed by lightning are very small (an average of about one in 300 000 in any one year), but this still suggests that about 25 000 people in the world are victims of lightning every year.

Examples of quantum tunnelling

- *Ionization* – some electrons can escape from the attraction of a nucleus without having to acquire the full ionization energy.
- *Alpha decay* – after alpha particles are created by nuclear decays in the nuclei of unstable atoms, calculations predict that they do not have enough energy to overcome the strong nuclear forces. But some do escape to become nuclear radiation.
- *Fusion in the Sun* – some hydrogen nuclei can fuse together with less than the kinetic energies that calculations suggest. This means that some nuclear fusion can occur at lower temperatures than would otherwise be predicted. For example, the absolute temperature in the core of the Sun is about 1000 times lower than calculations suggest is necessary for nuclear fusion.
- *Scanning tunnelling microscope (STM)* – see Figure 12.28. A very sharp tip is brought very close to the surface that is to be examined and a voltage is applied between the two to provide a 'barrier' for electrons moving from the surface to the tip. Some electrons tunnel across this barrier and a small current can be measured. The size of the current is very sensitive to the nature of the electron distribution in and around the surface, so it can be used to produce an

image of the surface if the tip is moved (scanned) across it. This type of microscope is capable of resolving individual atoms. Figure 12.29 shows an image from an STM.

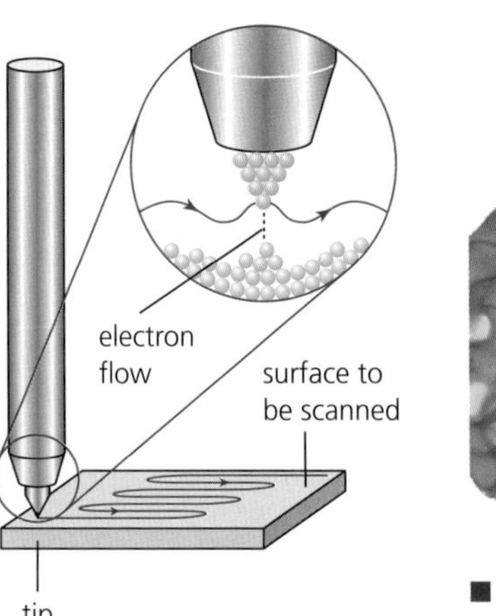

■ **Figure 12.28** The tip scans across the surface and may move up and down to keep the current constant

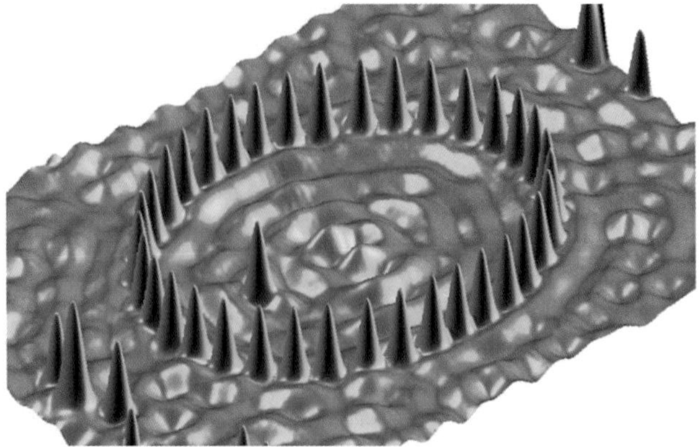

■ **Figure 12.29** Image of a copper surface produced by a scanning tunnelling microscope

The resolutions of various kinds of microscope are limited by the wavelengths of the radiation they use (remember Rayleigh's criterion from Chapter 9). Because electrons have a much shorter wavelength than light, they are capable of providing much higher resolution.

46 Redraw Figure 12.27 to represent a thicker barrier.

47 Use Heisenberg's uncertainty principle to estimate an order of magnitude for the maximum time for an electron to tunnel out of a hydrogen atom.

48 Use the internet to find more STM images.

49 Redraw Figure 12.25 to more realistically represent the potential (or potential energy) well for an electron in a hydrogen atom.

ToK Link

Technological advances can initiate new ways of thinking

The duality of matter and tunnelling are cases where the laws of classical physics are violated. To what extent have advances in technology enabled ***paradigm shifts*** *in science?*

As another example, the invention and use of the telescope was a major factor in the introduction of a sun-centred model of the universe and subsequent developments. Significant discoveries arising from new inventions or technological advances often precede new theories.

12.2 Nuclear physics – *the idea of discreteness that we met in the atomic world continues to exist in the nuclear world as well*

In this section we will consider nuclear size and structure in more detail, beginning with a closer look at the results of Rutherford's nuclear scattering experiments, previously discussed in Chapter 7 (Geiger and Marsden's experiments).

■ Rutherford scattering and nuclear radius

The graph shown in Figure 12.30 represents the typical results obtained from this experiment. The most unexpected observation was that some alpha particles were scattered through large angles. In fact, about one in every 10 000 alpha particles incident on the gold foil was deflected by an angle larger than 90°.

In order for an alpha particle to be scattered straight back (with very little loss of kinetic energy), it must have 'collided' with a much larger mass (see discussion of elastic collisions in Chapter 2). Because most of the alpha particles are *not* deflected, the mass of each atom must be concentrated in a very small centre (the nucleus), such that most particles do not collide with it. Rutherford further realized that the *pattern* into which the alpha particles were scattered could only be explained by the action of large electrical repulsions between the positive alpha particles and tiny positively charged

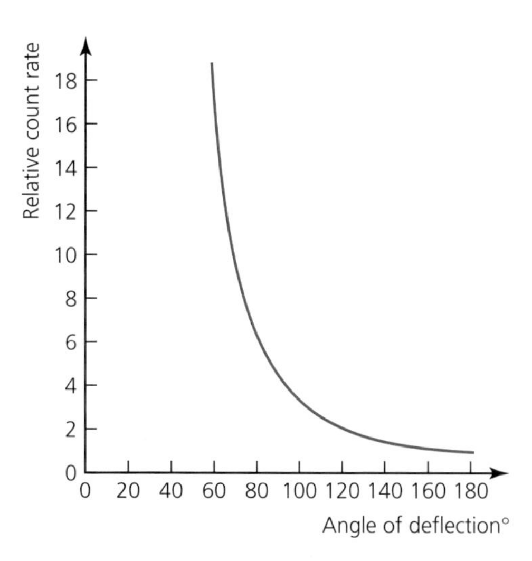

■ **Figure 12.30** The rate of scattering of alpha particles at different angles

nuclei. He proposed that each nucleus has a charge much higher than that of an alpha particle, and that this is effectively concentrated at a *point* at the centre of the atom. The size of the forces, F, between two charges, q_1 and q_2, is described by Coulomb's law (Chapter 5), where r is the distance between them:

$$F = \frac{kq_1q_2}{r^2}$$

This equation is given in the *Physics data booklet*.

Rutherford was able to estimate the radius of a gold nucleus by calculating how close an alpha particle got to the nucleus during its interaction. At its nearest distance, an alpha particle moving directly towards the nucleus stops moving, so all its original kinetic energy is then in the form of electric potential energy. It is as if an 'invisible spring' is being squeezed between the alpha particle and gold nucleus as they come closer and closer. See Figure 12.31.

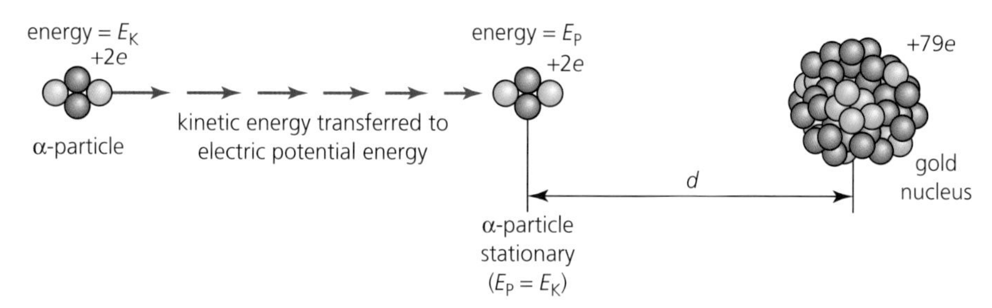

■ **Figure 12.31** At closest approach in a head-on collision, the electric potential energy stored in the electric field is equal to the initial kinetic energy of the alpha particle, $E_P = E_K$

The equation for electric potential energy (E_P) depends on the separation, r, of the two charges (see Chapter 10; it is also given in the *Physics data booklet*):

$$E_P = \frac{kq_1q_2}{r}$$

In the case of the alpha particle, at the distance of closest approach to the nucleus, all its kinetic energy, E_K, will have been transferred to electric potential energy, $E_P = E_K$, so:

$$E_P = E_K = \frac{kq_1q_2}{r}$$

The charge of the alpha particle is $+2e$ and the charge of the nucleus is $+Ze$, ($Z = 79$ for gold), so if the kinetic energy of the alpha particles is known (they are all the same), then r can be calculated as follows:

$$r = \frac{kq_1q_2}{E_K} = \frac{k(2e \times Ze)}{E_K}$$

If the target nucleus is gold and the incident alpha particles have a kinetic energy of 4.0 MeV (which is typical for alpha particles), the distance of closest approach is:

$$r = \frac{(8.99 \times 10^9) \times (2 \times 79) \times (1.6 \times 10^{-19})^2}{(4.0 \times 10^6) \times (1.6 \times 10^{-19})}$$

$$r = 5.7 \times 10^{-14}\,\text{m}$$

This is the separation of the alpha particle and gold nuclei at *closest* approach and gives an *upper limit* for the sum of their radii. Because alpha particles carry so much energy, from this calculation it may be assumed that the radius of a gold nucleus is of the order of 10^{-14} m.

The experiments can be repeated with different elements providing the target nuclei and by using alpha particles of different energies. However, much more energetic alpha particles may enter the nucleus, where strong nuclear forces will also interact with the alpha particle, such that the assumptions made above are no longer valid. The results of the Geiger and Marsden experiments were used to confirm if, for a particular nucleus and alpha particle energy, the scattering of the particles was dominated by an inverse square law of repulsion. If so, then the nuclear radius could be estimated by the theoretical calculations shown above. (No experimental data are involved apart from the energies of the alpha particles.)

50 Explain how results from Geiger and Marsden's experiments, such as those represented in Figure 12.31, indicate that an atom has a small, positive nucleus.

51 Determine the closest distance that an alpha particle of energy 1.37 MeV could approach to:
a a gold nucleus
b a copper nucleus.

Copper has a proton number of 29.

52 Calculate the velocity at which an alpha particle (mass of 6.64×10^{-27} kg) should travel towards the nucleus of a gold atom (charge +79*e*) in order to get within 2.7×10^{-14} m of it. Assume the gold nucleus remains stationary.

53 Explain why alpha particles were used in the Geiger and Marsden experiment.

Nature of Science

Theoretical advances and inspiration often precede discovery in particle physics

Rutherford's proposal of the existence of nuclei at the centre of atoms soon expanded to include the concepts of the proton and the neutron, but these were not actually discovered until years later. There are many examples in particle physics in which theories have been proposed long before the experimental expertise has developed enough for evidence to be found in support of the theory. Further examples include the proposals for the existence of antiparticles and the Higgs Boson, both of which were confirmed only many years after they were first proposed.

■ Electron scattering by nuclei

We have used Coulomb's law to *estimate* nuclear radii from alpha particle scattering experiments, but direct evidence is needed. Detecting what happens to beams of high-energy particles (of known properties, such as alpha particles) when they are directed at atoms and nuclei in targets is an obvious strategy. However, if the particles contain hadrons (as do alpha particles) their interactions with nuclei will become complicated if they experience nuclear strong forces when they are very close to, or entering, a nucleus (Figure 12.32).

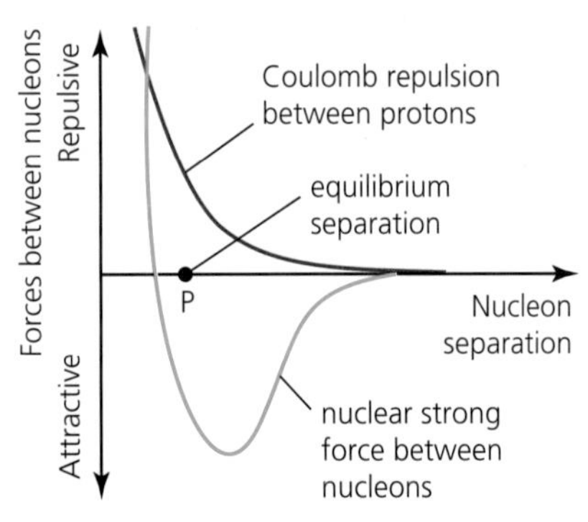

■ **Figure 12.32** Comparing Coulomb repulsion to the nuclear strong force

Electrons are leptons and do *not* experience nuclear strong forces (although they will still experience electromagnetic forces inside a nucleus). High-energy electron beams are also easily produced and controlled; this means that they are a good choice for investigating nuclei. Earlier in this chapter we saw that the wave properties of electrons can result in them being diffracted by atoms and the spacing between atoms, which are typically about 10^{-10} m in size. (Worked example 5 showed that an accelerating voltage of 1000 V produces electrons of wavelength 0.55×10^{-10} m.) Remember that diffraction effects are greatest if the wavelength and the object causing diffraction are about the same size. Much higher accelerating voltages can produce electrons with shorter wavelengths, as low as 10^{-15} m, which makes them suitable for diffraction by objects of that size, such as nuclei. It should be mentioned that, at very high speeds approaching the speed of light, relativity effects become important (Option A: Relativity) and the masses of accelerated electrons will increase, as well as their speeds.

We can use the equation $E = hf = hc/\lambda$ to determine the energy of electrons that will produce a required de Broglie wavelength of, for example, 5×10^{-15} m:

$$E = (6.63 \times 10^{-34}) \times \frac{3.00 \times 10^8}{5 \times 10^{-15}} \approx 4 \times 10^{-11}\,\text{J}$$

or about 2×10^8 eV (200 MeV).

That is, an accelerating voltage of the order of 200 MV is needed. (Electrons in beams produced by even higher accelerating voltages can penetrate deep into nuclei and provide information about the structure of nucleons.)

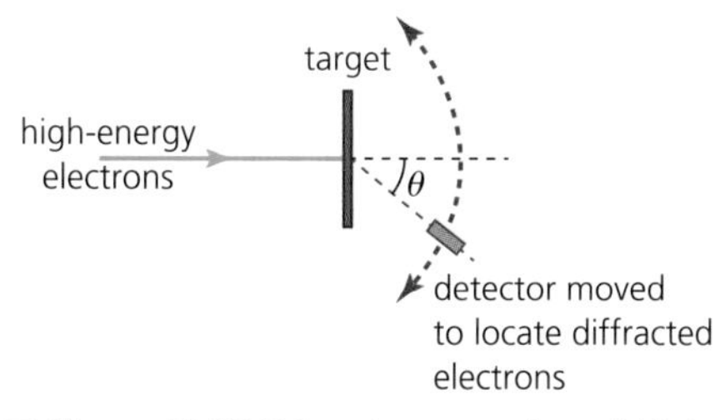

■ **Figure 12.33** Using the scattering of high-energy electrons to investigate nuclei

Figure 12.33 shows a simplified version of the experimental arrangement, excluding the apparatus designed to accelerate the electrons by about 200 MV or more.

The diffraction of electrons by a nucleus is similar in principle to the diffraction of light photons by a narrow slit.

In Chapter 9 we saw that the first diffraction *minimum* for light of wavelength λ passing through a narrow slit of width b occurred at an angle such that $\sin\theta = \lambda/b$. (Because the angles concerned with light were very small, the approximation $\sin\theta = \theta$ could be used, so that the equation became $\theta = \lambda/b$.) Similarly, for electron diffraction, the first diffraction *minimum* occurs at an angle such that:

$$\sin\theta \approx \frac{\lambda}{D}$$

where D is the width of the object causing the diffraction, the nuclear diameter. This equation is given in the *Physics data booklet*.

As the detector shown in Figure 12.33 is rotated from a central position, the intensity of the electron beam decreases. It reaches a diffraction minimum, then rises again. See Figure 12.34.

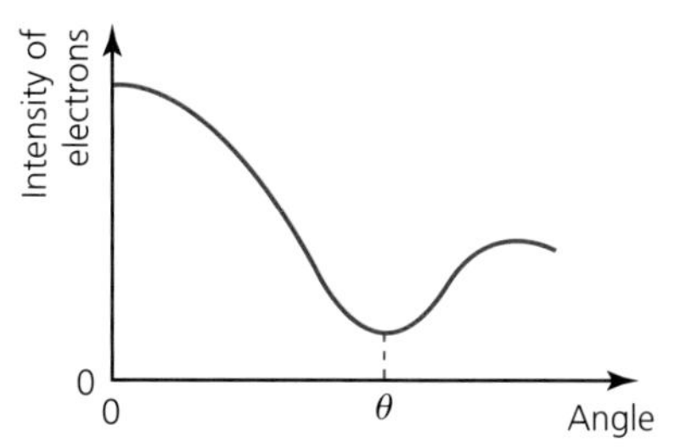

■ **Figure 12.34** Variation of electron intensity with angle

54 The nuclear diameter of carbon is 2.75×10^{-15} m. When carbon diffracts electrons of de Broglie wavelength 1.2×10^{-15} m, at what angle will the first diffraction minimum be detected?

55 Assuming that the maximum possible value of $\sin\theta$ is 1, what is the minimum electron energy (MeV) required to observe a diffraction minimum by a copper nucleus of diameter 4.8×10^{-15} m?

56 Consider Figure 12.32. Explain why the equilibrium separation occurs at point P.

Nuclear density

If we assume that, for a given separation, the strong nuclear force acts equally between all nucleons, so that the nucleons in a nucleus are packed close together, then it reasonable to suggest that the volume of a nucleus, V, is proportional to the number of nucleons, A.

$$V \propto A$$

Assuming that the nucleus is approximately spherical, its volume, $V = \frac{4}{3}\pi R^3$, so that its volume will be proportional to its radius, R, cubed:

$$R^3 \propto A$$

So that:

$$R \propto A^{1/3}$$

This is confirmed by electron scattering experiments.

Putting in a constant of proportionality:

$$R = R_0 A^{1/3}$$

This equation for nuclear radius is given in the *Physics data booklet*. The constant R_0 is called the **Fermi radius**, which is equivalent to the radius of a nucleus with only one proton ($A = 1$). From electron diffraction experiments it is found to have a value of 1.2×10^{-15} m (= 1.2 fm). This value is given in the *Physics data booklet*.

We can use $\rho = m/V$ to estimate **nuclear density**. The mass of a nucleus will be approximately equal to Au, where u is the unified atomic mass unit, giving:

$$\rho = \frac{Au}{\frac{4}{3}\pi A R_0^3} = \frac{3u}{4\pi R_0^3}$$

$$\rho = \frac{3 \times (1.661 \times 10^{-27})}{4\pi \times (1.2 \times 10^{-15})^3} \approx 2 \times 10^{17}\,\text{kg m}^{-3}$$

All nuclear densities are approximately the same, but it should be noted that it is an extremely large density!

If the electrons in an atom are considered to have negligible mass compared to the nucleons, and the radius of an atom is typically 10^5 times larger than a nucleus, then an order of magnitude density for atoms would be $(10^{17})/(10^5)^3 \approx 10^2\,\text{kg m}^{-3}$, which is comparable to everyday observations of the density of matter. The only macroscopic objects with densities comparable to nuclear densities are collapsed massive stars, known as *neutron stars* and *black holes* (Option D: Astrophysics).

57 Estimate the radius of:
 a a gold nucleus ($A = 197$)
 b an oxygen nucleus ($A = 16$)
 c The measured radius of a gold-197 nucleus is 6.87 fm. How does this compare with the value calculated in a?

58 Estimate a value for the following ratio: radius of largest possible nucleus to radius of smallest possible nucleus.

59 Gold is considered to be a dense element. Estimate what fraction of the volume of a gold ring is actually occupied by particles. (Density of gold = 19300 kg m^{-3})

60 The mass of the Earth is 6.0×10^{24} kg. Estimate the radius of a neutron star that had the same mass as the Earth.

61 Explain why the mass of a nucleus will be '*approximately* equal to Au'.

Nuclear energy levels

Like the atom, the nucleus is a quantum system and has discrete (individual and separate) energy levels.

The observation that the energies of alpha particles and gamma rays are *discrete* provides strong experimental evidence for the existence of nuclear energy levels. (This is in contrast to beta decays, in which the emitted electrons or positrons have a continuous range of energies.)

Alpha particle energies are discrete

During each alpha decay a single alpha particle is ejected from an unstable and larger nucleus. The smaller alpha particle has the higher speed and the nucleus moves backwards (recoils) at a lower speed, conserving momentum. Energy is released in alpha decay in the form of the kinetic energies of the alpha particle and the recoiling nucleus. Most of the energy is carried away by the alpha particle.

The alpha decays of any particular radionuclide will always produce the same daughter product, and the discrete energies of the emitted alpha particles will usually all be identical from nuclei of the same radionuclide. However, some radionuclides emit alpha particles with several different energies. This is because it is possible for the daughter product to be in one of several discrete *excited states*. Figure 12.35 shows the emission of alpha particles of five different energies from americium-241 to various nuclear energy levels of neptunium-239.

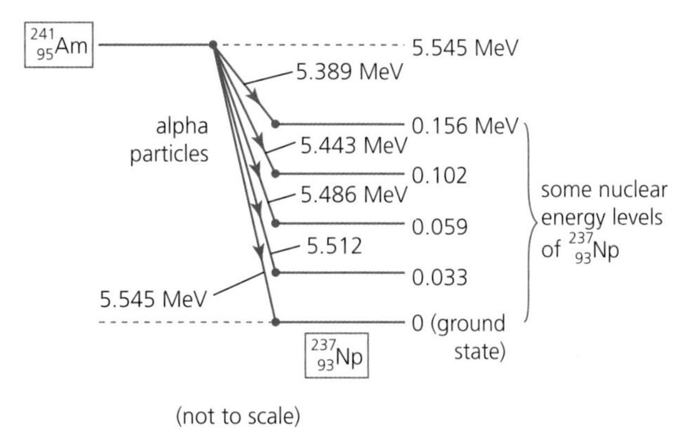

■ **Figure 12.35** Energies of alpha particles emitted from americium-241

The alpha decay of the same radionuclide always produces alpha particle(s) of the same discrete energies. This is predicted by the law of conservation of momentum because the *two* particles involved must always move in opposite directions.

Gamma ray energies are discrete

Evidence for the existence of nuclear energy levels also comes from studying gamma ray emissions from radioactive nuclei. These emissions do not change the numbers of protons and neutrons in the nucleus, but remove energy from the nucleus in the form of gamma rays (photons).

The gamma rays have energies that are discrete and unique for a particular nuclide. This suggests that the gamma rays are emitted as a result of **nuclear transitions** from an *excited state* with a higher energy level to a lower energy level, similar to electron transitions in the atom (Chapter 7). Nuclei are often left in excited states following the emission of alpha or beta particles, as was shown in Figure 12.35.

Another, simpler, example is shown in Figure 12.36. After carbon-15 has decayed by beta emission, some of the nitrogen-15 daughter nuclei are left in an excited state. These excited nuclei release gamma rays of a particular frequency, and hence energy, to reach the lowest energy level (the ground state).

The photons emitted by radioactive nuclei carry more energy, and therefore have higher frequencies, than photons emitted as a result of transitions involving electrons. This means that the energy changes involved in nuclear processes are much larger than those involved in transitions of electrons.

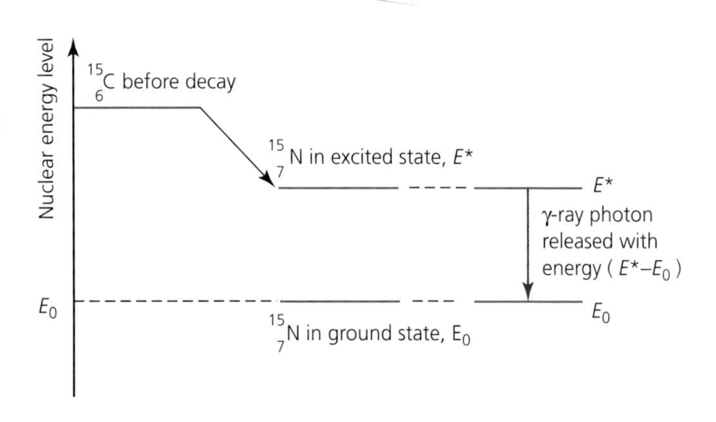

■ **Figure 12.36** The gamma rays released when nitrogen-15 nuclei fall from an excited state to the ground state have a characteristic frequency

62 Explain why measurements made of alpha particle and gamma ray energies suggests that there are discrete energy levels in nuclei.

63 Figure 12.37 shows the lowest two energy levels for the nucleus of a heavy atom.
a How does the size of nuclear energies compare with the size of electron energies?
b Calculate the wavelength of the electromagnetic radiation emitted in a nuclear change involving a transition from the first excited state to the ground state.
c To what part of the electromagnetic spectrum does this photon belong?

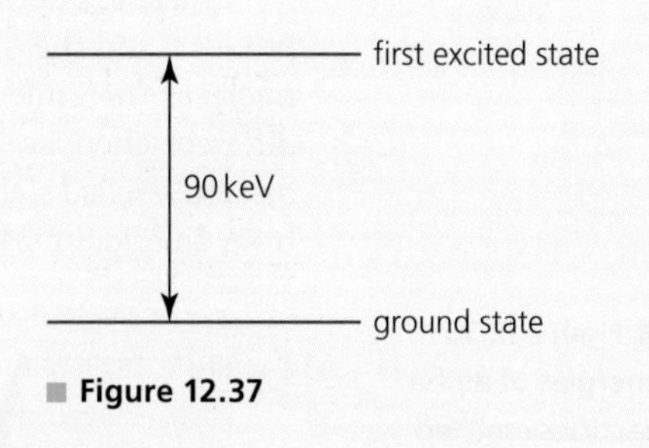

■ **Figure 12.37**

64 a Use Figure 12.35 to make a copy of five energy levels of a neptunium-237 nucleus. Draw arrows to represent the ten possible energy level transitions.
b What are the wavelengths of the largest and smallest gamma ray photons that could be emitted from these transitions?

65 Suppose that it is known that a particular nuclide emits gamma ray photons of energies 0.58 MeV, 0.39 MeV and 0.31 MeV. Determine the energies of another three different photons that might be emitted from the same nucleus.

Beta particle energies are *not* discrete

If beta particle emission involved only the nuclei and the beta particles, the particles emitted from a particular radionuclide should all have the same kinetic energy, in a similar way to alpha particle emission. However, measurements show that although the *maximum* kinetic energy of the beta particles is characteristic of the beta source, the particles are emitted with a *continuous range* of kinetic energies (Figure 12.38).

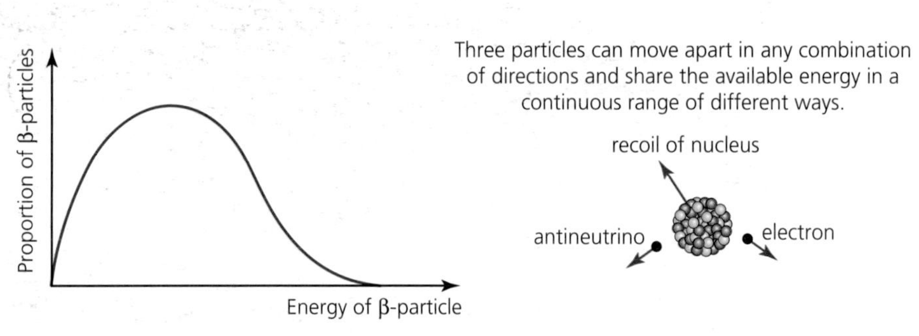

■ **Figure 12.38** Typical energy spectrum for beta decay

This unexpected observation was first recorded in 1928 and at first may have suggested that energy and momentum might not be conserved. However, in 1933 the Austrian physicist Wolfgang Pauli proposed that another undetected particle was also involved in beta decay. This particle was hypothesized to be neutral (otherwise the electric charge would not conserved) and have little or no mass (otherwise the energy curve of the beta particles would have a different shape). This new particle was named the **neutrino** (and its antiparticle the **antineutrino**). Because the neutrino has such a small mass and interacts very weakly with matter, it was not until the 1950s that evidence for its existence was obtained.

Reviewing information about beta decay (Chapter 7): there are two kinds: (a) *beta-negative decay,* in which a neutron changes into a proton with the emission of an electron and an (electron) antineutrino; and (b) *beta-positive decay,* in which a proton changes into a neutron with the emission of a positron (positive electron) and a (electron) neutrino. The full beta decay equations then become:

Beta-negative decay	*Beta-positive decay*
$n \rightarrow p + e^- + \bar{\nu}_e$	$p \rightarrow n + e^+ + \bar{\nu}_e$

Because each of the three particles may move away in any direction, applying the conservation of momentum, we see that the energy can be shared in any ratio, explaining the continuous spectrum of beta particle energies.

The neutrino

As we saw in Chapter 7, neutrinos (and their antiparticles) are elementary particles. There are three kinds of neutrino, each associated with either an electron, a muon or a tao particle. Neutrinos are uncharged and have very small masses. They travel at speeds close to the speed of light. It is now believed that they may be the second most common type of particle in the universe (after photons). It is estimated that over 10^{12} neutrinos pass through a finger nail every second, most of which were emitted in nuclear fusion reactions in the Sun. They usually pass through the entire Earth almost without being affected.

The detection and properties of neutrinos is becoming an increasingly important area of research in physics. It is hoped that deeper knowledge of neutrinos will increase our understanding of such matters as the origin of the universe, the ratio of matter/antimatter and the properties of 'dark matter' in the universe.

It is surprising that so little is known about such a common particle, but their properties and associated penetrating power make neutrinos very difficult to detect. The probability of them interacting with matter (in the form of sub-atomic particles) is very, very small, but finite. The probability is increased by having large detection chambers containing a lot of dense matter. Experiments are best located underground and/or in remote locations that are free from outside influences or the passage of less penetrating particles. See Figure 12.39.

Figure 12.39 The building at the top of the 'ice cube' neutrino detector, which is deep underground near the South Pole

Nature of Science

Modern computing power: analysing vast quantities of data

All particle physics and nuclear physics experiments require instruments to try to detect energetic sub-atomic particles. Generally, these detectors are required to measure various kinematic properties of each particle, such as its energy, momentum, the spatial location of its track and its time of arrival at the detector. Neutrinos are not detected directly but by the radiation emitted from the products of their interactions.

Scientific progress often emerges from advances in detector technology. Such advances include: enhanced precision in kinematic properties; the rate at which particles may be detected, leading to improved statistical precision; and in reduced costs, resulting in larger systems with greater sensitivity to rare processes.

All particle detectors ultimately produce information in the form of electric signals that must be processed by electronic circuits, digitized to produce numerical data, which in turn may be further processed in real time and then recorded for further analysis. All this data would be of less use without the internet and the data-handling abilities of computers around the world.

Additional Perspectives

The long-baseline neutrino experiment (LBNE)

This US/international project is intended to be completed by 2022. An artificial, high-intensity neutrino 'beam' will be created at Fermilab and detected by its rare interactions with target material 1300 km away. The beam will originate about 1.5 km under the Earth's surface at Fermilab, near Chicago, and travel through rock (at a depth of up to 35 km) to a detector in South Dakota. Because neutrinos are so penetrating no tunnel is required.

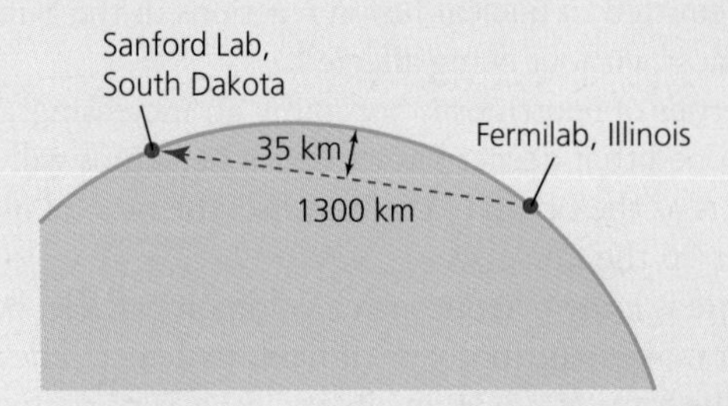

■ **Figure 12.40** Intended paths of the neutrinos

1 Find out:

a in what way this intended project will be able to provide information that is different from other neutrino detectors

b why these particular locations, and length of path, were chosen for the project.

2 If neutrinos are so difficult to detect/control, how is it possible to refer to a 'beam' of neutrinos?

ToK Link

We are all influenced and biased by our own previous experiences and backgrounds

Much of our knowledge about sub-atomic particles is based on the models one uses to interpret data from experiments. How can we be sure that we are discovering an 'independent truth' not influenced by our models? Is there such a thing as a single truth?

Of course, scientists involved in research into sub-atomic particles (such as the neutrino) will claim to be open-minded and receptive to new ideas. Undoubtedly, this is a primary aim of all scientists, but to some extent we are all inevitably influenced and restricted by our previous experiences. Expensive projects need to have specific aims (or governments would not provide funding), and these will focus the thinking of the scientists involved. However, it is clear from the development of science over the centuries that it is in the nature of science and scientists to use their imagination to formulate new and original ideas.

It is possible that no 'single truth' about neutrinos (in the sense of certain knowledge for all time) will emerge from the latest research, although it is to be hoped that our understanding of these elusive fundamental particles will expand.

■ The law of radioactive decay and the decay constant

Radioactive decay was discussed in Chapter 7 but the mathematics was restricted to examples involving only whole numbers of half-lives. In this chapter we will develop a deeper understanding of radioactive decay and, in particular, expand the theory to include calculations for any length of time.

In the dice experiment described on page 300, increasing the number of dice thrown increases the number of sixes that appear (this experiment provides a simple analogy for radioactive decay). Similarly, if the decay of a radioactive material is investigated, then it is found that the higher the number of radioactive nuclei in the sample, the greater the rate of decay, as measured by the count rate.

This can be described mathematically by the expression:

$$-\frac{\Delta N}{\Delta t} \propto N$$

where delta, Δ, represents 'change in' and N represents the number of undecayed nuclei in the sample at that time.

This relationship defines an **exponential decay**; that is, a change in which the rate of change (decrease) at any time is proportional to the value of the quantity at that moment.

(In Chapter 11 we met the idea of the exponential decrease of charge on a capacitor.)

$\Delta N/\Delta t$ represents the rate at which the number of nuclei in the sample is changing, and $-\Delta N/\Delta t$ represents the *rate of decay*.

Introducing a constant of proportionality, λ, we get:

$$-\frac{\Delta N}{\Delta t} = \lambda N$$

The constant λ is known as the **decay constant**. It has the units of reciprocal of time (for example s^{-1}). As we shall see later, the decay constant is closely linked to the half-life of the source. The higher the value of λ, the quicker the radionuclide decays and the shorter the half-life.

The decay constant is defined as the probability per unit time that any particular nucleus will undergo decay.

The **activity**, A, of a radioactive source is the number of nuclei decaying per second. Activity is the same as the rate of decay:

$$\text{Activity, } A, = -\frac{\Delta N}{\Delta t}$$

and so:

$$A = \lambda N$$

It should be noted that *none* of these equations are given in the *Physics data booklet*.

Activity is measured in **becquerels**, Bq, where one becquerel is defined as one decay per second. The becquerel is named after Henri Becquerel, who shared a Nobel Prize for Physics with Pierre and Marie Curie for their pioneering work on radioactivity.

Worked example

12 The activity of a radioactive sample is 2.5×10^5 Bq. The sample has a decay constant of $1.8 \times 10^{-16} s^{-1}$. Determine the number of undecayed nuclei remaining in the sample at that time.

$A = \lambda N$

$2.5 \times 10^{-5} = (1.8 \times 10^{-16}) \times N$

$N = \frac{2.5 \times 10^5}{1.8 \times 10^{-16}} = 1.4 \times 10^{11}$ undecayed nuclei

66 A radioactive sample emits alpha particles at the rate of 2.1×10^{12} per second at the time when there are 5.0×10^{20} undecayed nuclei left in the sample. Determine the decay constant of the radioactive sample.

67 A sample of a radioactive nuclide initially contains 2.0×10^5 nuclei. Its decay constant is $0.40 s^{-1}$. What is the initial activity?

68 Explain why the count rate detected from a radioactive sample reduces with time.

Solutions to the decay equation

Although the equation $A = -\Delta N/\Delta t = \lambda N$ defines the mathematics of radioactive decay, it does not directly provide us with what we are most likely to want to know – the value of A, or N, at any time, *t*. For that we can use the following equations, which may be described as *solutions* to the original equation. These equations are similar to those used in Chapter 11 for the exponential decrease in current (or p.d. or charge) for a discharging capacitor.

$$N = N_0e^{-\lambda t}$$

This equation is given in the *Physics data booklet*.

In this equation N_0 represents the number of undecayed nuclei in a sample at the beginning of a time t, and N represents the number of undecayed nuclei at the end of time t. Similarly:

$$A = A_0e^{-\lambda t}$$

where A_0 represents the activity from a sample at the beginning of a time t, and A represents the activity at the end of time t.

Since $A_0 = \lambda N_0$, the equation can also be expressed as:

$$A = \lambda N_0e^{-\lambda t}$$

This equation is given in the *Physics data booklet*.

The *count rate*, C (a non-standard symbol), measured by a radiation detector in a laboratory is *not* directly measuring the activity of the source, but it will usually be proportional to the activity, so that we can also write:

$$C = C_0e^{-\lambda t}$$

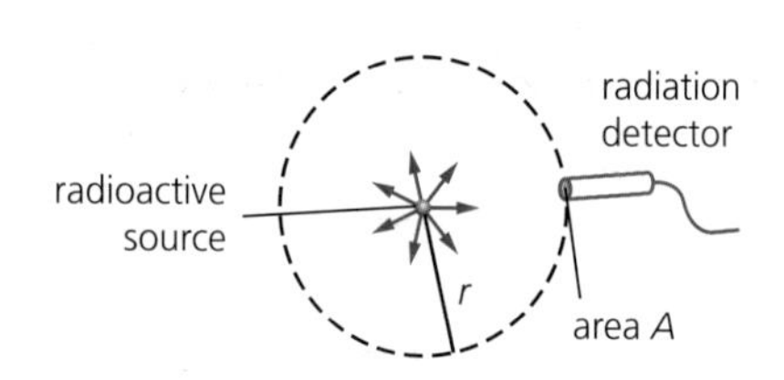

■ **Figure 12.41** A radiation detector receives only a fraction of the radiation emitted from a source

Figure 12.41 shows that a detector can only receive *some* of the total radiation from a source. We could use the ratio $A/4\pi r^2$ to determine the fraction arriving at the detector, but that would need to assume that no radiation was absorbed between the source and the detector, and also that the radiation was emitted equally in all directions. Determining the activity of the source would also be difficult because not all of the radiation passing into the detector will be counted.

Worked examples

13 The number of radioactive atoms in a sample decays by $\frac{7}{8}$ in 12 days. Calculate the fraction of radioactive atoms remaining after 24 days.

This can be done in a straightforward way *without* using exponentials: the sample passes through three half-lives to reduce to $\frac{1}{8}$, so the half-life is four days. In 24 days there are six half-lives, so the fraction reduces to $\frac{1}{64}$. We can get the same answer using the exponential equation $N = N_0e^{-\lambda t}$.

After 12 days:

$$\frac{N}{N_0} = \frac{1}{8} = e^{-\lambda(12)}$$

$$8 = e^{12\lambda}$$

$$\lambda = \frac{\ln 8}{12} = 0.173$$

After 24 days:

$$\frac{N}{N_0} = e^{-0.173(24)} = 0.0157 = \frac{1}{64}$$

14 The decay constant for a radioisotope is $0.054\,y^{-1}$. If the activity at the start of the year 2014 was 4.7×10^2 Bq, calculate the activity at the start of 2018.

$$A = A_0e^{-\lambda t}$$

$$A_{2018} = 470 \times e^{-0.054 \times 4}$$

$$\ln(A_{2018}) = \ln 470 - (0.054 \times 4)$$

$$A_{2018} = 379\text{ Bq}$$

69 A sample of radium contains 6.64×10^{23} radioactive atoms. It emits alpha particles and has a decay constant of $1.36 \times 10^{-11}\,s^{-1}$. How many atoms of radium are left after 100 years?

70 A radioactive nuclide has a decay constant of $0.0126\,s^{-1}$. Initially a sample of the nuclide contains 10 000 nuclei.
a What is the initial activity of the sample?
b How many nuclei remain undecayed after 200 s?

71 a The count rate from a radioactive source is measured to be $673\,s^{-1}$. If exactly three hours later the count rate has reduced to $668\,s^{-1}$, determine a value for the decay constant.
b Explain why the value may be unreliable.

72 A source has an activity of 4.7×10^4 Bq. If the background count average is $0.58\,s^{-1}$, what is the maximum possible count rate that could be recorded by a detector that has an effective receiving area of $0.85\,cm^2$ if it was placed:
a 50 cm from the source
b 5 cm from the source?
c Why are your answers *maximum* values?

73 The activity from a radioactive source is 8.7×10^5 Bq. If its decay constant is $6.3 \times 10^{-3}\,s^{-1}$, how many minutes will pass before the activity falls to 1.0×10^4 Bq?

74 A radioactive source of gamma rays, cobalt-60, is commonly used in school demonstrations. If the maximum allowable activity is 200 kBq, calculate the maximum mass of cobalt-60 in a school source. (Decay constant for cobalt-60 is $0.131\,y^{-1}$.)

Decay constant and half-life

The concept of the half-life of a radioactive nuclide was introduced in Chapter 7. Using the equation $N = N_0 e^{-\lambda t}$, we can easily derive an equation that relates the half-life, $T_{1/2}$, to the decay constant, λ.

For any radioactive nuclide, the number of undecayed nuclei after one half-life is, by the definition of half-life, equal to $N_0/2$, where N_0 represents the original number of undecayed nuclei. Substituting this value for N in the radioactive decay equation at time $t = T_{1/2}$ we have:

$$\frac{N_0}{2} = N_0 e^{-\lambda T_{1/2}}$$

Dividing each side of the equation by N_0:

$$\frac{1}{2} = e^{-\lambda T_{1/2}} \text{ or } 2 = e^{\lambda T_{1/2}}$$

Taking natural logarithms (to the base e):

$$\ln 2 = \lambda T_{1/2}$$

So that:

$$T_{1/2} = \frac{\ln 2}{\lambda}$$

This useful equation is *not* in the *Physics data booklet*. Alternatively, inserting a value for ln 2:

$$T_{1/2} = \frac{0.693}{\lambda}$$

Worked examples

15 A radioactive sample gives a count rate of $100\,s^{-1}$ at a certain instant of time. After 100 s the count rate drops to $20\,s^{-1}$. The background count rate is measured to be $10\,s^{-1}$. Calculate the half-life of the sample. Assume that the count rate is a measure of the activity.

initial count rate due to sample = $100 - 10 = 90\,s^{-1}$

count rate due to sample after 100 s = $20 - 10 = 10\,s^{-1}$

$$C = C_0 e^{-\lambda t}$$

$$10 = 90e^{-100\lambda}$$

$$\lambda = \frac{-\ln \frac{10}{90}}{100} = 0.022$$

$$T_{1/2} = \frac{0.693}{0.022} = 32\,\text{s}$$

16 The radioactive element A has 6.4×10^{11} atoms and a half-life of 2.00 hours. Radioactive element B has 8.0×10^{10} atoms and a half-life of 3.00 hours. Calculate how much time will pass before the two elements have the same number of radioactive atoms.

For A:

$$\lambda_A = \frac{0.693}{2.00} = 0.347\,\text{h}^{-1}$$

For B:

$$\lambda_B = \frac{0.693}{3.00} = 0.231\,\text{h}^{-1}$$

At time t:

$$N_A = (6.4 \times 10^{11})e^{-0.347t},\ N_B = (8.0 \times 10^{10})e^{-0.231t}$$

For $N_A = N_B$:

$$(6.4 \times 10^{11})e^{-0.347t} = (8.0 \times 10^{10})e^{-0.231t}$$

$$8 = \frac{e^{-0.231t}}{e^{-0.347t}}$$

$$8 = e^{0.116t}$$

$$\ln 8 = \ln(e^{0.116t});\ 2.079 = 0.116t$$

$$t = 18\,\text{h}$$

The two elements will have the same number of active atoms after 18 hours.

75 Radioactive carbon-14 in a leather sample decays with a half-life of 5730 years.
- a What is the decay constant?
- b Calculate the percentage of radioactive carbon remaining after 10000 years.

76 At a certain time a pure source contained 3.8×10^{15} radioactive atoms. Exactly one week later the number of radioactive atoms had reduced to 2.5×10^{14}. What is the half-life of this source?

77 The half-life of strontium-90 is 28.8 years. How long will it take for the count rate from a sample to fall by:
- a 1%
- b 99%?

78 A cobalt-60 source used in radiotherapy in a hospital needs to be replaced when its activity has reduced to 70% of its value when purchased. If it was purchased in mid-February 2011 and replaced in mid-November 2013, what was its half-life?

79 The count rate from a radioactive source is $472\,\text{s}^{-1}$. If its half-life is 13.71 minutes, what will be the count rate exactly 1 hour later?

80 Uranium-238 is the most common isotope of uranium. It has a half-life of 4.47 billion years. The age of the Earth is 4.54 billion years.
- a What percentage (to one decimal place) of the original uranium-238 is still present in the Earth's crust?
- b Suggest, in principle, how the 'age' of rocks may be determined by analysis of the radioisotopes they contain.
- c Why are such methods unlikely to be very accurate?

Measurement of half-life

The method used to measure the half-life of a radioactive element depends on whether the half-life is relatively long or short. If the activity of the sample stays approximately constant over a few hours then we can say that it has a relatively long half-life. However, if its activity decreases significantly during a few hours, then the radioactive element has a relatively short half-life.

Isotopes with long half-lives

$$A = \frac{\Delta N}{\Delta t} = -\lambda N$$

If the activity, A, of a source can be determined, then the decay constant (and therefore the half-life) can be calculated if the number of undecayed atoms at the time of the measurement, N, is known. Theoretically this is straightforward, but experimentally determining the number of atoms in a source is not easy, especially when the isotope in question is usually in a sample containing a mixture of other isotopes. This requires sophisticated equipment such as a mass spectrometer.

For a *pure* sample of mass m, the number of atoms of the isotope can be determined from the relative atomic mass, A_r, and Avogadro's constant, N_A, as follows:

$$N = \frac{mN_A}{A_r}$$

Therefore, if the activity ($A = \Delta N/\Delta t$) is measured, we can calculate the half-life, $T_{\frac{1}{2}}$, from the equation:

$$\frac{\Delta N}{\Delta t} = \frac{-\lambda m N_A}{A_r} = \frac{-0.693 m N_A}{T_{\frac{1}{2}} A_r}$$

Isotopes with short half-lives

If an isotope's half-life is a few hours or less, then it can be determined by recording the count rate (minus the background count) at regular time intervals. A radioisotope with a short half-life can used in a school laboratory by extracting it from the other radioisotopes in its decay chain *just* before the experiment is due to begin. The extraction process can be based on any chemical or physical property that is unique to the required isotope.

The adjusted count rate is assumed to be proportional to activity. If a graph of count rate, C, against time, t, is plotted, the half-life can be obtained directly from the graph as discussed on page 301.

Alternatively, a graph can be plotted of natural logarithm of the count rate, C, against time, t. This will give a straight line with a gradient of $-\lambda$ (Figure 12.42). Since $T_{\frac{1}{2}} = 0.693/\lambda$, the half-life can be calculated. This is a better method because the reliability of the data can be more readily assessed by seeing how close to the straight line of best-fit the data points lie.

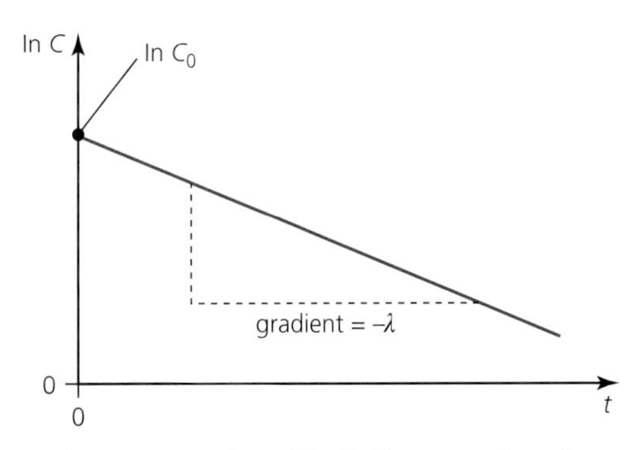

■ **Figure 12.42** A logarithmic–linear graph to show exponential decay of a radioactive nuclide

This method relies on transforming the equation $C = C_0 e^{-\lambda t}$, which describes the count rate in radioactive decay.

Taking natural logarithms:

$$\ln C = \ln C_0 - \lambda t$$

The equation can be compared to the equation for a straight line ($y = mx + c$), so the gradient is equal to $-\lambda$.

When the half-life of the isotope is very short, less than a second, then both of these methods are unsuitable. (Such half-lives may be found from tracks in a cloud or bubble chamber.)

Utilizations

Mass spectrometers

The first direct investigation of the mass of atoms (and molecules) was made possible by the development of the *mass spectrometer*. The first mass spectrometer was built in 1918 by William Aston, a student of J. J. Thomson. It provided direct evidence for the existence of isotopes. A more accurate mass spectrometer was developed by William Bainbridge in 1932.

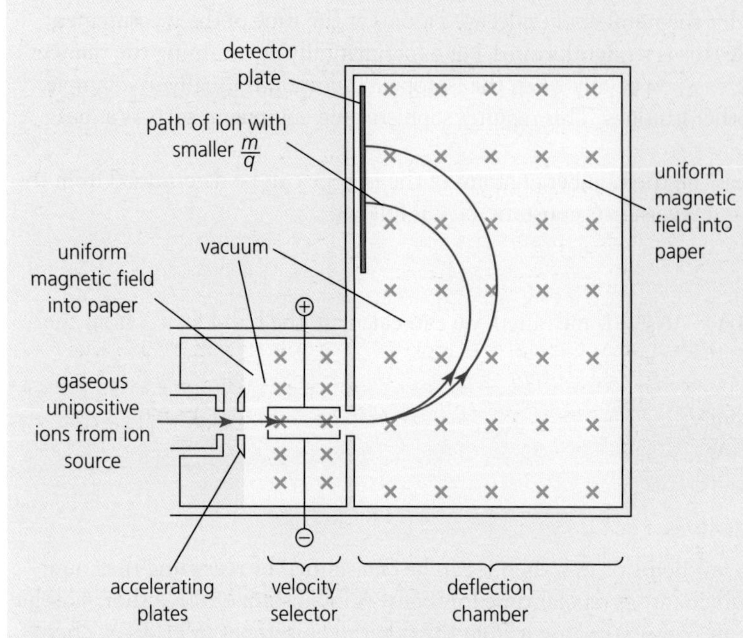

■ **Figure 12.43** Essential features of a Bainbridge mass spectrometer

This device uses the interaction of charged ions with electric and magnetic fields to measure the relative masses of atoms and to find their relative abundances. Figure 12.43 shows how a simple Bainbridge mass spectrometer works. There is a vacuum inside the machine; no air is allowed to enter. This is to avoid collisions with particles in the air that would disrupt the flight of the ions produced inside the instrument.

Positive ions are produced in the mass spectrometer by bombarding gaseous atoms or molecules (M in the following equation) with a stream of high-speed electrons:

$$M(g) + e^- \rightarrow M^+(g) + e^- + e^-$$

The ions are accelerated (by a pair of positive plates) and then pass through a *velocity selector*. Here electric and magnetic fields are applied to the ions so that only a narrow beam of ions travelling at the same velocity continue in a straight line to enter the next chamber. These ions then travel through a uniform magnetic field. This causes the ions to move in a circular path with radius, r, which depends on the ion's mass to charge ratio. For ions with mass, m, and charge, q, travelling with velocity, v, through the magnetic field, B:

centripetal force = force due to magnetic field

$$\frac{mv^2}{r} = Bqv$$

The radius of the circle for the ion is therefore:

$$r = \frac{mv}{Bq}$$

If the ions have the same charge, q (usually unipositive, i.e. with a single positive charge), and they are all selected to be travelling at the same velocity, v, then the radius of the circle of each ion's path will depend only on the mass of the ion. An ion with a large relative mass will travel in a wider circle (for the same magnetic field strength).

A number of vertical lines will be obtained on the detector plate, each line corresponding to a different isotope of the same element. The position of a line on the plate will allow the radius, r, to be determined. Because the magnetic field strength, B, the charge on the ion, e, and velocity of the ion, v, are all known, the mass of the ion, m, can be determined. The older type of Bainbridge mass spectrometer shown is actually a mass spectrograph, since the beam of ions is directed on to a photographic plate (Figure 12.44). The relative intensities of the lines allowed an estimate to be made of the relative amounts of isotopes.

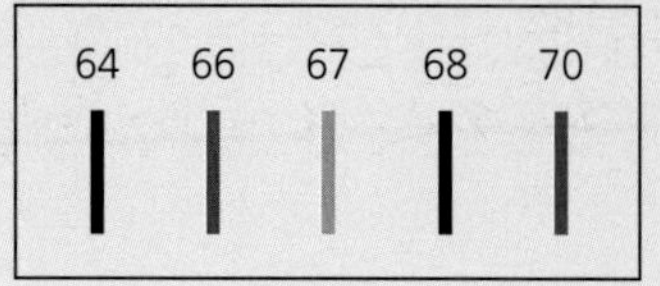

■ **Figure 12.44** A mass spectrum obtained by a mass spectrograph

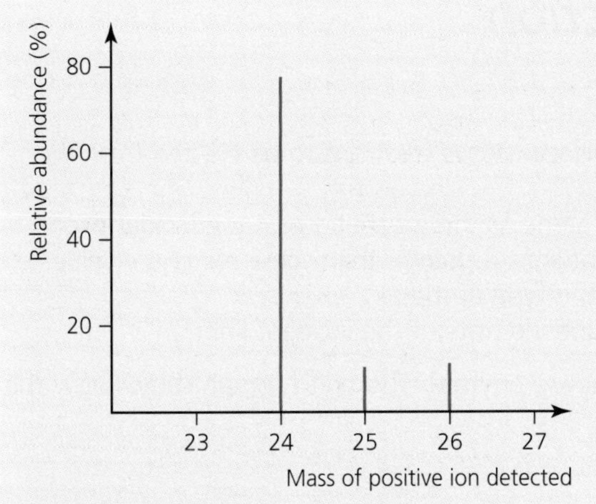

■ **Figure 12.45** The mass spectrum of naturally occurring magnesium atoms, showing the percentage abundances of three isotopes with their masses

Modern mass spectrometers neutralize the positive ions with electrons and count the number of ions directly, before amplifying the signal. The results are displayed on a computer screen in the form of a 'bar chart' (Figure 12.45).

Mass spectrometers are sensitive detectors of isotopes based on their masses. Some satellites and spacecraft are equipped with mass spectrometers to allow them to identify the small numbers of particles intercepted in space. For example, the SOHO satellite uses a mass spectrometer to analyse the solar wind. They are also used in carbon dating (page 305).

1 Find out how mass spectrometers on missions to Mars have been used to study the planet. What information have they provided to scientists?

2 Unipositive neon-21 ions, $^{21}Ne^+$, travelling with a velocity of $2.50 \times 10^5\,m\,s^{-1}$, enter a magnetic field of value 0.80 T, which deflects them into a circular path. Calculate the radius of the circular path.

■ Nuclear medicine

The use of radionuclides in hospitals for the diagnosis and treatment of disease is increasing rapidly worldwide. Any particular source of radiation has to be carefully selected for its purpose. This includes consideration of (interrelated) factors such as:

- types of radiation emitted
- energies of individual particles and rays
- length of time patients can be exposed to radiation
- activity of sources
- half-lives of sources.

81 The experimentally determined activity from 1.0 g of radium-226 was 1.14×10^{18} alpha particles per year. Calculate its half-life.

82 Radium-226 has a long half-life (>1000 years); radium-227 has a half-life of 42 minutes. Outline how the two half-lives of these radium isotopes can be determined experimentally.

83 The following data were obtained from the decay of caesium-130. Plot a graph of ln *A* against *t* to determine the decay constant, and hence the half-life. (The figures have been adjusted for background count.)

Time/s	Activity/s^{-1}
0	200
500	165
1500	113
2500	79
3500	54
4500	38
5500	26

84 A sample contains atoms of radioactive element A while another sample contains atoms of a radioactive element B. After a fixed length of time, it is found that ⅞ of atoms A and ¾ of atoms B have decayed. Calculate the value of the ratio: $\frac{\text{half-life of element A}}{\text{half-life of element B}}$.

85 Research the properties and applications of any one particular radionuclide that is used in hospitals.

Examination questions – a selection

Paper 1 IB questions and IB style questions

Q1 In the photoelectric effect, light incident on a clean metal surface causes electrons to be ejected from the surface. Which statement is correct?

A The de Broglie wavelength of the ejected electrons is the same as the wavelength of the incident light.
B Electrons are ejected only if the wavelength of the incident light is greater than some minimum value.
C The maximum energy of the electrons is independent of the type of metal.
D The maximum energy of the electrons is independent of the intensity of the incident light.

Q2 Ultraviolet light is shone on a zinc surface and photoelectrons are emitted. The sketch graph shows how the stopping potential V_s varies with frequency f.

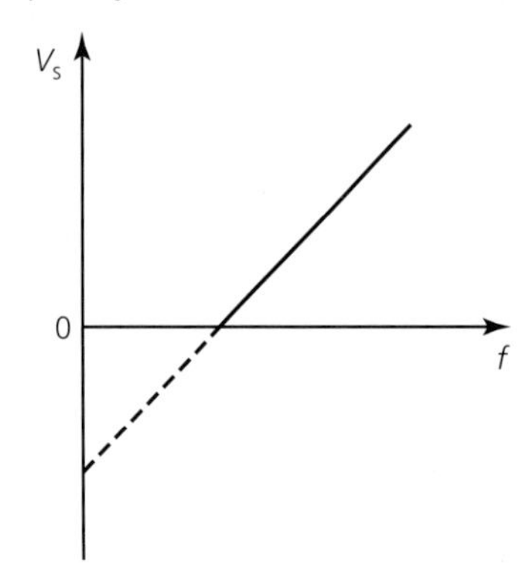

Planck's constant may be determined from the charge of an electron e multiplied by:

A the y-intercept
B the x-intercept
C the gradient
D the area under the graph.

Q3 In the Schrödinger model of the hydrogen atom, the probability of finding an electron in a small region of space is calculated from the:

A Heisenberg uncertainty principle (time–energy)
B de Broglie hypothesis
C (amplitude of the wavefunction)^2
D square root value of the wavefunction

Q4 A nucleus of potassium-40 undergoes β^+ decay to an excited state of a nucleus of argon-39. The argon-39 then reaches its ground state by the emission of a γ-ray photon. The diagram represents the β^+ and γ energy level diagram for this decay process.

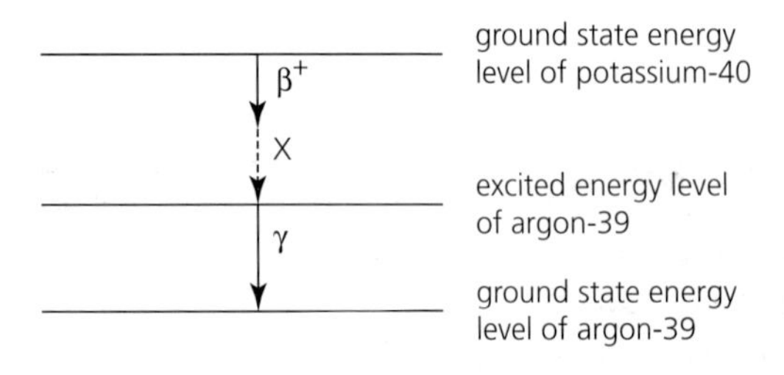

The particle represented by the letter X is

A an antineutrino
B a neutrino
C an electron
D a photon.

Q5 If an alpha particle and a proton have the same de Broglie wavelength, the ratio *speed of alpha particle/speed of proton* is approximately equal to:

A ¼
B ½
C 1
D 2

Q6 The diagrams show the variation with distance x of the wave function Ψ of four different electrons. The scale on the horizontal axis in all four diagrams is the same. For which electron is the uncertainty in the momentum the largest?

A
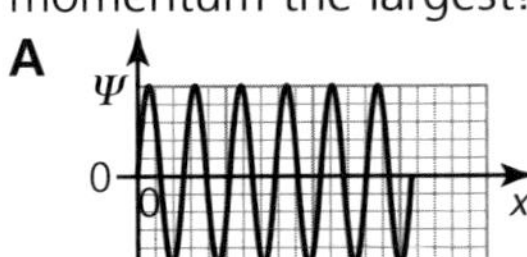

B
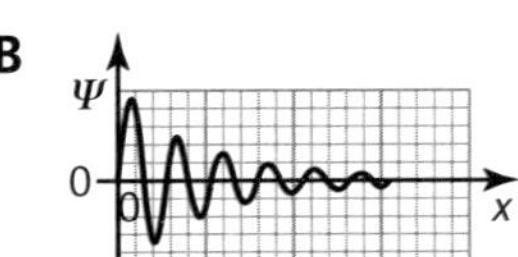

C
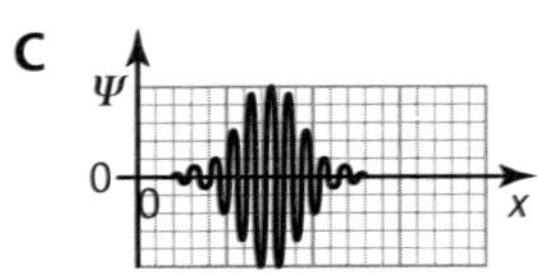

D
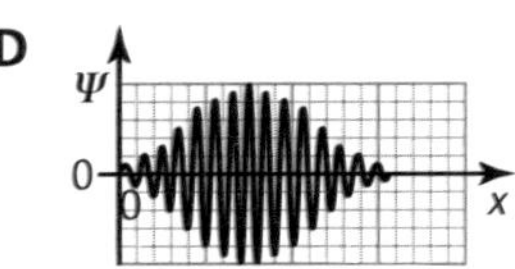

Q7 Which of the following provides evidence for the existence of discrete energy levels within the nucleus?

A Alpha particles are emitted with discrete energies
B Beta particles are emitted with discrete energies
C Neutrinos are emitted with discrete energies
D Light is emitted with discrete energies.

Q8 The equation $\Delta N/\Delta t = -\lambda N$ can be used to represent radioactive decay. In this equation:

A λ represents the half-life of the nuclide
B the negative sign represents a decay
C N is known as the activity of the source
D the magnitude of λ decreases with time.

Q9 Nuclear diameters can be determined from:

A mass spectrometer measurements
B gamma ray spectra
C decay constants
D the diffraction of high-energy electron beams.

Q10 The quantisation of angular momentum was used in the Bohr model of the atom to explain the emission of:

A line spectra
B alpha particles
C beta particles
D gamma rays.

Q11 The probability of a particle quantum tunnelling through a potential barrier does *not* depend on:

A the charge of the particle
B the mass of the particle
C the energy of the particle
D the width of the barrier.

Q12 A photon may be converted into a particle and its antiparticle. In order for this to be possible:

A the frequency of the photon must be below a certain threshold value
B another particle must be involved with the process
C the two created particles must move in the same direction
D momentum cannot be conserved in the interaction.

Q13 An electron is accelerated from rest through a potential difference V. Which *one* of the following shows the variation of the de Broglie wavelength λ of the electron with potential difference V?

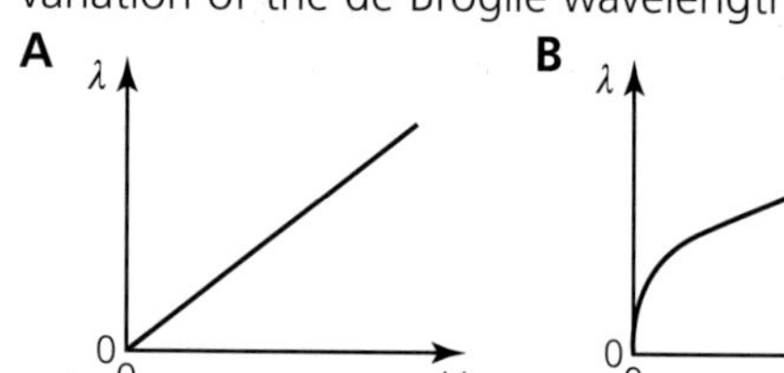

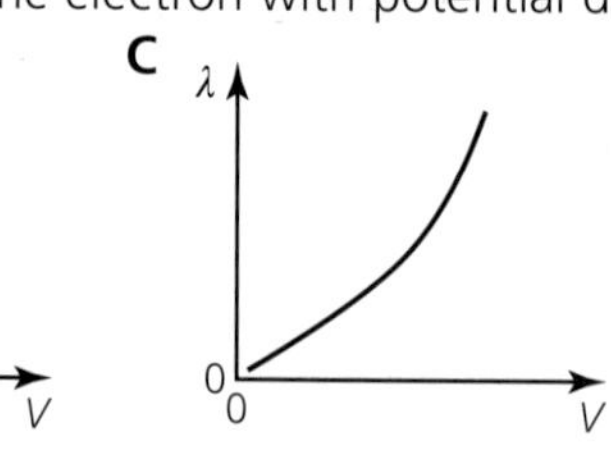

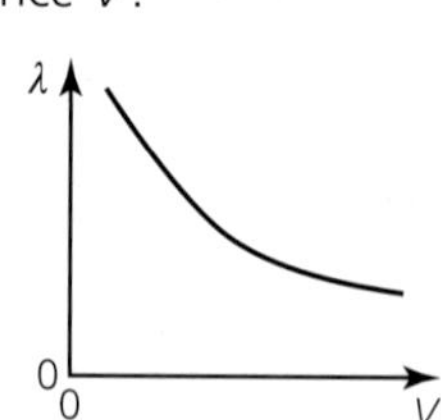

Paper 2 IB questions and IB style questions

Q1 **a** The diagram represents some of the energy levels of the mercury atom.

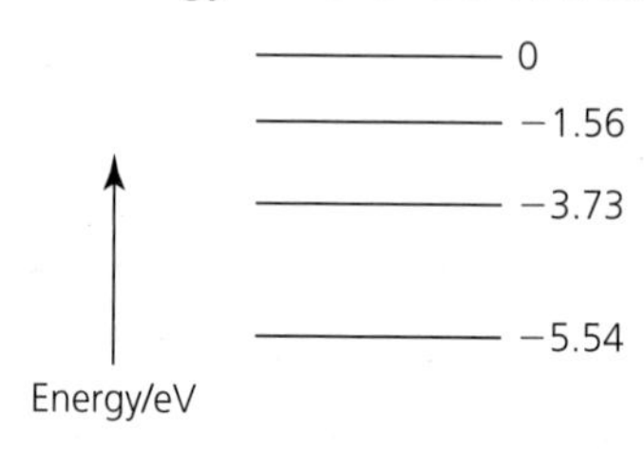

Photons are emitted by electron transitions between the levels. On a copy of the diagram, draw arrows to represent the transition, for those energy levels, that gives rise to:

i the longest wavelength photon (label this L) (1)

ii the shortest wavelength photon (label this S). (1)

b Determine the wavelength associated with the arrow you have labelled S. (3)

c A nucleus of the isotope bismuth-212 undergoes α-decay into a nucleus of an isotope of thallium. A γ-ray photon is also emitted. Draw a labelled nuclear energy level diagram for this decay. (2)

d The activity of a freshly prepared sample of bismuth-212 is 2.80×10^{13} Bq. After 80.0 minutes the activity is 1.13×10^{13} Bq. Determine the half-life of bismuth-212. (2)

Q2 A metal is placed in a vacuum and light of frequency f is incident on its surface. As a result, electrons are emitted from the surface. The graph shows the variation with frequency f of the maximum kinetic energy E_K of the emitted electrons.

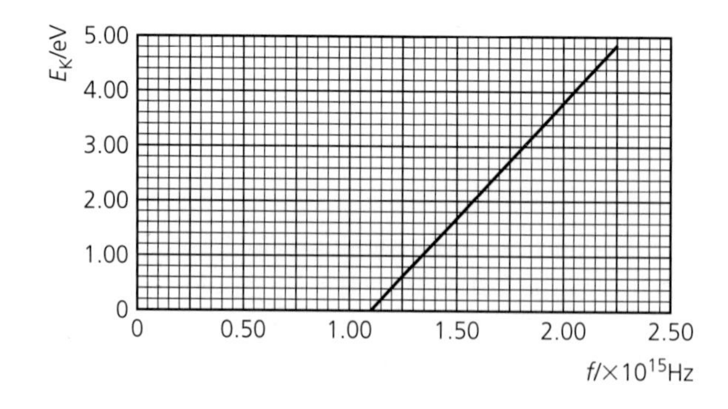

a The graph shows that there is a threshold frequency of the incident light below which no electrons are emitted from the surface. With reference to the Planck constant and the photoelectric work function, explain how Einstein's photoelectric theory accounts for this threshold frequency. (4)

b Use the graph in a to calculate the:

i threshold frequency (1)

ii Planck constant (1)

iii work function of the metal. (2)

© IB Organization

Q3 In a simple model of the hydrogen atom, the 'size' of the atom determines the kinetic energy of the electron. Its de Broglie wavelength is equal to the wavelength of the standing wave bounded by the nucleus and the 'edge' of the atom, as shown below.

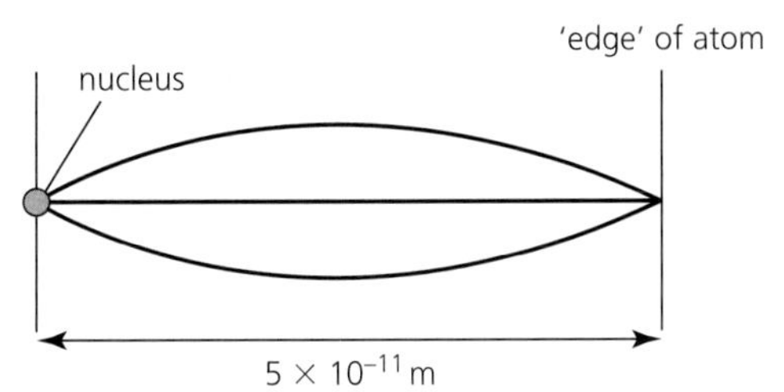

The 'edge' of the atom is 5×10^{-11} m from the nucleus.

a **i** State the de Broglie wavelength of the electron. (1)

ii The 'edge' of the hydrogen atom is moved closer to the nucleus. Describe what changes occur in the kinetic energy of the electron. (2)

A different model of the hydrogen atom takes into account the fact that the electrical potential energy of the electron depends on its distance from the nucleus.

b **i** Explain the variation with the distance from the nucleus of the electrical potential energy of the electron. (3)

ii Use your answer to b i to explain the variation with distance from the nucleus of the kinetic energy of the electron. (2)

iii Use your answer to b ii to suggest how the wavelength of the standing wave of the electron varies with distance away from the nucleus. (3)

© IB Organization

Answers to self-assessment questions in Chapters 1 to 12

1 Measurement and uncertainties

1 a 10^{-3} kg c 10 kg
  b 10^{-3} kg d 10^{21} kg
2 a 10 m c 10^6
  b 10^8 d 10^{-4} m
3 a 2×10^9s
  b 10^8 s
  c 1×10^{-8} s
4 a About 10 to 1; 1 order of magnitude
  b About 10^{25} to 1; 25 orders of magnitude
  c About 10^{-9} to 1; 9 orders of magnitude
5 a ± 3 g b 0.63%
6 a 4.2 (m) c 19%
  b *a*: 4.7%; *t*: 7.1% d ± 0.8 (m)
7 ± 0.03

2 Mechanics

1 a Acceleration for the first 100 s; then a constant speed until 200 s; followed by a deceleration to rest at 300 s
  b 3600 m
  c 20 m s^{-1}
  d 12 m s^{-1}
2 c Swimmers can accelerate by pushing on the end walls of the pool.
4 The runner starts from rest and accelerates towards a reference point (zero displacement); after that the runner has a constant speed away from the reference point.
5 a, c The object is moving (oscillating) backwards and forwards about a reference point; its greatest speed is in the middle of its motion and it decelerates as it moves away until its speed is zero at the maximum displacement; then it accelerates back.
  b Equal magnitude; opposite directions
  d 8 cm s^{-1} and 0 cm s^{-1}
  e A pendulum or a mass on the end of a spring
6 a Starts from rest, constant acceleration for 2 s; followed by constant velocity for 4 s; and finally a deceleration for 1 s to rest.
  b 1.5 m s^{-2}; 0 m s^{-2}; −3.0 m s^{-2}
  c 16.5 m
  d 2.4 m s^{-1}
7 2.8 m s^{-2} and 190 m
8 a The object experiences a constant acceleration for 8 s. In the first 4 s this decelerates the object to rest; then it accelerates in the opposite direction. Between 8 s and 12 s it has a constant velocity.
  b −3.0 m s^{-2}
  c 96 m
  d 48 m
  e 8 m s^{-1}
11 10 m s^{-1}
12 Graph should show a constant velocity changing very quickly to another constant velocity of the opposite sign. The acceleration should be zero except during the impact.
14 a 1.9 m s^{-1}
  b 2.7 m
  c 1.0 m s^{-2}
15 a 1540 m b 35.8 s
16 a 4100 m b 1600 s
17 a 3.0 m s^{-2} b 24 m s^{-1}
18 a 5.3 s after the police car started to move
  b No; private car has travelled 210 m; police car has travelled 192 m
  c After 12 s
19 a −13 m s^{-2} b 32 m s^{-1}
20 a 8.72 km s^{-1} b 3.01×10^6 km
23 a 0.65 s
  b 7.8 m s^{-1}
  c 7.8 m s^{-1}
24 After 1.2 s or 3.3 s
25 0.34 s later
26 a 3.77 s later
  b 24.2 m s^{-1}
  c 29.9 m
27 a 28.4 m
  b 39.6 m
28 5.5 m s^{-1}
29 The acceleration of the driver was twice the acceleration due to gravity, so the force acting on him was twice the driver's weight.
30 a The distance increases as they fall
  b Consider $s = ut + ½at^2$ for both stones from the moment the second stone is dropped; the first stone will always have travelled an extra distance of *ut*, which increases with time.
31 a 17 cm, ignoring air resistance
  b For example, a 60 cm height results from a take-off speed of 3.4 m s^{-1}.
  c For example, a 30 cm height reduction suggests an acceleration of about 20 m s^{-2}.
  d About 0.2 s
33 a 13.0 s b 127 m s^{-1}
34 b the acceleration will be equal to twice the gradient of an s–t^2 graph.
36 7.69 m s^{-1} and 27.3 m s^{-1}
37 a 17 km h^{-1} b 2 min
38 a 15 m s^{-1}
  b i Same ii Smaller
41 a 0.167 s
  b 0.137 m below the centre
42 a 46 m above sea
  b 30 m s^{-1} downwards
  c 4.7 s
  d 96 m
  e 36 m s^{-1} at 56° to horizontal
43 528 m
44 a 23°; $v_V = 1.5$ m s^{-1}
  $v_H = 3.5$ m s^{-1}
  b 98 cm
45 b There was no air resistance or friction (and the ball did not have any significant kinetic energy of rotation).
  c Larger masses will experience greater gravitational forces, but the same acceleration, $a = F/m$ (assuming there is no friction or air resistance).
46 28 m s^{-1}
47 a 1.23×10^4 N
  b 31.8 N
  c 2.43×10^{-3} N
48 1.33 N
49 a 5.60×10^6 N
  b Approximately 50 000 kg; about 9%
  c The forces stopping the motion of the plane when it lands are not as

large as the forces accelerating the plane when it is taking off.

d The mass has reduced by 180 tonnes; most of the fuel it was carrying has been burned.

50 There is less of the mass of the Earth beneath it, and some of the mass of the Earth is above it.

51 8.9 N kg^{-1}; this is about 10% less than on Earth.

52 A has twice the circumference, four times the surface area and eight times the volume, mass and weight.

55 0N, 2N, 4N, 6N

56 10.7 N at an angle of 44° to the 12 N force

57 9.1 N at an angle of 32° to the 7.7 N force

58 76 N at an angle of 14° to the 74 N force

59 b 37 N in the direction of the applied force

60 a 3.33 N

b 1.72 N

61 b 14 N parallel to the slope; 30 N perpendicular to the slope

62 a 6.68×10^4 N

b 1.74×10^4 N

c 0.045 m s^{-2}

d The component of weight (acting down the slope) of a heavy train is so large that it may be larger than the resultant forward force provided by the engine.

64 All arrows identical; pointing downwards

65 a A force vector of 300 N acting downwards from the middle of the suitcase, labelled weight; an equal and opposite arrow pushing up on the suitcase, labelled normal reaction force

b Add an arrow upwards from the handle of half the length of the previous vectors, labelled pull of hand, 150 N; the normal reaction force reduces to 150 N

66 The reading rises when a force is needed to accelerate the book upwards; and falls when the book is accelerated downwards.

67 a Not if the elevator is moving with constant velocity; this is because the forces acting on you would be the same in all three cases.

b In all diagrams the weight vector will be the same; if the person is accelerating, the force from the floor will be greater or less than the weight.

68 a i Weight and air resistance on the sky diver's body will be equal and opposite; then there will be an upwards force from the parachute on the skydiver.

ii The upwards forces from the parachute and air resistance on the skydiver will together be equal and opposite to the weight.

69 14 N

70 Weight acting downwards from climber's centre of mass; tension acting along the rope, away from the climber; push of rock acting from climber's feet to the point where the other two forces cross. (This could be resolved into a normal force and friction.)

71 a The weight of B will be eight times greater than A; the air resistance acting on B will be four times greater than A.

b A moves with constant velocity because the forces are balanced; B accelerates because there is a resultant force, since its weight is greater than the air resistance acting on it.

72 a Zero friction

b Yes. There is no reason why not, but it would mean that the object was easier to lift vertically than move horizontally. Which is unusual and may require further explanation.

73 a ≈ 90 N

b ≈ 20 N

74 a The angle of the slope can be slowly increased until the object *just* begins to slide down. At that angle the component of weight down the slope equals the frictional force up the slope.

b 0.71

75 a 9.2×10^3 N

b 1.1×10^4 N

c The extra weight will increase frictional forces between the tyres and the road and improve safety, but the extra mass in the car will mean that the car needs a greater force to stop it (than with only a driver).

76 Sand or very small stones ('grit') can be spread on the road. Salt can also be used to help melt the ice.

79 2.56×10^5 N

80 2.4 N

81 a 2.39 m s^{-2} b 1120 m

82 a −3.5 m s^{-2} b 4200 N

83 b 0.27 m s^{-2}

c A much greater force would be needed for the larger acceleration; the thin rope may not be strong enough, and may break.

84 a 0.44 b 0.47 m s^{-2}

85 a 25 degrees b 0.68 m s^{-2}

86 The hard ball cannot change shape so much and will therefore decelerate quicker. (It will have a greater negative acceleration.)

87 1.7×10^{-17} N

88 a 124 N b 933 N

90 a 1.6 m s^{-2}

b 3.3 N

c (Tension = 16.4 N)

91 a 4.58 m s^{-1}

b 1570 N

c 2310 N

d Pushed down hard on the ground

e The impact takes longer; reducing the size of the deceleration and the force needed to stop him.

92 a The impact takes longer, reducing the size of the deceleration and the force needed to stop.

b As a. The force is also spread over a large area, reducing the pressure.

93 a 230 m s^{-2}

b 0.14 s

94 The moving air from the fan will collide with the sail and exert a forward force. But, in order for the air to be pushed forward by the fan, the air pushes backwards with an equal force on the fan. The resultant force on the boat will also depend on how the air flows past the sail.

95 The two forces are acting on the same object and they are different types of force.

97 The weight will be the same.

98 a 59 J b 130 J

99 29 J

100 a 54 N b 140 J

101 0.11 J

102 a Your weight (say 600 N); 20 cm
b 60 J
b i 0.55 J ii ≈ 2.2 J
c The plastic gets warmer.

104 a 320 N m^{-1}
b 1.0 J
c 33 cm
d A mass of 10 kg would exert a force of approximately 100 N, which is well beyond the range shown on the graph; the spring may not continue to stretch proportionally to its load for such a large force.

106 a Loudspeaker
b Battery
c Nuclear power station
d Microphone
e Flame
f Car
g Bow and arrow
h Electrical generator
i Gas cooker
j Photovoltaic cell
(Other answers are possible.)

107 a 2.0×10^6 J
b The field strength is slightly less at the top of the mountain; the difference is too small to affect the answer.

108 a 1.5×10^7 J
b Energy is dissipated in the electric motor; because of friction and air resistance.

109 a 5.1×10^6 J
b As the elevator comes down a significant amount of its decrease in gravitational potential energy is transferred to the rising counterweight, rather than being dissipated as internal energy.

110 a i 130 J ii 2100 J
b 3000 J

111 15 m s^{-1}

112 4.1×10^{-18} J

113 900 kJ

115 a 7.5 cm
b 0.37 J
c 20.1 N, if it is not overstretched

116 About 100 J

117 1.27 m

118 10 m

119 a 22 m
b Measuring the time accurately; determining when the ball has reached the right height; the ball is not released from ground level.

120 a 2.8 m s^{-1}
b 2.8 m s^{-1}
c It moves a smaller distance at a greater average speed because it has a greater acceleration (the average component of its weight down the slope is greater).

121 a Elastic strain energy
b Gravitational potential energy to kinetic energy; to strain energy (and internal energy); to kinetic energy; to gravitational potential energy. This assumes that air resistance was negligible.
c 4.4 m and 3.4 m
d 22 cm; assuming that it loses the same fraction of its kinetic energy each time it bounces.

122 1.1 m s^{-1}

123 0.116 m

124 3.4 m

125 1.1×10^4 N m^{-1}

126 3.0×10^5 N

127 The average force exerted in the accident can be calculated from the kinetic energy before the collision divided by the distance the vehicles 'crumple'; the greater the deformation, the less the force.

128 a 16 000 J
b The work done in stretching the cord
c Equal increases in force produce smaller and smaller extensions (except for the first 250 N); the cord becomes stiffer.
d Approximately 25 m

129 a 2790 J
b Probably not; because this is quite a large mass falling a short distance
c 3400 N

130 To increase the length (and time) of the impact; and so reduce the force and prevent injury to the knees and legs.

131 a 6700 N
b The hammer comes to rest.

132 a 240 J
b 13 W

133 About 25 kW

134 1400 N

135 a 1.7×10^8 W
b 170 000 homes

136 900 MW

137 a 338 N
b 54 N; friction helps to stop the box sliding down the slope when the force is reduced.
c 830 J
d 0.86 (86%)

138 1.0×10^7 kg m s^{-1} to the west

139 3.3×10^3 N

140 a 8.5 kg ms^{-1}
b 8.5 kg ms^{-1}
c 26 m s^{-1}

141 a 6800 N s b 12 m s^{-1}

142 a 5 m s^{-1} to the left
b 2.5 kg m s^{-1} to the left
c 7.4 N to the left
d Force changing (linearly) from 0 to 14.8 N in 0.17 s; then back to zero in another 0.17 s
e Shorter times; greater forces
f Identical, except the forces are in opposite directions

143 a 2100 N
b 520 kg m s^{-1}
c 26 km h^{-1}

145 a 3.39×10^4 kg m s^{-1}
b 100 J
c 9.11×10^{-31} kg

146 0.45 m s^{-1} to the left

147 a 5.35 m s^{-1}
b Air resistance under these circumstances will be almost insignificant; so the actual speed will be close to the predicted value.
c About 5×10^{-23} m s^{-1}; cannot be measured

148 260 m s^{-1}

149 The ball is not an isolated system; it is acted on by the external force of gravity; the Earth loses an equal amount of momentum.

150 She moves in the opposite direction with a speed of 2.0 cm s^{-1}; an external force is needed to stop her motion.

151 22.8 cm s^{-1}; in the original direction of A

152 a 1.8 m s^{-1}
b The total (kinetic) energy of the balls would have increased.

153 At the 20 cm mark on the ruler

154 a 340 N
b 0.034 m s^{-2}
c 10 000 N s
d 1.0 m s^{-1}

156 a 11.9 m s^{-1} b 95%

157 a 0.6 m s^{-1} b 0.35 J

159 a 1.0 m s^{-1} b 99%

3 Thermal physics

1 a 331 K; 184 K b 38 K
2 a 127°C b 32 cm^3
3 a Probably not; they may have the same temperature, but they are different substances; there will probably be different numbers of molecules with different masses and different speeds.
 b Probably not; different substances require different amounts of energy for the same temperature rise.
 a i A relatively large attractive force pulls the molecules back together.
  ii A large repulsive force pushes the molecules apart.
 b Forces between gas molecules are much, much smaller; almost zero.
 c If the separation is ten times greater in each dimension, the average volume occupied by one molecule must be 10^3 times greater.
5 Each 'sparkle' is so small that it contains very little internal energy, although it has a high temperature. If a sparkle comes into contact with the skin it transfers only a tiny amount of thermal energy as it cools down rapidly.
6 9.5×10^4 J
7 1370 J kg^{-1} K^{-1}
8 4.8 K
9 34°C
10 4.8×10^4 J
11 40°C
12 128 s
13 23°C
14 41°C
15 980 W
16 a 3670 J
 b To reach thermal equilibrium with the oven
 c To reduce the thermal energy transferred to the surroundings
 d 313 J kg^{-1} K^{-1}
 e To make sure that it was all at the same temperature
 f Underestimate; the final temperature would have been higher if there had been no transfer of energy to the surroundings.
17 4.37×10^5 J kg^{-1}
18 a 38.3°C
 b All of the thermal energy that flowed out of the hot water went into the cold water; none went into the surroundings
19 17.3°C
20 320 s
21 3.4×10^5 J K^{-1}
22 840 s
23 a 1.80×10^5 J
 b 67°C
 c There would have been less time for energy to be transferred to the surroundings.
24 a 1.4°C
 b All the kinetic energy of the bullet is transferred to internal energy in the block and bullet.
 c Half the speed has a quarter of the kinetic energy; so the temperature rise would be 0.36°C.
25 a 4×10^4 J
 b 19.8°C
 c The temperature rises are too small to be easily measured with accuracy.
26 a 1.1°C
 b This assumes that all of the gravitational potential energy of the lead is transferred to internal energy in the lead; no energy is transferred to the surroundings.
 c Twice the potential energy has to be spread to twice the mass.
27 0.12°C
28 1600 J
29 2.56×10^6 J kg^{-1}
30 More bonds between molecules have to be broken during boiling than during melting.
31 a i 1.2×10^3 J ii 144 J
 b The energy transferred from the steam as it condenses is much greater than that from the water as it cools
32 a Some ice will melt because of thermal energy absorbed from the surroundings; by comparing the two masses of melted ice we can estimate the amount melted by the heater in A.
 b 4.0×10^5 J kg^{-1}
 c Some energy from the heater is transferred to the surroundings.
 d Carry out the experiment in a refrigerator; insulate the apparatus.
33 5.2×10^4 J
34 8.0×10^{10} J
35 10°C
36 a 0.1 N
 b 1.5×10^5 N
 c Because there is equal pressure inside the body.
37 Because of the extra weight of the water above them.
38 a Pressure under our feet is due to the force with which we push down – our weight, if we are standing still. The pressure in a gas is explained by random molecular collisions with surfaces, which can be in any direction.
 b The pressure in a liquid also acts in all directions.
39 a 6.21×10^{-21} J
 b 210°C
 c 1.20×10^{-19} J
40 a 461 m s^{-1} b 493 m s^{-1}
41 a 108 g
 b 5.7 g
 c 167 mol
 d 1.4×10^{25} molecules
42 a 0.062 mol
 b 1.9×10^{11} atoms per second
43 a 200 kg
 b 6700 mol
 c 4.0×10^{27} molecules
44 a 10 cm^3
 b 1.7×10^{-23} cm^3
 c 2.6×10^{-8} cm
45 143 g
46 1.71×10^{-2} m^3
47 39°C
48 130 mol
49 a 32 g b 0.60 m^3
50 2.1×10^6 Pa
51 a 2.51×10^5 Pa
 b Molecules collide with the walls more frequently.
52 a 7.56×10^6 Pa
 b On average the molecules are travelling faster; collide with the walls more frequently; and with greater force.
 c Helium may not behave like an ideal gas at low temperature and high pressure.
53 1.7 m^3
54 8/3
55 307°C
56 When the gas is burned, hot gases enter the balloon and displace cooler, denser gas out of the bottom; in this way the overall weight of the balloon can be adjusted to be less than the upthrust.

57 a The molecules are closer together.
b Electric forces act across space, without contact, and in theory they will never go completely to zero; but the forces become very small (negligible) if the separation of molecules is more than a few molecular diameters.

58 a When the molecules of the warmer gas hit the molecules of the colder surface, on average, kinetic energy will be transferred to the molecules in the wall.
b Internal energy and the temperature of the wall will increase; the internal energy and temperature of the gas will decrease.

59 a The molecules will have stopped moving, and stopped colliding with the walls
b A gas will condense to a liquid if its molecules do not have enough kinetic energy to overcome the electric forces between them when they get close together.

60 The graphs have the same shape; but for all volumes at the higher temperature, the pressures will be greater.

4 Waves

1 a 0.872 s b 1.15 Hz c 344
d The absolute uncertainty in measurement remains the same; but the percentage uncertainty decreases with larger measurements.

2 3.1×10^{-4} s

3 a 2.0×10^{8} Hz
b 200 MHz

4 a i 0 ii 2.4 cm to the right
b 24 cm
c Exactly out of phase

6 a Probably not
b Because the extension of the rubber is unlikely to be proportional to the force

8 a 4.0 cm c −1.4 cm
b 0.5 s d +1.4 cm

12 Each successive maximum would have less energy. The period would be unchanged.

14 a 0.155 m
b 4510 m s^{-1}
c The particles are closer together; there are much larger forces between them.

15 a 5.4×10^{5} Hz
b i 4.7×10^{-3} s
ii 0.11 s
iii Between 180 s and 1300 s

16 a 6.8×10^{13} MHz
b (Photons of) higher frequencies transfer more energy.

17 a 5.5×10^{-7} m
b Green
c Ultraviolet

18 3.6 s

20 a 0.29 s b Maybe 50%

21 b The distance needed for hand timing is much greater than can be found indoors. There are too many sound reflections indoors.

22 At the same temperature the molecules of different (ideal) gases have the same average kinetic energies. This means that molecules with less mass must have greater speeds (so that $\frac{1}{2}mv^2$ is the same).

24 a Sea water is denser than pure water.
b 65 s
c Noise pollution caused by human activities in recent times
d 31 m

25 3.2 MHz

26 a Increases by a factor of 4
b Energy is dissipated to the surroundings.

27 a 4.0×10^{4} W m^{-2}
b 370 W
c 6.3 minutes

28 a 18 W m^{-2} b 5.2 cm

29 a 5.0×10^{-7} W b 69 m

30 a 10 m
b No gamma rays are absorbed in the air.

32 a 200 kW
b Wave power is proportional to amplitude squared; efficiency of power station is independent of wave amplitude.
c 2.1 m
d 35 kW

37 a The resultant should have twice the amplitude of the individual waveforms.
b The resultant should always be zero.

38 Look through the sunglasses at light reflected from glass or water; if they are Polaroid, the intensity of the image will change as the glasses are rotated.

39 6.7%

40 63°

41 The sky appears blue because blue light is scattered from air molecules; simple scattering, like reflection, can result in polarization.

43 a 24 mW
b 21 mW
c 11 mW (or almost unaffected)

45 26°

46 Wavefronts in shallower water will be at 27° to the boundary.

47 a 1.35 b 35°

48 51°

50 Because that would mean that the waves travelled faster than 3×10^{8} m s^{-1} in the medium.

51 a 1.36 b 47.4°

55 Microwaves have a wavelength suitable for diffraction by gaps and obstacles with a size of a few centimetres.

56 60 cm and 600 Hz

57 Destructive interference

58 a When he is the same distance from both speakers, the waves he receives have both travelled the same distance and will interfere constructively. If he moves in any direction there will then be a path difference between the waves and they will no longer interfere perfectly constructively.
b 71 cm

59 The waves from the two sources of light are not coherent.

60 a To maximize diffraction of the waves emerging from the slits; so that they cross over each other and interfere.
b 12, 6, 4, 3 cm…
c Slowly move the receiver until it detects an adjacent maximum, the difference in the two path differences will equal the wavelength.

61 7.9 m

63 a 58.6 m s^{-1}
b 71.4 Hz
c 0.492 m

64 a i 0 ii π
d 21.6 cm; 127 Hz
e 27.5 m s^{-1}

65 a 2.0 Hz
b The chain has significant weight, so that the tension in it is not constant (greatest at the top).

66 a The wave speed will increase because of the larger forces in the system.

b The fundamental frequency increases because $f = v/l$ and v is higher but l has not changed.
c The oscillating string will accelerate more slowly because it is more massive.
d The fundamental frequency decreases because $f = v/\lambda$ and v is smaller but λ has not changed.

68 360 Hz

69 338 m s^{-1}

70 0.94 m

71 a 1.49 m
b 342 Hz
c 0.745 m
d To produce the same fundamental frequency, they can be half the length of pipes closed at both ends.

72 a 346 m s^{-1} b 256 Hz

5 Electricity and magnetism

2 a 1.4×10^3 N C^{-1}
b 4.0×10^{-4} N

3 a 2.4×10^{-17} N upwards
b Force of gravity is 8.9×10^{-30} N (there is a factor of about 10^{13} difference).

5 a 1.7×10^6 NC^{-1} to the south
b 3.9×10^6 NC^{-1}

6 a 1.5 cm
b Decrease. Because the paper of the book is not as good at transferring the electric field as air.
c About 0.2 N
d The book will not fill the space between the charges, it does not surround the charges; the exact properties of the material of the book are not known.

7 a Three forces are acting: weight, tension and the repulsive force between the charges.
b 4.3×10^{-4} N
c 54 nC

8 3.7×10^{-6} N to the left

9 5.4×10^{-3} N, diagonally away from the square

10 8.2×10^{-8} N

12 2.1×10^{-11}

13 Conductors would allow charges to move and so change the situation being considered.

14 a 0.25 C b 1.6×10^{18}

15 a 430 C b 830 s

20 6.3×10^{-4} m s^{-1}

21 a Speed must increase in order for the same number of electrons to pass any point every second in a thinner conductor.
b If the conductors had the same dimensions and the size of the charges are the same, in the semiconductor they would have to travel one million times faster for the same current. A typical drift speed would then be about 100 ms^{-1}.

22 a 8.5×10^{28} m^{-3}

23 a +44 V
b The magnitude of the p.d. would be the same, but its sign would be negative, meaning that, in the second example, the energy would be transferred *from* the electric field.

24 a 517 J
b 9100 C
c 5.7×10^{22}

25 a 220 V
b 40 C

26 The current would fall to a very low value because the voltmeter is a poor conductor. The value shown on the voltmeter would be almost equal to the p.d. across the battery.
a i 7.7×10^{-17} J ii 480 eV
b No; its kinetic energy is not large enough to overcome the repulsive force.

28 a 3.5×10^{-13} J b 1.0×10^7 m s^{-1}

29 8.0×10^{-16} J

30 a 3.2×10^{-19} J
b 1×10^{19} photons every second

35 9 A to the right

36 0.1 A, 1.1 A, 2.0 A

37 15 V and 14 V

39 375 V

40 27 Ω

41 1.2×10^{-4} A (0.12 mA)

42 136 V

46 a 3.0 Ω b 4.8 V

48 In the 2 Ω resistor; 4 Ω and 6 Ω in parallel have a total resistance of 2.4 Ω, so that the voltage across them is (only) slightly higher than the voltage across the 2 Ω resistor; because $P = V^2/R$, the lower resistance of the 2 Ω is the dominant factor.

49 a Low: two in series; medium: only one; high: two in parallel
b 605 W, 1210 W, 2420 W

50 a The ammeter will show a very low reading because the large resistance of the voltmeter prevents a greater current; the voltmeter will show (almost) 12 V because it has a resistance very much larger than 30 Ω; in a series circuit the voltages are in the same ratio as the resistances.
b 4.0 V and 0.4 A
c Because of the resistance of the voltmeter; the total resistance between the terminals of the voltmeter is not just 10 Ω, it is calculated from the two resistances in parallel.
d 260 Ω
e Negligible resistance

51 a 4.3 V
b 5.2 Ω; 3.6 W
c The lamp becomes dimmer; the current reduces when the overall resistance of the circuit increases.
d 3.4 V (assuming R is constant)

53 Current through the 2 Ω is 0.39 A to the left; current through the 3 Ω is 1.08 A upwards; current through the 4 Ω is 0.69 A to the left.

54 a 1.17 m b 120 Ω

55 They are inversely proportional

56 5.5×10^{-4} m

57 2.1×10^{-6} Ω m

58 a i 0.05 A ii 0.60 W iii 72 J
b 120 Ω

59 a 9.07 A b 220 V

60 a 2×10^5 W
b A power 'loss' of only 10 W m^{-1} may seem quite low; but it can be considerable for long cables.

61 a 150 W
b Heater, ammeter and power supply connected in series; voltmeter connected across the heater

62 a 0.86 A
b Energy is dissipated; because of friction and resistive heating.

63 a 9.68 Ω b 11.4 A
c The heater's resistance is lower when it is colder.

67 a 6.0 V b 2.4 Ω
c The student does not understand that the value of 6.0 V will change when the bulb is connected because the total resistance between A and B is no longer 100 Ω.
d 0.28 V; no
e Connect the lamp in series with an equal resistance (2.4 Ω).

68 a The circuit should include a variable resistance and a thermistor connected in series

to a power source; the output to the refrigerator control circuit will be taken across the variable resistance, so that if the temperature rises the output voltage to the control circuit also rises and turns the refrigerator on.

b Refrigerators, water heaters, room heaters, irons, air-conditioners, ovens

69 a The energy of the light releases more free electrons.

b It would not be possible to plot or interpret the graph accurately for low values if a linear scale was used; (because of the very large difference in the magnitude of different values)

c $\log R = 5.2 - (0.8 \times \log I)$

d 2300 Ω

e 5200 Ω

70 a 2.02%

b 0.942 V, assuming the voltmeter has very large resistance

c 0.944 V

71 b 96 Ω

c 0–10 Ω

d 4 V

e The voltage would only be 6 V if the potentiometer was used on its own; the lamp is in parallel with half of the potentiometer, and their combined resistance is less than the other half of the potentiometer.

72 a 5.5 Ω

b 0.37 A

c Internal resistance is constant.

73 a 3.9 Ω b 3.0 V

c i 2.0 W ii 0.12 W

74 a 12.5 V

b 0.28 Ω

c 4.7 Ω

75 a To get a high power from a low voltage ($P = VI$).

b The voltage across the battery falls because of the 'lost volts' (Ir) due to the internal resistance of the battery.

76 a 48 A

b No resistance in the wires causing the short circuit; internal resistance constant.

c 580 W

d Rapid rise in temperature; the battery may be damaged.

77 Usually a voltmeter would be connected across the resistor; but a voltmeter connected across a battery does not measure the p.d. across its internal resistance.

78 a 4.5 V and 3.0 Ω

b 1.5 V and 0.33 Ω

86 a About 107 min

b About 37 000 J

c About 1000 J

91 a Q downwards; R to the right

b The field strength at R is one-third of the field strength at Q (it is three times further away).

92 a The needle will become more aligned with the axis of the solenoid.

b Towards the northwest

93 a 4.3×10^{-3} N

b 7.5×10^{-3} N

c 8.6×10^{-3} N

d 0

94 a 2.4×10^{-3} N m^{-1}

b From west to east

95 b 3.1 A

96 a 37° b 9.5×10^{-3} N

98 a Only if it is moving parallel to the magnetic field.

b Unlike magnetic forces, electric and gravitational forces exist regardless of the direction of motion (indeed, motion is not needed for the forces to exist).

99 a 6.8×10^{-13} N

b Helical

100 a 1.3 T

b 1.1×10^{-25} kg

c 70 cm

101 a i 7450 eV

ii 1.19×10^{-15} J

b 5.12×10^{7} m s^{-1}

c 1.97×10^{-3} T

d 10.5 cm

102 a 1.0×10^{7} m s^{-1}

b 2.0×10^{5} V m^{-1}

c The fields need to be perpendicular to each other; and to the direction of movement of the particle.

6 Circular motion and gravitation

1 a 3.14 (π) rad

b 1.57 (π/2) rad

c 0.785 (π/4) rad

d 1.95 rad

2 240 cm

3 a 0.8727

b 0.08716

c 0.1%

4 a 26 rad s^{-1}

b 4.2 Hz

5 a 0.361 Hz

b 2.27 rad s^{-1}

c 2.99 m s^{-1}

9 There is no force acting on it in the direction of its instantaneous velocity.

10 a 0.093 m s^{-2}

b 0.040 N towards the centre of the circle

c 17 s

11 a 215.6 m s^{-2}

b 862 N

12 a 10 rad s^{-1}

b 6.0 s

13 a 1010 m s^{-1}

b 2.7×10^{-3} m s^{-2}

14 a 4500 N

b Friction between the tyres and the road

c The force needed will be four times great and there may not be enough friction to provide the extra force.

d There will be less friction.

e The centripetal force needed will increase to 6400 N, but the extra weight will also increase the friction.

15 a 3.0 m s^{-1}

b 0.53 revolutions per second

16 a 2.1 N

b 637.7 N

c The normal reaction force acting upwards on the boy is 635.6 N.

17 b 0.68 N

c 1.3 m s^{-1}; 1.4 s

18 19 m s^{-1}

19 Around 4×10^{-11} N

20 6×10^{-4} N

21 3.6×10^{22} N

22 3.7×10^{-47} N

23 3.7 N kg^{-1}

24 a 8.9 N kg^{-1} b 91%

25 10.3 N kg^{-1}

26 9.81 N kg^{-1}

28 b 1.42 N kg^{-1}

31 6 N kg^{-1} towards A

34 a 3.0×10^{4} m s^{-1}

b 5.9×10^{-3} m s^{-2}

c 2.0×10^{30} kg

35 a 5400 s; 7.7×10^{3} m s^{-1}

b Because it is close to the Earth communication is easy; it requires less energy to put it in orbit; it is better placed to monitor the surface of the Earth. But it may be affected by a very small amount of air resistance; it does not remain above the same location on the Earth's surface.

36 1.5×10^{11} m

38 44 days

7 Atomic, nuclear and particle physics

1 a 2.4×10^{19} Hz b Gamma rays

2 12 cm

3 a 1.2×10^{-19} J b Infrared

4 3.88×10^{-19} J

5 about 1×10^{19}

7 a 1.61×10^{-7} m
 b Ultraviolet
 c From the ground state up to −5.55 eV
 d 1.24×10^{-4} m

8 a Gaseous atoms of different elements are absorbing specific frequencies of light from the solar emission spectrum.
 b Each element has its own unique emission spectrum.

10 Iodine-129: 53 protons, 53 electrons and 76 neutrons
 Caesium-137: 55 protons, 55 electrons and 82 neutrons
 Strontium-90: 38 protons, 38 electrons and 52 neutrons

11 $+3.20 \times 10^{-19}$ C

12 $^{32}_{16}S$

13 13

14 a $^{35}_{17}Cl$ and $^{37}_{17}Cl$
 b A typical sample of chlorine has a mixture of isotopes and 35.45 is an average nucleon number.

16 0.216 MeV

18 1.34×10^{7} m s^{-1}

19 Because there are three particles after the decay, each of which may move in any direction. The momentum of the beta particle must be equal and opposite to the vector sum of the momenta of the other two particles, which can have a range of possibilities.

20 $^{23}_{12}Mg \rightarrow {}^{23}_{11}Na + {}^{0}_{+1}e + {}^{0}_{0}\nu$

21 1 cm

22 Because of their relatively large mass, alpha particles transfer more energy (than beta particles) in collisions with atoms.

23 At lower pressure the number of particles per m^3 in the air is reduced, so that the number of collisions per centimetre is decreased.

24 Alpha particles cannot penetrate the outer layers of the skin, but when released inside the body they come into direct contact with organs and tissue.

25 a about 1500 s^{-1}
 b about 35 s^{-1}

26 a All radioactivity measurements show random variations because individual decays are unpredictable.
 b 0.63 s^{-1}
 c Low count rates have a greater (%) variation than high count rates, and are therefore more likely to be indistinguishable from the varying background count, which may be of similar magnitude.

30 The activity of a radioisotope with a short half-life will quickly decrease to a level that will be undetectable. This means that it must be made available only a short time before it is needed.

31 80 min^{-1}

32 a 1/32
 b 15/16
 c 6.25×10^{13} Bq
 d 160 min
 e i 7/8 ii 1/16

33 a 3%
 b 1.3 million years

34 4.5×10^{16} J

35 2.0×10^{-13} J

36 4.67×10^{-13} kg

37 4.8 MeV

38 7.3 MeV

39 101 MeV

40 8.5666 MeV

41 a 7.7 MeV
 b 1800 MeV

42 b 6.1479 MeV

44 a 1.63×10^{7} ms^{-1}
 b 2.94×10^{5} ms^{-1}

45 a 236.132 u; 235.918 u
 b 0.214 u
 c 199.3 MeV
 d 8.14×10^{9} J

46 $^{2}_{1}H + {}^{1}_{1}p \rightarrow {}^{3}_{2}He$; 4.96 MeV
 $2\,{}^{3}_{2}He \rightarrow {}^{4}_{2}He + {}^{1}_{1}p + {}^{1}_{1}p$; 13.92 MeV

50 Neutrons would not have been deflected because they are not charged. They may be involved with reactions within the nucleus.

52 a Two up antiquarks and one down antiquark ($\bar{u}\bar{u}\bar{d}$).
 b Two down antiquarks and one up antiquark ($\bar{d}\bar{d}\bar{u}$).

54 a $u\bar{u}$ (this is a neutral pion)
 b An up quark and an anti–up quark can join together (a neutral pion is unstable and quickly decays).

55 Neutron

58 a strong nuclear, weak nuclear, electromagnetic (+ gravity)
 b strong nuclear, weak nuclear (+ gravity)
 c weak nuclear, electromagnetic (+ gravity)
 d strong nuclear, weak nuclear, electromagnetic (+ gravity)
 e weak nuclear, electromagnetic (only charged leptons) (+ gravity)

59 electric force = 9×10^{-9} N and gravitational force = 4×10^{-48} N; ratio ≈ 10^{39} : 1

60 The strong nuclear force has a very short range and only acts between quarks (and gluons) within the nucleus. The electromagnetic force can act on *all* charged particles.

8 Energy production

1 a 1.21×10^{4} J cm^{-3}
 b Higher; some thermal energy transferred from the burning fuel was spread into the surroundings, rather than into the water.

2 a 1.1×10^{11} J
 b 3600 W
 c 27%
 d Factories, schools, offices, shops, transportation, etc.

3 a 20 g
 b 2.7 g s^{-1}
 c 710 J g^{-1}

4 a i 31 kg ii 1.9×10^{7} kg
 b 5.41 kg

6 Oil and renewables sources would have a lower percentage; the others would be higher.

9 About 35%

11 a 1 kg of water at 35°C
 b The source of energy must be hotter than its surroundings.
 c Because all the things that surround us are at similar temperatures; the temperature differences are not great enough for efficient transfers of energy to occur.

12 57 kg

13 a 6.2 m^3
 b 3.7×10^{5} N
 c 1.3×10^{5} m^3

14 42 MJ kg^{-1}

15 a 34%
b 6.5×10^4 kg s^{-1}
c 9300 kg
17 a 1.4×10^7 m s^{-1}
b 1.4×10^4 m s^{-1}
18 Their physical properties depend on their masses, but there is only about 1% difference between the masses of the isotopes. In any mixture there will be a considerable overlap of the ranges of kinetic energies and momenta of the isotopes.
19 a 7.6×10^{19}
b 2.6 kg
20 a 7.8×10^{13} J kg^{-1}
b 2.3×10^{12} J kg^{-1}
c Uranium releases about 10^5 times more energy from every kilogram than coal.
22 a 3.2×10^{10} J
b 0.4 g
23 The three naturally occurring isotopes of uranium have very long half-lives; the half-life of uranium–238 is comparable to the age of the Earth.
24 The answer, of course, depends on what level is considered to be safe; for example, after about 60 years the level will still be about double the acceptable safety level.
25 a 100 000 years is a long time and the area may become prone to earthquakes, volcanoes or other unexpected natural disasters or dramatic changes in climate.
b There would be a small risk that something might go wrong during the launch; resulting in the radioactive material being scattered over a wide area of the Earth.
26 a 9.2 N
b 2.8×10^{27} m s^{-2}
31 A large wind generator is required to produce a relatively small power.
32 2.4×10^5 W
33 a 10 m s^{-1}
b 320 kW
34 4.4 m
35 a Diameter = 140 m (assuming efficiency = 25%, effective average wind speed = 8 m s^{-1}, density of air = 1.3 kg m^{-3})
b A larger generator would be expected to be more efficient; but design and construction problems (and the cost) of such a large structure may be too great.
36 a 400
b 36 km^2
c So that each generator does not affect the flow of wind to the others.
37 Calculations involve the wind speed cubed, and the cube of the average wind speed is a lot less than the average of the speeds cubed.
39 2.2×10^6 W
40 1.7×10^{14} J
41 a 98 cm
b All the rain that falls over the area flows into the lake and there is no evaporation from the surface. These are not reasonable assumptions.
42 6.1 kg
46 a 5.9×10^{24} J
b The total amount received is of the order of 10^4 times greater than the energy used.
49 Variation in the Sun's activity; variation in the distance between the Sun and the Earth
50 The metal pipes are good conductors of thermal energy; the glass cover prevents convection currents; black surfaces are good absorbers of thermal radiation.
52 a 0.023 W
b 0.031 A
c 2200
d 0.39 m^2
55 a 9.4×10^7 J
b 33°C
c All of the incident energy was transferred to the water in the tank; none was transferred to the surroundings, or retained in the solar panel.
56 c Greater angle of incidence and more atmosphere to pass through
60 a 1.8×10^{25} W
b It was a perfect black body.
c 0.25 W m^{-2}
61 a 1.5 m^2
b 700 W
c Conduction from the skin into the air; convection of warm air currents away from the body
d The body will also receive thermal energy from the surroundings.
62 a 44%
b 75°C
63 2500°C
64 a 8.3×10^{-7} m
b The light from the incandescent lamp will appear more yellow/'warmer'.
66 a 593 W m^{-2}
b 460 W m^{-2}
67 About 1.4 billion km
68 0.49 W m^{-2}
69 a Difference in the growth of trees and other plants; snow and ice in winter; variation in cloud cover; variation in angle of incidence
b Variation in cloud cover; variation in angle of incidence
70 600 W m^{-2}
71 293 K
72 a 3.6 W m^{-2}
b 65 K
73 a If the emissivity and albedo were the same as for the Earth the temperature would be about 160 K.
74 The received power and the radiated power would *both* be four times greater.
75 Carbon dioxide: about 15%; nitrous oxide: about 9%; methane: about 14%
77 a 2.26×10^6 J are needed to turn 1 kg of water into steam at 100°C.
b 1.1×10^{12} J
c The gravitational energy is very much smaller.
d About an hour

9 Wave phenomena

1 a 59 cm
2 a 36 N m^{-1}
b 2.1 Hz
c Force may not still be proportional to displacement; the coils of the spring may come into contact with each other.
3 a $T - \sqrt{m}$
b k can be found from the gradient of the graph ($= 2\pi/\sqrt{k}$).
4 0.045 s; 140 rad s^{-1}
5 450 rad s^{-1}
6 a 7.3×10^{-5} rad s^{-1}
b 7.2 rad
7 0.131 s
8 14 cm
9 a 9.2 rad s^{-1}
b 0.68 s
10 a 3.9 m s^{-2}
b Mass was undergoing SHM
11 −3.5 cm and −11 cm s^{-1}
12 a 0.23 m s^{-1} and 1.2×10^{-3} J
b −2.9 mm
13 5.63 Hz

14 a $0.39\ m\ s^{-1}$ b 6.0 cm
15 1.61 m above the low tide level
16 The area under the graph for one quarter of an oscillation (from $v = 0$) is equal to the amplitude.
17 a 4.32 s
b The accelerations are equal because $a = F/m$; the heavier pendulum has twice the mass, but also twice the weight.
c Amplitude does not affect period; if the amplitude is doubled, the restoring force is also doubled (for small amplitudes).
d $1.45\ rad\ s^{-1}$
e 3.97 J
f $1.75\ m\ s^{-1}$
18 0.83 m and $4.44\ m\ s^{-1}$
19 a 0.58 s
b $2.2\ m\ s^{-1}$
20 Same time period; reducing amplitude
21 a Ultraviolet
b By fluorescence
c 5.04×10^{-3} rad
22 5.7×10^{-7} m
23 0.085 mm
24 a Like Figure 9.12 – minima should occur at angles of $\pm 7.8 \times 10^{-3}$ rad and $\pm 15.6 \times 10^{-3}$ rad.
25 a 4.5×10^{-3} rad
c 1.4 cm
26 The light that would fall on the slits would not be coherent, monochromatic or intense.
27 1.54 mm
28 a 4.8×10^{-7} m
b The wavelength would be smaller than in air, so the fringe separation would decrease.
29 a So that the gaps were approximately the same size as the wavelength, to achieve maximum diffraction at each slit.
b About 1 m
30 a The spacing of the fringes will increase.
b The spacing of the fringes will increase.
c The spacing of the fringes will increase.
d The fringes will have coloured edges.
31 $260\ m\ s^{-1}$
33 16.0°
34 $80\ lines\ mm^{-1}$
35 1.6 m
36 a About 1 m
b Red light has a greater speed in glass than blue light, so that it is refracted less by a prism; but its greater wavelength means that a greater angle is needed to introduce the path difference of a whole wavelength that is needed for constructive interference.
37 Two
38 Because three times the wavelength of violet light is less than two times the wavelength of red light.
39 a 310 nm
b Ultraviolet
c By fluorescence or a suitable photovoltaic cell
40 This would have an effect similar to reducing the line separation, so that the pattern would broaden.
41 The line spacing is not comparable to the wavelength of X-rays; X-rays are too penetrating, so that they would pass through all of the grating largely unaffected.
42 Maxima at $\sin\theta = 0.272$, 0.544 and 0.816
43 0.267°
45 1.10×10^{-7} m
46 4.61×10^{-7} m or 6.92×10^{-7} m
47 1.4
50 a There is less diffraction with bigger lenses; they receive more light.
b It will be more difficult for a larger lens to focus all the light in the right places on the image.
51 Blue is near the short wavelength end of the visible spectrum and diffracts less.
52 A larger pupil at night means that diffraction is reduced, suggesting that resolution improves; but the much lower light intensity will reduce the quality of the image.
53 12 km
54 0.15 m
55 a 1.4×10^{14} m
b A line joining the stars is perpendicular to a line joining them to Earth.
56 Yes; the angle subtended at the telescope by the writing is 1.6×10^{-5} rad and this is much bigger than $1.22\lambda/b$ (about 5×10^{-6}).
57 About 100 km
59 a 610
b No
c Increase width of beam
60 413 Hz
61 $338\ m\ s^{-1}$
62 $31\ m\ s^{-1}$
63 a The sound will become louder as the train gets closer to P (allowing for a time delay for the sound to reach the observer). A pitch higher than that emitted by the train will be heard, but it will gradually fall as the train approaches (because the component of velocity towards the observer is decreasing). These processes are reversed as the train moves past P.
b The pitch and loudness will remain constant.
64 59 Hz
65 a 3000 Hz
b More easily absorbed and scattered in air; diffract and spread out more; slower speed
66 $260\ m\ s^{-1}$; 12.6 km
68 Moving away with a speed of $8.45 \times 10^{6}\ m\ s^{-1}$

10 Fields

1 9.6 V
2 -4.1×10^{8} J
3 2.3×10^{10} J
4 a 9.8×10^{9} J
b towards
5 24 J
6 $240\ J\ kg^{-1}$
8 a 7.5 J
b By the charge
15 Field lines would have double the separation and would point in the opposite direction.
16 Field lines would be straight lines pointing radially outwards from the centre. Equipotentials would be circular with increasing separation.
19 −77 pC
21 a 4.9×10^{24} kg
b 1.1×10^{4} N
23 a Any location below the Earth's surface will have mass distributed all around it and not just 'below' it.
b Mass is equally distributed around the central point.
24 a 6.4×10^{23} kg
c −250 MJ
25 a +2.5 nC
b -3.5×10^{-8} J
26 a 2200 N
b -4×10^{8} J

27 8.4×10^{-10} C
28 5.3×10^{33} J
29 a 6.3×10^{7} J would have to be given to 1 kg to move it to infinity from the Earth's surface
b P: -1.6×10^{7} J
Q: -6.25×10^{6} J
R: -3.1×10^{6} J
c 7.6×10^{9} J
31 a 3.3×10^{6} J kg^{-1}
b 1.3×10^{7} J kg^{-1}
c 2.8×10^{7} J
d It ignores the gravitational fields of Jupiter and its other moons (and the effect of Io's thin atmosphere).
32 4.3×10^{6} V
33 14 V
34 68 cm
35 18 N kg^{-1}
40 a 570 V
b 26 nC
41 About 6000 V
42 a 5.0×10^{4} V m^{-1}
b 0.039 N
c Downwards
d 3000 V
e 2.3×10^{-3} J
f Gains kinetic energy
43 a 1.6×10^{-19} C
b 7.5×10^{-16} J
c 2.0×10^{5} m s^{-1}
45 2.0×10^{-9} C
46 a 15.3 N C^{-1} at 17.9° to the direction of the field due to q_B
b 8 V
c Positive
47 a 6×10^{7} m
b 1×10^{27} kg
c 7.6×10^{8} m
d 0.52 N kg^{-1}
49 a 2.4 km s^{-1}
b An object escaping from the Moon also has to escape the Earth's gravitational field.
c The Earth is nearer than infinity and will attract the vehicle.
50 10.9 km s^{-1}
51 6.3×10^{10} m
52 a 3.0×10^{3} m
b The Sun's radius is roughly 200 000 times bigger.
53 a The escape speed increases by about 1%.
b This may be considered insignificant.
54 4.1×10^{3} m s^{-1}
55 a 2370 m s^{-1}
b 2.56 h
56 7.15 days
57 9.56×10^{6} m
58 a 3.17×10^{7} s
b Orbit is not a perfect circle.
60 a Decreases
b $E_T = -½GMm/r$: if E_T is decreasing it must be changing to a larger negative number, so that r must be getting smaller.
c The satellite will gain kinetic energy as it loses gravitational potential energy; as it goes faster it will encounter greater air resistance.
61 a The work done against air resistance is transferred to internal energy of the satellite; which is destroyed because it gets so hot that it vaporizes and/or chemically reacts with the air.
62 a -4.9×10^{10} J
b $+2.4 \times 10^{10}$ J
c -2.4×10^{10} J
d 1.1×10^{4} m s^{-1}
63 35 800 km
64 a The periods for the rotation of the Moon on its axis and its rotation around the Earth are the same.
b 8.8×10^{7} m
65 The length of the day (24 hours) may be considered as the duration between successive times when the Sun is at its greatest elevation. This is a little more than 23 hours and 56 minutes that the Earth takes to complete one revolution because the Earth is also orbiting the Sun.
66 a The time for a signal to travel from the Earth's surface to a geostationary satellite and back again is about 0.24 s.
b For example, compare satellites at heights of 358 km and 358 000 km (a ratio of 1/1000): if a transmitter on Earth sends the same signal to both, the geostationary satellite would receive a signal strength 1000^2 (10^6) times smaller, which suggests that a much larger receiving aerial would be necessary to receive sufficient power.

11 Electromagnetic induction

1 Because it has no free electrons; that can move to cause a charge separation.
2 a Into, or out of, the plane of the paper.
b Perpendicularly between the poles of the magnet, or along the line of the wire.
c It is not in a circuit.
3 A coil connected to a galvanometer should be moved quickly close to, or surrounding, the magnet.
4 a Use a stronger magnet; move it quicker.
b Reverse the motion; reverse the polarity.
5 Current flows in the opposite direction.
6 The induced emf increases as the speed of the falling magnet increases and also as the magnetic field passing through the coil gets stronger. The emf reverses direction when the magnet leaves the coil. The second peak is higher and quicker than the first because the speed is greater.
7 0.35 m s^{-1}
8 0.028 T
9 a 0.19 V
b The vertical component of the Earth's magnetic field is greater.
c No; any connecting leads would have the same voltage induced across them.
10 b 3.9 m s^{-1}
c 99 V
d 198 V and 0 V
11 1.3×10^{-5} Wb
12 7.2×10^{-4} Wb
13 4.5×10^{-4} Wb
14 a i The pointer on the galvanometer will deflect and then quickly return to zero. At the moment the current in the solenoid is switched on, a changing magnetic field through the loops of wire induces a current. There is no induction when the current in the solenoid is steady.

- ii As in i, but the deflection on the meter is in the opposite direction.
- b i Induced current will be doubled.
  - ii Induced current will be much greater.
- c Zero; no deflection

15 Increase the number of turns; move the coils closer together; place an iron core through the coils; increase the magnitude of the current in P.

16 a 8.8×10^{-3} Wb
 b 3.0×10^{-2} V

17 The amplitude and frequency would double; the maximum induced emf doubles because the current in A (and the resulting magnetic flux) is changing at twice the rate.

18 a 4.6×10^{-4} Wb
 b The magnetic field strength at A is negligible compared with the field at B.
 c 0.12 Wb
 d 0.083 V

19 a The plane of the coil should be parallel to the direction of the Earth's magnetic field.
 b 9.0×10^{-5} V

20 a 760
 b The coil and the solenoid should have the same axis.

21 a The change of flux linkage
 b The areas are equal; the flux linkage when the magnet enters the coil is equal to the flux linkage as the magnet leaves the coil.

23 a 2.6×10^{-3} Wb
 b 125 V
 c The answer for Question 10 is the *maximum* induced emf as the sides of the coil cut the field perpendicularly. This answer is the *average* during each rotation of the coil.

24 a The oscillations will be approximately simple harmonic.
 b Alternating voltage with the same frequency as a; voltage peaks when the magnet is passing through the middle of an oscillation.
 c The motion will be damped because kinetic energy of the magnet will be transferred to the induced current.

25 Currents are induced in the tube that set up a magnetic field opposing the motion of the falling magnet.

26 a The changing magnetic field passing through the loop induces a current, which sets up a magnetic field in opposition to the original field; the resulting force on the loop causes it to decelerate.
 b North pole
 c Into the plane of the paper
 d The magnetic flux through the loop is not changing.
 e The directions of current and induced magnetic field are opposite from those when the loop entered the magnetic field.
 f Transferred to internal energy in the loop because of the current in it.

27 a Its movement is parallel to the magnetic field.
 b Faster movement; stronger field; wind into a coil of many turns; place coil on an iron core

30 a 7 W
 b 89 Ω

31 a 325 V
 b 13.2 A
 c 9.35 A d 492 W

32 240 V

33 a 469 W
 b 1880 W
 c The resistor was ohmic and did not overheat.

34 a Peak power = 4.0 kW; V_0 = 350 V; I_0 = 11 A
 b Similar to Figure 12.27; period of voltage and current variations = 0.0167 s

36 a 30 turns c 5.8 W
 b 25 Ω d 0.024 A

37 a Turns ratio = 22/1
 b 0.072 A

38 a 1640
 b 0.012 A; 60 W

39 Resistance of the wires; cost; magnetic field strength in the core

40 a 69
 b 0.15 A
 c 89%
 d 69 mW
 e It would get hot.

41 Power losses are a smaller percentage of the overall power; magnetic flux is transferred more efficiently through a larger core; wires have less resistance.

42 Copper wires can be thinner and more flexible, and the properties of copper are better when making electrical connections.

43 a 3.1×10^5 W
 b 4.9 W

44 More power is transferred (for a given current).

45 So that most of the transmission line is at high voltage.

46 a The cables will have less weight; are easier to support on pylons.
 b Does not react with air and/or water.

47 a 2280
 b 4500 V
 c No voltage drop along the transmission line; transformer is 100% efficient.

48 a Hysteresis effects increase
 b Lower rate of change of magnetic flux

49 a 14 400 V
 b 1.92×10^5 W
 c 13 A
 d 7700 W
 e 29 W

51 FBCDEA

54 5.6×10^{-3} C

55 58.4 V

56 54 pF

57 1.1×10^{-10} $C^2\,N^{-1}\,m^{-2}$

59 11

61 The effective area between the plates changes.

62 Capacitance is proportional to area, but resistance is inversely proportional to area.

63 It can deliver enough power for an intense flash of light in a short time.

64 Any imperfections in the insulating layer may result in a flow of charge and electrical breakdown.

65 2.2×10^{-8} J

66 a 36 V
 b The thin layer of insulation between the plates may break down.

67 a 3.9 J
 b 2300 W

69 6.4×10^{-3} J

71 a 6000 μF
 b 540 μF

73 450 μF

74 a 137 μF
 b 0.054 J and 0.042 J
 c Internal energy and dissipated thermal energy when the current flows between them.

75 2.7×10^{-2} F

77 1.3×10^5 Ω

80 About 10 V and 1000 Ω
82 b 77 s
d A graph is a good way of assessing uncertainties and improving accuracy.
e 77 mF
83 3.3×10^{-3} C
84 5.1 V
85 80 kΩ
86 a 1.3 V
b 1.3×10^{-4} C
c 2.1×10^{-6} A

12 Quantum and nuclear physics

1 1.38×10^{24}
2 2 : 1
3 100
4 a 5×10^{-15} J; 3×10^{4} eV
5 a 4.91×10^{14} Hz
b 3.26×10^{-19} J
c 3.54×10^{-19} J
d No
f 5.34×10^{14} Hz
7 2.0×10^{-19} J
8 a 3.62×10^{-19} J
b 5.5×10^{-7} m; yellow light
c Red
9 1.44×10^{15} Hz
11 3.8×10^{-19} J; 2.4 eV
13 6.81×10^{-34} J s
17 6.6×10^{-34} J s
18 3.55×10^{-11} m
19 4.95×10^{2} m s^{-1}
20 9.0×10^{-7} m
21 105 V
22 a The wavelength of the de Broglie wave associated with a particle is inversely proportional to the mass of the particle (for the same velocity). The mass of an electron is less than that of a proton, so the wavelength of the de Broglie waves associated with the electron would be higher.
b Electron, neutron, alpha particle, gold nucleus
23 An airplane has a relatively large mass; hence the wavelength of the de Broglie waves associated with it is too small to be observed and measured.
25 6.7×10^{-13} m
26 0.719 MeV
27 1.3×10^{-15} m
28 Momentum could not be conserved with only one photon.
32 $2.5h/\pi$; 5.3×10^{-34} kg m^2 s^{-1}
33 a 8.5×10^{-10} m
b Infinite
34 a 2.17×10^{-18} J
b -4.34×10^{-18} J
c -2.17×10^{-18} J
d -13.6 eV
e Energy would have to be added to the atom to separate the proton and the electron, and when they are separate they are said to have zero potential energy.
35 a 4.5×10^{14} Hz
b Red light
c 0.66 eV
37 1.1×10^{-34} kg m s^{-1}
39 5.8×10^{7} m s^{-1}
40 1.46×10^{-33} m
41 3.7×10^{-19} eV
42 5×10^{-25} kg m s^{-1}; 1 eV
43 a 1.3×10^{-8} J
b 1.2×10^{-18} m
c The uncertainty principle will only allow the exchange particle to have a short lifetime because of its large mass-energy.
44 10^{-17} s
47 10^{-17} s
51 a 1.66×10^{-13} m
b 6.09×10^{-14} m
52 2.0×10^{7} m s^{-1}
54 26°
55 65000 MeV
56 That is the separation at which the attractive nuclear strong force is equal and opposite to the electric repulsive force.
57 a 7.0×10^{-15} m
b 3.0×10^{-15} m
c They are in close agreement; within 2%.
58 6 : 1
59 About 10^{-13}
60 About 200 m
61 The masses of the A protons and neutrons are each approximately equal to u, but vary depending upon the binding energy of the nuclei in which they are located.
63 a Typical nuclear energies are about 10^5 times greater than typical electron energies.
b 1.4×10^{-11} m
c Gamma radiation
64 b 7.97×10^{-12} m and 4.78×10^{-11} m
65 0.27 MeV, 0.19 MeV and 0.08 MeV
66 4.2×10^{-9} s^{-1}
67 8.0×10^{4} Bq
69 6.36×10^{23} atoms
70 a 126 Bq
b 805 nuclei
71 a 2.49×10^{-3} h^{-1}
b The small difference between the two count rates will be affected by random fluctuations in the activity of the source and the background count.
72 a 1.9 s^{-1}
b 130 s^{-1}
c Radiation may be absorbed in the air; all radiation passing through the detector may not be counted.
73 12 minutes
74 4.8×10^{-9} g
75 a 1.210×10^{-4} yr^{-1}
b 30%
76 0.26 weeks
77 a 0.42 yr
b 190 yr
78 5.3 years
79 23 s^{-1}
80 a 49.5%
81 1620 years
84 2/3

Answers to examination questions in Chapters 1 to 12

1 Measurements and uncertainties

Paper 1

1 B 2 B 3 C 4 A 5 C
6 C 7 A 8 A 9 B 10 C

2 Mechanics

Paper 1

1 B 2 B 3 C 4 D 5 C
6 B 7 C 8 A 9 C 10 C
11 C 12 A 13 B 14 C 15 B
16 C 17 D 18 A 19 D

Paper 2

1 a The equation can only be used for motion that has constant acceleration; the force on the bullet changes, so the acceleration cannot be constant.
b i 6.6×10^4 m s^{-2}
c i 280 m s^{-1}
ii 0.26 MW
d Newton's third law states that for every force there is an equal and opposite force acting on a different body; the forces of the bullet and gun are a pair of Newton's third law forces; the forward force on the bullet is equal to the backwards force on the gun.

2 a 1.8 m
b The initial velocity remains horizontal; the curve is steeper than the original because the horizontal distance travelled at any particular height becomes less and less; compared to that without air resistance
a i zero
ii Diagram should show normal reaction forces acting vertically upwards from the ground on both wheels; weight of bicycle acting downwards from its (approximate) centre of mass; force of foot pushing down on pedal (a downwards push of the hands may also be included); the total lengths of the upwards and downwards vector arrows should be about the same.
iii The forwards force on the bicycle is equal and opposite to the resistive forces (air resistance/drag/friction); so the resultant force is zero and therefore there is no acceleration.
b 320 W
c i 0.57 m s^{-2}
ii 56 m
iii The total resistive force is *not* constant (as assumed in the calculation); air resistance decreases with speed (the braking force may also change).

4 a The coefficient of dynamic friction applies to surfaces when there is relative motion. The coefficient of static friction is the maximum value just before motion begins.
b 29°

3 Thermal physics

Paper 1

1 C 2 B 3 D 4 A 5 A
6 B 7 A 8 D 9 A 10 D

Paper 2

1 a 3.8×10^5 J kg^{-1}
b Thermal energy will be transferred to the calorimeter and the surroundings; so less energy is available to boil the liquid (than used in the calculation).
a i Internal energy is the total potential energy and kinetic energy of the copper atoms; heating is the transfer of thermal energy to the copper from something hotter.
ii 240 J kg^{-1} K^{-1}
b i The molecules gain energy and move faster.
ii On average, each molecular collision with the walls exerts a greater force; so for the pressure to remain constant, the frequency of collisions must go down; which means that the volume must increase.

3 a low pressure, moderate temperature and low density
b 1.72×10^5 Pa
c Some of the water would evaporate and the pressure and volume would be greater.

4 Waves

Paper 1

1 C 2 C 3 A 4 C 5 C
6 D 7 C 8 D 9 C 10 C
11 D 12 C 13 B

Paper 2

1 a i A wave that transfers energy away from a source.
ii amplitude = 4.0 mm; wavelength = 2.4 cm; frequency = 3.3 Hz; speed = 8.0 cm s^{-1}
b i 63°

ii The three wavefronts are parallel to each other, equidistant and closer together than in medium A. They should make an angle of (approximately) 27° with the boundary. Two of the wavefronts should be continuous with the two leading wavefronts in A.

2 a
- The graph is a straight line through the origin, indicating that acceleration is proportional to displacement.
- The graph has a negative gradient, indicating that acceleration and displacement are in opposite directions.

c i The wave is progressive because it is transferring energy; the oscillations of the medium are parallel to the direction of energy transfer.

ii 0.94 m

3 a No energy is transferred; all oscillations between nodes are in phase; amplitude remains constant at each position (neglecting energy dissipation).

b 450 Hz

5 Electricity and magnetism

Paper 1

1 B 2 B 3 A 4 D 5 D
6 D 7 D 8 C 9 D 10 B
11 C 12 A 13 B 14 A 15 C

Paper 2

1 a The cell should be connected across the ends of the variable resistor; the lamp and the ammeter should be connected in series between one end of the variable resistor and its sliding contact; the voltmeter should be connected in parallel across the lamp.

b A curve passing through the origin (may be straight at first), with equal increases in voltage producing smaller and smaller increases in current.

2 a i resistance = $\frac{\text{potential difference}}{\text{current}}$

ii 54 Ω

b There are many possibilities: for example, one heater in series with a switch connected to the supply, and the other two heaters in parallel with each other, connected across the supply with a switch in series with the combination.

3 a Metal electrodes of two different metals placed inside the potato, variable resistor and ammeter in series with electrodes. Voltmeter connected across the electrodes.

b Voltage (pd) and current, and how they vary with time.

4 a 2.5×10^{-14} N

b The force is always perpendicular to motion, so the path will be along the arc of a circle, curving downwards.

6 Circular motion and gravitation

Paper 1

1 B 2 A 3 A 4 D 5 C 6 D 7 D

Paper 2

1 a gravitational field strength = $\frac{\text{gravitational force}}{\text{mass}}$

b i $g = \frac{GM}{R^2}$

ii 1.9×10^{27} kg

2 a friction

b 17 m s^{-1}

c The normal reaction force of the surface on the car will have a component acting towards the centre of the circle.

7 Atomic, nuclear and particle physics

Paper 1

1 D 2 C 3 B 4 D 5 A
6 D 7 C 8 B 9 D 10 A
11 C 12 B 13 D

Paper 2

1 a i Natural radioactive decay is the emission of particles; and/or electromagnetic radiation from an unstable nucleus; it is not affected by temperature or the environment. Radioactive decay is a spontaneous random process with a constant probability of decay (per unit time); the activity/number of unstable nuclei in a sample reduces exponentially.

ii Ionising means that when the radiation collides with a neutral atom, it will cause one or more electrons to be released, leaving behind a (positively) charged ion.

b i fission

ii U shown near the right-hand end of the line; Sr and Xe shown between U and the peak, with Sr to the left of Xe.

iii 7.60 MeV (1.22×10^{-12} J)

iv The binding energy of the neutrons is zero because neutrons are separate particles.

2 a i ${}^{220}_{86}\text{Rn} \rightarrow {}^{216}_{84}\text{Po} + {}^{4}_{2}\text{He}$

ii 1.01×10^{-12} J

iii $E_K = \frac{1}{2}mv^2$; 1.01×10^{-12}
$= \frac{1}{2} \times 4 \times 1.66 \times 10^{-27} \times v^2$;
$v = 1.74 \times 10^7$ m s^{-1}

b i The polonium nucleus moves in the opposite direction from the α-particle

ii By the conservation of momentum,
$m_\alpha v_\alpha = m_p v_p$:
$v_p = \frac{4}{216} \times 1.74 \times 10^7 = 3.22 \times 10^5$ m s^{-1}

iii The polonium nucleus has forward momentum; if the α-particle is emitted along the line of motion, there is no change in direction; but if it is emitted in any other direction, the polonium nucleus will deviate from its original path.

c i Decay is a random process; so it is not possible to state when nuclei will decay.

ii Half-life is the time taken for half of the radioactive nuclei in a sample to decay.

8 Energy production

Paper 1

1 A 2 D 3 C 4 C 5 A
6 B 7 B 8 D 9 C 10 D

Paper 2

1 a coal, oil, gas
  b The natural rate of production of the fuels is much lower than the rate of usage; so they will run out.
  c Using the width of the arrows; efficiency = 5/14 = 0.36 or 36%.
  d High energy density; readily available; cheap source of electrical energy; used widely by transport, etc.
2 a i infrared
    ii Nitrogen dioxide absorbs infrared radiation radiated from the Earth's surface; which it then re-radiates in random directions; thus reducing the net radiated energy transferred away from the Earth.
  b emissivity $= e = \dfrac{\text{power radiated by a surface}}{\text{power radiated from a black body of the same temperature and area}}$

  albedo $= \dfrac{\text{total scattered power}}{\text{total incident power}}$

  d i 154 W m^{-2} ii 402 W m^{-2}
  e 2 K

9 Wave phenomena

Paper 1

1 A 2 D 3 C 4 C 5 D
6 B 7 A 8 D 9 a B b B
10 D 11 B 12 A

Paper 2

1 a The wavefronts should all be circular, with their centres moving between S and P.
  b The diagram should show that the wavelength moving towards P is shorter than that moving towards Q; so that P hears a higher frequency.
  c i Doppler effect
    ii Change in frequency of the sound from a car moving past an observer.
2 a diffraction
  b The central maximum of the diffraction pattern of one image coincides with the first minimum of the diffraction pattern of the other.
  c 9.4×10^{20} m
3 a l is equal to the path difference between the rays from the edges of the slit; if $l = \lambda$, these rays will interfere constructively, but rays from the top of the slit and rays from X will interfere destructively because their path difference will be half a wavelength; in a similar way, pairs of rays across the slit width can be paired off to produce destructive interference; so when $l = \lambda$ the first minimum of the diffraction pattern will be produced.
  b The waveform is identical to the first, but the central maximum of one coincides with the first minimum of the other.
  c The smaller the value of λ/b, the better the resolution. Since the radio wavelengths received are relatively large, the width of the receiving aperture (dish) needs to be as large as possible.
  d Calculations show that they cannot be resolved; the angle subtended by the sources at Earth is smaller than $1.22\lambda/b$.
4 a Force proportional to displacement and in the opposite direction.
  b i Same shape as the kinetic graph, but inverted.

10 Fields

Paper 1

1 C 2 B 3 B 4 C
5 B 6 A 7 A 8 C

Paper 2

1 a The work done per unit mass; in bringing a small test mass from infinity; to that point.
  b 7.8 N kg^{-1}
  c 7.9×10^3 m s^{-1}
2 a i The field strength is constant at A and decreasing at B.
    ii Because it would mean that the force on a charge at that point acted in two different directions.
  b (See the red field lines in Figure 10.10.)
3 a Because infinity is chosen to be the location for zero potential and gravitational forces are always attractive, so that work has to be done to move any mass to the position where it has zero potential.
b +31 nC
c 2800 V
d 2800 V

11 Electromagnetic induction

Paper 1

1 A 2 C 3 B 4 C 5 B 6 D
7 B 8 B 9 A 10 C 11 C 12 D 13 B

Paper 2

1 a Faraday's law states that an induced emf is proportional to the rate of change of flux (linkage); in this example, the alternating current in the cable produces a changing magnetic field around it; so that a (time) changing magnetic flux passes through the coil.
  b The graph should have the same shape and frequency as the current graph, but show a phase shift of $\pi/2$.

c The size of the induced emf is proportional to the rms value of the current if the coil is always placed the same distance from the cables; with the same orientation.

2 a i The electrons are moving perpendicularly across the magnetic field and experience forces along the rod, which results in some separation of charges. The emf arises because work has been done to move the electrons.

ii The rate at which the rod passes through an area of magnetic field.

b i 2.0 mT

ii Lenz's law states that the direction of the induced current will oppose the change that produced it. In this example there is a force on the induced current in the rod in the opposite direction from its motion.

3 a Capacitor, resistor and ammeter in series. Voltmeter connected across the capacitor.

b 100 s

c Should show exponential decrease, falling from an initial value of 12 V.

12 Quantum and nuclear physics

Paper 1

1 D 2 C 3 C 4 B 5 A 6 C
7 A 8 B 9 D 10 A 11 A 12 B 13 D

Paper 2

a i, ii

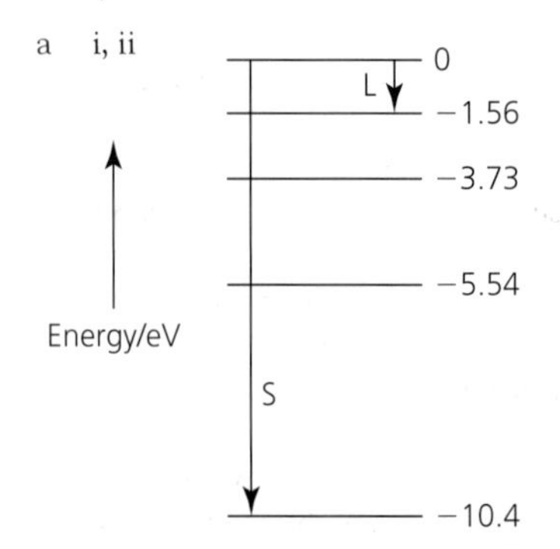

b $E = \frac{hc}{\lambda}$; $\lambda = \frac{6.6 \times 10^{-34} \times 3 \times 10^{8}}{10.4 \times 1.6 \times 10^{-19}} = 1.20 \times 10^{-7}$ m

c

bismuth

α

excited state of thallium

γ

ground state of thallium

Or

excited bismuth

γ

α

ground state thallium

d $A = A_0 e^{-\lambda t}$; $\frac{1.13}{2.80} = e^{-80\lambda} = 0.404$;

$\lambda = 0.0113$ (min^{-1});

$T_{½} = \frac{0.693}{0.0113} \approx 61$ min

2 a Light consists of photons; the energy of each photon is hf, where h is the Planck constant; a certain amount of energy, called the work function ϕ, is required to remove an electron from the metal surface; if f is less than ϕ/h, no electrons will be emitted.

b i 1.1×10^{15} Hz

ii $E_K = hf - \phi = Ve$, so slope of graph = h/e

slope $= 4.2\ (\pm 0.4) \times 10^{-15}$ giving h

$= 4.2\ (\pm 0.4) \times 10^{-15} \times 1.6 \times 10^{-19}$

$= 6.7\ (\pm 0.4) \times 10^{-34}$ J s

iii $\phi = hf$; using the value of h from *b* ii,

$\phi = 6.7 \times 10^{-34} \times 1.1 \times 10^{15} = 7.4 \times 10^{-19}$ J

Or

From the intercept on (the extended) E_K-axis,

$\phi = 4.5(\pm 0.2)$ eV

3 a i 1×10^{-10} m

ii The wavelength is smaller/frequency is higher; hence the kinetic energy is greater.

b i The electron is attracted to the nucleus, so work must be done (on the electron) to move it away from the nucleus; and so electrical potential energy increases as the distance from the nucleus increases

Or

The potential due to the nucleus is:

$V = k\left(\frac{Q}{r}\right)$

where Q is the nuclear charge; because $E_P = -V|e|$ the potential energy of the electron becomes less negative as the electron moves away.

ii The total energy of the electron is constant, so the kinetic energy decreases as the distance increases (because the potential energy increases)

iii Because the kinetic energy decreases, the wavelength must increase as the electron moves away from the nucleus.

Glossary

Standard level

The glossary contains key words, equations and terms from the IB Physics Diploma course. This section takes words from Chapters 1–8.

A

Absolute zero Temperature at which (almost) all molecular motion has stopped (0 K or −273°C).

Absolute temperature See *Kelvin scale of temperature*

Absorption When energy of incident particles or radiation is transferred to other forms within a material.

Absorption spectrum A series of dark lines across a continuous spectrum produced when light passes through a gas at low pressure.

Acceleration, *a* Rate of change of velocity with time, $a = \Delta v/\Delta t$ (unit: $\mathrm{m\,s^{-2}}$). Can be determined from the gradient of a velocity–time graph. Acceleration is a vector quantity.

Acceleration due to gravity, g Acceleration of a mass falling freely towards Earth. Numerically equal to the gravitational field strength. On or near the Earth's surface, $g = 9.8 \approx 10\,\mathrm{m\,s^{-2}}$.

Accuracy A single measurement is described as accurate if it is close to the correct result. A series of measurements of the same quantity can be described as accurate if their mean is close to the correct result. Accurate results have small systematic errors.

Air resistance Resistive force opposing the motion of an object through air. It depends on the cross-sectional area and shape of the object, and it is approximately proportional to the object's speed squared.

Albedo The total scattered or reflected power/total incident power (on part of a planet's surface). Albedo depends on the nature of the surface, cloud cover and inclination of the radiation to the surface. The Earth's annual mean albedo is 30%.

Alpha particle A fast-moving helium-4 nucleus emitted by a radioactive nucleus, consisting of two protons and two neutrons tightly bound together.

Alternating current (ac) A flow of electric charge that changes direction periodically.

Ammeter An instrument that measures electric current.

Ampere, A SI (fundamental) unit of electric current. One ampere is that current that, when flowing in two parallel wires 1 m apart in a vacuum, produces forces of $2 \times 10^{-7}\,\mathrm{N\,m^{-1}}$. $1\,\mathrm{A} = 1\,\mathrm{C\,s^{-1}}$

Amplitude, x_0 or A The maximum displacement of a wave or oscillation.

Analogue instrument Measuring instrument that has an indicator moving over a continuous scale. Compare with *digital instrument*.

Analogy Applying knowledge of one subject to another subject because of their similarities.

Angular velocity, ω Change of angle/change of time, $\omega = \Delta\theta/\Delta t = 2\pi/T = 2\pi f$ (unit: $\mathrm{rad\,s^{-1}}$). For regular motion in a circle, $\omega = 2\pi/T = 2\pi f$ and linear velocity = angular velocity × radius, $v = \omega r$.

Annihilation (pair) When a particle and its antiparticle interact, they annihilate and their mass is converted to electromagnetic energy.

Anode An electrode into which (conventional) current flows.

Antineutrino Low-mass and very weakly interacting particle emitted during beta-negative decay. Antiparticle of neutrino.

Antinodes The positions in a standing wave where the amplitude is greatest. See also *nodes*.

Antimatter Matter consisting of antiparticles.

Antiparticle Every particle has an antiparticle that has the same mass, but opposite charge and/or quantum numbers. (Some particles, for example photons, are their own antiparticle.)

Aperture A hole or gap designed to restrict the width of a beam of light (for example).

Atomic energy level One of a series of possible discrete (separate) energy levels of an electron within an atom.

Audible range Range of frequencies that can be heard by humans.

Avogadro's constant, N_A, The number of particles in 1 mole of a substance (the number of carbon atoms in 12 g of carbon-12).

B

Background radiation Radiation from radioactive materials in rocks, soil and building materials, as well as cosmic radiation from outer space and any radiation escaping from artificial sources.

Ballistics The study and use of projectiles.

Banking The use of sloping surfaces to enable faster motion around curves.

Barometer Instrument for measuring atmospheric pressure.

Baryons Particles made from the combination of three quarks – for example, protons and neutrons.

Battery One or more electric cells.

Beta particle A high-speed electron that is released from a nucleus during beta negative decay, or a high-speed positron released during beta positive decay.

Bias A preference for one opinion, one side of a discussion etc., often without fair consideration of the other.

Binding energy The energy released when a nucleus is formed from its constituent nucleons. Alternatively, it is equal to the work required to completely separate the nucleons. Binding energy is the energy equivalent of the *mass defect*.

Binding energy per nucleon (average) Binding energy of a nucleus divided by the number of nucleons it contains. It is a measure of the stability of a nucleus.

Biomechanics Application of the principles of mechanics to humans and animals.

Black body An idealized object that absorbs all the electromagnetic radiation that falls upon it. A perfect black body also emits the maximum possible radiation.

Black-body radiation Radiation emitted from a 'perfect' emitter. The characteristic ranges of different radiations emitted (*spectra*) at different temperatures are commonly shown in graphs of intensity against wavelength (or frequency).

Boil Change from a liquid to a gas at a well-defined temperature. Occurs throughout the liquid.

Boltzmann constant, k_B Important constant that links microscopic molecular energies to macroscopic temperature measurements. Linked to the universal gas constant, R, by the equation: $k_B = R/N_A$.

Bosons A class of sub-atomic particle that includes gauge bosons, the Higgs boson and mesons.

Boundary (between media) Surface where waves can be reflected or transmitted (and maybe refracted).

Boundary conditions The conditions at the ends of a standing wave system. These conditions affect whether there are nodes or antinodes at the ends.

C

Calibrate Put numbered divisions on a scale.

Caloric fluid An old, discredited theory about heat flow.

Calorimeter Apparatus designed for experiments investigating thermal energy transfers.

Carbon dating Using the radioactive decay of carbon-14 to estimate the age of once-living biological material.

Cathode An electrode out of which (conventional) current flows.

Cell (electric) Device that transfers chemical energy to the energy carried by an electric current. Also called an electrochemical cell or a voltaic cell. See also *primary cell* and *secondary cell*.

Celsius (scale of temperature) Temperature scale based on the melting point (0°C) and boiling point (100°C) of water.

Centre of mass Average position of all the mass of an object. The mass of an object is distributed evenly either side of any line through its centre of mass.

Centripetal acceleration and force Any object moving in a circular path has an acceleration towards the centre of the circle, called its centripetal acceleration. It can be calculated from the expression $a = v^2/r$. The force producing this acceleration is called a centripetal force, $F = mv^2/r$.

CERN European organization for nuclear research.

Chain reaction (nuclear) If, on average, one (or more) of the neutrons produced in a nuclear fission process causes further fission, the process will not stop and will become a self-sustaining chain reaction.

Charge Fundamental property of some sub-atomic particles that makes them experience electric forces when they interact with other charges. Charges can be positive or negative.

Charge density, n Number of mobile charges in unit volume of a material.

Charge (to) Add or remove electrons, so that an object acquires an overall net charge.

Charging characteristic of a secondary cell p.d.–time graph for the length of recharging process.

Chemical potential energy Energy related to the arrangement of electrons within the structure of atoms and molecules.

Circuit (electrical) A complete conducting path that enables an electric current to continuously transfer energy from a voltage source to various components.

Circuit breaker Electromagnetic device used to disconnect an electrical circuit in the event of a fault.

Classical physics Physics theories that pre-dated the paradigm shifts introduced by quantum physics and relativity.

Climate model A complex computerized model that attempts to predict the future climate of the planet, especially how it will be affected by global warming.

Coefficients of friction, μ Constants used to represent the amount of friction between two different surfaces. $F_f \leq \mu_s R$ (for static friction), $F_f \leq \mu_d R$ (for dynamic friction).

Coherent waves Waves that have the same frequency and a constant phase difference.

Collaboration (scientific) Two or more people sharing information or working together on the same project.

Collision Two (or more) objects coming together and exerting forces on each other for a short length of time. In an *elastic* collision the total kinetic energy before and after the collision is the same. In an *inelastic* collision the total kinetic energy is reduced after the collision. If the objects stick together it is described as a *totally inelastic* collision.

Combustion (of fuels) Burning. Release of thermal energy from a chemical reaction between the fuel and oxygen in the air.

Compass A device for determining the direction of north. Small *plotting* compasses are used to investigate the shapes of magnetic fields in the laboratory.

Components of a vector For convenience, a single vector quantity can be considered as having two parts (components), usually perpendicular to each other. The combined effect of these components is exactly the same as the single vector. See also *resolve*.

Composite particles Particles that have internal structure because they contain other particles.

Compression (force) Force that tries to squash an object or material.

Compressions (in a longitudinal wave) Places where there are increases in the density and pressure of a medium as a wave passes through it.

Condensation Change from a gas (or vapour) to a liquid.

Conduction (thermal) Passage of thermal energy through a substance as energy is transferred from particle to particle.

Conductor (electrical) A material through which an electric current can flow because it contains significant numbers of mobile charges (usually free electrons).

Confinement (quark) Term used to describe the fact that free individual quarks are never observed.

Conservation laws

Charge: the total charge in any isolated system remains constant.

Energy: the total energy in any isolated system remains constant. Energy cannot be created or destroyed.

Momentum: the total momentum in any isolated system remains constant. The total (linear) momentum of a system is constant provided that no external forces are acting on it.

Quantum numbers: in an equation describing a nuclear interaction, the total quantum numbers must be the same on both sides.

Contact (normal) forces Forces that occur because surfaces are touching each other. Contact forces are perpendicular (normal) to the surfaces.

Continuous spectrum A spectrum in which all possible wavelengths are present. A continuous visible spectrum shows a smooth and uninterrupted change from one colour to another.

Control rods Used for adjusting the rate of fission reactions in nuclear reactors by absorbing more or fewer neutrons.

Controlled and uncontrolled nuclear fission In a nuclear power station the number of neutrons in the reactor core is carefully *controlled* in order to maintain the rate of the nuclear reactions. In nuclear weapons the number of neutrons is *uncontrolled*.

Convection Passage of thermal energy through liquids and gases due to the movement of the substance because of differences in density.

Conventional current The direction of flow of a direct current is always shown as being from the positive terminal of the power source, around the circuit, to the negative terminal. Conventional current is opposite in direction from electron flow.

Correlation There is a correlation between varying data if they show similarities that would not be expected to occur because of chance alone.

Coulomb, C The derived unit of measurement of electric charge. 1 coulomb of charge passes a point in 1 second if the current is 1 amp.

Coulomb constant, *k* The constant that occurs in the Coulomb's law equation. $k = 8.99 \times 10^9\,\mathrm{N\,m^2\,C^{-2}}$. $k = 1/(4\pi\varepsilon_0)$, where ε_0 is the electrical *permittivity* of free space.

Coulomb's law There is an electric force between two point charges, q_1 and q_2 given by $F = kq_1q_2/r^2$, where r is the distance between them and k is the *Coulomb constant*. The law may also be applied to charged spheres that are relatively far apart.

Count rate (radioactivity) The number of nuclear radiation events detected in a given time (per minute or per second) by a radiation 'counter'.

Crest Highest part of a transverse mechanical wave.

Critical mass The minimum mass needed for a self-sustaining nuclear chain reaction.

Current (electric), *I* A flow of electric charge. Equal to the amount of charge passing a point in unit time: $I = \Delta q/\Delta t$. $1\,\mathrm{A} = 1\,\mathrm{C\,s^{-1}}$. See also *ampere*.

Cycle (oscillation) One complete oscillation.

D

Data logging Connecting sensors to a computer with suitable software to enable physical quantities to be measured and recorded digitally.

Daughter product The resulting nuclide when a radioisotope decays.

Decay series A series of nuclides linked in a chain by radioactive decay. Each nuclide in the chain decays to the next until a stable nuclide is reached.

Deceleration Negative acceleration. Reduction in the magnitude of a velocity (speed).

Deformation Change of shape.

Degraded energy Energy that has spread into the surroundings and cannot be recovered to do useful work.

Derived units Units of measurement that are defined in terms of other units.

Determinism The belief that future events are completely controlled by past events, so that full knowledge of the present can only lead to one outcome.

Diffraction The spreading of waves as they pass obstacles or travel through apertures (gaps).

Diffusion Movement of randomly moving particles from a place of high concentration to places of lower concentration.

Digital instrument Measuring instrument that displays the measurement only as digits (numbers). Compare with *analogue instrument*.

Dipole Two close electric charges (or magnetic poles) of equal magnitude but of opposite sign (or polarity).

Direct current (dc) A flow of electric charge that is always in the same direction.

Discharge Flow of electrons to or from an object that reduces the overall charge on it.

Discharge characteristic (of a battery) p.d.–time graph for the duration of the battery's use.

Discrepency An inconsistency, usually when some information is not as it was expected to be.

Disperse (light) Separate, usually into different wavelengths/colours (e.g. to form a spectrum).

Displacement, linear, *x* The distance from a reference point in a specified direction. A vector quantity.

Displacement, angular, θ The angle through which a rigid body has been rotated from a reference position.

Dissipate Spread out so that it cannot be recovered.

Drag Force(s) opposing motion through a fluid; sometimes called *fluid resistance*.

Drift speed The net speed of charges moving in an electric current.

Dynamo A type of electric generator that (usually) produces direct current.

E

Earth (ground) connection A good conductor connected between a point on a piece of apparatus and the ground. This may be part of a safety measure, or to ensure that the point is kept at 0 V.

Echo A reflected sound that is heard distinctly from the original sound.

Efficiency The ratio of the useful energy (or power) output from a device to the total energy (or power) input; often expressed as a percentage.

Elastic behaviour A material shows elastic behaviour if it regains its original shape after a force causing deformation has been removed.

Elastic strain potential energy Form of energy that is stored in a material that has been deformed elastically. The energy is transferred when the material returns to its original shape.

Electric field strength, *E* The electric force per unit charge that would be experienced by a small *test charge* placed at that point. $E = F/q$ (unit: $\mathrm{N\,C^{-1}}$).

Electric forces Fundamental forces that act across space between all charges. The forces between opposite charges are attractive. The forces between similar charges are repulsive.

Electrode Conductor used to make an electrical connection to a non-metallic part of a circuit.

Electrolysis Passage of an electric current through a substance in a liquid or molten state, which causes chemical changes.

Electromagnetic force Fundamental force acting between charged particles that is responsible for electric and magnetic forces. It reduces with an inverse square law with distance v**Electromagnetic induction** Process used by a generator to convert kinetic energy into electricity.

Electromagnetic spectrum Electromagnetic waves of all possible different frequencies, displayed in order. In order of increasing frequency: radio waves, microwaves, infrared, visible light, ultraviolet, X-rays, gamma rays. The visible spectrum in order of increasing frequency: red, orange, yellow, green, blue, indigo, violet.

Electromotive force (emf), ε The total energy transferred in a source of electrical energy per unit charge passing through it.

Electron Sub-atomic particle with a negative charge ($-1.6 \times 10^{-19}\,\mathrm{C}$) and mass of $9.110 \times 10^{-31}\,\mathrm{kg}$; present in all atoms and located in energy levels outside the nucleus.

Electronvolt, eV The amount of energy transferred when an electron is accelerated by a p.d. of 1 V. 1 eV = 1.6×10^{-19} J

Elementary particles Particles that have no internal structure. They are not composed of other particles. For example, electrons.

Elliptical In the shape of an ellipse (oval)

Emission spectrum Line spectrum associated with the emission of electromagnetic radiation by atoms, resulting from electron transitions from higher to lower energy states.

Emissivity, *e* The power radiated by an object divided by the power radiated from a black-body of the same surface area and temperature.

Emit To send out.

Empirical Based on observation or experiment.

Endoscope A medical device that uses the total internal reflection of light to obtain images from inside the body.

Energy density The energy transferred from a unit volume of fuel (units: $J\,m^{-3}$). See also *specific energy*.

Energy mass equivalence Any mass is equivalent to a certain amount of energy, according to the equation $E = mc^2$.

Equations of motion Equations that can be used to make calculations about objects that are moving with uniform acceleration.

$v = u + at$

$s = (u + v)t/2$

$v^2 = u^2 + 2as$

$s = ut + \frac{1}{2}at^2$

Equation of state for an ideal gas, $pV = nRT$ Describes the macroscopic physical behaviour of ideal gases.

Equilibrium position Position in which there is no resultant force acting on an object.

Equilibrium (translational) An object is in translational equilibrium if there is no resultant force acting on it, so that it remains at rest or continues to move with a constant velocity.

Error When a measurement is not exactly the same as the correct value.

Error bars Vertical and horizontal lines drawn through each data point on a graph to represent the uncertainties in the two values.

Evaporation The change from a liquid surface to a gas at any temperature below the boiling point of the liquid. Occurs at the liquid surface.

Excitation The addition of energy to a system, changing it from its *ground state* to an *excited state*. Excitation of a nucleus, an atom or a molecule can result from absorption of photons or from inelastic collisions with other particles.

Exchange particles Also known as *gauge bosons*. The exchange of these particles is used to explain fundamental forces (interactions). Photons, gluons and bosons.

Excited state When an electron in an atom is at a higher energy level than its ground state.

Expansion Increasing in size.

Exponential change Occurs when the rate of change of the quantity at any time is proportional to the actual quantity at that moment.

Extension Displacement of the end of an object that is being stretched.

F

Feynman diagram A graphical means of representing particle interactions by the use of one or more vertices.

Field (gravitational, electric or magnetic) A region of space in which a mass (or a charge, or a current) experiences a force due to the presence of one or more other masses (charges, or currents – moving charges).

Field lines and patterns Representation of fields in drawings by a pattern of lines. Each line shows the direction of force on a mass (in a gravitational field), of force on a positive charge (in an electric field), or on a north pole (in a magnetic field). A field is strongest where the lines are closest together. See also *uniform field* and *radial field*.

Filament lamp Lamp that emits light from a very hot metal wire. Also called incandescent lamp.

Finite Limited.

Fission fragments The nuclei produced in a fission reaction.

Flavours (of quark) There are six different kinds ('flavours') of quark: up, down, strange, charm, bottom, top. The different types of lepton are also called flavours.

Fluid Liquid or gas.

Fluid resistance Force(s) opposing motion through a fluid; sometimes called drag.

Fluorescent lamp Lamp that produces light by passing electricity through mercury vapour at low pressure.

Force constant, *k* Ratio of force to extension for a stretched material or spring. $k = \Delta F/\Delta x$. (Sometimes called the *spring constant*.) Unit: Nm^{-1}

Force meter Instrument used to measure forces. Also called a newton meter or spring balance.

Fossil fuels Naturally occurring fuels that have been produced by the effects of high pressure and temperature on dead organisms (in the absence of oxygen) over a period of millions of years. Coal, oil and natural gas are all fossil fuels.

Foucault's pendulum Very large pendulum designed to show the rotation of the Earth.

Frame of reference (for motion) Location to which observations and measurements of motion are compared. For example, the speed of a car might be $10\,m\,s^{-1}$ compared to the Earth's surface.

Free electrons Electrons (most commonly in metals) that are not attached to individual atoms. Also called *delocalized electrons*. They provide the mobile charges that are needed for an electric current to flow in solid conductors.

Free-body diagram Diagram showing all the forces acting *on* a single object, and no others.

Free fall Motion through the air under the effects of gravity but without air resistance. In common use *free fall* can also mean falling towards Earth without an open parachute.

Free space Place where there is no air (or other matter). Also called a vacuum.

Freeze Change from a liquid to a solid. Also called *solidify*.

Frequency, *f* The number of oscillations per unit time, or number of waves passing a point per unit time (usually per second). $f = 1/T$ (unit hertz, Hz)

Friction Resistive forces opposing relative motion, particularly between solid surfaces. *Static friction* prevents movement, whereas *dynamic friction* occurs when there is already motion.

Fuel A store of energy (chemical or nuclear) that can be transferred to do useful work (for example, generate electricity or power vehicles).

Fuel enrichment Increasing the percentage of ^{235}U in uranium fuel in order to make it of use in a nuclear power station or for a nuclear weapon.

Fundamental Having no simpler explanation.

Fundamental forces (interactions) Strong nuclear, weak nuclear, electromagnetic and gravitational forces.

Fundamental units Units of measurement that are not defined as combinations of other units.

Fuse A device used to disconnect an electrical circuit in the event of a fault. It comprises a thin wire that melts if the current gets too large.

Fusion (thermal) Melting.

G

Gamma radiation/ray Electromagnetic radiation (photons) emitted from some radionuclides and having an extremely short wavelength.

Gas laws Laws of physics relating the temperature, volume and pressure of a fixed mass of an ideal gas: Boyle's law, Charles's law and the pressure law.

Gauge bosons See *exchange particles*

Geiger–Marsden experiment The scattering of alpha particles by a thin sheet of gold foil, which demonstrated that most particles passed through the foil completely undeflected, while a few were deflected at extremely large angles. This demonstrated that atoms consist of mostly empty space with a very dense positively charged core (the nucleus).

Geiger–Muller tube Apparatus used to detect the radiation from a radioactive source.

Generator (electrical) Device that converts kinetic energy into electricity.

Global warming Increasing average temperatures of the Earth's surface, atmosphere and oceans.

Gluon Exchange particle for the strong nuclear force.

Gravitational field strength, g The gravitational force per unit mass that would be experienced by a small *test mass* placed at that point. $g = F/m$ (unit: N kg^{-1})

Gravitational forces Fundamental attractive forces that act across space between all masses. Gravitational force reduces with an inverse square law with distance between point masses.

Gravitational potential energy, E_P Energy that masses have because of the gravitational forces between them. Changes in gravitational potential energy in a uniform field can be calculated from $\Delta E_P = mg\Delta h$.

Greenhouse effect The natural effect that a planet's atmosphere has on reducing the amount of radiation emitted into space, resulting in a warmer planet (warmer than it would be without an atmosphere).

Greenhouse effect (enhanced) The reduction in radiation emitted into space from Earth due to an increasing concentration of greenhouse gases in the atmosphere (especially carbon dioxide) caused by human activities; believed by most scientists to be the cause of *global warming*.

Greenhouse gases Gases that absorb and emit infrared radiation and thereby affect the temperature of the Earth. The principal greenhouse gases are water vapour, carbon dioxide, methane and nitrous oxide. Atmospheric concentrations of the last three of these have been increasing significantly in recent years.

Ground state The lowest energy state of an atom/electon.

H

Hadrons Particles made from the combination of quarks – for example, baryons and mesons.

Half-life (radioactive) The time taken for the activity or count rate from a source, or the number of radioactive atoms, to be reduced by half; the half-life is constant for a particular radioisotope.

Harmonics Different frequencies (modes) of vibration of a given system. They are mathematically related and numbered as first, second, third etc. Wavelengths are related to the length of the system.

Heat engine Device that uses the flow of thermal energy to do useful work.

Heat exchanger Equipment designed to efficiently transfer thermal energy from one place to another.

Hertz, Hz Derived unit of measurement of frequency. 1 Hz = one oscillation per second.

Higgs boson The exchange of Higgs bosons is responsible for giving particles the properties of mass.

Holistic theory A theory that is more concerned with the whole of an issue rather than concentrating on parts.

Hooke's law The force needed to deform a spring is proportional to the extension (or compression).

Human science The study of how people interact with each other and the world around them

Hydroelectric power (HEP) The generation of electrical power from falling water.

I

***I*–*V* characteristic** Graph of current–p.d., representing the basic behaviour of an electrical component.

Ideal gas The kinetic model of an ideal gas makes the following assumptions. **i** The molecules are identical. **ii** The molecules are point masses with negligible size or volume. **iii** The molecules are in completely random motion. **iv** There are negligible forces between the molecules, except when they collide. **v** All collisions are elastic, that is, the total kinetic energy of the molecules remains constant.

Ideal meters Meters that have absolutely no effect on the electrical circuits in which they are used. An ideal ammeter has zero resistance, and an ideal voltmeter has infinite resistance.

Imagination Formation of new ideas that are not related to direct sense perception or experimental results.

Immersion heater Heater placed inside a liquid or object.

Impact Collision involving relatively large forces over a short time. The effect of such an impact may be greater than from the same *impulse* (Ft) delivered by a smaller force over a longer time.

Impulse The product of force and the time for which the force acts. It is equal to the change of momentum. (Unit: N s)

Incidence, angle of The angle between an incident ray and the normal (or between the incident wave and the boundary).

Incident wave or ray Wave (or ray) arriving at a boundary.

Inclined plane Flat surface at an angle to the horizontal (but not perpendicular). A simple device that can be used to reduce the force needed to raise a load; sometimes called a *ramp*.

Infinite Without limits.

Inspiration Stimulation (usually to be creative).

Insulator (electrical) A non-conductor. A material through which a (significant) electric current cannot flow, because it does not contain many mobile charges. See also *conductor*.

Insulator (thermal) A material that significantly reduces the flow of thermal energy.

Intensity, I Wave power/area: $I = P/A$ unit: Wm^{-2}. The intensity of a wave is proportional to its amplitude squared, $I \propto A^2$.

Interaction Any event in which two or more objects exert forces on each other.

Interference Superposition effect that may be produced when similar waves meet. Most important for waves of the same frequency and similar amplitude. Waves arriving in phase will interfere *constructively* because their *path difference* = $n\lambda$. Waves completely out of phase will interfere *destructively* because their *path difference* = $(n + \frac{1}{2})\lambda$.

Internal energy Total potential energy and random kinetic energy of the molecules of a substance.

Internal resistance, r Sources of electrical energy, for example, batteries, are not perfect conductors. The materials inside them have resistance in themselves, which we call internal resistance. This results in energy dissipation in the battery and a reduction in the useful voltage supplied to the circuit. See *lost volts*.

Intuition Immediate understanding, without reasoning.

Inverse square law For waves/energy/particles/fields spreading equally in all directions from a point source without absorption or scattering the intensity is inversely proportional to the distance squared. $I \propto x^{-2}$

Ionization The process by which an atom gains or loses one or more electrons, thereby becoming an *ion*. The required energy is called the *ionization energy*.

Ionizing radiation Radiation with enough energy to cause ionization.

Isochronous Describing events that take equal times.

Isotopes Two or more atoms of the same element with different numbers of neutrons (and therefore different masses). A *radioisotope* is unstable and will emit radiation.

J

Jet engine An engine that achieves propulsion by emitting a fast-moving stream of gas or liquid in the opposite direction from the intended motion.

Joule, J Derived SI unit of work and energy. $1\,J = 1\,N\,m$

Joulemeter Electrical meter that displays the energy transferred in a circuit.

K

Kelvin scale of temperature Also known as the *absolute temperature scale*. Temperature scale based on absolute zero (0 K) and the melting point of water (273 K). The kelvin, K, is the fundamental SI unit of temperature. $T/K = \theta/°C + 273$. The kelvin (absolute) temperature is a measure of the mean random translation kinetic energy of one molecule of an ideal gas: $\bar{E}_K = \frac{3}{2}k_BT$. The total translational kinetic energy of one mole of an ideal gas = $\frac{3}{2}RT$.

Kilogram, kg SI unit of mass (fundamental).

Kilowatt hour, kWh The amount of electrical energy transferred by a 1 kW device in 1 hour.

Kinematics The study of moving objects

Kinetic energy, E_K Energy of moving masses; translational KE is calculated from $\frac{1}{2}mv^2$.

Kinetic model of an ideal gas The idealized motions of the molecules in a gas used to predict the macroscopic behaviour of gases. See also *ideal gas*.

Kirchoff's first circuit law $\Sigma I = 0$ (junction).

Kirchoff's second circuit law $\Sigma V = 0$ (loop).

L

Lagging Thermal insulation.

Latent heat Thermal energy that is transferred at constant temperature during any change of physical phase. See *specific latent heats of fusion and vaporization*.

Left-hand rule (Fleming's) Rule for predicting the direction of the magnetic force on moving charges, or a current in a wire.

Lens Curved transparent surfaces used to refract waves to a focus.

Leptons Elementary particles of low mass (electrons, taus and muons and their neutrinos).

Light-emitting diodes (LEDs) Small semiconducting diodes that emit light of various colours at low voltage and power.

Light-dependent resistor (LDR) A resistor that has less resistance when placed in light of greater intensity.

Linear relationship One which produces a straight line graph.

Line of action (of a force) A line through the point of action of a force, showing the direction in which the force is applied.

Liquid crystal display (LCD) A display that uses liquid crystal layers between polarizing filters. Voltages applied to the liquid crystal change its plane of polarization.

Longitudinal wave A wave in which the oscillations are parallel to the direction of transfer of energy, for example sound waves. Sometimes called a compression wave.

Lost volts Term sometimes used to describe the voltage drop (below the emf) that occurs when a source of electrical energy delivers a current to a circuit.

M

Macroscopic Can be observed without the need for a microscope.

Magnetic field strength, B The force acting per unit length on unit current moving across the field at an angle θ: $B = F/(IL\sin\theta)$. (Unit: tesla; $1\,T = 1\,N\,A^{-1}\,m^{-1}$)

Magnetic forces Fundamental forces that act across space between all moving charges, currents and/or permanent magnets. The forces are perpendicular to the direction of the current.

Magnitude Size.

Mains electricity Electrical energy supplied to homes and businesses by cables from power stations.

Malus's law Used for calculating the intensity of light transmitted by a polarizing filter: $I = I_0\cos^2\theta$. I_0 is incident intensity and θ is the angle between the polarizer axis and the plane of polarization of the light.

Manometer A U-tube containing liquid that is used for measuring differences in gas pressure.

Mapping Representing the interrelationships between ideas, knowledge or data by drawing (graphically).

Mass defect The difference in mass between a nucleus and the total masses of its constituent neutrons and protons when separated.

Mechanics Study of motion and the effects of forces on objects.

Medical scanners/imagers Equipment used in hospitals for obtaining images of the internal structure of the body.

Medium (of a wave) Substance through which a wave is passing (plural: media).

Meltdown (thermo-nuclear) Common term for the damage to the core and reactor vessel that results from overheating following some kind of accident at a nuclear power station.

Mesons Particles that combine a quark and an antiquark.

Metre, m SI unit of length (fundamental).

Metholodology An outline of the way in which a study, investigation or project is carried out.

Microscopic Describes anything that is too small to be seen with the unaided eye.

Modelling A central theme of science that involves representing reality with simplified theories, drawings, equations etc., in order to achieve a better understanding and make predictions.

Moderator Material used in a nuclear reactor to slow down neutrons to low energies.

Modes of vibration The different ways in which a standing wave can be set up in a given system. See also *harmonics*.

Molar mass The mass of a substance that contains 1 mole of its defining particles.

Mole, mol SI unit of amount of substance (fundamental). Defined as the amount of a substance that contains as many of its defining particles as there are atoms in exactly 12 g of carbon-12.

Momentum (linear), p Mass times velocity: $p = mv$ (unit: kg m s^{-1}). A vector quantity.

Monochromatic Containing only one colour/frequency/ wavelength (more realistically, a narrow range).

Moral and ethical issues Changes to the world, often brought about by scientific and technological developments, that some individuals and societies believe to be wrong.

Motor effect Magnetic force on a current in a magnetic field, as used in electric motors.

N

Natural gas Naturally occurring mixture of gases (mainly methane) that can be used as a fuel. May be either a fossil fuel, or produced more recently by biological processes.

Natural philosophy The name used to describe the (philosophical) study of nature and the universe before modern science

Negligible Too small to be significant

Neutral Uncharged, or zero net charge.

Neutrino Low-mass and very weakly interacting particle emitted during beta-positive decay. Antiparticle of *antineutrino*.

Neutron Neutral sub-atomic particle with a mass of 1.675×10^{-27} kg. The number of neutrons in a nucleus is called the *neutron number* (N).

Neutron capture Nuclear reaction in which a neutron interacts with a nucleus to form a heavier nucleus.

Newton, N Derived SI unit of force. $1\text{ N} = 1\text{ kg m s}^{-2}$

Newton's laws of motion

First law: an object will remain at rest, or continue to move in a straight line at a constant speed, unless a resultant force acts on it.

Second law: acceleration is proportional to resultant force; $F = ma$ or $F = \Delta p/\Delta t$.

Third law: whenever one body exerts a force on another body, the second body exerts exactly same force on the first body, but in the opposite direction.

Newton's universal law of gravitation There is a gravitational force between two point masses, M and m, given by $F = GMm/r^2$, where r is the distance between them and G is the universal gravitation constant. The law may also be applied to spherical masses that are relatively far apart.

Nodes The positions in a standing wave where the amplitude is zero. See also *antinodes*.

Non-renewable energy sources Energy sources that take a very long time to form and which are being rapidly used up (depleted).

Normal Perpendicular to a surface.

Nuclear potential energy Energy related to the forces between particles in the nuclei of atoms.

Nuclear equation An equation representing a nuclear reaction. The sum of nucleon numbers (A) on the left-hand side of the nuclear decay equation must equal the sum of the nucleon numbers on the right-hand side of the equation. Similarly with proton numbers (Z).

Nuclear fission A nuclear reaction in which a massive nucleus splits into smaller nuclei whose total binding energy is greater than the binding energy of the initial nucleus, with the simultaneous release of energy.

Nuclear fusion Nuclear reaction in which two light nuclei form a heavier nucleus whose binding energy is greater than the combined binding energies of the initial nuclei, thereby releasing energy.

Nuclear waste Radioactive materials associated with the production of nuclear power that are no longer useful, and which must be stored safely for a long period of time.

Nucleon A particle in a nucleus, either a neutron or proton. The *nucleon number* (A) is the total number of protons and neutrons in the nucleus.

Nucleus The central part of an atom containing protons and neutrons (except for hydrogen). A nucleus is described by its atomic number and nucleon number. See *nuclide*.

Nuclide Term used to identify one particular species (type) of atom, as defined by the structure of its nucleus. A *radionuclide* is unstable and will emit radiation.

O

Objective Free from bias and emotion. Compare with *subjective*.

Observer effect When the act of observation, or measurement, has an effect on the phenomenon being observed.

Ohm, Ω The derived unit of measurement of electric resistance. $1\,\Omega = 1\text{ V}/1\text{ A}$

Ohmic (and non-ohmic) behaviour The electrical behaviour of an ohmic component is described by *Ohm's law*.

Ohm's law States that the current in a conductor is proportional to the potential difference across it, provided that the temperature is constant.

Optically active substance Substance that rotates the plane of polarization of light that is passing through it.

Optical fibre Thin, flexible fibre of high-quality glass that uses total internal reflection to transmit light along curved paths and/or over large distances.

Orbit The curved path (often assumed circular) of a mass, or charge, around a larger central mass, or charge.

Order of magnitude When a value for a quantity is not known precisely, we can give an approximate value by quoting an order of magnitude. This is the estimated value rounded to the nearest power of ten. For example, 400 000 has an order of magnitude of 6 (10^6) because $\log_{10} 400\,000 = 5.602$, which is nearer to 6 than 5.

Oscillation Repetitive motion about a fixed point.

P

Pair production Conversion of photon energy into a particle and its antiparticle. The opposite of annihilation.

Parabolic In the shape of a parabola. The trajectory of a projectile if air resistance is negligible.

Paradigm (physics) A widely accepted model and way of thinking about a particular aspect of the physical world.

Parallax error Error of measurement that occurs when reading a scale from the wrong position.

Parallel connection Two or more components connected between the same two points, so that they have the same p.d. across them.

Particle accelerator Apparatus designed to produce particle beams.

Particle beams Streams (flows) of very fast-moving particles, most commonly charged particles (electrons, protons or ions), moving across a vacuum. Properties of the individual particles can be investigated by observing the behaviour of the beams in electric and/or magnetic fields.

Passive safety standards Requirements for the design and construction of vehicles that are aimed at limiting the risks to drivers and passengers in accidents.

Path difference The difference in distance of two sources of waves from a particular point. If the path difference between coherent waves is a whole number of wavelengths, constructive interference will occur.

Peer review Evaluation of scientific results and reports by other scientists with expertise in the same field of study.

Penetrating power The penetrating power of nuclear radiation depends upon the ionizing power of the radiation. The radiation continues to penetrate matter until it has lost all of its energy. The greater the ionization per cm, the less penetrating power it will possess.

Period (time period), *T* The time taken for one complete oscillation, or the time taken for one complete wave to pass a point.

Permanent magnet Magnetized material that creates a significant and persistent magnetic field around itself. Permanent magnets are made from ferromagnetic materials, like certain kinds of steel. *Soft iron* cannot be magnetized permanently.

Permeability Constant that represents the ability of a particular medium to transfer a magnetic force and field.

Permeability of free space, μo Fundamental constant that represents the ability of a vacuum to transfer a magnetic force and field, $\mu o = 4\pi \times 10^{-7}\,T\,mA^{-1}$

Permittivity of free space, ε_0 Fundamental constant that represents the ability of a vacuum to transfer an electric force and field, $\varepsilon_0 = 8.85 \times 10^{-12}\,C^2 N^{-1} m^{-2}$.

Permittivity of a medium, ε Constant that represents the ability of a particular medium to transfer an electric force and field. Often expressed as *relative permittivity*: $\varepsilon_r = \varepsilon/\varepsilon_0$ (no units), which is also sometimes called *dielectric constant.*

Perpetual motion machine Machine that can continue to move without any energy source. An impossible dream of scientists from the past. We now know that dissipation of energy will cause all motion, which is not powered, to slow down and stop (unless moving through space).

Phase (of matter) A system (substance) in which all the physical and chemical properties are uniform.

Phase (oscillations) Oscillators are in phase if they have the same frequency and they are doing exactly the same thing at the same time.

Phase difference When oscillators that have the same frequency are out of phase with each other, the difference between them is defined by the angle (usually in radians) between the oscillations. Phase difference can be between 0 and 2π radians.

Phlogiston An old, discredited theory about combustion.

Photon A quantum ('packet') of electromagnetic radiation, with an energy (E) given by $E = hf$. Exchange particle for the electromagnetic interaction.

Photosynthesis Chemical process that produces plant sugars (chemical energy) from carbon dioxide and water using the radiant energy from the Sun.

Photovoltaic cell Device that converts electromagnetic radiation (mainly light) into electrical energy. Also called a solar cell.

Planck relationship The frequency of electromagnetic radiation, f, emitted or absorbed when an electron undergoes a transition between two energy states, is given by $\Delta E = hf$, where h represents *Planck's constant* and ΔE is the difference in energy levels.

Plane waves Waves with parallel wavefronts, which can be represented by parallel rays.

Plane of polarization The plane in which all oscillations of a plane-polarized wave are occurring.

Plasma State of matter that is similar to a gas, but which contains a significant proportion of charged particles (ions).

Point particle, mass or charge Theoretical concept used to simplify the discussion of forces acting on objects (especially in gravitational and electric fields).

Polarity Separation of opposite electric charges or opposite *magnetic poles*, which produces uneven effects in a system.

Polarization (plane) A property of some transverse electromagnetic waves in which the electric field (and magnetic field) oscillations are all in the same plane.

Polarizing filter A filter that transmits light that is polarized only in one plane. A filter used to produce polarized light from unpolarized light is called a *polarizer.* A polarizing filter that is rotated in order to analyse polarized light is called an *analyzer. Crossed filters* prevent all light from being transmitted.

Positron Antiparticle of the electron; released during beta-positive decay.

Potential difference (electric), p.d. The electrical potential energy transferred (work done) as a unit charge moves between two points, $V = W/q$. Commonly referred to as *voltage.*

Potential divider Two resistors used in series with a fixed potential difference across them. When one resistance is changed, the p.d.s across each resistor will change; this can be used for controlling another part of the circuit.

Potential energy Energy that is stored. See *chemical potential energy, elastic strain potential energy, gravitational potential energy* and *nuclear potential energy.*

Potentiometer Variable resistor (with three terminals) used as a *potential divider.*

Power, *P* Energy transferred/time taken ($P = \Delta E/\Delta t$) or, for mechanical energies, work done/time taken ($P = \Delta W/\Delta t$). (Unit: watt; $1\,W = 1\,J\,s^{-1}$).

Power (electrical) The rate of dissipation of energy in a resistance. $P = VI = I^2R = V^2/R$

Precision A measurement is described as precise if a similar result would be obtained if the measurement was repeated. Precise measurements have small random errors.

Prefixes (for units) Used immediately before a unit to represent powers of ten. For example, 'milli-' in front of a unit represents $\times 10^{-3}$, as in millimetre.

Precautionary principle The idea that scientific research should not continue unless scientists are able to confirm that the research will not be harmful.

Pressure, *P* Defined as force acting normally per unit area: pressure = force/area. The unit is the pascal, Pa (1 Pa = 1 N m^{-2}.) *Atmospheric pressure* is the pressure in the air due to molecular motions.

Pressure gauge An instrument for measuring gas pressure.

Primary cell An electric cell that cannot be used again after the chemical reactions have finished.

Primary energy source Natural source of energy that has not been converted to another form (for example, not to electricity or hydrogen).

Prism A regularly shaped piece of transparent material (such as glass) with flat surfaces, which is used to refract and disperse light.

Projectile An object that has been projected through the air and which then moves only under the action of the forces of gravity and air resistance.

Propagation (of waves) Transfer of energy by waves.

Propulsion Method by which force is provided to produce motion.

Proton Sub-atomic particle with a positive charge ($+1.6 \times 10^{-19}$ C) and mass of 1.673×10^{-27} kg. The number of protons in a nucleus is called the *proton number* (Z).

Pulse (wave) A travelling wave of short duration.

Pumped storage (HEP) Large quantities of water are pumped up to a higher location using excess electrical power. When the water is allowed to fall down again, electricity can be regenerated.

Q

Qualitative Involving qualities, rather than quantities.

Quantitative Involving quantities, measurements

Quantized Can only exist in certain definite (discrete) numerical values.

Quantum The minimum amount of a physical quantity that is quantized (can only have discrete values).

Quantum numbers Used to describe the quantized properties of sub-atomic particles, which are conserved in interactions. (Charge, baryon number, lepton number and strangeness.)

Quantum physics (mechanics) Study of matter and energy at the sub-atomic scale. At this level quantities are *quantized*.

Quarks Elementary particles that cannot exist as individual particles. There are six kinds (flavours). Quarks have a charge of $\pm\frac{1}{3}$e or $\pm\frac{2}{3}$e. Combinations of quarks are called *hadrons*.

R

Radial field Field that spreads out from a point equally in all directions.

Radian Unit of measurement of angle. There are 2π radians in 360°.

Radiation Particles, or waves, that radiate away from a source. Principally this refers to various kinds of electromagnetic radiation and nuclear (ionizing) radiation.

Radiation sickness The condition associated with intense exposure to *ionizing radiation*.

Radioactive decay (radioactivity) Spontaneous *transmutation* of an unstable nucleus, accompanied by the emission of ionizing radiation in the form of alpha particles, beta particles or gamma rays.

Random Without pattern or predictability.

Random errors Measurements of any quantity that are bigger or smaller than the correct value and are scattered randomly around that value (for various reasons).

Range (of a projectile) Horizontal distance travelled before impact.

Rarefactions (in a longitudinal wave) Places where there are reductions in the density and pressure of a gas as a wave passes through it.

Raw data Measurements made during an investigation.

Ray A line showing the direction in which a wave is transferring energy. Rays are perpendicular to wavefronts.

Ray diagrams Drawings that represent the direction of waves or particles as they pass through a system.

Reaction force See also *Newton's third law*: forces *always* occur in pairs and these forces are sometimes described as *action* and *reaction*. For example, a person's force pressing down on the ground could be described as the action force and the normal contact force pushing up could be described as the reaction force.

Reaction time The time delay between an event occurring and a response. For example, the delay that occurs when using a stopwatch.

Real gases Modeling of gas behaviour is idealized. Real gases do not behave exactly the same as the model of an ideal gas

Recharging a cell Passing a current through a secondary cell to reverse the chemical reactions, so that it can be reused.

Reflection (waves) Change of direction that occurs when waves meet a boundary between two media such that the waves return into the medium from which they came.

Reflection, law of Angle of incidence = angle of reflection.

Refraction Change of direction that can occur when a wave changes speed (most commonly when light passes through a boundary between two different media).

Refractive index, *n* The ratio of the speed of waves in vacuum to the speed of waves in a given medium. n = sine of angle in air/sine of angle in medium

Regenerative braking Decelerating a vehicle by transferring (some of) its kinetic energy into a form that can be of later use (rather than dissipating the energy into the surroundings). For example, by generating an electric current that charges a battery.

Renewable energy sources Sources that will continue to be available for our use for a very long time. They cannot be used up (depleted), except in billions of years, when the Sun reaches the end of its lifetime.

Resistance (electrical) Ratio of p.d. across a conductor to the current flowing through it. $R = V/I$ Unit: ohm, Ω

Resistive force Any force that opposes motion, for example friction, air resistance, drag.

Resistivity, *ρ* Resistance of a specimen of a material that has length of 1 m and cross-sectional area of 1 m^2. $R = \rho L/A$

Resistor A resistance made to have a specific value or range of values.

Resistors in series Resistors connected one after another so that the same current passes through them all. $R_{total} = R_1 + R_2 + \ldots$

Resistors in parallel Resistors connected between the same two points so that they all have the same potential difference across them. $1/R_{total} = 1/R_1 + 1/R_2 + \ldots$

Resolve (vector) To express a single vector as components (usually two perpendicular to each other). See also *components*.

Rest mass Mass of an isolated particle that is at rest relative to the observer.

Restoring force Force acting in the opposite direction from motion, returning an object to its equilibrium position.

Resultant The single vector that has the same effect as the combination of two or more separate vectors.

Resultant force The vector sum of the forces acting on an object, sometimes called the unbalanced or net force.

Rheostat Variable resistance used to control current.

Right-hand grip rule Rule for determining the direction of the magnetic field around a current.

Ripple tank A tank of water used for investigating wave properties.

Rocket engine Similar to a jet engine, but there is no air intake. Instead an oxidant is carried on the vehicle, together with the fuel.

Rotational kinetic energy Kinetic energy because of rotation (spin).

S

Sankey (energy flow) diagram Diagram representing the flow of energy in a system. The widths of the arrows are proportional to the amounts of energy (or power). Degraded energy is directed downwards in the diagram.

Satellite Object that orbits a much larger mass. Satellites can be natural (like the Earth or the Moon), or artificial (as used for communication, for example).

Satellite footprint The area on the Earth's surface that receives radio (micro) waves from a satellite that are intense enough for the effective transmission of signals.

Scalars Quantities that have only magnitude (no direction).

Scattering Irregular reflection of waves from their original path by interactions with matter.

Schrödinger's theory A mathematical description of the probability of locating electrons within atoms as forms of standing waves.

Scientific notation Every number is expressed in the following form: $a \times 10b$, where a is a decimal number larger than 1 and less than 10 and b is an exponent (integer).

Second, s SI unit of time (fundamental).

Secondary cell Electric cell that can be recharged.

Secondary energy source Source that has been converted from a primary energy source, for example electricity.

Self-sustaining nuclear chain reaction Occurs when enough of the neutrons created during fissions then go on to cause further fissions.

Semiconductor Material (such as silicon) with a resistivity between that of conductors and insulators. Such materials are essential to modern electronics.

Sense perception How we receive (become aware of, perceive) scientific information, using the five human senses (most commonly sight).

Sensor An electrical component that responds to a change in a physical property with a corresponding change in an electrical property (usually resistance). Also called a *transducer*.

Serendipity Unplanned and unexpected good luck.

Series connection Two or more components connected such that there is only one path for the electrical current, which is the same through all the components.

Short circuit An unwanted (usually) electrical connection that provides a low resistance path for an electric current. It can result in damage to the circuit, unless the circuit is protected by a fuse or circuit breaker.

Simulation Simplified visualization (imitation) of a real physical system and how it changes with time. Usually part of a computer modelling process.

SI system of units International system of units of measurement (from the French 'Système International') which is widely used around the world. It is based on seven fundamental units and the decimal system.

SI format for units For example, the SI unit for momentum is kilogram metre per second and should be written as kg m s^{-1} (not kg m/s).

Signal generator Electronic equipment used to supply alternating currents of a wide range of different frequencies.

Significant figures (digits) All the digits used in data to carry meaning, whether they are before or after a decimal point.

Simple harmonic motion (SHM) Oscillations in which the acceleration, a, is proportional to the displacement, x, and in the opposite direction, directed back to the equilibrium position. $a \propto -x$

Sinusoidal In the shape of a sine wave (usually equivalent to a cosine wave).

Slow neutrons Low-energy neutrons (typically less than 1 eV) that are needed to sustain a chain reaction. They are slowed down in a nuclear reactor by the use of a *moderator*. They are sometimes called *thermal neutrons* because they are in approximate thermal equilibrium with their surroundings.

Snell's law (of refraction) Connects the sines of the angles of incidence and refraction to the wave speeds in the two media (or the refractive indices). $n_1/n_2 = \sin\theta_2/\sin\theta_1 = v_2/v_1$

Soft iron Form of iron (pure or nearly pure) that is easily magnetized and demagnetized. Soft iron cores are used in a wide variety of electromagnetic devices.

Solar constant Intensity of the Sun's radiation arriving perpendicularly to the Earth's upper atmosphere.

Solar heating panel Device for transferring radiated thermal energy from the Sun to internal energy in water.

Solenoid Long coil of wire with turns that do not overlap (helical). Solenoids are often used because of the strong uniform magnetic fields inside them.

Sound Longitudinal waves in air, or other media, that are audible (can be heard).

Specific energy Amount of energy that can be transferred from unit mass of an energy source (unit J kg^{-1}).

Specific heat capacity, c The amount of energy needed to raise the temperature of 1 kg of a substance by 1 K. $c = Q/m\Delta T$

Specific latent heat, *Lf* or *Lv* The amount of energy needed to melt (fusion) or vaporize (vaporization) 1 kg of a substance at constant temperature. $L = Q/m$ (unit: J kg^{-1})

Spectroscopy The production and analysis of spectra using instruments called spectroscopes or spectrometers.

Spectrum, continuous The components of radiation displayed in order of their wavelengths, frequencies or energies (plural: spectra).

Spectrum, line A spectrum of separate lines (rather than a continuous spectrum), each corresponding to a discrete wavelength and energy.

Speed, *v* Average speed is defined as distance travelled/time taken, $v = \Delta s/\Delta t$ (unit m s^{-1}). Instantaneous speed is determined over a very short time interval, during which it is assumed that the speed does not change. It can also be determined from the gradient of a distance–time graph. Speed is a scalar quantity; compare with *velocity*, a vector quantity.

Standard (particle) model Summary of the quarks, leptons and bosons that are believed to be the elementary particles of the universe.

Standing wave The kind of wave that can be formed by two similar travelling waves moving in opposite directions. The most important examples are formed when waves are reflected back upon themselves. The wave pattern does not move and the waves do not transfer energy.

Stefan–Boltzmann law An equation that can be used to calculate the total power radiated from a surface, $P = e\sigma AT^4$. σ is known as the *Stefan–Boltzman constant* and *e* is the *emissivity*.

Strain If a material has a strain, it has been deformed.

Strangeness Property of some quarks that was introduced to explain their unexpectedly long lifetimes. Quantum number of strangeness is conserved in some particle interactions.

State of a gas Specified by quoting the pressure, *P*, temperature, *T*, and volume, *V*, of a known amount, *n*, of gas.

Streamlined Having a shape that reduces the resistive forces acting on an object that is moving through a fluid.

Strong nuclear force Fundamental force acting on quarks that is responsible for attracting nucleons together. It is a short-range attractive force (the range is about 10^{-15} m), but for smaller distances it is repulsive, and hence it prevents the nucleus from collapsing. Exchange particle is the *gluon*.

Subjective Describes an opinion based on personal experiences and emotions.

Superconducting Without significant electrical resistance; only occurring at very low temperatures.

Supernumary An amount greater than normal.

Superposition (principle of) The resultant of two or more waves arriving at the same point can be determined by the vector addition of their individual displacements.

Superstring theory A theory of 'everything' that tries to explain all fundamental forces and particles with one theory involving vibrations of 'superstrings'.

Surroundings Everything apart from the *system* that is being considered; similar to the 'environment'.

Synchronization Arrangement of events so that they occur at the same time.

System The object(s) being considered.

Systematic error A reading with a systematic error is always either bigger or smaller than the correct value by the same amount, for example a zero-offset error.

T

Technology The application and use of scientific knowledge for practical purposes.

Temperature Determines the direction of thermal energy transfer. It is a measure of the average random *kinetic energy* of the molecules of a substance.

Tension (force) Force that tries to stretch an object or material.

Terminal potential difference The potential difference across the terminals of a battery (or other voltage supply) when it is supplying a current to a circuit.

Terminal speed (velocity) The greatest downwards speed of a falling object that is experiencing resistive forces (for example, air resistance). It occurs when the object's weight is equal in magnitude to the sum of resistive forces (+ any upthrust).

Tesla, *T* Unit of measurement of magnetic field strength. $1\,\text{T} = 1\,\text{N A}^{-1}\,\text{m}^{-1}$

Test charge (or mass) Idealized model of a small charge (or mass) placed in a field in order to determine the properties of that field, but without affecting those properties.

Theory A term that can have many different interpretations. A scientific theory provides a fully tested and checked explanation of particular observations.

Thermal capacity The amount of energy needed to raise the temperature of something by 1 kelvin.

Thermal contact Objects can be considered to be in thermal contact if thermal energy (of any kind) can be transferred between them.

Thermal energy (heat) The (non-mechanical) transfer of energy between two or more bodies at different temperatures (from hotter to colder).

Thermal equilibrium All temperatures within a system are constant.

Thermistor (negative temperature coefficient) A resistor that has less resistance when its temperature increases. Also called a temperature-dependent resistor.

Thermodynamics Branch of physics involving transfers of thermal energy to do useful work.

Thermostat Component that is used with a heater or cooler to maintain a constant temperature.

Tidal water storage Hydroelectric power generation involving water stored behind artificial dams (barrages) at locations where there is a large difference in water level between the high tide and low tide.

Total internal reflection Occurs when a wave meets a boundary with another medium with a lower refractive index (in which it would travel faster). The angle of incidence must be greater than the *critical angle*.

Tracer (radioactive) Radioisotope introduced into a system (for example, a human body) to track where it goes by detecting the radiation that it emits.

Trajectory Path followed by a projectile.

Translational Moving from place to place.

Translational equilibrium Remaining at rest or continuing to move with constant velocity.

Transmission Passage through a medium without absorption or scattering.

Transmutation When a nuclide changes to form a different element during radioactive decay.

Transition (between energy levels) A photon is emitted when an electron makes a transition to a lower energy level. The energy of the photon is equal to the difference in energy of the levels involved.

Transmit To send out (usually a signal).
Transparent Describes a medium that transmits light without scattering.
Transverse wave A wave in which the oscillations are perpendicular to the direction of transfer of energy, for example light waves.
Trough Lowest point of a transverse mechanical wave.
Turbine Device that transfers the energy from a moving fluid (gas or liquid) to do mechanical work and cause (or maintain) rotation.

U

Ultrasonic Relating to frequencies of sound above the range that can be heard by humans (approximately 20 kHz).
Uncertainty (random) The range, above and below a stated value, over which we would expect any repeated measurements to fall. Uncertainty can be expressed in absolute, fractional or percentage terms.
Unified atomic mass unit, u A unit of mass used to express the mass of atoms and molecules equal to one-twelfth of the mass of the nucleus of a carbon-12 atom (at rest and in the ground state).
Uniform field Field of constant strength, represented by parallel field lines.
Universal (molar) gas constant The constant, R, that appears in the equation of state for an ideal gas ($pV = nRT$). $R = 8.31\,\mathrm{J\,K^{-1}\,mol^{-1}}$
Upthrust A force exerted vertically upwards on any object that is in a fluid.
Universal gravitation constant, G The constant that occurs in the Newton's universal law of gravitation, $G = 6.67 \times 10^{-11}\,\mathrm{N\,m^2\,kg^{-2}}$.

V

Vaporization Change from a liquid to a gas by boiling or evaporation.
Vapour Gas (which can be condensed by pressure to a liquid).
Variable Quantity that can change during the course of an investigation. It can be measurable (*quantitative*) or just observable (*qualitative*). A quantity being deliberately changed is called the *independent variable* and the measured or observed result of those changes occurs in a *dependent variable*. Usually all other variables will be kept constant (as far as possible); they are called the *controlled variables*.
Variable resistor A resistor (usually with three terminals) that can be used to control currents and/or p.d.s in a circuit.
Vector A quantity that has both magnitude and direction.
Velocity, linear, *v* Rate of change of displacement with time, $v = \Delta s/\Delta t$ (unit: $\mathrm{m\,s^{-1}}$). Velocity is a vector quantity and can be considered as speed in a specified direction. Velocity can also be determined from the gradient of a displacement–time graph. If the velocity (speed) of an object changes during a period of time t, the initial velocity (speed) is given the symbol u and the final velocity (speed) is given the symbol v. Instantaneous velocity is determined over a very short time interval.
Verify To show that something is true or accurate.
Vibration Mechanical oscillation.
Vibrational kinetic energy Kinetic energy due to vibration/oscillation.
Video analysis Analysis of video recordings of moving objects by freeze-frame or slow-motion replay.
Virtual particles Particles that are exchanged during fundamental interactions, but the speed of the exchange makes their lifetimes so short that they are impossible to observe.
Visualization Helping understanding by using images (mental or graphic).
Volt Derived unit of measurement of potential difference. $1\,\mathrm{V} = 1\,\mathrm{J\,C^{-1}}$
Voltage See *potential difference*
Voltmeter An instrument used to measure *potential difference*.

W

Watt, W Derived SI unit of power. $1\,\mathrm{W} = 1\,\mathrm{J\,s^{-1}}$
Wave (electromagnetic) A transverse wave composed of perpendicular electric and magnetic oscillating fields travelling at $3 \times 10^8\,\mathrm{m\,s^{-1}}$ in free space.
Wave (mechanical) A wave involving oscillating masses.
Wave (travelling) A wave that transfers energy away from a source. Sometimes called a progressive wave.
Wavefront A line connecting adjacent points moving in phase (for example, crests). Wavefronts are one wavelength apart and perpendicular to the rays that represent them.
Wavelength, λ The shortest distance between two points moving in phase (for example, the distance between adjacent crests).
Wave–particle duality Some properties of light need a wave theory to explain them; other properties can only be explained in terms of particles (photons).
Wave speed, c The speed at which energy is transferred by a wave. $c = f\lambda$
Weigh Determine the weight of an object. In everyday use the word 'weighing' usually means quoting the result as the equivalent mass: 'my weight is 60 kg' actually means I have the weight of a 60 kg mass.
Weak nuclear force Fundamental force acting between quarks and between leptons. It is involved with radioactive decay. Exchange particles are **W** or **Z bosons.**
Weight, W Gravitational force acting on a mass. $W = mg$
Wien's displacement law Relationship between absolute temperature and the wavelength emitted with maximum power by a black body at that temperature. $\lambda_{max} = 2.90 \times 10^{-3}/T$
Work, W The energy transfer that occurs when an object is moved with a force. More precisely, work done = force × displacement in the direction of the force: $W = Fs\cos\theta$, where θ is the angle between the direction of movement and the direction of the force.

Z

Zero offset error A measuring instrument has a zero offset error if it records a non-zero reading when it should be zero.

Higher level

The glossary contains key words, equations and terms from the IB Physics Diploma course. This section takes words from Chapters 9–12.

A

Activity (radioactivity), A The number of decays per second that occur in a radioactive source. Unit: bequerel, Bq.

Alternator See *generator (ac)*

Angular separation Angle subtended at the eye (or an optical device) by two points or objects.

Anti-reflection coating A very thin layer of a transparent material that is coated onto glass in order to increase the amount of light transmitted. The process is called *blooming.*

B

Beams of charged particles Created and controlled by strong electric and magnetic fields. Used to investigate the atomic and subatomic structure of matter.

Becquerel, Bq SI unit of (radio)activity, equal to one nuclear decay every second.

Beta particle spectra Beta particles emitted by nuclei of the same radioisotope have a range of different energies. This is evidence that a third particle is involved with the decays (neutrino or antineutrino).

Bohr model A theory of atomic structure that explains the spectrum from hydrogen atoms. It assumes that the electron orbiting around the nucleus can exist only in certain energy states at specific radii. In this model the electron could only have values of angular momentum (mvr) that fitted the equation $mvr = nh/2\pi$, where n is an integer, known as the principal quantum number.

Breakdown (electrical) Flow of current through an insulator when a fault occurs, or if the voltage rises too high.

Bridge circuit Circuit in which two pairs of components are connected in parallel across a power source, with another component connected between their mid-points. Diode bridges are used for rectification. (Wheatstone and Wien bridge circuits are used for precise electrical measurements.)

C

Capacitance, C The ratio of the charge on a component to the potential difference across it: $C = q/V$ (unit: farad, F; μF and pF are also in common use).

Capacitor An electric circuit component designed to store small amounts of energy temporarily, often consisting of two parallel metallic plates separated and insulated from each other.

Capacitor discharge When the charge on a capacitor is allowed to flow around a circuit (equations: $I = I_0e^{-t/\tau}$, $q = q_0e^{-t/\tau}$, $V = V_0e^{-t/\tau}$).

Capacitor, energy storage Capacitors can store only relatively small amounts of energy, but they are able to transfer the energy quickly, and so can deliver high power for a short time. $E = \frac{1}{2}CV^2$

Capacitors in parallel Two or more capacitors connected so that they have the same potential difference. $C_{parallel} = C_1 + C_2 + \dots$

Capacitors in series Two or more capacitors connected so that the same charge flows onto each of them. $1/C_{series} = 1/C_1 + 1/C_2 + \dots$

Combination of fields Requires vector addition to determine the resultant.

Combination of potentials Requires scalar addition to determine the resultant.

Communication Vital ingredient of scientific research and in providing the general public with information.

Conjugate quantities A pair of physical variables describing a quantum-mechanical system such that either of them, but not both, can be precisely specified at any given time. Also called *linked variables*. See *Heisenberg's uncertainty principle*.

Curiosity (scientific) The desire to learn and understand – an essential characteristic of successful scientists.

D

Davisson–Germer experiment Experiment that verified the wave properties of matter by showing that a beam of electrons is diffracted by a crystal at an angle dependent upon the velocity of the electrons.

De Broglie's hypothesis All particles exhibit wave-like properties, with a de Broglie wavelength, $\lambda = h/p$, where p is the momentum of the particle.

Decay constant, λ The probability of decay of a nucleus per unit time: $\lambda = (-\Delta N/N)/\Delta t$ (unit: s^{-1}). The decay constant is linked to the half-life by the equation: $T_{\frac{1}{2}} = \ln(2/\lambda)$.

Dielectric materials Insulators used between the plates of capacitors in order to increase their capacitance.

Diffraction grating A large number of parallel slits very close together. Used to disperse and analyse light. Angles for constructive interference are predicted by the equation $n\lambda = d\sin\theta$.

Discrete energy level Energy that does not vary continuously but that is restricted to a specific value.

Diode Semiconducting component that allows the passage of electric current in in one direction, when it is then described as *forward biased.*

Diode bridge Four diodes connected in a bridge circuit to provide full-wave rectification of an alternating current.

Dish aerial Aerial placed at the focus of a parabolic reflector, typically for receiving or transmitting radio waves.

Doppler effect When there is relative motion between a source of waves and an observer, the emitted frequency and the received frequency are not the same. Sometimes called a *Doppler shift*. The frequency received from a moving source can be determined from the equation $f' = vf/(v \pm u_s)$; the frequency received from a moving observer can be determined from the equation $f' = (v \pm u_o)f/v$.

Doppler effect with electromagnetic waves If the relative speed between the source and the observer, v, is significantly less than the speed of the wave, c, we can use the following approximation: $\Delta f/f = \Delta\lambda/\lambda \approx v/c$.

E

Eddy currents Circulating currents induced in solid pieces of metal when changing magnetic fields pass through them, for example in the iron core of a transformer.

Edge effects The electric field between parallel plates is assumed to be uniform except at the edges – but these effects are not considered in this course.

Electromagnetic braking See *Regenerative braking*

Electromagnetic induction Production of an emf across a conductor that is experiencing a changing magnetic flux. This may be as a result of moving a conductor through a magnetic field, moving a magnetic field through a conductor, or a time-changing magnetic flux passing from one circuit to another (without the need for any physical motion).

Electromagnetic induction in a straight conductor When a straight wire moves perpendicularly across a magnetic field the induced emf can be determined from the equation $\varepsilon = Bvl$ (or $\varepsilon = BvlN$ if there are N turns).

Electron scattering (by nuclei) High-energy electrons have a wavelength comparable to nuclear diameters, D, so that they will be diffracted in a similar way to light by narrow slits. The first diffraction minimum will occur at angle θ such that $\sin\theta \approx \lambda/D$. This equation can be used to determine nuclear diameters.

Energy levels of hydrogen Because hydrogen is the atom with the simplest structure, scientists were very interested in determining the energy levels of the electron within the atom by examining hydrogen's line spectrum. They were able to show that the energy levels could be predicted by the empirical equation $E = 13.6(\text{eV})/n^2$. This equation was explained later by using the *Bohr model* of the hydrogen atom.

Electrostatic field Electric field that is constant, not changing.

Equipotential line (or surface) Line (or surface) joining points of equal potential. Equipotential lines are always perpendicular to field lines.

Escape speed (velocity) Minimum theoretical speed that an object must be given in order to move to an infinite distance away from a planet (or moon, or star): $v_{escape} = \sqrt{(2GM/R)}$. This assumes that air resistance is not significant.

Expansion of the universe The redshift of light (similar to the Doppler effect) from distant galaxies provides evidence of an expanding universe.

Exponential radioactive decay Represented by the equation $N = N_0e^{-\lambda t}$. N_0 is the number of undecayed nuclei at the start of time t and N is the number remaining at the end of time t. Alternatively, equations of the same form can be used with activity, A, or the count rate, C. Activity is linked to the initial number of atoms by the equation $A = \lambda N_0e^{-\lambda t}$.

F

Farad, F The derived unit of capacitance in the SI system equal to the capacitance of a capacitor having an equal and opposite charge of 1 coulomb on each plate when there is a potential difference of 1 volt between the plates.

Faraday's law of electromagnetic induction The magnitude of an induced emf is equal to the rate of change of magnetic flux linkage, $\varepsilon = -N\Delta\Phi/\Delta t$. For an explanation of the negative sign, see *Lenz's law.*

Fermi radius, R_0 Constant in the equation for nuclear radius ($R = R_0A^{1/3}$). Equal to the value of R for $A = 1$.

Ferromagnetic material Material containing iron that has excellent magnetic properties (high permeability).

Fourier analysis Any periodic waveform can be reproduced (analysed) as the addition of a series of trigonometrical functions.

G

Galvanometer Ammeter that measures very small currents.

Generator (ac) A device containing coils that rotate in a magnetic field (or a field that rotates within coils), transferring kinetic energy to the energy carried by an alternating electric current.

Geophone Instrument that converts movements of the ground into voltages.

Geostationary orbit A satellite is described as geostationary if it appears to remain 'above' the same location on the Earth's surface. This can be very useful for communications. Geostationary satellites follow a type of *geosynchronized orbit*, which is in the same plane as the equator, and with exactly the same period as the Earth's oscillation on its axis (1 day).

Geosynchronized orbit Any satellite orbit that has the same period as the Earth spinning on its axis. The orbit must have exactly the correct radius.

Global positioning system (GPS) A navigation system that provides accurate information on the location of the GPS receiver, which continually communicates with several orbiting satellites.

H

Heisenberg uncertainty principle A fundamental principle of quantum mechanics, which states that it is impossible to measure simultaneously the momentum and the position of a particle with infinite precision: $\Delta x\Delta p \geq h/4\pi$. The principle also applies to measurements of energy and time: $\Delta E\Delta t \geq h/4\pi$.

Hollow charged sphere There is a constant potential within the sphere, which means that the electric field is zero. Faraday's cages are an example.

Hysteresis (magnetic) The changing magnetic properties of a ferromagnetic material depend on what has happened to it before and how quickly the changes take place. These effects are called hysteresis.

I

Insight (scientific) Ability to achieve a deep understanding of a complex situation.

Iridescence The property of certain surfaces to produce variable coloured effects depending on the angle at which they are viewed. (The feathers of some birds and insect wings are common examples.)

Iteration Repetitive mathematical procedure that calculates the changes that occur in small increments in order to determine an overall result.

J

Joule heating Transfer of electrical energy to thermal energy as a current passes through a resistance.

L

Laminations (iron core) Alternate layers of iron and insulation in a core of an electromagnetic device, which are designed to limit energy dissipation due to eddy currents.

Law of radioactive decay The number of nuclei that decay per second, $\Delta N/\Delta t$ (= the activity, A, of the source) is proportional to the number of radioactive atoms still present that have not yet decayed,
N: $-\Delta N/\Delta t = \lambda N$, where λ represents a constant, known as the *decay constant.*

Lenz's law (of electromagnetic induction) The direction of an induced emf is such that it will oppose the change that produced it. This is represented mathematically by the negative sign in the equation representing *Faraday's law.*

Lightning Discharge to the ground, or another cloud, of the charge that has accumulated in a cloud.

M

Magnetic flux, Φ Defined as the product of an area, A, and the component of the magnetic field strength perpendicular to that area, $B\cos\theta$. $\Phi = BA\cos\theta$ (unit: Weber, Wb)

Magnetic flux density, B The term commonly used at Higher level for magnetic field strength.

Magnetic flux linkage, $N\Phi$ The product of magnetic flux and the number of turns in a circuit (unit: Wb).

Mass on a spring oscillator Can be used as a simplified visualization of many oscillators. $T = 2\pi\sqrt{m/k}$.

Mass spectrometer A device that can measure the masses and relative abundances of gaseous ions.

Matter waves Waves that represent the behaviour of an elementary particle, atom or molecule under certain conditions. See *De Broglie's hypothesis*.

Modulation Changing the amplitude (or frequency) of a wave according to variations in a secondary signal or effect.

Multiple slits By increasing the number of parallel slits (of the same width) on which a light beam is incident, it is possible to improve the resolution of the fringes/spectra formed.

N

Nuclear energy levels The emission of alpha particles and gamma rays with discrete energies during radioactivity indicates that nuclei have discrete energy levels.

Nuclear medicine Use of radioisotopes in the diagnosis and treatment of disease.

Nuclear density Assuming that the nucleus is spherical, nuclear density can be determined from nuclear mass ($\approx Au$) divided by the volume of a sphere having the appropriate nuclear radius. All nuclear densities are similar in magnitude, which is extremely large.

Nuclear radius, R R is proportional to the cube root of the nucleon number. $R = R_0A^{1/3}$, where R_0 is called the *Fermi radius*.

Nuclear transition A change in nuclear energy level that results in the emission (or absorption) of a high-energy photon.

O

Orbital speed For a satellite in a circular orbit, its speed must have the correct value for the chosen radius: $v_{orbit} = \sqrt{(GM/r)}$.

Orbital energy An orbiting satellite has both gravitational potential energy and kinetic energy: $E_P = -GMm/R$; $E_K = +\frac{1}{2}GMm/R$. Adding these together gives the total energy: $E_T = -\frac{1}{2}GMm/R$.

Oscilloscope An instrument for displaying and measuring voltages that change with time.

P

Peak values (electrical) The maximum values of an alternating current, I_0, voltage, V_0 or power, P_0 (compare with *rms values*). $P_0 = I_0V_0$. Average power, $P = \frac{1}{2}I_0V_0$.

Pendulum A weight suspended from a point so that it can oscillate freely. There are many designs. A simple pendulum consists of a small spherical mass on the end of a string ($T = 2\pi\sqrt{l/g}$). A *torsion pendulum* involves twisting in a horizontal plane.

Phase change Waves undergo a phase change of π when they reflect off the boundary with a medium in which they would travel slower.

Pitch The sensation produced in the human brain by sound of a certain frequency.

Photoelectric effect Ejection of electrons from a substance by incident electromagnetic radiation, especially by ultraviolet radiation. Sometimes called *photoemission*. The ejected electrons are called *photoelectrons*.

Photoelectric equation The maximum kinetic energy of an emitted photoelectron is the difference between the photon's energy and the work function: $KE_{max} = \frac{1}{2}mv_{max}^2 = hf - \Phi$.

Polar orbit Descriptive of the path of a low-orbit satellite that passes over the poles of the Earth and completes many orbits every day, usually passing over many different parts of the planet.

Potential difference ΔV The p.d. between two points is the work done when moving unit mass (or unit positive charge) between those points. $\Delta V_e = W/q$; $\Delta V_g = W/m$

Potential, V The potential at a point is equal to the work done per unit mass (or unit positive charge) in bringing a small test mass (or positive charge) from infinity to that point. $V_g = E_p/m$ and $V_e = E_p/q$ (units: J kg^{-1} or J C^{-1}, which is more commonly called volts, V.)

Potential barrier Graphical representation of the potential (energy) that a bound particle needs to overcome in order to escape from the forces that are confining it. Potential barriers can also be described as potential hills or potential wells.

Potential energy in fields, E_P Work has to be done to move masses in gravitational fields, or charges in electric fields. Potential energy is defined as the work done when bringing a mass (or charge) to its present position from infinity. Potential energy equals the area under a force-distance graph between the point and infinity.

Potential gradient $\Delta V/\Delta r$ Equals the magnitude of the field strength. $g = -\Delta V_g/\Delta r$ and $E = -\Delta V_e/\Delta r$

Primary and secondary coils (transformers) The coils to which the input and output of a transformer are connected.

Probability density $P(r)$ The probability of finding a particle in unit volume at a distance r from a reference point. Related to the wave function by the equation: $P(r) = |\Psi|^2\Delta V$.

Q

Quantum tunnelling Because of uncertainty, there is the possibility that a particle can pass through a *potential (energy) barrier* and thereby create effects that would not be considered possible using the principles of classical physics.

Radar Using the reflection of microwaves to locate the position and speed of planes and other vehicles.

Radial fields (gravitational equations) $F_G = GMm/r^2$; $E_p = GMm/r$; $g = GM/r^2$; $V_g = -GM/r$

Radial fields (electric equations) $F_E = kq_1q_2/r^2$; $E_p = kq_1q_2/r$; $E = kq/r^2$; $V_e = kq/r$

Radioastronomy Study of space utilizing the detection of radio waves emitted by astronomical sources.

Rayleigh's criterion A guide to resolution: two point sources can *just* be resolved if the first minimum of the diffraction pattern of one occurs at the same angle as the central maximum of the other. This means that if the sources are observed through a narrow slit, they will just be resolved if they have an angular separation of $\theta = \lambda/b$. For a circular aperture, $\theta = 1.22\lambda/b$.

Rectification Changing *alternating current* (ac) into *direct current* (dc). Rectification can be *full wave* or *half wave*.

Resistor-capacitor (RC) circuit Circuit with a resistor and a capacitance with values selected so that the capacitor charges or discharges at a required rate. See *time constant*.

Resolution The ability of an instrument (or an eye) to detect (resolve) separate details.

Resolvance, R A measure of the ability of a diffraction grating with N slits to resolve separate wavelengths. $R = \lambda/\Delta\lambda = mN$, where m is the diffraction order.

Retina Surface at the back of the eyeball where light is detected.

Root mean squared (rms) value The effective value of an alternating current (or voltage), also called its *rating*. It is equal to the value of the direct current (or voltage) that would dissipate power in a resistor at the same rate. $I_{rms} = I_0/\sqrt{2}$ and $V_{rms} = V_0/\sqrt{2}$, where I_0 and V_0 are the peak values.

Rutherford scattering Sometimes called Coulomb scattering. The scattering of alpha particles by nuclei, which can be explained by the action of (only) an inverse square law of electric repulsion. For particles that are scattered through 180°, their initial kinetic energy can be equated to the electric potential energy when closest to the nucleus. This provides an estimate for the radius of the nucleus. When high-energy particles are used they might enter the nucleus, so that strong nuclear forces are also involved and then the scattering will no longer follow the same pattern.

S

Schrödinger wave equation An equation that mathematically represents the Schrödinger model of the hydrogen atom by describing electrons using *wavefunctions*. The square of the amplitude of the wavefunction gives the probability of finding the electron at a particular point.

Secondary waves The propagation of waves in two or three dimensions can be explained by considering that each point on a wavefront is a source of secondary waves.

Seismology The study of earthquakes.

Simple harmonic motion (SHM) Defining equation: $a = -\omega^2 x$. Equations for displacement: $x = x_0 \sin \omega t$ (for oscillations that start with zero displacement at time $t = 0$); $x = x_0 \cos \omega t$ (for oscillations that start with displacement x_0 at time $t = 0$). Equations for velocity: $v = \omega x_0 \cos \omega t$ (for oscillations that start with zero displacement at time $t = 0$); $v = -\omega x_0 \sin \omega t$ (for oscillations that start with displacement x_0 at time $t = 0$). The velocity can also be determined from the displacement and amplitude: $v = \pm\omega\sqrt{(x_0^2 - x^2)}$.

Simple harmonic energy transfers All mechanical oscillators continuously interchange energy between potential and kinetic forms. $E_K = \frac{1}{2}m\omega^2(x_0^2 - x^2)$; $E_P = \frac{1}{2}m\omega^2 x^2$; $E_T = \frac{1}{2}m\omega^2 x_0^2$

Single-slit diffraction The simplest diffraction pattern is that produced by wavefronts interfering after they have passed through a narrow, rectangular slit. Minima occur at angles such that $\theta = n\lambda/b$.

Slip rings and brushes In an ac generator these are used for connecting the rotating coil to the external circuit.

Smoothing (capacitor) Use of a capacitor to make the output of a diode rectifier 'smoother' – less variable.

Stopping potential The minimum voltage required to reduce a photoelectric current to zero.

T

Tangible and discernable Recognized by the senses as being real and definite (not vague or imaginary).

Terminology The words and phrases used in a particular area of study.

Threshold frequency, f_0 The minimum frequency of a photon that can eject an electron from the surface of a metal.

Thin-film interference Interference that occurs after a wavefront is split by reflection off two surfaces of a very thin transparent medium. Constructive interference occurs if $2dn = (m + \frac{1}{2})\lambda$. Destructive interference occurs if $2dn = m\lambda$.

Time constant, τ Value that characterizes the rate at which an RC circuit discharges (or charges). $\tau = RC$ (unit: s). It is the time taken for the current (or charge, or p.d. of capacitor) to fall to $1/e$, or 37% of previous value.

Transducer Device that converts one form of energy to another. The word is most commonly used with devices that convert to or from changing electrical signals.

Transformer A device that transfers electrical energy from one circuit to another using electromagnetic induction between coils wound on an iron core. Transformers are used widely to transform one alternating voltage to another of different magnitude. *Step-up transformers* increase voltages; *step-down transformers* decrease voltages.

Transformer (ideal) An *ideal* transformer has no energy dissipation and is 100% efficient, so that $\varepsilon_p I_p = E_s I_s$.

Transmission of electrical power Electrical power is sent (transmitted) from power stations to different places around a country along wires (cables), which are commonly called *transmission* (or power) *lines*. These lines are linked together in an overall system called the *transmission grid*.

Tuning fork Vibrating instrument of a simple shape used to produce a sound of a single, precise frequency.

Tunnelling electron microscope Microscope that uses quantum tunnelling of electrons between a pointed electrode and the surface being scanned.

Tunnelling probability Probability that a particle can tunnel through a potential barrier.

Turns ratio (transformer) The ratio of turns in the primary and secondary coils of a transformer controls the output voltage. For an ideal transformer: $N_P/N_S = \varepsilon_P/\varepsilon_S$.

U

Uniform electric field Created between parallel charged plates. $E = V/d$ (unit: $V\,m^{-1}$)

W

Wavefunction, $\Psi(x,t)$ Mathematical function of space and time that describes the quantum state of a subatomic particle, such as an electron. It is a solution to the Schrödinger wave equation.

Weber, Wb Unit of magnetic flux. 1 Wb = 1 T m^2

Work function, Φ The minimum amount of energy required to free an electron from the attraction of atoms in a metal's surface. Since the energy of the incident photons is equal to hf, $hf_0 = \Phi$, where f_0 represents the threshold frequency.

X

X-Ray diffraction (crystallography) Investigating the atomic and molecular structure of matter by detecting how X-rays are diffracted by crystalline materials. X-ray wavelengths are comparable to atomic dimensions.

Y

Young's double-slit experiment Classic physics experiment that demonstrated the wave properties of light by producing an interference pattern.

Acknowledgements

Photo credits

p.4 *l* Courtesy Zettl Research Group, Lawrence Berkeley National Laboratory and University of California at Berkeley. All rights reserved © 2009, *r* © NASA, ESA, HEIC, and The Hubble Heritage Team (STScI/AURA); **p.8** © sciencephotos/Alamy; **p.22** © ImageZoo/Alamy; **p.31** © Markus Schreiber/AP/Press Association Images; **p.32** © Edward Kinsman/Science Photo Library; **p.34** © Hoch Zwei/Corbis; **p.38** *l* © ESA–AOES-Medialab, *r* © The Photos/Fotolia; **p.39** *tl* © Photodisc/Getty Images, *tr* © REX/Red Bull Content Pool, br © The Art Archive/Alamy; **p.40** *t* © NASA Goddard Space Flight Center, *b* © Andy Lyons/Getty Images; **p. 42** © Dr. Loren Winters/Visuals Unlimited/Corbis; **p.43** *l* © Royal Astronomical Society/Science Photo Library, *r* © Edward Kinsman/Science Photo Library; **p.46** © icsnaps/Fotolia; **p.47** *t* © Peter Parks/AFP/Getty Images, *b* © NASA Marshall Space Flight Center (NASA-MSFC); **p.49** © KeystoneUSA-ZUMA/Rex Features; **p.50** *b* © 1996 R. Strange/Photodisc/Getty Images; **p.52** *t* © Tomohiro Ohsumi/Bloomberg via Getty Images, *c* © tingimage/Alamy, *b* © Takeshi Takahara/Science Photo Library; **p.59** © Argus/Fotolia; **p.62** © shefkate/Fotolia.com; **p.63** © The Art Archive/Alamy; **p.65** © Ian Walton/Getty Images; **p.66** © imagebroker/Alamy; **p.68** © Vasily Smirnov/Fotolia; **p.69** © Yoshikazu Tsuno/AFP/Getty Images; **p.70** © ITAR-TASS Photo Agency/Alamy; **p.73** © Cameron Spencer/Getty Images; **p.78** © Photolink Ltd/Alamy; **p.81** © Julija Sapic/Fotolia; **p.86** *t* © fStop/Alamy, *b* © David Gee/Alamy; **p.87** *t* © Andrew Jankunas/Alamy; **p.93** © JGI/Tom Grill/Blend Images/Getty Images; **p.97** © Stephen J. Boitano/Lonely Planet Images/Getty Images; **p.98** *t* © RSBPhoto/Alamy, *b* © Koji Sasahara/AP/Press Association Images; **p.100** *tl* © Tim Graham/Alamy, *b* © Johner Images/Alamy, *tr* © chalabala/Fotolia; **p.101** © kasiastock/Fotolia; **p.102** © Jerry Taylor/Alamy; **p.109** © Tony McConnell/Science Photo Library; **p.112** © SuperStock/Alamy; **p.113** Courtesy of Kelson/Wikimedia Commons (http://commons.wikimedia.org/wiki/File:SS-joule.jpg); **p.118** © Bert de Ruiter/Alamy; **p.119** © Siemens AG, *r* © 2003 by Daniel Tückmantel/Fotolia; **p.120** © SSPL/Getty Images; **p.126** *l* © Railpix/Fotolia, *cl* © Deborah Benbrook Photography, *cr* © arquiplay77/Fotolia, *r* © Luis Viegas/Fotolia; **p.130** © Anne Trevillion; **p.133** *t* © Mark Thomas/Alamy, *c* © Photodisc/Getty Images/World Commerce & Travel 5, *b* © Photodisc/Getty Images/Business & Industry 1; **p.141** © natureshots/Fotolia; **p.142** © okinawakasawa/Fotolia; **p.151** © Laurence Cartwright/Fotolia; **p.154** © sciencephotos/Alamy; **p.155** © Alistair Cotton/Fotolia; **p.156** © INTERFOTO/Alamy; **p.158** © alexandre zveiger/Fotolia; **p.159** © Konstantin Sutyagin/Fotolia; **p.162** © David Fleetham/Visuals Unlimited/Corbis; **p.168** © Jerome Wexler/Science Photo Library; **p.170** © Pasieka/Science Photo Library; **p.172** *t* © Deklofenak/Fotolia, *b* © Christopher McGowan/Alamy; **p.175** *l* © Robert Fried/Alamy, *r* © Imagestate Media (John Foxx); **p.176** © sandy young/Alamy; **p.179** © Horizon International Images Limited/Alamy; **p.180** Courtesy of Wikipedia Commons (http://creativecommons.org/licenses/by-sa/3.0/); **p.181** © Art Photo Picture/Fotolia; **p.182** *t* © Edward Kinsman/Science Photo Library, *b* © jarous/Fotolia; **p.183** *t* © APTN/AP Photo/Press Association, *b* © Utah Images/NASA/Alamy; **p.185** © sciencephotos/Alamy; **p.188** © Fred Hirschmann/Superstock; **p.190** © Edward Kinsman/Science Photo Library; **p.192** © Mary Evans/Grenville Collins Postcard Collection; **p.193** © 1997 Doug Menuez/Photodisc/Getty Images; **p.197** Courtesy of Steen Jeberg/Wikimedia (http://creativecommons.org/licenses/by-sa/2.5/deed.en); **p.203** © Leslie Garland Picture Library/Alamy; **p.204** © Ted Kinsman/Science Photo Library; **p.206** Courtesy of Artmechanic/Wikimedia; **p.208** © REX/DECC; **p.209** © Eye Ubiquitous/Alamy; **p.212** © niroworld/Fotolia; **p.217** *t* © Gerrit de Heus/Alamy, *b* © Photo Researchers/Mary Evans Picture Library; **p.220** © David J. Green - electrical/Alamy; **p.222** *t* © Alx/Fotolia, *c* © Dmitry Kolmakov/Fotolia, *bl* © Alx/Fotolia, *br* © Alx/Fotolia; **p.224** © Sheila Terry/Science Photo Library; **p.230** © bloomua/Fotolia; **p.233** *l* © David J. Green - electrical/Alamy, *c* Martyn F. Chillmaid/Science Photo Library, *r* © David J. Green - technology/Alamy; **p.234** © Trevor Clifford Photography/Science Photo Library; **p.239** © Liu Xiaokun/Xinhua Press/Corbis; **p.240** © Xuejun li/Fotolia; **p.243** © Phil Degginger/Alamy; **p.244** © Phil Degginger/Alamy; **p.246** © OJO Images Ltd/Alamy; **p.252** © CERN; **p.253** © CERN;

p.254 © Joshua Strang, USAF, Wikipedia; **p.263** *t* © magann/Fotolia, *b* © PCN Photography/Alamy; **p.264** © Valeriya Potapova/Alamy; **p.267** *t* © Alban Egger/Fotolia, *b* © Sergio Azenha/Alamy; **p.274** © sciencephotos/Alamy; **p.278** © Universal History Archive/Getty Images; **p.279** © European Space Agency, S. Corvaja/Science Photo Library; **p.306** © BSIP SA/Alamy; **p.319** © Keystone Pictures USA/Alamy; **p.325** © Shelley Gazin/Corbis; **p.327** © Denis Balibouse/Reuters/Corbis; **p.334** © lifeonwhite.com/Fotolia; **p.336** © RIA Novosti/Alamy; **p.339** © Jeremy Walker/The Image Bank/Getty Images; **p.340** *t* © U.S. Coast Guard/Science Photo Library, *b* © alohaspirit/iStockphoto.com; **p.341** © Sipa Press/Rex Features; **p.343** © Ria Novosti/Science Photo Library; **p.345** © GAMMA/Gamma-Rapho via Getty Images; **p.346** *t* © Mere Words/Science Photo Library, *b* © David Boily/AFP/Getty Images; **p.347** © Barbara Sax/AFP/Getty Images; **p.348** © EFDA (www.efda.org); **p.349** *l* © Birute Vijeikiene/Fotolia, *r* © seraphic06/Fotolia, *b* © Visions of America, LLC/Alamy; **p.352** © Mike Goldwater/Alamy; **p.355** © Friedrich Saurer/Science Photo Library; **p.356** *t* © F1online digitale Bildagentur GmbH/Alamy, *b* © LMR Media/Alamy; **p.359** © Laurence Gough/Fotolia; **p.361** © SuperStock/Alamy; **p.365** © POOL/Reuters/Corbis; **p.367** © Kevin Schafer/Alamy; **p.372** *t* © Ian Patrick/Alamy, *b* © Li jianshu/Imaginechina/Corbis; **p.376** © REX/Hannu Vallas; p.377 © Glyn Thomas Photography/Alamy; **p.389** *t* © The University of Texas at Austin, *b* © sciencephotos/Alamy; **p.394** © Dietrich Zawischa; **p.397** © NASA Goddard Space Flight Center; p.403 © Action Press/Rex Features; **p.404** © Stocktrek Images, Inc./Alamy; **p.405** *l* © Ria Novosti/Science Photo Library, *r* © marcel/Fotolia, *b* © siamphoto/Fotolia; **p.408** © Mark Williamson/Fotolia; **p.409** © Louisa Howard; **p.414** © Christopher Pasatieri/Reuters/Corbis; **p.417** © Danita Delimont/Alamy; **p.418** © dave stamboulis/Alamy; **p.432** © Art Directors & TRIP/Alamy; **p.433** © Philippe Wojazer/Reuters/Reuters/Corbis; **p.441** © Matt Writtle/PA Archive/Press Association Images; **p.446** *l* Courtesy NSSDC, NASA, *r* G AFP/Getty Images; **p.447** © KeystoneUSA-ZUMA / Rex Features; **p.449** © NASA; **p.454** *t* © Jenny Thompson/Fotolia, *b* © NASA; **p.467** © oslobis/Fotolia.com; **p.468** Courtesy of Wikimedia Commons (http://en.wikipedia.org/wiki/Michael_Faraday); **p.471** © Lucky Dragon/Fotolia; **p.480** *b* © Galina Semenko/Fotolia; **p.481** © Lester Lefkowitz/Taxi/Getty Images; **p.482** © REX/Image Broker; **p.486** *l* © Himmelssturm/Fotolia, *r* © Islemount Images/Alamy; **p.487** © Cn Boon/Alamy; **p.489** © UrbanZone/Alamy; **p.493** © David J. Green - electrical/Alamy; **p.494** yuri4u80/Fotolia; **p.519** © Archive Pics/Alamy; **p.522** © INTERFOTO/Alamy; **p.529** © Joseph A. Stroscio, Robert J. Celotta, Steven R. Blankenship and Frank M. Hess/NIST; **p.535** © NSF/Xinhua Press/Corbis

t = top, *c* = centre, *b* = bottom, *l* = left, *r* = right

Computer hardware and software brand names mentioned in this book are protected by their respective trademarks and acknowledged.
Every effort has been made to trace all copyright owners, but if any have been inadvertently overlooked the Publishers will be pleased to make the necessary arrangements at the first opportunity.

Artwork credits

p.112 Fig 3.8, **p.180** Fig 4.56 and **p.468** Fig 11.19 all Jon Homewood.

Examination questions

Examination questions have been reproduced with kind permission from the International Baccalaureate Organization.

Index

D

E

F

S